Investigating Chemistry

Second Edition

M. J. Denial, B.Sc.,
Deputy Headmaster,
King Edward VII School, Sheffield

in conjunction with

L. Davies, B.Sc.,
Formerly of Whitehaven Grammar School

A. W. Locke, B.Sc., Ph.D.,
Head of Chemistry, Whitehaven Grammar School

M. E. Reay, B.Sc.,
Formerly of Whitehaven Grammar School

Heinemann Educational Books

Heinemann Educational Books Ltd
22 Bedford Square, London WC1B 3HH

LONDON EDINBURGH MELBOURNE AUCKLAND
HONG KONG SINGAPORE KUALA LUMPUR NEW DELHI
IBADAN NAIROBI JOHANNESBURG
PORTSMOUTH (NH) KINGSTON

First published 1973
Reprinted 1973, 1974, 1976, 1978
Second edition published 1981
Reprinted 1982, 1985, 1986

British Library Cataloguing in Publication Data

Investigating chemistry. – 2nd ed.
Pupil's book
1. Chemistry
I. Denial, M. J.
540 QD33

ISBN 0 435 64166 2

Filmset in 'Monophoto' Times and Univers
by Eta Services (Typesetters) Ltd., Beccles, Suffolk

Printed and bound in Great Britain
by Richard Clay (The Chaucer Press) Ltd., Bungay, Suffolk

Introduction

This second edition of *Investigating Chemistry* is in many respects a new book. It has been produced in response to syllabus changes and in the light of the experience of practising teachers who successfully use the first edition.

The second edition

The main changes from the first edition are as follows.

1. The subject matter has been changed to conform with new syllabuses produced by the major examining boards.
2. The style and language have been considerably simplified to allow the book to be used with a wider ability range than is normally possible with the first edition.
3. Much more guidance has been given to help pupils to deduce the appropriate conclusions from their experiments. For example, they are given more help about what to look for in writing 'results' and 'conclusions' to experiments, even though the *actual* results are often not given directly. Where it is felt that the average pupil cannot be expected to appreciate all of the principles introduced by an experiment, the appropriate conclusion has been made in full. Information which is needed to make a conclusion has been given under the heading 'Notes about the Experiment'.
4. Questions which are used to provide support or further experience have been separated from those connected with an important principle or experimental conclusion, so as not to confuse the main issues. Such questions appear under the heading 'Points for Discussion'; they are not directly related to the main experimental conclusions. We have also included short examination questions chosen to test understanding of the preceding work.
5. End-of-chapter summaries make the book easier to use as a reference or revision text. The summaries are sometimes used to indicate the main points which the student needs to know rather than to restate facts and principles in full, which could produce some very long summaries, especially for Chapters such as 19, 21, and 22. The student thus has a check list of what is to be learned for each chapter, and it is a simple matter for the teacher to indicate which of the areas, if any, are not appropriate to the syllabus being followed. The summaries have been made as general as possible, so that student and teacher are not affected by choosing experimental approaches which are not described in the book. Summaries of this kind should be particularly useful in revision, when the syllabus is broken down into small areas of teaching material, in a way that is not possible if a student works from an official syllabus produced by an Examining Board.
6. We do not believe that teaching should be dominated by examination requirements, but nevertheless we feel that it would be quite wrong to ignore the problems which some pupils face when preparing for an examination. Many pupils find it difficult to dissociate experiments which are used 'as a means to an end' (e.g. to teach a principle) from those which are more likely to be examined in full (e.g. chemical techniques). The end-of-chapter summaries draw attention to the *important experiments* in an attempt to avoid unnecessary learning during revision, and more attention has been given to the problems which pupils so frequently encounter during a chemistry course.

7. More consideration has been given to the relation between chemistry and the outside world. There is increased emphasis on the industrial scene, and energy problems, pollution, recycling, and the conservation of raw materials are considered in much greater detail.

All the successful features of the first edition have been retained. We have continued to use the personal approach to our readers. We have interspersed the theoretical material with numerous experiments so that almost every aspect of the work is approached through the laboratory. Our aim has been to provoke thought and discussion, and to foster the investigational spirit by withholding the experimental results and conclusions in some cases until later in the chapter.

We have made every effort to present chemistry as a living and developing science. It is a subject of such importance in our rapidly developing technological society that it is essential that all its members should understand enough chemistry to assist in making some of the vital decisions facing us in the future.

Safety

Safety measures have been stressed throughout. We have suggested whether each experiment could be a demonstration, classwork, or both, but it is impossible to be dogmatic about this because such decisions depend upon the experience of the teacher, the equipment that is available, the age and ability of the pupils, and so on. If an experiment is given 'one star' we consider that the experiment is suitable for classwork; two stars indicate that the work could be either (or both) classwork or demonstration; an experiment marked with three stars would normally be a demonstration.

Teachers are responsible for the safety of pupils within a laboratory, and although the experiments used in the text have been conducted many times, it cannot be overemphasized, particularly to inexperienced colleagues, that even the most traditional or apparently trivial experiment is still a potential source of danger. It is essential that teachers familiarize themselves with reactions and manipulations before a lesson takes place, and that careful thought is given to laboratory management and potential hazards before *each* practical session. More details about safety factors and hints on practical work are given in *Investigating Chemistry, Notes for Teachers*. This also includes answers to the numerical problems set in the main book. Teachers may have to deal with a wide ability range in one class, and it is important to approach every lesson by assuming that even very simple instructions can be misunderstood. It is much better to assume that a pupil may do something which could result in an accident than to ignore the possibility, and there is no doubt that the eyes are extremely vulnerable if a chemistry experiment goes wrong. (Note that goggles are not included in the list of apparatus given for each experiment, nor are they always referred to in the procedure, because we have assumed that they will always be worn during practical work.)

Teaching sequence

It is not intended that the material in the book has to be taught chapter by chapter and in the sequence presented. No two people would ever teach the same syllabus in the same way. The chapter sequence does constitute a logical progression, in which the basic aim has been to introduce as many principles as possible early in the course, and then to use these in order to establish the more factual aspects. In each topic we have tried to reserve a little 'depth' for later work, partly because it may be more easily appreciated with maturity and also to provide a final stage which contains only a relatively small amount of new material, all of which requires a revision of earlier work before it can be established. However, many such sequences are possible, and these are likely to be further modified depending on the nature of preceding elementary science courses, the time allocated to the teaching of chemistry, the resources and manpower available, and so on. On some occasions it is better to think of sections (rather than chapters) as modules of teaching material, e.g. it is not intended that Section 5.5 (More about acids and alkalis) should be taught at the same time as the rest of Chapter 5.

Contents

PERIODIC TABLE OF

1	2							
1 **H** Hydrogen **1** 1								
7 **Li** Lithium **3** 2)1	9 **Be** Beryllium **4** 2)2							
23 **Na** Sodium **11** 2)8)1	24 **Mg** Magnesium **12** 2)8)2							
39 **K** Potassium **19** 2)8)8)1	40 **Ca** Calcium **20** 2)8)8)2	45 **Sc** Scandium **21** 2)8)9)2	48 **Ti** Titanium **22** 2)8)10)2	51 **V** Vanadium **23** 2)8)11)2	52 **Cr** Chromium **24** 2)8)13)1	55 **Mn** Manganese **25** 2)8)13)2	56 **Fe** Iron **26** 2)8)14)2	59 **Co** Cobalt **27** 2)8)15)2
85.5 **Rb** Rubidium **37** 2)8)18)8)1	88 **Sr** Strontium **38** 2)8)18)8)2	89 **Y** Yitrium **39** 2)8)18)9)2	91 **Zr** Zirconium **40** 2)8)18)10)2	93 **Nb** Niobium **41** 2)8)18)12)1	96 **Mo** Molybdenum **42** 2)8)18)13)1	[99] **Tc** Technetium **43** 2)8)18)14)1	101 **Ru** Ruthenium **44** 2)8)18)15)1	103 **Rh** Rhodium **45** 2)8)18)16)1
133 **Cs** Caesium **55** 2)8)18)18)8)1	137 **Ba** Barium **56** 2)8)18)18)8)2	139 **La** Lanthanum **57** 2)8)18)18)9)2	178.5 **Hf** Hafnium **72** 2)8)18)32)10)2	181 **Ta** Tantalum **73** 2)8)18)32)11)2	184 **W** Tungsten **74** 2)8)18)32)12)2	186 **Re** Rhenium **75** 2)8)18)32)13)2	190 **Os** Osmium **76** 2)8)18)32)14)2	192 **Ir** Iridium **77** 2)8)18)32)15)2
[223] **Fr** Francium **87** 2)8)18)32)18)8)1	[226] **Ra** Radium **88** 2)8)18)32)18)8)2	227* **Ac** Actinium **89** 2)8)18)32)18)9)2						

Key
Atomic weight Symbol Name Atomic number
Electronic structure

139 **La** Lanthanum **57** 2)8)18)18)9)2	140 **Ce** Cerium **58** 2)8)18)20)8)2	141 **Pr** Praseodymium **59** 2)8)18)21)8)2	144 **Nd** Neodymium **60** 2)8)18)22)8)2	[147] **Pm** Promethium **61** 2)8)18)23)8)2	150 **Sm** Samarium **62** 2)8)18)24)8)2
227* **Ac** Actinium **89** 2)8)18)32)18)9)2	232 **Th** Thorium **90** 2)8)18)32)18)10)2	[231] **Pa** Protactinium **91** 2)8)18)32)20)9)2	238 **U** Uranium **92** 2)8)18)32)21)9)2	[237] **Np** Neptunium **93** 2)8)18)32)23)8)2	[242] **Pu** Plutonium **94** 2)8)18)32)24)8)2

[ ] This is the mass number of the isotope with the longest known half life of the element indicated.
* This is the mass number of the most stable or best known isotope of the element indicated.

ELEMENTS

			3	4	5	6	7	0
								4 **He** Helium 2 2
			11 **B** Boron 5 2)3	12 **C** Carbon 6 2)4	14 **N** Nitrogen 7 2)5	16 **O** Oxygen 8 2)6	19 **F** Fluorine 9 2)7	20 **Ne** Neon 10 2)8
			27 **Al** Aluminium 13 2)8)3	28 **Si** Silicon 14 2)8)4	31 **P** Phosphorus 15 2)8)5	32 **S** Sulphur 16 2)8)6	35.5 **Cl** Chlorine 17 2)8)7	40 **Ar** Argon 18 2)8)8
59 **Ni** Nickel 28 2)8)16)2	64 **Cu** Copper 29 2)8)18)1	65 **Zn** Zinc 30 2)8)18)2	70 **Ga** Gallium 31 2)8)18)3	73 **Ge** Germanium 32 2)8)18)4	75 **As** Arsenic 33 2)8)18)5	79 **Se** Selenium 34 2)8)18)6	80 **Br** Bromine 35 2)8)18)7	84 **Kr** Krypton 36 2)8)18)8
106 **Pd** Palladium 46 2)8)18)18	108 **Ag** Silver 47 2)8)18)18)1	112 **Cd** Cadmium 48 2)8)18)18)2	115 **In** Indium 49 2)8)18)18)3	119 **Sn** Tin 50 2)8)18)18)4	122 **Sb** Antimony 51 2)8)18)18)5	128 **Te** Tellurium 52 2)8)18)18)6	127 **I** Iodine 53 2)8)18)18)7	131 **Xe** Xenon 54 2)8)18)18)8
195 **Pt** Platinum 78 2)8)18)32)17)1	197 **Au** Gold 79 2)8)18)32)18)1	201 **Hg** Mercury 80 2)8)18)32)18)2	204 **Tl** Thallium 81 2)8)18)32)18)3	207 **Pb** Lead 82 2)8)18)32)18)4	209 **Bi** Bismuth 83 2)8)18)32)18)5	[210] **Po** Polonium 84 2)8)18)32)18)6	[210] **At** Astatine 85 2)8)18)32)18)7	[222] **Rn** Radon 86 2)8)18)32)18)8

152 **Eu** Europium 63 2)8)18)25)8)2	157 **Gd** Gadolinium 64 2)8)18)25)9)2	159 **Tb** Terbium 65 2)8)18)27)8)2	162.5 **Dy** Dysprosium 66 2)8)18)28)8)2	165 **Ho** Holmium 67 2)8)18)29)8)2	167 **Er** Erbium 68 2)8)18)30)8)2	169 **Tm** Thulium 69 2)8)18)31)8)2	173 **Yb** Ytterbium 70 2)8)18)32)8)2	175 **Lu** Lutetium 71 2)8)18)32)9)2
[243] **Am** Americium 95 2)8)18)32)25)8)2	[247] **Cm** Curium 96 2)8)18)32)25)9)2	[249] **Bk** Berkelium 97 2)8)18)32)27)8)2	[251] **Cf** Californium 98 2)8)18)32)28)8)2	[254] **Es** Einsteinium 99 2)8)18)32)29)8)2	[253] **Fm** Fermium 100 2)8)18)32)30)8)2	[256] **Md** Mendelevium 101 2)8)18)32)31)8)2	254* **No** Nobelium 102 2)8)18)32)32)8)2	[257] **Lw** Lawrencium 103 2)8)18)32)32)9)2

Acknowledgements

Some of the questions used in the book are reproduced by kind permission of the following examining boards:

The Associated Examining Board (A.E.B.)
University of Cambridge Local Examinations Syndicate (C.)
The Joint Matriculation Board (J.M.B.)
The Welsh Joint Education Committee (W.J.E.C.)
The Oxford and Cambridge Schools Examination Board (O. & C.)

The origin of such questions is acknowledged separately in the text.

We should also like to thank the many firms, organizations, and individuals who have so generously provided information and photographs and allowed us to quote data and other material from their own publications. In particular our thanks are due to:

Friends of the Earth Ltd (especially Chris Jordan) for allowing use of their publication 'Material Gains';
British Steel Corporation Ltd;
Imperial Chemical Industries Ltd (especially Bob Finch, Schools Liason Officer and Mr A. J. Williams, Planning Group Manager, Fertilizer Marketing Department);
The Fertilizer Manufacturers Association Ltd, especially the Director General, Mr H. S. S. Few;
Esso Petroleum Co. Ltd;
Severn Trent Water Authority, especially Christine Moseley;
Mr A. F. Maule, Senior Adviser, Sheffield and District Clean Air Committee;
Housemann (Burnham) Ltd, Permutit Standard Division;
Shell International Petroleum Co. Ltd;
Elga Products Ltd;
The Controller of Her Majesty's Stationary Office, for permission to reproduce data from 'Energy—a key resource';
Fisons Ltd, particularly Mr J. M. Lukey, Divisional Public Relations Manager, Fertilizer Division;
British Petroleum Co. Ltd;
Unilever Ltd;
The Marchon Division of Albright and Wilson Ltd;
British Nuclear Fuels Ltd;
United Kingdom Atomic Energy Authority;
British Sulphur Corporation, London;
National Sulphuric Acid Association, London.

Photographs and other material provided by these organizations have also been acknowledged separately in the text. If any names of organizations, individuals, or books have been inadvertently omitted from the above list we extend our sincere apologies.

We would also like to thank Martyn Berry, Chris Knowles and Colin Johnson for their detailed reading of the draft material and their many helpful comments, Geoff Salter for translating our ideas into such excellent cartoons, and Bob Grier for the cover photographs. Finally we would like to thank our Publishers, particularly Graham Taylor, for their advice, and also the wives and husbands who have shown so much patience during the preparation of the book.

1 The particles which make up matter (atoms, molecules, ions)

1.1 HOW DO WE KNOW THAT THESE PARTICLES EXIST?

Chemistry is mainly concerned with what substances are made of, and how and why these different substances react with each other. We now understand a great deal about the way in which chemicals react together. This understanding has allowed chemists to make new substances which are even better suited for a particular purpose than some of those found in nature. Before you can begin to understand how and why chemicals react together, you must have a simple understanding about the 'building blocks' from which *all* substances are made, i.e. atoms, molecules, and ions.

You will learn much more about these particles later. For the moment you must understand the differences between them, and how to recognize (from a formula) which type of particle we are referring to. Most of you probably believe that atoms and molecules exist, because you have heard about them from an early age. Unfortunately, it is almost impossible to *prove* that these particles exist by doing experiments at school, because the particles are so small that they cannot be seen even with the most powerful light microscope.

You may have taken part in a game, or seen an experiment, in which you had to guess what object is inside a sealed box. By tilting and shaking the box, it is possible to decide whether the object inside is round or square, hard or soft. By using a magnet, it is possible to find out whether the object inside is attracted by a magnet. These and other tests sometimes make it possible to build up a fairly good picture of what is inside the box, even though you cannot see it. In a similar way, some simple experiments which you can do will help to build up evidence for the theory that everything is made of particles. Each experiment, taken on its own, does not contribute very much to the evidence. However, the results of many experiments can only be explained satisfactorily if we believe that all substances are made up of very small particles, either atoms, molecules, or ions. Some of the other ideas referred to in Chapter 2, such as diffusion, Brownian movement, and change of state, are easy to explain if we believe everything is made of particles, but they are impossible to explain if we do not believe this.

Note: It is most important that you understand that the experiments described in this book are typical only. There may be occasions when, to illustrate a particular point, you will use a different experiment from the one we have chosen. When this does occur you will find it a valuable exercise to link your work on the alternative experiment with that given for a similar experiment in the book. A comparison of the methods used and results obtained will often lead to a better understanding of the principles involved. Although the individual experiments may differ, the conclusions reached from them will be the same.

Experiment 1.1
Two into one will go*

Apparatus
Small gas jar, spatula
Sodium chloride (common salt)

Procedure
(a) Pour water into the gas jar until it is *almost* overflowing (see Figure 1.1), i.e. the surface of the water can be seen to be above the rim of the gas jar.
(b) Guess how many spatula measures of sodium chloride you will be able to add to the gas jar before the water overflows.
(c) Carefully add a spatula measure of sodium chloride to the water in the gas jar. Continue to add sodium chloride by spatula measures until the liquid overflows. Count the number of measures of sodium chloride you add.

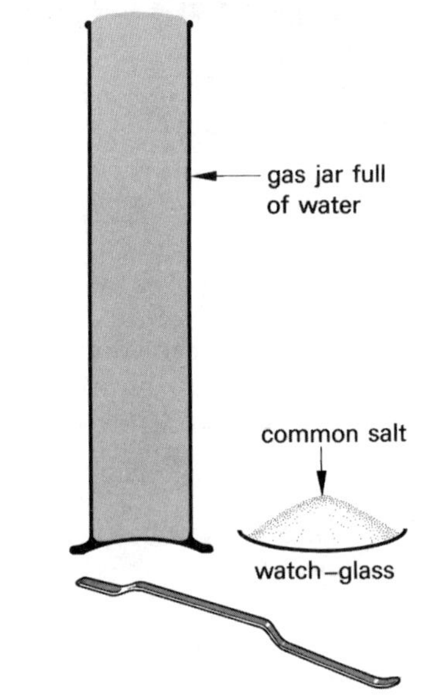

Figure 1.1 Is the gas jar as full as it appears?

Results
Write down (i) your forecast of the number of spatula measures you thought you would use, and (ii) the number used, to make the liquid overflow.

Conclusion
Read the following points. If you think carefully about them, you ought to be able to write a conclusion about the experiment, in your own words. If you believe that the water and the sodium chloride are made of very tiny particles, it is possible to explain what happened. Can you explain it in any other way?

1. If the gas jar was *really* full, the liquid would have overflowed when only a few spatula measures of sodium chloride were added. (Don't imagine that some sodium chloride 'no longer exists' when it dissolves; it is still there!)
2. If the gas jar was not really full, the water must have had some 'spaces' in it.
3. It might help you to understand how water can have 'spaces' in it if you think of a gas jar full of marbles. Although the jar is full of marbles, you could still pour rice into it.
4. Do not imagine that *crystals* of sodium chloride fit into gaps in the water. Gaps between atoms and molecules are *extremely* small, far too small for even tiny crystals to fit into.
5. Do not state that sodium chloride particles go between water particles, because it could be the other way round. It is better to state that the particles of water and sodium chloride rearrange themselves to make more use of the spaces between them.

Points for discussion
1. If everything is made of particles, we can begin to understand why solids can dissolve and why liquids can evaporate. If a solid already consists of particles, could it be that when it is added to a liquid, the liquid causes the particles to separate from each other, and this is how a solid dissolves?
2. If 25 cm^3 of water are shaken up with 25 cm^3 of ethanol, the total volume is found to be less than 50 cm^3, even though nothing has escaped. Can you explain what has happened?
3. A few cm^3 of a certain liquid, when left on a watch-glass, evaporated into the air in such a way that it spread throughout the room and could be detected by its smell in every part of the room. Why should a liquid 'come apart' in this way, spread so widely, and do so without any help from us?

1.2 THE DIFFERENT KINDS OF PARTICLE

Atoms

The 'units' which make up a chemical are rarely atoms, although this is the word which most young chemists expect to use. Free atoms are very rare in chemistry, and most substances are made up of either molecules or ions. However, it is important to know what atoms are before we discuss molecules and ions, as these are produced from atoms.

Free atoms

Imagine that you could cut a small piece of an element such as gold into two even smaller pieces. In theory, you could then cut each new piece into two, and so on, so that you would produce smaller and smaller pieces of gold. Eventually you would *in theory* produce a piece so small that it would be an atom of gold, but then you could not make that piece any smaller and still expect it to have the properties of gold.

An atom is the smallest particle into which an element can be divided without losing the properties of the element.

There are only about 100 different kinds of atoms, but all the substances around us are made from various combinations of these atoms. The differences between substances depend on which particular atoms are used to make them, how many atoms are used, and the order in which they are joined together. In a similar way, there are only 26 letters in the English alphabet but these can be grouped in different ways to make many thousands of different words. Also, you will learn that certain combinations of atoms make more 'sense' than others, in the same way that letters can form real words or 'nonsense' words.

Molecules

The word molecule is used to describe a group or cluster of atoms which are joined together. We need this special word because it is unusual for a substance to consist of single atoms. Most atoms cannot exist on their own but must join with one or more other atoms to form a group called a molecule. (Or they become ions, see p. 4.) Molecules can exist on their own.

A molecule is a group of atoms which are bonded together and which can exist as a separate unit.

The atoms in a molecule may be of the same kind, or of several different kinds. Figure 1.2 illustrates the ways in which atoms join to form molecules, spheres being used to represent different atoms. Note that the figure is not representing scale models of actual molecules, it is indicating the kinds of molecules which exist.

An atom can be given a *symbol* (e.g. Cl for a chlorine atom) and a molecule can be given a *formula*. The atoms within a molecule can be of the same type, e.g. Cl_2 is the formula for a chlorine molecule. (Note that we do not write 2Cl for a chlorine molecule. 2Cl means two free atoms of chlorine, but Cl_2 means two chlorine atoms joined together.) A molecule may contain two or more different atoms, e.g. H_2O is the formula of a water molecule. You will learn later why a molecule of a certain substance always has the same formula, e.g. why a chlorine molecule is always Cl_2 and cannot be Cl_4.

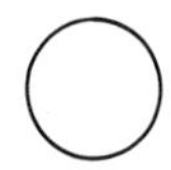

An atom of hydrogen, symbol H, which does not exist on its own at ordinary temperatures and pressures

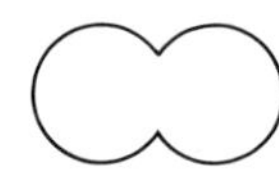

A molecule of hydrogen, formula H_2, which can exist on its own

A simple way of representing a molecule of hydrogen (i.e. two hydrogen atoms bonded together)

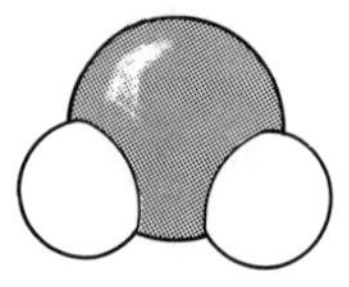

A molecule of water, formula H_2O. This molecule contains two atoms of hydrogen and one atom of oxygen, i.e. three atoms altogether. In simple form this is shown

H—O—H

A molecule of sulphuric acid, formula H_2SO_4. This molecule contains seven atoms, four of oxygen, two of hydrogen, and one of sulphur. In simple form this is shown as

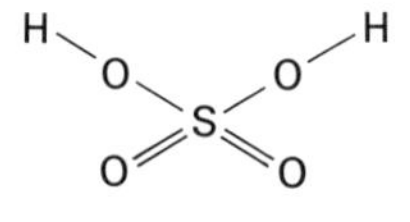

(You will understand what a 'double stick' means later)

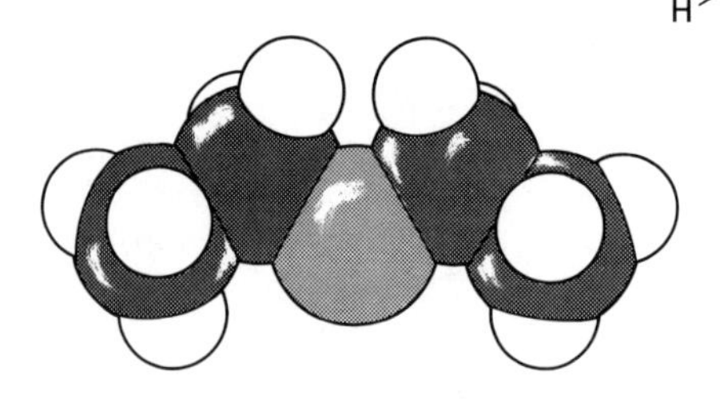

A molecule of ethoxyethane, formula $C_4H_{10}O$. This contains fifteen atoms, four of carbon, ten of hydrogen, and one of oxygen. In simple form this is shown as

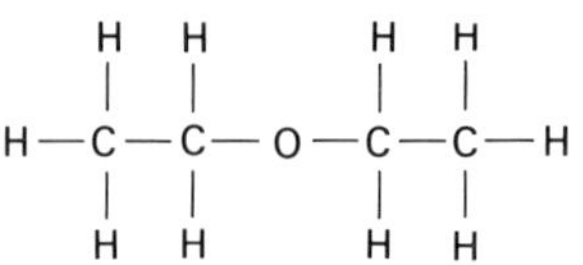

Figure 1.2 Some simple representations of atoms and molecules

Ions

You will also need to understand about a third type of particle called an ion. You will only understand ions properly when you have studied ionic bonding, but for the moment you can imagine that an ion is an electrically *charged* particle formed from an atom or group of atoms.

Atoms and molecules are always electrically neutral, but ions are either negatively charged or positively charged. The symbol for an ion always has the charge written alongside it, e.g.

Cl	Cl_2	Cl^-
chlorine atom	chlorine molecule	chloride ion
(neutral)	(neutral)	(negatively charged)

Mg	Mg^{2+}
magnesium atom	magnesium ion
(neutral)	(positively charged)

As you will learn later, the degree of charge on the ion (i.e. whether it is 1+, 2+, 3+, or 1−, 2−, 3−) depends on the *valency* of the parent atom (p. 31), and the *type* of charge (negative or positive) usually depends upon whether the ion is formed from an atom of a non-metal or from an atom of a metal.

Using the correct name for a particle

It is important to understand that each of the terms atom, molecule, and ion has a very special meaning. It is quite wrong to describe something as an atom when you really mean a molecule, and so on. (The noble gases such as neon and argon are the only substances which exist naturally as individual atoms.) Students are often very careless about using these words. As your knowledge of chemistry and structure grows, you will find it easier to use these words correctly. If in doubt, for the moment use the word *particle*, which is a general term covering atoms, molecules, and ions. For example, you may know that pieces of iron filings fizz (effervesce) when they are added to a dilute acid. You may not know, at this stage, whether the actual reacting 'pieces' of iron are atoms, molecules, or ions, so do not use these words in describing the reaction. You are perfectly safe, however, to say 'particles of iron react with the dilute acid'.

Make sure that you understand the differences between the following ways of writing symbols and formulas.

I	I_2	I^-
This means an atom of the element iodine.	This means a molecule (in this case two atoms joined together) of iodine.	This means an ion of iodine (with a single negative charge).
$2I$	$2I_2$	$2I^-$
This means two *separate* atoms of iodine.	This means two *separate* molecules of iodine.	This means two *separate* iodide ions.

Check your understanding

1. Which of the following stand for one or more separate atoms?

$I \quad Cl_2 \quad I_2 \quad Br \quad Br_2$

$2Br \quad SO_4 \quad NO_3 \quad 2NO_3 \quad 2I$

2. Which of the following are molecules?

$H_2O \quad SO_3 \quad O \quad Pb \quad Cl_2$

$Cl \quad I_2 \quad Br \quad Br_2$

3. Using the table of symbols on page 31, write the correct symbol or formula for each of the following:

(a) an atom of zinc;
(b) two atoms of copper;
(c) a molecule of nitrogen (which contains two nitrogen atoms);
(d) a molecule of carbon monoxide (which contains one carbon atom and one oxygen atom);
(e) two molecules of nitrogen.

How small are these particles?

The diameter of a typical atom is about 10^{-8} cm. Even if you understand what this means, it is almost impossible to appreciate such a very small size because you have never 'seen' such a small object, and you cannot compare it with anything.

Look at a 1.0 cm length on your ruler. If we divide this centimetre into 10 equal parts, each of the parts will have a length of $\frac{1}{10}$ cm, or 0.10 cm, or 10^{-1} cm (three different ways of saying the same thing). This particular length is commonly called a millimetre, because it is one-thousandth of a metre.

Imagine dividing one of the parts just obtained (i.e. a millimetre) again into 10 equal parts. Each new part is now so small that you would need some magnifying aid to see it properly. This tiny distance is now $\frac{1}{100}$ cm, or 0.010 cm, or 10^{-2} cm.

Imagine taking that tiny distance, and again dividing it into 10 equal parts. We would need a microscope to see each of these new lengths, which are now $\frac{1}{1000}$ cm, or 0.0010 cm, or 10^{-3} cm long.

If we continue to take one length from each division, and divide it into 10 equal small parts, we would eventually have a length about the same as the diameter of an atom. We would then have made eight of these divisions, starting from 1 cm, and the length would be $\frac{1}{100\,000\,000}$ cm or 0.000 000 01 cm or 10^{-8} cm long. Such a length is far too small to be seen by even the most powerful light microscope.

You may wonder how we can be so sure of such very small sizes, especially as we cannot even see them. In fact, it is not very difficult to show, even in a school laboratory, that particles have diameters of about this length, and you may do such an experiment later in the course. For the moment, the following comparisons may help you to appreciate just how small these particles are.

If a drop of water could be magnified to the size of the earth, then on the same scale the molecules within it would be as large as golf balls.

Two million hydrogen atoms would cover an average full stop.

In 1000 cm^3 of water there are as many atoms as there would be grains of sand if the whole of the earth's surface (continents and oceans) were covered with a layer of sand 30 cm thick.

1.3 A SUMMARY OF CHAPTER 1

The main points covered in this chapter are as follows:

1. Definitions

(a) An atom (p. 3)
(b) A molecule (p. 3)

2. Other ideas

(a) A simple understanding of what an ion is
(b) All substances are made of particles
(c) These particles are very small
(d) How symbols and formulae are used to describe atoms, molecules, and ions

2 Particles in motion: the kinetic theory

2.1 PARTICLES IN MOTION

It will be obvious from some of the ideas in Chapter 1 (e.g. the spreading out of a liquid when it evaporates) that the particles of a gas are moving. The work in this section provides answers to questions such as the following. Are the particles of all substances constantly moving? Do we have to do something to make them move? Do the particles of different substances move in different ways and at different speeds?

Diffusion

Experiment 2.1
The movement of bromine molecules in air***

Apparatus
Two gas jars, white paper, cover glass, teat pipette. Bromine. (This chemical must only be used by the teacher. Gloves and goggles must be used when working with bromine, and a fume cupboard is essential.)

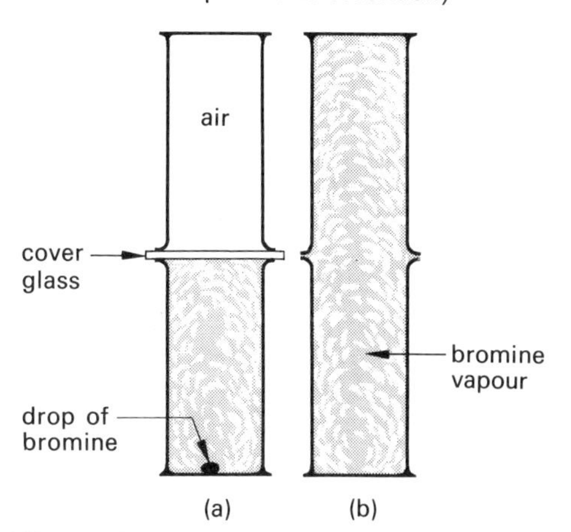

Figure 2.1 The movement of bromine particles in air (a) with cover glass separating gas jars and (b) five minutes after cover glass is removed

Procedure
(a) Using the teat pipette, place two or three drops of bromine at the bottom of a gas jar, and immediately cover the jar.
(b) Place a piece of white paper behind the jar, and observe what happens.
(c) When the gas jar is full of bromine vapour, quickly remove the cover glass and put a second gas jar on top of the first, as in Figure 2.1. Note how long it takes for the bromine vapour to fill the gas jar.
(d) Again watch what happens.

Results
Write down (a) what you saw in the first gas jar, (b) how long it took for the second jar to fill with bromine vapour, and (c) what you saw when the two gas jars were used.

Conclusion
Use words or sentences of your own to complete the following:

Liquid bromine ________ when a few drops of it are placed in a large container. This happens because bromine consists of lots of small ________, and these separate when given a chance to do so. Bromine vapour is more dense than air, but in spite of this it moved ________ until it filled the gas jar. This movement took place without any help (e.g. heating), and this suggests that the ________ of bromine are moving at room temperature.

Diffusion is the movement of particles from a region of high concentration to a region of low concentration (i.e. from a region in which there are many of the particles in a given volume, to a region where there are only a few particles in the same volume).

In this experiment, bromine particles diffused into the ________ in the gas jar. I think that bromine would have diffused (more quickly, less quickly, at the same rate—insert whichever is best) if the second gas jar had been completely empty, i.e. had contained a vacuum, because ________.

Experiment 2.2
The movement of copper(II) sulphate particles in solution**

Apparatus
Gas jar or measuring cylinder, pipette, polystyrene sphere. Concentrated copper(II) sulphate solution.

Procedure
(a) Pour copper(II) sulphate solution into the gas jar or measuring cylinder until it is about half full.
(b) Float the sphere on top of the solution, and slowly run water from the pipette on to the top of the sphere so that the water forms a separate upper layer.
(c) Remove the sphere when the container is about two-thirds full, and then fill the container by adding water more quickly, but still carefully.
(d) Leave the container in a place where it can remain undisturbed for several weeks.

Results
Describe what the liquids looked like at the beginning of the experiment, then how they appeared one week later, and also two weeks later.

Conclusion
Write your own conclusion to the experiment, making sure that you include the following points:

1. Explain what happened in terms of particles, using the scientific word diffusion. (Both the water and the copper(II) sulphate solution diffuse, and you should explain this.)
2. Does the experiment suggest that particles are moving in the liquid state? If so, how does their speed of movement compare with the movement of particles in the gas state, e.g. in the previous experiment? Explain your comments.

Brownian movement

Experiment 2.3 will enable you to understand what Brownian movement means. The effect was first *explained* by C. Wiener in 1863, but it is named after the person who first recorded it, the botanist Robert Brown. Brown made his discovery by accident in 1827, when he was looking at some pollen grains in water under a microscope.

Experiment 2.3
The smoke cell**

Apparatus
Smoke cell, microscope, low voltage d.c. supply, milk straw.
Note: the procedure may have to be modified for certain kinds of smoke cell.

Procedure
(a) Make sure that you understand the various parts of the apparatus.
(b) Carefully slide the cover from the smoke cell unit. Notice the very small 'beaker'.

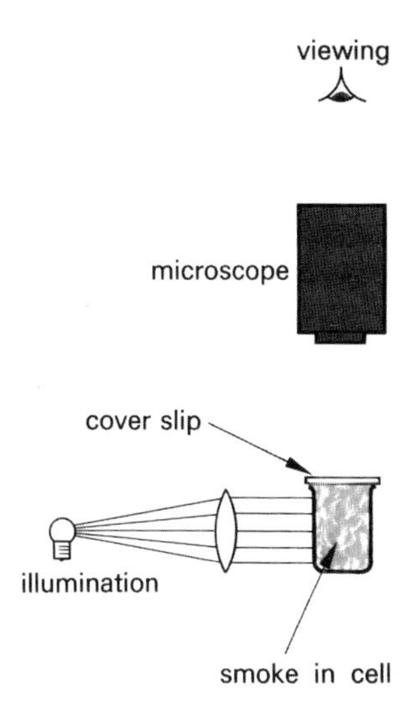

Figure 2.2 The smoke cell

(c) Light one end of the straw, and tilt the lit end downwards for a few seconds.
(d) Blow out the flame and place the opposite end of the straw just inside the top of the small 'beaker' in the smoke cell. Smoke will fall from the end of the straw and into the 'beaker'.
(e) When the 'beaker' is full of smoke remove the straw and place a cover slip over the 'beaker' to trap the smoke.
(f) Place the smoke cell unit on the stage of the microscope, with the 'beaker' below the lens.
(g) Switch on the low voltage supply to the smoke cell so that the lamp lights. Light passes through the smoke from the side of the 'beaker'. (The apparatus at this stage is shown in a simplified form in Figure 2.2.)
(h) Carefully adjust the microscope (you will be told which lens to use) until you see something moving inside the 'beaker'.
Note: (i) You must be extremely careful to make sure that you do not cause the lens of the microscope to touch the cell.
(ii) The movement you are looking for is not easy to see. Be patient; if necessary, ask for help.

Results
Try to describe in your own words what you saw under the microscope.

Notes about the experiment

You will need help to explain what happens in this experiment. Robert Brown noticed that when pollen grains which were floating on water were examined under a microscope, the pollen grains were moving about in the water in an irregular manner (see Figure 2.3). There were no currents in the water and no draughts in the room, and pollen grains cannot move by themselves, but the tiny pollen grains were nevertheless darting about in the water. If pieces larger than pollen grains were used, they no longer showed this kind of movement.

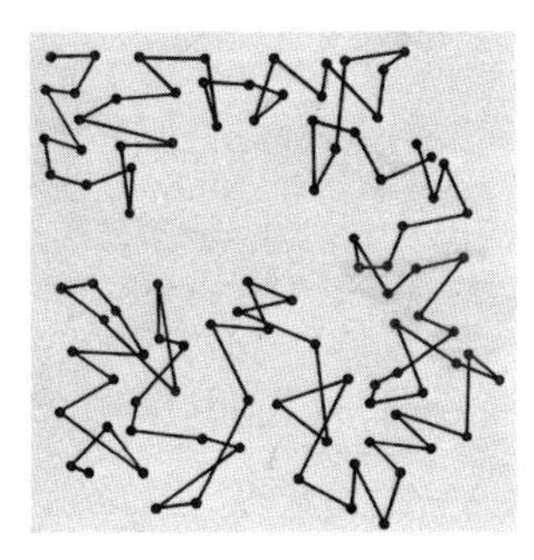

Figure 2.3 Brownian movement. This shows the path taken by a single particle suspended in water. The position of the particle was recorded at ten second intervals

We now know that this strange effect is caused by the bombardment of the pollen grains by the much smaller and rapidly moving water molecules which surround them (see Figure 2.4). Note that the figure is not to scale; a pollen grain is much, much larger than a water molecule.

Many other examples of this type of behaviour are now known. Whenever small pieces of solid, each too small to be seen with the naked eye but big enough to be seen under a microscope, are 'suspended' in a liquid or gas, they can be seen to be moving about in an irregular manner. This movement is caused by the rapid bombardment by the (invisible) liquid or gas molecules which surround the small pieces of solid, and is called Brownian movement.

In Experiment 2.3, the smoke was made of very small pieces of carbon suspended in air. The flashes of light you saw were caused by light being reflected off the pieces of carbon in the smoke. The pieces of carbon (and therefore the flashes of light) were moving in a random way because they were being bombarded by (invisible) air molecules. This experiment thus provides further evidence that the particles of gases and liquids are moving about; it is impossible to explain the observations in any other way.

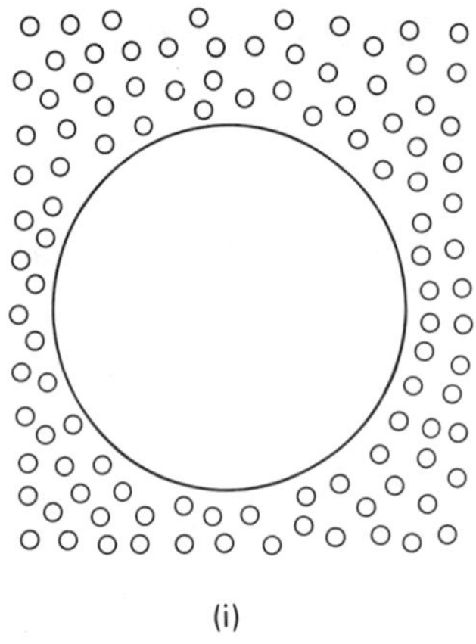

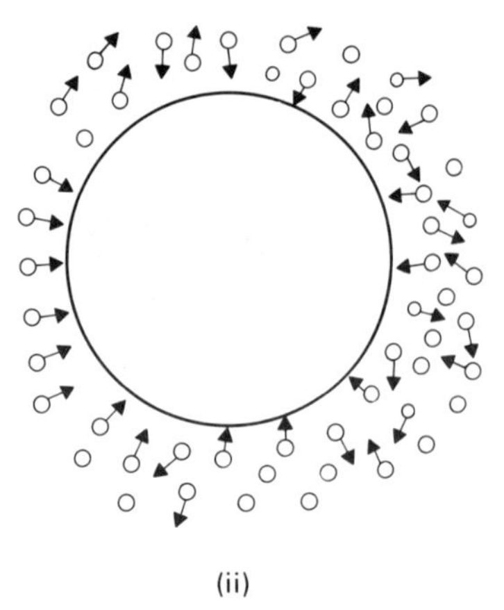

Figure 2.4 An illustration of Brownian movement. (i) Pollen grain surrounded by water molecules. (ii) At any given moment, it is unlikely that *exactly* equal numbers of water molecules hit the pollen grain on opposite sides. The pollen grain thus moves away from the greatest number of 'hits'. A split second later it is sent in another direction. This produces a constant but irregular movement as in Figure 2.3

Brownian movement is the random (irregular) movement of microscopic (i.e. very small) pieces of solid caused by irregular bombardment from the molecules which surround them in the gas or liquid state.

Conclusion
Write a conclusion to Experiment 2.3 *in your own words*, including an explanation of what you were looking at, and say why you see silvery specks and why they move. Include in your answer the term Brownian movement. Remember that you did not see any actual molecules in the experiment, but the observations are difficult to explain unless you believe that the invisible air molecules are constantly moving. Can you also explain why Brownian movement only works if the solid pieces are very small?

Comparing the movement of particles in liquids and gases

Experiments 2.1, 2.2, and 2.3 have suggested that the particles in liquids and gases are constantly moving. They also suggest that particles move more rapidly in a gas than in a liquid, because bromine vapour diffused much more quickly than the copper(II) sulphate solution.

Your next question could be 'how fast do the particles in gases move?', or 'do all gases diffuse at the same speed?'. The next two experiments will help to answer questions such as these.

Experiment 2.4
To compare the rate of movement of two gases***

Apparatus
Glass tube about 90 cm long and 2.5 cm diameter, supported horizontally by two retort stands, two rubber bungs to fit the tube (Figure 2.5), two test-tubes (150 × 25 mm), two test-tube stands, cotton wool.

Concentrated hydrochloric acid, concentrated ammonia solution.

Procedure
(a) Almost fill one test-tube with the acid and stand it at one end of the bench. Make a small ball of cotton wool, and using tweezers dip it in the acid. Remove the cotton wool and hold it over the test-tube so that the excess acid drips back into the tube. When it has stopped dripping, insert the cotton wool into one end of the long tube and stopper it with a bung.

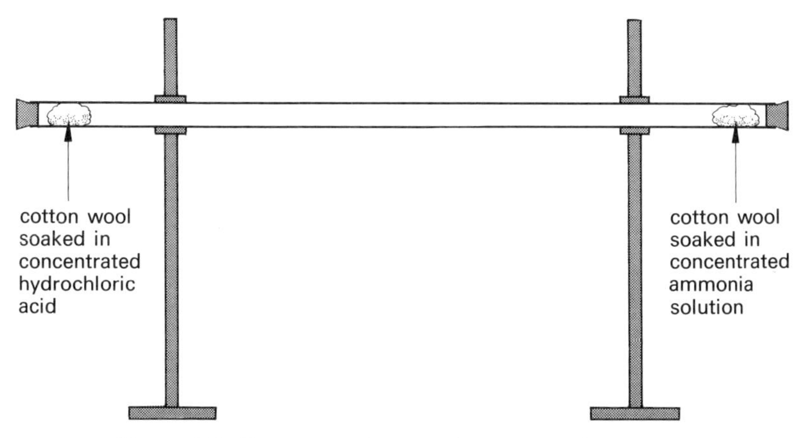

Figure 2.5 To compare the rate of movement of two gases

(b) Repeat the procedure as quickly as possible with another ball of cotton wool, and another pair of tweezers, but dipping the cotton wool this time into the *other* test-tube which has been almost filled with ammonia solution and placed at the other end of the bench. Insert the ammonia-soaked piece of cotton wool into the *other* end of the tube, and stopper with another bung. (It is an advantage to insert both pieces of cotton wool at the same time, so that both gases start to diffuse at the same instant.) Stopper the two test tubes.
(c) Leave the tube undisturbed for some time, but watch for any developments. Note the time taken for any change to occur.

Results
Write down what you see happen inside the tube, how long it takes before you see anything happen, and how far each gas has moved.

Notes about the experiment

(i) The liquids themselves are not important in this experiment; it is the gases which come from them that we are concerned with. Concentrated ammonia solution is a concentrated solution of the *gas* ammonia (NH_3) dissolved in water, and ammonia gas is constantly escaping from the solution. Similarly, concentrated hydrochloric acid is a concentrated solution of the *gas* hydrogen chloride (HCl) in water, and hydrogen chloride gas is constantly escaping from the solution. When these two gases meet a chemical reaction takes place, as shown by the equation:

$$NH_3(g) + HCl(g) \rightarrow NH_4Cl(s)$$

(See notes about equations and state symbols on page 12.) The product of the reaction is called ammonium chloride and this forms as tiny pieces of white solid in the air, which looks like a white smoke. In the experiment the two gases diffuse along the tube (why?) and the point where they meet will be marked by the formation of a ring of white smoke.
(ii) If both gases diffuse at the same rate they will meet in the middle of the tube. Where did the gases meet, and what does this tell you about their rates of diffusion?
(iii) Ammonia gas (relative molecular mass 17, see p. 301) is less dense than hydrogen chloride gas (relative molecular mass 36.5); how does the relative molecular mass of a gas affect its rate of diffusion?
(iv) The experiment also shows another important point. It took some time before the gases met, and from this you might conclude that gas molecules do not move particularly quickly. In actual fact, molecules of ammonia are known to move at an

Ammonia gas is constantly escaping from concentrated ammonia solution

average speed of 550 metres per second at 0 °C (i.e. about 1200 mph) and molecules of hydrogen chloride also move very quickly. How is it, then, that gas molecules travelling at hundreds of metres per second take several minutes before they 'travel' 0.5 metres or so and meet? To answer the question, we need to imagine what is happening as the molecules move along the tube. In the tube there are also millions of molecules of air, moving in all directions at enormous speeds, constantly bumping into each other and rebounding. Each ammonia molecule, for example, is constantly bumping into other ammonia molecules and into molecules of air as it moves down the tube. Sometimes it moves backwards, sometimes sideways, but always at great speeds. Although each molecule moves very quickly, it does not make much progress down the tube because of the collisions. This is rather like having to make your way down the subway of a tube station in the rush hour in the opposite direction to the majority of the crowd of people. An average ammonia molecule *will* travel 550 metres in every second (at 0 °C), but not in a straight line.

It has been calculated that in air at room temperature, each molecule collides with another molecule about 10^9 times in every second. This is an enormous number of collisions, and it is easy to understand how gas molecules do not make progress as quickly as we might imagine from their actual speeds. Make sure that you understand why gases diffuse more slowly than would be expected from the speed of their molecules.

Conclusion

Write a conclusion to the experiment, using the following questions as a guide. Do not just answer the questions.

(a) Which gases were diffusing along the tube?
(b) What happens when these gases meet? (Give an equation.)
(c) Where did they meet?
(d) Do all gases diffuse at the same speed? If not, which gas diffused more quickly in this experiment? Is this gas the less dense of the two?
(e) Gas molecules actually move at great speeds, so why did it take several minutes before they met?

Points for discussion

1. Knowing the time it took for the gases to meet (in Experiment 2.4) and also the distance that each gas had moved, calculate the apparent speed of each gas in metres per second.
2. Why is it important to use the word 'apparent' in the first question?

Some notes about equations and state symbols

The equation used in Experiment 2.4 is the first equation in the book, and so a brief explanation may be helpful. For the moment, think of equations simply as the chemist's way of summarizing a reaction by using symbols and formulae for the chemicals taking part in the reaction. The equation for the reaction in Experiment 2.4 summarizes the fact that each ammonia molecule which reacts combines with one molecule of hydrogen chloride to form one unit of a compound called ammonium chloride.

Not all chemicals react in 'equal proportions'. For example, when hydrogen and oxygen combine to form steam, two molecules of hydrogen react with one molecule of oxygen to form one molecule of steam, and we write:

$$2H_2(g) + O_2(g) \rightarrow 2H_2O(g)$$

At this stage, you will not understand fully why chemicals react as they do, and you will need to learn the equations you are given. Putting the correct numbers in front of formulae in an equation is called *balancing an equation*. The equation used in Experiment 2.4 is already balanced when one unit of each substance is used in the equation, and so no further numbers are required in front of the formulae. A properly balanced equation provides a great deal of information, and you will learn how to balance and work out equations later. For the moment, remember that any numbers in a correctly balanced equation are important, and the equation is quite wrong if other numbers are inserted in it.

You will have noticed that the symbols (g) and (s) were added to the formulae in the equation for Experiment 2.4, and in the equation for the reaction of hydrogen and oxygen. These symbols are called *state symbols* and they are used to describe whether the chemicals taking part in the reaction are solids, gases, pure liquids, or in solution. There are four state symbols, and the appropriate one is placed *after* the formula of the substance it refers to:

(g) means that the substance is in the gas state in this particular reaction;
(l) means that the substance is in the liquid state in this reaction;
(aq) means that the substance is dissolved in water, i.e. in *aqueous* solution, in this reaction; and
(s) means that the substance is in the solid state in this reaction.

Many substances can be used or produced in all four 'forms', i.e. as a solid, or as a liquid (when the solid is melted), or as a gas (when the liquid evaporates) or when dissolved in water. (Note that many students confuse a pure liquid with a solution; they are *not* the same thing.) Different forms of the same substance often react in different ways, and so state symbols are important; they tell us the form in which each substance is reacting.

Experiment 2.5
Comparing the rates of diffusion of hydrogen and air***

Apparatus
Demonstration apparatus as in Figure 2.6 (the gas jar can be replaced by a suitable beaker, if more convenient), hydrogen generator or cylinder.

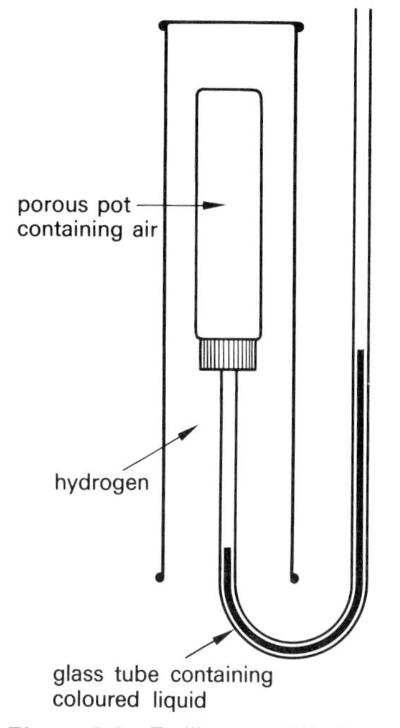

Figure 2.6 To illustrate diffusion

Procedure
(a) Collect a sample of hydrogen by upward delivery into a beaker or small gas jar, and quickly place it over the porous pot as in Figure 2.6. Note any changes. (A porous pot allows gases to move through it, and liquids to seep through it.)
(b) When no further change takes place, remove the gas jar or beaker and again note what happens.

Results
Write down what happened to the liquid levels after both procedures (a) and (b).

Notes about the experiment
(i) At the beginning of the experiment the porous pot was full of air. When the sample of hydrogen was placed over it, we had a region in which there was a high concentration of hydrogen and a low concentration of air (the gas jar or beaker), next to a region in which there was a high concentration of air and a low concentration of hydrogen (inside the porous pot). Gases can move through the sides of the porous pot. From what you know of diffusion, and given that the density of air is greater than that of hydrogen, what will happen?

(ii) If the hydrogen diffuses into the porous pot at the same rate as the air diffuses out, there will still be the same number of gas molecules per cm^3 inside the porous pot as there are gas molecules per cm^3 outside the pot. If this happened, the pressure would be the same as before (i.e. ordinary air pressure) and the liquid levels would not change. If the two gases diffuse at different rates, more gas will enter the pot than leaves it (or the other way round) and there will be for a short time more molecules of gas per cm^3 on one side of the porous pot than on the other. This will cause the pressure to increase on one side of the pot. In order to return the pressure to 'normal' the molecules on this side spread out more to make the number of molecules per cm^3 the same as it was before. The liquid level therefore moves down on the side which briefly had a higher pressure, so that the increased number of molecules are no more concentrated than they were before. From what you saw in the experiment, which gas won the diffusion race?

(iii) In order to explain what happened when the gas jar was removed, you must consider what would be inside the porous pot and what would be outside the porous pot just after the removal of the gas jar.

Conclusion
Explain what happened in the experiment in your own words, including a statement comparing the rates of diffusion of air and hydrogen.

Points for discussion
Hydrogen is the least dense of all gases. Remembering what you learned in Experiment 2.4, what do you think would happen if Experiment 2.5 was repeated but using carbon dioxide (a gas which is more dense than air) instead of hydrogen? It would also be necessary to modify the apparatus slightly; can you suggest how and why?

2.2 THE KINETIC THEORY

The kinetic theory as a model for the behaviour of particles

Types of particle movement
In order to move, particles must have energy, and this 'energy of movement' is called *kinetic energy*; particles which move about very quickly have a large amount of kinetic energy. The movement shown by a particle depends upon two opposite effects. It is moving because of its kinetic energy, but at the same time there are forces *between* particles which have the effect of holding the particles together. In a solid, the forces between the particles are not completely overcome by the kinetic energy of the particles, and so the particles of a solid can vibrate but do not move from place to place. If the solid is heated until it becomes a liquid, the particles have more kinetic energy and they can now move around from place to place, but they are still not completely free from the forces which hold them together. In a gas the particles have so much kinetic energy that they almost completely overcome the forces trying to hold them together, and they separate from each other and are free to move from place to place, in any direction. These ideas are summarized in Figure 2.7.

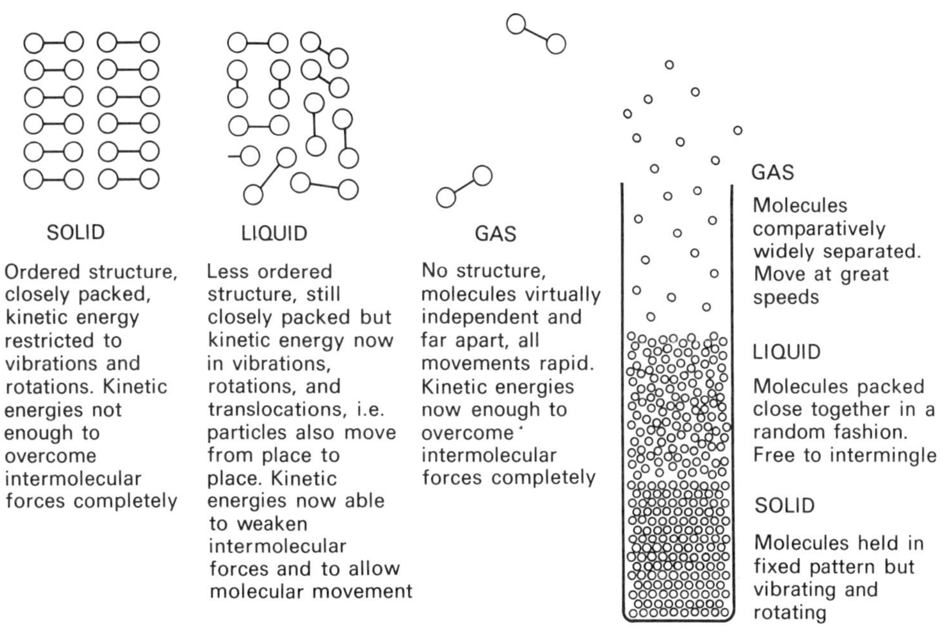

Figure 2.7 The states of matter

Change of state

Most substances can exist in three states of matter, as a gas, a liquid, or a solid. These states are connected by the following physical changes.

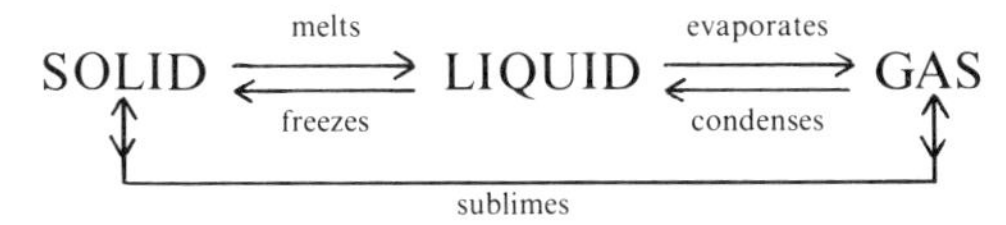

In a solid the particles are packed closely together and are held fairly tightly together. The particles in the solid do not move from place to place, they only vibrate. If the solid is heated, the particles gain more kinetic energy and their vibrations increase until eventually they have enough energy to escape from each other. The particles then move from place to place, and the solid melts as the particles come apart from each other.

In the liquid the particles are still fairly close together, but they now have more kinetic energy than they had in the solid state and are able to move about within the liquid. It is easy to understand how the liquid form of water can be so different from the solid form of water, ice, even though both consist of exactly the same particles, i.e. molecules of H_2O. It is not possible to pour a block of ice, but water can be poured because its molecules are free to move. Similarly, a block of ice keeps its own shape because its molecules are 'locked together'; water has no shape, it simply takes up the shape of whatever it is put in because its molecules are free to move.

If a liquid is given heat energy, the kinetic energy of its particles increases and some escape from the surface of the liquid as vapour. We say that the liquid is evaporating. It is important to realize that even a cold liquid evaporates, i.e. starts to change into the gas state, long before it is hot enough to boil. Many students think that liquids only evaporate at their boiling point, but the 'disappearance' of a puddle of water on a hot day is not due to the water boiling. However, the rate of evaporation of a liquid increases with temperature, so that at low temperatures the rate of evaporation is very slow.

All of the particles in a substance do not have the same kinetic energy. Some particles always have more kinetic energy than others. Some of the fast particles in a liquid, for example, have so much kinetic energy that they can escape from the liquid as vapour. As the temperature is increased, more molecules gain enough kinetic energy to escape in this way, and the rate of evaporation increases.

Evaporation occurs only at the *surface* of a liquid. The boiling point is a special situation, as explained later, in which liquid changes to vapour *inside* the liquid as well as at the surface. Each liquid has its own temperature (at ordinary pressure) at which this special situation is reached, i.e. its own boiling point, but it evaporates at temperatures below this point.

In a gas, molecules have sufficient kinetic energy to overcome the forces which attract them to each other and are moving very quickly indeed. They are very far apart compared with the distances between particles in a solid or liquid. A big change of volume therefore takes place when a liquid changes to a gas, e.g. one kettleful of water will produce about 1700 kettlesful of steam. It is easy to understand why we cannot see the gas form of water, and why a gas spreads quickly to 'fill' a room.

Absolute zero and the Kelvin scale

We have seen that when a substance is heated the kinetic energy of its particles increases and they move more quickly. On the other hand, when a substance cools, the particles lose kinetic energy and their movement slows down. If this cooling is continued a substance will reach a state when its particles will have lost *all* of their kinetic energy and are no longer moving. No more energy can be given out and so the substance cannot lose any more heat energy. It has reached the lowest possible temperature. This temperature, which is −273 °C, is called *absolute* zero, and it forms the lowest point on a scale of temperature known as the Kelvin or Absolute scale. 0 K is −273 °C. To convert °C to K, simply add 273. For example, 20 °C is 293 K.

Note that a change of temperature from 20 °C to 40 °C is not doubling the temperature, for this is only a change of temperature from 293 K to 313 K.

Check your understanding

1. What is the temperature on the Kelvin scale of the following: 100 °C, 273 °C, 0 °C, −100 °C, −273 °C?
2. What is the temperature in °C of the following: 303 K, 273 K, 100 K, 0 K?

Points for discussion

1. Have you ever seen steam? Explain your answer.
2. Why are gases so easily compressible whereas it is practically impossible to compress a liquid or a solid?

Applications of the kinetic theory

Gas pressure

You will have used the idea of gas pressure in daily life, e.g. in checking the air pressure in a bicycle tyre. You can now understand what causes this pressure. If a gas is trapped inside a container, the molecules of the gas, as they move around, are constantly hitting the walls of the container, and these hits cause the pressure.

For example, the wall of a paper bag which is being inflated is being bombarded by air molecules both inside and outside the bag. However, during the inflation, more molecules are hitting a certain area per second from the inside than are hitting the same area from the outside (see Figure 2.8) because the air inside is more 'concentrated'. The pressure is thus greater inside than outside for a short period, and the bag 'expands' until the pressures inside and outside are again the same. The expansion allows the pressure inside the bag (which has suddenly increased) to fall by giving the molecules of air more room to move around in. The expansion continues until the molecules make the same number of hits per cm^2 on the inside of the bag as they do on the outside. When the bag stops 'expanding' the pressures inside and outside are again equal.

Points for discussion

Explain what is wrong with each of the following statements:

1. There are more molecules of air inside a balloon than there are outside the balloon whilst it is being inflated.
2. An inflated paper bag has a greater pressure inside it than outside it.

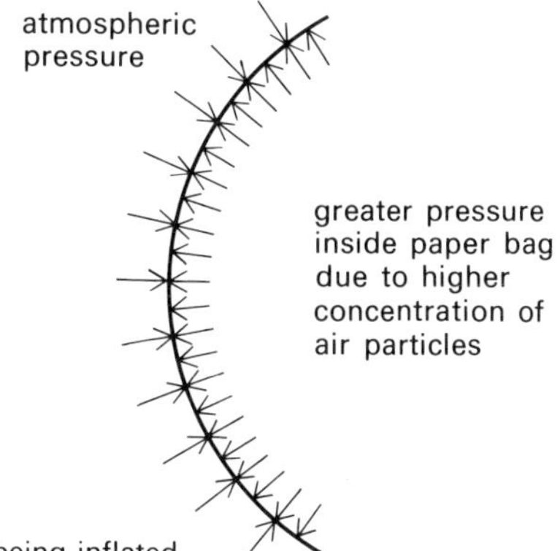

Figure 2.8 Why a paper bag expands whilst it is being inflated

The effect of temperature changes on gas pressure

You may have noticed that when the velocity of hydrogen molecules was quoted earlier, it was given at 0 °C. It is important to state the temperature at which the velocity is measured, because a change of temperature can make a big difference to the velocity of gas molecules (see Table 2.1).

If the temperature of a gas is increased, the gas molecules move more rapidly and this means they will also collide with the walls of their container more often and with more force. An increase in temperature of a gas in a closed container thus results in an increase in pressure. There are many simple examples of this effect, such as when an inflated balloon is left too near a fire, or when the pressure of a car tyre increases on a hot day.

In 1782 J. A. Charles discovered the law which tells us exactly how much the pressure is affected by a change in temperature.

Charles' Law states that the volume of a fixed mass of gas is directly proportional to its temperature (on the Kelvin scale) if the pressure remains constant.

It can also be shown that the pressure of a fixed mass of gas is directly proportional to its temperature (on the Kelvin scale), i.e. $P \propto T$, providing that T is measured in K, not °C. This means, for example, that if the Kelvin temperature of a gas is doubled, the pressure of the gas will be doubled providing the volume is kept constant.

Table 2.1 The effect of temperature on the average velocity of molecules of gases

	0 °C	*100 °C*	*820 °C*
Hydrogen	1823	2430	3626
Oxygen	460	538	920
Nitrogen	493	576	986
Carbon dioxide	390	456	780

Velocity in metres per second

The effect of volume changes on gas pressure

If the volume of a gas is halved, the same number of molecules will be contained in a smaller volume. The molecules are then more concentrated and make twice as many collisions with the walls of the container as before, so that the pressure is doubled (see Figure 2.9). Similarly, if the volume increases the pressure decreases. This fact was first discovered by Robert Boyle in 1662.

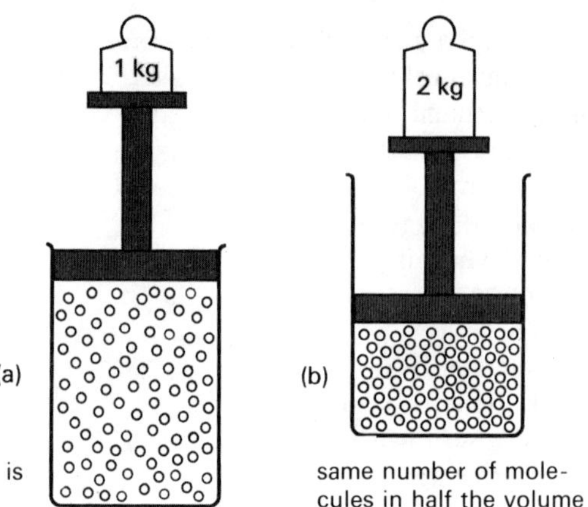

Figure 2.9 Effect on pressure when volume is decreased

Boyle's Law states that the volume of a fixed mass of gas is inversely proportional to the pressure, providing the temperature remains constant.

Thus $P \propto \frac{1}{V}$. This means, for example, that if the pressure is doubled the volume is halved.

Remember, if the temperature goes UP the pressure goes UP (if volume is kept constant), but if volume goes UP the pressure goes DOWN (if temperature is kept constant).

The liquefaction of gases

The main differences between the gas state and the liquid state are that in the gas state particles have more kinetic energy and that they are also farther apart. All gases can be changed into liquids, but some of them only with great difficulty. To change a gas into a liquid, it must be cooled down (which reduces the kinetic energy of the particles so that they become closer together, like the particles in a liquid) and/or compressed (which brings the particles closer together, as they are in a liquid). Some gases can be liquefied by cooling alone (e.g. steam) or by applying pressure at room temperature (e.g. sulphur dioxide), but some need to be both cooled and compressed before they change to a liquid, e.g. air.

Boiling point

When a pure liquid is heated its temperature rises until it reaches the boiling point. After this the temperature of the liquid will not rise any further (unless the pressure is changed) no matter how much heat energy is given to it. Each pure liquid has its own boiling point, and this varies from one liquid to another.

The boiling point of a liquid is the temperature at which its vapour pressure (i.e. the pressure caused by its vapour) is equal to the atmospheric pressure.

At this temperature, vapour can be formed *inside the liquid* as well as at the surface, so that bubbles of vapour appear within the liquid and cause the bubbling appearance we associate with boiling.

You can also think of the boiling point of a liquid as the point at which its molecules cannot receive any further energy (e.g. heat) without the liquid changing into a gas. The heat energy being supplied at the boiling point is being used to separate the liquid molecules from each other so that they can enter the gas state, but is *not* being used to raise their temperature. Energy is always needed to make molecules separate from each other.

You have probably felt the effect of a drop of ethoxyethane (ether) evaporating on the back of your hand. It feels very cold. This is because the particles of ethoxyethane need energy in order to separate from each other and evaporate. In this case the energy is the heat taken from your skin. Energy is always needed to make some of the particles separate from each other and escape from the surface of the liquid, as they do when the liquid evaporates. As this effect happens *throughout* a liquid when it boils, it requires a lot of energy to change a liquid into gas at its boiling point.

When a gas condenses into a liquid, the opposite happens. As the molecules of a gas change into the liquid state, the molecules come close together again and give back the energy which was needed to separate them. This is why a scald from steam is more serious than a scald from boiling water, even though both are at 100 °C. In effect, the steam scalds you twice. It scalds because it is at 100 °C, and it also scalds because extra heat energy is given out as it condenses on your skin. This 'hidden heat' or extra heat is called *latent heat*, and you may study this further in your Physics lessons.

The effect of air pressure on boiling point

Imagine a beaker of water being heated until it boils. As explained earlier, it boils when its vapour pressure is equal to the atmospheric pressure. If the pressure of the air above the beaker is higher than normal, then the molecules of water have to be heated to a higher temperature before their vapour pressure is equal to the atmospheric pressure. This is why the boiling point of a liquid depends upon the atmospheric pressure; if the atmospheric pressure is increased, so is the boiling point.

The following experiments can be used to demonstrate this point.

Experiment 2.6
The effect of air pressure on boiling point***

Apparatus

Set up the apparatus as shown in Figure 2.10. The thermometer should read to 110 °C. Pieces of broken porcelain or anti-bumping granules should be added to the water in the flask. All connections should be made with pressure tubing, and the bungs must fit tightly. The apparatus should be surrounded by a safety screen.

Procedure

(a) Heat the water until it boils. When the temperature of the boiling water is steady, record it, and also note the atmospheric pressure.
(b) Remove the Bunsen flame and connect the pump to the apparatus. Use the pump to reduce the pressure by about 3 cm of mercury.
(c) Again boil the water in the flask and note the steady temperature and the difference in the levels of the mercury in the manometer.
(d) Continue to reduce the pressure in the apparatus in stages of about 3 cm and note the boiling point and pressure difference each time. Four readings at reduced pressure are sufficient.
(e) Plot a graph of 'boiling point of water' against 'pressure'.

Results

(a) Record the results in a table.
(b) Fasten the graph into your notebook.

Notes about the experiment

(i) The shield was placed between the flask and the bottle so that heat energy from the Bunsen does not reach the bottle. If this did happen, the heat energy would further change the pressure of the gas.
(ii) The filter pump is used to remove some of the air from the apparatus, i.e. to reduce the number of molecules of air present and so to decrease the pressure.

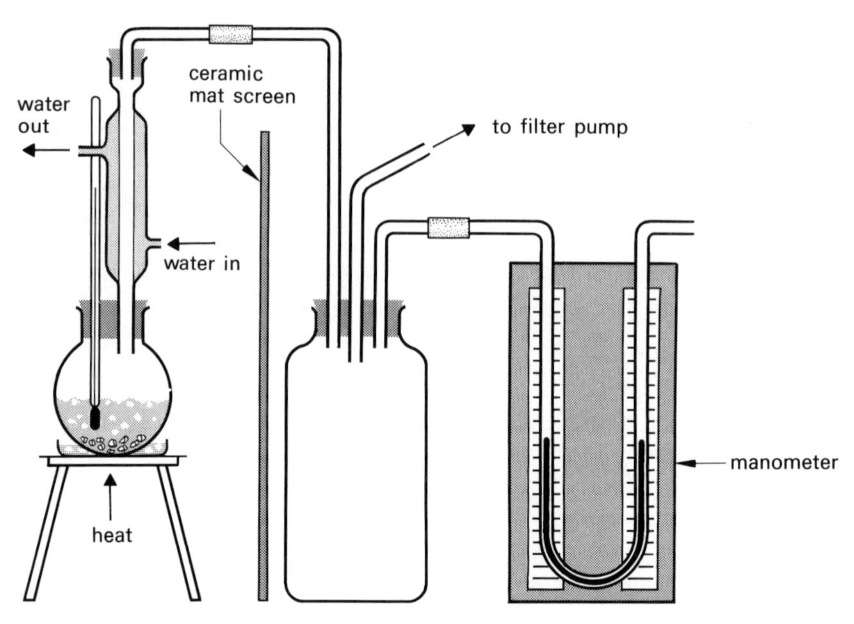

Figure 2.10 The effect of air pressure on the boiling point of water

(iii) The large bottle is there so that the pressure changes can be easily controlled (and made in small steps) by using the pump. If the pump had been connected directly to the condenser and heating flask, there would be a *large* decrease in pressure each time the pump was used. This would happen because there is only a small volume of air in the flask, and even if only a little of this was removed by the pump, there would still be a large reduction in pressure.

Conclusion
State how (not why) the boiling point of a liquid is affected by air pressure.

Points for discussion

1. Why does water boil at about 70 °C near the top of Mount Everest?
2. How do you think a pressure cooker cooks food more quickly than it can be cooked by using boiling water in an ordinary open pan?

Experiment 2.7
The mysterious flask of water***

Apparatus
As in Figure 2.11(a), plus a cloth and a safety screen.

Procedure
(a) Set up the apparatus as in Figure 2.11(a), with the clip open. Place the safety screen around the flask.
(b) Heat the flask until the water boils. Note the temperature.
(c) Allow steam to escape through the rubber tube for a minute or so, then quickly close the clip and *immediately* turn off the Bunsen burner.
(d) Turn the flask over and re-clamp it as in Figure 2.11(b).
(e) Soak the cloth in cold water and hold the cloth on 'top' of the flask. Observe what happens inside the flask and note the temperature of the water.
(f) Remove the cloth, again soak it in cold water, and repeat step (e).
(g) Repeat this stage for a third time.
(h) Turn the flask over to the original position, open the clip and listen carefully.

Results
Write down what you saw, and the temperature of the water, each time the cloth was used. Also record any other observations.

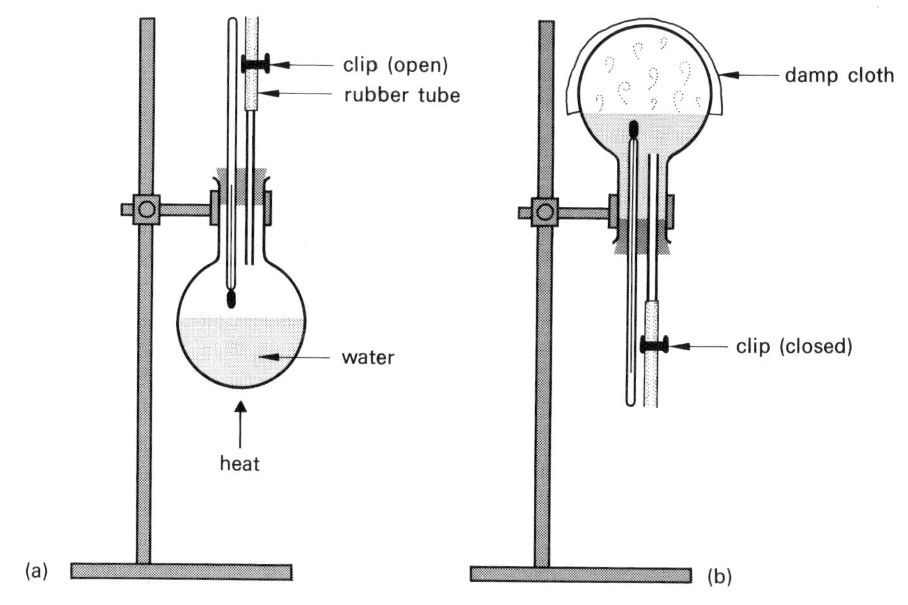

Figure 2.11 The mysterious flask of water

Notes about the experiment

You may need some help before you can explain what happened. It should help if you think about the following questions:

1. What was in the space above the water when the clip was closed?
2. What would happen to this substance when the cold cloth cooled down the glass around it?
3. When a gas changes to a liquid, it takes up much less volume. What will happen to the gas pressure in the space above the liquid when the cloth is used?
4. As the water is still hot when the cloth is applied (e.g. about 97 °C), what effect will the pressure change have on the water?
5. Why can the effect be repeated several times?
6. Can you explain what happened in procedure (h)?

Conclusion

If you have understood all the work on kinetic theory, the conclusion to Experiment 2.6, and have answered the questions above, you ought to be able to write a conclusion to Experiment 2.7 in your own words.

2.3 A SUMMARY OF CHAPTER 2

The following 'check list' should help you to organize the work for revision.

1. Definitions

(a) Diffusion (p. 8)
(b) Brownian movement (p. 10)
(c) Charles' Law (p. 17)
(d) Boyle's Law (p. 18)
(e) Boiling point (p. 18)

2. Other ideas

(a) At all temperatures above Absolute zero (−273 °C), molecules, atoms, and ions are moving; they have kinetic energy.
(b) Gases diffuse more quickly than liquids (e.g. bromine vapour diffuses more rapidly than copper(II) sulphate solution).

(c) Diffusion, change of state, and Brownian movement all suggest that particles have energy and are moving.
(d) All gases diffuse rapidly, but some diffuse more rapidly than others. The rate of diffusion depends on the density of the gas, e.g. ammonia gas diffuses more rapidly than hydrogen chloride, and hydrogen more rapidly than air.
(e) Gas molecules move at high speeds but only make relatively slow progress because of the enormous number of collisions they make.
(f) The Kelvin scale or Absolute scale as a scale of temperature. To change °C into K, add 273.
(g) The three states of matter, solids, liquids, and gases. The changes which take place (in terms of the kinetic energy of the particles and their distance apart) when a solid is heated until it melts and eventually boils. How the states of matter are interconnected, e.g. by temperature changes and (in the case of gases), by pressure changes.
(h) The difference between evaporation and boiling. A liquid evaporates at any temperature, but at a particular pressure it only boils at one temperature.
(i) Gas pressure is caused by gas molecules hitting the walls of a container. Pressure is increased by a rise in temperature, and by a decrease in volume.
(j) The boiling point of a substance depends upon the atmospheric pressure; the higher the air pressure, the higher the boiling point.

3. *Important experiments*

(a) An experiment to show diffusion in gases (e.g. Experiment 2.1).
(b) An experiment to show diffusion in liquids (e.g. Experiment 2.2).
(c) An experiment to show Brownian movement (e.g. Experiment 2.3).
(d) An experiment to compare the rates of diffusion of two gases (e.g. Experiment 2.4 or 2.5).
(e) An experiment to show how boiling points depend upon atmospheric pressure (e.g. Experiment 2.6 or 2.7).

QUESTIONS

Multiple choice and short answer questions

1. The table below gives the melting and boiling points of five pure substances.

Substance	*Melting point (°C)*	*Boiling point (°C)*
A	−100	−35
B	−7	58
C	−6	225
D	44	280
E	328	1751

Which substance, A, B, C, D, or E, is a volatile liquid at room temperature? (C.)

2. When fine chalk is suspended in water and viewed through a microscope, the chalk particles appear to move in a random fashion. This motion is the result of

A bombardment of water molecules by chalk particles
B bombardment of chalk particles by water molecules
C chemical reaction between chalk particles and water molecules
D a change in pressure caused by the addition of chalk particles to the water
E convection currents in the water (C.)

3. A crystal of the purple compound potassium manganate(VII) is placed at the bottom of a beaker of water and the beaker left until there is no further change. One may then observe a

A uniformly purple solution
B colourless liquid with the purple crystal unchanged
C purple layer below a colourless layer
D colourless layer below a purple layer
E deep purple layer below a pale purple layer (C.)

4. Which one of the following is an example of Brownian motion?

A the slow settling of a precipitate
B the movement of coloured substances in paper chromatography

C the movement of smoke particles in a container of still air
D the bubbling in a heated liquid
E the passage of a gas through a porous pot (C.)

5. An increase in temperature causes an increase in the pressure of a gas because

A it increases the mass of the molecules
B it causes the molecules to combine together
C it decreases the average velocity of the molecules
D it increases the average velocity of the molecules
E it increases the number of collisions between the molecules

6. A liquid Z was found to have a boiling point of 75 °C at 760 mm pressure, and of 85 °C at 800 mm pressure. At a pressure of 820 mm the boiling point of Z is likely to be

A 86 °C
B 88 °C
C 90 °C
D 92 °C
E 94 °C

7. If an experiment is conducted using the apparatus shown in Figure 2.6, but with nitrogen in the beaker and carbon monoxide in the porous pot, it is found that the water level does not change. Can you suggest an explanation for this?

Questions requiring longer answers

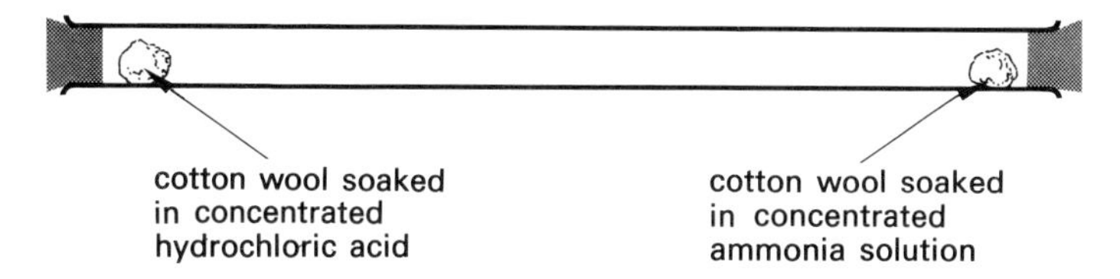

8. (a) Two pieces of cotton wool, one soaked in concentrated hydrochloric acid and the other in concentrated ammonia solution, were put at the ends of a glass tube of length 2 m which was then stoppered, as shown.
(i) Describe what you would see happen in the tube.
(ii) Give reasons for your observations.
What measurements would you make to determine the apparent velocity in metres per second of an ammonia molecule? Such experiments give results of approximately 0.001 metres per second whilst the velocity of an ammonia molecule in a vacuum is approximately 600 metres per second.
Suggest a reason for the difference.

(b) Suggest a method, based on diffusion, by which a gas jar containing hydrogen could be distinguished from one containing carbon dioxide. State what you would observe, giving reasons.

(c) Water is carefully placed on top of a blue solution of copper(II) sulphate to form two layers. After some days the solution is uniformly coloured. Suggest a reason for this, and comment on the time involved. (A.E.B. 1979)

9. A strong, round-bottom flask was half filled with water which was then boiled vigorously for five minutes. Heating was discontinued and the flask was closed with a well-fitting rubber stopper. The flask was inverted and cold water was poured on the outside. (a) What would be seen in the flask? (b) Give the reasons for your expected observations. (J.M.B.)

3 Elements, compounds, and mixtures
Symbols and formulae

Over three million pure substances have now been discovered. Try to imagine what it would be like to have to learn about three million different chemical substances without any way of classifying them, i.e. sorting them out into groups. It would be a frightening and certainly an almost impossible task. Fortunately we are able to break down this enormous number of substances by classifying them into various groups. One of the simplest classifying systems used in chemistry is the division into elements, compounds, and mixtures. These particular groupings do not directly help us to simplify the problem of studying three million chemicals, but they are important in helping us to understand the classification systems you will use later, such as the Periodic Table, the Activity Series, and the use of homologous series in organic chemistry.

3.1 ELEMENTS

Copper(II) sulphate is a substance which can be split up into substances simpler than itself, one of which is copper. For example, if electricity is passed through a solution of copper(II) sulphate a layer of copper is formed on one of the rods which pass the current into the solution. We can 'simplify' many chemicals by using forms of energy such as heat and electricity, but no matter what we do to copper we cannot simplify it any further. This is because copper is an element; the atoms within it are all copper atoms, and no other kinds of atom are present. We can define an element as follows (you will soon understand what the term atomic number means, page 97).

It would be impossible to learn about 3 million chemicals

An element is a substance which cannot be simplified by any chemical process. All of the atoms within it have the same atomic number.

Over a hundred elements are now known, though only ninety or so occur naturally. These elements are the 'building bricks' of all substances. Every substance on this planet consists of at least one element, and the differences between substances are caused by the fact that they contain different elements or have different proportions of elements within them. A full alphabetical list of the elements and their symbols is given at the end of the book. You might like to find out how some of them came to be named as they are.

Table 3.1 Some physical differences between metals and non-metals

Metals	*Non-metals*
1. Good conductors of electricity	Poor conductors of electricity, except carbon in the form of graphite
2. Good conductors of heat	Poor conductors of heat
3. When freshly cut, have a shiny surface	Are often dull, even when freshly cut
4. Can be beaten into shape (are malleable) and drawn into wire (are ductile)	Usually shatter or break when treated in this way, i.e. are brittle

The elements can be divided into two main groups, metallic elements and non-metallic elements, although some elements have the properties of both. There are far more metallic elements than there are non-metallic elements.

You have probably done some simple experiments which show the main physical properties of metallic elements and non-metallic elements (e.g. testing for the conduction of electricity and heat). Metallic elements often have different physical properties from non-metallic elements, and these differences are summarized in Table 3.1.

Physical properties (like physical changes) are concerned with what a substance looks like or with facts about it such as its melting point, boiling point, density, whether it is magnetic, etc. Physical properties (or changes) do not affect the formula of the chemical, but chemical properties refer to the ways in which the substance reacts and changes into something else. For example, the melting of ice into water is a physical change because both ice and water are the same substance, i.e. H_2O. The heating of copper in air produces a chemical change because the copper forms copper(II) oxide, which is a completely different substance from copper.

Metallic and non-metallic elements also differ in their chemical properties. You will not understand some of these until later in your course, and they are summarized in Table 21.1 (p. 368).

It is important to understand that if an element has just one or two of the 'metallic' properties you cannot conclude that it is a metal. On the other hand, an element need not have *all* the 'metallic' properties to be classified a metal. You can only come to a meaningful conclusion if an element has many of the metallic properties or many of the non-metallic properties.

Points for discussion

1. Four of the following substances have something in common. Which is the odd one?

A Nickel
B Sulphur
C Sodium
D Copper
E Silver

2. Can you name a non-metallic element which is shiny in appearance and is therefore an exception to point 3 in Table 3.1?

3. When asked to suggest other exceptions to the general points given in Table 3.1, a class suggested that an exception to point 3 is glass (which is shiny, but not a metal) and an exception to point 4 was plastic (which can be drawn into thin strands but is not a metal). The teacher said that these substances were not exceptions because they did not even belong in the table. Who was right, and why?

4. Is it true that most metals are magnetic? Explain your answer.

3.2 COMPOUNDS AND MIXTURES

A compound is a substance composed of two or more elements chemically joined together, and which cannot be separated into its elements by physical processes. (Physical processes include melting, boiling, dissolving, using a magnet, and all of the purification techniques described in Chapter 4).

A mixture consists of two or more different substances which are not chemically joined together, and which can therefore be separated by physical processes (e.g. purification techniques).

Most common chemicals contain more than one kind of atom joined together and therefore are compounds. Remember that bronze, brass, and other alloys are mixtures, not elements.

A mixture can consist of (i) two or more elements (e.g. brass contains zinc and copper), (ii) two or more compounds (e.g. copper(II) sulphate solution contains water and copper(II) sulphate), or (iii) a combination of one or more elements and compounds (e.g. a mixture of zinc and sodium chloride).

These points are summarized in Figure 3.1, in which atoms of four different elements are represented by ○, □, ● or ⊗. The 'sticks' between the atoms in the figure are simply used to show which atoms are joined together to form molecules.

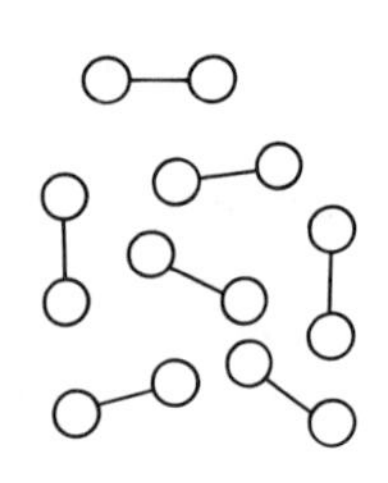

Element.
All atoms are of the same kind

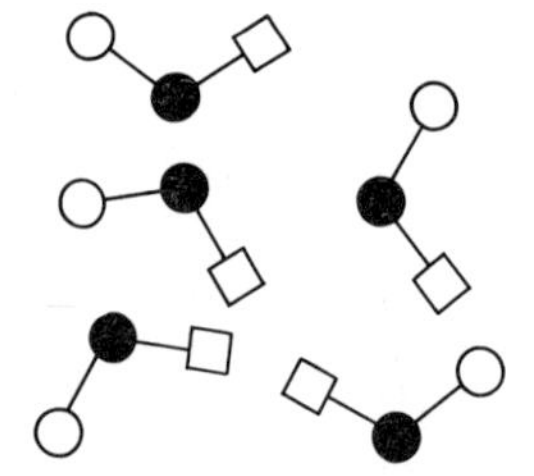

Molecules of a compound. Two or more elements joined together

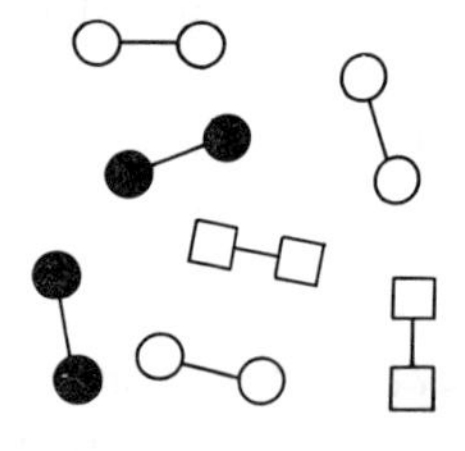

Mixture of elements. Different molecules, but each molecule contains only one kind of atom. The different chemicals are not joined together

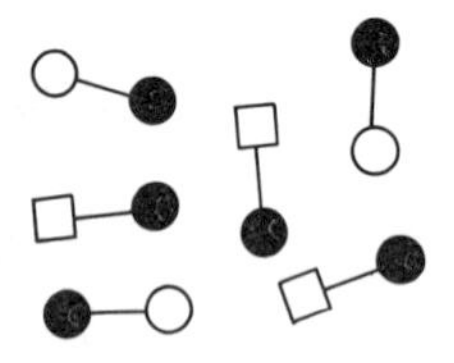

Mixture of compounds. Each molecule contains more than one kind of atom

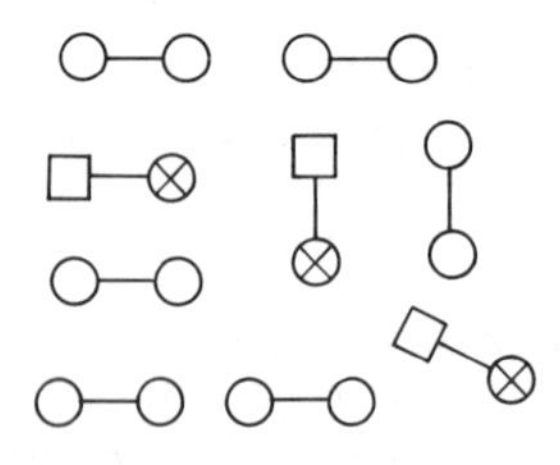

Mixture of an element and a compound. You should be able to decide which is the element

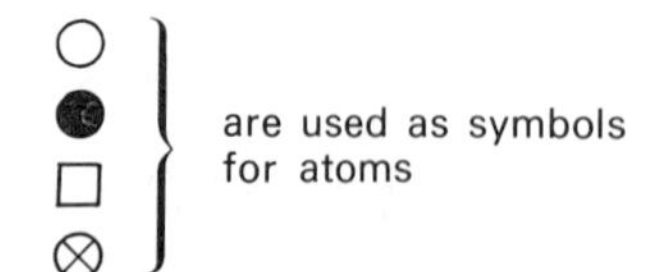

Atoms which are joined together are connected by a 'bond',

e.g. 

Figure 3.1 A simple representation of elements, compounds, and mixtures

Note that these very simple diagrams do not correspond to actual molecules, nor to their shapes. Also, some compounds do not consist of molecules.

You will be expected to know whether each chemical you use is an element, a compound, or a mixture.

Check your understanding

Note: In the examples (a) to (e), letters are used to represent different atoms (rather than their normal symbols) and 'sticks' are used to show which atoms are joined together to form molecules.

Say whether each of the following 'substances' is an element, a compound or a mixture. If the answer is a mixture, say whether it is a mixture of elements, or of compounds, or of elements and compounds.

(a) A–A
A–A A–A
A–A
A–A

(b) B–C–B
B–C–B
B–C–B

(c) D–D
E–E
D–D D–D
E–E

(d) F–G
G–H F–G
G–H F–G

(e) I–J
I–I
J–J
I–J

(f) water

(g) air

(h) sea water

(i) steel

(j) a drink of tea

(k) chromium

(l) milk

(m) magnesium chloride

(n) iodine

3.3 SOME IMPORTANT DIFFERENCES BETWEEN ELEMENTS, COMPOUNDS, AND MIXTURES

Some very important changes take place when elements *combine* together to form a compound. These changes are summarized in Table 3.2, and the next experiment will help you to understand these changes.

Table 3.2 Some differences between compounds and mixtures

Compounds	*Mixtures*
1. Always homogeneous, i.e. each sample of the compound looks exactly the same, and it is not possible to distinguish the separate elements within it	May be homogeneous (e.g. water and ethanol) or heterogeneous (e.g. iron and sulphur)
2. Cannot be split up by physical processes	Separation is possible by physical processes, e.g. purification techniques
3. The properties are totally unlike those of the elements which combine together to make it	The substances in the mixture retain their own properties, and the mixture has all of these properties
4. Always contains the same elements in a fixed and definite ratio, i.e. has a constant composition (see the Law of Constant Composition, page 306)	Can have any composition, e.g. 1% sulphur and 99% iron, 1% iron and 99% sulphur
5. The melting points and boiling points are fixed although the boiling point will vary with atmospheric pressure	The melting and boiling points are not fixed; they depend upon the composition of the mixture, which varies

Experiment 3.1
To show some differences between compounds and mixtures **

Apparatus
Teat pipette, magnet, pieces of scrap paper, 4 watch-glasses, rack of test-tubes, spatula, access to fume cupboard. The separate elements, iron (in the form of iron filings) and sulphur; a mixture of iron and sulphur; a compound of iron and sulphur called iron(II) sulphide; carbon disulphide; dilute hydrochloric acid.
Note: Your teacher may allow you to make your own iron(II) sulphide by gently heating the mixture of iron and sulphur on a tin lid in the fume cupboard, in which case you will be given separate instructions. When iron and sulphur are heated together, they *combine* to form the compound. The carbon disulphide is to be used only by the teacher, and only in a fume cupboard.

Procedure
(a) Make a full-page table for your results, using the headings given at the bottom of the page. You will add further points to the first column as you do the various steps in the experiment.
(b) Carefully examine the elements, the mixture, and the compound and fill in the columns in the table of results.
(c) Add the statement to column 1 in your table '2. The effect of a magnet, a physical change'.
(d) Place a small sample of iron filings on a piece of paper, and stroke the *underside* of the paper with the magnet. Do not allow iron filings to touch the magnet. Record your observations (e.g. 'not magnetic') in the table. Repeat with separate samples of sulphur, the mixture, and the compound.
(e) Add the statement to column 1 in your table '3. Effect of carbon disulphide, a physical change'.
(f) Your teacher will have done this stage for you. Separate samples of iron, sulphur, the mixture, and the compound have been stirred in test-tubes with the liquid carbon disulphide in the fume cupboard. Excess solid in each tube was allowed to settle, leaving a clear liquid above. To find out if the liquid had dissolved some of the solid, a few drops of the clear liquid were withdrawn from each test-tube using a teat pipette, and placed on separate, labelled watch-glasses. These were left in the fume cupboard for the liquid to evaporate. This procedure is illustrated in Figure 3.2.

Examine the watch-glasses carefully. If any iron (or sulphur) can be seen on one of them, the carbon disulphide must have dissolved some iron (or sulphur) in the original test-tube and it has reappeared (crystallized) when the liquid evaporated. Make sure that you understand this.

Complete your table, using expressions such as 'dissolved the iron', 'did not dissolve anything'; 'dissolved only the iron in the mixture', etc.
(g) Add the statement to column 1 of your table '4. The effect of a dilute acid, a chemical change'.
(h) Place a small sample of iron in a test-tube. Add a few cm^3 of dilute hydrochloric acid, and warm (*not* boil) the mixture. If the mixture fizzes (effervesces), it means that a gas is being formed. If gas is being formed, remove the tube from the flame and trap the gas by holding your thumb over the mouth of the test-tube. Do not hold your thumb over the tube whilst it is being heated or if effervescence is rapid! Then test the gas for hydrogen (as on page 450). Complete the table by writing expressions such as 'no reaction'; 'hydrogen given off' etc.
(i) Repeat step (h) using separate samples of sulphur, the mixture, and the compound, but using the compound only in the fume cupboard. You may be shown how to test any gas given off from the compound.

Notes about the experiment
(i) The results of the experiment should help to convince you about points 1 and 2 in Table 3.2. Point 2 is often confused. Young pupils sometimes think that fish

Test	The elements: Iron	The elements: Sulphur	Mixture	Compound
1. Appearance (colour, homogeneous or heterogeneous)				

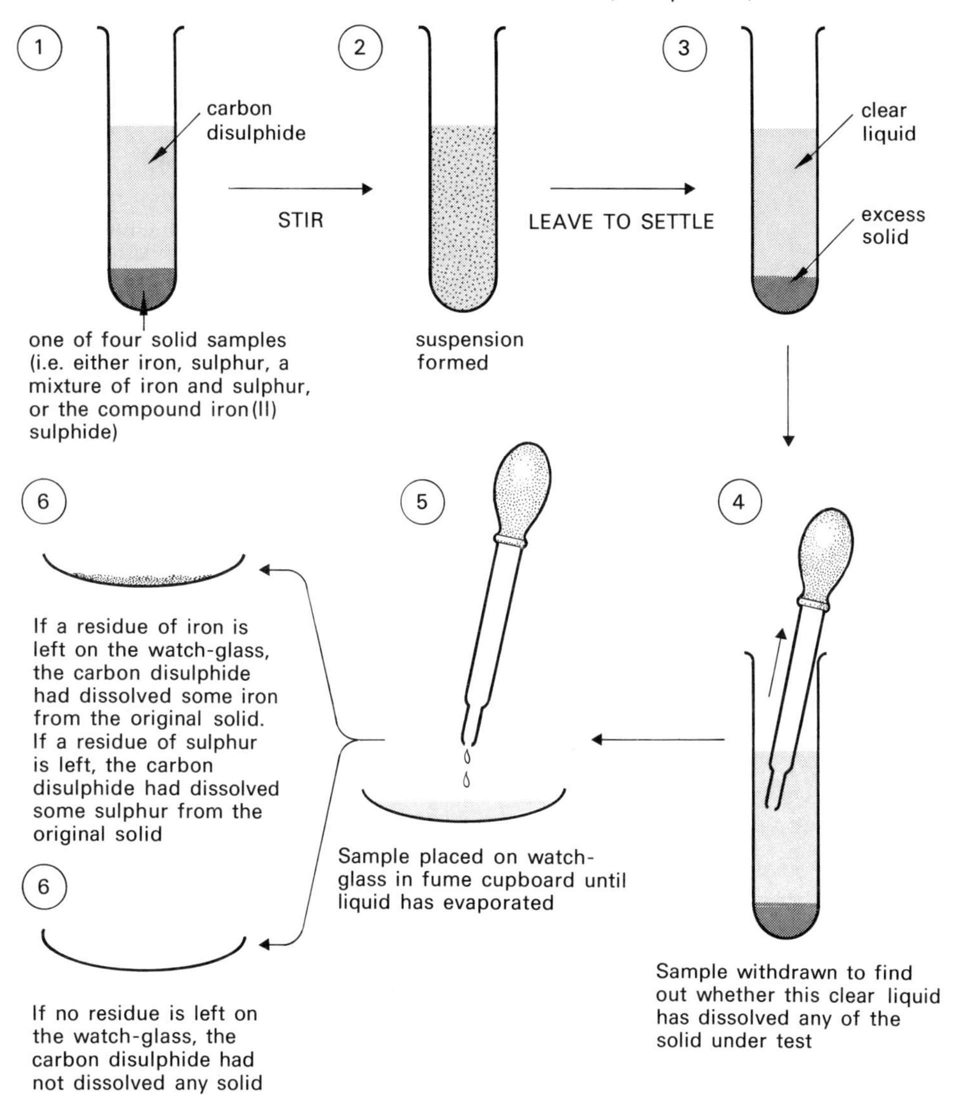

Figure 3.2 Finding out whether carbon disulphide is a solvent for various solids used in Experiment 3.1

'breathe' by taking the oxygen out of the water molecules, H_2O. The oxygen in water molecules is, of course, 'locked together' with hydrogen atoms because water is a compound. This oxygen cannot be separated from the hydrogen by a physical process such as breathing. Fish obtain oxygen from air *dissolved* in the water.

(ii) The results should also help you to understand point 3 in Table 3.2, but this is so important that it needs further explanation. The experiment should have convinced you that the mixture still keeps all the separate properties of iron and sulphur, but not only has the compound *lost* the properties of iron and sulphur, it has also *gained* new ones of its own. Which property of iron(II) sulphide was not given by the elements or the mixture?

This complete change always takes place when a compound is formed. The elements are 'locked together' in the compound, and are no longer able to show their individual properties. Instead, the compound has a completely new set of reactions. A more common example is the compound formed from the elements sodium and chlorine, i.e. common salt, sodium chloride. Sodium is a soft, grey metal which reacts violently with water and needs to be used with great caution. Chlorine is a non-metallic element, a poisonous, green, unpleasant gas which reacts with water to form an acidic solution. However, the compound formed from these two elements, sodium chloride, is a white solid, shows no metallic properties, does not react with water, and in small quantities is perfectly safe to eat with your food.

(iii) Although the experiment does not confirm point 4 in Table 3.2, you should realize that all pure samples of the compound iron(II) sulphide will have the same formula (FeS) and will always consist of iron and sulphur particles joined in the proportions one particle of iron to one particle of sulphur. If the composition of a compound *could* change, we could not give it a single formula. You know, for example, that the formula of water is H_2O; you have never seen any other formula for water, because a compound with a different formula (e.g. H_2O_2) is no longer water. (This idea is further emphasized by the Law of Constant Composition, page 306). The composition of a mixture, however, can be anything you wish; a mixture cannot be given a single formula, or indeed *any* chemical formula.

(iv) The experiment does not confirm point 5 in Table 3.2. However, you will carry out an experiment to prove this point later (p. 49).

Conclusion

In your own words, describe how the experiment has demonstrated some of the differences between compounds and mixtures.

Check your understanding

1. How would you obtain a pure sample of iron from a mixture of iron and sulphur? (You should be able to suggest two ways.)

2. How would you obtain a pure sample of sulphur from a mixture of iron and sulphur? (Again you might be able to suggest two ways, one of which involves a chemical change.)

3. A pupil was asked to prove that hydrochloric acid, HCl, contains hydrogen and suggested testing a sample of the acid for hydrogen by using the 'pop' test as described on page 450. What is wrong with this suggestion?

4. Water is a compound of formula H_2O, formed from the elements hydrogen and oxygen. Find out some of the properties of hydrogen, oxygen, and water and then compare these properties in table form. Use your information to show how a compound has different properties from those of the elements within it.

3.4 SYMBOLS, VALENCIES, AND FORMULAE

Now that you have a simple understanding of what symbols and formulae are (pp. 3 and 5), you can learn some of the common symbols and also how to use them to work out simple formulae. You will understand much more about formulae later, but as chemists constantly use them in chemical equations and as you see formulae throughout any chemistry textbook, it is useful to have a 'working understanding' of formulae at this stage.

Symbols and valencies

Each element has a chemical symbol. A full list of the elements and their symbols is reproduced at the end of the book, and some of the more common ones are given in Table 3.3. Note that sometimes several elements have the same first letter (e.g. phosphorus and potassium, or sodium and sulphur) and so some of them have been given symbols which do not appear to be connected with their name. For example, the symbol for phosphorus is P, but a different symbol (K) has to be used for potassium, which was discovered after phosphorus. The 'second' symbols are often (but not always) derived from the Roman names for the elements. Thus I is the symbol for iodine, and cannot also be used for iron; the Roman name for iron was *ferrum*, hence the symbol Fe for iron.

Table 3.3 Symbols and combining powers of the common elements and radicals

	Symbol	Stable ion formed
Elements and radicals with a combining power of 1		
chlorine (chloride)	Cl	Cl^-
bromine (bromide)	Br	Br^-
iodine (iodide)	I	I^-
hydroxide	OH	OH^-
hydrogencarbonate	HCO_3	HCO_3^-
hydrogensulphate	HSO_4	HSO_4^-
nitrate	NO_3	NO_3^-
manganate(VII) (permanganate)	MnO_4	MnO_4^-
copper(I)	Cu	Cu^+
silver	Ag	Ag^+
sodium	Na	Na^+
potassium	K	K^+
ammonium	NH_4	NH_4^+
hydrogen (hydride)	H	H^+ or H^-
Elements and radicals with a combining power of 2		
oxygen (oxide)	O	O^{2-}
sulphate	SO_4	SO_4^{2-}
sulphite	SO_3	SO_3^{2-}
carbonate	CO_3	CO_3^{2-}
sulphur (sulphide)	S	S^{2-}
dichromate(VI)	Cr_2O_7	$Cr_2O_7^{2-}$
lead(II)	Pb	Pb^{2+}
zinc	Zn	Zn^{2+}
tin(II)	Sn	Sn^{2+}
magnesium	Mg	Mg^{2+}
calcium	Ca	Ca^{2+}
copper(II)	Cu	Cu^{2+}
iron(II)	Fe	Fe^{2+}
Elements and radicals with a combining power of 3		
phosphate	PO_4	PO_4^{3-}
aluminium	Al	Al^{3+}
iron(III)	Fe	Fe^{3+}
Elements and radicals with a combining power of 4		
carbon	C	no ion
silicon	Si	no ion
lead(IV)	Pb	Pb^{4+}
tin(IV)	Sn	Sn^{4+}

Note: The elements in italic have more than one combining power

Many compounds consist of a metallic part and a non-metallic part. These 'parts' may be elements, or *radicals*. Table 3.3 also includes the common radicals. A radical is a group of atoms which can remain intact during a chemical reaction, behaving in many ways like a single atom. The sulphate group, SO_4, is a radical. Most radicals, if present, act as the non-metallic part of a compound. It is important to understand the difference between a molecule and a radical. A radical cannot exist on its own, whereas a molecule can. For example, it is not possible to have a bottle of just 'sulphate'; the sulphate radical has to have a partner, e.g. copper(II) sulphate. On the other hand, it is possible to have just 'water' in a bottle because water consists of molecules.

Elements and radicals can be given one or more numbers which are called their *valencies* or *combining powers*. For example, sodium has a valency of one. You cannot understand what this number really means until you study atomic structure and bonding (p. 95). Then you will realize that the valency of an element is usually the number of electrons an atom of the element must lose or gain, either completely or by sharing, in order to obtain a stable atomic structure. This explanation is not the complete answer, however, for some elements have more than one valency.

Formulae

For the moment, imagine that the valency number is a guide as to how atoms and radicals combine with other atoms and radicals. It might help to think of the 'sticks' used in Figure 3.1 to show which atoms join together. An atom with a valency of one can be imagined as having one 'stick' for joining with other atoms, and an atom with

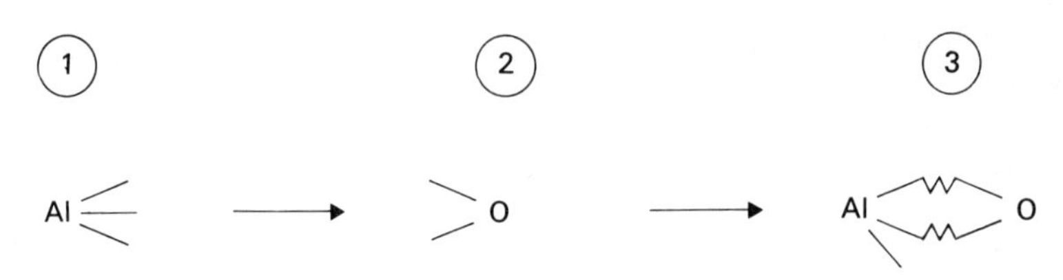

An atom of aluminium has the symbol Al and a valency of 3; the valency is shown by three sticks

An atom of oxygen has the symbol O and a valency of 2

The two atoms are joined together using the valencies they each have. In this case a spare valency is left on the aluminium

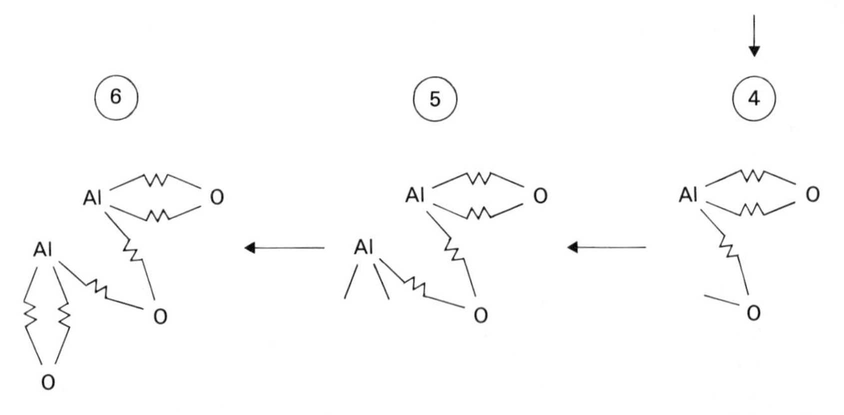

A third 'atom' of oxygen is added, with its two valencies, and this completes the compound because all the atoms have used up their valencies in combining together

An 'atom' of the other kind (this time aluminium) is added, with its normal valencies. This satisfies the two oxygen 'atoms', but leaves the second aluminium 'atom' with two spare valencies

An 'atom' of the other kind is added, with its normal valencies. This satisfies the aluminium 'atom' but leaves the second oxygen 'atom' with a spare valency

7

The final structure shows that the formula unit consists of two atoms of aluminium combined with three atoms of oxygen, so the formula of the compound is Al_2O_3

Figure 3.3 A simple but unscientific way of working out a formula a valency of two has two such 'sticks' etc. It does not matter at this stage where you put the 'sticks', as long as you use the correct number. Examples of using sticks in this way are given in Figure 3.3.

Note: It is vital that you understand that this system has no real scientific basis. The idea is used only to give you a way of working out a formula.

The valencies of the common elements and radicals are given in Table 3.3. Suppose that we need to work out the formula of aluminium oxide. The table shows us that the symbol for aluminium is Al, and its valency is three. The symbol for oxygen or oxide is O, and its valency is two. Figure 3.3 shows how the 'stick idea' can be used in stages to work out the formula, which is Al_2O_3.

This method provides a simple picture of how atoms bond together. There is a quicker way of producing the same result, and again this is simply a series of steps which provide the right answer but do not explain what is happening. The explanation will come with further experience. This alternative method can be summarized by the following 'rules'.

A chemical formula can be worked out if the appropriate symbols and valencies are known. The steps (using aluminium oxide as an example) are as follows:

1. Write down the symbols involved, putting the metal (if present) first, and leaving a small gap between them:

$$Al \quad O$$

2. Write the valency of the *first* symbol *after* the *second* symbol. In the example, the valency of aluminium is 3 so we write the figure $_3$ after the symbol for the *other* element or radical, which in this case is the symbol O (for oxide). Similarly, the valency of the second symbol is written after the first symbol. The valency of oxygen (or oxide) is 2, and so the figure 2 is added after the symbol for aluminium, Al. The formula of aluminium oxide is thus Al_2O_3. In effect, the two valencies have been 'swapped over'.

Note

(a) If you use either of these methods to work out a formula, remember that it has no scientific basis; you will understand what is really happening later. Only the final answer should appear in your book.

(b) It is not necessary to write the number 1 in any chemical formula; just leave it out. For example, write $MgCl_2$ rather than Mg_1Cl_2.

(c) If one or more radicals are present, brackets may be needed so as to avoid confusion between any numbers already present and those numbers being added. Thus it would be quite wrong to write Al_2SO_{43} for aluminininium sulphate; the correct formula is $Al_2(SO_4)_3$. Similarly, the correct formula for magnesium hydroxide is $Mg(OH)_2$, not $MgOH_2$.

Incorrect use of brackets is very common when you first start working out formulae. You need much practice in this.

The incorrect use of brackets is very common

(d) When writing the name of a chemical, remember that when a metal is combined only with oxygen, or chlorine, or bromine, or iodine, the compounds are called oxides, chlorides, bromides, and iodides respectively.

(e) If a metal has more than one valency, the *name* must clearly show which valency is being used in the compound. This is done by placing the appropriate valency (in Roman numerals) inside a bracket immediately after the element to which it refers, e.g. copper(II) sulphate, iron(III) chloride, etc.

It is not necessary to show the valency of such an element when writing its *formula*, however, because the formula itself enables us to work out the valency being used. For example, $FeCl_3$ must mean iron(III) chloride and cannot mean iron(II) chloride.

Check your understanding

1. Copy Table 3.4 into your notebook. You should have 48 spaces in which to write the formulae of different substances, e.g. the first space is for sodium sulphate. Use your knowledge of symbols, valencies, and formulae to work out the formulae of the 48 substances, and write your answers in the appropriate spaces.

2. What is wrong with each of the following formulae?

(a) $AlOH_3$
(b) $Fe(III)Cl_3$
(c) OMg
(d) $(Na)_2O$

Table 3.4 Typical table used to show the formulae of common substances

Metal or radical	*Sulphate*	*Chloride*	*Oxide*	*Nitrate*	*Hydrogen-carbonate*	*Hydroxide*
Sodium						
Iron(II)						
Aluminium						
Magnesium						
Calcium						
Potassium						
Copper(II)						
Iron(III)						

3.5 A SUMMARY OF CHAPTER 3

The following 'check list' should help you to organize the work for revision.

1. Definitions

An element (p. 24)
A compound (p. 26)
A mixture (p. 26)

2. Other ideas

(i) The division of the elements into metals and non-metals, and the important physical and chemical differences between them (Table 3.1 on p. 25, and Table 21.1 on p. 368).
(ii) The difference between physical properties and chemical properties (p. 25).
(iii) The important differences between a mixture and a compound (Table 3.2, p. 27)
(iv) The symbols and valencies of the common elements and radicals (Table 3.3, p. 31).
(v) What radicals are (p. 31).
(vi) How to work out a chemical formula in a rather unscientific way.
(vii) The correct use of Roman numerals and brackets in writing names and formulae.

4 Purification techniques: finding out whether a substance is pure

4.1 THE METHODS CHEMISTS USE TO PURIFY SUBSTANCES

When chemicals are made in the laboratory they are likely to be mixtures rather than pure substances, because they will probably contain one or more impurities. Chemists have many ways of removing impurities from substances. The purification methods mentioned in this section are all physical methods, so they can be used to separate or purify mixtures but cannot be used to break down compounds. Much more drastic methods are needed to break down a compound, and these involve chemical changes.

The experiments carried out in this section are to illustrate chemical *techniques* which are frequently used in laboratories to purify compounds. It is important that you fully understand these techniques and can describe how to use them, and that you can recognize which of them would be useful in a particular situation. The actual examples of the compounds purified and impurities separated are only *some* examples of how the techniques can be used.

Solution, filtration, and crystallization

It is likely that you have done experiments of this type at an earlier stage, and so the ideas are only summarized here. When describing processes of this type, try to use as many of the key words (given in italic) as possible. Make sure that you understand these words.

This method is normally used to separate a mixture of two or more solids. A *solvent* is chosen which (when hot, if necessary) will *dissolve* one of the solids (the *solute*) but not the others, to form a *solution*. The *insoluble* solids are at first left *suspended* in the solution. The suspension is filtered, when the insoluble material is trapped on the filter paper as the *residue* and the clear *filtrate* (the solution) is collected. (*Centrifuging* is an alternative to filtration, especially if small volumes are being used.)

The filtrate is heated in an evaporating basin to *evaporate* some of the solvent (not all of the solvent), and when the concentrated solution is allowed to cool in a crystallizing dish the solute *crystallizes* out, separated from the original mixture.

Note that this technique can be used to obtain a pure sample of the soluble substance, or a pure sample of the insoluble substance (if only one substance in the mixture is insoluble in the solvent), or both. It cannot be used, however, to recover the solvent unless distillation is also used. If the residue is required it is washed and dried. If the solute is required, it also is washed (with a *little* solvent) and dried.

The sequence of events is summarized in Figure 4.1.

Note: To find out whether enough solvent has been evaporated so that crystallization can take place, keep transferring a few drops of solution on to a microscope slide

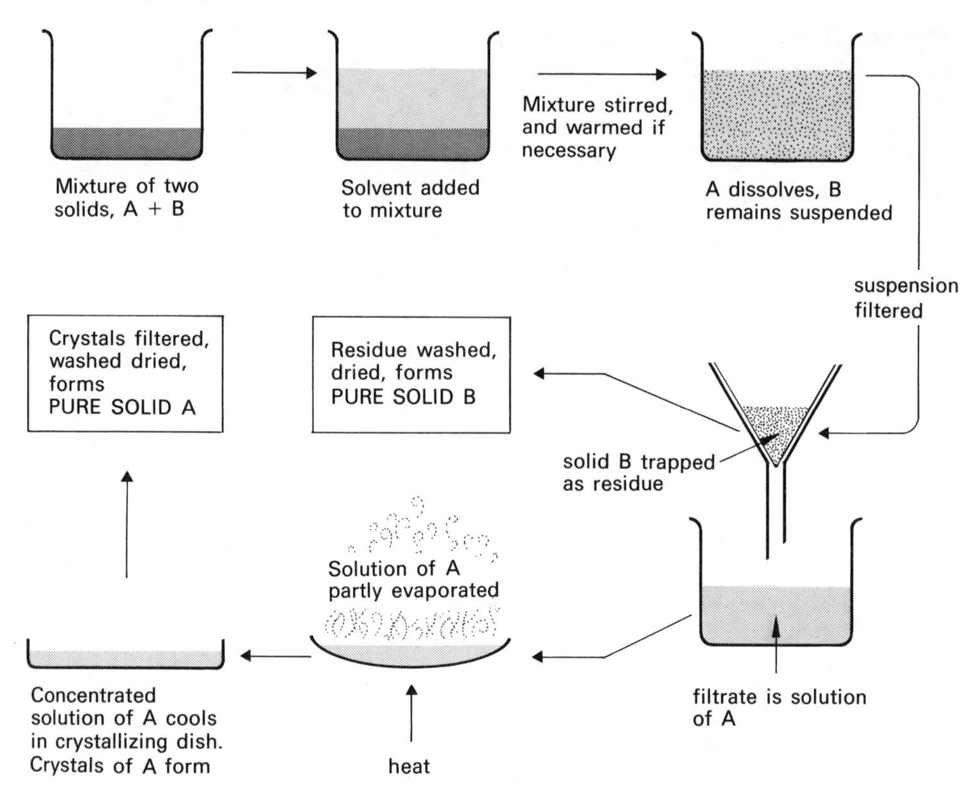

Figure 4.1 One way of separating a solid from a mixture of solids

using a teat pipette. When crystals form on the slide as the drops of solution cool, heating can be stopped. Usually, the evaporation of about half the volume of solvent is enough.

Typical examples of this method are the purification of rock salt (using water as the solvent, when the salt is the solute and the sand forms the residue), and the preparation of many soluble salts as in Chapter 6.

It is important to remember that this is a *general* method and it has many applications. The solvent does not have to be water. There are some mixtures where water will not dissolve any of the components, but another liquid might be suitable as an alternative solvent. If you use a liquid other than water it may be flammable, and great care will be needed when it is heated (e.g. in the evaporation). In these cases a method such as direct electrical heating or heating on a water bath is advisable, so as to avoid naked flames and to reduce the risk of fire. A flammable liquid should never be evaporated into an open laboratory; ideally, it should always be distilled (p. 37).

More to do

1. Write down as many examples as possible in which 'filtering' takes place in everyday life.

2. Although water has often been described as the 'universal solvent', some substances such as sulphur, oil, and tar are almost insoluble in water. Name suitable solvents, one in each case, for these substances.

The separation of immiscible liquids

In a mixture of ethanol and water it is impossible to distinguish one liquid from the other because they mix completely. Such liquids are said to be *miscible*. Liquids which do not mix are said to be *immiscible*. You must never describe *one* liquid as being immiscible; the word only applies to a named pair of liquids. For example, water is miscible with ethanol but is immiscible with petrol.

A separating funnel (Figure 4.2) is used to separate immiscible liquids. A typical procedure would be as follows, and you may be allowed to practise this.

A separating funnel?

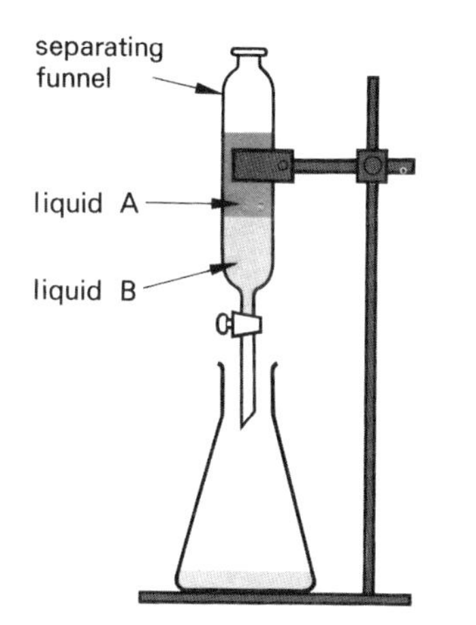

Figure 4.2 Separation of two immiscible liquids

(a) Pour the mixture (e.g. oil and water) into the funnel, after first making sure that the tap is closed.
(b) Clamp the funnel over a clean conical flask, as in Figure 4.2.
(c) Allow the two liquids to separate.
(d) Make sure that you know which liquid forms the lower layer. This is sometimes obvious from the colour, but it may be necessary to use a book of data to find out which is the more dense of the two. Alternatively, you could add a further volume of one of the liquids, watching carefully to see which of the two layers increases in volume.
(e) Open the tap on the funnel and allow the lower layer to run into the conical flask. Close the tap again just *before* the upper liquid reaches it.
(f) Unclamp the funnel and hold it over a beaker. Open the tap and allow a little liquid to run out until some of the upper liquid has passed into the beaker. (This small volume of liquid is waste.) Close the tap.
(g) You should now have a reasonably pure sample of one liquid in the conical flask, and a reasonably pure sample of the other liquid in the funnel. This can be carefully poured out of the *top* of the funnel if any traces of the other liquid are left in or below the tap.

Simple distillation

This technique is used to separate a pure solvent from a solution (e.g. pure water from copper(II) sulphate solution) or one liquid from a mixture of *miscible* liquids (e.g. ethanol from a mixture of ethanol and water) providing their boiling points differ by more than about 20 °C.

Distillation is the boiling of a mixture in an apparatus so that the vapour of one of the substances in the mixture is condensed and collected in a container separate from the original solution; the liquid collected in this way is called the distillate.

The basic principle is that the different substances in a mixture have different boiling points. When a mixture is heated, molecules of the most volatile substance in the mixture (i.e. the one with the lowest boiling point) will evaporate more readily than molecules of other substances. The vapour given off will therefore contain more molecules from one liquid than from the other(s). If the boiling points of the substances are far apart, the vapour will consist almost entirely of molecules from the substance with the lowest boiling point. If this vapour is allowed to leave the distillation flask before being condensed back into liquid in a separate container, it is separated from the mixture.

Experiment 4.1
The use of simple apparatus to distil copper(II) sulphate solution*

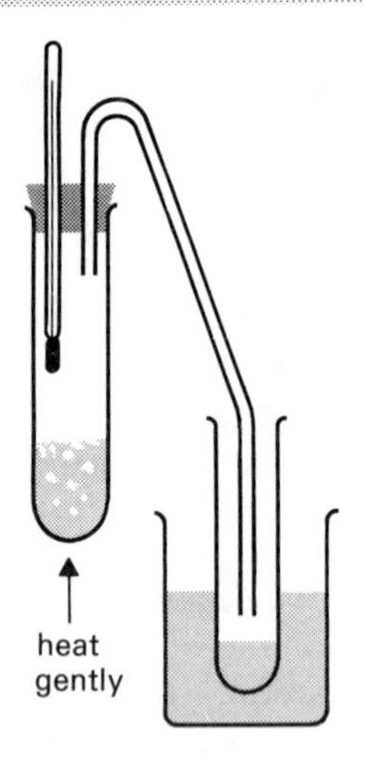

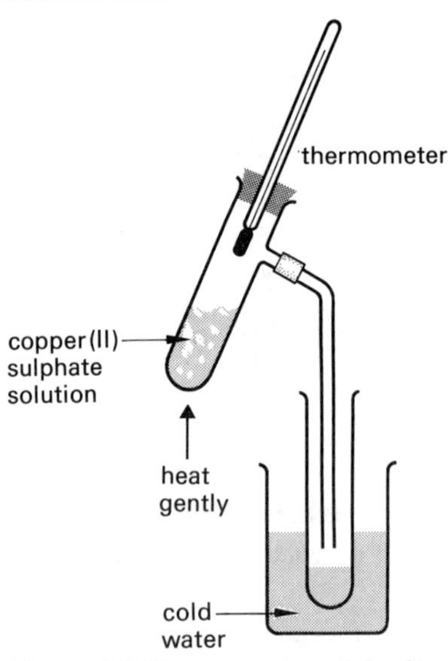

Figure 4.3 Two examples of simple apparatus which can be used to obtain pure water from copper(II) sulphate solution

Apparatus
One of the alternatives shown in Figure 4.3. Bunsen burner, stand and clamp. (The thermometer is for −10 °C to 110 °C.) Copper(II) sulphate solution, antibumping granules.

Procedure
(a) Set up the apparatus in one of the ways shown in Figure 4.3.
(b) Pour a few cm^3 of the solution to be distilled into the test-tube, and add a few antibumping granules.
(c) Heat the solution gently until it boils and note the steady temperature.
(d) Continue to boil the mixture gently until several cm^3 of the distillate have been collected.

Results
Describe what you see happening, and record the temperature of the vapour and the colour of the distillate.

Notes about the experiment

The thermometer is placed where shown and not in the liquid, because we are interested in the temperature of the vapour (which is the same as the boiling point of the *distillate*) and not that of the mixture. The temperature of the boiling liquid is not necessarily the same as that of the vapour being condensed, as you will understand later (p. 41). Compare the position of the thermometer with that used in a boiling point test on page 52.

Conclusion
Has distillation separated a pure liquid from the mixture? Your conclusion should include an answer to this question, together with an explanation of what happened and an explanation of how you knew whether the distillate was pure or not.

Points for discussion

1. What was the purpose of the beaker of water in Figure 4.3?
2. What would be left in the distillation tube if the heating was continued until distillation stopped?

Experiment 4.2
Distillation with a Liebig condenser***

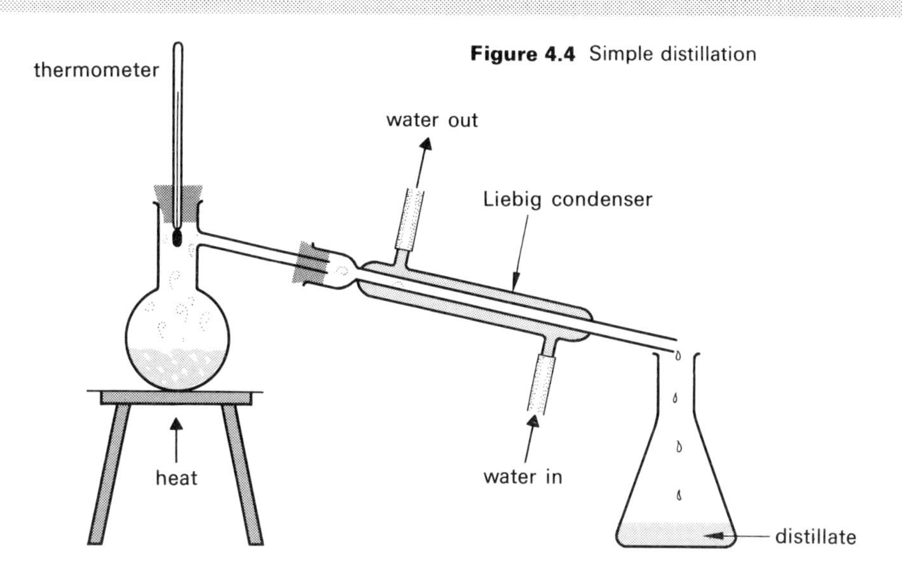

Figure 4.4 Simple distillation

Apparatus
As in Figure 4.4, retort stand and clamp, muddy water, antibumping granules.

Procedure
(a) About half fill the distillation flask with muddy water, add a pinch of antibumping granules, and set up the apparatus as in Figure 4.4.
(b) Make sure that cooling water passes through the condenser.
(c) Heat the mixture to boiling point, and keep it boiling steadily. Note the temperature of the vapour. Collect a few cm^3 of distillate.

Results
Write down what you see and record the temperature of the vapour, and the appearance of the distillate.

Conclusion
Has the distillation separated a pure distillate from the mixture? Again explain how you know whether the distillate is pure or not, and also explain why you think the use of a Liebig condenser results in a more efficient distillation than the use of apparatus as in Figure 4.3.

Points for discussion

1. Whenever a Liebig condenser is used, it is important that the water supply enters the condenser at the lower opening, and leaves at the upper opening. Why is this so?
2. Suppose that you were distilling a mixture of water and another liquid, X, and that you know the boiling point of pure X is 55 °C. What should the first sample of

distillate be, and what is the temperature of the vapour which condenses to give this distillate?
3. Why would you not separate cooking oil from water by distillation?
4. How could you produce some distilled water by using only common kitchen equipment?
5. Imagine that you are camping by a muddy river which is your only source of water. Describe how you would improvise from your camping equipment a method of obtaining (a) clear water for washing and (b) pure water for drinking.

Fractional distillation

Simple distillation is very useful, particularly for obtaining the pure solvent from a dissolved solid, but it is not always efficient if a mixture of *liquids* is being separated. The problem arises because it is quite possible for molecules of *both* liquids to evaporate when the mixture is heated, for a liquid is evaporating long before it boils. The vapour will always contain more molecules from the more volatile liquid (i.e. the liquid with the lower boiling point) but it may also contain molecules of the other liquid. This is particularly important if the two liquids have boiling points which are fairly close together, for then the vapour will contain a large proportion of both types of molecules and the distillate will certainly not be pure. In these cases, and whenever efficient separation is needed, the distillation is carried out with a *fractionating column* (Figure 4.5) and is then called *fractional distillation.*

The fractionating column is usually packed with glass beads or some other unreactive substance which provides a large surface area. When a mixture is boiled in the distillation flask, the vapour coming off reaches the fractionating column and

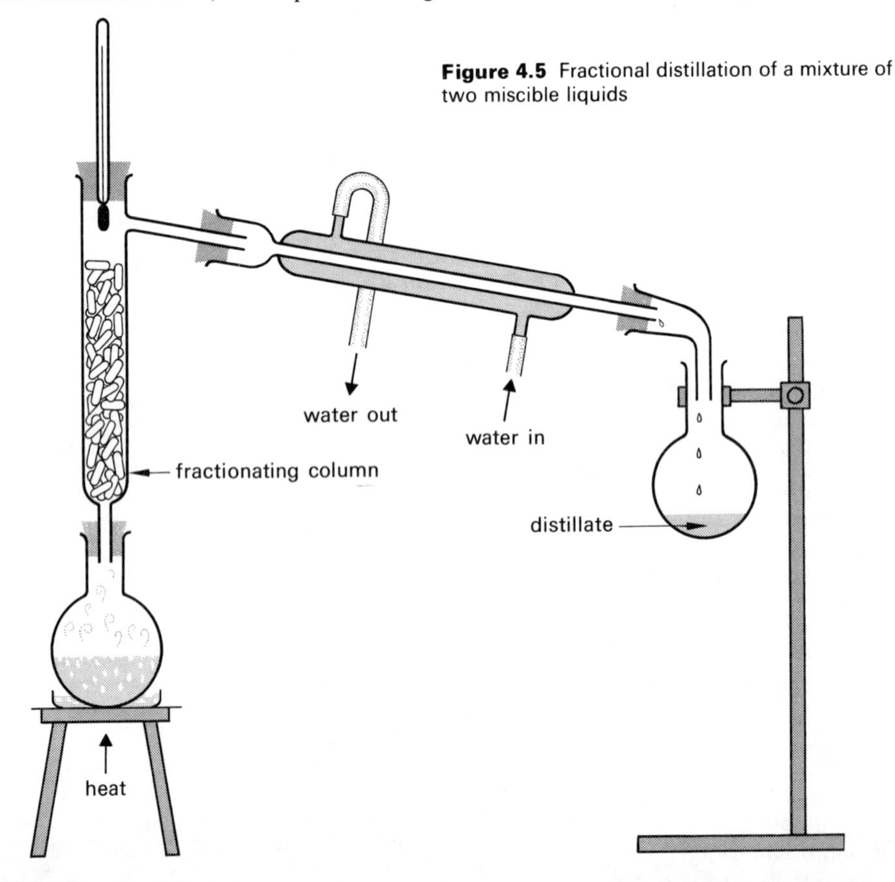

Figure 4.5 Fractional distillation of a mixture of two miscible liquids

begins to warm it up. A temperature gradient is set up in the column, which means that one end of the column is hotter than the other. In this case, the lower part (which is nearer to the source of heat) is warmer than the top.

At first the vapour evaporating from the mixture simply condenses on the surfaces inside the column (which are at a temperature lower than its boiling point), gives up heat to the column as it does so, and trickles back into the flask. As the column gets hotter, the molecules in the vapour state rise higher before condensing. Eventually the top of the column reaches the temperature at which one of the liquids in the mixture boils (the one with the lower boiling point).

Fractional distillation?

The vapour coming off from the mixture will contain molecules of each of the substances in the mixture, although it will be richer in molecules of the liquid with the lowest boiling point. When the temperature at the top of the column reaches the boiling point of one of the substances in the mixture, molecules of that substance can now 'survive' as vapour right to the top of the column. This happens because all parts of the column have a temperature either equal to or greater than its boiling point. This substance therefore passes through the column as vapour and enters the condenser where it is condensed and runs down to be collected in a separate container as a pure liquid. The temperature at the top of the column remains steady until all that particular substance has 'boiled off', because a liquid cannot normally be heated to a temperature higher than its boiling point.

Molecules of other substances will also rise up the column, but they have little or no chance of surviving as vapour for the length of the column because they must keep touching the surfaces within the column, some of which are at a temperature below their boiling point. Such molecules thus condense before they reach the top of the column, and trickle back down as a liquid.

When the first liquid has boiled off completely, the temperature at the top of the column rises quickly until it reaches the boiling point of another liquid in the mixture, and then it remains steady again until that, too, has boiled off. If, for example, there are four different components in the mixture, you will collect four separate *fractions*. Note once again the importance of placing the thermometer in the *vapour* passing into the condenser and not in the liquid.

The description above is not a full explanation of what happens inside a fractionating column, but it is adequate for an understanding of the process. It is very unlikely that you will ever be asked, at an elementary level, to *explain* the process. It is important, however, to understand what it achieves and when it is better to use it rather than simple distillation.

Experiment 4.3
The laboratory distillation of crude oil*

Apparatus

Apparatus as in Figure 4.3 (page 38). The thermometer range is from 0 °C to 360 °C. Test-tubes in rack, glass wool, tongs, evaporating basin. Crude oil.

Procedure

(a) Copy out the table headings at the top of p. 42 for your results.

Fraction number	*Boiling range*	*Colour*	*Viscosity*	*Flammability*

(b) Place a small piece of glass wool in the tube (do not touch the glass wool with your hands), and pour in about 6 cm^3 of crude oil.
(c) Carefully heat the tube with a small flame and collect the distillate which boils off between room temperature and 75 °C in a test-tube.
(d) Allow the distillation tube to cool for a minute and then repeat procedure (c). As soon as the temperature rises above 75 °C start collecting the distillate in a second test-tube and continue until the temperature reaches 120 °C.
(e) Again allow the distillation tube to cool for a minute and reheat up to 120 °C, collecting any more distillate produced.
(f) Repeat this procedure two more times, but this time collecting the distillate which boils off between 120 °C and 170 °C, and then between 170 °C and 220 °C in two more test-tubes, producing four distillates altogether.
(g) Note the colour and viscosity of each sample. A thick, treacle-like liquid which only flows very slowly is described as being very *viscous*, or as having a high viscosity.
(h) Using tongs, soak a piece of glass wool in one of the samples and place it in a dry evaporating basin. Try to ignite the sample on the glass wool with a lighted taper. Observe carefully how easy it is to light the sample, the colour of the flame, and whether any smoke is formed.

Figure 4.6 Columns of distillation units at BP's Europort refinery, Rotterdam, Holland

(i) Repeat step (h) using the other samples in turn, cleaning out (e.g. wiping with a paper towel) and drying the evaporating basin each time.

Results
Complete the table.

Notes about the experiment

This experiment is often described as the fractional distillation of crude oil, but as no fractionating column is used it is better to think of it as a simple distillation in which separate 'fractions' distil off as the temperature is raised. We are not particularly concerned with an efficient separation of the mixture in this case, because crude oil consists of so many compounds that the fractions are themselves mixtures. Distillation of crude oil at a refinery involves a fractionating column or tower (Figure 4.6) and this is correctly described as fractional distillation. Even the fractions obtained at the refinery are still mixtures, but the use of the fractionating tower enables the boiling range of each fraction to be carefully controlled.

Conclusion
Write a conclusion to the experiment, based upon the following points. Do not just answer the questions.

1. Did the experiment separate crude oil into simpler parts?
2. Were each of the fractions pure liquids?
3. Which column in your results table helps to confirm your answer to the previous question? (Explain.)
4. Your fractions probably correspond roughly to liquids used in everyday life, such as paraffin, light oil, petrol, and heavy oil. Which fraction corresponds to each of these names?
5. The dark mass left in the distillation tube is also important. What is it, and what is it used for?

Fractional distillation of crude oil at a refinery
Crude oil is a mixture of many hydrocarbons; separation is a difficult process, and some of the products are further processed on the site to obtain a wide variety of important organic chemicals (p. 413). An oil refinery is thus a large, complicated site (Figure 4.7). At first, oil refineries were built quite close to the oil fields, but it is now more convenient and economical to build them in the area where the products will be distributed.

Figure 4.7 Aerial view of Stanlow oil refinery (Shell)

The crude oil is first treated to remove sulphur, for if this is present in any of the fuels made from oil, it burns to form sulphur dioxide, which is a toxic and corrosive gas and pollutes the atmosphere. Separation into main fractions is achieved by using a kind of fractionating column, but the design is more complicated than the simple glass columns used in a laboratory. The vapours of the various chemicals pass into the fractionating tower, which contains trays and bubble caps as in Figure 4.8. Each tray is a little cooler than the one below it because it is further away from the source of heat. The primary fractionating column (i.e. the one used for the first stage of the processing) contains more trays and bubble caps than those shown in the simplified figure.

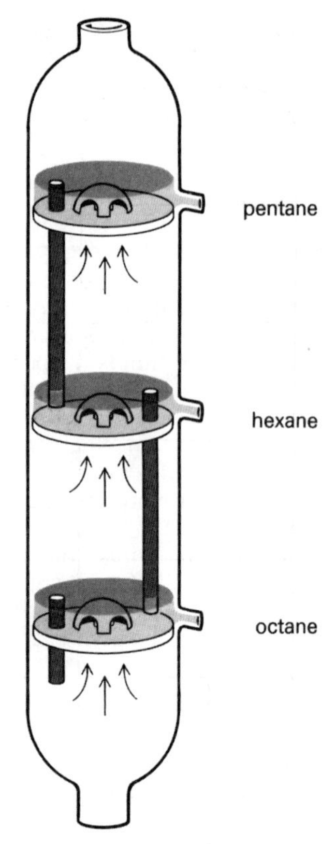

Figure 4.8 Secondary fractional distillation of a fraction from a primary distillation

As in any type of fractionating column, the proportion of molecules from liquids with low boiling points increases as the vapours pass up the tower. However, instead of allowing the fractions to emerge one by one from the top of the tower as the temperature is increased (as you did in the laboratory experiment), the fractions are collected at different heights (temperatures) within the tower, as shown in Figure 4.9.

Each fraction from crude oil is not a pure substance, but simply a mixture of chemicals which boil within a chosen range of temperatures. For most purposes, these mixtures are adequate. Considerable use is made of these carefully controlled mixtures as different fuels for use in homes and industries, and for transport (e.g. petrol, paraffin). If required, however, each fraction from this 'primary' distillation can be given further (secondary) fractional distillations and eventually pure chemicals can be separated, as in Figure 4.8.

Chromatography

The name of this technique comes from the Greek 'chromos' which means colour, because the early work was done with coloured substances. You will almost certainly use only coloured substances to illustrate the technique in elementary work, but it is now possible to separate many mixtures by this method, whether they are coloured or not.

Experiment 4.4
To separate the components of black ball point ink*

Apparatus

Teat pipette with finely drawn jet, Petri dish or evaporating basin, filter paper. Solvent, e.g. propanone (acetone) or ethanol, black ball point pen.

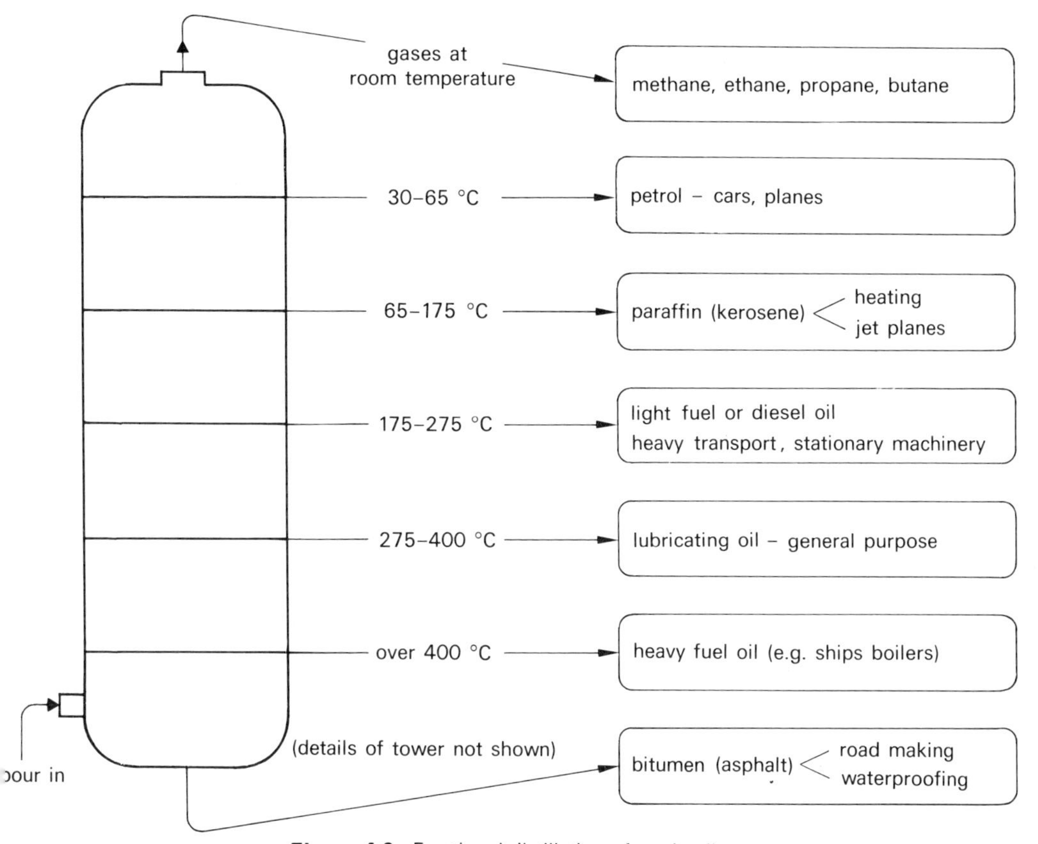

Figure 4.9 Fractional distillation of crude oil

Procedure

(a) Place the filter paper on a clean surface, e.g. a new page in your notebook. Use the ball point pen to make a black dot of ink in the centre of the paper, about 2 mm in diameter. Add more black ink to the same spot, but do not make the spot any bigger and do not damage the paper.

(b) Place the filter paper over the top of the Petri dish, and using the teat pipette carefully add a few drops of solvent on to the centre of the paper as in Figure 4.10. Allow the solvent to spread out as far as possible.

(c) Add more solvent in the same way and allow it to spread out. Repeat this procedure until the solvent has almost spread to the edge of the paper.

Results

Describe what happened, and stick the dried paper (now called the chromatogram) in your book.

Notes about the experiment

The principle of the technique is the same in all its many forms. The mixture to be separated is placed on a suitable surface

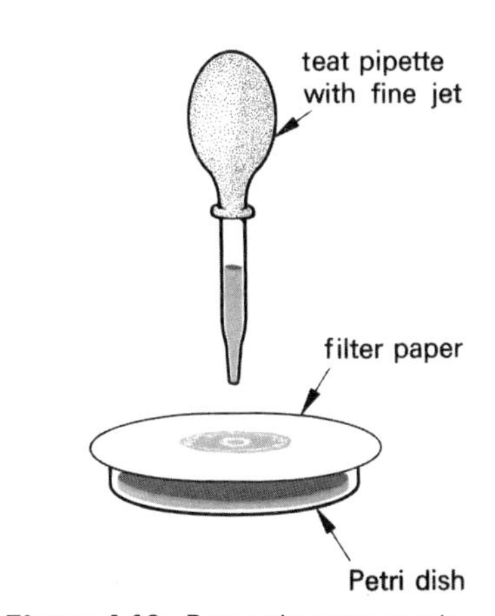

Figure 4.10 Paper chromatography

(e.g. paper, chalk, or aluminium oxide) over or through which a solvent is allowed to flow. The flowing solvent can be a liquid or a gas. The material of the stationary surface is such that the substances in the mixture can cling to it (i.e. be *ad*sorbed on its surface). At the same time the moving liquid or gas has a tendency to 'dissolve' the substances and carry them with it. Each substance in the mixture will have different tendencies to be adsorbed on the surface and to move with the solvent, and so they will be separated. The substances which are both easily adsorbed and poor at dissolving will hardly move from their starting point, whereas the 'poor clingers and good dissolvers' will travel much further with the solvent.

Conclusion

Write your own conclusion to the experiment, saying whether black ink is a single substance or a mixture, and explaining what happened in the experiment.

Experiment 4.5
Is the green colour in grass a single substance?**

Apparatus

Pestle and mortar, chromatography paper, small screw-top jar, capillary tubing, teat pipette. Grass, propanone (acetone), clean sand.

Procedure

(a) Place some chopped grass in the mortar and just cover it with a little propanone. Add a little sand. Crush and stir the grass by means of the pestle.

(b) Take a piece of chromatography paper about 1.5 cm wide and slightly less than the height of the jar in length. Fold in two along its length as in Figure 4.11.

(c) Pour propanone into the jar to a depth of about 0.4 cm.

(d) Make a line in pencil about 1 cm from one end of the paper as in Figure 4.11. Use a small piece of capillary tubing to transfer a small amount of the grass 'solution' on to the line, as in the figure. For good results, keep each spot small, although more solution can be added to each spot after it has dried. Make about three applications to each spot so that the solution is concentrated on the paper.

(e) Stand the paper in the jar with the marked end to the bottom of the jar. Screw on the top of the jar, and leave the jar to stand undisturbed until the level of the solvent is about 1 cm from the top. Remove the paper and allow it to dry.

Note: The coloured substances sometimes fade very quickly, especially in sunlight. If the laboratory is well lit, the jar should be placed in a dark cupboard while the solvent is rising, and the completed chromatogram should be kept away from bright light.

Results

Describe what happened, and/or fasten the chromatogram into your notebook.

Notes about the experiment

(i) This is an alternative way of producing a paper chromatogram, and it is sometimes called ascending chromatography because the solvent moves up the paper rather than across it. The same effect can be achieved by using a piece of blackboard chalk. If the stick of chalk is dipped into ink and then stood upright in a small beaker containing a little water, the water rises up the chalk and separates the ink into its various components. As always in chromatography, the separation depends upon the differing adsorption and dissolving properties of the various substances in the mixture.

Figure 4.11 Another method of preparing a paper chromatogram

(*Note*: For a particular surface and a particular solvent, each substance being used in a chromatography experiment always 'travels' a certain proportion of the distance travelled by the solvent. This is described by its R_f value. The R_f value of a particular substance =

$$\frac{\text{distance moved by substance (from base line)}}{\text{distance moved by solvent (from base line)}}$$

and it is fixed for a given set of conditions.)
(ii) The individual separated substances can be obtained by cutting out the 'spots' on the final chromatogram, and adding a little solvent to each spot in a separate beaker. The pure substance dissolves in the solvent. If the piece of paper is then removed and the solvent evaporated off, the pure substance is obtained. Note that chromatography can be used both for analysis (i.e. finding out what is in a mixture) *and* for separating the components of a mixture into pure samples.

Conclusion
The green colour in leaves is due to a substance called chlorophyll. Say whether you think chlorophyll is the only coloured chemical in the leaves, and again explain what happened in the experiment.

Applications of chromatography

Remember that chromatography is also used to identify substances as well as to separate them. For example, a dye used in food colourings can be identified by chromatographing it alongside known dyes. In this way a public analyst can find out whether a dye used in food is a permitted one or an 'illegal' one. The idea is illustrated in Figure 4.12, from which it can be concluded that dye A contains at least substance C and substance D. Similarly, the police may use gas chromatography to establish the alcohol content of blood taken from a driver suspected of drinking too much alcohol.

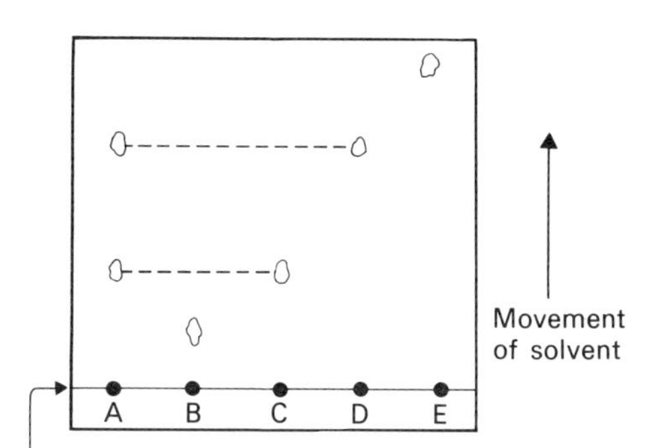

Figure 4.12 Analysing a dye by paper chromatography

Sublimation

Experiment 4.6 introduces another method of separation. An important part of this experiment involves testing for the presence of sulphate, chloride, and ammonium ions. If a chemist suspects that a chemical is ammonium chloride, the substance can be tested separately for ammonium ions and chloride ions. If it gives both tests and no others, then the substance is ammonium chloride. The tests for ammonium ions, sulphate ions, and chloride ions are given on pages 450–1. If you have not tried these tests before, you should practise them on other known compounds which you are given, before starting Experiment 4.6. Barium chloride and silver nitrate solutions are poisonous, silver nitrate solution can stain the skin and clothing, and sodium hydroxide solution can damage the eyes so follow the instructions for the tests very carefully.

Experiment 4.6
The effect of heat on a mixture of ammonium chloride and sodium sulphate*

Apparatus
Hard glass test-tubes and holder, spatula, Bunsen burner, protective mat.

Mixture of ammonium chloride and sodium sulphate (solids), and the chemicals for testing for chloride, sulphate, and ammonium ions.

Procedure
(a) Place two spatula measures of the mixture in a test-tube, and heat the mixture, gently at first and then more strongly.
(b) Watch what happens in the test-tube while you heat the mixture for a minute or two, and then allow the tube to cool on the mat.
(c) When the tube has cooled, use a clean spatula to scrape some solid from the top of the tube into a clean, second test-tube.
(d) Add a few cm^3 of water to the second tube and shake to dissolve the sample of solid.
(e) Pour roughly one-third of the solution from (d) into each of two more tubes, so that you have a small sample of solution in three separate tubes.
(f) Test one sample for ammonium ions, one for chloride ions, and one for sulphate ions (page 451).

Results
(a) Describe what you saw when the first tube was heated (a simple diagram might help).
(b) Describe the results of the tests made on the sample taken from the top of the tube.

Notes about the experiment
It may not have occurred to you, but when you heated the mixture of solids something rather unusual happened. A glance at the change of state 'table' on page 15 should make you realize what this was. The process is called *sublimation*, and the solid which reforms from the vapour in such a process is called the sublimate. Only one of the solids sublimed in the experiment, and the results of the tests should tell you which one it was.

Another common way of collecting a sublimate is to put a glass funnel upside down over the top of the test-tube. The sublimate collects on the inside of the funnel.

This method of separation is not very useful because only a few solids sublime when heated. Apart from the one which sublimed in the experiment, other chemicals which give this effect are iodine, benzene carboxylic acid, anhydrous aluminium chloride, and anhydrous iron(III) chloride.

Conclusion
Copy out and complete the following.
When a solid mixture of ammonium chloride and sodium sulphate is heated, the ________ stays as a solid because it has a high ________. The ________ misses out the ________ stage when heated, and changes straight from a ________ to a ________. This vapour then reforms the ________ in a cooler part of the tube, and is thus separated from the mixture. The direct change from solid to vapour is called ________.

Check your understanding
1. You are given a mixture of dimethylbenzene (xylene) and a dilute aqueous solution of potassium chloride. Dimethylbenzene is a liquid which boils at 140 °C and is immiscible with water. (a) Which of the following methods would you first use to separate the mixture as far as possible?

A filtration
B use a separating funnel
C distillation
D sublimation
E chromatography

(b) What would you then do to complete the separation of pure water?

2. From the following methods of separation (filtration, evaporation, sublimation, fractional distillation, chromatography) select the one most suitable for separation of each of the following mixtures.
A petrol and paraffin
B the various colours in red rose petals.
C sea water, to obtain salt

3. Name the method of purification you would use in each of the following:
(a) to separate a mixture of red and blue inks
(b) to remove the 'cloudiness' from lime water
(c) to obtain pure water from sea water
(d) to separate a mixture of paraffin and water

4.2 FINDING OUT WHETHER A SUBSTANCE IS PURE

When a chemical is first made in a laboratory, it is rarely pure; it usually contains one or more impurities. The substance prepared is therefore a mixture and we can use physical methods such as those just described in Section 4.1 to separate the impurities from the chemical we are interested in. The next problem is, how do we know whether the purification has worked? Is the chemical really pure or is it just less contaminated than before?

To find out if a substance is pure, we always check its physical properties. Any pure substance has its own special physical properties, such as its boiling point and melting point. It also has its own special chemical properties, but these will still be given if the chemical is impure. For example, the mixture of iron and sulphur in Experiment 3.1 still showed the chemical properties of iron although it was impure iron, i.e. iron mixed with sulphur. Chemical properties do not tell us whether something is pure, they only give an idea of the kind of substance present. Physical properties such as melting point, however, are altered if an impurity is present. These properties therefore can not only be used to help to identify a substance, but also to tell us how pure it is.

There are many physical properties which could provide information about a substance, apart from its melting point and boiling point. For example it is sometimes possible to recognize something by its taste, appearance, or smell, but it would be *most unwise* to taste any unknown substance, and probably unwise to smell it. Even if we did recognize a substance in one of these ways, we cannot tell whether it is *pure* or not.

Usually melting point, boiling point, and perhaps density are the physical properties we measure in deciding whether a substance is pure or not, or in helping to identify a substance. Experiments conducted to find these values are usually safe and provide very useful information. If, say, the melting point of an unknown substance is checked against the lists of melting points in a book of data, we get a useful clue as to the identity of a substance, and we get further clues from measuring its boiling point or density.

Alternatively, if we *know* what the chemical is, the values can be compared with those given for the pure chemical, and therefore the degree of purity can be found. The next three experiments illustrate this point, and also show what happens if a substance is impure. If a solid is being tested, it is usual to take its melting point, and the boiling point is determined for a liquid. Ideally, to find the identity of a substance *both* the melting point and boiling point should be determined.

Experiment 4.7

To find the melting point of (a) pure naphthalene, and (b) impure naphthalene (a mixture of naphthalene and camphor)*

Apparatus

Test-tube (125 × 25 mm), capillary tube, thermometer 0–110 °C, wire stirrer, bung with V notch for the stirrer, clamp and stand, Bunsen burner, white tile, spatula, fuse wire.

Naphthalene and camphor.

Procedure

(a) Place a few crystals of naphthalene on the white tile and crush these to a fine powder by means of the spatula. Keep some of the powder until later.

(b) Seal the end of a capillary tube in a Bunsen flame and allow it to cool. Push the open end of the tube through the powder and up against the flat end of the spatula. Push the naphthalene to the bottom of the tube using the fuse wire. Repeat if necessary until about 0.5 cm length of tube is filled.

(c) Attach the tube to the bulb of the

thermometer by means of two small pieces of rubber cut from the end of a piece of small bore rubber tubing (Figure 4.13).
(d) Insert the bung carrying the thermometer and stirrer into the test-tube, which is half filled with water. Clamp the tube upright in the stand. The open end of the capillary tube should not be in the water.
(e) Carefully and *slowly* heat the test-tube with a small flame. Stir the water constantly and note the temperature at which the solid melts. It is best not to heat the water continuously, so as to ensure a very slow temperature rise.

If you have time, use this first result as a rough reading. Allow the water to cool until the naphthalene resolidifies and then repeat the heating and melting. This time, as the temperature approaches the melting point, heat *very slowly* and stir constantly to obtain a more accurate melting point.
(f) Prepare a sample of impure naphthalene by mixing the rest of the crushed naphthalene with a little crushed camphor. Find the melting point of the impure naphthalene by the same method as before; one melting point is sufficient.
(g) You may be asked to add your results to a table on the blackboard, so that the results obtained by other groups can be compared. If so, find the class average for the melting point of *pure* naphthalene.

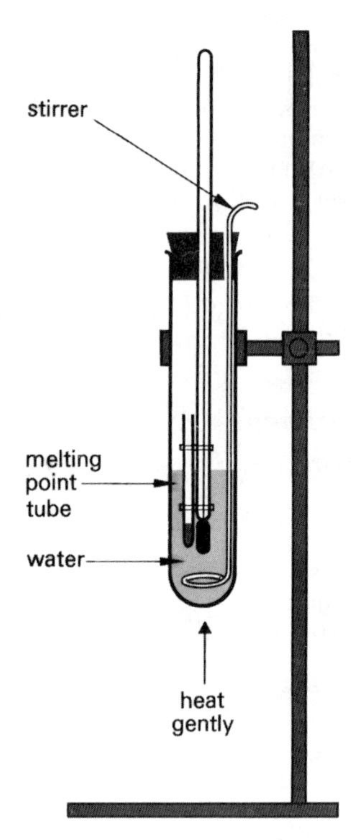

Figure 4.13 Melting point determination

Results
Write down *your* results for (a) the melting point of pure naphthalene, and (b) the melting point of impure naphthalene. Also write down: (c) the class average for the melting point of pure naphthalene; (d) the class *range* (i.e. the lowest and highest values) for the melting point of impure naphthalene; and (e) the melting point of pure naphthalene obtained from a book of data.

... not yet experienced in the technique

Notes about the experiment

(i) The water is used in the test-tube so that the temperature of the capillary tube can be increased *slowly*. If the solid under test has a melting point greater than 100 °C the water can be replaced by a liquid with a higher boiling point.
(ii) This method is particularly useful if only a small amount of solid is available. The next experiment gives a more accurate value for a melting point, but it needs a much larger amount of solid. Melting points are now often found by using an electrically heated apparatus, which also gives accurate results.
(iii) Pure solids have sharp and fixed melting points, and the class average for the melting point of pure naphthalene should be very close to the value given in a book of data. Individual groups may have found different melting points, but this is to be expected because you are not yet experienced in the technique.

Conclusion
Make your own conclusion to the experiment, pointing out that the melting point of a pure solid is sharp and fixed, and comparing this with what happens when the solid is impure. (The effect you have observed in the experiment always happens whenever a solid is impure.)

Points for discussion

1. You will have noticed that you were not asked to calculate a class average for the melting point of impure naphthalene. Why do you think that there was such a wide range of results for this value, and why would it be pointless to calculate the average?

2. What two changes would you make to the apparatus in Figure 4.13 if you were checking the melting point of a solid which melts around 140 °C?

Experiment 4.8
A more accurate method of determining a melting point**

Apparatus
As in Figure 4.14 (the thermometer is for 0 °C to 100 °C), stop-clock, access to fume cupboard.
Naphthalene.

Procedure
The experiment should be carried out in the fume cupboard.
(a) Put about 2 cm depth of naphthalene into the test-tube and clamp it in a beaker which is half filled with water (Figure 4.14). Place the thermometer in the tube.
(b) Heat the beaker on a gauze supported on a tripod. When the naphthalene melts note the temperature, and continue heating until the temperature of the melt rises another 10 degrees. Use the thermometer to stir the liquid.
(c) Lift the clamp and test-tube free of the water, and wipe the tube.
(d) Stir the molten naphthalene with the thermometer, record the temperature and at the same time start the stop-clock. Record the temperature every half minute. When the naphthalene is beginning to solidify, continue to stir the pasty mass, but when it becomes stiff leave the thermometer embedded in it. Continue to record the temperature every half minute until it has dropped a further 10 degrees.
(e) Remove the thermometer by replacing the tube in the beaker of hot water and remelting the naphthalene.
(f) Draw a graph of your results, bearing in mind that the factor you control (time) is always plotted on the horizontal (x) axis, and the dependent variable (temperature) along the vertical axis (y).

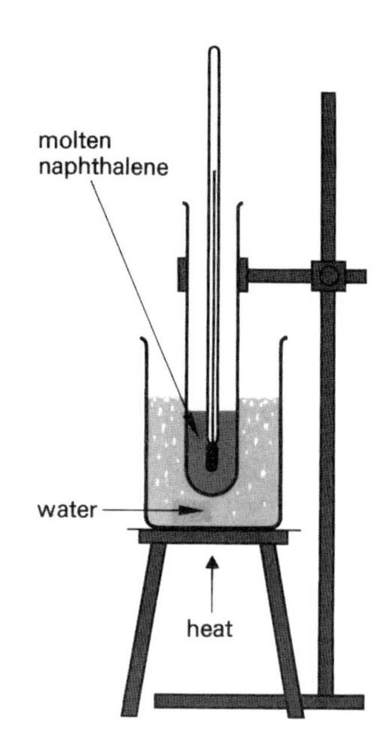

Figure 4.14 Apparatus for determining the cooling curve for naphthalene

Results
Fasten the graph into your notebook. Make sure that it is fully labelled.

Notes about the experiment

The point of interest on your graph is where the naphthalene stayed at the same temperature for a little while. You might have expected the temperature of the naphthalene to fall steadily at all temperatures above room temperature, but it did not always do so during the experiment.

The steady temperature is the melting point of naphthalene, and it should be a simple matter to decide what it is. Add this to your results.

The situation which causes the temperature to remain steady at the melting point is very similar to that which causes the temperature of a liquid to remain steady at its boiling point, even though energy is then being *supplied*. In Experiment 4.8 the change of state is from liquid to solid (rather than that which happens on boiling), and the temperature remains constant even though heat is being *lost* to the atmosphere. The 'hidden' extra energy (latent heat) which prevents the temperature from falling at the melting point comes from the attractive forces which form between the particles as they come back together in the solid state. If this puzzles you, read the account of why steam 'scalds twice' on page 19, and remember that the temperature of the naphthalene would be *expected* to fall steadily at all temperatures above room temperature.

The mixed melting point test

A melting point test is also useful in *identifying* an unknown solid, as well as finding out whether it is pure or not. Suppose that an unknown solid, Z, is suspected of being naphthalene because it melts at 81 °C. Some pure naphthalene is obtained from a laboratory bottle, and a mixture of pure naphthalene with a little Z is made. The melting point of the mixture is found. If the melting point is below 81 °C, Z must be a chemical other than naphthalene and it has acted as an impurity to the naphthalene, lowering the melting point. If the melting point of the mixture is still 81 °C, then Z and naphthalene must be the same substance.

Experiment 4.9
To determine the effect of impurities on the boiling point of water*

Apparatus

Boiling tube, clamp stand with two bosses and clamps, spatula, Bunsen burner, thermometer (0 °C to 110 °C) mounted in bung, antibumping granules, sodium chloride.

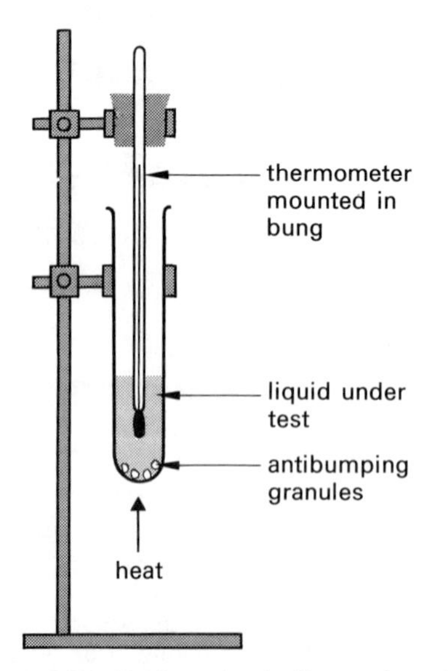

Figure 4.15 Finding the boiling point of a liquid

Procedure

(a) Pour water into the boiling tube to a depth of about 3 cm. Add a pinch of antibumping granules.
(b) Set up the apparatus as in Figure 4.15, making sure that the thermometer is clamped firmly but not too tightly.
(c) Heat the liquid gently to boiling, and note the steady boiling point of the water.
(d) Remove the burner temporarily from under the clamped tube.
(e) While the water is hot but *not* boiling add two heaped spatulas of sodium chloride, and stir gently to dissolve.
(f) Replace the burner, and note the steady boiling point of the solution.

Results

Write down the boiling point of (a) pure water and (b) impure water (sodium chloride solution).

Notes about the experiment

(i) If you found the boiling point of pure water to be different from 100 °C, this might be because you were not reading the thermometer correctly (e.g. not reading it at eye level), or there might be a fault in the thermometer. You should also be able to suggest another explanation.

(ii) If a liquid is flammable, it is not advisable to find its boiling point using an open tube and a naked flame as in this experiment. A simple and safer alternative would be to use electrical heating. Another (better) method would be to use a distillation apparatus as in Figure 4.4, but placing the thermometer bulb in the liquid rather than in the vapour. The vapour is kept separate from the source of heat, and it is safely condensed to a liquid again.

Conclusion
Copy out and complete the following.
When soluble impurities are added to a liquid, the boiling point of the solution is ______ than that of the pure liquid. This test is not quite as reliable as the melting point test, because a change in ______ also affects the boiling point.

Points for discussion

1. A student did a boiling point test on a liquid, and found the boiling point to be 102 °C. The student concluded that the liquid was impure water, but the teacher said that there *could* be two other explanations. What are they? (The thermometer was not faulty, and the student had used it correctly.)

2. Which one of the following statements about salt water is false?
A It boils at a higher temperature than pure water.
B It freezes at a higher temperature than pure water.
C Its density is greater than that of pure water.
D Its appearance is similar to that of pure water.

3. Antifreeze is added to the water in a car cooling system to reduce the risk of the water in the radiator freezing and thereby bursting the radiator. Why does the addition of antifreeze have this effect?

4. Explain the chemical principle behind the reason for adding salt to roads in the winter months.

4.3 A SUMMARY OF CHAPTER 4

The following 'check list' should help you to organize the work for revision.

1. Definitions

(a) Immiscible liquids (p. 37)
(b) Distillation (p. 38)

2. Other ideas

(a) How and when to use the following techniques.

Technique	*What it separates*
Solution and filtration	An insoluble substance from a soluble one, e.g. salt from sand
Crystallization	A solid from its solution, e.g. copper(II) sulphate crystals from copper(II) sulphate solution
Separating funnel	Immiscible liquids, e.g. petrol and water
Distillation (simple)	A liquid from a solution containing a solid (e.g. pure water from sea water), or two liquids with well-separated boiling points
Fractional distillation	Liquids with boiling points close together, and a better separation of other liquid mixtures
Chromatography	Mixtures of substances which are adsorbed to different extents on (e.g.) paper and which also differ in their solubilities in certain solvents. Easier to use with coloured substances
Sublimation	Solids, one of which sublimes on heating

(b) Precautions to be taken in working with flammable liquids.
(c) The main fractions obtained by the fractional distillation of crude oil, and their different properties (see Figure 4.9 and also Table 4.1, p. 54).

Table 4.1 Typical fractions obtained from the laboratory distillation of crude oil

Fraction	*Boiling range/°C*	*Colour*	*Viscosity*	*Flammability*
Petrol	30–75	Colourless	Very mobile (i.e. flows easily)	Very flammable, clean flame
Paraffin	75–120	Colourless or pale yellow	Mobile	Flammable, yellow flame, some smoke
Light oil	120–170	Yellow	Slightly viscous	More difficult to ignite, yellow flame, more smoke
Lubricating oil	170–220	Dark yellow/brown	More viscous	Difficult to ignite, very smoky flame, carbon residue

Note: The residue usually consists of heavy fuel oil and bitumen, which are difficult to separate in a school laboratory. The number of fractions obtained, and their properties, will depend upon the sample of oil used and experimental technique.

(d) A simple understanding of the distillation of crude oil at a refinery.
(e) The use of chromatography to identify chemicals, as well as to separate them.
(f) Impurities make a melting point go down; the fall in melting point depends upon the amount of impurity.
(g) Impurities make a boiling point go up; the rise in boiling point depends upon the amount of impurity.
(h) Atmospheric pressure also affects boiling points (p. 19).
(i) While a pure liquid is boiling, or while a pure solid is melting, or while a pure liquid is freezing (solidifying), the temperature remains steady.

3. *Important experiments*

(a) The experimental details of all of the separation (purification) techniques used in the chapter.
(b) An experiment to determine the melting point of a solid (e.g. Experiment 4.7 and/or Experiment 4.8).
(c) An experiment to determine the boiling point of a liquid (e.g. Experiment 4.9).

QUESTIONS

Multiple choice and short answer questions

1. Which one of the following processes would you use to separate the dyes in universal indicator solution?
A chromatography
B distillation
C use of a separating funnel
D filtration
E crystallization (C.)

2. Copper(II) sulphate crystals can be separated from sand using the four processes listed below. The order in which these processes should be used is

	1st	2nd
A	filtration	solution
B	solution	evaporation
C	filtration	solution
D	solution	filtration
E	solution	evaporation

	3rd	4th
A	crystallization	evaporation
B	filtration	crystallization
C	evaporation	crystallization
D	evaporation	crystallization
E	crystallization	filtration

(C.)

3. Crude oil can be separated by fractional distillation because its constituents
A are compounds
B are chemically similar
C have different boiling points
D have different densities
E do not mix with water (C.)

4. Which one of the following is an *incorrect* statement about the boiling point of water?
A It rises if the atmospheric pressure increases
B It rises if a solid is dissolved in the water
C It is lower at high altitudes
D It increases when water is heated by a powerful energy source
E It is 373 K at 760 mm pressure

5. A student measured the boiling point of deionized water and found it to be 100.5 °C. The thermometer was accurate and he concluded that the water was impure in some way. Can you suggest an alternative explanation?

Structured questions and questions requiring longer answers

6. The use of chemicals to colour foodstuffs has to be carefully controlled to ensure that no possibly harmful chemicals are used. Below is some information concerning two brown dyes which are manufactured chemically for use in food.

Brown NG—composed of only one brown, water soluble dyestuff which is thought to be a health hazard.
Brown G—composed of three harmless, water soluble dyes—yellow, orange, and black.

What process would you use in the laboratory to distinguish between the two dyes?

Show diagrammatically the result you would expect to obtain on subjecting each dye to the process. (A.E.B. 1978)

7. The table below gives some details about the properties of three compounds.

Compound	*Heat*	*Cold water*	*Hot water*
Naphthalene	Sublimes	Insoluble	Insoluble
Calcium fluoride	No effect	Insoluble	Insoluble
Potassium chloride	No effect	Fairly soluble	More soluble

Use this information to devise a scheme for obtaining pure dry samples of the compounds from a mixture of the three, and carefully describe the procedure.

Give one property of calcium fluoride in which it differs from calcium chloride.

How could you prepare a dry sample of calcium fluoride from potassium fluoride solution? (J.M.B.)

5 Acids, bases, and pH

5.1 COMMON LABORATORY ACIDS

Most people have heard of acids, and to a beginner the term may create impressions of highly dangerous, corrosive, and fuming liquids. Although this often is not the case, acids must always be treated with great care and handled sensibly, even when dilute. If possible, you should wear goggles and a laboratory coat whenever acids are being used. You may be able to name some of the common acids used in the laboratory, and you may know some of their formulae. Two acids which are familiar to you in everyday life are hydrochloric acid which is produced by the stomach as an aid to digestion, and ordinary vinegar which contains ethanoic (acetic) acid. Acids can be obtained as pure substances, although they are normally used as *concentrated* or *dilute* solutions, according to whether they have been dissolved in a small or large volume of water respectively. The names, formulae, and appearances of some common laboratory acids are summarized in Table 5.1.

The first three acids in the table are often called the mineral acids because they were first obtained from minerals. Acids used rather less frequently in the laboratory include citric acid, tartaric acid, and oxalic acid, all of which are white solids when pure. Citric acid occurs in many fruits, especially those of the citrus variety, and lemon juice may contain up to 10 per cent of the acid. Tartaric acid is found in

Table 5.1 Some common laboratory acids

Acid	*Formula*	*Appearance of pure substance*	*Appearance of concentrated solution*	*Appearance of dilute solution*
Sulphuric	H_2SO_4	Dense, oily liquid	Colourless, oily liquid	Looks like water
Hydro-chloric	HCl	A gas called hydrogen chloride	Resembles water, often fumes, sharp smell	Looks like water
Nitric	HNO_3	Dense, fuming, oily liquid	Colourless when pure, often yellow-brown due to impurities, usually kept in brown bottles	Looks like water
Ethanoic (Acetic)	CH_3COOH	Colourless liquid called *glacial* ethanoic acid because it easily freezes to an ice-like solid, smells very strongly of vinegar	Colourless liquid, rarely used	Looks like water, smells of vinegar

grapes, and small quantities of the very poisonous oxalic acid occur in rhubarb leaves and sorrel. All three solids are soluble in water.

As the properties of the dilute and concentrated forms of the same acid are often very different, it is important to state the concentration of acid you are referring to. This is either done in words (in written statements) or by using state symbols (behind the formula). Thus $H_2SO_4(aq)$ means the dilute acid (dissolved in water) and $H_2SO_4(l)$ means the pure (liquid) acid. Note that there is a very important difference between a solution and a pure liquid. Similarly, remember that HCl(g) refers to the gas hydrogen chloride whereas HCl(aq) means an aqueous solution of hydrochloric acid; students sometimes read HCl(g) to mean hydrochloric acid solution.

Before you investigate the properties of acids, you must learn an important rule. *Never* dilute concentrated sulphuric acid by adding water to it. The heat produced may cause the water to turn to steam and the acid to spray out in a very dangerous manner. It is possible to dilute concentrated sulphuric acid by adding the acid to water (i.e. mixing it the other way round), but this still requires great care and you will be shown how to do it if ever the occasion arises.

Properties of acidic solutions

When an acid is dissolved in water the solution is described as being acidic, and such solutions all have certain properties in common, as the next few experiments will show. Most dilute acids also have sharp, sour tastes but the tasting of materials in a chemistry laboratory is not recommended under any circumstances, and this is not a 'property' which should be investigated.

'But it looked *like water Sir, and he* was *thirsty'*

How dilute acids affect indicators

It has been known for a long time that acidic solutions can change the colour of certain substances, especially some occurring naturally as the colouring matter of plants. Red cabbage, for example, changes from its natural blue-purple colour to bright red when placed in vinegar, which is an acidic solution. Substances which change colour when solutions of acids or alkalis are added are called *indicators*; many indicators are extracted from plants.

Experiment 5.1
To investigate how acids and alkalis affect indicators*

Note: (i) Alkalis are studied in more detail later in the chapter, but two alkalis (ammonia solution and sodium hydroxide solution) are included in this experiment so that the full colour changes can be investigated.
(ii) Goggles are particularly important when acids and alkalis are being used.

Apparatus
Solutions of methyl orange, neutral litmus, and phenolphthalein. Red and blue litmus papers. Rack of test-tubes, teat pipette, spatula, 100 cm^3 beaker.
Dilute solutions of hydrochloric acid, nitric acid, sulphuric acid, ammonia solution, sodium hydroxide.

Procedure

(a) Make a table for your results as follows.

Indicator	*Colour in nitric acid*	*Colour in sulphuric acid*	etc.

(b) Place about 2 cm^3 of one of the three acid solutions into a test-tube. Add two drops of one of the indicators.

Note: Do not allow a teat pipette containing indicator to touch an acid solution, or even the sides of the test-tube. If this does happen, and the pipette is then placed back in the indicator, all the indicator in the bottle may change colour and be of no further use in the experiment.

(c) Shake the test-tube gently to mix the solutions and note any colour change.

(d) Repeat (b) and (c) using the other acids, alkalis, and indicators until each acid and alkali has been tested with each indicator. (Thoroughly rinse out the test-tubes after each test.) The red and blue litmus papers are used by simply touching the test solution with a small strip of litmus paper. Remember that sodium hydroxide solution and ammonia solution are alkalis, not acids.

Results

Complete the table.

Conclusion

Is it true to say that all of the acid solutions used have the same effect with a particular indicator? Is this true of the alkalis? If so, summarize your conclusion in a simple table as follows.

	Colour of methyl orange in	*Colour of litmus in*	*Colour of phenolphthalein in*
Acidic solutions			
Alkaline solutions			

The action of dilute acids on carbonates and hydrogencarbonates

Most metals form compounds called carbonates, which contain the carbonate ion, CO_3^{2-}; e.g. copper(II) carbonate, $CuCO_3$. A similar range of compounds is the hydrogencarbonates, which contain HCO_3^- ions; e.g. sodium hydrogencarbonate, $NaHCO_3$. The hydrogencarbonates are less important than the carbonates and only a few of them are stable in the solid state.

In the next experiment you will add dilute acids to several carbonates and hydrogencarbonates and try to find out whether the same thing happens each time, whether there are any exceptions to a general pattern, or whether there is no pattern at all. If a gas is formed in any of the reactions it will be necessary to identify it.

If you consider the formulae of the carbonate radical (CO_3) and the three mineral acids (HCl, HNO_3, H_2SO_4), it is apparent that several gases *could* be formed no matter which acid and carbonate are mixed together. The carbonate radical could always produce carbon dioxide, CO_2, or oxygen, O_2, and any of the acids could produce hydrogen, H_2. Make sure that you know the tests for these gases (p. 450) before you begin the experiment, and if you have not done them before, practise them. Certain combinations (but not all combinations) of acid and carbonate could produce other gases, and if there is no *general* pattern of results you may have to test for other gases. For example, hydrochloric acid and a carbonate could produce chlorine, Cl_2, but sulphuric and nitric acids could not do so. Sulphuric acid and a carbonate could produce sulphur dioxide, SO_2, but hydrochloric and nitric acids could not do so.

Experiment 5.2

To investigate the reactions of dilute acids on carbonates and hydrogencarbonates*

Apparatus
Rack of test-tubes, teat pipette, splints, Bunsen burner, spatula.
A selection of solid carbonates and hydrogencarbonates, a solution of sodium carbonate, dilute hydrochloric, nitric, and sulphuric acids, calcium hydroxide solution (lime water).

Procedure
(a) Make a table for your results as follows, leaving two lines between the name of each carbonate.

	Hydrochloric acid	*Sulphuric acid*	*Nitric acid*
Sodium carbonate(s) etc.			

(b) Place a small quantity of a solid carbonate or hydrogencarbonate in a test-tube and cover it with a few cm^3 of a dilute acid. Record your observations and test any evolved gases for hydrogen, oxygen, and carbon dioxide.
(c) Repeat (b) using the other solid samples and the other acids until each acid has reacted with each carbonate and hydrogencarbonate.
(d) Pour a few cm^3 of sodium carbonate solution into a test-tube and add a few cm^3 of a dilute acid. Test any gas evolved as before. Repeat with the other acids.

Results
Record your results in the table, using expressions such as 'effervesced; hydrogen gas formed', or 'no effervescence', etc. The word effervesce is the scientific way of saying 'fizz', and it is used to describe the appearance of gas bubbles in a solution where there is a chemical reaction. Whenever effervescence takes place a gas is being formed; usually there is no need to test for gases from a solution if there is no effervescence.

Conclusion
Make your own conclusion after thinking about the following questions. (i) Do *all* acids appear to react in the same way with *all* carbonates and hydrogencarbonates? (If so, how do they react?) (ii) Do *most* combinations of acid and carbonate react in the same way? (iii) Do they all react differently? (iv) Does dissolving a carbonate in water appear to make any difference to the reaction?

You may be given some equations to be included in your conclusion. Some typical examples are given in Table 6.3 on page 94.
Note: The following points will only be understood when you have done more chemistry. When you revise this work at the end of your course, you must be able to construct a balanced equation for the reaction between an acid and any common carbonate. At this stage this may be difficult, as you are not familiar with some of the other products. If, at the end of the course, you prefer to use ionic equations, remember that just one ionic equation applies to any reaction between a carbonate and an acid:

$$CO_3^{2-}(s) \text{ (or aq)} + 2H^+(aq) \rightarrow CO_2(g) + H_2O(l)$$

The other ions from the acid and the carbonate do not take part in the reaction and so do not appear in the equation; they are spectator ions (page 91). Similarly, for a hydrogencarbonate:

$$HCO_3^-(aq) + H^+(aq) \rightarrow CO_2(g) + H_2O(l)$$

Spectator ions

The reactions between dilute acids and metals

Acidic solutions can also react with metals to produce gases. Look again at the argument used to decide which gases could be formed by reacting acids with carbonates, on page 58. If you apply the same idea to the reactions of acids with metals you will see that the only gas which could be formed from *any* combination of acid and metal is hydrogen, H_2. This does not mean to say, however, that other gases cannot be formed from some combinations of acid and metal. An important part of the next experiment is to find out which, if any, combinations of acid and metal *do* produce hydrogen. As these reactions are a little more difficult to classify, you are given some guidance on how to 'sort out' your results under the heading 'Conclusion'.

Experiment 5.3
To investigate the reactions of dilute acids with metals*

Apparatus

Rack of test-tubes, splints, Bunsen burner, spatula, protective mat. Access to fume cupboard.

A range of metals such as calcium, zinc, iron, aluminium, magnesium, and copper; dilute solutions of the three mineral acids.

Procedure

(a) Make a table for your results as shown below.

(b) Place a small sample of magnesium metal in a test-tube and cover it with a few cm^3 of dilute hydrochloric acid. Observe what happens and test any evolved gas for hydrogen. Write your observations in the table, using words such as 'effervescence; hydrogen formed' etc.

(c) Repeat (b) with the other metals and acids until each acid has been added to each metal. If no reaction takes place with an acid in the cold, you may *warm* the mixture but do not boil it. Distinguish between different reactions by stating 'very rapid reaction' or 'needed to be warmed' etc. The reactions with nitric acid should be performed in the fume cupboard.

Conclusion

In order to make it easier for you to classify your results, these have been summarized in Table 5.2, but letters have been used instead of the names of the metals and acids. For your conclusion, make a table like this but replace each letter with a chemical name. Study the table and your results carefully; the exercise needs thought. You may also be asked to write some typical equations for the reactions, examples of which are given in Table 6.3 on page 94. When you come to revise this work at the end of the course you must be able to construct balanced equations for the reactions between acids and metals which produce hydrogen. (It is not necessary to be able to construct balanced equations for the reactions between metals and nitric acid.)

Metal under test	*Reaction with hydrochloric acid*	*Reaction with sulphuric acid*	*Reaction with nitric acid*

Other properties of acidic solutions

Acids also react with bases, and this is studied in Section 5.4. Acidic solutions are also electrolytes, as you will learn in Chapter 13.

Tests for acid solutions

The most reliable tests for an acidic solution are the colour changes with indicators and the reaction with a carbonate or hydrogencarbonate. For example, dilute hydrochloric acid will give an indicator test (e.g. blue litmus goes red) and it will

Table 5.2 The results of Experiment 5.3

Metals	*Dilute acids X and Y*	*Dilute acid Z*
Sodium, Potassium	THESE METALS ARE VERY REACTIVE AND MUST NEVER BE ADDED TO AN ACID	
A, B, C, D	These fairly reactive metals react with these dilute acids, and appear to 'dissolve'. The mixtures effervesce to produce hydrogen, although A hardly reacts with X. D often contains impurities, which produce an unpleasant smell	These fairly reactive metals react with Z, 'dissolving' and producing gases which do not usually contain hydrogen. The main gases formed are oxides of nitrogen, which you were unable to test for but may have seen as brown fumes
E	THIS METAL SHOULD BE FAIRLY REACTIVE BUT IT IS PROTECTED BY A LAYER OF ITS OXIDE, WHICH PREVENTS REACTION WITH MOST DILUTE ACIDS	
F	This unreactive metal does not react with these acids	This unreactive metal reacts with this acid and 'dissolves'. The mixture effervesces and produces brown oxides of nitrogen

effervesce with a carbonate to form carbon dioxide. These tests show that it is an acid, but do not show *which* acid. Individual acids can also be identified; e.g. hydrochloric acid will, in addition to the 'acidity' tests, also give a *chloride* test (p. 451) because it contains chloride ions. How would you recognize sulphuric acid or nitric acid?

The properties of acids are summarized at the end of the chapter and also in Table 6.3 on page 94.

Points for discussion

1. Name the acids found in lemons, milk, and vinegar.
2. Name three mineral acids. Which of these play a part in digestion?
3. What is meant by an indicator? Name two common indicators and give their colours in acid solution.
4. Name two metals which react with dilute hydrochloric acid. What is the name of the gas evolved? Describe how you would test for this gas.
5. Name the gas evolved when a dilute acid reacts with a carbonate or hydrogencarbonate. Describe carefully how you would test for this gas.

5.2 BASES AND ALKALIS

You may already be familiar with the terms base and alkali. The 'parent' term is in fact base, for an alkali is a special kind of base. Acids and bases may be regarded as 'chemical opposites' and when they react together each destroys the other's characteristic properties, forming a neutral substance. You will be studying this opposing action in more detail in Section 5.4, but before that you will work with a number of the common laboratory bases and learn some of their characteristic properties.

The relationship between bases and alkalis

A compound which consists of only an element and oxygen is called an oxide, e.g. copper(II) oxide, CuO. Many elements form compounds containing hydrogen and oxygen in which the hydrogen and oxygen atoms are joined together to form OH groups, and such compounds are called hydroxides, e.g. sodium hydroxide, NaOH.

As a general guide, ammonia (NH_3) and most oxides and hydroxides of *metals* are bases, but whereas all the acids you studied in Section 5.1 are soluble in water, many bases are insoluble. Bases which do dissolve in or react with water form solutions which are given the special name alkali. All alkalis are thus automatically bases and have the same properties as bases and in addition they dissolve in water. All alkalis are bases but not all bases are alkalis (cf. all buses are vehicles but not all vehicles are buses).

A base is usually defined as a substance which can neutralize an acid to form a salt and water (and nothing else).

Note that some substances which are not metal oxides or hydroxides can also neutralize acids to from salts, but there is another product as well as water. For example, calcium carbonate is used to neutralize soil acidity (p. 395) and metals neutralize acids (p. 81), but these are not acting as bases in such reactions because the products are not *a salt and water only*.

BASES

Usually oxides or
hydroxides of metals

↓

Some are soluble in water,
others are not, so there is
a subdivision into

INSOLUBLE BASES | SOLUBLE BASES (ALKALIS)

For many pupils the word 'base' remains a word which they do not use properly and never really understand until late in the course. There is no excuse for this because it is very easy to recognize a base. Remember that any metal oxide or hydroxide (and also ammonia) can act as a base. This means that if you are working with, for example, copper(II) oxide, you should automatically think 'copper(II) oxide is an oxide of a metal, so it is a base'. Similarly, if you are asked to name some bases you could safely list copper(II) oxide, copper(II) hydroxide, calcium oxide, calcium hydroxide, etc.

Common laboratory alkalis

There are very many bases (any metal oxide or hydroxide) but only a few alkalis. An alkali is a *solution* of a base in water, and most oxides and hydroxides of metals are *insoluble* in water. However the alkalis are frequently used and so it is important to learn the names and formulae of the common laboratory alkalis, which are listed in Table 5.3. Calcium hydroxide is rarely used as an alkali in elementary chemistry because its solubility in water is very small, so most of your work with alkalis is likely to be restricted to sodium hydroxide, potassium hydroxide, and ammonia solution.

An alkali is a base which dissolves in water.

Precautions in using alkalis

Most students do not realize that alkaline solutions are more dangerous in the laboratory than dilute acids. The so-called caustic alkalis (sodium hydroxide and potassium hydroxide) rapidly attack the cornea if they splash into the eye. They also attack clothing. It is therefore particularly important that goggles should *always* be worn when using these chemicals.

Table 5.3 Laboratory alkalis

Chemical name	*Formula*	*Common name*	*Appearance before dissolving in water*	*Appearance of solution*
Sodium hydroxide	NaOH	Caustic soda	White solid as pellets, sticks or flakes	Resembles water
Potassium hydroxide	KOH	Caustic potash	White solid as pellets, sticks or flakes	Resembles water
Calcium hydroxide	$Ca(OH)_2$	Slaked lime (solid) lime water (solution) milk of lime (suspension)	White powder	Resembles water
Ammonia solution	$NH_3(aq)$	'Ammonia solution'	A colourless pungent gas ammonia, NH_3	Both concentrated and dilute solutions resemble water but the concentrated solution has a very powerful smell of ammonia gas—care!

Concentrated solutions of the caustic alkalis will attack the skin, and even in dilute solution they feel 'soapy' on the skin because of this reaction. This slipperiness also means that if a caustic alkaline solution is spilled down the outside of a beaker, for example, it is easy for the beaker to slip out of the hand.

Ammonia solution must also be used with care; the gas coming off from it is unpleasant and in high concentration can be dangerous, so bottles should be held well away from the face.

Properties given by all bases (i.e. insoluble bases and alkalis)

Neutralization of acidic solutions

Bases and alkalis will neutralize acidic solutions to form salts. This is studied in more detail in Section 5.4.

The reaction with ammonium compounds

When a base is heated with an ammonium compound, a reaction takes place (Experiment 5.4). This reaction is useful to identify a base (or an ammonium compound) and for preparing an important gas. However, insoluble bases only react slightly in this way, and so only alkalis (when solid or in solution) are normally used for the test.

Experiment 5.4

The action of alkalis on ammonium compounds*

Apparatus

Rack of test-tubes, Bunsen burner, test-tube holder, spatula, teat pipette.
Ammonium chloride, ammonium sulphate, calcium hydroxide, dilute sodium hydroxide solution, red litmus paper.

Procedure

(a) Place a spatula measure of ammonium chloride into a test-tube. Add 2 cm³ of sodium hydroxide solution.
(b) Heat the mixture *gently* until it is almost boiling. (When you heat a solid mixture, later, heat it more strongly.)

(c) *Carefully* sniff the gas evolved in the reaction, but do not do so while the mixture is being heated. The best way to do this is to hold the tube some distance from your face, and then direct some of the gas to your nose by waving a hand over the tube towards your face.
(d) Test the gas with moist red litmus paper. (Do not touch the tube with the litmus paper, for the glass may have sodium hydroxide solution on it, and this will make the paper turn blue anyhow.)
(e) Repeat procedures (a) to (d) but using ammonium sulphate instead of ammonium chloride.
(f) Repeat steps (a) to (d), but using solid calcium hydroxide *instead* of sodium hydroxide solution. Make sure that the two solids in the tube are thoroughly mixed. Then try a solid mixture of ammonium sulphate and calcium hydroxide.

Notes about the experiment

(i) Even if you do not recognize the gas given off by its smell, its effect on litmus paper should prove its identity. Ammonia gas is the *only* common alkaline *gas*.
(ii) When you understand ionic equations you will be able to use the following for the reaction between any alkali and an ammonium compound:

$$OH^-(aq) + NH_4^+(aq) \rightarrow NH_3(aq) + H_2O(l)$$

Conclusion
Copy out and complete the following.

Sodium hydroxide and calcium hydroxide are both ______ of metals, and so they are ______. They both also dissolve in water, so they belong to the subgroup of bases called the ______. Whenever an alkali is heated with an ammonium compound, ______ is formed and this is the only common gas which turns ______ litmus ______. This reaction can be used to identify a ______, or an ______, and also for the preparation of the gas ______.

Properties given only by alkalis (because of their solubility)

Colour changes with indicators
Solutions of alkalis (but not insoluble bases) will change the colour of indicators. Your results for Experiment 5.1 included the colour changes caused by alkalis. You should learn the various colour changes of the common indicators as they provide a simple way of deciding whether a solution is acidic or alkaline. When you do such a test, always wash out test-tubes thoroughly before using them.

Alkalis as electrolytes
Solutions of alkalis conduct electricity and are therefore electrolytes. You will learn more about this kind of reaction in Chapter 13.

Alkalis as sources of hydroxide ions
Alkalis are used as a laboratory source of hydroxide ions, $OH^-(aq)$, e.g. in order to make metallic hydroxides by precipitation. You may make use of this when you study metal compounds in Chapter 19. Remember that all solutions of alkalis contain hydroxide ions.

A summary of the properties of bases
The reactions involving acids, bases, and salts are interconnected and are summarized in Table 6.3 (p. 94) and at the end of this chapter.

5.3 THE pH SCALE

All acidic solutions contain hydrogen ions, $H^+(aq)$, and all alkaline solutions contain hydroxide ions, $OH^-(aq)$. If a solution contains a high concentration of $H^+(aq)$

ions it is very acidic, and if a solution contains a high concentration of $OH^-(aq)$ ions it is very alkaline. The degree of acidity depends not only upon the volume of water present but also on the substance dissolved. You will learn later (p. 74) that some acids (the strong acids) very easily form $H^+(aq)$ ions when they dissolve, but some (the weak acids) are very reluctant to form $H^+(aq)$ ions.

You may see a simple demonstration of the fact that some acidic solutions are stronger than others. If equal volumes of nitric, hydrochloric, and ethanoic acids, each containing the same number of dissolved acid molecules, are added to a standard coil of magnesium ribbon in a test-tube, it can be seen that the acids take different times to react with the metal. The volume of water is the same in each case, the number of molecules of acid dissolved is the same in each case, and yet some solutions are more acidic than others. This is because some acids (strong acids) form $H^+(aq)$ ions more easily than do other acids, and it is these $H^+(aq)$ ions that are responsible for the acidity of the solution.

The pH scale and universal (or full range) indicator

We obviously need some way of measuring, even if only approximately, the concentration of $H^+(aq)$ or $OH^-(aq)$ ions in a solution, so that we know *how* acidic or alkaline the solution is.

In order to compare acidity and alkalinity in a scientific way, a scale of numbers is used called the pH scale (pH is derived from a German expression for the 'power of hydrogen'). This scale ranges approximately from 0 to 14. If a solution has a pH of less than 7 it is acidic. Neutral liquids have a pH of 7, and alkaline solutions have a pH of more than 7. An acidic solution with a pH of 0 or 1 is very acidic, and a strongly alkaline solution has a pH of 13 or 14.

The difficulty with an ordinary indicator such as litmus is that it only shows whether a solution is acidic, neutral, or alkaline; it does not show how acidic or alkaline a solution is. However, *universal indicator* (or full range indicator) is a special mixture of indicators, which changes to different colours at different concentrations of $H^+(aq)$ and $OH^-(aq)$ ions. Universal indicator can therefore change to a number of different colours, each of which tells us something about the degree of acidity or alkalinity in the solution being tested. The pH scale, and the colours given by universal indicator at different pH values, are illustrated in Figure 5.1.*

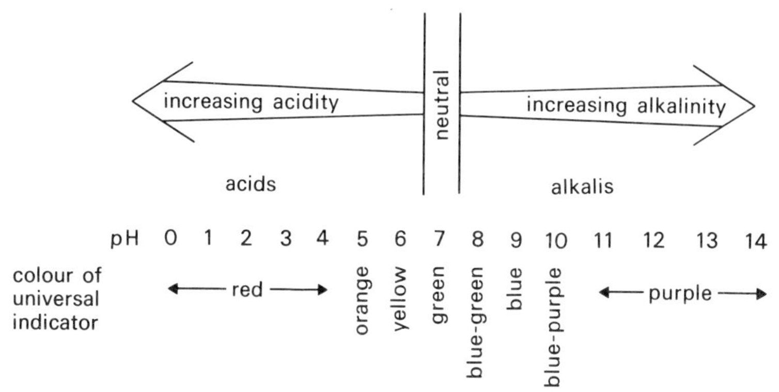

Figure 5.1 The pH scale and the colours of universal indicator

* It is useful to remember that the pH scale changes in units of powers of 10, i.e. a solution of pH 1 has a concentration of $H^+(aq)$ ions 10 times greater (*not* 2 times greater) than that in a solution of pH 2, and 100 times greater than that in a solution of pH 3.

The term pH is being used increasingly in everyday life, e.g. gardeners and farmers talk of the pH of soils and hairdressers of the pH of shampoos. You may be asked to do some experiments in which you find the pH of common substances such as toothpaste, bleach, vinegar, stomach powder, and solutions of laboratory chemicals, using either universal indicator solution or universal indicator paper. Remember that it is very important that all tubes used for such tests should be absolutely clean, and any water used should be as near neutral as possible.

There are much more accurate ways of finding the pH of a solution than the simple use of universal indicator. It is very difficult to estimate the pH of a strongly acidic or alkaline solution using universal indicator, because 'red' corresponds to several pH units, as does 'purple'. This is not particularly important in elementary chemistry, where the properties of a solution of pH 3 are virtually the same as those of a solution of pH 4. For more advanced work or processes where an accurate value of pH is required, instruments such as a pH meter (Figure 5.2) allow a very accurate determination of pH to be made directly.

Figure 5.2 Portable pH meter (Electronic Instruments Ltd)

5.4 NEUTRALIZATION

Introduction

As acids and bases are chemical opposites, when they are mixed they 'cancel out' each other's properties. Substances which are acidic or alkaline can cause problems in everyday life, and they often have to be neutralized. Sufferers from indigestion often have too much acid formed in the stomach, farmers may be unable to grow certain crops if the soil pH is unsuitable, and you probably know that bacteria feeding on food particles trapped between teeth produce acids which are partly responsible for tooth decay. Bases can be used to overcome difficulties caused by too much acidity. Similarly, acids can be used to neutralize bases, but it is important not to add too much of the 'neutralizing' substance or a new problem may be created. For example, if too much acid is added in an attempt to neutralize a base, the final substance will be acidic, and this may be just as inconvenient as the starting situation.

You may have done a very simple 'experiment' to illustrate what is meant by the term neutralization. If a piece of an acid drop is placed in the mouth and allowed to remain there until the flavour develops, the taste of the acid drop 'disappears' if a *little* sodium hydrogencarbonate or stomach powder is placed on the tongue. This happens because the taste of the acid drop is due to a mildly acidic solution which is formed as it dissolves. (You can show that a crushed acid drop stirred with a little water causes the water to become acidic, e.g. by testing the solution with universal indicator.) Sodium hydrogencarbonate and stomach powder both form weakly alkaline solutions, and they neutralize the acidity of the acid drop and cause the taste to disappear. If you use too much of these mild 'alkalis', you will have an unpleasant taste caused by an excess of alkali—another illustration of the importance of 'getting the quantities right' when you perform a neutralization.

The use of indicators to illustrate neutralization

The neutralization of an acid drop showed how an alkali such as a stomach powder can overcome problems due to excess acidity. No attempt was made then to measure the amount of alkali used as an excess of such a mild alkali is unimportant and will result in no major discomfort. If a farmer were to add too much alkali to an acid soil he would obtain an alkaline soil which may be just as difficult to produce crops from as the original. He would have swung the pendulum from the acid side of the pH scale *through* the neutral point and into the alkaline side. This may or may not be important to a farmer but in chemical reactions it is sometimes essential to be able to mix just the right amounts of acid and base to produce a neutral solution.

The pH changes which take place when a substance is being neutralized can be seen by adding milk of magnesia (a safe, mild 'alkali') to vinegar (a weak acid). Typically one milk of magnesia tablet is ground up with 10 cm^3 of water to provide the alkali, and a very dilute ethanoic (acetic) acid solution (e.g. 0.5 mol dm^{-3}) can be used for the 'vinegar'. If 5–10 drops of universal indicator are added to 10 cm^3 of the acid solution in a beaker, and the alkali suspension added dropwise from a teat pipette, with constant stirring and with pauses between additions, the indicator changes from red, to orange, to yellow as the acid is neutralized by the alkali and the pH of the solution slowly rises. You may see the green colour when the solution is neutral, but this will almost certainly quickly change to the 'alkaline colours' of blue-green, blue, and then purple as the pH continues to rise and the solution becomes alkaline due to excess alkali.

To summarize, when an alkali is added slowly to an acid, the pH of the solution gradually rises as the solution becomes less acidic and the acid is neutralized. If the addition of alkali continues after the neutral point (pH 7), the solution then becomes alkaline. The opposite happens when an acid is added to an alkali. It is important, therefore, to have a method by which we can add just the right amount of acid or alkali needed to neutralize something, so that the final pH is 7. The next experiment shows a simple way of achieving this.

Experiment 5.5
Neutralizing a solution of hydrochloric acid with sodium hydroxide solution*

Apparatus

Teat pipette, 25 cm^3 measuring cylinder, white tile, 250 cm^3 beaker, 250 cm^3 conical flask.

Solutions of sodium hydroxide and hydrochloric acid each of concentration 0.1 mol dm^{-3}; phenolphthalein indicator.

Procedure

(a) Measure out 25 cm^3 of the hydrochloric acid solution into the conical flask and stand it on the white tile.

(b) Add 2–3 drops of phenolphthalein to the acid in the flask.

(c) Rinse out your measuring cylinder *several times*, and then use it to measure 25 cm^3 of sodium hydroxide solution.

(d) Pour about 15 cm^3 of the sodium hydroxide solution into the acid, and carefully swirl the mixture. If the indicator changes to a scarlet colour and the colour remains after swirling the mixture, you have passed the neutral point and made the solution too alkaline. If a scarlet colour does not stay after swirling, continue to step (e).

(e) Add approximately 2 cm^3 of the alkali to the acid, and again swirl the contents of the flask. Do this several times, if necessary, until the scarlet colour of the indicator remains even on swirling. Note the *total* volume of alkali you have added at this point.

(f) The first result is only a rough guide to show you *approximately* how much alkali is needed to neutralize 25 cm^3 of the acid. Now you will repeat the operation more carefully in order to gain a more accurate neutralization.

(g) Thoroughly rinse out both the measuring cylinder and the conical flask. Pour a fresh 25 cm^3 sample of the acid into the conical flask, and add 2–3 drops of indicator.

(h) Thoroughly rinse out the measuring cylinder and use it to measure 25 cm^3 of sodium hydroxide solution.

(i) Pour sodium hydroxide solution from the cylinder into the acid until you have used about 3 cm^3 *less* than the volume you used in the first experiment. Swirl the mixture; it should still be colourless.

(j) You now know that you are very near to the neutral point. The next stage must be done carefully, and with patience.

(k) Add more sodium hydroxide solution, *one* drop at a time, from the teat pipette, swirling after each addition. Repeat the addition and swirling until the solution is very pale scarlet. You now have a solution which is as near to neutral as you can get using this indicator. *Keep this solution (stoppered) for Experiment 6.1.*

Notes about the experiment

(i) There are no results or conclusion to write, as this is only a *very simple technique*. You will learn in the next chapter that this technique can be further improved upon when the situation needs it. For example, a measuring cylinder is not a very accurate way of measuring volume, and the final solution, although neutral, is contaminated with the indicator.

(ii) The white tile is used to provide a white background so that colour changes can be easily seen. The conical flask is used for the mixing, rather than a beaker, so that the contents can be swirled without spilling.

(iii) Make sure that you understand the need for the 'rough' estimate of how much alkali to add; if you had added the alkali drop by drop right from the beginning, you would have had to be very patient in finding the neutral point. The rough estimate allowed you to take a short cut; you were near the neutral point before you started using *drops* in the second (more accurate) part of the experiment.

(iv) This problem of adding the right amount does not arise in the special case of neutralizing an acid by adding an *insoluble* base, and it is important that you understand why. Once enough insoluble base has been added to neutralize the acid, adding more base (i.e. an excess) will *not* make the product alkaline, for only soluble bases (alkalis) can affect pH directly. In other words, as soon as the acid has been neutralized, the pH is 7 and stays at 7 no matter how much excess insoluble base is added. This particular way of neutralizing an acid is simple, and is often used. For example, in Chapter 6 you will neutralize acids by this method in order to make salts. Similarly, farmers may prefer to use calcium carbonate (limestone) to neutralize an acid soil, rather than calcium hydroxide (slaked lime, or hydrated lime). This is because an excess of the calcium carbonate only neutralizes the acid; it is itself insoluble, and will not make the soil alkaline. An excess of calcium hydroxide, on the other hand, can make parts of the soil alkaline because it is slightly soluble. These changes are summarized in Table 5.4. The other possibilities for neutralization, i.e. by adding an acid to an insoluble base, or by adding an acid to an alkali, or by adding an alkali to an acid, need to be controlled in some way (as in the last experiment) so that they can be stopped at the neutral point and so avoid problems caused by adding an excess.

Table 5.4 The neutralization cycle

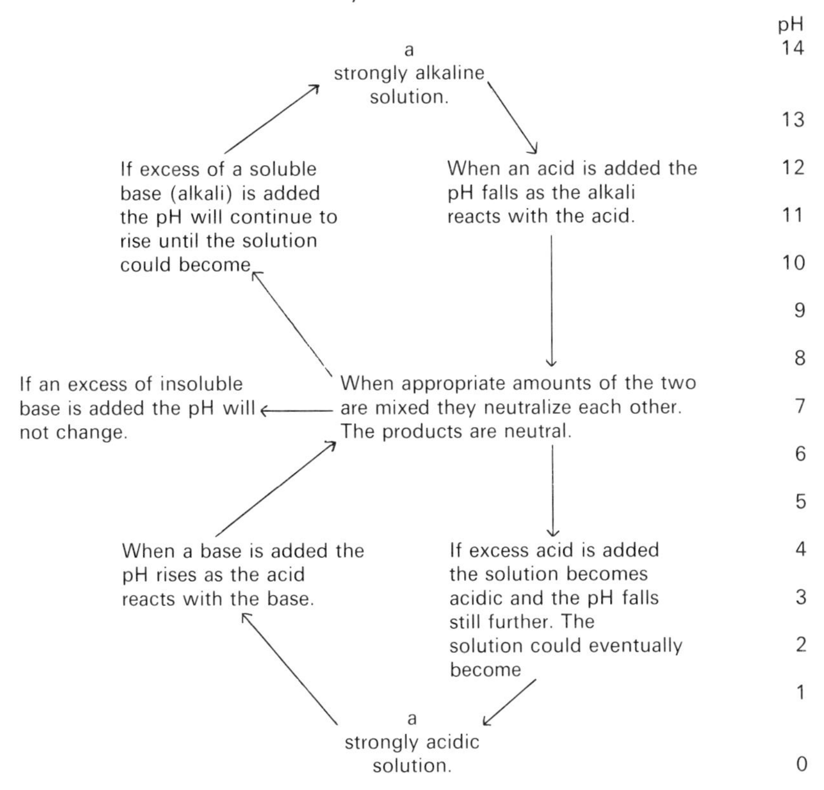

Check your understanding

1. A solution of sulphuric acid
A is a poor conductor of electricity
B has a pH below 7
C will react with copper to produce hydrogen
D has no reaction with zinc carbonate

2. What do we mean by an *alkali*? Name two common alkalis and say what effect they have on two named indicators.

3. Sodium hydroxide and potassium hydroxide solutions both turn red litmus blue and have a soapy feel. What other properties would you expect these solutions to have in common?

4. Select from the pH values 1, 5.5, 7, 8, and 11 the one you consider most applicable to each of the following solutions: lime water, household soap, hydrochloric acid, lemon juice, sodium chloride. (W.J.E.C.)

5. Explain what is wrong with each of the following statements.
(a) When universal indicator is added to an acid solution, the acid goes red.
(b) Zinc reacts with nitric acid to produce hydrogen gas.
(c) If bottles of dilute sulphuric and hydrochloric acids have lost their labels, it is not possible to tell the difference between them, because the two acids have the same properties.
(d) Sand is an insoluble base because when it is added to water containing universal indicator, there is no colour change.
(e) A solution of pH 2 is twice as acidic as a solution of pH 4.
(f) Litmus is a good indicator for distinguishing between a strong acid and a weak acid.
(g) A student was adding the base copper(II) oxide to an acid in order to neutralize it. The teacher said that the student had wasted some of the base by adding too much of it, but the student showed that the solution was neutral and this proved that an excess of base had not been added.

5.5 MORE ABOUT ACIDS AND ALKALIS

Note: You will probably study this section at a fairly late stage in your course. It is not intended that you should follow straight from Section 5.4 to this section, but it is included in the same chapter as the other work on acids and alkalis so that the material is all together for revision purposes.

What do all acids have in common?

You know that acidic solutions all show similar properties, but what is it that makes them behave in this way? You may have noticed that the only obvious link is that all of the common acids contain the element hydrogen. You may think that acids could be defined as substances which contain hydrogen, but this is not the case. Methane (CH_4), glucose ($C_6H_{12}O_6$), and many other substances contain hydrogen but they are not acids.

Acids were once defined as substances containing hydrogen which can be replaced by a metal. This works nicely for sulphuric, hydrochloric, and many other common acids, each of which contains hydrogen which can be replaced by a metal to form a salt. This definition at least helps us to think of the hydrogen part of an acid as being something special (not 'any old hydrogen'). In other words, some of the hydrogen in a chemical must be capable of being replaced by a metal before we can say that the substance is an acid. For example, the formula of ethanoic (acetic) acid is CH_3COOH, but it is only the 'odd' hydrogen atom which can be replaced by a metal; the other three atoms of hydrogen are not 'acidic'. This is still not a full explanation, however, for iron(III) chloride solution, $FeCl_3(aq)$, is acidic (as litmus will show) but the formula of the solute does not contain any hydrogen. The next few experiments will help to explain what it is that makes an acid an acid, and why some hydrogen atoms can fulfil this requirement whereas others cannot.

Experiment 5.6
To test the acidity of pure acids**

Apparatus

Rack of *dry* test-tubes, spatula.

Glacial ethanoic (acetic) acid, solid tartaric acid, dilute ethanoic acid, *dry* indicator papers, magnesium ribbon and sodium hydrogencarbonate (all dry reagents kept in a desiccator); universal indicator solution.

Procedure

Note: Glacial ethanoic acid must be used with great care; it is important to wear goggles, and do not allow the acid to come near your face.

Test a sample of each of the acids with *dry* indicator paper, magnesium ribbon and sodium hydrogencarbonate. Use dry containers for the glacial ethanoic and tartaric acids. Also test a sample of each acid with a little water followed by a few drops of universal indicator solution.

Results

Record your results in a simple table.

Notes about the experiment

Only the dilute ethanoic acid behaved as a typical acid solution with the dry or solid reagents. The glacial acid and tartaric acid did not give these properties with the dry or solid reagents. The only difference between the acid which showed acidic properties and those which did not, was that the 'typical' acid was dissolved in water. The other acids are *potential* acids in this experiment; they were not able to show their acidity until water was added. As soon as water is added they then show acidic properties.

The part played by the solvent in acidic solutions

You may have made two rather surprising conclusions so far in this section. You have found that some substances labelled as acids do not *always* act as acids (e.g. tartaric acid); they only do so if mixed with something else, e.g. water. As a complete contrast, you have also learned that some substances not labelled as acids can form acidic solutions when dissolved in water (e.g. iron(III) chloride). Note that a solvent (water) is needed in both cases.

There is nothing wrong with calling solid tartaric acid an acid; it *is* an acid, but the properties you think of as 'acidic properties' are those of acidic *solutions*. An acid can only give these reactions when it is dissolved in a solvent such as water; just being a pure solid or liquid is not enough. Pure acids are still called acids, however, because they have the potential to show acidic properties.

Experiment 5.7

How does water give an acid its 'acidic' properties?***

Apparatus

Hydrogen chloride generator for the *dry* gas, dry boiling tubes, thermometer (0–110 °C) mounted in bung to fit boiling tube, bungs, access to fume cupboard, filter paper, rack of test-tubes.

Dry methylbenzene (toluene), dry indicator paper, dilute hydrochloric acid, solution of dry hydrogen chloride in dry methylbenzene, magnesium ribbon, sodium hydrogencarbonate.

Procedure

(a) Fill two boiling tubes with dry hydrogen chloride in the fume cupboard, insert bungs, and keep the tubes in a rack.

(b) Moisten the bulb of the thermometer with dry methylbenzene, and note the temperature.

(c) Remove the bung from one of the tubes of hydrogen chloride, and quickly replace it with the bung holding the thermometer. Does the temperature go up?

(d) Remove the thermometer and bung, wipe off the methylbenzene using filter paper, and dip the bulb in water. Note the temperature.

(e) Remove the bung from the other tube of hydrogen chloride and replace it with the bung holding the thermometer, making sure that the bulb of the thermometer is still wet. Does the temperature go up?

(f) Test the solution of hydrochloric acid (hydrogen chloride in water) separately with indicator paper, magnesium ribbon, and sodium hydrogencarbonate.

(g) Repeat (f) with the solution of dry hydrogen chloride in methylbenzene, using *dry* test-tubes and making sure that no moisture comes into contact with any of the reagents.

(h) Repeat (g) but this time add a few drops of water to the solution *after* adding the appropriate reagent.

Results

Record your results in a table as follows.

	Hydrogen chloride and water (hydrochloric acid)	*Hydrogen chloride in dry methylbenzene*
1. Acidic properties (e.g. observations with indicators, magnesium, sodium hydrogencarbonate)		
2. Temperature change on adding gas to solvent		

Notes about the experiment

The experiment should have further convinced you that water is needed before hydrogen chloride can show the properties you think of as 'acidic'. The solvent, water in this case, is very important. Not every liquid can act as a solvent in which a substance can form an acidic solution. Hydrogen chloride can dissolve in methylbenzene, for example, but the solution does not then have the properties of acidic solutions. There is clearly something special about water, and you will remember that solid tartaric acid and glacial ethanoic acid also gave typical acidic properties when dissolved in water, as did iron(III) chloride.

The final key to the problem is therefore to find how water makes these solutions all show the same acidic properties. (Water is not the only solvent which can bring out acidic properties, but it is the only one you need to know about at this level.) Part of the answer is given by the temperature changes in the experiment. They should show that there is no significant temperature change when hydrogen chloride dissolves in methylbenzene, but there is a rise in temperature when hydrogen chloride dissolves in water. This is because there is a chemical *reaction* (which is exothermic) between hydrogen chloride and water. One of the products of this reaction is the actual 'chemical' which causes acidic behaviour.

You will remember from your work on electrolysis (p. 209) that the only solutions or pure liquids which can conduct electricity are ionic. Glacial ethanoic acid, solid tartaric acid and other *pure* acids are covalently bonded and will not conduct electricity. As soon as they are dissolved in water, they become electrolytes. This is because they *react* with water to form ions (and at the same time show acidic properties if given a chance to do so.) They must all form the same cation, because when any dilute solution of an acid is electrolysed, hydrogen gas is always formed at the cathode; the solution contains no other type of cation. This gas comes from hydrated hydrogen ions, $H^+(aq)$:

$$2H^+(aq) + 2e^- \rightarrow H_2(g)$$

It is the $H^+(aq)$ ions which make acidic solutions behave in the way they do. All acidic solutions contain these ions, and they are formed only when the pure acid is dissolved in water. The acid then *reacts* with the water to form these $H^+(aq)$ ions. Only solutions which contain these ions can show acidic properties. A solution of iron(III) chloride is acidic because it also contains these ions, which are formed when the solute reacts with water molecules.

A liquid which contains *covalently* bonded hydrogen (e.g. pure ethanoic acid, CH_3COOH, or dry hydrogen chloride dissolved in methylbenzene) will not affect indicators, etc., because the acid molecules cannot form $H^+(aq)$ ions on their own. They will act as acids, however, if they can react with another substance in such a way that the other substance gains a H^+ ion from the acid. Water is just one of the chemicals which can react with acids in this way, i.e. it accepts an H^+ ion from them, e.g.

$$HCl(g) + \text{water} \rightarrow H^+(aq) + Cl^-(aq)$$

covalent molecules	ions responsible for acidic behaviour

Modern definitions of acids and bases

Hydrogen ions, which as we have seen can be donated by acids, are the same as protons. This is because when a hydrogen atom loses its one electron, it becomes a hydrogen nucleus which contains a single proton. We can therefore define an acid as follows.

An acid is a substance which is capable of DONATING a proton to another substance.

This definition of an acid was first introduced by J. N. Brønsted and T. M. Lowry in 1922. In many of the examples encountered at elementary level, the 'other substance' is water, and so solutions of acids in water are acidic because the acids have donated protons to water molecules, to form hydrated protons. Pure sulphuric acid, nitric acid, and hydrogen chloride are all covalently bonded, but they react with water and give up protons to form hydrated protons and the appropriate hydrated anions (SO_4^{2-}(aq), NO_3^-(aq) and Cl^-(aq) respectively). Their solutions all have similar properties because they all contain the ions responsible for the behaviour of acidic solutions, H^+(aq); the hydrated anions play no part in acidic behaviour.

Other substances, as well as water, can make acids give up protons. For example, when hydrogen chloride gas reacts with ammonia gas,

$$HCl(g) + NH_3(g) \rightarrow NH_4Cl(s) \text{ [which can be shown as } NH_4^+Cl^-(s)],$$

the hydrogen chloride is again given a chance to act as an acid by donating a proton, this time to an ammonia molecule.

Ethanoic acid is a monobasic acid (i.e. it contains only *one* hydrogen atom per molecule which is replaceable by a metal ion) even though it contains four hydrogen atoms per molecule of acid, CH_3COOH. This is because only one of the hydrogen atoms can be donated as a proton to another substance, e.g.

$$CH_3COOH(l) + \text{water} \rightarrow CH_3COO^-(aq) + H^+(aq)$$

The other three hydrogen atoms are not capable of giving rise to acidic properties. Similarly, many substances contain hydrogen which cannot be donated as protons to another chemical, e.g. methane, CH_4, and glucose, $C_6H_{12}O_6$, and so such substances show no acidic behaviour.

Bases, according to the Brønsted–Lowry theory

Brønsted and Lowry defined a base in terms similar to those used in their definition of an acid.

A base is a substance which is capable of ACCEPTING a proton.

All alkalis contain OH^- ions, and their solutions are bases because the OH^- ions can accept hydrated protons (from solutions behaving as acids) to form water:

$$\underset{\text{from base}}{OH^-(aq)} + \underset{\text{from acid}}{H^+(aq)} \rightarrow H_2O(l)$$

Similarly, many metal hydroxides and oxides are basic because the oxide or hydroxide ions can accept protons, e.g.

$$CuO(s) + 2H^+(aq) \rightarrow Cu^{2+}(aq) + H_2O(l)$$

Ammonia gas is a further example of a base:

$$NH_3(g) + HCl(g) \rightarrow NH_4Cl(s) \text{ [which can be shown as } NH_4^+Cl^-(s).]$$

Ammonia is a base because it can accept a proton, in this case from hydrogen chloride, to form the NH_4^+ ion.

Whenever there is a 'battle for protons' in a chemical reaction, the substance which 'wins' is a base and the substance which 'loses' is an acid. The terms acid and base are to some extent relative (like oxidizing agents and reducing agents, p. 294); a compound usually described as an acid can be made to act as a base if it is dissolved in a solvent which makes the 'acid' accept a proton. Substances such as nitric, sulphuric, and hydrochloric acids can be safely called acids, however, as in routine situations at this level there are no common chemicals which can make them *accept* a proton.

Neutralization in terms of the Brønsted–Lowry theory

We can now regard neutralization as the reaction between H^+ ions (the acidic part) and a substance capable of accepting these ions, i.e. usually compounds containing O^{2-} or OH^- ions (the basic part). The hydrogen ions are usually the hydrated ions, $H^+(aq)$, found in acidic solutions, e.g.

$$H^+(aq) + OH^-(aq) \rightarrow H_2O(l)$$

though they may come from other sources, e.g.

$$HCl(g) + NH_3(g) \rightarrow NH_4Cl(s)$$

As long as there are no excess $H^+(aq)$ ions or $OH^-(aq)$ ions in a solution, the solution is neutral because the acidic and basic units have neutralized each other. If there are more $H^+(aq)$ ions than there are $OH^-(aq)$ ions, the solution is acidic and has a pH of less than 7. Similarly, alkaline solutions contain more $OH^-(aq)$ ions than $H^+(aq)$ ions and have a pH greater than 7.

Note that the equation above can be used to summarize any neutralization between a solution of a strong acid and a strong alkali, because the other ions present do not join together unless the water is removed by evaporation, as in salt preparations. Thus if a solution of sodium hydroxide is neutralized by hydrochloric acid, and the products are left in solution, it is correct to write

$$H^+(aq) + OH^-(aq) \rightarrow H_2O(l)$$

because the other ions are spectator ions. If, after the same reaction, the water is evaporated by heating, the sodium and chloride ions then come together to form the solid salt sodium chloride. This second 'stage' is not a chemical reaction in the normal sense, but we can then summarize the complete reaction by using the 'full' equation:

$$NaOH(aq) + HCl(aq) \rightarrow NaCl(s) + H_2O(l)$$

Strong and weak acids and bases

We can now use the theory of acidity to explain why some acids are stronger than others (p. 65). Remember that the expressions 'strong acid' and 'concentrated acid' do *not* mean the same thing. A strong acid is one in which most of the molecules react with water to donate protons and form $H^+(aq)$ ions; a solution of a strong acid thus contains a high concentration of $H^+(aq)$ ions. Similarly, a strong alkali forms a high concentration of $OH^-(aq)$ ions when added to water.

A concentrated acid, on the other hand, means an acid with hardly any water present. It so happens that as there are comparatively few water molecules present to *react* with the acid molecules, a concentrated acid quite often does not contain a high concentration of $H^+(aq)$ ions. This is why concentrated sulphuric acid does not show the properties of typical acidic solutions; it has a high percentage of H_2SO_4 molecules but has a low percentage of $H^+(aq)$ ions.

Think about the terms 'strong acid' and 'concentrated acid' very carefully; these terms are frequently misunderstood.

A weak acid has molecules which are reluctant to react with water molecules to form $H^+(aq)$ ions; many of its molecules in a solution do not react to form these ions, they simply stay dissolved as 'whole molecules'. A solution of a weak acid therefore contains a low proportion of $H^+(aq)$ ions, even if the acid itself is concentrated. (Think about it!) Similarly, a weak alkali is one in which only a small proportion of its molecules provide $OH^-(aq)$ ions.

These ideas are summarized in Table 5.5. Examples of strong acids include the mineral acids sulphuric, hydrochloric, and nitric. Weak acids include most organic acids such as ethanoic (acetic) acid, and the so called 'carbonic acid' (carbon dioxide dissolved in water, sometimes given the formula H_2CO_3).

Strong alkalis include sodium hydroxide solution and potassium hydroxide solution. Weak alkalis include ammonia solution; the solution contains a high proportion of dissolved ammonia molecules, but only a few of these react with water to produce OH^-(aq) ions.

Table 5.5 Strong and weak acids

(a) Dilute solutions of strong acids, (e.g. HCl, HNO_3, H_2SO_4)	Out of every 100 molecules of the acid, perhaps 99 react with water to form H^+(aq). EASILY FORM H^+(aq), PROPORTION OF H^+(aq) USUALLY HIGH
(b) Concentrated solutions of the strong acid sulphuric acid	Out of every 100 molecules, only a few react with water to form H^+(aq) ions, not because they are unable to form these ions but because there are not enough water molecules to react with. READILY FORMS H^+(aq) IONS, BUT PROPORTION OF H^+(aq) IONS ONLY LOW, OXIDIZING PROPERTIES DOMINATE ACIDIC PROPERTIES (p. 462)
(c) Concentrated solutions of the strong acid nitric acid	The *pure liquid* contains only a very low proportion of H^+(aq) ions for the same reasons as in (b), but the concentrated acid contains more water (about 30%) than concentrated sulphuric acid does. The proportion of H^+(aq) in concentrated nitric acid is thus greater than that in concentrated sulphuric acid, but the solution does not appear to behave as a 'more active' acid because it is also a powerful oxidizing agent (p. 381). READILY FORMS H^+(aq) IONS, PROPORTION OF H^+(aq) QUITE HIGH, BUT ACIDIC PROPERTIES OBSCURED BY ITS BEING A POWERFUL OXIDIZING AGENT
(d) Concentrated solutions of the strong acid hydrochloric acid	Concentrated hydrochloric acid contains a fairly large proportion of water (about 60%) and the concentration of H^+(aq) ions is high. This is why the concentrated form of this acid *does* show the properties of acidic solutions. READILY FORMS H^+(aq) IONS, PROPORTION OF H^+(aq) HIGH, BEHAVES AS A MORE ACTIVE FORM OF A STRONG ACID
(e) Dilute solutions of a weak acid, e.g. ethanoic acid, CH_3COOH	Out of every 100 molecules, perhaps only 2 react with water molecules to form H^+(aq) ions, even though there are lots of water molecules present. ONLY FORM H^+(aq) IONS WITH DIFFICULTY. PROPORTION OF H^+(aq) LOW
(f) Concentrated solutions of weak acids	Out of every 100 molecules perhaps only 1 reacts with water molecules to form H^+(aq) ions. The concentration of these ions is even lower than in (e) because there are fewer water molecules with which the acid can react. ONLY FORM H^+(aq) IONS WITH DIFFICULTY. PROPORTION OF H^+(aq) IONS VERY LOW

Note: The same ideas are true of alkalis if OH^- ions are considered.

5.6 A SUMMARY OF CHAPTER 5

The following 'check list' should help you to organize the work for revision.

1. Definitions

(a) A base (p. 62)
(b) An alkali (p. 62)
The following definitions may not be studied until later in the course.
(c) An acid (according to the Brønsted–Lowry theory), page 72.
(d) A base (according to the Brønsted–Lowry theory), page 73.

2. Other ideas

(a) The names and formulae of the common acids and alkalis (Tables 5.1 and 5.3).
(b) Acidic solutions turn blue litmus red, cause methyl orange to turn pink, produce a red colour with universal indicator, and make phenolphthalein stay colourless.
(c) Acidic solutions react with most carbonates and hydrogencarbonates; the mixtures effervesce to produce carbon dioxide. A salt and water are also formed.
(d) The reactions of dilute acids with metals are more complicated, and individual details should be learned (see Table 15.2 on page 251).
(e) Hydrochloric acid, sulphuric acid, and nitric acid, in addition to giving the usual tests for acidity, also give tests for chloride ions, sulphate ions, and nitrate ions respectively.
(f) Oxides and hydroxides of metals can act as bases. Bases which dissolve in water are given the special name alkali.
(g) Solutions of alkalis (but not insoluble bases) make red litmus go blue, make methyl orange go yellow-orange, change phenolphthalein to scarlet, and universal indicator to purple.
(h) When alkalis are warmed with ammonium compounds, the gas ammonia is formed; this is the only common alkaline gas. An example of this reaction is:

$$Ca(OH)_2(s) + 2NH_4Cl(s) \rightarrow CaCl_2(s) + 2NH_3(g) + 2H_2O(l)$$

(i) The pH scale, and the colour of universal indicator at different pH values (Figure 5.1, page 65).
(j) Acids and bases can neutralize each other to form a salt and water only; what neutralization means.
(k) The need to *control* the addition of an acid to an alkali or insoluble base, or the addition of an alkali to an acid, in order to produce a neutral solution.
(l) The neutralization of an acid by adding an excess of an *insoluble* base does not produce an alkaline solution.
(m) The need to first obtain a 'rough' reading when trying to neutralize a solution accurately.

The remaining points may not be studied until later in the course.
(n) All acidic solutions show the same properties because they contain $H^+(aq)$ ions.
(o) Pure acids do not contain free ions; they are covalently bonded and cannot act as acids on their own.
(p) Acids can only behave as acids when they can donate a proton to another substance (often water) which is capable of accepting a proton. A substance which accepts a proton is acting as a base.
(q) A strong acid forms a high concentration of $H^+(aq)$ ions when dissolved in water; a weak acid only forms a low concentration of these ions when dissolved in water.
(r) A strong alkali forms a high concentration of $OH^-(aq)$ ions when dissolved in water; a weak alkali only forms a low concentration of these ions.

3. Important experiments

Learn the results and conclusions of the experiments in this chapter rather than the experimental details.

6 Salts

6.1 WHAT ARE SALTS?

Introduction

In the previous chapter you learned of the chemical families called acids and bases. Another family of chemicals is called the salts, and there are many more salts than there are acids and bases. Most students fail to realize how widely this term is used; many of the chemicals studied in elementary courses are salts, and this point is further emphasized in the introduction to Chapter 19. Although you have learned to tell whether a substance is an acid or base by looking at its properties, you can easily decide whether a substance is a salt or not just by knowing its name. This point will become clearer as you work through the chapter.

Experiment 6.1
Is water the only product formed when an acid neutralizes a base?*

Apparatus
The neutral solution from Experiment 5.5. Bunsen burner, tripod, gauze, beaker, watch-glass suitable for resting on top of the beaker.

Procedure
(a) Half fill the beaker with water and place it on the tripod and gauze.
(b) Pour a few cm^3 of the neutral liquid on to the watch-glass and rest the glass over the beaker as shown in Figure 6.1.
(c) Boil the water in the beaker until the liquid on the watch-glass has evaporated.

Results
Draw the appearance of the watch-glass (as seen from above) at the end of the experiment.

Notes about the experiment

Some people imagine that when an acid and a base neutralize each other, water is the only product. You should now be able to agree or disagree with this idea. If there is a product besides water, can you recognize what it might be?

Conclusion
Make your own conclusion after looking again at the title of the experiment. You will understand the significance of this when you read the next section.

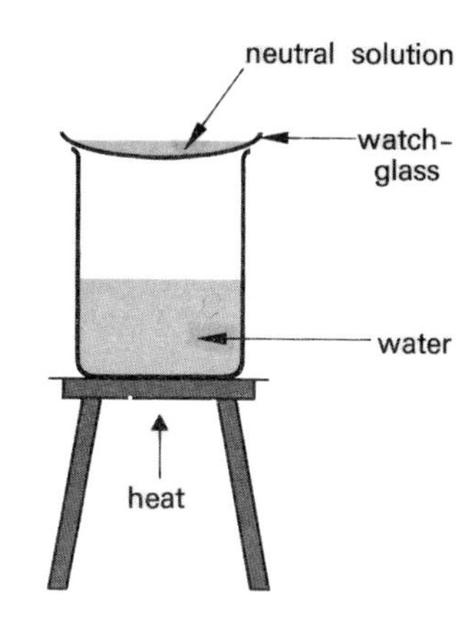

Figure 6.1 A simple evaporation technique

What are salts?

When hydrochloric acid and sodium hydroxide neutralize each other the particles 'rearrange themselves' and can form water and dissolved sodium chloride (or, if the water is evaporated away, solid sodium chloride).

$$NaOH(aq) + HCl(aq) \rightarrow H_2O(l) + NaCl(aq)$$

Sodium chloride is sometimes called common salt, and this is a rather unfortunate term as it is just one member of a large family called 'salts'.

The parent compound of any salt is an acid. All acids form salts. Sodium chloride is a salt formed from the acid hydrochloric acid. In fact hydrochloric acid forms lots of salts, but they are always called chlorides (e.g. magnesium chloride, sodium chloride). All metal chlorides are salts.

If you consider the formulae of hydrochloric acid (the 'parent' acid, HCl), and of sodium chloride (one of its salts, NaCl), you will see that the salt is formed by replacing the hydrogen part of the acid by a metal (Figure 6.2). This idea leads to the definition of a salt.

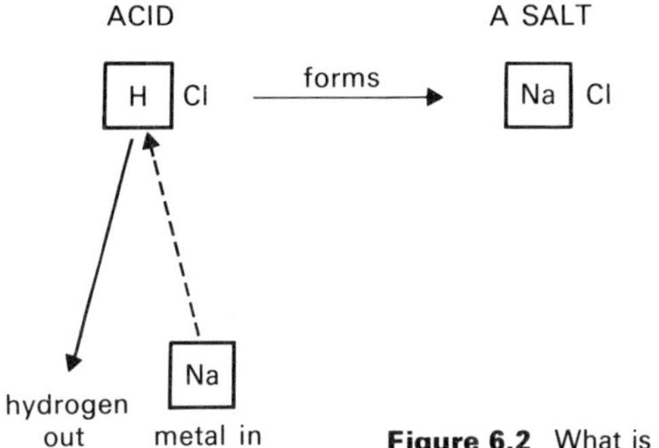

Figure 6.2 What is a salt?

A salt is a substance formed when the 'acidic hydrogen' of an acid is partly or completely replaced by a metal ion or an ammonium ion.

(You will understand the need for the word 'partly' when the special case of sulphuric acid is considered on page 79.) Note also that some acids contain hydrogen atoms which are not 'acidic' and cannot be replaced by a metal ion or an ammonium ion. This is explained in more detail on page 73, but for the moment this aspect need not concern us because all of the hydrogen atoms in molecules of the mineral acids are 'acidic'.)

In the example above, the replacement of hydrogen by the metal was not done directly (i.e. by using the metal itself), for the 'metal part' came from the base sodium hydroxide. Sometimes the replacement can be done directly by a metal, but the important idea in making a salt is to use some substance which can provide a 'metal part' when it reacts with an acid. You will learn later in the chapter that the choice of metal or metal compound to be used in making a salt must be made with care; not all metals or metal compounds can be used to make salts directly. In the example above, for instance, sodium would react in a dangerously violent way with hydrochloric acid.

Nitric acid, HNO_3, is similarly the parent acid of a group of salts called the nitrates. If the hydrogen part of this acid is replaced by a metal part (ion) or ammonium ion (NH_4^+), a salt is formed which is called a nitrate, e.g. sodium nitrate, $NaNO_3$, or potassium nitrate, KNO_3. Nitric acid always forms salts called nitrates because the part that is 'left behind' when the hydrogen part has been replaced is the nitrate ion, $NO_3^-(aq)$. All metal nitrates are salts.

Other, less common acids, also form salts in this way, so that all iodides, all bromides, and all sulphides are salts, e.g. sodium iodide (NaI), sodium bromide (NaBr), and iron(II) sulphide (FeS).

The special case of sulphuric acid as the 'parent' of salts

Hydrochloric and nitric acids each have only one hydrogen atom per molecule which is replaceable by a metal ion (i.e. one 'acidic' hydrogen atom per molecule). Such acids are called *monobasic acids*, and these acids can each form only one group of salts, i.e. chlorides from hydrochloric acid and nitrates from nitric acid.

Sulphuric acid is a *dibasic acid* because it has two hydrogen atoms per molecule which are replaceable by a metal ion, i.e. two 'acidic' hydrogen atoms per molecule (H_2SO_4). This acid can form two different types of salts. If *one* of the two hydrogen atoms in a molecule of sulphuric acid is replaced by a metal ion, the salt formed still contains one acidic hydrogen atom. Such salts are called *acid salts* because some of the hydrogen of the original acid is still present and such salts are usually acidic in solution. This particular acid salt will be a hydrogensulphate (see Table 6.1), e.g. sodium hydrogensulphate, $NaHSO_4$, and its acidic nature is shown by the fact that it is sometimes supplied in chemistry sets as a safer alternative to a mineral acid.

If both of the acidic hydrogen atoms in a molecule of sulphuric acid are replaced by a metal ion or ions, a *normal salt* is produced. In this case the normal salt will be a sulphate (Table 6.1), e.g. sodium sulphate, Na_2SO_4. Hydrochloric acid and nitric acid, being monobasic, can only produce normal salts.

Table 6.1 Acid and normal salts

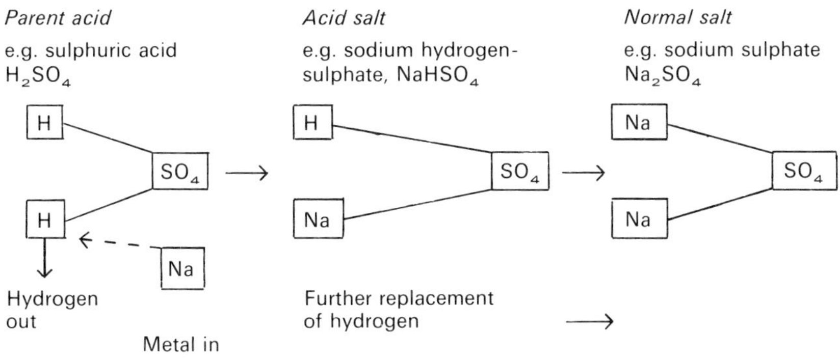

Parent acid		*Acid salt*		*Normal salt*
e.g. sulphuric acid H_2SO_4		e.g. sodium hydrogen-sulphate, $NaHSO_4$		e.g. sodium sulphate Na_2SO_4
H, H — SO_4	→	H, Na — SO_4	→	Na, Na — SO_4
H ← - - Na; Hydrogen out; Metal in		Further replacement of hydrogen	→	

An acid such as phosphoric(V) acid, H_3PO_4, is *tribasic*; it contains three atoms of hydrogen which can be replaced by metal ions and forms three different kinds of salt, e.g. NaH_2PO_4, Na_2HPO_4, and Na_3PO_4. The first two of these are acid salts, and the third one is a normal salt. *Note*: some acids may have more than one hydrogen atom per molecule but are still called monobasic, because only *one* of these hydrogen atoms can be replaced by a metal ion, i.e. only one of them is 'acidic', see ethanoic acid, page 73.

When carbon dioxide is dissolved in water, an acidic solution is formed. This solution is sometimes called carbonic acid, and given the formula H_2CO_3, although in actual fact the acid is very unstable and cannot be separated in the pure state from the solution. The solution does *behave* as if it is a dibasic acid of formula H_2CO_3, and it gives rise to two types of salt. These are the (often unstable) acid salts called hydrogencarbonates, in which one of the two hydrogen atoms per molecule of 'carbonic acid' is replaced by a metal ion, e.g. $NaHCO_3$, sodium hydrogencarbonate, and the usually stable normal salts called carbonates, e.g. Na_2CO_3, sodium carbonate. Some of these ideas are summarized in Table 6.2 and at the end of the chapter.

Table 6.2 Naming common salts

Sulphuric acid, H_2SO_4		*Hydrochloric acid, HCl*	*Nitric acid, HNO_3*	*Carbonic acid, H_2CO_3*	
Part of hydrogen replaced by a metal	All of hydrogen replaced by a metal	Only one series of salts, called chlorides, e.g. NaCl, sodium chloride	Only one series of salts, called nitrates, e.g. $NaNO_3$, sodium nitrate	Part of hydrogen replaced by a metal	All of hydrogen replaced by a metal
Acid salts called hydrogen-sulphates, e.g. $NaHSO_4$, sodium hydrogen-sulphate	Normal salts called sulphates, e.g. Na_2SO_4, sodium sulphate			Acid salts called hydrogen-carbonates, e.g. $NaHCO_3$, sodium hydrogen-carbonate	Normal salts called carbonates, e.g. Na_2CO_3, sodium carbonate

Check your understanding

1. Name 5 salts of copper.
2. Name an acid salt of magnesium.
3. Why is sulphur chloride not a salt?
4. What are salts of citric acid called?
5. A liquid turns universal indicator red, attacks some metals to liberate hydrogen, and liberates carbon dioxide when added to hydrogencarbonates. Which of the following is the most reasonable conclusion about the liquid?

A It is water
B It is a normal salt
C It is a base
D It is an acid
E It is an alkali

More to do

Salts are important substances in everyday life. Epsom salts, washing soda, and limestone are salts. Can you find their chemical names?

6.2 METHODS OF PREPARING SALTS

It is important to learn the *principle* of each method rather than the particular examples used in the experiments, for by simply changing the chemicals hundreds of different salts can be made.

1. By the action of an acid on a metal

In this method a metal is used directly to replace the hydrogen of an acid. Always think carefully before you describe a salt preparation involving this reaction, for not all combinations of acid and metal will react, and some will react too violently (see page 251). Nitric acid can react with *most* metals to form a solution of the nitrate, but this acid often needs to be warmed and the reaction should be conducted in a fume cupboard as poisonous oxides of nitrogen will also be produced. The sequence outlined in Figure 6.3 is typical of the preparation of a metal nitrate by this method, and you should make sure that you understand the principle before you actually

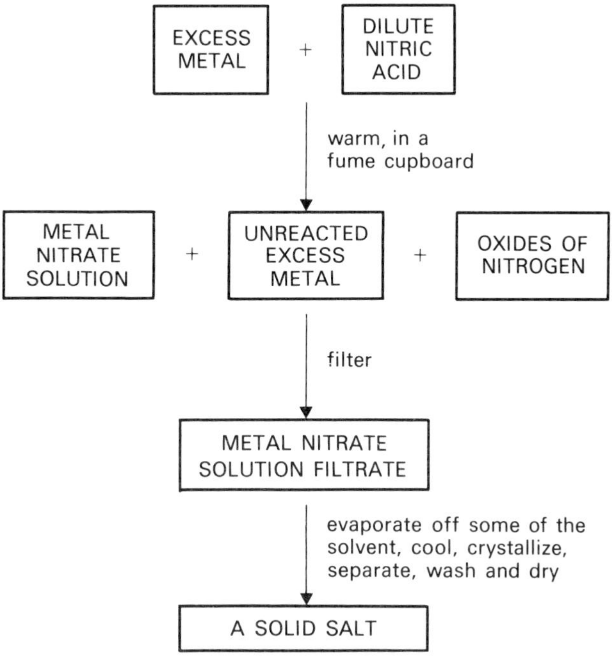

EXCESS METAL + DILUTE NITRIC ACID

warm, in a fume cupboard

METAL NITRATE SOLUTION + UNREACTED EXCESS METAL + OXIDES OF NITROGEN

filter

METAL NITRATE SOLUTION FILTRATE

evaporate off some of the solvent, cool, crystallize, separate, wash and dry

A SOLID SALT

Figure 6.3 The preparation of a salt by the action of an acid on a metal

begin Experiment 6.2. If hydrochloric or sulphuric acids are used, the principle is the same but fewer metals are suitable and the fume cupboard is not always necessary. Note that in all these cases *excess* metal is used (i.e. until no more will 'dissolve') so that the acid is completely neutralized. The excess metal does not cause any problems, or change the pH of the solution, because it is insoluble in water (the acid has now been 'used up') and can be filtered off.

Experiment 6.2

To prepare a salt by the action of an acid on a metal*

Apparatus

250 cm³ beaker, stirring rod, filter paper, tripod, gauze, Bunsen burner, Petri dish or crystallizing dish, measuring cylinder, spatula, filter funnel, access to fume cupboard.

Iron filings, sulphuric acid. (*Care*: the acid may be more concentrated than usual.)

Procedure

(a) Measure out 30 cm³ of the sulphuric acid, pour it into the beaker. Heat it *gently*. Add a spatula measure of iron filings to the acid (preferably in a fume cupboard) and stir. Vigorous frothing will occur.

(b) Continue to add portions of iron filings with stirring until effervescence ceases and there is a slight excess of unreacted metal.

(c) Set up filtration apparatus and filter the liquid into a clean Petri dish or crystallizing dish. You should obtain a clear (not necessarily colourless) filtrate. Repeat the filtration if necessary.

(d) The salt has already been made but it is dissolved in water. It remains to isolate and purify the solid salt. If the laboratory is warm, put the dish to one side until crystals form. In general good crystals are obtained only if evaporation is slow but under the conditions used in the experiment crystals usually form readily. If left for even a short time these particular crystals are likely to go brown.

Note: one way of ensuring slow evaporation for crystallizations is to cover the dish with a piece of filter paper and then pierce some holes through the paper. If crystals do not form after a day or two your solution is not sufficiently concentrated and you will have to evaporate off more of the water by boiling before repeating the crystallization procedure. In this particular experiment boiling will cause the salt to change chemically.

(e) Pour off any remaining liquid (the 'mother liquor') from the crystals and rinse them several times with a little cold water. Use only a little water with each rinse as some of the crystals will redissolve. Press the crystals gently between sheets of filter paper in order to dry them.

Results

Write down all your observations (including colour changes) and describe the appearance of the crystals.

Notes about the experiment

(i) The acid and iron react to form a salt and hydrogen. Excess metal is used to neutralize all the acid. When the excess metal is filtered off it leaves only the salt dissolved in water, because the hydrogen escapes. If excess metal was not used, then it would be the acid which was in excess and this could not be separated easily from the salt solution.

(ii) If some of the water is evaporated from the hot solution of the salt, the solution becomes more concentrated. When this hot solution is left to cool in a crystallizing dish, the salt starts to come out of solution (crystallizes).

(iii) The crystallizing dish allows a large surface area of the liquid to be exposed to the air, so that the solvent easily evaporates.

(iv) The salt crystals are separated from the liquid, washed with a small amount of distilled water to remove any solution still on them, and dried to remove water.

(v) If you understand this procedure, you will realize that all other substances should have been separated from the salt (i.e. acid, metal, hydrogen, water) and the crystals should be reasonably pure.

(vi) When writing about a salt preparation you may not have time or space to *explain* why each of the steps is done, but you should point out that after the initial reaction most of the steps are designed to remove other chemicals from the salt.

(vii) The particular salt made in this experiment is called iron(II) sulphate, $FeSO_4$, and the equation for the reaction is

$$Fe(s) + H_2SO_4(aq) \rightarrow FeSO_4(aq) + H_2(g)$$

(viii) You must *always* write an equation when describing a chemical reaction of this kind, unless nitric acid is being reacted with a metal. Nitric acid usually reacts with metals to produce a mixture of gases (oxides of nitrogen) and the equations can be quite complicated, so a *word equation* is acceptable in these cases, e.g.

nitric acid + zinc →
 zinc nitrate + oxides of nitrogen + water

Conclusion

State what you have succeeded in making, and give the equation.

Points for discussion

1. Do you think that this method of making salts (i.e. acid plus metal) is a *general* method? (See Experiment 5.3, page 60 and Table 15.2, page 251). Illustrate your answer by giving examples of other metal and acid combinations which you think *will* make salts (naming the salt in each case), and some which would *not* make salts, with reasons.

2. Try to construct an equation for some of the combinations which you think would work, but do not write equations for those reactions involving nitric acid.

2. Preparing a salt by direct combination between elements

This is the only other method of salt preparation in which a metal is used directly. The method is not a general method because it is used only for making those simple salts which contain *two* elements only; such salts are called *binary salts*. However, although the method is restricted, there are one or two reactions of this type which are important. For example, metal chlorides (which are salts) can be made by joining together the two elements, the metal and chlorine. Metal sulphides can be made in a similar way, using sulphur.

$$\text{METAL} + \text{CHLORINE} \rightarrow \text{METAL CHLORIDE (a salt)}$$

e.g. $$2Na(s) + Cl_2(g) \rightarrow 2NaCl(s)$$

$$\text{METAL} + \text{SULPHUR} \rightarrow \text{METAL SULPHIDE (a salt)}$$

e.g. $$Fe(s) + S(s) \rightarrow FeS(s)$$

The method cannot be used to prepare salts which contain three or four elements, such as the sulphates (e.g. $CuSO_4$), the hydrogensulphates (e.g. $NaHSO_4$), the nitrates (e.g. $NaNO_3$), the carbonates (e.g. $CaCO_3$), and the hydrogencarbonates (e.g. $NaHCO_3$).

The method is completely different from the other methods considered in this section because (i) no other product is formed, and (ii) it involves only pure elements; no water or solutions are used as is the case with the other methods. For this reason the method is often called a 'dry method' (contrasting with the other 'wet' methods) and is useful for preparing *anhydrous* salts, particularly those such as anhydrous iron(III) chloride and aluminium chloride, which cannot be prepared by heating aqueous solutions.

For example, it is possible to heat or burn the metal in chlorine, or to heat the metal with sulphur. You have probably already prepared the salt iron(II) sulphide by this method (p. 28). You may see demonstrations of the reactions of chlorine with metals, and these are considered in more detail in Chapter 24.

3. The action of an acid on an insoluble carbonate

In this kind of reaction the 'metal part' of the salt comes from a metal carbonate. You have met this reaction before, e.g. in studying the properties of acids and in testing for a carbonate. You will remember that a mixture of an acid and a carbonate

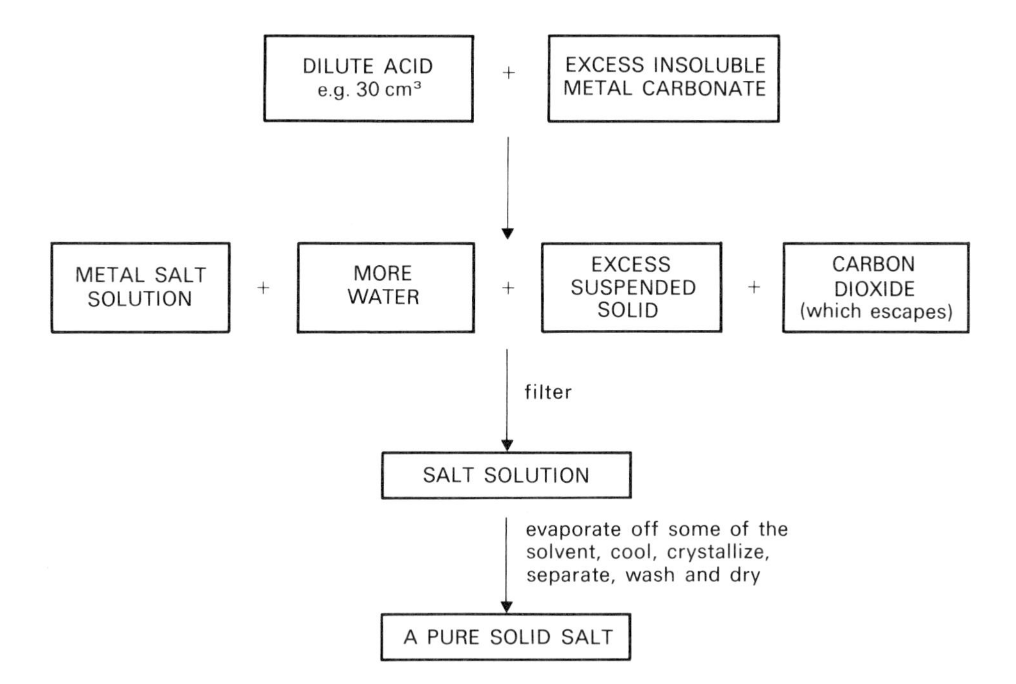

Figure 6.4 The preparation of a salt from an acid and a metal carbonate

effervesces because carbon dioxide gas is produced. There are also two other products, a salt and water:

CARBONATE + ACID → A SALT + WATER + CARBON DIOXIDE

This method is a very useful way of making a salt; it is more general than the previous methods because almost any combination of acid and carbonate will react in this way. Combinations which are not convenient are (a) calcium carbonate and sulphuric acid and (b) lead(II) carbonate with either sulphuric acid or hydrochloric acid. In each of these cases the salt formed (calcium sulphate, lead sulphate, or lead chloride) is *insoluble* in water and forms a coating over the carbonate, preventing further reaction between acid and carbonate. Yields of the salt are therefore low in these particular cases.

Apart from its general nature, the method is also very simple. All carbonates, except those of sodium and potassium, are insoluble in water and so an excess of the carbonate can normally be added (to neutralize all of the acid); the excess of insoluble carbonate will not dissolve and is simply filtered off. The gas carbon dioxide escapes, and so it only remains to evaporate off some of the water, and to crystallize, wash, and dry the salt. The steps are summarized in Figure 6.4; make sure that you understand the principle of the method before you do Experiment 6.3.

Experiment 6.3
To prepare a salt by the action of an acid on a carbonate*

Apparatus

250 cm³ beaker, evaporating basin, crystallizing dish, stirring rod, Bunsen burner, tripod, gauze, filtration apparatus, microscope slides, spatula.
Zinc carbonate (or any other insoluble carbonate), dilute nitric acid.

Procedure

(a) Pour about 30 cm³ of the nitric acid into the beaker and add spatula measures of the metal carbonate, with stirring, until the reaction has stopped and there is a slight excess of metal carbonate.
(b) Filter off the excess metal carbonate and collect the filtrate in the evaporating basin. Boil the solution in order to concentrate it.
(c) When the solution is sufficiently concentrated (check in the usual way from time to time) transfer it to a crystallizing dish and allow to crystallize as in the previous experiment.
(d) Collect, wash, and dry the crystals as before.

Results

State what you saw during the procedure (e.g. colour changes) and describe the appearance of the salt.

Conclusion

State the name of the salt you made, and give an equation for the reaction. Comment on whether the method is a very useful way of preparing salts.

Points for discussion

1. Which chemicals would you use in order to make the following salts by the same method as that used in Experiment 6.3? (a) magnesium sulphate, (b) calcium nitrate, (c) copper(II) chloride.
2. Construct equations for each of the reactions in 1.
3. How would you modify the method if you were using either sodium or potassium carbonates? Explain your suggestions.

4. The neutralization of a base by an acid

There are three ways in which this can be done, but all three are based on exactly the same reaction:

ACID + BASE → A SALT + WATER

The 'metal part' comes from the base.

(a) *Using an insoluble base and an acid*

This method is very similar in principle to methods 1 and 3 described earlier, because in all three cases an excess of an insoluble solid is added to neutralize the acid, and the excess solid is then filtered off. An important difference, however, is that in this case there is no effervescence because no gas is formed, and it is therefore a little more difficult to decide when the acid has been neutralized. The usual method is to add *small* quantities of the solid to the *warm*, dilute acid, and to stir for a few minutes after each addition. When the solid no longer reacts and 'dissolves', the reaction has finished and the acid has been neutralized. As an extra check, a piece of blue litmus paper can be dipped into the solution at this stage; it should not turn red. The principle of the method is summarized in Figure 6.5.

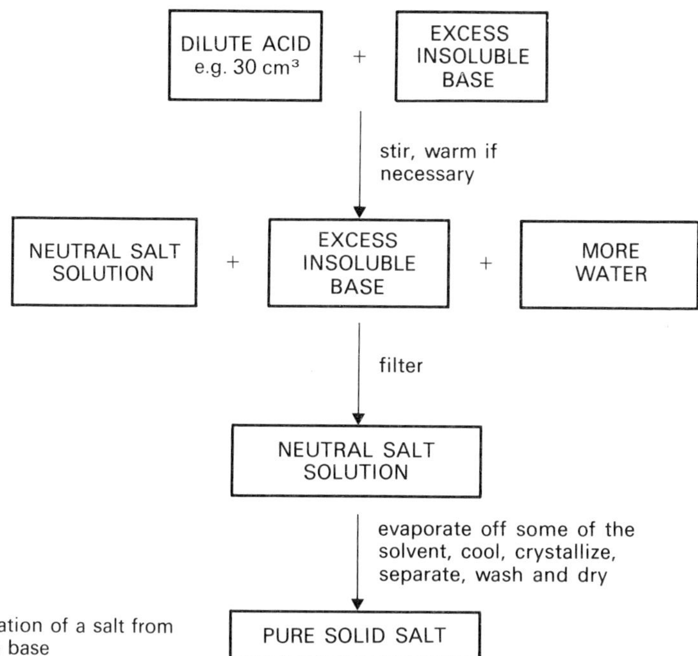

Figure 6.5 The preparation of a salt from an acid and an insoluble base

Experiment 6.4
To make a salt by the action of an acid on an insoluble base*

As this method is similar to some earlier ones, you should be able to use it to prepare a salt such as copper(II) sulphate without any further detailed instructions. Make sure that you understand the principle of the experiment before you begin. Start with the insoluble base copper(II) oxide and prepare some pure, dry crystals of copper(II) sulphate. Use 25 cm^3 of the acid solution.

Write up your experiment using headings such as apparatus, procedure, results, and conclusion. In your conclusion, include an equation for the reaction and comment on the general usefulness of the method.

(b) Using a soluble base (alkali) and an acid to make a normal salt
The reaction taking place is the same as in (a):

$$\text{ACID} + \text{BASE} \rightarrow \text{A SALT} + \text{WATER}$$

Note that there is no visible sign of reaction (e.g. effervescence) to indicate when the acid has been 'used up'. Also, unlike method (a), it is not possible to add an excess of one of the reagents and then filter off the excess, because both reagents are soluble. The method involves the exact and complete neutralization of two *solutions*, and is therefore a more accurate version of Experiment 5.5 (p. 67).

A fixed volume of *one* solution is used throughout the experiment (e.g. 10.0 cm^3 or 25.0 cm^3) and this is carefully measured out by a piece of apparatus called a *pipette*. (A measuring cylinder, which you used in Experiment 5.5, is not as accurate as a pipette.) A *burette* is used to find how much of the second solution is needed to neutralize the fixed volume of the first solution. It is important to practise using a pipette, a pipette filler, and a burette before starting the next experiment. Make sure that you can control the volume of liquid in either piece of apparatus with confidence, and that you take readings only when the apparatus is vertical and the top of the liquid is at eye level. Allow the pipette to drain naturally and do not be tempted to shake out or blow out the last drop or so of liquid from the pipette; pipettes are made to *deliver* the stated volume, not to hold it.

As in Experiment 5.5, the first stage of the experiment is to find a 'rough reading' of the volume of acid needed for the neutralization, so that more accurate readings can then be obtained quickly. Also as in Experiment 5.5, an indicator is used to show when neutralization is complete. An important difference, however, is that the final neutralization is done *without* the indicator, so that a pure salt, uncontaminated by indicator, can be made. It is possible to do this because as soon as the volumes needed for neutralization have been found (using the indicator), the same volumes can be used again to form a neutral solution but without using the indicator. The principle is summarized in Figure 6.6.

The method has only a few applications because only two alkalis are commonly used in this way, i.e. sodium hydroxide and potassium hydroxide. Typical salts which can be made by the method therefore include KCl, NaCl, Na_2SO_4, K_2SO_4, $NaNO_3$, and KNO_3.

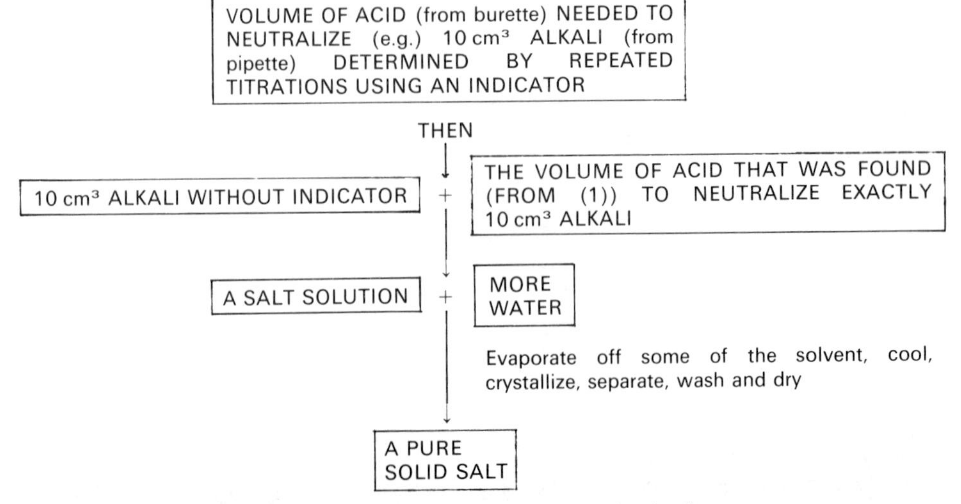

Figure 6.6 The preparation of a normal salt by titration

Experiment 6.5

Using a titration to make a normal salt**

Note: A titration is the name given to an experiment in which a burette and pipette are used to find the exact volume of one solution which reacts with a given volume of another solution.

Apparatus
10.0 cm^3 pipette and pipette filler, burette and stand, funnel, white tile, two 250 cm^3 beakers, a conical flask, Bunsen burner, tripod, gauze, evaporating basin, microscope slide, teat pipette. Solutions of sodium hydroxide and hydrochloric acid, each of concentration approximately 0.1 mol dm^{-3} (0.1 M page 303), phenolphthalein indicator.

Procedure
(a) Make out a table for your results as follows:

	Rough titration	*1*	*2*
Final burette reading (cm^3)			
Initial burette reading (cm^3)			
Volume of acid used (cm^3)			

(b) Label one of the beakers 'acid' and the other 'alkali'. Wash out each beaker with water and then with the appropriate liquid. Pour about 100 cm^3 of each liquid into the corresponding beaker.
(c) Pipette exactly 10.0 cm^3 of the sodium hydroxide solution into a clean conical flask. (Use a pipette filler if possible, but if these are not available take great care not to suck any liquid into your mouth.) A little water in the flask will not affect the result. Add two drops of indicator.
(d) Clean and check the burette, clamp it in position and fill it up with the acid solution. Run a little acid through to fill the tip of the burette. Adjust the level of the meniscus and record the level in your notebook. Place the conical flask on a white tile or piece of white paper below the tip of the burette.
(e) Run the acid into the alkali fairly quickly, shaking the flask all the time, until the colour of the indicator just permanently changes from pink to colourless. Close the tap. Note the new level of the acid in the burette.
(f) The first result is only approximate as there was no drop-by-drop control near the end of the reaction, so that a slight excess of acid was probably added. Repeat with a further 10.0 cm^3 of alkali after washing out the conical flask. This time run in the acid quickly until you have added about 1 cm^3 less than the volume used in the rough titration. Swirl the contents of the flask and add one drop of acid. Swirl again. Repeat this dropwise addition until the indicator *just* changes colour. Record the readings as before. The volume of acid used in this reaction should represent the volume needed to neutralize accurately the fixed volume of alkali.
(g) As a check on your own technique it is advisable to repeat (f) until two or more titration results agree within 0.1 cm^3.
(h) The normal titration is now over (i.e. you know the volumes which react together) but as you need to obtain a pure salt it is necessary to repeat the operation by mixing the dilute acid and alkali in the same proportions as those given by the titration results, but without the indicator. Use a fresh 10.0 cm^3 portion of alkali in a cleaned conical flask, and add exactly the same volume of acid as was used in your accurate titration(s). Swirl thoroughly.
(i) Transfer the neutral liquid to an evaporating basin, evaporate some of the water and crystallize the salt in the usual way. Wash and dry the sample.

Results
Complete the table of results. Describe the appearance of the salt.

Conclusion
Give an equation for the reaction, and comment on the usefulness of the method.

Use a pipette filler if possible

Points for discussion

1. Why is it important to swirl the contents of the flask during the addition of the acid?
2. Why does it not matter if there is water in the flask when you add the 10.0 cm^3 of alkali?
3. Why is a conical flask used and not a beaker?
4. If the burette and pipette have been washed out with water, they must be rinsed out with the appropriate solution before being used. Why is this necessary?

(*c*) *Using a soluble base* (*alkali*) *and an acid to form an acid salt*

(You will probably understand this technique better when you have studied equations in more detail.)

This is even more restricted than method (b) because the only dibasic acid normally used in an elementary course (i.e. an acid which can form both normal and acid salts) is sulphuric acid. The only acid salts you are likely to prepare by this method are therefore the hydrogensulphates of sodium and potassium.

The equation for the preparation of a typical normal salt might be

$$2NaOH(aq) + H_2SO_4(aq) \rightarrow Na_2SO_4(aq) + 2H_2O(l)$$

The equation for the preparation of the corresponding acid salt would be

$$NaOH(aq) + H_2SO_4(aq) \rightarrow NaHSO_4(aq) + H_2O(l)$$

The equations show that in order to make the acid salt, half as much alkali (or twice as much acid) must be used compared to that used for the normal salt.

The first stage of the method is similar to that used in Experiment 6.5. As before, the volume of acid required to neutralize completely the 10.0 cm^3 of alkali (and form the normal salt) is determined. (This stage can be omitted if you have just done Experiment 6.5, and more of the same solutions are available, for you will then already know the required volume.)

Stage 2 of the experiment is then done in which the same volume of alkali (10.0 cm^3) is used, but *twice* the volume of acid needed for complete neutralization is used. (No indicator is needed in this stage.) The solution obtained contains the acid salt, which can be crystallized in the usual way.

You may be asked to prepare an acid salt by this method, in which case you should write up your experiment in the usual way.

Preparing insoluble salts

Salts which are insoluble in water cannot be prepared by methods 1, 3, or 4, all of which involve the crystallization of a *soluble* salt from solution.

Relatively few insoluble salts are encountered in elementary chemistry courses, so be quite sure that you know which they are. The common insoluble salts are listed below. Each of these is usually prepared by the principle which is developed in this section.

Insoluble chlorides: silver chloride, AgCl, and lead(II) chloride, $PbCl_2$. (*Note*: Lead(II) chloride is soluble in *hot* water.)

Insoluble sulphates: barium sulphate, $BaSO_4$, calcium sulphate, $CaSO_4$ (slightly soluble), and lead(II) sulphate, $PbSO_4$.

Insoluble carbonates: *all* carbonates of metals are insoluble *except* for those of sodium and potassium.

Insoluble nitrates: NONE; all nitrates are soluble.

Others: lead(II) iodide, PbI_2, is insoluble in *cold* water.

It is fairly common in chemistry to mix two solutions, which then react together to form a *precipitate*. A precipitate is the name given to an insoluble solid which *forms* when solutions react together as in the next experiment. When some calcium carbonate is added to water it is insoluble in the water and remains suspended in the liquid, but such a solid must *not* be called a precipitate because it was formed *before* it was added to the liquid.

The following experiment enables you to make some precipitates.

Experiment 6.6
Making precipitates*

Apparatus

Rack of test-tubes.

Solutions of sodium carbonate, sodium chloride, copper(II) nitrate, magnesium nitrate, lead(II) nitrate.

(*Care*: lead and copper salts are poisonous.)

Procedure

(a) Make a table for your results. You will be adding (i) sodium carbonate and (ii) sodium chloride solutions separately to each of the other solutions.

(b) Pour a few cm^3 of copper(II) nitrate solution into two separate test-tubes.

(c) Add a little sodium carbonate solution to one of the samples of copper(II) nitrate, and a little sodium chloride solution to the other. Record your results in the table, using expressions such as 'pink precipitate formed' or 'no precipitate formed', but leaving enough space to add the names of the precipitates later.

(d) Wash out the test-tubes and repeat (b) and (c) but using magnesium nitrate solution instead of copper(II) nitrate solution.

(e) Wash out the test-tubes and repeat (b) and (c) but using lead(II) nitrate solution instead of copper(II) nitrate solution.

Results

Complete the table.

Notes about the experiment

(i) When a solid substance is dissolved to make a solution, the particles in the solid separate from each other and are free to move about in the solution. A solution of copper(II) nitrate contains copper(II) ions, $Cu^{2+}(aq)$, and nitrate ions, $NO_3^-(aq)$, which are moving about freely in the water. Similarly sodium carbonate solution contains sodium ions ($Na^+(aq)$) and carbonate ions ($CO_3^{2-}(aq)$) moving about in the water.

(ii) The first mixture you made in the experiment was between copper(II) nitrate solution and sodium carbonate solution. At the very instant of mixing, the mixture contained free copper(II) ions, free nitrate ions, free sodium ions, and free carbonate ions ($Cu^{2+}(aq)$, $NO_3^-(aq)$, $Na^+(aq)$, and $CO_3^{2-}(aq)$). These ions will keep bumping into each other, making millions of collisions per second. If any of these collisions cause ions to *combine* and make a chemical which is *insoluble*, then the insoluble compound forms as a *precipitate*. A reaction of this type is sometimes called *ionic association*, or *double decomposition*. A blue precipitate should have formed when you mixed copper(II) nitrate solution and sodium carbonate solution, and so some of the particles in the mixture must have combined to form an insoluble compound.

The various 'combinations' of collisions in the mixture are:

1. SODIUM IONS + NITRATE IONS
   = sodium nitrate if they combine
2. SODIUM IONS + CARBONATE IONS
   = sodium carbonate if they combine
3. COPPER(II) IONS + NITRATE IONS
   = copper(II) nitrate if they combine
4. COPPER(II) IONS + CARBONATE IONS
   = copper(II) carbonate if they combine

One of these possible combinations must have produced an insoluble blue solid, which formed a precipitate. You started with *solutions* of sodium carbonate and copper(II) nitrate, so these two compounds must be soluble and the precipitate must be either sodium nitrate or copper(II) carbonate. Look again at the information about the solubility of common salts on page 88 and you should be able to name the blue precipitate. Put this name in your table of results. You should also realize that this chemical is a salt, so you have made an insoluble salt in the experiment.
(iii) The second mixture you made was between copper(II) nitrate solution and sodium chloride solution. Work out the names of the four chemicals which could be made by collisions between ions in this mixture. Did you see a precipitate form from this mixture? If the answer is yes, name the precipitate; if no, can you explain why no precipitate formed?
(iv) Look at the other mixtures you made. Work out the names of any precipitates that formed, and make sure that you understand why a precipitate does not form in some cases. Write the names of the precipitates in your table, and say whether each precipitate is a salt or not.

How the formation of precipitates is used in some chemical tests

You may have already used the tests for soluble chloride ions and soluble sulphate ions (p. 451), e.g. in Experiment 4.6 on page 48. The chloride test depends upon the fact that silver chloride is insoluble, and so when $Ag^+(aq)$ ions (from silver nitrate solution) are added to a solution containing $Cl^-(aq)$ ions, a white precipitate (of silver chloride) is formed, and this indicates the presence of the $Cl^-(aq)$ ions. You should now be able to suggest why silver nitrate is used for this test, rather than any other silver compound.

$$\underset{\text{(from silver nitrate solution)}}{Ag^+(aq)} + \underset{\text{(from soluble chloride)}}{Cl^-(aq)} \rightarrow AgCl(s)$$

Similarly, the sulphate test depends upon the fact that barium sulphate is insoluble in water, and so when $Ba^{2+}(aq)$ ions (from barium chloride solution) are added to a solution containing $SO_4^{2-}(aq)$ ions, a white precipitate (of barium sulphate) forms, and this indicates the presence of the $SO_4^{2-}(aq)$ ions.

$$\underset{\text{(from barium chloride solution)}}{Ba^{2+}(aq)} + \underset{\text{(from soluble sulphate)}}{SO_4^{2-}(aq)} \rightarrow BaSO_4(s)$$

Note that although a salt can be recognized by its *name* (e.g. sodium sulphate is a salt because all metal sulphates are salts), an 'unknown' substance can be identified as a salt by performing chemical tests. If the unknown compound gives a sulphate or chloride test as described above, and it is a metal or ammonium compound, then it is a salt. Other tests do not depend upon the formation of precipitates (e.g. those for carbonate ions and for nitrate ions, as on page 451) but if a metal or ammonium compound gives any of the tests for chloride, sulphate, carbonate, or nitrate it is automatically a salt. (There are also less common salts such as the acid salts and the bromides.) You may be allowed to practise these tests by being given some 'unknown' solutions and being asked to recognize which of them are salts.

Point for discussion

Silver chloride is not the only insoluble chloride; lead(II) chloride, $PbCl_2$, is also insoluble in cold water. Can you suggest why (a) *silver* nitrate solution is used to detect the presence of chloride ions rather than *lead(II)* nitrate solution, (b) an acid must be added during the test, and (c) this acid must be *nitric* acid? (These are difficult questions!)

The principles used in preparing insoluble salts

The principle of the method is: (i) mix a *solution* containing the required metal ions (i.e. positive ions, or cations) with a *solution* containing the required 'non-metal' ions (i.e. negative ions, or anions); (ii) filter or centrifuge off the precipitate; (iii) wash it, and dry it.

Note that both starting materials must be soluble, so choose them carefully. As *all* metal nitrates are soluble, you can be *certain* that the metal ions can be provided by a solution of the appropriate metal nitrate. Similarly, as *all* sodium and potassium compounds are soluble, a solution of the appropriate sodium compound will safely provide the anion. This is why, in Experiment 6.6, the *nitrates* of the metals were used to provide copper(II) ions, magnesium ions, and lead(II) ions. Similarly, *sodium* carbonate solution was used to provide carbonate ions (anions) and *sodium* chloride solution to provide chloride ions (anions).

An example should make this clear. Suppose we need to make the insoluble salt, silver chloride. We need to mix a *solution* containing 'SILVER something' with a *solution* containing 'something CHLORIDE'. We can safely pick silver nitrate and sodium chloride for these solutions. Any other *soluble* silver compound would be equally satisfactory, as would any other *soluble* chloride. Lead(II) chloride would not be a suitable source of chloride ions because lead(II) chloride is insoluble (see list on page 88).

If you understand the principle, and you have learned the list of insoluble salts, you are now in a position to make any pure, dry, insoluble salt.

Experiment 6.7
To make the insoluble salts lead(II) chloride and lead(II) iodide by precipitation*

Apparatus

Teat pipette, stirring rod, two small beakers, centrifuge, centrifuge tubes, filter paper.

Dilute solutions of sodium chloride, potassium iodide, and lead(II) nitrate. (*Care*: lead salts are poisonous.)

Procedure

Devise your own way of preparing pure, dry samples of the two insoluble salts, using only the apparatus listed above. Write up your own procedure.

Results

Write down what you saw during the experiment, and the colours of the insoluble salts.

Notes about the experiment

Precipitation experiments (including the test for chlorides and sulphates, page 451) are best shown by *ionic equations*. The basic idea of an ionic equation is simple enough. When you made a precipitate of lead(II) chloride by mixing lead(II) nitrate solution with sodium chloride solution, you left two other types of ions free to move in the liquid. These were the sodium ions (left from the sodium chloride solution) and nitrate ions (left from the lead(II) nitrate solution). As these ions were free in the solution before mixing and are still free in solution after mixing, they take no part in the reaction and may be called *spectator ions*. An ionic equation shows only those ions which actually join together and therefore end up in a situation

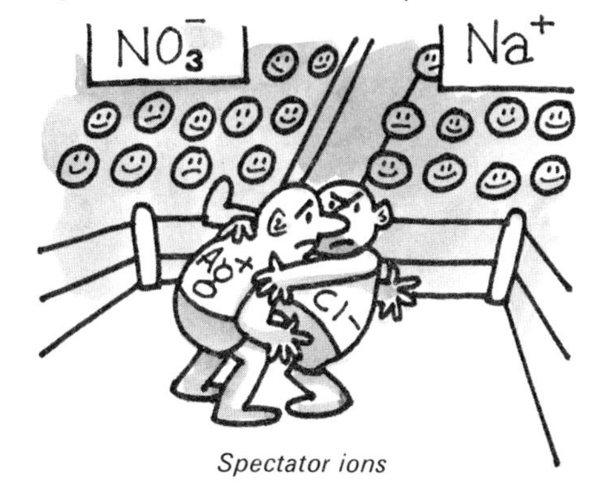

Spectator ions

different from that they were in at the beginning. The ionic equations for the two precipitations in the last experiment are:

$$Pb^{2+}(aq) + 2Cl^-(aq) \rightarrow PbCl_2(s)$$
$$Pb^{2+}(aq) + 2I^-(aq) \rightarrow PbI_2(s).$$

(Table 3.3 on page 31 shows the common ions and their charges.)
The corresponding full equations would be

$$Pb(NO_3)_2(aq) + 2NaCl(aq) \rightarrow PbCl_2(s) + 2NaNO_3(aq)$$
$$Pb(NO_3)_2(aq) + 2KI(aq) \rightarrow PbI_2(s) + 2KNO_3(aq).$$

Conclusion
Explain the *principle* of making insoluble salts, and give equations (ionic and full) for the reactions in the experiment.

Points for discussion

1. Why is it that salts made by earlier methods were washed with a little water, whereas it is safe to wash the precipitates obtained in Experiment 6.7 with lots of water?
2. It is possible to make the 'spectator ions' in these reactions combine together to form a solid compound. If this happens *two* salts are being formed by one reaction. How can this be done?
3. Could you have used potassium chloride solution instead of sodium chloride solution in Experiment 6.7? Explain your answer.
4. The following statements are frequently made by young students of chemistry, and there is a mistake in each of them. Explain what is wrong in each case.
(a) 'Sir, this solution of copper(II) sulphate isn't clear as you said it should be; it's blue.'
(b) 'I added the copper carbonate until the fizzing stopped, and then filtered off the excess copper.'
(c) 'I added the copper(II) oxide to the acid until effervescence stopped.'
5. A pupil complained that an examination question was unfair because it had asked for a description of the preparation of the salt zinc sulphate, and the class had not prepared this particular salt. The teacher said that the question was perfectly fair, and that the class should know several ways of making zinc sulphate even though they had not actually made it. What did the teacher mean?

Check your understanding

1. Name suitable starting solutions for making the following insoluble salts by precipitation: (a) lead(II) sulphate, (b) calcium carbonate, (c) barium sulphate.
2. Which of the following combinations of reactants would NOT be used to produce magnesium chloride?
A Magnesium carbonate and dilute hydrochloric acid
B Magnesium oxide and dilute hydrochloric acid
C Magnesium nitrate and dilute hydrochloric acid
D Magnesium and dilute hydrochloric acid
E Magnesium hydroxide and dilute hydrochloric acid
3. When sodium hydroxide solution is added to dilute hydrochloric acid in a beaker, which of the following is NOT happening in the beaker?
A The pH of the solution increases
B The hydrogen ion concentration falls
C The hydroxide ions neutralize some of the hydrogen ions
D The volume of water increases
E The reaction $Na^+(aq) + Cl^-(aq) \rightarrow NaCl(s)$ takes place
4. When potassium chloride solution is mixed with silver nitrate solution,
A a salt cannot be obtained from the mixture
B one salt only can be obtained from the mixture
C two soluble salts can be obtained
D two insoluble salts can be obtained
E one insoluble salt and one soluble salt can be obtained
5. Write the equations in words for the neutralization of an acid by a base. Write equations for the following neutralizations: sulphuric acid by potassium hydroxide, and hydrochloric acid by calcium hydroxide.
6. What do we mean by an acid salt? Give the names and formulae of two acid salts.
7. Write ionic equations for reactions between solutions of (a) silver nitrate and hydrochloric acid, (b) barium chloride and copper(II) sulphate.
8. Complete and balance the following equations:
$NaOH(aq) + H_2SO_4(aq) \rightarrow$
$CaCO_3(s) + HCl(aq) \rightarrow$
$Mg(OH)_2(s) + HNO_3(aq) \rightarrow$
$Mg(s) + H_2SO_4(aq) \rightarrow$
$Fe(OH)_3(s) + H_2SO_4(aq) \rightarrow$
$Cu(OH)_2(s) + HNO_3(aq) \rightarrow$

9. Identify the following. Explain the reactions taking place.
(a) A green liquid which turns purple when added to sodium hydroxide solution and red when added to dilute hydrochloric acid.
(b) A white precipitate formed by mixing silver nitrate and sodium chloride solutions.
(c) A white precipitate formed by mixing barium chloride solution with dilute sulphuric acid.
(d) A dilute acid which only rarely produces hydrogen when added to fairly reactive metals.

6.3 A SUMMARY OF CHAPTER 6

The following 'check list' should help you to organize the work for revision.

1. Definitions

(a) A salt (p. 78)
(b) A precipitate (p. 89)

2. Other points

(a) Sulphuric acid is a dibasic acid (its molecules each contain two hydrogen atoms which can be replaced by a metal, page 79) and forms salts called sulphates (its normal salts, page 79) and also hydrogensulphates (its acid salts, page 79).
(b) Carbonic acid (H_2CO_3, from CO_2+H_2O) cannot be isolated, but its solution acts as a dibasic acid and forms normal salts called carbonates, and acid salts called hydrogencarbonates.
(c) Nitric acid, HNO_3, is a monobasic acid which forms salts called nitrates.
(d) Hydrochloric acid, HCl, is a monobasic acid which forms salts called chlorides.
(e) All metal (and ammonium) chlorides, sulphates, nitrates, hydrogensulphates, carbonates, hydrogencarbonates, iodides, bromides, and sulphides are salts, although you cannot assume that all of the salts of a particular metal are stable or can be obtained in the solid state.
(f) The names of the common insoluble salts (p. 88).
(g) How precipitates of insoluble salts are formed, and how to choose appropriate solutions with which to make them. (A solution of the appropriate *sodium* salt will provide the anion in solution, and a solution of the appropriate metal nitrate will provide the metal ion in solution.)
(h) The main reactions of acids and bases, and the preparations of salts, are interconnected and they are summarized in Table 6.3 (p. 94).

3. Important experiments

You must learn the *principles* (summarized under the appropriate headings in the chapter, and in Figures 6.3 to 6.6) and the *experimental details* of the five methods of preparing salts which are listed below. Before you revise these points, however, it is important that you remember the following.

(i) A particular soluble salt can usually be prepared by several methods, e.g. zinc chloride by zinc metal and hydrochloric acid (metal + acid), zinc oxide and hydrochloric acid (base + acid), zinc hydroxide and hydrochloric acid (base + acid), and zinc carbonate and hydrochloric acid (carbonate + acid). You will have to decide which of the methods is appropriate or convenient for a particular situation.
(ii) Do not worry if you are asked to describe the preparation of a particular salt and you cannot recall ever having made it; you should use the principles given in this section to plan a suitable experiment. Remember that you may not be told that a particular chemical is a salt—it is up to you to recognize a salt from its name or formula.
(iii) You must always write equations when describing salt preparations.

The methods of preparation to be learned are:

(a) the action of an acid on a metal (and its limitations), page 80.
(b) direct combination between elements (and its limitations), page 82.
(c) the action of a carbonate on an acid, page 83.
(d) the neutralization of an acid by an insoluble base, and also by soluble bases (alkalis) to form both normal and acid salts (pages 85–8).
(e) preparing insoluble salts by precipitation, page 91.

Table 6.3 Acids, bases, and salts—a summary of the main reactions

Reactions		*Gas evolved (laboratory preparation)*	*Salt formed*	*Any other product*
1. *Reactions of acids*				
(a) *Acid + carbonate*	⟶	carbon dioxide	+ salt	+ water
e.g.				
$2HCl(aq) + Na_2CO_3$(aq or s)	⟶	$CO_2(g)$	$+ 2NaCl(aq)$	$+ H_2O(l)$
$H_2SO_4(aq) + CuCO_3(s)$	⟶	$CO_2(g)$	$+ CuSO_4(aq)$	$+ H_2O(l)$
Exceptions: calcium carbonate + sulphuric acid (very slow)				
(b) *Acid + metal*	⟶	hydrogen	+ a salt	———
e.g. $Zn(s) + 2HCl(aq)$	⟶	$H_2(g)$	$+ ZnCl_2(aq)$	———
$Mg(s) + H_2SO_4(aq)$	⟶	$H_2(g)$	$+ MgSO_4(aq)$	———
Exceptions: not given by unreactive metals; usually no hydrogen from nitric acid; very reactive metals are dangerous to use				
(c) *Acid + base*	⟶	———	a salt	+ water
e.g.				
$HCl(aq) + NaOH$(aq or s)	⟶	———	$NaCl(aq)$	$+ H_2O(l)$
$2HNO_3(aq) + CuO(s)$	⟶	———	$Cu(NO_3)_2(aq)$	$+ H_2O(l)$
Exceptions: none				
2. *Reactions of bases*				
(a) *Base + acid* (see acid + base)				
(b) *Base + ammonium salt*	⟶	ammonia	+ a salt (not usually used as a preparation)	+ water
e.g.				
$CuO(s) + 2NH_4Cl(aq)$	⟶	$2NH_3(g)$	$+ CuCl_2(aq)$	$+ H_2O(l)$
$NaOH(aq) + NH_4Cl(aq)$	⟶	$NH_3(g)$	$+ NaCl(aq)$	$+ H_2O(l)$
Exceptions: none				

QUESTIONS

1. Explain what is meant by the term 'salt'. Describe how you would prepare (a) a dry, crystalline sample of the soluble salt lead (II) nitrate starting from lead(II) carbonate, (b) a dry sample of insoluble barium sulphate from barium nitrate.

Include in your answers the names of the substances used, the conditions required, the technique by which the sample is separated in a pure condition, and an equation for the reaction. Outline how the method described in (a) would be altered if you were asked to make a sample of sodium nitrate starting from sodium carbonate. (J.M.B.)

2. Give four general methods of preparing salts. Name the starting materials you would use to make four named salts, each one prepared by a different method, and describe *one* of the methods in detail.

3. Describe with full experimental details how you would prepare (a) pure dry crystals of magnesium sulphate starting from magnesium oxide, (b) a pure dry specimen of zinc carbonate (insoluble) starting from zinc chloride.

4. Suppose that you have to prepare fairly pure specimens of (a) solid calcium sulphate, and (b) crystalline copper(II) nitrate, starting from marble chips (calcium carbonate) in each case. Suggest how you would proceed to prepare the two samples.

5. The following is an extract from a pupil's notebook. 'Solutions of sulphuric acid and hydrochloric acid are equally concentrated because they each have a pH of 3. They are also as strong as each other for the same reason.' These statements are inaccurate and show a lack of understanding of the terms 'strength' and 'concentration'. Describe how you would try to help the pupil to understand the terms.

7 Atomic structure
Radioactivity and nuclear power

7.1 ATOMIC STRUCTURE

Sub-atomic particles

Atoms are small, but they contain even smaller particles called sub-atomic particles. There are three main kinds of sub-atomic particle, two of which are found in the nucleus. The nucleus is a very small, extremely dense central portion of the atom in which there are particles called *protons* (positively charged) and *neutrons* (neutral – no charge). Outside the nucleus is a much larger region of the atom in which *electrons* (negatively charged) occur in *energy levels*. Electrons can be imagined to be orbiting the nucleus at great speeds.

'We can only take two of you on this floor, the others will have to go upstairs'

Many energy levels exist within an atom although they do not all contain electrons, in the same way that a house may have many rooms but it does not mean that they all contain people. Each level can take up to a certain maximum number of electrons. Most basic chemistry courses are concerned only with the atomic structures of the first twenty elements, and for this purpose we can assume a maximum of 2 electrons in energy level one (the lowest energy level), 8 electrons in energy level two, and 8 electrons in energy level three.

The electrons in a particular energy level all have the same amount of energy associated with them. Electrons always occupy the lowest available energy level, i.e. the energy level nearest to the nucleus, providing it still has space available. When an energy level is full, further electrons occupy the next energy level until that is filled, and so on.

The main characteristics of the sub-atomic particles are shown in Table 7.1. Their masses are so very small that they are measured in atomic mass units (amu). One atomic mass unit is approximately the same mass as that of the lightest atom, a hydrogen atom.

Table 7.1 Sub-atomic particles

Particle	*Approximate mass (atomic mass units)*	*Relative charge*
Proton	1.0	+1
Neutron	1.0	0
Electron	1/1840	−1

Atomic structure diagrams

We usually represent the arrangement of the particles within an atom by drawing a labelled diagram which shows its *atomic structure*. Such diagrams are not to scale, nor do they represent a true picture of what an atom is like. For example, in reality the nucleus of an atom is *extremely* small, and if we represented it on a diagram as a circle of diameter 1 cm, electrons occupying the first energy level, on the same scale, would be more than 100 metres away from the nucleus.

Another simplification is that in atomic structure diagrams it is convenient to show an energy level as if it were a 'region' in the atom, represented by a circle. The various energy levels are shown by a series of concentric circles, separated from each other by roughly equal distances. An electron does not, in fact, follow a fixed path as such diagrams suggest. It is possible to think of each circle as indicating a 'region' of the atom where a particular electron is most likely to be found. These 'regions' are sometimes called orbitals, or shells of electrons. It is far better, however, to think of each circle as meaning an energy level (rather than a region), in which electrons all have the same energy. The electrons in the first energy level spend most of their time fairly close to the nucleus. Electrons in the second energy level have more energy than those in the first energy level, and spend most of their time further away from the nucleus. This is why the circle representing the second energy level is drawn outside the circle representing the first energy level.

Do not imagine that the energy 'jumps' between different energy levels are all equal, as is suggested by the spacing of the circles in these diagrams. For all these reasons, atomic structure diagrams are not accurate pictures of what atoms are like (in fact we are not really sure what they *are* like!), but they are useful representations which can be used to explain many fundamental chemical principles, e.g. bonding.

Showing sub-atomic particles on bonding diagrams

As atoms are electrically neutral, the number of protons and the number of electrons in any given atom must always be the same. The differences between atoms of one element and those of another are due to the differing numbers of electrons, protons, and neutrons they contain.

You will learn later that chemical reactions involve rearrangements of the *outer electrons* in atoms, ions, or molecules, and it is therefore particularly important to understand the arrangement of the *electrons* in an atom, ion, or molecule. Chemistry deals with the outer electrons only of atoms; neutrons and protons play no part at all in chemical reactions.* If you know the number of protons and hence electrons in an atom, you can work out (using the information given under the next heading) the number of electrons in its outer shell, and therefore, as you will soon see, you will be able to predict what sort of reactions it will perform.

The number of electrons occupying a particular energy level in an atom is normally indicated by a number on the appropriate 'circle'. Thus Figure 7.1(a) shows that an atom of chlorine has 17 electrons altogether, 2 in energy level one (the maximum number this level can contain), 8 in energy level two (which is also full), and 7 in the third (incomplete) energy level.

Sometimes it is more convenient (particularly in bonding diagrams) to show *individual* electrons in energy levels, especially in outer energy levels. In these cases each electron is represented by some symbol, e.g. ⊗ or ⊖, etc. Figure 7.1(b) is a slightly different way of showing a chlorine atom, using this idea.

The number of protons or neutrons in the nucleus of an atom can be shown as in Figure 7.1(a). Often, however, it is sufficient just to put the symbol of the element inside the nucleus, as in Figure 7.1(b), because the particles inside a nucleus do not change during a chemical reaction and so we do not need to show them individually.

* Neutrons play an important part in holding the nucleus together. The protons in the nucleus are all positively charged, and like charges try to push each other apart (repel each other); the neutrons help to overcome these repulsive forces.

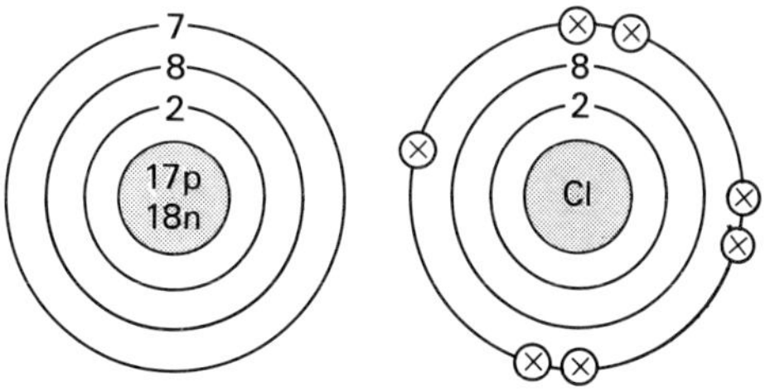

Figure 7.1 Atomic structure diagrams for chlorine

p = protons in nucleus ⊗ = electron in
n = neutrons in nucleus outer energy level

(a) (b)

It is important, however, that you should know how to calculate the number of protons and neutrons inside a nucleus.

In order to draw the atomic structure of a given atom (and to work out the numbers of protons and neutrons within its nucleus) we need to know two pieces of information, its atomic number and its mass number.

Atomic number and mass number

The atomic number of an atom is defined as the number of protons present in that atom.

This information is often provided as a number written as a subscript in front of the symbol for the element, e.g. ${}_{2}He$ means that the atomic number of helium is 2. (This means that an atom of helium contains 2 protons.) The atomic number of an element is often given the symbol Z.

The mass number of an atom is defined as the sum (i.e. total) of the numbers of neutrons and protons in the atom.

The mass number is normally given as a number written as a superscript in front of the symbol for the element, e.g. ${}^{4}He$ means that the mass number of a helium atom is 4 (and that the helium atom contains a number of protons and neutrons which add up to four altogether). The mass number of an atom is often given the symbol A.

Usually both numbers are provided at the same time, e.g. ${}^{16}_{8}O$ means that the atomic number of this atom of oxygen is 8, and that its mass number is 16. Where the two numbers are different, the larger of them is always the mass number.

Sometimes information is provided only for the mass number, A, or the relative atomic mass, A_r (which is explained on page 100), in the form:

$$A_r(H) = 1,\ A_r(Na) = 23,\ A_r(Cl) = 35.5, \text{ or simply } H = 1,\ Na = 23, \text{ etc.}$$

The information provided by these two numbers is extremely important and can be summarized as follows.

ATOMIC NUMBER = NUMBER OF PROTONS
(also = number of electrons in a neutral atom, from which the number of electrons in the *outer* shell can be determined.)

MASS NUMBER = NUMBER OF PROTONS + NUMBER OF NEUTRONS

NUMBER OF NEUTRONS = MASS NUMBER − ATOMIC NUMBER

An atom of $^{35}_{17}Cl$ thus consists of 17 electrons and 17 protons (because the atomic number is 17) and also 18 neutrons (because mass number − atomic number = 18). As the first energy level can contain a maximum of 2 electrons, the second level can contain up to 8 electrons, and the third up to 8 electrons, the arrangement of the sub-atomic particles in this atom of chlorine is summarized by the atomic structure diagrams shown in Figure 7.1. *Note*: When drawing atomic structures always label your diagram or use a key.

Check your understanding

1. Table 7.2 shows the number of protons, neutrons, and electrons in typical atoms of the first twenty elements. The table is incomplete. Copy the table into your book and complete it.

2. Using information from your completed table, draw atomic structures for each of the first twenty elements.

Table 7.2 Incomplete table showing the number of protons, electrons, and neutrons in atoms of the first twenty elements (most abundant isotopes only)

Atom and symbol		*Number of protons*	*Number of electrons*	*Number of neutrons*	*Mass number*	^A_Z*Symbol*
Hydrogen	H	1	1	0	1	^1_1H
Helium	He	2	2	2	4	
Lithium	Li	3	3	4	7	
Beryllium	Be	4	4	5	9	
Boron	B	5	5	6	11	
Carbon	C	6		6		
Nitrogen	N	7			14	
Oxygen	O	8			16	
Fluorine	F		9	10		
Neon	Ne		10		20	
Sodium	Na		11	12	23	
Magnesium	Mg	12		12		
Aluminium	Al	13			27	
Silicon	Si		14	14		
Phosphorus	P		15		31	
Sulphur	S			16	32	
Chlorine	Cl			18	35	
Argon	Ar		18	22		
Potassium	K	19		20		
Calcium	Ca		20	20		

Isotopes

Atoms of the same element *always* contain the *same* number of protons and the *same* number of electrons, but the number of neutrons in such atoms may vary. Most elements therefore consist of more than one kind of atom. The variation in the number of neutrons only affects the mass of the atom and has no influence on the chemical properties, which depend largely on the number of electrons in the outer shell of the atom. All of the different atoms of the same element therefore have identical chemical properties because they have the same number of electrons and protons, i.e. the same atomic number.

Isotopes are atoms of the same element which contain different numbers of neutrons, i.e. they have the same atomic number but different mass numbers.

Isotopes of the element chlorine include $^{35}_{17}Cl$ and $^{37}_{17}Cl$, and their atomic structures are illustrated in Figure 7.2. These isotopes of chlorine occur in the ratio of approximately three atoms of $^{35}_{17}Cl$ to one atom of $^{37}_{17}Cl$, and this is always the situation no matter how much chlorine is made, or how it is made, or where the starting materials come from. Many elements have isotopes (e.g. tin has ten), but a few elements have no *naturally occurring* isotopes. These include aluminium, fluorine, iodine, and sodium, but it is now possible to produce radioactive isotopes of all the elements, including those that have no naturally occurring isotopes.

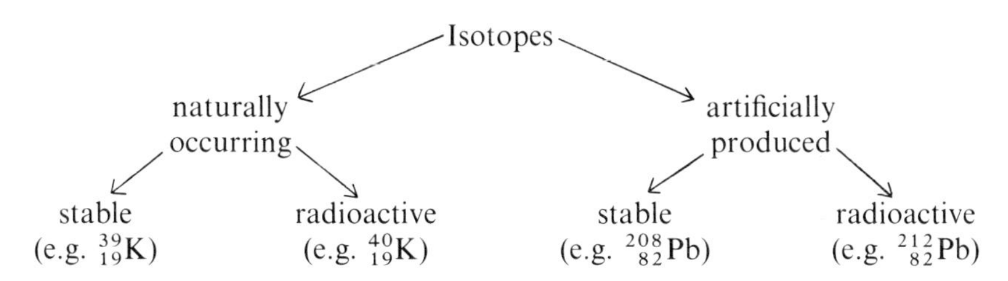

Nowadays an enormous number of isotopes is known, both naturally occurring and artificially produced.

In most cases where an element has several isotopes, one of the isotopes makes up a high percentage of the atoms present, e.g. carbon consists of 98.89 per cent $^{12}_{6}C$, 1.1 per cent $^{13}_{6}C$, and minute traces of $^{14}_{6}C$. The chlorine isotopes are much used as examples in elementary chemistry, and this element is unusual in having relatively high proportions of both isotopes, i.e. 75 per cent $^{35}_{17}Cl$ and 25 per cent $^{37}_{17}Cl$.

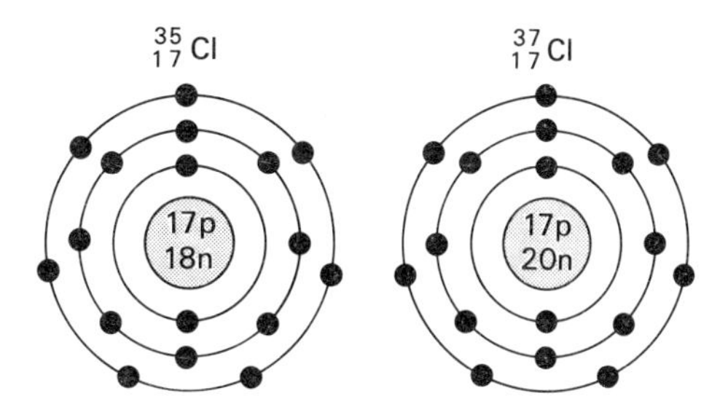

Figure 7.2 Two isotopes of chlorine
p = proton n = neutron ● = electron

Nuclides and isotopes

A particular kind of atom is identified by its symbol, mass number, and atomic number, and is known as a nuclide, e.g. $^{35}_{17}Cl$ is a nuclide. The word isotope should only be used when *two or more* nuclides with the same atomic number are being compared, e.g. $^{35}_{17}Cl$ is a nuclide, but $^{35}_{17}Cl$ and $^{37}_{17}Cl$ are isotopes.

It is possible that a given nuclide could be described by giving *up to* four items of information in addition to its symbol or formula, although all four items (numbers) would not be used at the same time.

e.g. $^{\text{mass number}}_{\text{atomic number}}\text{SYMBOL}^{\text{charge on ion}}_{\text{number of atoms or ions joined}}$

e.g. $^{35}_{17}Cl$, $^{35}_{17}Cl^{-}$, H_2, Mg^{2+}, etc.

Relative atomic mass and relative molecular mass

In the nineteenth century it was impossible to find the mass of individual atoms, and so it was decided to fix a scale for atomic masses by *comparing* the masses of atoms. Hydrogen was the lightest element, and so the mass of a hydrogen atom was fixed as 'one unit'. The masses of the atoms of other elements were then found (on this relative atomic mass scale) by comparing their masses with the mass of hydrogen.

On this scale, the relative atomic mass of oxygen was 16, which means that an atom of oxygen weighed sixteen times as much as an atom of hydrogen. Many changes have occurred since this scale of atomic mass was introduced. For example, hydrogen was not a good standard because it does not combine directly with many elements, and oxygen (which does combine with most elements) replaced it as the standard. A more serious problem was raised when isotopes were discovered. If chlorine contains two isotopes, for example, what was meant by the 'atomic mass of chlorine'? Did this mean the atomic mass of the ${}^{35}_{17}Cl$ nuclide, the atomic mass of the ${}^{37}_{17}Cl$ nuclide, or the 'average' atomic mass which the mixture of isotopes appears to have? And which oxygen nuclide are we comparing the mass with, anyway?

The unsatisfactory situation was clarified in 1961 by the recommendations of the International Union of Pure and Applied Chemistry. The standard comparison for all atomic masses was changed to one particular *nuclide* (rather than the vague 'an atom of oxygen', which could have several different masses, according to which nuclide was being considered). In addition, the standard for comparison was changed from oxygen to carbon, because carbon is a particularly convenient standard to use in very accurate modern instruments such as the mass spectrometer. The actual nuclide chosen as the standard is the most abundant of the carbon isotopes, ${}^{12}_{6}C$.

An amicable agreement

The relative atomic mass of an element is defined as the mass of 'an average' atom of the element compared with the mass of an atom of ${}^{12}_{6}C$, the mass of which is taken as exactly 12 units.

Relative atomic mass is normally given the symbol A_r.

Note that the 'relative atomic mass' of a particular nuclide would be more or less the same as its mass number, because electrons have virtually no mass and so the mass of an atom depends almost entirely on the number of neutrons and protons present. However, the term relative atomic mass *must not* be used with reference to a particular nuclide. It is only used to refer to the *mixture* of isotopes which many elements naturally consist of. This is why the term 'average atom' is used in the definition above. The term 'average atom' allows for all the isotopes present in a mixture, and for the proportions in which they occur. For example, as chlorine always contains two isotopes in the ratio of nearly three parts of ${}^{35}_{17}Cl$ to one part ${}^{37}_{17}Cl$, the relative atomic mass of an 'average atom' of chlorine is

$$\frac{(3 \times 35) + 37}{4} = 35.5$$

35.5 is the relative atomic mass of the element chlorine, and is *not* the mass number.

A mass number refers only to one particular nuclide, and is always a whole number because it is the sum total of the numbers of neutrons and protons present.

Elements may have more than one mass number, e.g. chlorine has two mass numbers, 35 and 37.

A relative atomic mass refers to an 'average atom' of the element, and individual atoms of this description do not exist, e.g. there is no real atom of chlorine which has a relative atomic mass of 35.5, in the same way that there is no real family like the 'average family' of 2.2 children. A relative atomic mass is often not a whole number (because of the presence of isotopes) and an element has only *one* relative atomic mass.

Check your understanding

1. Work out the relative atomic mass of bromine, assuming that it is made up of equal proportions of the two isotopes $^{79}_{35}Br$ and $^{81}_{35}Br$.

2. What would be the new relative atomic mass of bromine if the isotopes occurred in a 2:1 ratio, the lighter isotope being more abundant?

3. Isotopes are atoms of the same element which have

A the same mass number but different atomic numbers

B the same mass number and the same atomic number

C different atomic numbers and different mass numbers

D the same number of neutrons but different numbers of protons

E the same number of protons but different numbers of neutrons

4. The number of neutrons present in an atom of $^{81}_{35}Br$ is

A 46 B 35 C 81 D 116 E 17.

5. What information regarding the atomic structure of a nuclide is given by its mass number and its atomic number? Illustrate your answer by reference to $^{16}_{8}O$, $^{32}_{16}S$, $^{7}_{3}Li$, and $^{12}_{6}C$.

Relative molecular mass

The relative molecular mass of a substance (M_r) is defined as the mass of an 'average formula unit' of the element or compound relative to the mass of an atom of $^{12}_{6}C$, which is taken as exactly 12 units.

The term 'formula unit' is used rather than molecule so that the definition can be applied to both molecular and ionic compounds, even though the latter do not contain molecules. It is perfectly correct to use the term relative molecular mass, or the symbol M_r, when working with *ionic* compounds as long as you do not refer to the particles inside the structure as molecules. For example, it is correct to state that the relative molecular mass (or molar mass) of sodium chloride (which is ionic) is 58.5. It would be incorrect to say that the relative molecular mass of sodium chloride *molecules* is 58.5, because sodium chloride consists of ions, not molecules.

A relative molecular mass is obtained by adding together the individual relative atomic masses of the atoms or ions present in the formula unit. For example, if the relative atomic masses are $A_r(H) = 1$, $A_r(S) = 32$, and $A_r(O) = 16$, then the relative molecular mass of sulphuric acid, H_2SO_4, is $(2 \times 1) + 32 + (4 \times 16) = 98$. This means that an average molecule of sulphuric acid would 'weigh' 98 units on a scale where an atom of $^{12}_{6}C$ 'weighs' 12 units.

Note

(i) Relative atomic masses and relative molecular masses do not have units as they are ratios. We can say that the relative atomic mass of chlorine is 35.5, but not that it is 35.5 g.

(ii) In obtaining relative molecular masses, we add together the individual relative atomic masses but *not* mass numbers; if the latter were used we would not be referring to an '*average* formula unit'.

Check your understanding

1. Work out the relative molecular masses of the following substances: (a) Cl_2, (b) H_2O, (c) $MgCl_2$, (d) HNO_3, (e) CH_3COOH, (f) $C_6H_{12}O_6$, (g) $FeCl_3$.

2. Can you explain note (ii) above?

7.2 RADIOACTIVITY AND NUCLEAR POWER

The discovery of radioactivity

Radioactivity was discovered in 1896 by a French scientist, Henri Becquerel. He had been investigating some uranium salts and was trying to find out what made them phosphoresce (glow). He found that they were giving out some type of rays, which he thought were similar to X-rays.

The next development was due to the work of Becquerel's pupil, Marie Curie (Figure 7.3). She found that a number of uranium and thorium compounds were able to ionize gases, i.e. they were able to 'remove' electrons from atoms or molecules of the gas and change them into positive ions. They did this by sending out 'rays' which in turn hit the particles of the gas, and Marie Curie described substances which gave out rays of this kind as being *radioactive*. Her most memorable achievement was while working with her husband Pierre, when they discovered a new element, very much more radioactive than any others known at the time. The Curies called the new element *radium*.

For their work on radioactivity the Curies (and Becquerel) were jointly awarded the Nobel Prize for Physics in 1903, and in 1911 Marie also won the Nobel Prize for Chemistry.

Figure 7.3 Marie Curie in her laboratory (*Radio Times Hulton Picture Library*)

Different kinds of radioactivity

Although only very small amounts of radium had been isolated by the Curies, Ernest Rutherford was able to obtain some for use in his laboratory at Cambridge. From his experiments the true nature of radioactivity was established.

He found that the rays given off from radium could be divided into three main kinds. Some were attracted towards a negatively charged plate, and these were named *alpha rays*. Some were attracted towards a positively charged plate, and were named *beta rays*. Rays which were undeflected by charged plates were called *gamma rays*.

Beta rays consist of streams of electrons. Alpha rays consist of streams of positively charged helium ions, $^{4}_{2}He^{2+}$. (A helium ion is the same as a helium nucleus, because the two electrons present in the neutral atom have both been removed.) An alpha particle has a mass about 7000 times greater than that of a beta 'particle' (i.e. an electron); consequently they travel at slower speeds, and have less 'penetrating' power. Gamma rays are waves similar to light waves and X-rays, but have a much shorter wavelength. They have great 'penetrative' power. Some of the important properties of these different types of radioactivity are summarized in Figure 7.4 and on page 114.

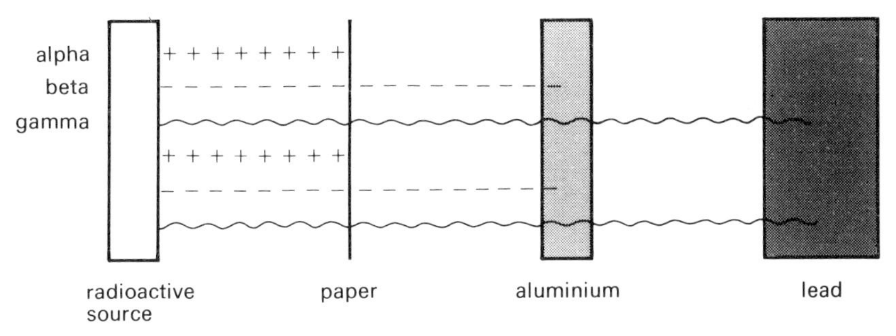

Figure 7.4 Illustrating the penetrating power of the different rays

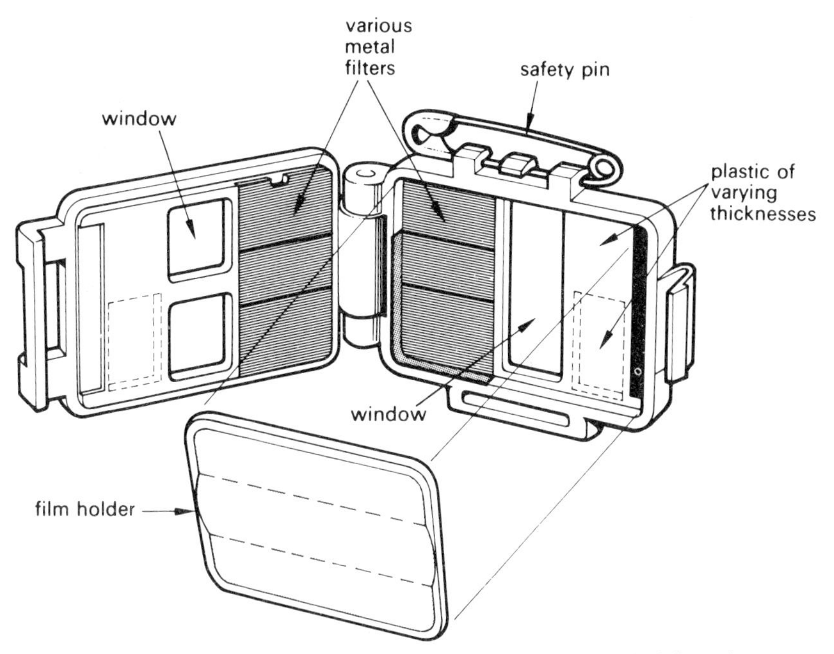

Figure 7.5 Personal radioactivity detector (*Radiochemical Centre*)

Detecting radioactivity

Becquerel found that radioactivity fogged a photographic plate. This method of detecting radioactivity is used today in works and factories where there is any radioactive hazard. Everyone entering the premises carries a portion of special film in a small container (Figure 7.5). This is developed on leaving, or after a certain period of time, and it shows whether the wearer has been exposed to excess radiation and also of which type. This method is too slow for experimental work, and more rapid and sensitive methods are supplied by the gold leaf electroscope, the ionization tube, the Geiger-Müller tube and counter, the cloud chamber, and the scintillation counter. The most familiar of these methods is the Geiger-Müller tube and counter, often called simply a Geiger counter (Figure 7.6). Many schools now possess apparatus for demonstrating the properties of radioactive substances. You may see some of these demonstrations, using a Geiger counter.

Before doing an experiment on radioactivity, it is usual to make a *background count*. This is necessary as a control, because there is always a certain amount of radioactivity in the air. These natural radiations arise from naturally-occurring radioactive substances and also from the ionizing of atoms in the upper atmosphere by cosmic rays. This atmospheric radiation is allowed for in any experimental results.

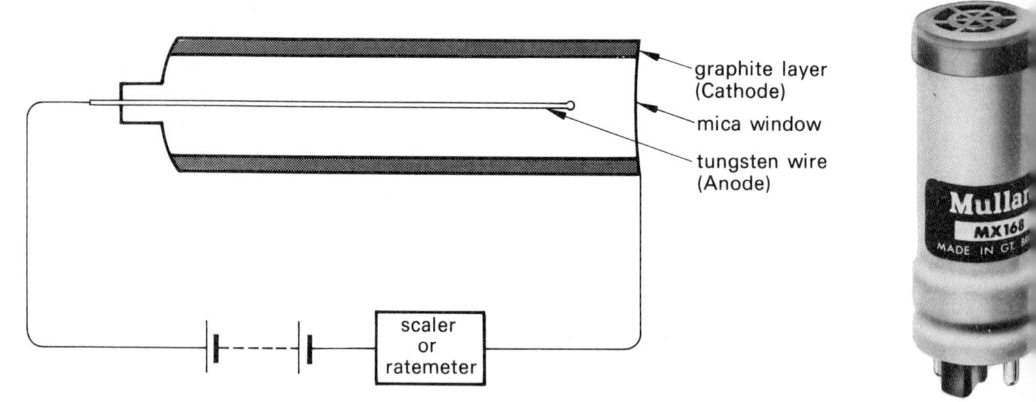

Figure 7.6 A Geiger-Müller tube, (left) the structure of the tube and (right) an example of a commercially available tube

What makes some atoms radioactive?

The nuclei of some atoms are more stable than others. The forces which act within a nucleus are extremely complicated and not yet fully understood, but to some extent the stability of a particular nucleus depends on the ratio of protons to neutrons. If a nucleus is unstable, it sometimes 'decays' by losing alpha particles (each consisting of two neutrons and two protons), or beta particles, or both. It is easy to understand how the loss of one or more alpha particles causes a nucleus to 'rearrange' itself, but less obvious how the loss of one or more electrons (beta particles) can cause changes in the *nucleus*. A beta particle does not come from one of the energy levels outside the nucleus, but is instead formed by converting a neutron inside the nucleus into a proton and an electron; the newly formed proton stays in the nucleus and the electron is ejected as a beta particle. Some changes inside a nucleus also produce gamma rays.

The important thing to understand is that an unstable nucleus can change into a more stable one by 'rearranging' itself, and that it gives out radiation (sub-atomic particles and/or energy) in the process. All the naturally occurring elements from bismuth to uranium are radioactive.

Radioactive decay

The change which occurs when an atom loses particles from the nucleus is known as *radioactive decay*. Unlike chemical changes (Chapter 20), the rate of decay is unaffected by temperature, pressure, or chemical action.

Some radioactive materials decay into other substances which are themselves radioactive. This process continues until an element is produced which is not radioactive and has a stable nucleus. Such a chain of reactions is known as a *decay series*. Loss of beta particles is known as beta decay and the emission of alpha particles as alpha decay.

Decay by beta emission

In this type of decay a neutron changes spontaneously into a proton and an electron. The electron is emitted from the nucleus at great speed and, with millions of other electrons produced in the same way from other atoms, forms the beta rays.

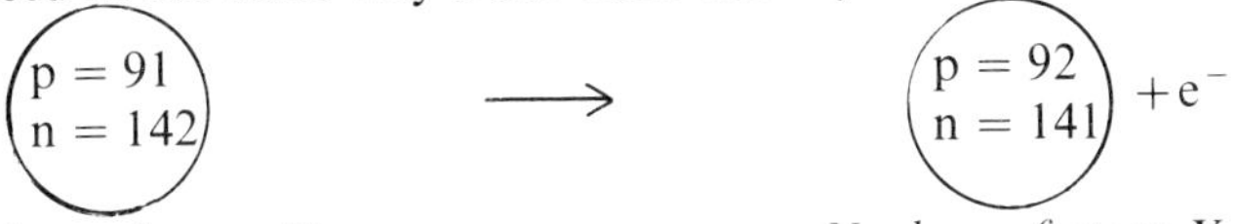

(p = 91, n = 142) ⟶ (p = 92, n = 141) + e^-

Nucleus of atom X
Atomic number = 91
Mass number = (91 + 142) = 233

Nucleus of atom Y
Atomic number = 92
Mass number = (92 + 141) = 233

Result Mass number is unchanged.
Atomic number increases by 1.
Electron emitted (beta ray).

Decay by alpha emission

An alpha particle is a stable particle consisting of two protons and two neutrons, i.e. the nucleus of a helium atom. If an alpha particle is emitted from the nucleus both the atomic number and the mass number of the atom are changed.

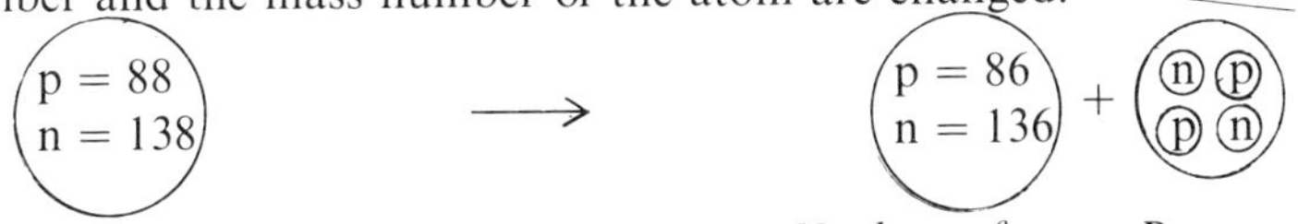

(p = 88, n = 138) ⟶ (p = 86, n = 136) + (n p p n)

Nucleus of atom A
Atomic number = 88
Mass number = (88 + 138) = 226

Nucleus of atom B
Atomic number = 86
Mass number = (86 + 136) = 222

Result Mass number decreases by 4.
Atomic number decreases by 2.
Alpha particle emitted.

Reactions of this kind are normally shown by equations such as:

$$^{226}_{88}A \rightarrow {}^{222}_{86}B + {}^{4}_{2}He^{2+}$$

In most decay series isotopes of some of the elements appear. You will remember that isotopes are atoms which have the same atomic number, i.e. the same number of protons, but different numbers of neutrons. The production of radioactive isotopes will be better understood if we consider a natural decay series, e.g. that of uranium.

An atom of $^{238}_{92}U$ emits an alpha particle and becomes an atom of $^{234}_{90}Th$. (The particle emitted is shown above the arrow.)

$$^{238}_{92}U \xrightarrow{\alpha} {}^{234}_{90}Th$$

The thorium then loses a beta particle which, having negligible mass, does not affect the atomic mass. In the process, a neutron becomes a proton (i.e. neutron → proton + β particle) and so the atomic number is increased by one and the new atom is an atom of protactinium.

$$^{234}_{90}\text{Th} \xrightarrow{\beta} {}^{234}_{91}\text{Pa}$$

Protactinium decays by emitting a beta particle and by the same process becomes $^{234}_{92}$U (an isotope of $^{238}_{92}$U)

$$^{234}_{91}\text{Pa} \xrightarrow{\beta} {}^{234}_{92}\text{U}$$

This process continues until the final product, in this case a non-radioactive isotope of lead, is formed. The start of the decay series can be written:

$$^{238}_{92}\text{U} \xrightarrow{\alpha} {}^{234}_{90}\text{Th} \xrightarrow{\beta} {}^{234}_{91}\text{Pa} \xrightarrow{\beta} {}^{234}_{92}\text{U} \xrightarrow{\alpha} {}^{230}_{90}\text{Th} \xrightarrow{\alpha} {}^{226}_{88}\text{Ra} \xrightarrow{\alpha} {}^{222}_{86}\text{Rn} \rightarrow \text{etc.}$$

Rate of decay

We have said that the rate of radioactive decay cannot be affected by any chemical or physical change. The other important point about radioactive decay is that the rate of decay is not constant, i.e. so many atoms per second. Instead the rate depends on the number of atoms present at any one time. The more radioactive atoms there are present, the more will decay, and as these break up, the rate at which the remaining atoms decay gradually decreases.

It will help you to understand this better if we take an example of a similar type of rate of change from everyday life. Suppose a number of people, say thirty-two, start out on a cross country run. After the first hour's running sixteen have dropped out, leaving sixteen still in the race. By the end of another hour, eight more have given up and eight remain; after three hours four of these have dropped out and four remain and at the end of four hours only two runners are left. Note that the number dropping out is always the same as the number remaining and after the same period of time, one hour in this case, the number of people left is always half the number present at the beginning of the period.

This is what happens with radioactive substances. The time taken for the number of atoms to be reduced to half in a sample of a particular nuclide is always the same and is characteristic of the nuclide. It is known as its *half life*. Half lives range from small fractions of a second for some man-made elements to millions of years in the case of some which occur naturally (Figure 7.7).

The half life of a nuclide of thorium is twenty-four days. This means that if we start with eight million atoms of the thorium nuclide we would be left with four

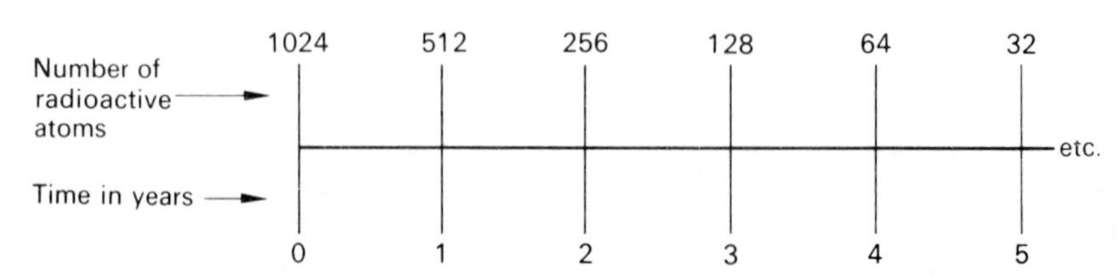

Figure 7.7 To show how a radioactive element, with a half life of one year, would decay

million atoms after twenty-four days. After the next twenty-four days (forty-eight altogether) there would be two million atoms left and after the next twenty-four days, one million, and so on.

What is the 'half life' of the people on the cross country run?

Check your understanding

1. $^{232}_{90}Th$ gives out alpha particles and forms an element X which decays by emitting beta particles to form element Y. Y decays by beta emission to form element Z.

(a) Complete the following by writing the mass numbers and the atomic numbers in front of the elements X, Y, and Z.

$$^{232}_{90}Th \xrightarrow{\alpha} X \xrightarrow{\beta} Y \xrightarrow{\beta} Z$$

(b) How are Z and thorium related?

2. A radioactive substance has a half life of fifteen minutes. If 2 grams of the substance are available at the start of an experiment, how much will remain at the end of an hour?

Power from the atom

Fission

$$^{235}_{92}U + ^{1}_{0}n \rightarrow ^{89}_{36}Kr + ^{144}_{56}Ba + 3^{1}_{0}n$$

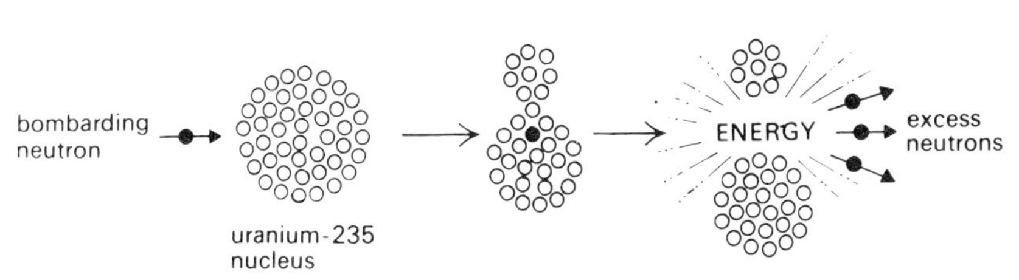

Figure 7.8 The fission of a $^{235}_{92}U$ nucleus

When a 'heavy' nucleus splits to form one or more smaller, more stable nuclei, energy and sub-atomic particles are released. The energy changes are tremendous, many million times greater than the energy produced by chemical reactions. An example of *fission* (i.e. the splitting up of a nucleus into two or more smaller nuclei) occurs when an atom of $^{235}_{92}U$ is bombarded by a neutron (Figure 7.8). The neutron is absorbed by the nucleus which then splits in two forming, for example, nuclides of barium and krypton and releasing further neutrons.

$$^{235}_{92}U + ^{1}_{0}n \rightarrow ^{89}_{36}Kr + ^{144}_{56}Ba + 3^{1}_{0}n.$$

During this process there is a slight loss in mass, i.e. the total mass of the products is slightly less than the total mass of the reactants. The 'lost' mass is converted into energy—about sixty million times the energy produced by the combination of one carbon atom and two oxygen atoms during the combustion of coal.

The extra neutrons given out during the fission of a uranium atom might cause further fissions in neighbouring atoms, with the production of still more neutrons, all capable of causing fission. If this process were to continue, the fission process would accelerate rapidly and a self-sustaining or *chain reaction* would result (Figure 7.9).

A chain reaction cannot occur in *natural* uranium because very few of the neutrons which are given out can cause further fissions. Uranium consists of two main isotopes, $^{238}_{92}U$ (99.3 per cent) and $^{235}_{92}U$ (0.7 per cent). $^{235}_{92}U$ will undergo fission by

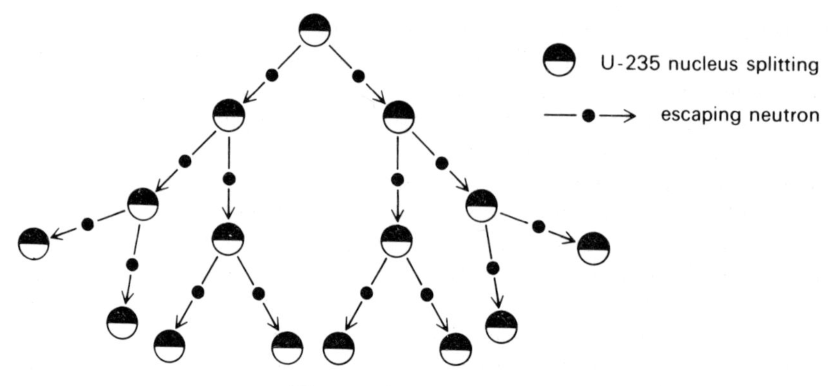

Figure 7.9 A chain reaction

both fast and slow neutrons, but more easily by slow neutrons. $^{238}_{92}U$, which forms the greater part of natural uranium, is fissile by fast neutrons only, and very few of the neutrons produced by the fission process are capable of producing fission in $^{238}_{92}U$. In fact, a neutron is much more likely to be absorbed by a $^{238}_{92}U$ nucleus than to split it. Consequently, most of the neutrons produced by fission of $^{235}_{92}U$ nuclei are either absorbed by the $^{238}_{92}U$ nuclei or escape into the air, and no chain reaction is possible.

Uncontrolled chain reactions. The atomic bomb

If the two isotopes of uranium are separated (this has to be done by physical means as the isotopes are chemically identical) and a fairly pure sample of $^{235}_{92}U$ is used, an uncontrolled chain reaction can be produced. The factor controlling this possibility

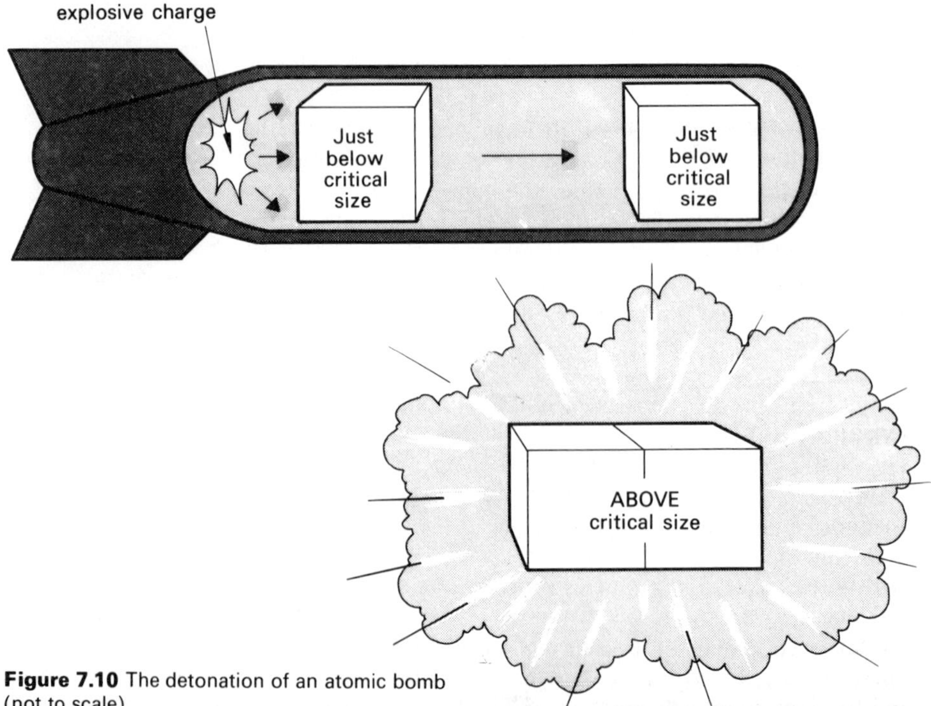

Figure 7.10 The detonation of an atomic bomb (not to scale)

is the size and shape of the piece of uranium. A small block of uranium with a relatively large surface area will lose many of its neutrons to the surrounding air before they are able to produce further fissions (remember that ${}^{235}_{92}U$ is more easily affected by slow neutrons). As the size of the piece of uranium is increased, a smaller proportion of neutrons will escape and more will be slowed down by collision and be available for fission. A point will be reached where the number of neutrons which fail to escape and are slowed down is just sufficient to form a chain reaction, each fission producing one neutron available for a further fission. The size of the piece of uranium when this state of affairs is reached is called the *critical size*.

In the simplest type of atomic bomb, two pieces of ${}^{235}_{92}U$, each just under the critical size, are used (Figure 7.10). These are kept apart until the bomb is to be detonated. They are then brought together and the result is a block of uranium of above critical size. A sufficient number of neutrons is now retained and slowed down within the mass to produce an accelerating chain reaction. The consequent release of energy is enormous and produces an explosion many thousands of times more violent than that made by a more conventional explosive such as TNT. This is what occurred when the atomic bomb destroyed Hiroshima in 1945.

Controlled chain reactions. The nuclear reactor

The principle

A more constructive use of nuclear fission is the nuclear reactor which uses nuclear energy to produce electricity. One of the earliest types of reactor consisted of a huge cylinder of graphite into which were inserted fuel rods of natural uranium and control rods of boron steel.

The graphite acts as a *moderator*, i.e. it cuts down the speed of the neutrons escaping after the fission of a ${}^{235}_{92}U$ nucleus to such an extent that most of them are able to produce fission in other ${}^{235}_{92}U$ nuclei. Thus the use of the graphite moderator makes a chain reaction possible in *natural* uranium.

The boron steel rods are used to control the reaction and prevent it getting out of hand. Boron is able to absorb neutrons and so prevent them producing further fissions. The rods are moved in or out of the reactor until a balance is obtained in which each fission produces, on average, just one neutron capable of causing a further fission, and thus a controlled chain reaction takes place.

The energy is released inside the reactor in the form of heat, and this is removed by a coolant such as carbon dioxide gas, which is forced through ducts in the pile. The hot gas changes water into steam in a heat exchanger at some distance from the pile, and this steam is used to drive electricity generators as in a conventional power station. The principle of the process is summarized on page 115.

Types of nuclear reactor

The first batch of commercial-size nuclear power stations used in Britain are known as *magnox stations* because the fuel rods are placed in a magnesium alloy (magnox) container. More modern types of reactor show various improvements.

During the 1960s a phase of *advanced gas-cooled reactors* (AGR) was introduced in Britain. These are very similar to the magnox stations, but the fuel is 'enriched' and encased in stainless steel tubes (Figure 7.11). The enriched fuel contains a greater proportion of ${}^{235}_{92}U$ (i.e. the fissile part) than occurs in natural uranium. These reactors operate at higher temperatures than earlier types, are more compact, and have an increased power output.

Figure 7.11 Loading the uranium fuel elements into a reactor (*United Kingdom Atomic Energy Authority*)

A further stage in development is the *fast reactor* or *self-breeder* reactor. In theory, these reactors are very attractive propositions because they produce more fuel than they consume, but they are still handicapped by technical problems. The reactor core has no conventional moderator and is therefore very small. It consists of natural uranium enriched with extra $^{235}_{92}U$ and $^{239}_{94}Pu$. Plutonium is not a naturally occurring element but is produced when $^{238}_{92}U$ is bombarded by fast neutrons. It is therefore formed in the magnox and AGR reactors. $^{239}_{94}Pu$ is fissile and can be used in the same way as $^{235}_{92}U$. In the early reactors, $^{239}_{94}Pu$ is produced at about the same rate as $^{235}_{92}U$ is used up, but the fast reactors are designed to produce excess plutonium. As plutonium is a different element from uranium the two can easily be separated by chemical means, and the separated plutonium (from both the early types of reactor and the fast reactors) is used to enrich the fuel in the fast reactors.

The fast reactors produce excess plutonium because the core is surrounded by a blanket of uranium, which is mostly the $^{238}_{92}U$ nuclide. This acts as a kind of 'moderator' in that it absorbs neutrons, and in so doing $^{238}_{92}U$ is converted into plutonium at a rate faster than the fuel is consumed. The vast quantities of heat produced in these reactors is removed by liquid sodium (which is pumped around the core) and then transferred to steam generators in the usual way.

In theory, fast reactors can utilize all the uranium which is given to them, whereas the early reactors utilize only a small fraction of their fuel (see page 504). This is because the fast reactors can convert $^{238}_{92}U$ (which is non-fissile, and forms by far the greatest part of natural uranium) into nuclear fuel, whereas the early types of reactor are unable to do this. It is claimed that the 20 000 tonnes of $^{238}_{92}U$ which are stored in Britain would be equivalent to more than 40 000 million tonnes of coal if 'burned' in fast reactors.

In summary, all of the various reactors in current use have essentially the same parts: the fuel, some form of moderator, control rods, and a coolant. They may also have a neutron reflector (a graphite casing which acts as a reflector, propelling some of the escaping neutrons back into the fuel), and the whole system is enclosed in a thick protective shield of concrete and lead or steel.

Fusion

This is considered on page 503, as a possible source of energy in the future.

Uses of radioactive isotopes

The first use outside the laboratory of a radioactive substance came when it was discovered that radium stopped the growth of certain tumours in the human body. Since that time the use of radioactive substances has spread into many types of work. This is due mainly to the discovery of artificial radioactive isotopes. These are man-made isotopes which are produced by bombarding the nuclei of stable atoms with sub-atomic particles travelling at high speeds.

Tracers

If a small quantity of a radioactive element or compound is mixed with a non-radioactive form of the same substance, then the movement or behaviour of the whole mass can be followed with some form of radiation detector. Radioisotopes which are used for this purpose are called tracers. They can be detected in extremely small concentrations and they are indistinguishable chemically from stable isotopes of the same element. The following are just a few of the many examples of the use of tracers.

Radioactive sodium ions can be introduced into the body in common salt and their path followed in the blood stream. Mosquito larvae at a particular place are made radioactive and later checks on the adults show how they migrate. Radio-phosphorus introduced into the soil with normal ammonium phosphate tells us exactly how a plant picks up and uses phosphorus in its growth. The path of the same element can be traced from a hen to the eggs she lays. Radio-iodine can be used to check the activity of the thyroid gland. Oil distributors use radioactive tracers to follow the paths of different grades of oil which flow in sequence through the same pipe.

One of the most interesting uses of a radioisotope as a tracer is that of radio-carbon, $^{14}_{6}C$. This isotope has been valuable for following the mechanism of various organic reactions which involve isomers (p. 403) and particularly for tracing the complex series of events which occur during photosynthesis.

Naturally occurring isotopes also have their uses. In atmospheric carbon dioxide there is a very small fraction of the combined carbon which is radio-carbon. The ratio of this $^{14}_{6}C$ to its stable isotope $^{12}_{6}C$ is constant and its value is known. Suppose it were in the ratio of eight to ten million. As carbon dioxide is utilized by plants, which are then eaten by animals, all *living* things contain the two carbon isotopes in this ratio. When the plant or animal dies no more carbon is taken in but the $^{14}_{6}C$ in the body continues to decay. It has a half life of about 5600 years. If now the radioactivity of the carbon in an ancient piece of wood, say from a prehistoric canoe, is measured and the ratio of the $^{14}_{6}C$ to the $^{12}_{6}C$ is only four to ten million, this would show that the wood was over 5600 years old. Using similar methods, scientists are able to date objects which were in existence up to fifty thousand years ago.

Penetration

As you already know, the beta and gamma rays emitted by radioactive substances can penetrate to varying distances in different materials. The degree to which the radiation is absorbed depends on the thickness and density of the substance. Thus radioisotopes can be used to measure and control the thickness of metals, paper, plastics, etc. (Figure 7.12).

The radioactive source is placed below the material which is to be tested and the detector, e.g. a Geiger tube, above it. The material is gradually unwound and then rewound on another roller after passing under the detector. Variations in the thickness of the strip produce variations in the reading of the counter. The current variations operate a control which automatically adjusts the thickness to the correct measurement. Other examples of the same type are radioactive density gauges which

Figure 7.12 Use of isotopes in industry to measure thickness of a pipe wall (*United Kingdom Atomic Energy Authority*)

control the packing of tobacco in cigarettes, as well as the filling of various packets, cans and other containers. Again gamma rays can be used like X-rays to 'photograph' metals; hidden flaws show up as dark spots on the film.

Shortly after the isolation of radium it was discovered that radiations could destroy cancerous tissue and for a short time radium was used for this purpose. Today, radioisotopes such as radio-cobalt are used instead of radium. Radio-cobalt is particularly effective as it is a strong source of gamma rays and these can penetrate deep into the body (Figure 7.13). Of course these radiations can harm healthy cells and cause genetic changes, so great care must be exercised when they are used. A new development in this field is the use of neutron generators, and it is hoped that neutron bombardment will overcome some of the most stubborn and, at present, incurable forms of cancer.

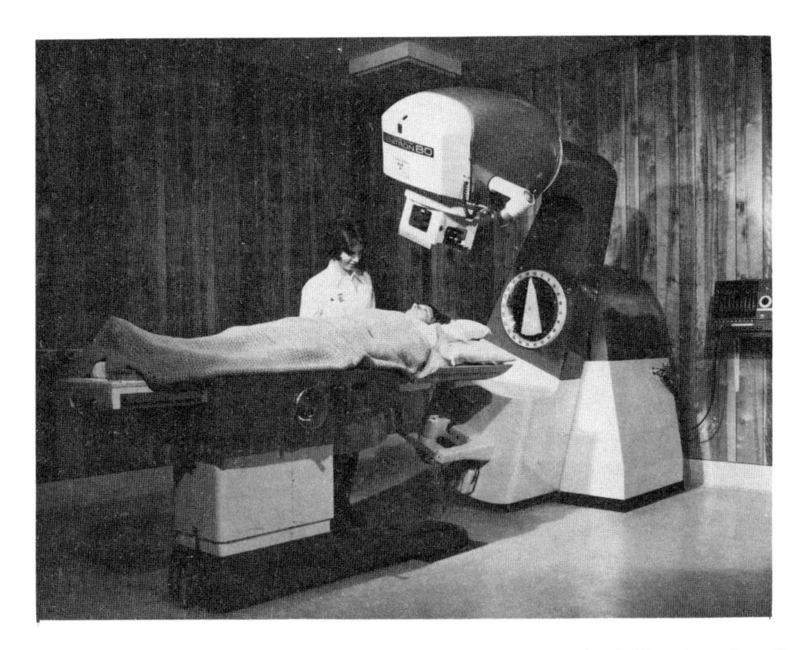

Figure 7.13 Theratron-80 unit at Churchill Hospital, Oxford (*United Kingdom Atomic Energy Authority*)

Sterilization

Medical instruments are sterilized by strong doses of gamma rays from $^{60}_{27}Co$, which kill bacteria.

Provision of artificial light

Some lights in coal mines are made of tubes coated with zinc fluoride. The tubes contain $^{85}_{36}Kr$ which emits beta rays and causes the zinc fluoride to fluoresce. The tubes can produce light for over two years before renewal, but the light is rather dim.

Summary

The subject is of absorbing interest and has grown to such an extent as to be far beyond the confines of this book. If you are interested and would like to follow it further you will be sure to find books dealing with radioactivity in more detail in either your school or local library. You will perhaps want to know more about nuclear reactors, atomic power, and the hydrogen bomb, and to discuss the merits and demerits of their effect on modern civilization. The work of the researchers on radiation is moving at a rapid pace. Work that started with the Atomic Theory of John Dalton may one day progress to make use of all the enormous store of energy trapped in the atomic nucleus.

7.3 A SUMMARY OF CHAPTER 7

The following 'check list' should help you to organize the work for revision.

1. *Definitions*

(a) Atomic number (p. 97)
(b) Mass number (p. 97)
(c) Isotopes (p. 98)
(d) Relative atomic mass (p. 100)
(e) Relative molecular mass (p. 101)

2. *Other points*

(a) Details about sub-atomic particles (electrons, protons, and neutrons), and where they are found in an atom (p. 95). The maximum number of electrons which each of the first few energy levels can contain (p. 95).
(b) How to construct atomic structure diagrams, given the atomic number and mass number of a nuclide, e.g. $^{37}_{17}Cl$ (p. 96).
(c) The correct use of the terms isotope and nuclide ($^{37}_{17}Cl$ and $^{35}_{17}Cl$ are isotopes, but $^{37}_{17}Cl$ is a nuclide).
(d) The difference between relative atomic mass and mass number. (A mass number refers to one particular nuclide. A relative atomic mass refers to a mixture of isotopes, and may not be a whole number.)
(e) How to determine relative molecular masses from relative atomic masses, by adding together the relative atomic masses of each of the atoms in a molecule.
(f) *Radioactive emissions*

	β rays	*α particles*	*neutrons*	*γ rays*
Nature	Fast electrons. Originate in nucleus (where a neutron → proton + electron) *not* electron shells. Same as cathode rays	A combination of 2 neutrons and 2 protons, i.e. a helium nucleus	Neutrons	Electromagnetic radiation of very short wavelength
Charge	Negative (−1)	Positive (+2)	0	0
Mass cf. hydrogen atom	$\frac{1}{1837}$	4	1	0
Approximate speed ($m\,s^{-1}$)	Up to 2.5×10^8	3×10^6 to 3×10^7	3×10^3 to 9×10^5	3×10^8
Penetrating power	Paper has little effect. Slower ones stopped by thin sheets of aluminium, but sheets of 0.35 cm thickness needed to stop most	Poor, e.g many stopped by sheet of paper	—	Great. Paper and aluminium have little effect. Considerable thickness of lead needed to stop this radiation

(g) *Radioactive decay* is when atoms give off one or more types of radioactive emission in order to become more stable. The rate of radioactive decay cannot be influenced by any of the methods used to change the rates of chemical reactions.
(h) Each radioactive nuclide has a *half life*, which is the time taken for half of the atoms present at any moment to decay. Half lives vary from fractions of a second to millions of years.
(i) *What happens to an atom when it decays*

	Mass number of atom	*Atomic number of atom*
By loss of a β particle	Unchanged	Increases by 1
By loss of an α particle	Decreases by 4	Decreases by 2
By loss of a neutron	Decreases by 1	Unchanged
By γ rays	Unchanged	Unchanged

(j) *Nuclear power (simplified version of early reactors)*

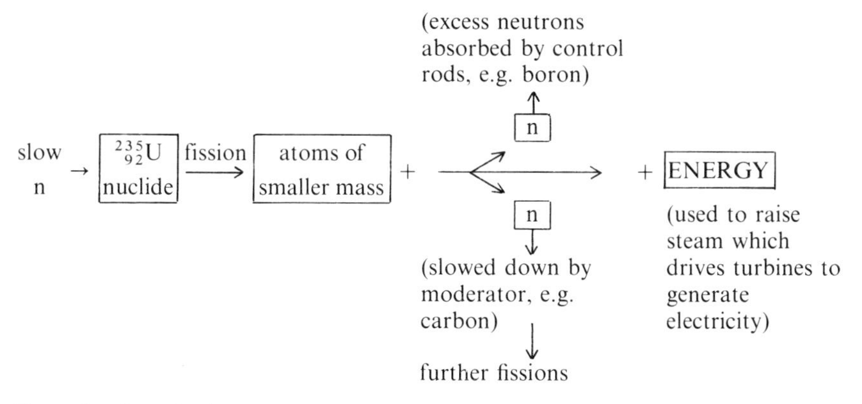

(k) *Uses of radioactive isotopes.*
(i) Medical, e.g. controlling some kinds of cancer.
(ii) Tracers. A radioactive nuclide is chemically identical to a stable nuclide of the same element, but is easily detected even in very small concentrations because of its radioactivity. Tracers are used to follow the path of chemicals in plant and animal cells etc.
(iii) Radio-carbon dating. The ratio of non-radioactive carbon:radioactive carbon in an object made of once-living material (e.g. wood) depends upon the date when the material 'died', and this is used to date objects which were in existence up to 50 000 years ago.
(iv) Penetration. Radioactivity penetrates varying distances in different materials. Radioactive isotopes can be used to measure the thickness of sheet materials such as metals, paper, and plastics when they are being produced automatically.

QUESTIONS

Some questions relevant to this chapter are included at the end of Chapter 8.

8 Bonding

8.1 INTRODUCTION

You now know that chemicals each have a formula, e.g. sodium chloride has the formula NaCl. You must now learn the answers to three very important questions.

1. Why do sodium and chlorine join together (when given a chance to do so) to form sodium chloride, instead of staying separately as sodium and chlorine?
2. Why do sodium and chlorine always combine in the ratio of one particle of sodium to one particle of chlorine, so that the formula of the product is always NaCl and never $NaCl_2$, Na_2Cl, etc?
3. When sodium and chlorine do combine, what holds the different particles together in the compound sodium chloride?

The structure and stability of the noble gases

If you look back to your drawings of atomic structures for elements with atomic number 2 (helium), 10 (neon), and 18 (argon), you will notice that these three elements all have atoms which contain only fully filled electron shells. These elements belong to a family called the *noble gases*. This family used to be called the inert gases, and their old name gives a clue to their chemical properties; these elements are very unreactive. The noble gases do not take part in normal chemical reactions, and only the ones with large atoms (krypton, xenon) form a few compounds. Their almost unique lack of chemical reactivity is shown by the fact that they do not even join with other atoms of their own kind to form molecules (unlike elements such as hydrogen, in which two atoms join together to form a H_2 molecule). Atoms of the noble gases are the only *atoms* to be found *free* in nature, e.g. Ar, He. Atoms of all other elements are not found free; they always join with other atoms to form larger units. These other atoms can be of the same type (e.g. in Cl_2) or be atoms of a different type (e.g. in H_2O).

Chemists soon realized that atoms of noble gases are stable (i.e. they will not easily react or change) because they have fully filled electron shells. Atoms of other elements can become stable by losing, gaining, or sharing electrons until they also have fully filled electron shells. You must understand this very important point, because chemical reactions involve sharing or exchanging electrons in order that atoms can

'He never joins with us'

obtain a more stable arrangement of electrons. The most common stable electron arrangement is that of an atom of a noble gas (i.e. fully filled electron shells) and the 'atoms' in most elementary reactions end up having this arrangement. They achieve a stable arrangement of electrons by joining (bonding) with other atoms, even if this means atoms of their own kind.

There are two main types of bonding, *ionic* (or *electrovalent*) and *covalent*. When different atoms form ions (in order to achieve a stable electron arrangement), these ions then attract each other and this force of attraction is called ionic bonding. When atoms share electrons in joining together (and achieving a stable electron arrangement), the bonding is called covalent.

8.2 IONIC BONDING

The formation of positive ions

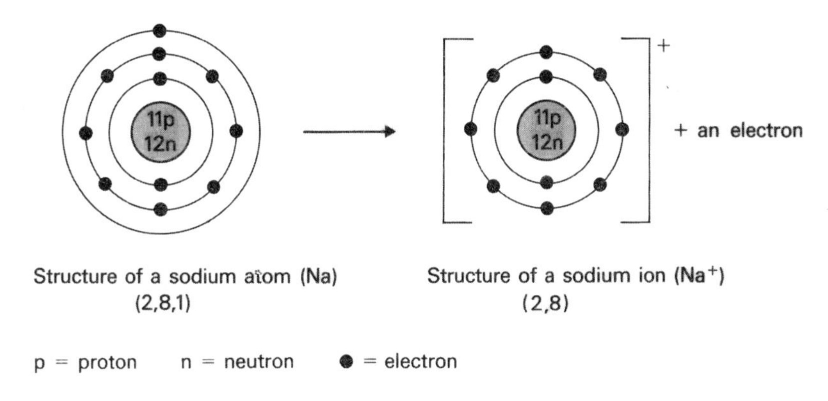

Figure 8.1 The structure of a sodium atom and a sodium ion

Figure 8.1 shows the electron structures of a sodium atom and a sodium ion. The sodium atom has one electron in its outer shell of electrons. If this atom is to achieve the stable electron structure of a noble gas, it must either *lose* the outer electron (in which case it will be left with two fully filled electron shells) or *gain* seven electrons (in which case it will have three fully filled electron shells). It requires less energy to lose one electron than to gain seven, and so a sodium atom does this in order to achieve a stable structure. This produces a sodium ion.

Figure 8.1 shows that when a sodium atom loses an electron, it is left with 11 protons and only 10 electrons. The particle formed is no longer neutral because it has more positive protons than negative electrons, and so has an overall positive charge of one unit. This is shown by using the square bracket and the charge symbol as in Figure 8.1, or in simple form as Na^+. This particle must no longer be called an atom because it is no longer neutral, and although it has the same nucleus as the 'parent atom' it has the electron arrangement of a *different* element, a noble gas. The ion takes its name from the parent atom and *not* from the atom whose electron arrangement it now has. Thus in this example an ion of *sodium* is formed and not an ion of *neon*. This is because the sodium ion still has the *nuclear structure* of sodium, even though it has the electron arrangement of neon. The ion is completely different in behaviour when compared with the parent atom because of its stable arrangement of electrons. You will now begin to understand why sodium chloride (which contains sodium *ions* and chloride *ions*) is so different from the reactive (and potentially dangerous) elements sodium and chlorine (p. 29).

All metals form *positive* ions, because metal atoms usually have 1, 2, or 3 electrons in their outer shells, and it is easier for them to become stable by losing electrons than by gaining electrons.

The formation of negative ions

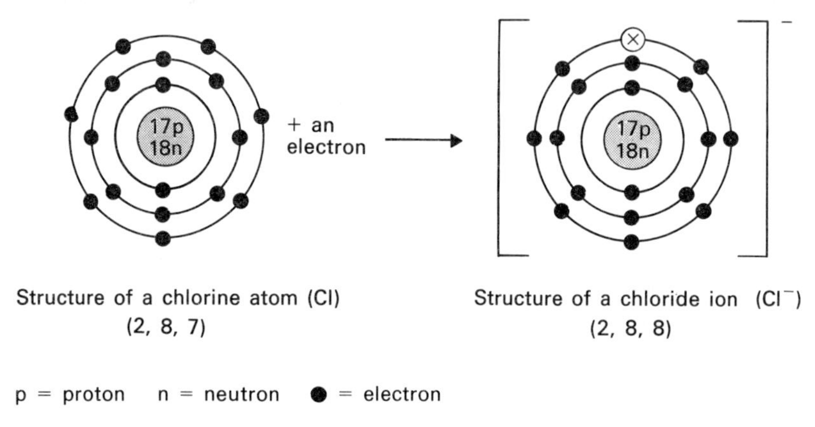

Figure 8.2 The structure of a chlorine atom and a chloride ion

Figure 8.2 shows the electron structures of a chlorine atom and a chloride ion. The atom has 7 electrons in its outer shell of electrons, and can either *gain* one electron or *lose* seven in order to obtain a stable electron structure. It requires less energy to gain one electron than to lose seven, and the ion formed is shown in Figure 8.2.

The chloride ion has 17 protons but 18 electrons, and so it has one more negative electron than positive protons, and therefore an overall negative charge of one unit. The ion is shown by using the square bracket and charge symbol as in the figure, or in simple form as Cl^-. This ion is completely different in behaviour from the parent chlorine atom.

All non-metals can form *negative* ions because their atoms often have 5, 6, or 7 electrons in their outer electron shells, and it is easier for them to become stable by gaining electrons than by losing them.

How sodium and chlorine combine by ionic bonding

Sodium atoms and chlorine atoms combine because they can help each other to become stable. The one electron which each sodium atom *loses* can be *gained* by a chlorine atom, so that both atoms become ions with stable electron structures. We can now answer the three questions posed in the introduction to this chapter.

1. Sodium and chlorine atoms join together because by doing so their 'atoms' become stable. (Stability, of course, is relative (see page 475). Thus Na^+ and Cl^- ions are not stable with respect to electricity, and revert to the elements when molten sodium chloride is electrolysed, page 533.)
2. They join (in this case) in the ratio 1:1 because this combination provides the correct number of electrons for atoms of *both* elements to become stable. If they had joined in the ratio 2Na:1Cl, the two sodium atoms which react with one chlorine

atom would each lose one electron, providing two electrons altogether. One chlorine atom, however, can only take *one* of these electrons (it only has 'room' for one) and so the other electron could not be accepted.
3. The ions are held together in the compound because the positively charged sodium ions are attracted to the oppositely charged chloride ions. The force which holds them together is called electrostatic attraction.

We are saved!
A satisfactory conclusion

Note: You must not imagine that sodium ions and chloride ions form simple 'ion pairs' of sodium chloride, NaCl. The attractions between all the negative and positive ions in crystals of sodium chloride produce a 'giant structure' which consists of a three-dimensional arrangement of millions of ions, called a *crystal lattice*. You will understand this better after reading Chapter 16.

How to work out a bonding diagram for any example of ionic bonding

If you understand how and why sodium and chlorine atoms become ions, and how the ions bond together, you will be able to work out how other ionic compounds are formed. It is easy to decide whether a chemical is ionically bonded, because *whenever a metal* (*or the ammonium radical*) *is combined with a non-metal, the bonding will be ionic*. (There are a few exceptions to this rule, but these can easily be learned if you come across them.) This is because metals form positive ions (by losing electrons) and non-metals can form negative ions (by gaining electrons), so that they can help each other by exchanging electrons, forming ions, and bonding together.

In order to show how ions combine by ionic bonding, use the steps which are set out below. An example, using magnesium chloride, is shown in Figure 8.3, and this should be consulted at each stage. You may feel that you could combine the atom and ion drawings to make one diagram only, but this would be unwise. The separate drawings show clearly what happens, and in answering a question it is important to show that you fully understand this.

The instructions may seem complicated at first, but they should enable you to explain how atoms combine by ionic bonding for any common example, and they will soon seem easy if you apply them to examples other than magnesium chloride. Note that all electrons are identical, but those present in different atoms should be given different symbols in bonding diagrams in order to show electron transfers more clearly.

1. Determine the formula of the substance, e.g. $MgCl_2$.
2. Work out the atomic structures (excluding neutrons) of the atoms concerned from their atomic numbers.
3. Draw the atomic structure of each of the atoms in the formula (e.g. for $MgCl_2$ draw one atom of magnesium and two atoms of chlorine), keeping the metal atoms

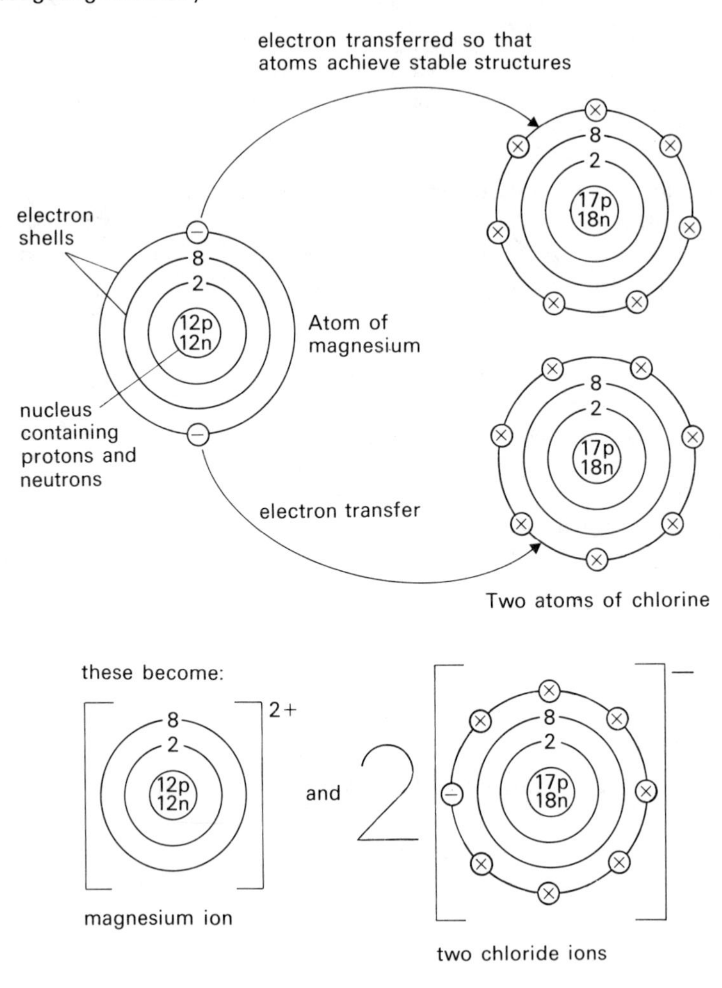

Figure 8.3 The formation of magnesium chloride

on the left hand side. (If more than one atom of the same kind is needed, draw the additional atoms in line below the first one of the same type.) Draw the *outer* electrons in the atoms individually, using different colours or different symbols for those in the metal atom(s) and those in the non-metal atom(s).

4. Show by means of one or more arrows how the outer electrons in the metal atoms move over to the outer energy levels of the non-metal atoms, so as to leave the original outer shells of the metal atoms empty and to fill the outer shells of the non-metals. (If you have worked out the formula correctly, then the transfer of the appropriate number of electrons should be automatic.)

5. Label fully as shown in Figure 8.3.

6. Redraw the 'atoms' as they will be after the electron transfer, showing the individual electrons in the outer shells of the non-metal ions with their different colours or symbols. Only one example of each type of ion need be drawn, but the numbers of each should be clearly shown.

7. Put a bracket and charge symbol around each ion. Remember that if an atom loses electrons it becomes a positively charged ion, with the same number of positive charges as the number of electrons lost. This is because the protons now outnumber the electrons. The opposite is true of non-metals, which gain electrons. Metals always form positive ions.
8. Write the name and formula of the compound formed, and label as shown; it is not necessary to label the nucleus or electrons in the second drawing as they were labelled in the first.
9. Check that the *total* number of positive charges is equal to the *total* number of negative charges, e.g. Mg^{2+} and $2Cl^-$.

Check your understanding

1. What do a chloride ion and an argon atom have in common? Why is a chloride ion not called an argon ion?

2. Work out bonding diagrams for the following compounds *if they are ionic*. Two of them will not have ionic bonding; you must recognize which two these are, and then ignore them.
(a) potassium chloride, (b) potassium fluoride, (c) sulphur chloride, (d) sodium oxide, (e) calcium fluoride, (f) magnesium oxide, (g) carbon dioxide, and (h) aluminium fluoride.

3. (a) The atomic number of potassium is 19. Give three deductions which can be made from this statement.
(b) The following symbols represent atoms of elements showing their mass numbers and atomic numbers.

$^{40}_{18}Ar \quad ^{39}_{19}K$

(i) What is the electronic structure of the argon atom?
(ii) How many neutrons are there in the nucleus of the argon atom?
(iii) Explain why the argon atom has a lower atomic number but a greater mass number than the potassium atom.
(c) $^{32}_{16}S \quad ^{35}_{17}Cl \quad ^{40}_{18}Ar \quad ^{39}_{19}K \quad ^{40}_{20}Ca$
Atoms and ions which have the same number of electrons are said to be *isoelectronic*. From the elements given above, write the formulae for three ions which are isoelectronic with the argon atom. (J.M.B.)

4. The table shows the mass numbers and atomic numbers of atoms labelled T to Z.

	Mass number	*Atomic number*
T	2	1
V	3	1
W	3	2
X	6	3
Y	9	4
Z	11	5

(a) How many protons are there in an atom of Y?
(b) How many electrons are there in an atom of W?
(c) How many neutrons are there in an atom of Z?
(d) Which atoms are isotopes of the same element?
(e) Which atom would readily form an ion with a single positive charge?
(f) Which is an atom of a noble gas? (J.M.B.)

5. The table shows the numbers of fundamental particles in some atoms and/or ions. The letters shown are not the symbols for the elements.

	Protons	*Neutrons*	*Electrons*
A	10	10	10
B	11	12	10
C	12	12	12
D	17	18	17
E	17	20	17

Select the particles which best fit the descriptions given below.
(a) a metal atom
(b) a metal ion
(c) a noble gas
(d) an atom of a divalent element
(e) atoms which are isotopes of one element (J.M.B.)

6. Sodium atoms and sodium ions
A are chemically identical
B have the same number of electrons
C have the same number of protons
D both react vigorously with water
E react and form compounds with covalent bonds

7. When magnesium combines with chlorine
A each magnesium atom gains two electrons
B each chlorine atom loses one electron

C a covalent bond is formed
D the compound formed contains equal numbers of magnesium and chlorine ions
E the compound formed will be a solid at room temperature

8. An ionic bond is often formed when
A the combining atoms need to lose electrons in order to attain a noble gas configuration
B the combining atoms both need to gain electrons in order to attain a noble gas configuration
C two non-metallic elements react together
D a metallic element combines with a non-metallic element
E two metallic elements react together

Properties of ionic compounds

You will understand these better when you have studied structure (Chapter 16) and electrolysis (Chapter 13), but they are given here so that all aspects of bonding can be revised together.

1. Ionic compounds are composed of two or more different kinds of oppositely charged ions.
2. These oppositely charged ions attract each other and form a large three-dimensional lattice, called a giant structure, which is held together by electrostatic attractions in all directions. Ionic compounds are thus usually crystalline solids.
3. Because of the great attraction between the ions, a large amount of energy has to be used to separate them, and ionic compounds usually have high melting points and boiling points.
4. Ionic compounds, when molten or in aqueous solution, are electrolytes (see page 210).
5. Ionic compounds are often soluble in water and do not usually dissolve in organic solvents such as ethanol or benzene, but see page 169.
6. Once ions are formed they show none of the chemical properties of the parent atoms.

8.3 COVALENT BONDING

You will realize that there are many compounds which clearly do not have the properties of ionic compounds, for example those which are gases at room temperature.

When two atoms of non-metals join together, the bonding is likely to be covalent.

This happens because non-metallic atoms (which usually have 5, 6, or 7 electrons in their outer shells) need to *gain* electrons in order to obtain a stable electron configuration. If two such atoms combine, they cannot transfer electrons (as in ionic bonding) because they *both* need to gain electrons, and there are no metal atoms to provide them. In cases like this, the atoms *share* electrons; in a sense, they 'help each other'. A group of two or more atoms held together by shared electrons (i.e. by covalent bonding) is called a molecule, and you should now understand why it is important to use the correct word (atom, ion, or molecule) when describing the particles of a substance.

The chlorine molecule, Cl_2

Figure 8.4 shows how two chlorine atoms are held together by covalent bonding in a chlorine molecule, Cl_2. It shows that a pair of electrons is shared by *both* atoms, so that in effect each atom appears to have a fully filled outer shell of electrons. The atoms are held together by the shared pair of electrons, and this is the 'bond'. As

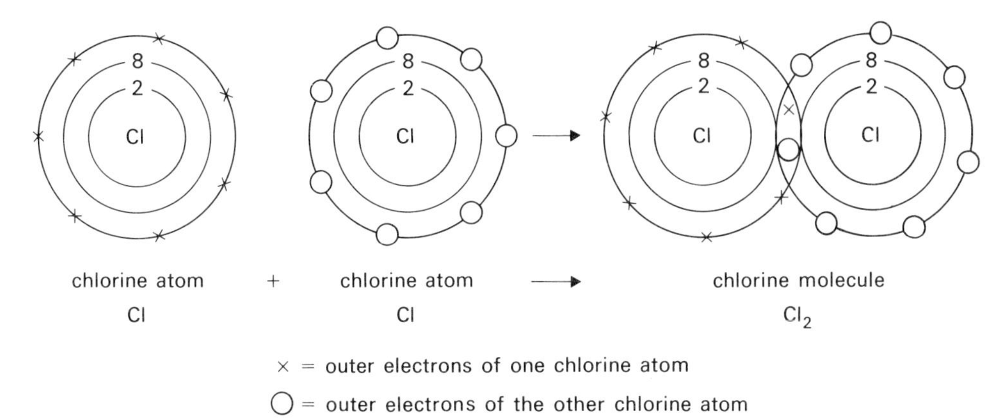

Figure 8.4 The formation of a chlorine molecule, Cl_2 (Cl–Cl)

neither atom has lost or gained an electron, there are no positive or negative ions formed.

A pair of electrons (i.e. one electron from each atom) which bonds two atoms together is called a *single covalent bond*. A single covalent bond is sometimes shown in simple form by a single line between the symbols of the atoms concerned. Thus in Figure 8.4 the chlorine atoms are joined by a single covalent bond in forming a chlorine molecule, and we can also show this by Cl–Cl.

When you work out examples of covalent bonding, you will find that two atoms sometimes have to share two pairs of electrons (two electrons from each atom) in order that they each have stable electron structures. The bond formed by sharing two pairs of electrons is called a *double covalent bond*, and is sometimes shown as X=X.

Similarly, two atoms occasionally form *triple covalent bonds* by sharing three pairs of electrons, and this is sometimes shown as X≡X.

Whenever you draw diagrams to show covalent bonding, you must always show the *type* of covalent bonding in each case, i.e. whether it is single, double, or triple.

How to work out a bonding diagram for any example of covalent bonding

The example given so far, the chlorine molecule, has a covalent bond between two atoms of the same kind. There are a number of other molecules like this, such as H_2, O_2, N_2, etc., although the covalent bond between the atoms is not always a single one as it is with chlorine. It is equally possible for covalent bonds to form between atoms of different non-metallic elements, e.g. in H_2O, NH_3, etc. The rules which follow should enable you to work out a bonding diagram for any common, covalently bonded molecule. At first these steps may seem complicated, but you should follow each step using a simple example, such as the chlorine molecule already given. Two further examples, for an ammonia molecule (NH_3) and an ethyne molecule (C_2H_2), are given in Figure 8.5. With practice, you will soon be able to combine some of the steps.

1. Determine the formula of the molecule in question.
2. Draw atomic structures for each of the atoms in the molecule, setting them out as follows. (Inner electron shells can be shown by numbers on the circles, but at this

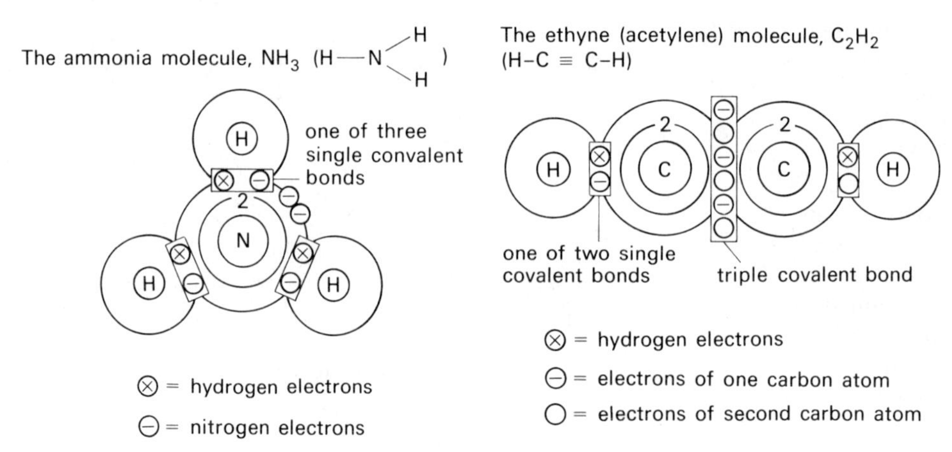

Figure 8.5 The bonding in ammonia and ethyne molecules

stage draw only a circle for the outer shell of each atom.) If there are only two atoms in the molecule draw them side by side, with outer shells overlapping. If there are three or more atoms in the molecule, place the atom with the largest combining power (i.e. the one with the smallest number of atoms in the formula) in the centre of the page, and distribute the other atoms evenly around it so that their outer shells overlap the outer shell of the central atom. Thus for an ammonia molecule, NH_3, the atoms would be spaced like this:

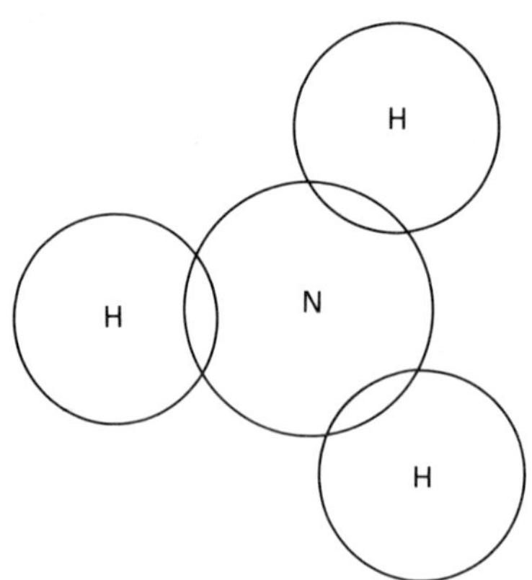

If the atom with the largest combining power occurs twice, draw the two atoms alongside each other (with outer shells overlapping) in the centre of the page and divide the other atoms evenly around them, not touching each other. Thus for ethane, C_2H_6, the atoms would be spaced like this:

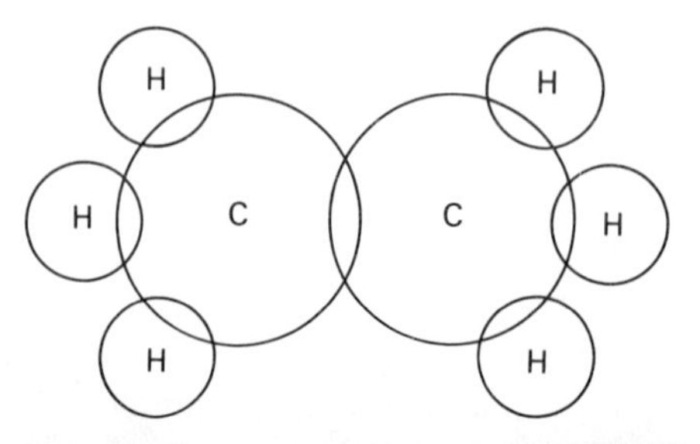

3. Using a pencil, lightly mark in the individual electrons in each outer energy level, using different symbols or colours for each atom and keeping them away from areas where atoms touch.
4. Draw a box around each of the areas where outer shells touch (see diagrams).
5. Move to each box (by erasing and redrawing) one electron from each of the atoms which form the box. Each box will now contain a pair of electrons shared by both atoms touching the box. (Retain your original symbols for the electrons.)
6. If necessary move more electrons (always working in pairs, one electron from each atom) until each outer shell is filled with electrons. (The electrons in each box are counted twice, i.e. they are included for counting purposes in the outer shells of both atoms.)
7. Label each atom as shown in the completed examples, and write 'single covalent bond', 'double covalent bond', or 'triple covalent bond' alongside each box according to the number of shared electrons.
8. Write the formula of the compound together with a summary of the bonding arrangement using single lines (single bonds), double lines, or triple lines between the atoms according to the bonding.

Check your understanding

1. Work out bonding diagrams for the following substances *if they are covalently bonded*. Two of them have ionic bonding; you must recognize which two these are, and then ignore them.
(a) calcium fluoride, (b) a hydrogen molecule, H_2, (c) oxygen, O_2, (d) nitrogen, N_2, (e) potassium oxide, (f) hydrogen chloride, (g) water, (h) methane, CH_4, (i) ethane, C_2H_6, (j) ethene, C_2H_4, (k) carbon dioxide, CO_2, (l) trichloromethane ($CHCl_3$).
2. The elements W, X, Y, and Z have atomic numbers respectively of 7, 9, 10, and 11. Write the formula for the compound you would expect to form between each of the following pairs of elements, and indicate the type of bonding present.
(a) W and X, (b) X and Z, (c) X and X, (d) Y and Y, (e) Z and Z.

Point for discussion

A carbon atom has four electrons in its outer shell, and it might appear that it could gain four electrons or lose four electrons with equal ease. In fact, carbon atoms never lose or gain electrons, they always form covalent bonds. Can you suggest a reason for this?

Properties of covalent compounds

1. Covalent compounds consist of two or more atoms linked together by covalent bonds.
2. Although the covalent bonds *inside* each molecule (the intramolecular bonds) are very strong and cannot easily be broken, the forces acting *between* the molecules (the intermolecular forces) are weak and relatively easy to break, e.g. hydrogen bonds (p. 287), or van der Waals forces (p. 287).

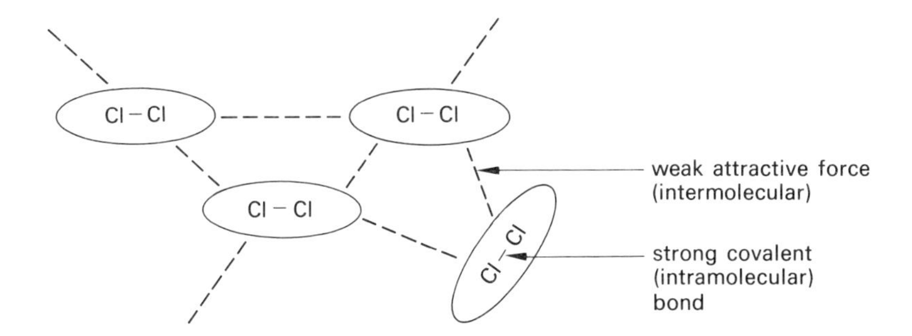

This means that molecules are easily separated, i.e. *molecular* substances have low boiling points and melting points. In the example above, chlorine is a gas at room temperature because it has 'already' melted and boiled at this temperature, because the weak intermolecular forces have become small when compared with the increased kinetic energy of the molecules. The molecules have thus separated and enjoy the independence and freedom of movement of the gas state.

Note that silicon oxide, diamond, and graphite are covalently bonded but they exist as *giant structures* rather than molecules. (This is explained in Chapter 16.) The points about boiling points, melting points, intermolecular forces, and intramolecular bonds apply only to those covalent substances which are molecular, i.e. to most covalent materials but not all.

3. Covalent compounds are often insoluble in water, and dissolve more readily in organic solvents, but see page 169.
4. Covalent compounds do not conduct electricity when molten or dissolved in water.
5. Covalent bonds are directional (i.e. they 'lock' atoms in certain positions and at certain angles) unlike ionic bonds, where the positive or negative charges are distributed evenly around the whole ion. This means that covalent molecules have a fixed shape. (See Figures 16.7 and 16.11, pp. 276 and 282.)

Bonding by the hydrogen atom

Of all the elements, hydrogen is unique in that its atom has only one electron, and it can become stable in three different ways.

1. It can lose the only electron, forming a positive ion, H^+. This ion is in fact a simple proton. It cannot exist on its own, and is usually found hydrated (i.e. joined to water molecules), when it is given the symbol $H^+(aq)$. This ion is responsible for the properties of acidic solutions.
2. It can gain an electron, thus making a complete outer shell and forming the H^- ion. This ion is relatively rare, and it only forms when hydrogen combines with very reactive metals such as sodium, potassium, and calcium.
3. It can share an electron to form a single covalent bond, as in several examples of covalent bonding you have worked out in this chapter.

8.4 A SUMMARY OF CHAPTER 8

The following 'check list' should help you to organize the work for revision.

(a) The significance of the stable electron structure of atoms of noble gases; other atoms are more stable if they can also attain fully filled electron shells.
(b) When a metal combines with a non-metal, the bonding is probably ionic. When two non-metal atoms combine, the bonding is covalent.
(c) Metal atoms normally lose electrons to form positive ions; the charge on the ion is equal to the number of electrons lost, which is usually the same as the normal valency of the atom.
(d) Non-metal atoms form ions by gaining electrons to form negative ions; the charge on the ion is equal to the number of electrons gained, which is equal to the normal valency of the atom.
(e) Oppositely charged ions are held together in ionic compounds by electrostatic attraction to form a crystal lattice. A crystal lattice has no separate units or molecules.
(f) How to work out a bonding diagram for common examples of ionic bonding (p. 119).
(g) The properties shown by typical ionically bonded compounds (p. 122).
(h) The difference between single covalent (one shared pair of electrons), double covalent (two shared pairs of electrons), and triple covalent bonds (three shared pairs of electrons).
(i) How to work out bonding diagrams for common examples of covalent bonding (p. 123).
(j) The properties shown by typical covalent compounds, with particular reference to the difference between intermolecular and intramolecular bonds (p. 125).

QUESTIONS

1. Chlorine has an atomic number of 17 and exists mainly in isotopic forms, A and B, of relative atomic masses 35 and 37 respectively. State the number of
(a) electrons in each atom of A
(b) protons in each atom of B
(c) neutrons in each atom of B
(d) electrons in each $^{35}_{17}Cl^-$ ion.
Calculate the A:B ratio in ordinary chlorine gas (relative molecular mass = 71). (J.M.B.)

2. (a) The following symbols refer to non-radioactive isotopes of the element boron: $^{10}_{5}B$ $^{11}_{5}B$.
(i) Explain the term *isotopes*.
(ii) Give the composition of the nucleus of each of the boron atoms.
(iii) How do you account for the fact that the relative atomic mass of naturally occurring boron is 10.8?
(b) The atomic numbers of the elements chlorine and argon are 17 and 18 respectively.
(i) Give the electronic structures of chlorine and argon.
(ii) How do you account for the fact that chlorine is a diatomic gas but argon is a monatomic gas? (W.J.E.C.)

3. The following symbols refer to atoms of sodium, fluorine, and neon: $^{23}_{11}Na$, $^{19}_{9}F$, $^{20}_{10}Ne$, $^{22}_{10}Ne$.
Using the above information, answer the following questions.
(a) What are the electronic structures of the sodium and fluorine atoms?
(b) Sodium and fluorine combine to give the ionic compound sodium fluoride. Explain, with the aid of a diagram, the changes in electronic structures which take place in this reaction.
(c) State two characteristic properties that would be expected of the ionic compound, sodium fluoride.
(d) In what respects do the neon atom, the sodium ion, and the fluoride ion (i) resemble each other, (ii) differ from each other?
(e) By means of a diagram, give the electronic structure of the covalent fluorine molecule, F_2.
(f) What can you deduce about the composition of the nuclei of the neon atoms shown above?
(g) How do you account for the fact that the relative atomic mass of naturally occurring neon is 20.2? (W.J.E.C.)

4. For *three* of the following (gaseous hydrogen, liquid water, metallic sodium, crystalline sodium chloride)
(a) draw a labelled diagram indicating the distribution of electrons in the molecule or lattice involved. Clearly show the electrons associated with each atom/ion.
(b) State the types of bonding involved and show how the bonding relates to (i) *two* physical properties, (ii) *two* chemical properties of the substance. (A.E.B. 1979)

5. Fluorine, F, (9); Neon, Ne, (10); Sodium, Na, (11); Magnesium, Mg, (12).
The following refer to the above four elements, the atomic numbers of which are shown in brackets.
(a) What *two* facts about the structure of the atom of any *one* of these elements can be deduced from its atomic number?
(b) State the numbers of electrons in successive electron shells of the magnesium atom.
(c) If the symbol for the sodium ion is written Na^+, write similar symbols for the ions of fluorine and magnesium.
(d) Write the chemical formula for (i) magnesium fluoride, (ii) sodium fluoride.
(e) Name and explain briefly, with the aid of a diagram, the type of chemical bond linking atoms of fluorine in the molecule F_2.
(f) Caesium and sodium are both alkali metals. In what respect do the structures of atoms of these two elements resemble each other? (W.J.E.C.)

9 Working with gases: the air, air pollution

9.1 WORKING WITH GASES

You will prepare a number of gases in your chemistry course, and it is important that you understand the various pieces of apparatus which are used in the preparations. Sometimes several alternative arrangements are possible, and the actual choice may simply depend upon how much gas is needed (e.g. do we use a small scale apparatus or a larger scale version?), or simply on which is the most convenient to use.

In general, a gas is made in some form of *generator*, and leaves the generator by a *delivery tube*. This may lead directly to a *collecting system* or it may first pass through a *purification system* (e.g. to remove water from it, or to remove another gas which may be mixed with it). You will be told if a purification stage is necessary for a particular preparation, but the collecting system always needs careful thought as the choice of system depends upon the physical properties of the gas.

Gas generators

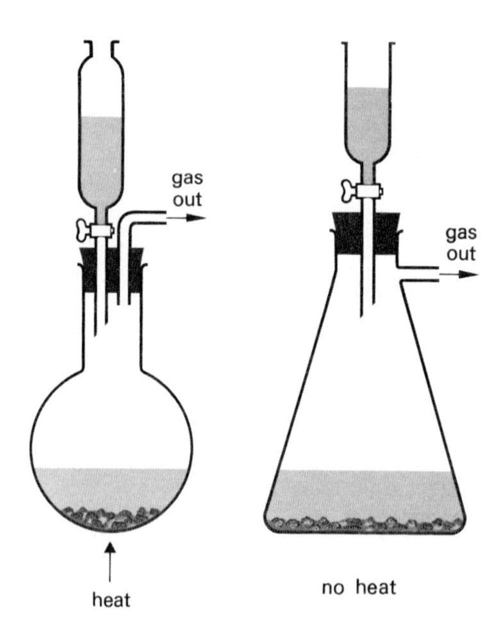

Figure 9.1 Two types of gas generator. These are suitable for preparing gas jars of gas

Typical examples are shown in Figure 9.1. Note the following points. They may seem obvious, but many students forget about one or more of them.

1. If the generator flask is to be heated, a round bottomed flask should be used to reduce the risk of cracking.
2. Always think of the scale of preparation; do not use a test-tube generator if large quantities of gas are needed.
3. The gas outlet tube (to the delivery tube) must always be well *above* the level of the reacting chemicals.
4. Any tube used to supply a liquid reagent must dip *below* the level of the reacting chemicals unless it has a tap on it, for otherwise gas can escape through the supply tube instead of through the delivery tube.

Purification systems

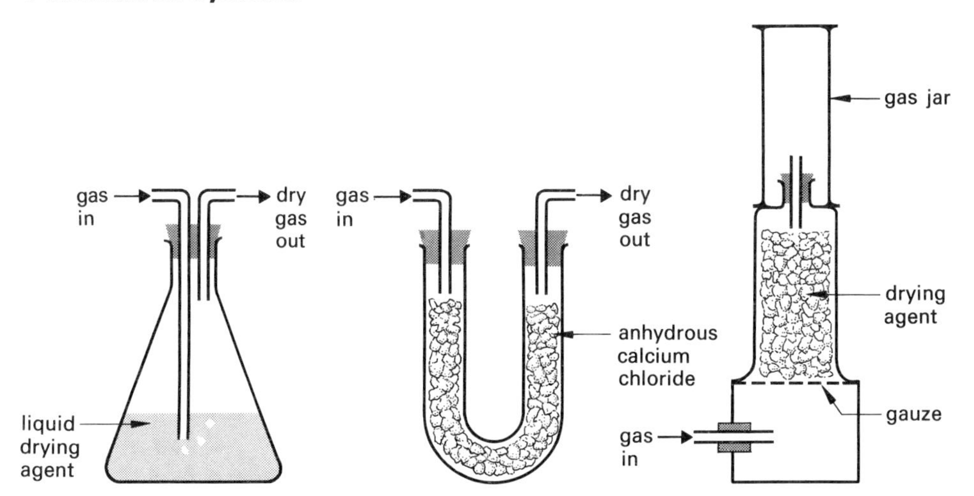

Figure 9.2 Drying or purifying systems

Many of the gases we prepare at an elementary level are pure enough for our purposes, and we do not then use any purification stage. One common exception is the preparation of carbon dioxide from the reaction between calcium carbonate and dilute hydrochloric acid (p. 370). In this reaction, there can be a significant amount of acid spray which is carried over with the carbon dioxide, and so the gas is passed through water to 'wash out' the acid spray. In industry a series of purification systems may be needed in order to produce a pure gas (such purifications are often important, because even slight traces of impurities can poison a catalyst), or in order to make it safe to be expelled into the air as a waste product. Occasionally we need to dry a gas, particularly if the gas is very soluble in water (e.g. sulphur dioxide, ammonia, and hydrogen chloride), but more sophisticated purification stages are rarely used at this level. Typical methods used to dry a gas in the laboratory are shown in Figure 9.2. Note the following points.

1. Make sure that when tubes deliver a gas *into* a liquid, they dip well below the surface of the liquid, and that those tubes which take a gas away from a liquid *do not enter* the liquid.
2. The only common liquid drying agent is concentrated sulphuric acid, but this must not be used for drying ammonia gas (a base) which would react with it.

3. There are several solid drying agents. Anhydrous calcium chloride, or silica gel are commonly used, but they are used in a U-tube or tower rather than a conical flask (can you understand why?).
4. Anhydrous calcium chloride cannot be used to dry ammonia gas because, like concentrated sulphuric acid, it also reacts with ammonia. Ammonia is an odd one out; it is the only common alkaline gas and is normally dried by using calcium oxide. A drying tower is often used as in Figure 9.2.

The collecting system

Two typical ways of collecting a gas *over water* are shown in Figure 9.3. The gas jar or tube must be *filled* with water first, and although there are disadvantages of the method (see points 1 and 2, page 131), useful advantages are that we can always see when a gas jar or tube is full of gas, and gas collected should contain very little air.

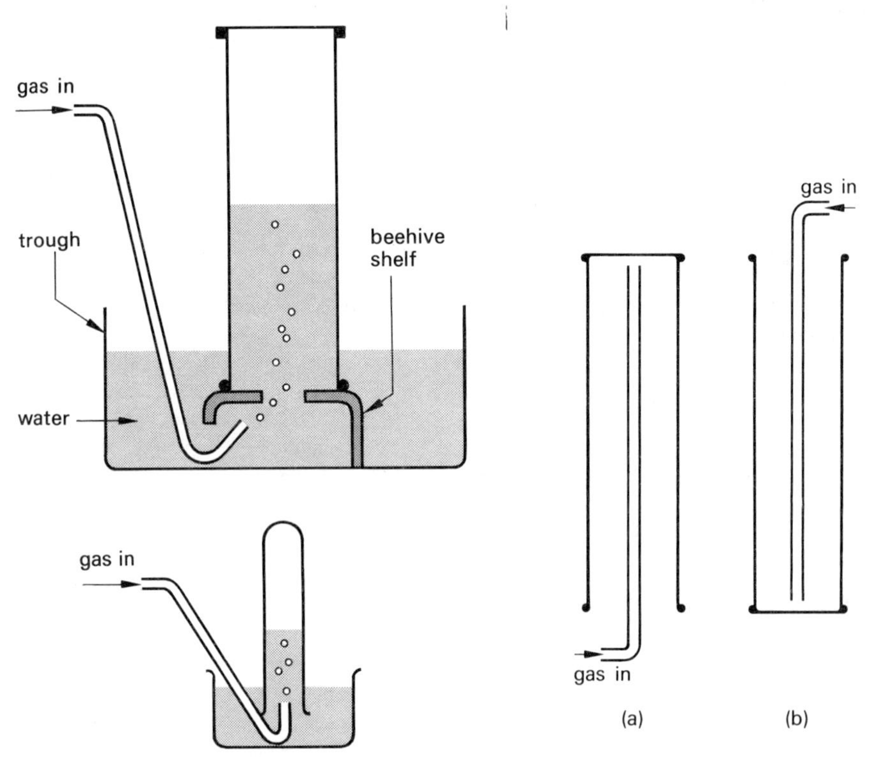

Figure 9.3 The collection of gases which are insoluble in water

Figure 9.4 The collection of gases which are soluble in water, (a) upward delivery and (b) downward delivery

Gases which are less dense than air can be collected by *upward delivery*, and gases more dense than air can be collected by *downward delivery* (Figure 9.4). The main disadvantages of these two collecting methods are that gases diffuse so quickly that the gas jar will almost certainly contain some air, and unless the gas is coloured it may not be easy to decide when the gas jar is full. Notice the length of the delivery tubes—shorter tubes would result in even more mixing with air.

A third method of collecting a gas is to use a syringe as in Figure 9.5. Obvious advantages are (i) the collected sample of gas is then easy to carry around, (ii) it is possible to eject a known volume of gas into a separate container simply by pushing

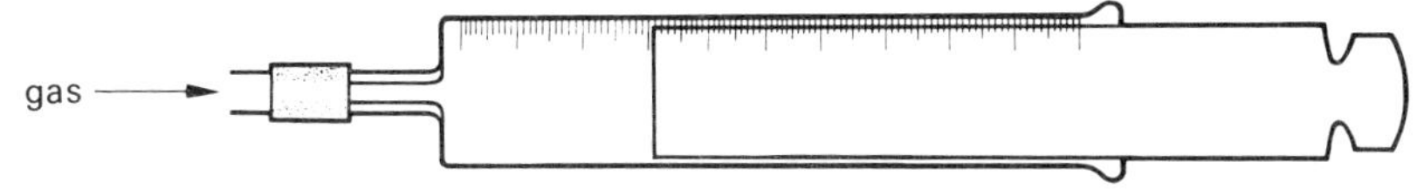

Figure 9.5 Collection of a gas by syringe

the plunger in to the required mark, (iii) air is unlikely to enter and mix with the gas even when some of it has been used, and (iv) we know when the syringe is full. Unfortunately there are three disadvantages. Glass syringes are expensive and easily broken, they are inconvenient to change when more than one syringe is to be filled, and if gas is produced rapidly the syringes are filled quickly and the collection is then difficult to control.

If a gas can be condensed to a liquid at a convenient temperature (i.e. one reached by simple cooling), it can be dried and then liquefied, and at the same time automatically separated from gaseous impurities. A typical arrangement for the collection of liquid dinitrogen tetraoxide is shown in Figure 9.6. Note that such systems should be kept open, so that gaseous impurities can escape.

Note the following points about collecting systems.

1. Never dry a gas and then collect it over water.
2. Never collect a gas over water if it is fairly soluble in water or reacts with it.
3. As the gas generator is full of air at the start of the experiment, the first sample of gas collected should be discarded as it will be mainly air.

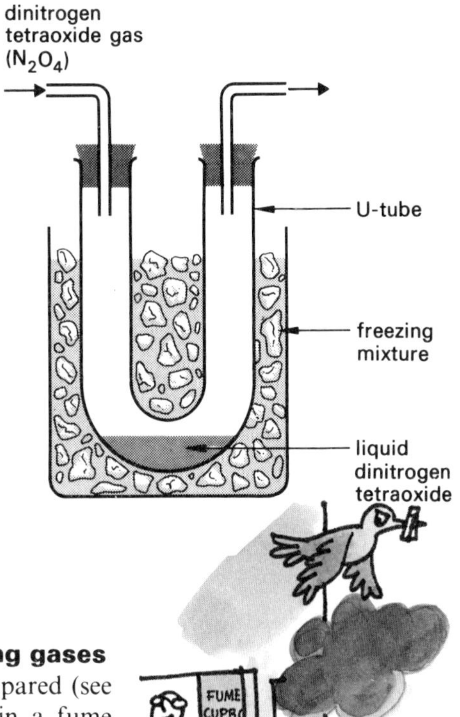

Figure 9.6 The liquefaction of dinitrogen tetraoxide gas

Some general points about preparing gases

1. If a toxic or unpleasant gas is being prepared (see Table 9.1), the apparatus must be used in a fume cupboard, and a note to this effect should be added to the diagram.
2. When drawing diagrams to show the preparation and collection of a gas, try to make sure that the various parts of the apparatus are drawn to approximately the same scale. Too many students draw, for example, a tiny gas jar next to a large conical flask.
3. Table 9.1 gives those facts about common gases which influence how they are made and collected.

Use a fume cupboard for toxic gases

Table 9.1 Physical properties of some common gases

Name and formula	*Odour*	*Physiological effect*	*Colour*	*Relative density (H = 1)*	*Solubility in water (cm^3 100 cm^{-3}) (room temperature and pressure)*	*Melting point (°C)*	*Boiling point (°C)*
Air	—	—	—	14.4	5.0	—	—
Carbon monoxide, CO	—	Very dangerous	—	14.0	2.5	−207	−190
Carbon dioxide, CO_2	Faint, pleasant	Poisonous only at high concentrations	—	22	100	Sublimes at −78	
Chlorine, Cl_2	Characteristic choking smell	Poisonous	Yellow-green	35.5	263	−101	−34
Hydrogen, H_2	—	—	—	1.0	2.1	−259	−253
Hydrogen chloride, HCl	Characteristic acrid smell	Corrosive, poisonous	—	18.25	46 100	−114	−85
Nitrogen, N_2	—	—	—	14.0	1.6	−210	−196
Ammonia, NH_3	Characteristic choking smell	Poisonous	—	8.5	80 670	−78	−33
Dinitrogen tetraoxide, N_2O_4	Characteristic, unpleasant	Poisonous, corrosive	Brown (colourless only when *pure*)	46 (varies)	Soluble, decomposes	−11	21
Oxygen, O_2	—	—	—	16	3.4	−219	−183
Sulphur dioxide, SO_2	Pungent, acrid	Corrosive, poisonous	—	32	4730	−73	−10
Sulphur trioxide, SO_3	Sharp, acrid	Corrosive, poisonous	Solid is white	—	Reacts violently	17	44
Hydrogen sulphide, H_2S	Rotten eggs	Very poisonous	—	17	260	−86	−60

Points for discussion

1. Which gases could you collect over water in an open laboratory?
2. Which gases could you collect over water in a fume cupboard?
3. Which gases could you collect by upward delivery in an open laboratory?
4. Which gases could you collect by downward delivery in an open laboratory?
5. Which gases could you collect by upward delivery in a fume cupboard?
6. Which gases could you collect by downward delivery in a fume cupboard?
7. Which gases could you easily collect as solids?
8. Which gases could you easily collect as liquids?

9.2 THE AIR

Heating substances in the air

You have probably done experiments at an earlier stage in which you found out what can happen when a substance is heated in air. If the substances are weighed before and after heating, it can be shown that one of three things normally happens.

1. There is no change in mass. This happens with substances such as pure, dry sand, which neither reacts with air nor decomposes on heating.
2. There is a loss in mass. This occurs when a substance decomposes on heating and gives something off, e.g. a gas or water vapour. Detailed examples of this type are studied as the properties of individual compounds later in the book, e.g. when heated, hydrated copper(II) sulphate loses water vapour to become the anhydrous salt, and copper(II) carbonate loses carbon dioxide gas on heating.
3. There is an increase in mass. This happens when a substance reacts with something in the air, e.g. when magnesium metal burns to form magnesium oxide, it gains (combines with) oxygen from the air and increases in mass. Some reactions of this type *appear* to produce a loss in mass if one or more of the products is allowed to escape, e.g. as a gas. Thus when a candle burns away completely, there appears to have been a loss in mass. However, if all the products from the burning candle (carbon dioxide and water vapour) had been collected and weighed, there would have been an increase in mass when compared with the mass of candle burned.

Oxygen, the active gas in air

The substances which increase in mass when heated in air usually do so because they have combined with oxygen from the air. Oxygen is the active gas in the air, and combines with many substances. You may have seen a simple experiment in which a piece of copper is weighed and then heated in the air. It becomes covered with a black powder, and increases in mass. If a similar piece of copper is heated in the absence of air, e.g. by using a simple piece of apparatus incorporating a Bunsen valve (Figure 9.7, p. 134), or by wrapping it in an airtight sandwich of metal foil, there is no increase in mass and the copper does not become coated with the black powder.

Experiments like this prove that something in the air is combining with the copper when it is heated, and this is of course oxygen. The black powder is copper(II) oxide:

$$2Cu(s) + O_2(g) \rightarrow 2CuO(s)$$

This very simple reaction is used in the next experiment.

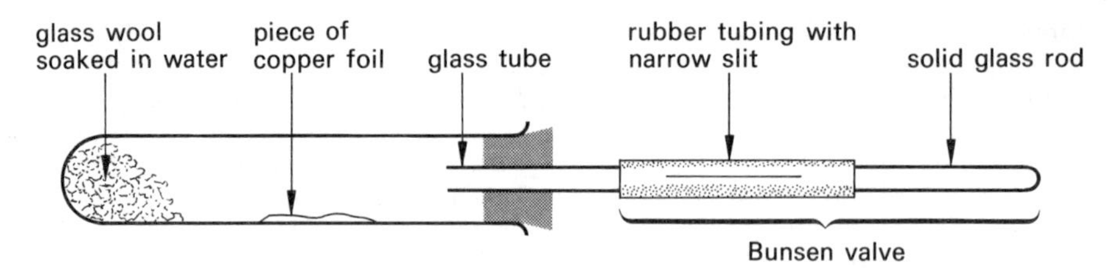

The glass wool is heated, and the tube fills with steam. Excess steam forces out through the slit under pressure (taking the air from the tube with it). When the tube is full of steam instead of air, the copper is heated. Air cannot *enter* the tube through the slit because its pressure (atmospheric pressure) is less than that of the escaping steam

Figure 9.7 Heating copper in the absence of air by using a Bunsen valve

Experiment 9.1
Finding how much of the air is oxygen**

Apparatus
As in Figure 9.8.
Copper powder or wire.

Procedure
(a) Place some copper in the silica tube and connect the tube to the two syringes. Make sure that there is as little air space as possible in the tube and connectors. One syringe should be full of air (100 cm^3 or 50 cm^3) and the other empty.
(b) Test the apparatus for leaks by pulling air through from one syringe to the other. The syringes should now read 0 cm^3 and 100 cm^3 or 50 cm^3.
(c) Heat the silica tube containing the copper vigorously and at the same time continually pass air slowly from one syringe to the other through the hot copper. After about three minutes stop heating and cool the silica tube by applying a cold damp cloth. When cold, record the volumes of gas in the syringes.
(d) Repeat procedure (c) until no further change of volume takes place.

Results
Write down the original and final volumes of the gases in the syringes. Also note any other changes which you observe.

Notes about the experiment
(i) The silica tube is used so that it can be heated strongly and then cooled rapidly (e.g. by using a damp cloth) without risk of cracking. Ordinary glass would crack if treated in this way.
(ii) Note that the tube is cooled each time before a reading is taken. This is because the hot gases take up a larger volume than they would at room temperature, and this could cause a false reading. When the gases are cooled to room temperature, their volume can then be fairly compared to the original volume (e.g. 100 cm^3).
(iii) Notice the important scientific principle of continuing the experiment until there is no further change; only then can we be sure that the reaction has finished. You will probably use this idea

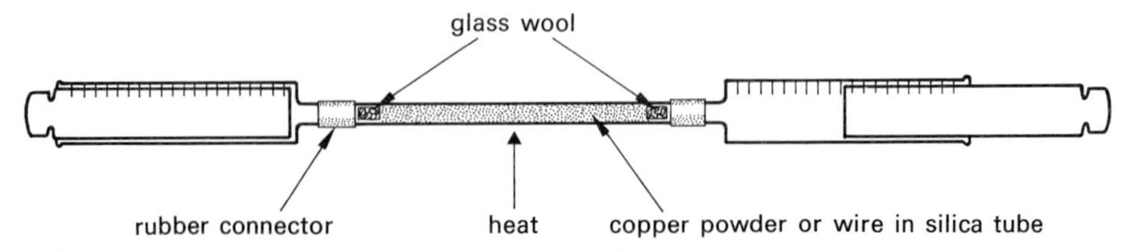

Figure 9.8 To determine the percentage of oxygen in the air

in other experiments, when *mass changes* are being recorded. In such cases the experiment is repeated until there is no further change in mass, which shows that the reaction has finished.

Conclusion
Write your own conclusion, based on answers to the following questions.
(a) If there was a change in volume, what has caused it and why did this happen?
(b) If there was a change in volume, what does it tell you about the composition of the air?

Carbon dioxide and water vapour in the air

You probably know that air also contains carbon dioxide and water vapour, and the next experiment demonstrates this.

Experiment 9.2
Two further components of the air**

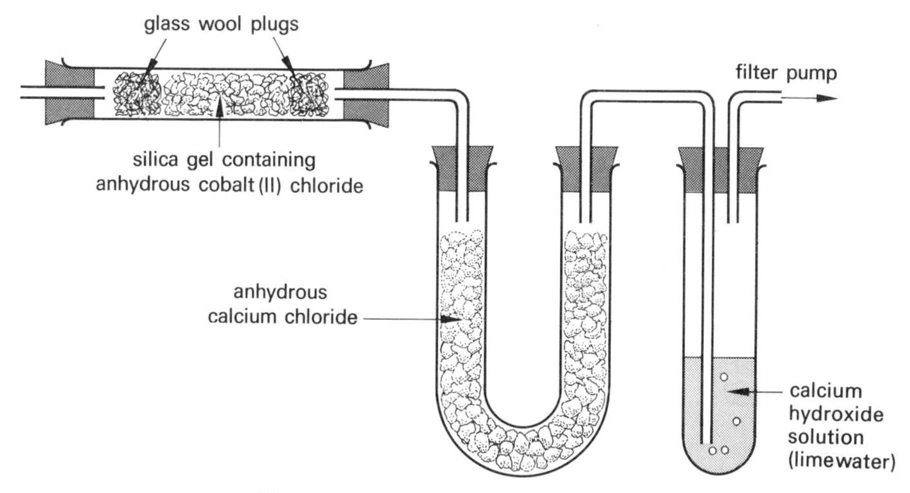

Figure 9.9 Two further components of air

Apparatus
As in Figure 9.9.

Procedure
Assemble the apparatus as shown in Figure 9.9 and turn on the filter pump. This draws a stream of air through the apparatus.

Results
Write down all the changes you observe during the experiment.

Notes about the experiment

Silica gel is a drying agent, because it readily absorbs moisture from the air. You may have seen small bags of silica gel crystals placed inside a container to keep the air dry, e.g. a case for binoculars. Silica gel crystals often contain anhydrous cobalt(II) chloride, which acts as an indicator of the presence of water, appearing blue when dry and pale pink when hydrated. (When this happens the crystals are no longer useful as a drying agent.)

Conclusion
Explain your observations in the experiment, and state what the experiment shows about gases in the air. Include an explanation as to why the calcium chloride tube was used.

Nitrogen in the air

The other main component of the air is nitrogen. As this gas is very unreactive, it is fairly easy to remove most of the other (more active) gases from the air so as to leave fairly pure nitrogen. You may have seen a demonstration of this kind, and been shown that the nitrogen collected will extinguish a lighted taper and other burning substances. Remember that this is *not* a test for nitrogen; other gases such as carbon dioxide also do this.

Check your understanding

1. Figure 9.10 is a young student's plan for an experiment to obtain fairly pure nitrogen from the air. Study it carefully; it contains six important mistakes. Try to explain what is wrong with the plan, or redraw the apparatus correctly, with notes to explain what you have done. Note that an aspirator is a useful way of 'pulling' (or 'pushing' if used slightly differently) a slow stream of air through an apparatus. You may need to revise Section 9.1 before you can answer the question fully.

2. Two other students chose solid drying agents instead of concentrated sulphuric acid when they designed their apparatus, and placed these in a U-tube instead of a conical flask. One labelled the drying agent 'calcium chloride', and the other chose 'calcium carbonate'. One of these is completely incorrect, and the other is incompletely labelled. What is wrong?

3. Another student wanted to draw air through the apparatus by using a filter pump. Why would this not be useful in preparing a sample of nitrogen?

4. Yet another student said that it would be better to have two combustion tubes containing copper to make sure that all of the oxygen is removed (remember Experiment 9.1?). The teacher said that this was a sensible suggestion, but if the aspirator was used correctly and the combustion tube filled with copper, there ought to be no neeed for a second combustion tube. Why?

5. Only one student obtained full marks for designing the experiment, because not only was the apparatus chosen correctly and in the appropriate sequence, but the student was the only one to also state that it was important not to *collect* the gas immediately. Why?

6. The nitrogen obtained in such an experiment is not absolutely pure. What other substances are present? Are these impurities likely to make the collected gas reactive?

7. How is *pure* nitrogen obtained from the air?

8. In order to determine the proportion by volume of one of the major components of air, the air was passed in turn through sodium hydroxide solution, through concentrated sulphuric acid, and into a glass syringe. The volume of remaining gas was measured and the gas was passed repeatedly over red hot copper until no further contraction in volume occurred. The gas was then allowed to cool and its volume was measured.

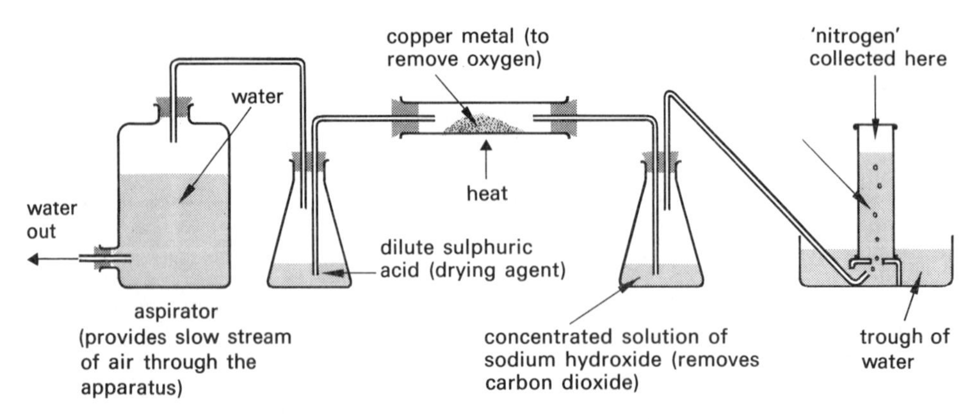

Figure 9.10 An incorrect way of preparing a sample of nitrogen from the air

Volume of gas before passing over hot copper: $90.0\ cm^3$

Volume of gas after passing over hot copper: $70.2\ cm^3$

(a) Why was the air passed through sodium hydroxide solution?

(b) Why was it passed through concentrated sulphuric acid?

(c) How would the appearance of the copper before heating and after cooling differ?

(d) Which gas was removed by the copper?

(e) Name the main gas remaining in the syringe at the end of the experiment and calculate the approximate percentage of this gas in the air from the data provided.

(f) Give the name of another element which would still be present in the residual gas.

(J.M.B.)

Point for discussion

You may be interested to know that an experiment similar to the one you have just 'corrected' led to the discovery of the noble gas argon. In 1892 Lord Rayleigh repeatedly observed that samples of nitrogen prepared chemically (i.e. as the result of a chemical reaction) always had a slightly lower density compared with samples of nitrogen extracted from the air. He reported the respective values of $1.2505\ g\ dm^{-3}$ and $1.2572\ g\ dm^{-3}$. Small though the differences were, they were consistent and Sir William Ramsay thought the discrepancy was due to an impurity in the atmospheric nitrogen. He examined the spectrum of 'nitrogen' obtained from the air and was able to detect the presence of a hitherto unknown element, which he and Rayleigh were later to isolate in small quantities. They called the gas argon (from the Greek argos, 'idle'), for they were unable to remove it from the air by any chemical means. Is argon likely to be more or less dense than nitrogen?

The composition of the air

The air is a mixture of gases and not a compound. Its composition varies slightly from time to time and from place to place. For example, the water vapour content varies with temperature, the carbon dioxide level is higher inside a poorly ventilated crowded classroom than it is outside, and pollutant gases are more obvious near industrial cities.

The main gases in the air are nitrogen, oxygen, argon, and carbon dioxide. The proportion of water vapour in the air varies much more than that of the other gases. The following is a typical analysis of *dry* air.

Component	*Composition by volume (%)*
Nitrogen	78.08
Oxygen	20.95
Argon	0.93
Carbon dioxide	0.03
Neon	0.002
Other noble gases	0.0006

The air also contains small traces of methane and ozone, and solids such as soot, bacteria, and pollen. In addition it is likely to contain pollutant gases such as hydrogen sulphide and sulphur dioxide (especially over industrial areas) and carbon monoxide (especially over towns and cities). The air will also contain traces of compounds of heavy metals, such as lead compounds formed by the combustion of petrol (which contains a lead-based additive, see page 146).

9.3 COMBUSTION, RESPIRATION, PHOTOSYNTHESIS

Combustion and respiration normally take oxygen *out* of the air and photosynthesis helps to balance these effects by putting oxygen back into the air. Oxygen is therefore a common factor in all three processes.

Combustion

Combustion is the chemical reaction of a substance with a gas (usually oxygen) so that heat (and sometimes light) is produced.

Although most of the combustions you will study involve the gas oxygen, the term can be used for reactions in other gases. For example, sodium metal will burn in chlorine to form sodium chloride.

As explained earlier, the products of a combustion process (if they are all collected) will weigh more than the material which has been burned, because atoms have *combined* with another gas such as oxygen.

The combustion of *elements* is considered in detail in other sections of the book, but many of the important combustion processes involve the burning of *compounds* in the air. Most of our fuels are *hydrocarbons* (i.e. compounds containing hydrogen and carbon only), such as petrol, paraffin, and natural gas. The *complete* combustion of a sample of hydrocarbon fuel always produces carbon dioxide (from the carbon in the fuel), water vapour (from the hydrogen in the fuel), and energy, e.g. heat. This energy is used to propel cars, heat our homes, produce electricity, and so on.

e.g.

$$\underset{\text{Methane from natural gas}}{CH_4(g)} + 2O_2(g) \rightarrow CO_2(g) + 2H_2O(g) + \text{energy}$$

$$\underset{\text{Pentane from petrol}}{C_5H_{12}(l)} + 8O_2(g) \rightarrow 5CO_2(g) + 6H_2O(g) + \text{energy}$$

You may have seen simple experiments which demonstrate that carbon dioxide and water vapour are formed when a hydrocarbon fuel is burned. For example, candle wax is a solid hydrocarbon. If a candle is burned under a large inverted beaker, the flame goes out when most of the oxygen is used up. A mist can be seen on the inside of the beaker, and this moisture will turn anhydrous copper(II) sulphate blue, proving the presence of water. The condensed liquid is also slightly acidic when tested with universal indicator solution because carbon dioxide has been formed, and some of this dissolves in the water formed to produce an acidic solution (the so-called carbonic acid).

Respiration

This is not the same as breathing. All living things respire, but all do not breathe.

Respiration is the process by which a living organism obtains its energy from food substances, and it can be considered as a kind of combustion process.

Breathing is the process used by organisms to ensure that oxygen enters the body (so that respiration, the 'combustion' of food, can take place) and to remove some of the products of combustion (e.g. carbon dioxide) from the body.

Some living organisms respire without using oxygen (e.g. yeast when it is used for fermentation, page 418) in which case we describe the process as *anaerobic respiration* (without air) to distinguish it from the more usual *aerobic respiration.*

You have probably done simple experiments, either earlier in your chemistry course or in biology lessons, which show that *inspired air* (i.e. normal air, that which is breathed in) differs from *expired air* (that which is breathed out) in the following ways:

Inspired air	*Expired air*
1. Temperature may vary, but is usually less than blood heat (37 °C)	Temperature usually around blood heat (37 °C) which is normally warmer than inspired air
2. Contains approximately 20% oxygen and a little carbon dioxide	Contains rather less than 20% oxygen, and the proportion of carbon dioxide is increased
3. Contains some water vapour	Contains much more water vapour

The foods we eat (e.g. carbohydrates and fats) always contain hydrogen and carbon, but they are not hydrocarbons because they also contain other elements such as oxygen. Respiration is very similar to combustion in that the carbon and the hydrogen in the food are oxidized by oxygen to carbon dioxide and water vapour respectively, and energy is released. However, respiration occurs in a series of small steps and the food is not actually 'burned'. Equations such as the following summarize these many small steps:

$$\underset{\text{(a sugar)}}{C_6H_{12}O_6(aq)} + 6O_2(g) \rightarrow 6CO_2(g) + 6H_2O(l) + \text{energy}$$

Photosynthesis

Photosynthesis is the process by which green plants produce carbohydrate foods (sugars and starches) from carbon dioxide and water, with the liberation of oxygen. Chlorophyll (a green pigment) and energy from sunlight are needed for the process.

A summary equation for the process is

$$\underset{\text{(from sunlight)}}{\text{energy}} + 6CO_2(g) + 6H_2O(l) \xrightarrow{\text{chlorophyll}} C_6H_{12}O_6(aq) + 6O_2(g)$$

In effect this is the 'opposite' of respiration. It is one of the most important chemical reactions in the world, for two reasons. All of the food we eat is produced through photosynthesis, for even meat is produced by animals which have themselves eaten plant sugars and starches. It is also obvious that without photosynthesis our oxygen supply would have been exhausted by the processes of combustion and respiration. Similarly, photosynthesis by aquatic plants (i.e. plants living in water) helps to 'aerate' the water and provides some of the necessary oxygen for animal life in the water.

The differences between combustion, respiration, and photosynthesis are summarized in Table 9.2.

Table 9.2 A comparison between combustion, respiration, and photosynthesis

Component	*Combustion*	*Respiration*	*Photosynthesis*
Oxygen	Often needed, although some combustions proceed in other gases	Needed for aerobic respiration	Liberated
Carbon dioxide	Nearly always formed if the 'fuel' contains carbon	Liberated	Needed
Water	Always formed if the 'fuel' contains hydrogen	Liberated	Needed
Energy	Always liberated	Liberated	Needed

9.4 AIR POLLUTION

The problem

The earth's atmosphere is the mixture of gases which surrounds the earth. It is kept 'in position' by gravity, and the lower zone of the atmosphere (which will support life) extends upwards for about 8 miles. The atmosphere is absolutely vital to us; we need its oxygen for respiration, to help supply most of our energy requirements by the combustion of fuels, and for many of the chemical reactions which are important in industry. The atmosphere is one of our most important natural resources, and we must preserve it in a clean and acceptable state.

On page 137 we gave you the composition by volume of air. If the air in a country district is analysed it will probably have this same approximate composition although the proportion of carbon dioxide in the atmosphere is increasing at a very slow rate (see page 141). Unfortunately, the problem of controlling the composition of the atmosphere does not just depend upon the carbon dioxide/oxygen balance. The air over and near large centres of population or industrial areas is very different from pure air as it contains a number of highly undesirable substances. Chemical pollution means the introduction to the environment of substances which have harmful effects on the ability of the environment to support life and/or make the environment 'dirty'. Some of the pollutants in air certainly fit this description.

When man and other living things breathe, they take in with the air any pollutants that are mixed with it. The human respiratory system can filter out much of the solid material (e.g. dust and dirt) which may be present so that it does not reach the lungs. This filtering system is not completely effective, however, and it has little or no effect on any pollutant *gases* which may be in the air.

Atmospheric pollution endangers health and well-being. For example, it causes respiratory diseases and adds carcinogenic (i.e. cancer producing) substances to the air. It also promotes and stabilizes fog, it accelerates corrosion in metal structures and in buildings, it blackens paint, and it affects soil pH, crop production, and the health of animals. It is obviously extremely important that we should do everything possible to ensure that the atmosphere is kept clean.

Sources of air pollution

Atmospheric pollution is caused by the discharges from domestic and industrial chimneys, from chemical factories and other industrial processes, and by exhaust gases produced by vehicles, etc. Many of the reactions which produce pollution are normal (i.e. planned) operations, although the situation is occasionally made more serious by accidents or by unusual operating conditions.

Industry has a bad image with respect to pollution, but in many cases this is not deserved. For example, by far the greatest amount of smoke and dirt in the atmosphere over Britain was brought about by the burning of coal in open fires in domestic houses, until the Clean Air Act of 1956 (see Figure 9.11). Figure 9.11 also shows that another significant contribution to the problem in Britain (in the days of steam locomotives) was the railway system. Again, many people fail to realize that enormous amounts of pollutants enter the atmosphere by *natural* processes such as those of decay, volcanic action, and wind erosion. It is quite

Many of the actions which produce pollution are 'normal'

unfair, therefore, to lay the whole problem on industry's doorstep. You will learn later in this section about successful methods which are being developed in industry to further reduce pollution, but we must remember that we live in an age when the demand for energy is greater than ever before, and much of this energy is provided by the combustion of fuels. Wherever raw fuels are burned, pollutants are inevitably formed. Any improvements in pollution control cost vast amounts of money, and we must be prepared for this cost to be passed on to the customer who uses the products, i.e. to us.

As an effect of the Clean Air Act in Britain, the amount of smoke in the atmosphere is being progressively reduced. The figures are in million metric tons.

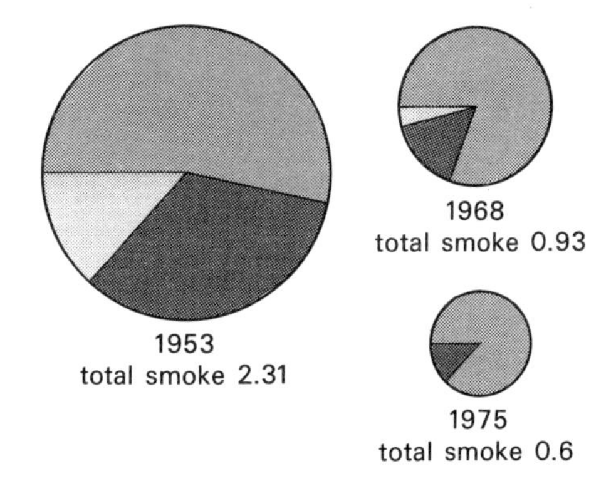

Figure 9.11 The effect of the Clean Air Act in Britain (*Reproduced from Man and Industry by permission of Shell International Petroleum Ltd*)

Clearly, the nature of the problem is likely to change because of other factors, some of which are beyond our control. In Chapter 26 emphasis is placed on the problems caused by the dwindling stocks of fossil fuels, and on the need to conserve energy sources and to develop new ones. Some alternative fuels which may be developed could be virtually pollution-free, e.g. hydrogen. The combustion of pure hydrogen would produce only water vapour in addition to energy. The increased use of other fuels, on the other hand, may produce very different pollution risks, such as the safe disposal of radioactive waste from nuclear power stations.

The main atmospheric pollutants are considered separately under the following headings.

Carbon dioxide as a pollutant

The main way in which carbon dioxide is removed from the atmosphere is by photosynthesis. Until fairly recently the carbon dioxide/oxygen ratio was more or less constant because photosynthesis removed as much carbon dioxide from the air as was added to it, and replaced it with a quantity of oxygen roughly equal to that which was being used up.

The volume of carbon dioxide removed from the atmosphere by photosynthesis remains almost constant, but the quantity of the gas produced by the combustion of fossil fuels has increased steadily. At the moment, carbon dioxide is being added to the atmosphere faster than it can be removed, and so the concentration of the gas in the air is increasing at about 0.2 per cent per year. (For a more detailed consideration of the carbon dioxide balance, see the carbon cycle on page 373.)

Carbon dioxide is not toxic at the low concentrations which occur in the air, and as a *local* pollutant it is therefore harmless. The gas is, however, one of the main factors which help to control the temperature of the earth's surface due to the so-called 'greenhouse effect' (see below). This in turn affects the climate. Carbon dioxide is thus considered to be a pollutant on the *global* scale because of the way its increasing concentration in the atmosphere could alter the world's climate.

The 'greenhouse effect'

A greenhouse allows sunlight to enter it, and the infra-red and ultraviolet parts of the light warm up the air and materials inside the greenhouse. The hot contents of the greenhouse then radiate heat energy, mainly as infra-red light rather than ultraviolet light. As glass does not readily allow infra-red radiation to pass through it, heat energy is trapped inside the greenhouse and it escapes only slowly by conduction through the glass and/or by convection through open windows.

In a similar way, the earth's surface is warmed by ultraviolet light from the sun, and much of this heat is lost as infra-red radiation escaping from the earth and through the atmosphere. Carbon dioxide (like the glass in a greenhouse) does not readily allow infra-red radiation to pass through it, and so carbon dioxide in the atmosphere helps to prevent the escape of infra-red radiation from the earth's surface. As the concentration of carbon dioxide in the atmosphere increases, less heat energy is lost from the surface of the earth and the surface temperature gradually increases. It has been calculated that doubling the concentration of carbon dioxide in the atmosphere will increase the average surface temperature by between 2 and 3 °C. Some people believe that if the concentration of carbon dioxide continues to rise over the next 100 years or so there will be perhaps catastrophic effects on our climate. However, the concentration of the gas in the air may not continue to rise in this way, as explained under the carbon cycle on page 373.

Point for discussion

Can you explain why there is more likely to be a severe ground frost in winter on a clear night rather than on a cloudy night?

Toxic gases in the air

Carbon monoxide

If a fuel is burned in an inadequate supply of oxygen (e.g. because of poor ventilation or because of blockages or dirt in the equipment), any carbon in the fuel is not completely oxidized to carbon dioxide, and instead forms some carbon monoxide and even elemental carbon (soot). Most fuels (e.g. coal, coke, natural gas, paraffin, oil, and petrol) contain a high proportion of carbon, and so the problem is potentially a serious one.

Incomplete combustion occurs in homes (e.g. where gas fires have poor ventilation or have not been serviced regularly), and in internal combustion engines where petrol is the fuel and carbon monoxide is one of the components of the exhaust gases. This can be a dangerous form of pollution because carbon monoxide has no odour, is invisible (and thus cannot be detected easily), and is toxic. Carbon monoxide is poisonous because the red blood cells combine with it in preference to oxygen, so a person poisoned by the gas is, in effect, suffering from a lack of oxygen.

It is obvious that if poor ventilation is responsible for the formation of the gas, this will also cause it to build up in concentration in an enclosed space (e.g. the room of a house), with potentially fatal consequences. Even in the open air, concentrations of carbon monoxide have built up to such an extent in the centres of certain cities during the rush hour that policemen on traffic duty have had to wear gas masks.

Carbon monoxide is only a *local* pollutant, i.e. it may build up to a concentration higher than a safe limit in certain areas and at certain times, but there is no global increase in the proportion of the gas in the atmosphere. This is because natural processes soon convert it into other compounds (e.g. more carbon dioxide!). Recent figures even suggest that more carbon monoxide enters the atmosphere through natural processes than from car exhausts, but every effort is being made to ensure that fuels are now burned more efficiently to avoid forming products of incomplete combustion. Car engines are more efficient than they used to be, and methods have been introduced to reprocess the products of incomplete combustion, e.g. to oxidize carbon monoxide into carbon dioxide before it is allowed to enter the atmosphere.

Oxides of sulphur

Most fossil fuels contain sulphur compounds. Sulphur is present in all living organisms (e.g. in proteins), and as coal, oil, and natural gas are formed by the decay of dead organisms it follows that they will contain sulphur compounds. The proportion of sulphur (by mass) in raw coal varies between about 0.2 per cent and 7.0 per cent, and averages about 1.5 per cent. The proportion of sulphur in crude oil varies between 0.4 per cent and 5.0 per cent, and a typical heavy fuel oil may contain 3.5 per cent of sulphur. Crude oil from the Middle East contains a much higher proportion of sulphur than that from the USA, which may contain less than 1 per cent of sulphur. The extent of the problem therefore varies from fuel to fuel, and from one sample to another.

Whenever such raw fuels are burned, the sulphur compounds form sulphur dioxide and smaller quantities of other gases such as sulphur trioxide (SO_3) and hydrogen sulphide (H_2S). These gases are serious pollutants. They are toxic, they cause respiratory disease, and they form acidic solutions which corrode metals and buildings.

It is difficult to appreciate just how vast this particular problem is. It has been estimated that in just one year 5.7 million tonnes of sulphur dioxide have been added to the atmosphere in Britain alone. This quantity of the gas would occupy about 2 000 000 000 000 dm^3 at room temperature and pressure! The scale on which atmospheric pollution occurs is brought home to us when we learn that a typical coal-fired power station may burn over 4000 tonnes of coal in a single day, and Britain's power stations can consume 70 million tonnes of coal each year, all of which contains an average of 1.5 per cent sulphur. The coke ovens which produce both the coke needed for steel making and the solid smokeless fuels used in domestic heating also consume vast amounts of coal. The sources of some of these pollutant gases are summarized in Table 9.3.

Pollution of this kind may have its greatest effect in an area far away from its source. The 5.7 million tonnes of sulphur dioxide referred to above could produce 9×10^6 tonnes of sulphuric acid, and most of the gas does indeed dissolve in the rain, which falls back on the earth as an acidic solution. The pH of normal rain water is about 5.3 (it is acidic because of the presence of dissolved carbon dioxide from the air). In early spring, some rivers in Scandinavia contain water with a pH as low as 3.0, and it is now known that this acidity is almost entirely formed by the sulphur dioxide liberated by the power stations in Britain. The gas is carried by the

Table 9.3 Sources of pollutant gases in the atmosphere

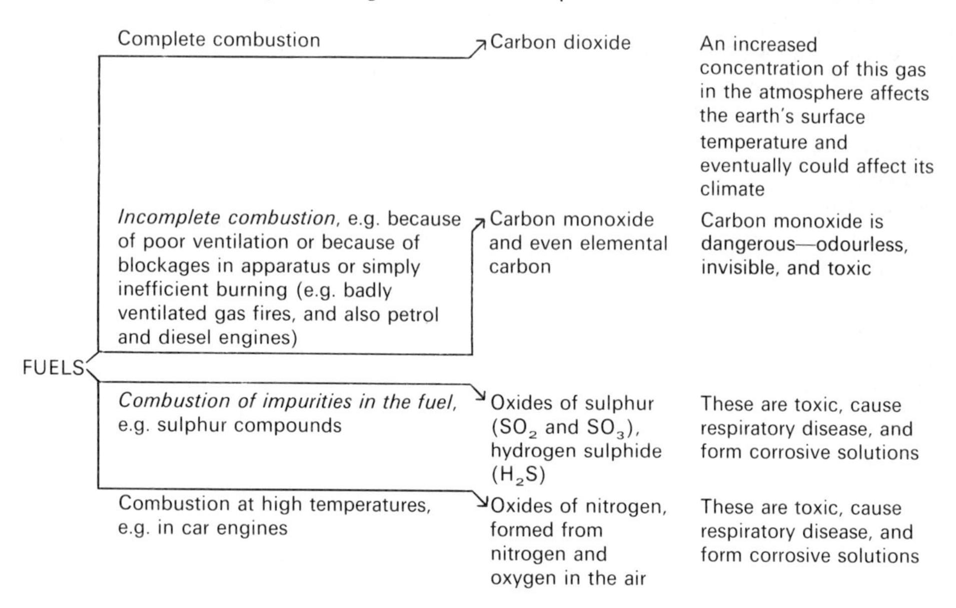

FUELS		
Complete combustion	Carbon dioxide	An increased concentration of this gas in the atmosphere affects the earth's surface temperature and eventually could affect its climate
Incomplete combustion, e.g. because of poor ventilation or because of blockages in apparatus or simply inefficient burning (e.g. badly ventilated gas fires, and also petrol and diesel engines)	Carbon monoxide and even elemental carbon	Carbon monoxide is dangerous—odourless, invisible, and toxic
Combustion of impurities in the fuel, e.g. sulphur compounds	Oxides of sulphur (SO_2 and SO_3), hydrogen sulphide (H_2S)	These are toxic, cause respiratory disease, and form corrosive solutions
Combustion at high temperatures, e.g. in car engines	Oxides of nitrogen, formed from nitrogen and oxygen in the air	These are toxic, cause respiratory disease, and form corrosive solutions

prevailing winds to Scandinavia, where it falls as dilute sulphuric acid in rain or snow.

It is interesting to note that a 'biological' estimate of the sulphur dioxide concentration in the air can be made by studying the mosses and lichens in a given area. A sulphur dioxide pollution scale has been worked out which depends on the lichen community growing on deciduous trees with rough bark.

Desulphurization and other remedies

The fuels burned in a modern home (e.g. smokeless solid fuels, natural gas, and paraffin) are reasonably pure and contain very little sulphur. This is because sulphur compounds have been almost completely removed from the fuels (i.e. they have been desulphurized) by preliminary processing. Desulphurization is important not only in helping to prevent atmospheric pollution, but also in providing a valuable supply of sulphur which can be burned in *controlled* conditions to produce sulphur dioxide which is then used to make sulphuric acid (p. 545).

However, industry uses fuels on an enormous scale and often has to use coal and other fuels which are (given present day economic thinking) either too difficult or too expensive to purify completely before use. Here again, great improvements have been made. A move over to burning natural gas (instead of solid fuel, or oil) has helped because natural gas often has a naturally low sulphur content, and is anyway relatively easy to desulphurize. Very efficient systems have been devised for washing and filtering waste gases, and most of the fuels burned on an industrial scale are at least partially desulphurized. Very tall chimneys (up to 280 metres high) are now used to pass waste gases into the atmosphere, so that any pollutants which do escape are then diluted by the atmosphere before they reach ground level where plants and people live. In addition, the Alkali Inspectorate and Local Authorities insist upon high standards of pollution control and have powers to penalize any companies not complying with their requests.

Most companies now regularly monitor the environment of the works to ensure that pollution problems are strictly controlled. An example of the successful way in which industry is tackling the problem is the £3 million antipollution programme initiated by ICI at their Billingham, Teesside, complex. Dust emission has been cut by 98 per cent, sulphur dioxide emission by 97 per cent, and 'Teesside fogs' are a thing of the past.

Oxides of nitrogen

Nitrogen is the main gas in the air, but under normal conditions it is almost inert. If sufficient energy is supplied, however, it will react with oxygen in the air to form oxides of nitrogen. These conditions are created by an electric discharge (e.g. lightning, as in the nitrogen cycle, page 383) and also by the high temperatures at which hydrocarbon fuels (petrol, diesel oil) burn in internal combustion engines. Cars thus add oxides of nitrogen to the atmosphere from their exhaust gases, in addition to carbon monoxide and lead compounds (see later).

The first compound of nitrogen formed under these conditions is usually nitrogen monoxide, NO.

$$N_2(g) + O_2(g) \rightarrow 2NO(g)$$

This gas normally reacts with more oxygen to form nitrogen dioxide, NO_2.

$$2NO(g) + O_2(g) \rightarrow 2NO_2(g)$$

This gas is toxic, promotes respiratory diseases and also dissolves in water to form a solution of nitric acid. (These last two reactions are important steps in the manufacture of nitric acid, as explained on page 552.)

Solid particles in the air

Very large quantities of smoke (which is mainly small particles of carbon suspended in the air) were produced by the burning of raw coal on domestic fires. The old-fashioned open grates were not only responsible for a great deal of pollution; they were also a very inefficient way of using the fuel, because over 70 per cent of the heat energy escaped through the chimney. Many industrial processes also produce a great deal of smoke, and such smoke may contain metallic particles and particles of silicon oxide.

Smoke, be it tobacco smoke or any other form, consists of carbon, tar, salts, other solid particles, and often carcinogenic substances. The particles are very small (they may be less than 7 μm in diameter; 1 $\mu m = 10^{-6}$ m) so that they can enter the lungs and be retained there. Smoke therefore promotes respiratory problems, and increases the risk of lung cancer. It also produces a dirty environment. In the London smog of 1952 there were over 4000 'extra' deaths (i.e. compared to the average number of deaths) in a period of just five days. Most of these deaths were caused by various respiratory problems induced by the smoke and the sulphur dioxide which it also contained.

In the UK the introduction of the Clean Air Act in 1956 enabled Local Authorities to introduce *smokeless zones* in which the burning of solid fuels was not permitted unless they were smokeless, i.e. had been pretreated as described on page 502. Similarly, industry had to conform to strict regulations about the emission of smoke. Modern heating systems also burn solid fuels more efficiently, and use far more of the heat energy produced.

Industry, which cannot always afford to use vast quantities of smokeless solid fuel, has developed very efficient smoke and dust extraction systems, and again the use of tall chimneys helps to dilute whatever smoke still escapes. The effect of the Clean Air Act can be appreciated by looking again at Figure 9.11.

Compounds of heavy metals

Ions of heavy metals (e.g. lead and mercury) are usually toxic, and have the added disadvantage of often being *cumulative* poisons, i.e. they build up in the body and are only very slowly excreted. There is concern over the way compounds of this type are sometimes found in liquid industrial waste (which eventually pollutes the rivers and seas) and they are also found in the air.

Lead is the worst example of this kind of atmospheric pollution. It occurs in the air as lead compounds formed by the combustion of petrol. Petrol contains several additives, one of which is tetraethyllead(IV), $Pb(C_2H_5)_4$, which is important in preventing premature explosion of the petrol/air mixture in the cylinders (i.e. the process often called 'knocking'). Approximately 0.4 g of lead have been present as this compound in 1 dm^3 of petrol in the UK, but this will soon be reduced by about 75%. Other compounds are added (e.g. 1,2-dibromoethane) to ensure that lead compounds do not build up in the engine but are instead discharged as volatile lead bromide in the exhaust gases. This results in a build up of lead compounds in the atmosphere and on surfaces, particularly in densely populated areas where traffic flow is heavy.

Over 94 per cent of the lead which is found in the air and on surfaces in urban areas has been derived from petrol, and this enters the body by direct inhalation and also through food. The levels of lead in the atmosphere and in human blood are measured regularly (see Table 9.4). (The World Health Organization recommended limit for inhalation by people is 2.0 $\mu g\ m^{-3}$.) Blood levels of lead are below those known to cause problems, but they are nearer to the permitted levels than those of any other pollutant. This is of particular concern with respect to young children, because lead compounds are known to cause brain damage in young children at levels lower than those which affect adults.

Most governments are taking action to ensure that lead levels in petrol are being reduced, and it could well be that petrol will become 'lead-free'.

Table 9.4 Concentrations of lead in different parts of the atmosphere

Area	*Concentration of lead* ($\mu g\ m^{-3}$)
South Pole	0.004
Rural areas of Britain	0.01 to 0.1
Small town in Britain	0.1 to 0.5
Large city in Britain	0.1 to 2.0
Fleet Street, London	up to 5.4
Los Angeles, USA (city centre)	up to 7.6
Motorways in Los Angeles	20 to 70
Busy road tunnels	up to 200

Radioactive substances in the air

There has always been some form of natural radioactivity in the air, but man has added to this. Nuclear power inevitably results in the production of radioactive waste materials which are either solid or in solution. These are not likely to affect the atmosphere in any way unless an accident occurs, although there is a potential risk to the sea and to the land. The explosion of any form of atomic weapon does, however, add to the natural radioactivity in the atmosphere, and successful attempts are being made internationally to reduce or stop nuclear experiments of this type. The Nuclear Test Ban Treaty of 1963, which has been signed by all the nuclear powers except France and China, has greatly reduced the 'fall-out' of radioactive material from the atmosphere.

Check your understanding

1. Combustion of the hydrocarbons in petrol should produce carbon dioxide and water only. However, petrol, on combustion in the internal combustion engine, produces many other compounds. Complete the following table which relates to such products of combustion.

Substance in the exhaust of car engine	*Reason for that substance being in exhaust*	*Pollutant effect of the substance*
Carbon monoxide		
Carbon		
Lead compounds		
Oxides of nitrogen		

(A.E.B. 1976)

2. Complete the following table which is concerned with pollutants in the atmosphere.

Pollutant	*One cause of this pollutant being in the atmosphere*	*One effect of this pollutant*
Sulphur dioxide		
Carbon dioxide		
Oxides of nitrogen		
Radioactive particles (i.e. dust)		

(A.E.B. 1978)

9.5 A SUMMARY OF CHAPTER 9

The following 'check list' should help you to organize the work for revision.

1. Definitions

(a) Combustion (p. 138)
(b) Respiration (p. 138)
(c) Photosynthesis (p. 139)

2. Other ideas

(a) The principles used to plan a suitable set of apparatus to prepare, dry and collect common gases, and the mistakes to be avoided in drawing the apparatus.
(b) The common drying agents are concentrated sulphuric acid (a liquid), anhydrous calcium chloride, soda lime, and silica gel (all solids). Calcium oxide (a solid) is used to dry the gas ammonia.
(c) Why the mass of a substance may decrease, increase, or remain unchanged when it is heated in the air (with examples).

(d) Anhydrous cobalt chloride is blue, hydrated cobalt chloride is pink, and this colour change is sometimes used to detect the presence of (not necessarily pure) water. Similarly anhydrous copper(II) sulphate (white) changes to the blue hydrated form when water is added.
(e) The typical composition of dry air (p. 137).
(f) The difference between hydrocarbons (compounds consisting only of the elements carbon and hydrogen) and carbohydrates (sugars and starches, which contain carbon and hydrogen and also oxygen).
(g) When hydrocarbons or carbohydrates are burned *completely*, carbon dioxide and water vapour are the only chemical products.
(h) The differences between inspired and expired air (p. 139).
(i) The difference between aerobic (using oxygen) and anaerobic (without oxygen) respiration.
(j) The importance of photosynthesis in helping to maintain the oxygen/carbon dioxide balance in the air and in providing food, and how it compares with combustion and respiration.
(k) The problem of atmospheric pollution. The main pollutants, carbon dioxide, carbon monoxide, sulphur dioxide, oxides of nitrogen, solid particles, compounds of heavy metals and radioactivity. How these pollutants are formed, why they are harmful, and what is being done in each case to solve the problem.

3. *Important experiments*

(a) An experiment to show that oxygen is about 21 per cent of the air by volume (e.g. Experiment 9.1).
(b) An experiment to prepare a partially purified sample of nitrogen from the air (e.g. a corrected version of the apparatus shown in Figure 9.10).

QUESTIONS

Multiple choice questions

1. Photosynthesis and respiration both
A release oxygen to the atmosphere
B release energy to the atmosphere
C take place only in light
D result in carbon dioxide being absorbed from the atmosphere
E occur in living material (C.)

2. Ordinary air from the atmosphere contains about 21 per cent of oxygen, whereas the proportion of oxygen in the mixture released by boiling river water is about 30 per cent. The best explanation of the increase in the percentage of oxygen is that
A oxygen is more soluble in water than is nitrogen
B carbon dioxide is more soluble in water than is oxygen
C nitrogen reacts with water
D the noble gases are insoluble in water
E the gases from boiled water contain no water vapour (C.)

3. Which of the following does *not* reduce the amount of oxygen in the air?
A Aerobic respiration
B Rusting
C Photosynthesis
D The use of petrol as a fuel
E The combustion of natural gas

Questions requiring longer answers

4. Discuss the various methods available for the generation, purification, drying, and collection of gases. Point out the advantages and disadvantages of each of the methods you mention.

5. Compare and contrast the processes of combustion, respiration, and photosynthesis.

6. Explain why each of the following gases are pollutants if present in air: (a) carbon monoxide, (b) carbon dioxide, (c) nitrogen dioxide, (d) sulphur dioxide.

7. Explain the meaning of each of the following terms, and explain how each is connected with air pollution: (a) incomplete combustion, (b) desulphurization, (c) carcinogenic, (d) the 'greenhouse effect', (e) smokeless zone, (f) cumulative poison.

10 Oxygen and hydrogen

10.1 OXYGEN

Obtaining oxygen

Oxygen is by far the most abundant element on Earth. About 49 per cent by mass of the Earth's crust and oceans consists of combined oxygen (as silicates, carbonates, water, etc.) and the free element makes up about 21 per cent by volume of the atmosphere.

Pure oxygen can be *separated* from the air if the air is liquefied and then fractionally distilled. The cylinders of pure, compressed oxygen obtained in this way are a convenient source of the gas. It is also useful to be able to *prepare* the gas in the laboratory. The gas was first prepared by Carl Wilhelm Scheele about 1772, and named by Antoine Lavoisier from the Greek *oxys genon* (acid former) because he thought (incorrectly) that all oxides were acidic.

The laboratory preparation of oxygen

A typical laboratory preparation of the gas is shown in Figure 10.1. The dilute hydrogen peroxide solution, H_2O_2(aq), decomposes into oxygen and water, but the rate of this reaction is very slow under ordinary conditions. The catalyst

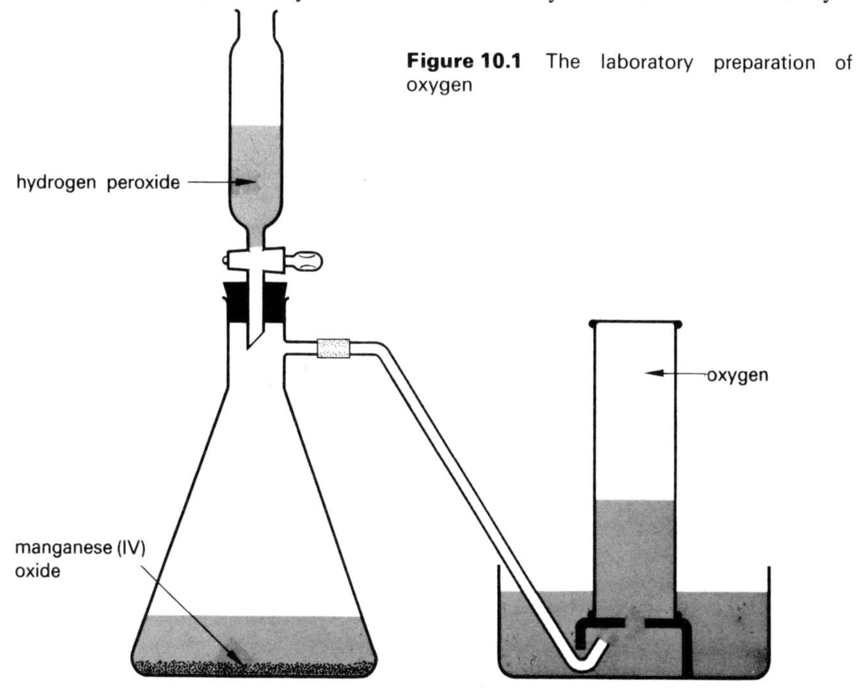

Figure 10.1 The laboratory preparation of oxygen

manganese(IV) oxide greatly increases the rate at which the dilute hydrogen peroxide solution decomposes, and no heating is required.

$$2H_2O_2(aq) \xrightarrow{MnO_2 \text{ catalyst}} 2H_2O(l) + O_2(g)$$

If you record the preparation in your book, include a note about what you see during the preparation, e.g. any effervescence or colours.

Oxygen can also be prepared, in small volumes, by the electrolysis of several different solutions, as described in Chapter 13, but these reactions are not normally used to prepare the gas.

Physical properties of oxygen

Solubility in water	*Colour*	*Odour*	*Density relative to air*	*Toxicity*
Very slight	None	None	About the same	Not toxic

Chemical properties of oxygen

General

Oxygen is a very reactive element, and reacts with most other elements. Occasionally an element may appear reluctant to react with oxygen, but this may be because it has *already reacted* to a small extent and is coated with a thin layer of oxide, which then protects the element from further reaction with oxygen. The best example of this is aluminium. Similarly, many elements do not actually *burn* when heated in the gas, because they quickly form a layer of oxide which protects the rest of the element from oxidation. The great reactivity of oxygen is shown by the large amounts of energy (e.g. as heat or light) which are often given out when it reacts with other substances.

When oxygen combines with other elements, a compound called an oxide is formed; even water, H_2O, is an oxide. Many very common processes involve oxygen, and it is easy to forget these, e.g. respiration, combustion, and photosynthesis.

Reactions of oxygen with metallic elements to form metallic oxides

Experiment 10.1
The formation of metallic oxides**

Apparatus

Oxygen generator and collection apparatus. Rack of test-tubes (150 × 25 mm) with bungs to fit, combustion spoon mounted in cork or bung (see Figure 10.2), tongs, Bunsen burner, gas jars and covers, deflagrating spoons.

Magnesium ribbon, steel wool, sodium, and a selection of oxides of other metals (i.e. some of the oxides which are difficult to prepare by direct combination, such as iron(III) oxide and lead(II) oxide).

Procedure

Note: Appropriate safety precautions should be observed throughout the experiment. Sodium must be used only by the teacher, as a demonstration. The gas jars and deflagrating spoons are for demonstrations. If students are allowed to use some metals, the boiling tubes and combustion spoons can be used, but as these tubes will be wet it is particularly important that goggles must be worn when hot metals are held inside the tubes. Every effort should be made to avoid touching the sides of the

glass with burning metal. In all the tests in this and the next experiment, the heated metals must be inserted into the gas jars or boiling tubes whilst they are on the bench or in a rack, *not* while held in the hand.
(a) Make out a table for your results as shown below.
(b) Collect the appropriate number of gas jars and/or tubes of oxygen, sealing each with a greased cover (gas jars) or tight-fitting bung (test-tube).
(c) Wind a small length of magnesium ribbon around the bottom of the combustion spoon so that a straight piece about 1.5 cm long hangs free. Ignite the ribbon and immediately insert it into a sample of oxygen. *Warning*: do not stare *directly* at the burning magnesium.

Add a few cm^3 of deionized water to the product in the tube and then add a few drops of universal indicator solution.
(d) Place some iron wool on the end of the spoon, heat it to red heat and *quickly* plunge it into another sample of oxygen. Test the product with deionized water and indicator as before.
(e) Place a small piece of sodium on a dry deflagrating spoon. Hold the spoon in the flame until the sodium begins to burn and then transfer it to a gas jar of oxygen. Carefully test the product with water and indicator as before.
(f) Add small samples of the metal oxides provided to water in separate test-tubes, and add a drop or two of universal indicator to each tube.

Results
Complete the table.

Metal	*Observations when hot metal placed in oxygen*	*Colour and state of oxide*	*Effect of oxide on universal indicator*

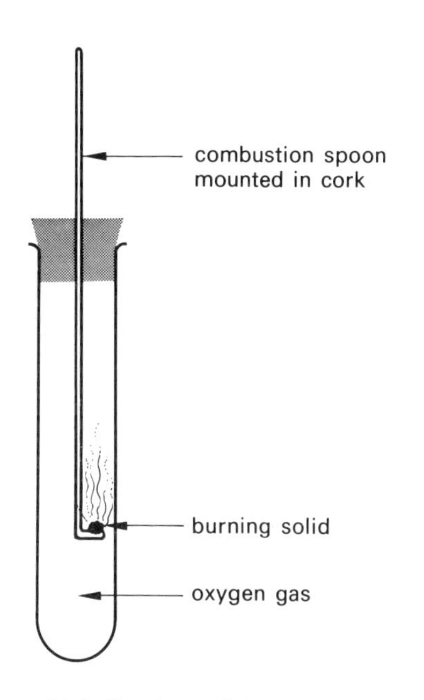

Figure 10.2 Burning solids in oxygen (small scale)

Notes about the experiment

Some elements can form more than one oxide, in which case the product of combustion will often be a mixture. Sodium is a typical example of this. Equations for some of the reactions are:

$$2Mg(s) + O_2(g) \rightarrow 2MgO(s)$$
$$3Fe(s) + 2O_2(g) \rightarrow Fe_3O_4(s)$$
$$4Na(s) + O_2(g) \rightarrow 2Na_2O(s) \quad \text{(trace only)}$$
$$2Na(s) + O_2(g) \rightarrow Na_2O_2(s) \quad \text{(main product)}$$

You may think that some of the metal oxides are neutral. The term 'neutral oxide' can be confusing. Many oxides are basic or acidic, but this is only obvious if they happen to dissolve in water and then affect indicators. It is possible for a basic oxide or an acidic oxide to be insoluble in water and therefore not to affect universal indicator. It would be quite wrong to conclude that an oxide which does not affect universal indicator *must* be a neutral oxide. The only test which will definitely show whether such an oxide is basic or acidic is to see if it will react with (and dissolve in) a known acid or alkali. If it is insoluble in water but does dissolve in an acid, the oxide is basic. If it is insoluble in water but dissolves in an alkali, the oxide is acidic. (If it dissolves in *both* acids and alkalis, it is a special type of oxide called an *amphoteric* oxide). It is only a neutral oxide if it fails to react with water, acids, and alkalis. These points are referred to again on page 153.

In your experiment we are not looking for this kind of detailed chemistry. It should be possible to make a fairly definite conclusion using the following as a guide. It is up to you to choose the appropriate alternative words from those given in brackets.

Conclusion

(Make sure that you can write equations for the combustion of the metals you used.)

Most metals, especially if heated, will react with oxygen to form oxides. Metal oxides are usually in the (solid, liquid, gas) state at room temperature, because they are (ionically, covalently) bonded and have fairly (high, low) melting points. Metal oxides are usually (acidic, basic) although they may be neutral. Common ones are never (acidic, basic).

Reaction of oxygen with non-metallic elements

Experiment 10.2
The formation of non-metallic oxides**

Apparatus

As in Experiment 10.1, but using sulphur, carbon, and red phosphorus instead of the metals. Sample of nitrogen dioxide gas (in fume cupboard), silicon dioxide.

Procedure

Note the safety precautions mentioned for Experiment 10.1; the same general principles must be used in this experiment. Red phosphorus and nitrogen dioxide must only be used for demonstrations. All of the reactions can be demonstrated (on the 'gas jar' scale) if preferred.

(a) Make a table for your results, similar to that used for Experiment 10.1.

(b) Working in a fume cupboard, place a small quantity of sulphur on the spoon, ignite it in the Bunsen flame and then plunge it into a sample of oxygen. When the reaction has finished, open the tube or jar (still in the fume cupboard) and add a few cm³ of deionized water followed by a few drops of universal indicator solution.

(c) Place a small quantity of carbon on the spoon, heat to redness and *quickly* plunge it into a sample of oxygen. Test the product with deionized water and indicator as in (b).

(d) With care, ignite a *little* red phosphorus on the spoon and place it in a sample of oxygen in the fume cupboard. Test the product with deionized water and universal indicator as in (b).

(e) Add a sample of silicon dioxide to a few cm³ of water in a test-tube, and add a few drops of universal indicator solution.

(f) You will see a demonstration of the effect of adding water and indicator to a sample of nitrogen dioxide gas (in the fume cupboard).

Notes about the experiment

As with metals, some non-metals burn to form more than one oxide. Some typical equations are:

$$S(s) + O_2(g) \rightarrow SO_2(g),$$
sulphur dioxide, main product

$$2S(s) + 3O_2(g) \rightarrow 2SO_3(g),$$
sulphur trioxide, trace only

$$C(s) + O_2(g) \rightarrow CO_2(g)$$

$$P_4(s) + 5O_2(g) \rightarrow P_4O_{10}(s)$$

Remember that pure water is also a non-metallic oxide; what is its pH?

Conclusion

Make your own conclusion, using the following statements as a guide. It is up to you to choose the correct alternative words from those given in brackets.

Most non-metal oxides are usually in the gas or liquid state, although a few are solids at room temperature. This is because they are formed by bonding two non-metals together and are therefore (ionically, covalently) bonded and often have molecular structures. Non-metal oxides are often (acidic, basic) although they can be neutral. They are never (acidic, basic).

The test for oxygen gas

If a *glowing* wooden splint is placed in a sample of oxygen, the splint relights. Another gas also gives this test, but you are unlikely to come across it at this level.

Oxygen as an oxidizing agent

The term oxidizing agent now has several meanings, but in its simplest (and earliest) form, an oxidizing agent is defined as a substance which can add oxygen to another chemical. The substance which gains the oxygen is said to be oxidized.

Obviously, the simplest oxidizing agent is oxygen itself. In Experiments 10.1 and 10.2, the metals or non-metals were being oxidized to their oxides by the oxygen.

You will study this idea in Section 10.2, and, in a more advanced way, in Chapter 17.

Uses of oxygen

1. When mixed with ethyne (acetylene), oxygen forms a mixture which burns at a very high temperature (the oxyacetylene flame) and this is used in welding and cutting metals.
2. Vast quantities are used in steel making (p. 527).
3. In rocket fuels.
4. Respiratory aid (hospitals, diving, high altitude flying).

Oxides

Oxides of elements can be classified as acidic, basic, neutral, or amphoteric, according to their chemical properties. Sometimes, however, it is quite difficult to place an oxide in one of these classes because its properties may not be obvious. The various types of oxide are discussed under the following headings.

Acidic oxides

An acid oxide will react with a base so that they are neutralized and form a salt.

Acidic oxides which 'dissolve' in water always form acidic solutions, which can be detected by using an indicator. Those acidic oxides which do not dissolve in water will not affect indicators, but they will always dissolve in hot alkali, e.g. silicon dioxide, SiO_2. Sand is impure silicon dioxide.

Acidic oxides are often oxides of non-metals, although not all non-metal oxides are acidic.

The reaction of an acidic oxide with water is conveniently summarized by an equation, e.g. sulphur trioxide dissolves in (and reacts with) water to form a solution of sulphuric acid:

$$SO_3(g) + H_2O(l) \rightarrow H_2SO_4(aq)$$

The oxide in this example is like 'sulphuric acid without water' and it is called an *acid anhydride* (acid without water). Details of the formation of some common non-metal oxides are given in Table 10.1, and the properties of some common non-metal oxides are given in Table 10.2. Acidic oxides are usually liquids or gases at room temperature because although they have strong covalently bonded molecular structures there are only weak forces between the molecules and so they have relatively low boiling points and melting points. Those which are solids either have low melting points or exist as giant covalent structures (macromolecules, page 272).

Table 10.1 The combustion of some common non-metals in oxygen

Element	*Reaction details*
Sulphur	The pale yellow solid melts when heated (amber liquid) then burns with a bright blue flame to form a misty gas with a choking smell, sulphur dioxide $S(s) + O_2(g) \rightarrow SO_2(g)$ (a little sulphur trioxide, SO_3, is also formed) The oxides are acidic, turning damp blue litmus red
Red phosphorus	The red powder rapidly reacts when heated, burning with a white flame to form a white smoke: $P_4(s) + 5O_2(g) \rightarrow P_4O_{10}(s)$ The oxide is acidic, and turns damp blue litmus red
Carbon	The black powder (or granules) only reacts when red hot, and then smoulders (or burns with a white flame) producing an invisible gas: $C(s) + O_2(g) \rightarrow CO_2(g)$ The gas formed is weakly acidic, turning damp blue litmus a purple-red colour

Table 10.2 Some common non-metal oxides

Oxide, formula, and state at room temperature	*Effect on water, and type of oxide*
Sulphur dioxide, SO_2, gas	Some dissolves, some reacts: $SO_2(g) + H_2O(l) \rightarrow H_2SO_3(aq)$ Acidic oxide, the anhydride of sulphurous acid
Sulphur trioxide, SO_3, white solid usually seen as a smoke	Violent reaction: $SO_3(s) + H_2O(l) \rightarrow H_2SO_4(aq)$ Acidic oxide, the anhydride of sulphuric acid
Nitrogen dioxide, NO_2, gas	Reacts to form two acids: $2NO_2(g) + H_2O(l) \rightarrow HNO_3(aq) + HNO_2$ Acidic oxide, a mixed anhydride (i.e. forms *two* acids when it reacts with water)
Silicon dioxide, SiO_2, solid	Insoluble. Neutral to indicators, but is an acidic oxide which dissolves in hot alkali
Carbon dioxide, CO_2, gas	Some dissolves and some reacts: $CO_2(g) + H_2O(l) \rightarrow H_2CO_3(aq)$ Weakly acidic, the anhydride of the unstable carbonic acid
Carbon monoxide, CO, gas	Insoluble; neutral oxide
Water, H_2O, liquid	Neutral oxide
Phosphorus(V) oxide, P_4O_{10}, solid	Reacts violently to form a solution of phosphoric(V) acid, H_3PO_4. Acidic oxide

Basic oxides

A basic oxide will react with an acid so that they are neutralized and form a salt.

Basic oxides which dissolve in (and react with) water always form alkaline solutions, which can be detected by using an indicator, e.g.

$$CaO(s) + H_2O(l) \rightarrow Ca(OH)_2(aq)$$

$$Na_2O(s) + 2H_2O(l) \rightarrow 2NaOH(aq)$$

Those basic oxides which do not dissolve in water will not affect indicators, but they will always dissolve in a hot solution of an acid.

Basic oxides are often (or, at an elementary level, always) oxides of metals, and as these are ionically bonded they are usually solids at room temperature, with fairly high melting points and boiling points. The formation of metal oxides from oxygen and some of their properties is summarized in Table 15.1 (p. 250) (in connection with the activity series) and in Table 19.2 (p. 328) (in the chapter on metal compounds). Typical basic oxides include magnesium oxide and copper(II) oxide, and these and other similar examples are frequently used to neutralize acids in salt preparations.

Remember that metal *hydroxides* are also bases.

Amphoteric oxides

An amphoteric oxide will neutralize (and dissolve in) both acids and strong alkalis to form salts.

The only examples you need to learn are zinc oxide, ZnO, and aluminium oxide, Al_2O_3. Zinc oxide, for example, will react with a typical acid such as sulphuric acid to form a salt, zinc sulphate.

$$ZnO(s) + H_2SO_4(aq) \rightarrow ZnSO_4(aq) + H_2O(l)$$

It will also dissolve in a typical strong alkali solution such as sodium hydroxide to form a salt such as sodium zincate. Aluminium oxide forms salts called aluminates when reacted with strong alkalis. Equations are not normally required for the formation of zincates and aluminates.

'Ampho's great; he can deal with acids and alkalis'

Neutral oxides

A neutral oxide will not affect the pH of water, nor neutralize either an acid or an alkali.

The only common examples are water (H_2O), nitrogen monoxide (NO), and dinitrogen monoxide (N_2O). Under most conditions, carbon monoxide (CO) also behaves as a neutral oxide, but at high pressures and temperatures it will react with the alkali sodium hydroxide to form sodium methanoate (formate) HCOONa, and so may be considered to be an acidic oxide.

The main types of oxide are summarized in Table 10.3.

Table 10.3 The main types of oxide

	Acidic oxide	*Basic oxide*	*Neutral oxide*	*Amphoteric oxide*
Will it affect the colour of indicators?	If soluble, will form an acidic solution which affects indicators	If soluble, will form an alkaline solution which affects indicators	NO	NO
Will it neutralize an acid?	NO	YES	NO	YES
Will it neutralize a base?	YES	NO	NO	YES

10.2 HYDROGEN

Obtaining hydrogen

Hydrogen is one of the ten most common elements on Earth, although it is usually found only in combined form. It has been estimated that approximately 15.4 per cent of all the atoms in the Earth's crust are hydrogen atoms. The gas was named by Antoine Lavoisier from the Greek *hydro genon* (water former).

Hydrogen is used on a large scale in industry. It is obtained by processing natural gas and petroleum, and also by the electrolysis of solutions such as sodium chloride solution.

The laboratory preparation of hydrogen

In Experiment 5.3 (p. 60) you learned that some dilute acids and some metals react to produce hydrogen and a salt. This general reaction can be used to prepare the gas, but the metal and the acid need to be chosen carefully. Typically, the metal used is zinc and the acid is dilute hydrochloric acid, as in Figure 10.3, in which case the hydrogen will contain some hydrogen chloride vapour unless it is collected over water, which dissolves the hydrogen chloride.

$$Zn(s) + 2HCl(aq) \rightarrow ZnCl_2(aq) + H_2(g)$$

or ionically

$$Zn(s) + 2H^+(aq) \rightarrow Zn^{2+}(aq) + H_2(g)$$

(The reaction with dilute sulphuric acid tends to be slow unless a little copper(II) sulphate solution is added.)

Note: Mixtures of hydrogen and air or oxygen are very dangerous because they can explode if ignited. A flame should never be used near a hydrogen generator, and a large sample of the gas (e.g. a gas jar full) must never be ignited.

If hydrogen is needed in a laboratory it is often obtained from a cylinder of the compressed gas. This is safer to use and more convenient than a generator, because the gas is released under pressure and air does not normally get a chance to mix with the hydrogen to form explosive mixtures. However, large samples of the gas, even from a cylinder, should not be ignited.

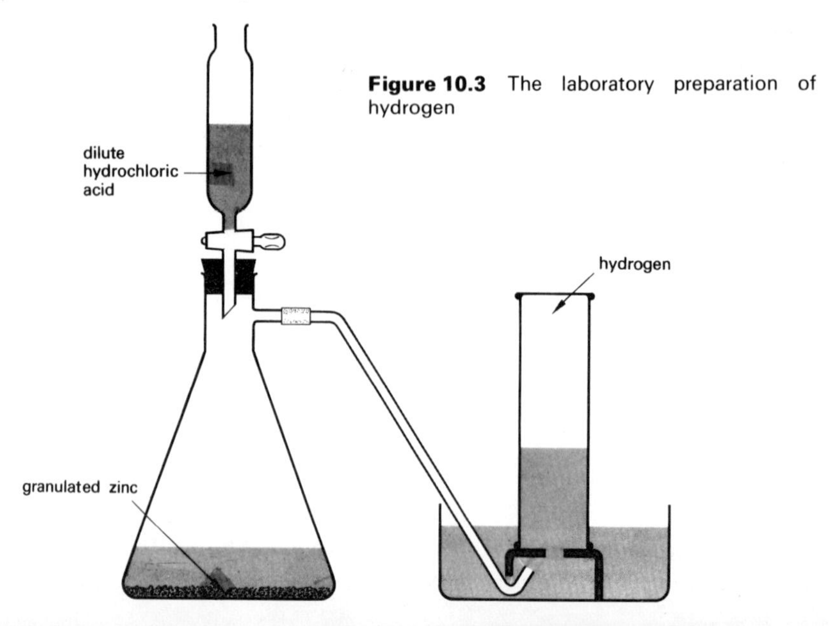

Figure 10.3 The laboratory preparation of hydrogen

Physical properties of hydrogen gas

Solubility in water	*Colour*	*Odour*	*Density relative to air*	*Toxicity*
Very slight	None	None	Much less dense	None

Hydrogen is the lightest gas, and so diffuses very rapidly (Experiment 2.5, page 13).

Points for discussion

1. A test-tube of hydrogen was placed below a test-tube of air, the bungs removed, and the tubes held mouth to mouth. After a few seconds the tubes were separated and each quickly tested for hydrogen. The experiment was repeated but with the tube of hydrogen on top of the tube of air. Which of the four tubes would give a hydrogen test after a few seconds? Explain your answer.

2. If the experiment was repeated by leaving the tubes mouth to mouth for two minutes rather than a few seconds, which of the four tubes would you expect to give a hydrogen test? Again explain your answer.

Chemical properties of hydrogen

Combustion

A stream of *pure* hydrogen burns in air or oxygen with a blue flame to form steam.

$$2H_2(g) + O_2(g) \rightarrow 2H_2O(g)$$

A *mixture* of hydrogen with air or oxygen explodes when a flame is applied, and again steam is formed. A small scale version of this explosion is used as a test for the gas.

You may see a demonstration of the fact that water is the product when hydrogen burns in air, although such reactions are potentially dangerous and must be conducted with great care.

The test for hydrogen gas

If a gas is liberated in a test-tube reaction it can be tested for hydrogen as follows. Trap some of the gas by holding the thumb tightly over the mouth of the test-tube for a few seconds (unless effervescence is rapid). Remove the thumb and almost at the same time apply a lighted splint to the mouth of the tube. If the evolved gas is hydrogen, a 'squeaky pop' is heard as the mixture of hydrogen and air reacts with a miniature explosion.

The formation of hydrides

Just as a compound formed from an element and oxygen only is called an oxide, so a compound formed from an element and hydrogen only is called a hydride. Hydrogen forms hydrides with the most reactive metals (e.g. MgH_2, CaH_2, NaH), but the only hydrides you are likely to study are the non-metal hydrides such as H_2O, and NH_3, which are discussed in later chapters of the book.

Hydrogen as a reducing agent

Oxidation was defined in very simple terms on page 153. Reduction can be regarded as the opposite of oxidation, and two simple definitions are given (p. 158). You will also need to use this term in a different way (p. 290).

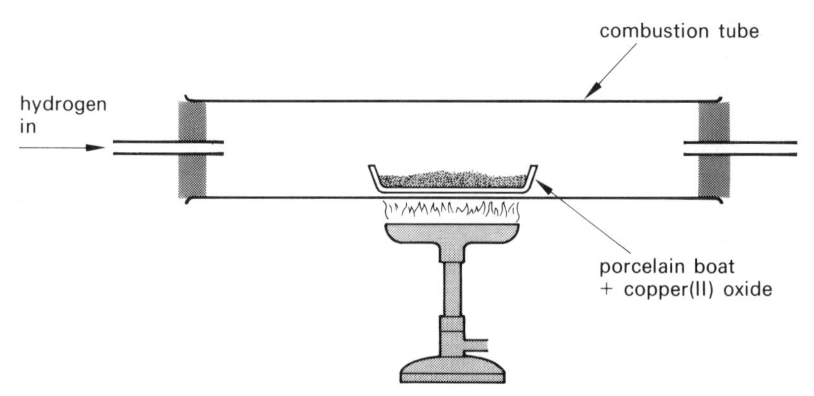

Figure 10.4 The reduction of copper(II) oxide by hydrogen

A substance is reduced when it has hydrogen added to it.
A substance is reduced when oxygen is removed from it.

It should be obvious that hydrogen must be a good *reducing agent* (from the first definition) and as it forms a strong bond with oxygen it also removes oxygen from substances and reduces according to the second definition. For example, when it reacts with chlorine

$$Cl_2(g) + H_2(g) \rightarrow 2HCl(g)$$

we can say that the chlorine has been reduced to hydrogen chloride (because it has gained hydrogen; first definition above). When hydrogen gas is passed over a heated metal oxide, e.g. in a porcelain boat in a combustion tube as in Figure 10.4, there is a 'battle for oxygen'. The metal already has the oxygen (because we start with a metal oxide) but the hydrogen could combine with the oxygen from the metal oxide to form water. If the metal 'wins the battle' it keeps the oxygen and there is no reaction. If hydrogen wins, it takes the oxygen from the metal oxide (to form water) and therefore reduces the metal oxide to the metal (second definition above).

You will learn later that reactive metals form strong bonds with oxygen, and hydrogen cannot take away the oxygen from (i.e. cannot reduce) their heated oxides. Metal oxides formed from less reactive metals cannot hold on to their oxygen so well, and hydrogen can reduce them to the metal.

Reactive metals: hydrogen cannot reduce their oxides, e.g. Na, K, Ca, Mg, Al, Zn.
Less reactive metals; hydrogen can reduce their oxides, e.g. Cu, Pb.
Hydrogen can reduce one of the oxides of iron, Fe_3O_4, but the reaction is not a simple one.

A typical equation for the reduction of a metal oxide is

$$CuO(s) + H_2(g) \rightarrow Cu(s) + H_2O(g)$$

The copper(II) oxide has been reduced to copper by losing oxygen.

Point for discussion

When a metal oxide is reduced to the metal as shown in Figure 10.4, it is usual to allow the metal formed to cool down in a stream of hydrogen before the combustion tube is opened. Why?

Uses of hydrogen gas

1. The manufacture of ammonia by the Haber process (p. 548). Some of the ammonia is then made into nitric acid.
2. The manufacture of many organic chemicals such as methanol and nylon.
3. The manufacture of cooking fats and margarine. Most animal fats such as mutton fat and pork fat are solids at room temperature. Most vegetable fats are liquids at room temperature and are called oils, e.g. olive oil. The differences in melting points are largely due to the presence of C═C double covalent bonds in the oils. If the comparatively cheap and readily available oils have their C═C double bonds changed into C—C single bonds (i.e. if they are changed from being *unsaturated* compounds into *saturated* compounds) they become solids at room temperature, and can be easily converted into solid fats for use in cooking or as a butter substitute. The process is called 'hardening' the oil, and it is achieved by adding hydrogen to the unsaturated molecules. The oil is heated and mixed with a finely divided nickel catalyst, and then hydrogen is blown through the mixture under pressure.

Some points of general interest

Many people believe that hydrogen atoms are the 'building bricks' of all substances in the universe. It is probable that over 90 per cent of all matter in the known universe is hydrogen, and although the planet Earth has a comparatively small amount of free hydrogen our own sun consists almost entirely of hydrogen and helium. At the enormous temperatures generated in the sun and other stars, fusion reactions take place (p. 503), resulting in the formation of larger atomic nuclei. All other elements could *theoretically* be produced from hydrogen by such reactions.

Hydrogen is not particularly reactive under ordinary conditions; much energy is needed to break open the H—H bond

The two atoms in a hydrogen molecule are joined by a single covalent bond, H—H. When hydrogen reacts this bond must be broken so that the free atoms formed can make new bonds with the other reactant(s). For example, in the reaction

$$2H_2(g) + O_2(g) \rightarrow 2H_2O(g)$$

the hydrogen atoms are at first joined together, but in the product they are separate, i.e.

$$\text{H—H} + \text{H—H} + \text{O=O} \rightarrow \text{H—O—H} + \text{H—O—H}$$

A great deal of energy is needed to break open the H—H bond and so hydrogen is not particularly reactive under ordinary conditions although it forms a highly explosive mixture with air.

Once the bond is broken the hydrogen atoms are exceptional in that they can then do one of three things in order to gain stable electronic configurations (p. 126). This is why hydrogen is studied separately; it does not conveniently fit into any one group of the Periodic Table.

'Ordinary' hydrogen ($^{1}_{1}H$) is sometimes given the special name *protium* to distinguish it from two other isotopes of hydrogen, $^{2}_{1}H$ (deuterium) and $^{3}_{1}H$ (tritium).

Atoms of deuterium occur naturally in the proportion of one part in 6400 parts of hydrogen. Thus water, although always assumed to contain hydrogen (^1_1H) and oxygen only, actually contains a small amount of deuterium as well. If water is electrolysed for a long period the residual liquid becomes richer and richer in D_2O (deuterium oxide), as the lighter protium ions are discharged about six times more readily than are deuterium ions. This process is not feasible on a large scale unless cheap electricity (e.g. hydroelectric power) is available, and so large quantities of 'heavy water' are made in Norway. Heavy water is used as a moderator in some nuclear reactors. Tritium is radioactive.

Some physical properties of H_2O and D_2O are as follows:

Property	H_2O	D_2O
Boiling point, °C	100	101.5
Melting point, °C	0	3.8
Density at 20 °C (g cm^{-3})	0.998	1.1

Points for discussion

1. Draw the structure of an atom of (a) deuterium, (b) tritium.

2. Why do you think there is a more marked difference in physical properties between atoms of ^2_1H and ^1_1H than there is between, say, the two chlorine isotopes $^{37}_{17}Cl$ and $^{35}_{17}Cl$? The difference in boiling point of these two hydrogen isotopes is, for example, 3.3 °C.

10.3 A SUMMARY OF CHAPTER 10

1. Definitions

(a) Acidic oxides (p. 153)
(b) Basic oxides (p. 154)
(c) Amphoteric oxides (p. 155)
(d) Neutral oxides (p. 155)

2. Other ideas

The following 'check list' should help you to organize the work for revision.

(a) How to prepare oxygen in the laboratory by the decomposition of hydrogen peroxide solution, using manganese(IV) oxide as a catalyst.

$$2H_2O_2(aq) \xrightarrow{\text{catalyst}} 2H_2O(l) + O_2(g)$$

(b) The physical properties of oxygen (p. 150).
(c) The colour changes and other changes observed when common metals are heated in oxygen to form their oxides, and the equations for the reactions (see Table 15.1 on page 250).
(d) The colour changes and other changes observed when common non-metals are heated in oxygen to form their oxides, and the equations for the reactions (see Tables 10.1 and 10.2).
(e) Two ways in which oxygen can oxidize substances (by adding oxygen or removing hydrogen), two ways in which hydrogen can reduce substances (by adding hydrogen or removing oxygen), and why some metal oxides can be reduced by hydrogen (oxides of less reactive metals) but others cannot (oxides of reactive metals).
(f) Some uses of oxygen, and the test for the gas.
(g) The differences between acidic, basic, neutral, and amphoteric oxides, and examples of each type. See also Tables 10.1, 10.2, 10.3, 15.1 (p. 250) and 19.2 (p. 328).
(h) How to prepare hydrogen gas in the laboratory from zinc metal and dilute hydrochloric acid (p. 156).

$$Zn(s) + 2HCl(aq) \rightarrow ZnCl_2(aq) + H_2(g)$$

(i) The physical properties of hydrogen, including its rapid rate of diffusion.
(j) What hydrides are (binary compounds of hydrogen and one other element), that water is the only product of the combustion of hydrogen, and how to test for hydrogen.
(k) Some uses of hydrogen.

QUESTIONS

1. Which of the following statements about hydrogen is *untrue*?
A It is a neutral gas, almost insoluble in water
B It is a reducing agent
C It will burn in air to form steam
D It diffuses more rapidly than carbon dioxide
E It is prepared by the action of dilute nitric acid on zinc

2. Explain the meaning of each of the following terms, and give an example in each case: (a) neutral oxide, (b) amphoteric oxide, (c) acidic anhydride, (d) basic oxide.

3. (a) (i) Name *three* substances, other than oxygen and nitrogen, that are always present in the atmosphere.
(ii) Describe, with the aid of a diagram of the apparatus, an experiment by which you could demonstrate the presence in the atmosphere of *one* of these three substances.
(b) Describe, with the aid of a diagram of the apparatus, the preparation and collection of a sample of oxygen, starting with hydrogen peroxide.
(c) Outline briefly how air is liquefied. How is oxygen obtained from liquid air?
(d) State the approximate percentage of oxygen by volume in the atmosphere, and explain briefly why the figure does not vary very much. (C.)

4. Draw a labelled diagram of the apparatus you would use to prepare and collect gas jars of *either* oxygen or hydrogen (not by electrolysis). Describe briefly *three* experiments you have seen that demonstrate chemical or physical properties of the gas that you have chosen above. In any chemical reaction mentioned, name and describe the product(s). Give two important different uses of *each* gas (excluding balloons). (C.)

11 Water (1): pure water, solubility, water of crystallization

11.1 WATER AS A CHEMICAL

Introduction

We tend to take water for granted, and yet our lives depend very much upon a good supply of pure water. Although water is by far the most common compound found naturally on earth, we are sometimes faced with a shortage of *pure* water in the summer months. This is partly because we have so little respect for pure water, and we tend to imagine that we can keep on using it (and wasting it) without thinking about it. The purification of water and its supply to our homes etc. is an expensive business. Some 4000 million tonnes of water are processed annually by the water authorities in the United Kingdom alone for human, industrial, and animal use. We are very fortunate that most people in Britain have piped water which is fit to drink directly, although it is ironical that we then use large volumes of this clean water for purposes which do not require such a degree of purity, e.g. for flushing the toilet.

The following are examples of just a few of the ways we use water without perhaps thinking of the part which water plays. We need enormous volumes of water in industry for generating electricity, cooling, transporting things, cleaning, and as a liquid medium for reactions. We need water to irrigate crops. It is possible to live for five weeks without food, but most people could live for only seven days without water. We need water to wash away waste, from inside our bodies, our homes, and factories. Many of the chemical reactions you study in school only take place if the chemicals are dissolved in water, and many important electrolysis reactions take place in solution in water.

Some physical properties of water

1. Pure water has no smell, no taste, and no colour.
2. Pure water boils at 100 °C (at a pressure of 760 mm mercury) and freezes at 0 °C.
3. The solid form of water (ice) is unusual because it is *less* dense than the liquid form, so that ice floats on water; this is because water expands when it freezes, unlike most other liquids. This property can cause pipes to burst in winter.
4. The intermolecular forces between water molecules (hydrogen bonds, page 287) are quite strong compared with those in other molecular liquids, and so water has a high latent heat of vaporization (page 478).
5. Water is a very good solvent, especially for ionically bonded substances.

If you understand your work on the kinetic theory, you should be able to explain why most liquids do not behave as water does in point 3.

Some chemical properties of water

Water can *react* with other chemicals. You will, for example, study the way water (or steam) reacts with metals. You already know that some non-metal oxides and some metal oxides react with water to form acidic or alkaline solutions, e.g.

$$SO_3(g) + H_2O(l) \rightarrow H_2SO_4(aq)$$

and you may know that acids such as hydrogen chloride can only show the properties of acidic solutions after they have *reacted* with water (p. 72).

The only reliable *test* for pure water is to check its boiling point and its freezing point. Some so-called tests for water are only tests for the *presence* of water, not necessarily pure water. For example, in the presence of water, anhydrous copper(II) sulphate will change from white to blue and anhydrous cobalt(II) chloride will turn from blue to pink. A drop of liquid withdrawn from a cup of tea, or a glass of beer, would also give these tests, however, but tea and beer would not boil at 100 °C (at 760 mm mercury pressure) or freeze at 0 °C.

The fact that water is formed by the combination of two volumes of hydrogen gas and one volume of oxygen gas is shown by the so-called 'electrolysis of water', page 218.

Points for discussion

1. Can you suggest which of the following statements is/are true?
(a) It can take up to 1 160 000 litres of water to grow 0.5 kg of cotton.
(b) It can take up to 116 000 litres of water to produce one tonne of steel.
(c) An average person uses approximately 160 litres of water each day, only 9 litres of which is used for drinking or cooking.

2. Why do you think a person with a piped water supply uses more water per day than a person who obtains water from a well?

3. Suggest reasons why we use far more water now than we did fifty years ago.

11.2 OBTAINING PURE WATER

The water cycle

Most of the water in, on, and around the Earth has been here since the Earth began, and we keep using the same water over and over again. Only a small proportion of this water is 'pure' at any moment; the rest, mainly in rivers or seas, contains dissolved substances.

The heat from the sun evaporates pure water from the seas, and this water vapour forms clouds and then condenses, falls as rain and eventually flows back into the sea. Your Water Authority traps *some* of this water on its way back to the sea, treats it as on page 164 and sends it as drinking water to your home. The water which evaporates from the sea is *pure* water, but by the time it has reached reservoirs etc. most of it has dissolved some carbon dioxide from the air, picked up bacteria, and dissolved substances from rocks and the soil, and so it must be treated before we drink it.

Your Water Authority also treats *used* water from homes and industry so that it is safe to put back into the rivers and seas. Such water is often polluted by chemicals added as waste or by accidental leakage. The treatment of used water occurs at a sewage works. Some of this treated, used water can be further purified at a water works so that it is fit to drink. These processes are summarized in Figure 11.1.

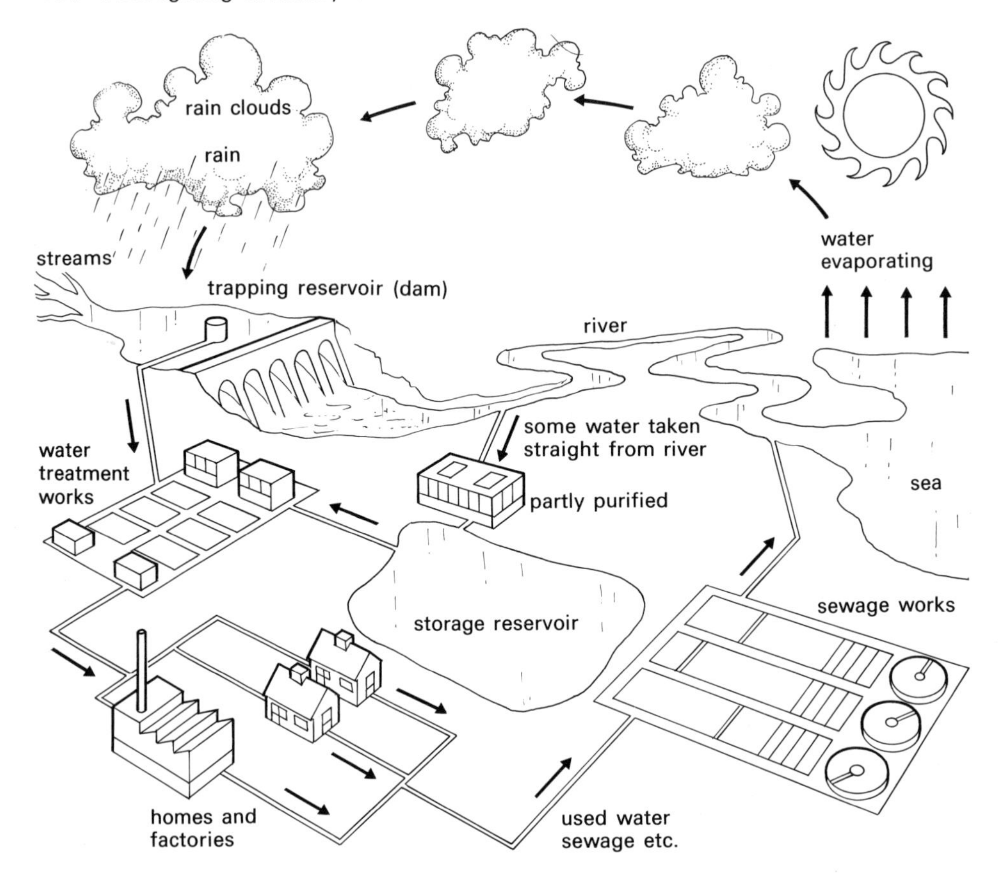

Figure 11.1 The Water Cycle (*Modified from material supplied by Severn Trent Water Authority*)

The purification of water at a treatment works

The processing of water at the treatment works depends on the impurities it contains. For example, water which has flowed over moorland peat will be acidic, and such water may be partly or completely neutralized at the treatment works. Water which has flowed over limestone will contain dissolved chemicals which make it hard (this is studied in Chapter 12), and such water may be softened at the treatment works. Again, if the water has a lot of suspended solids in it, chemicals may be added to cause the particles of suspended solid to join together to form larger pieces which then settle out in settling tanks. Some Water Authorities add fluoride ions to the water to help prevent tooth decay. However, certain processes *always* occur at a treatment works, no matter what the water contains.

Filtration

Large floating objects such as pieces of wood are removed when the water passes through wire mesh screens. Smaller solid particles are then filtered out in *filter beds* (Figure 11.2) which often consist of layers of sand and gravel. As water trickles through the beds, most of the solid particles are trapped on the sand and gravel. The filter beds are cleaned when necessary.

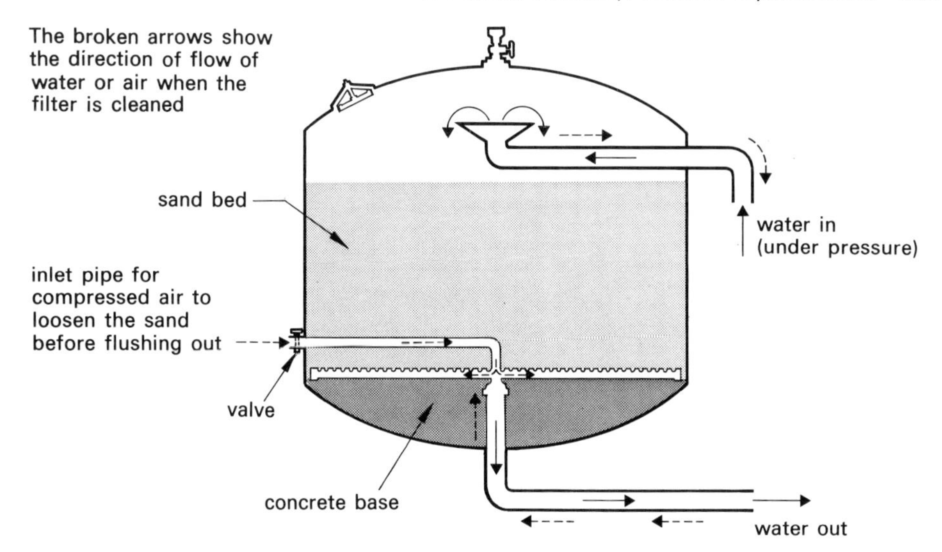

Figure 11.2 Typical pressure filter (2.7 m diameter) at a water treatment works (Output approximately 30 000 dm³ per hour)

Chlorination

Chlorine gas is poisonous, and was used in gas warfare in World War I. In very *small* quantities it is harmless to us, especially if dissolved in water, but it is still poisonous to smaller forms of life such as the bacteria which cause disease. Water which is to be used for drinking is therefore *chlorinated* at the treatment works, and this process kills bacteria in the water. Chlorine gas is passed into the water from cylinders of the compressed gas, and it dissolves in the water. By the time the water reaches our homes the smell and taste of the gas have almost disappeared, although the taste may be more obvious if the water has been more heavily chlorinated, e.g. if there is an increased risk of bacteria being present. This may happen after heavy rainfall, which may wash a variety of undesirable things into a reservoir.

Water used in swimming pools is also chlorinated, but slightly higher concentrations of chlorine are used because the water is not used for drinking and because the bacterial population in the water is likely to be greater.

In Europe, diseases such as cholera and typhoid, which can be caused by bacteria living in dirty water, have been almost stamped out by chlorinating the water we drink. (Typhoid epidemics occurred in London until the 1860s.) It is important to remember, however, that we do not chlorinate the water in rivers etc., and this means that bathing in polluted rivers and seas can still bring us into contact with dangerous bacteria which can cause problems ranging from mild stomach upsets to potentially fatal diseases such as cholera and typhoid.

Some of the processes which take place at a water treatment works are summarized in Figure 11.3. Methods used to soften water are discussed in Chapter 12.

Points for discussion

1. What does the statement 'This water has probably been through a few kidneys before mine' mean, when it is applied to a glass of water about to be drunk?

2. Why do some people object to the deliberate addition of fluoride ions to drinking water?

3. How do you think filter beds are cleaned at a water treatment works?

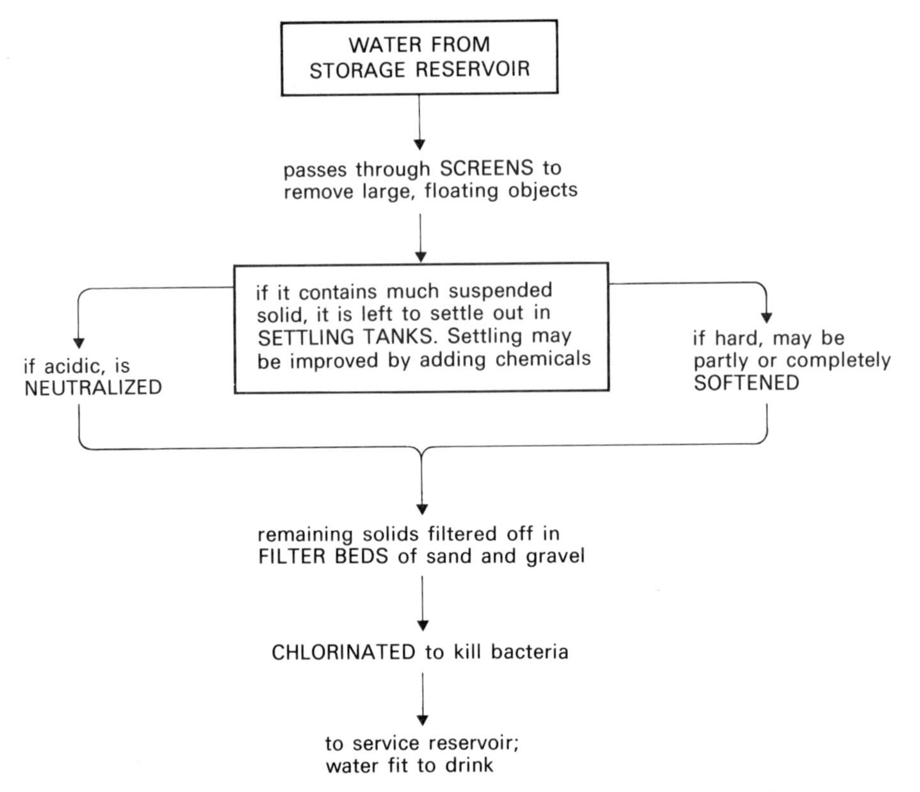

Figure 11.3 Typical sequence of operations at a water treatment works

Obtaining pure water in the laboratory

Tap water is safe to drink, but it is not really pure. It contains dissolved air, dissolved chlorine (it sometimes gives a positive test for chloride), and other chemicals such as those which cause the water to be hard. Chemicals such as these are *harmless* to health, and so they are not always removed. Some of the dissolved chemicals are *beneficial*, for example fluoride ions help to prevent tooth decay, calcium and magnesium ions help bone formation, and the small amounts of iodine which we all need are often supplied by very small traces of dissolved iodides in drinking water. However, some of these dissolved chemicals can interfere with industrial processes and chemical reactions, and so we sometimes need to take the purification of water a stage further. Drinking water is far too impure, for example, to use in medicine. A modern hospital needs large volumes of ultra-pure water for use in pathological laboratories, for making up solutions, for using in the operating theatre, and for use in kidney machines. Factories producing food, pharmaceuticals, transistors, cosmetics, and paint (to name just a few) also need very pure water.

The ions which most commonly occur in drinking water are $Na^+(aq)$, $K^+(aq)$, $Ca^{2+}(aq)$, $Mg^{2+}(aq)$, $Fe^{2+}(aq)$, $Fe^{3+}(aq)$, $CO_3^{2-}(aq)$, $HCO_3^-(aq)$, $SO_4^{2-}(aq)$, $Cl^-(aq)$, and $NO_3^-(aq)$.

The concentrations of these ions are only very low, and it is usual to express concentrations of dissolved solids as *parts per million* (ppm). *Deionized* water (see later) should have less than 1 ppm of dissolved solids.

There are two main ways in which this purification can be achieved, although the first of these is being used less extensively.

Distillation

You will remember from your study of purification techniques that usually only pure water vapour evaporates from impure water when it is heated to its boiling point. This is the same physical process as the natural one by which pure water vapour evaporates from the seas and rivers and forms clouds. If this water vapour is condensed separately from the original solution, we have distilled the water and the distillate will be pure water (distilled water). There are many varieties of distillation apparatus or stills, but on a small scale the standard Liebig condenser etc. can be used.

Distilled water is tasteless and is not so pleasant to drink as tap water, in which small quantities of dissolved air and solids produce a liquid which is slightly more pleasant for most people to drink. Distillation is rarely used for the purification of water because it is now very expensive to use any kind of fuel, and a considerable amount of fuel is needed to produce even a small volume of distilled water. For example, an ion exchange unit which can supply 400 litres of pure water before its resin is discharged weighs typically 12 kg, and yet it may need 1000 kg of equipment and fuel to make the same volume of water by distillation. Even the so-called 'distilled water' used by garages for topping up car batteries is likely to be deionized water rather than distilled water.

Deionized water

You may have a deionizing system in your laboratories (see Figure 11.4) or in your home if you live in a 'hard water' area. The essential part of the system is an *ion exchange resin*, which is usually made from an unreactive, covalent 'backbone' of atoms (e.g. a polymer such as polystyrene) on to which ionic groups are weakly attached. These ions may be positively charged such as $H^+(aq)$, or $Na^+(aq)$, negatively charged such as $OH^-(aq)$, or a mixture of both positive and negative ions. An ion exchange resin which contains only positive ions is called a cation exchange resin. Such resins can be used to soften water, but in order to completely purify water a mixed resin containing both types of ion is used. Such a resin removes all dissolved ions from water, i.e. deionizes it.

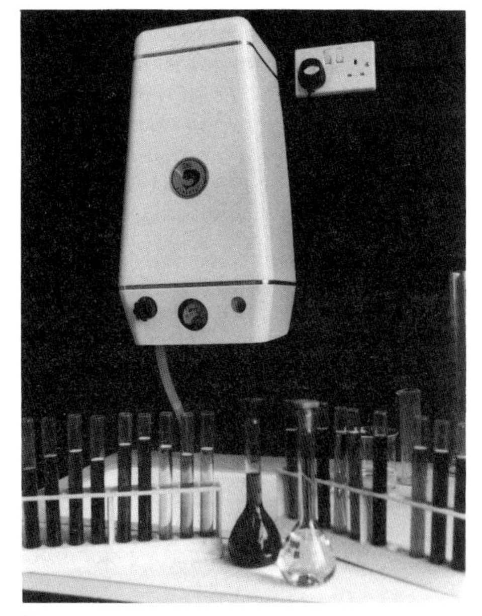

Figure 11.4 A typical laboratory deionizer (*Elga*)

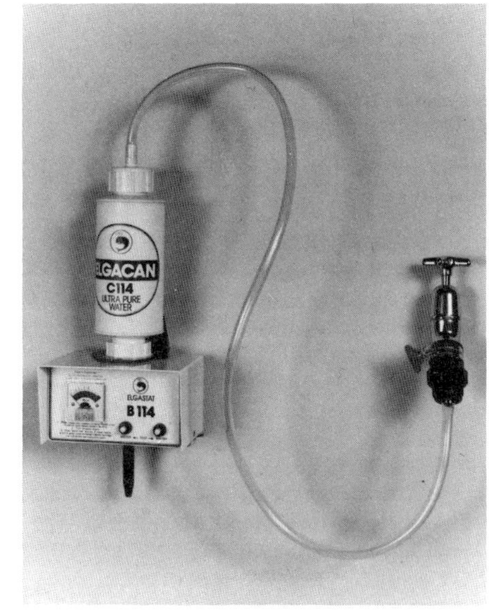

Figure 11.5 A small deionizer (*Elga*)

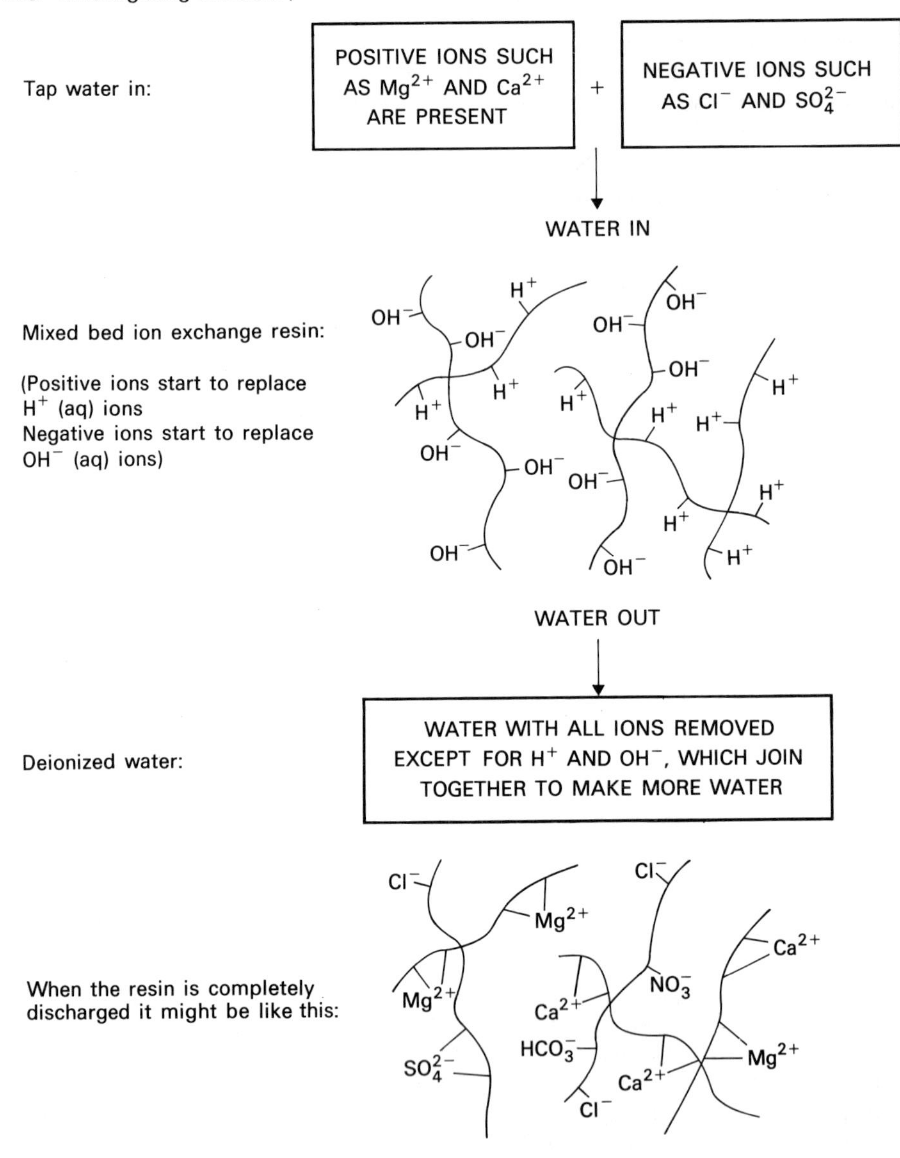

Figure 11.6 An illustration of the deionization process

When tap water passes through an ion exchange resin of this type, positive ions in the water (i.e. cations such as Ca^{2+}(aq) and Mg^{2+}(aq)) are adsorbed by (i.e. they cling to the surface of) the resin in exchange for H^+(aq) ions already on the resin. All of the negatively charged ions in the water, i.e. anions such as Cl^-(aq) and SO_4^{2-}(aq), are adsorbed on the resin in exchange for OH^-(aq) ions already on the resin. In simple terms this means that *any* positive ions are 'swapped' for H^+(aq) ions, and any negative ions are 'swapped' for OH^-(aq) ions. This is summarized in Figure 11.6. The water leaving the resin in this example thus contains H^+(aq) and OH^-(aq) ions only, and as these combine to form water

$$H^+(aq) + OH^-(aq) \rightarrow H_2O(l)$$

the water leaving the resin is pure.

Modern deionizers are usually quicker, cheaper to run, easier to use, and far more portable than the traditional still which is used to provide distilled water. In addition, they can provide water with a greater degree of purity. Portable deionizers have been used in laboratories, hospitals, etc., since the 1940s.

As most of the dissolved substances in water provide free ions in solution, they increase the electrical conductivity of water. The conductivity of a water sample is therefore a guide to its purity. This provides a convenient method of monitoring (checking) the quality of water. Most deionizers have a built-in conductivity meter which continuously monitors the quality of the water leaving the unit, and indicates when a resin is no longer producing water of the required purity (see Figures 11.4 and 11.5). A typical *small* deionizer (Figure 11.5) has a disposable can containing 0.56 litres of resin and can deionize 45 litres of water containing 200 ppm of dissolved solids, at a maximum flow rate of 15 litres per hour.

Deionized water may be slightly acidic because it contains dissolved carbon dioxide from the air, but this does not interfere with most chemical reactions. Very pure, exactly neutral water is difficult to obtain, but is produced for research purposes etc.

An ion exchange resin eventually becomes *discharged* when the useful ions on it have been replaced by unwanted ions (Figure 11.6). Such resins can sometimes be recharged by passing very concentrated solutions containing the original ions through the resin. More usually a used cartridge of resin is replaced by a recharged one from the manufacturer.

There are many natural ion exchange systems as well as the synthetic ones used in laboratories. Clay-like materials, for example, cause ion exchange reactions to occur in the soil and such reactions are vital for plant life.

11.3 WATER AS A SOLVENT

Introduction

Water has been described as 'the universal solvent' because it dissolves most of the substances familiar to us, even if only to a slight extent. For example, each time we drink water from a tumbler we are also drinking a *minute* amount of dissolved glass.

You should know that water is said to be a particularly good solvent for ionically bonded solids, but it would be quite wrong to imagine that this is true of *all* such solids. In fact, the solubility of ionic substances in water is probably overstressed, because there are a large number of ionic solids which are insoluble in water. Calcium carbonate is just one of the common ionically bonded solids which will not dissolve in water. Others include barium sulphate and silver chloride, which are met in analysis (p. 451).

It is often stressed that water is a poor solvent for covalently bonded chemicals, but again there are many common exceptions. Substances such as chlorine (Cl_2) and bromine (Br_2) have true covalent bonding but they dissolve in water, as do sugars such as glucose, $C_6H_{12}O_6$. Solubility is a complicated subject which cannot be fully explained at this level. As a general guide it is perhaps useful to think of water as a good solvent for many ionic solids, and of molecular liquids such as petrol as solvents for molecular, covalent solids. These statements, however, are not *rules* for solubility, as there are too many exceptions.

The important point is that water dissolves *many* substances it comes into contact with, and this can be useful or a problem, depending on the circumstances. The solvent properties of water are sometimes very useful or even essential (e.g. in carrying dissolved food, oxygen, and waste to appropriate parts of our bodies), but they are sometimes very inconvenient or even dangerous, e.g. the dissolved chemicals which cause water pollution.

The presence of water can be detected

The solubility of solids in water

Solubility and saturated solutions

If a soluble solid is added in small quantities at a time (with stirring) to a *fixed volume* of water at a *fixed temperature*, eventually a point is reached when no more of the solid will dissolve.

A solution which contains as much dissolved solid as it can hold when in contact with undissolved solid at a certain temperature is called a saturated solution.

It is usual to measure and compare the solubilities of different chemicals by using the following definition.

The solubility of a solid in water at a particular temperature is the mass of solid which will dissolve in 100 g of water at that temperature to form a saturated solution.

The definition can also be applied to solvents other than water. To make sure that the solution really is saturated, measurements of solubility are usually taken when there is excess undissolved solid in contact with the solution. Experiment 11.1 enables the solubility of a solid (in water) to be found.

Sometimes, in rather exceptional circumstances, it is possible to make a solution dissolve more solid than would be expected from its solubility at that temperature. Such solutions are known as *supersaturated solutions* and they are very unstable; shaking or the addition of even the slightest trace of a solid (e.g. a further crystal of the solute or even dust) will make the 'excess' dissolved solid come out of solution. A common example of a supersaturated solution is that produced by gently warming sodium thiosulphate crystals, when they dissolve in their own water of crystallization. If the solution is allowed to cool and one crystal of sodium thiosulphate is then added, a mass of crystals forms rapidly with a considerable rise in temperature.

How to use experimental results to calculate the solubility of a substance

This kind of calculation is needed to work out the results of Experiment 11.1(b), and so it is important that you understand the method of working before you do the experiment.

Suppose that the following results have been obtained in connection with the solubility of a solid, using the procedure as in Experiment 11.1(b).

Temperature of sample solution: 20 °C

Mass of evaporating basin + sample of saturated solution	= 49.35 g
Mass of evaporating basin empty	= 28.63 g
∴ Mass of sample of saturated solution	= 20.72 g

Mass of evaporating basin + solid (after evaporation) = 30.31 g
Mass of evaporating basin empty = 28.63 g
∴ Mass of dissolved solid in sample = 1.68 g

Mass of sample of solution = 20.72 g
Mass of solid in sample = 1.68 g
∴ Mass of solvent in sample = 19.04 g

1.68 g of solid dissolved in 19.04 g of water at 20 °C

$$\therefore \quad \frac{1.68}{19.04} \text{ g solid dissolve in 1 g of water at 20 °C}$$

$$\therefore \quad \frac{1.68}{19.04} \times 100 \text{ g solid dissolve in 100 g of water at 20 °C}$$

$$= 8.82 \text{ g}$$

∴ the solubility of this solid in water at 20 °C is 8.82 g per 100 g water

Experiment 11.1
To determine the solubility of a solid in water at various temperatures*

(*a*) *Simple version*

Apparatus
Five boiling tubes in rack, 10 cm³ pipette or measuring cylinder, −10 °C to 110 °C thermometer, stirring rod, access to balance. Bunsen burner, test-tube holder, spatula.
Potassium chlorate or potassium nitrate.

Procedure
(a) Place 10 cm³ (10 g) of water into each of the boiling tubes, using either the pipette or the measuring cylinder.
(b) Label the tubes 1 to 5. Weigh out 1.0 g of potassium chlorate (or potassium nitrate) and add it to tube 1. Similarly, add 2.0 g of the solid to tube 2, 3.0 g to tube 3, 4.0 g to tube 4, and 5.0 g to tube 5.
(c) Warm test-tube 1 and stir its contents until all the solid dissolves. Place the tube in the rack, place the thermometer in the solution, and allow the tube to cool. Take the temperature of the solution when the *first* sign of crystallization occurs. At this point the solution is saturated.
(d) Repeat (c) using each of the other tubes in turn, so as to obtain five readings altogether.

Results
Plot the results on a graph, using solubility values (in g per 100 g of water) on the *y* (vertical) axis, and temperature on the *x* (horizontal) axis. Label your graph fully (include the name of the solute) and fasten it in your notebook. If other groups have used different solids, compare the various graphs.

For conclusion and notes about the experiment, see page 173.

(*b*) *Alternative version*

Apparatus
250 cm³ beaker, stirring rod, spatula, −10 °C to 110 °C thermometer, Bunsen burner, tripod, gauze, evaporating basin, 100 cm³ measuring cylinder, access to balance.
Suitable solids to act as solutes, e.g. potassium nitrate, potassium chloride, potassium chlorate.

Procedure
Note: If there is insufficient time to complete the experiment as instructed, each solid can be shared between say three groups, so that each group finds its solubility at two different temperatures. This can be organized so that the three groups can share solubility values at six *different* temperatures.

Table 11.1 Setting out the results for Experiment 11.1

	First temperature (°C)	*Second temperature (°C)*	*(etc. as necessary)*
Mass of evaporating basin + sample of saturated solution (g)			
Mass of evaporating basin empty (g)	______	______	______
Mass of sample of saturated solution =	______	______	______
Mass of evaporating basin + solid (after evaporation) (g)			
Mass of evaporating basin empty	______	______	______
Mass of dissolved solid in sample (g) =	______	______	______
Mass of sample of solution (g)			
Mass of solid in sample (g)	______	______	______
Mass of solvent in sample (g) =			

(a) Make out a table for your results as shown in Table 11.1.
(b) Weigh a clean evaporating basin and record its mass in *all* of the appropriate columns in your table.
(c) Pour approximately 100 cm^3 of water into the beaker, and heat the water until the temperature is about 85 °C. (If more than three samples are to be taken from the solution, the volume should be increased from 100 cm^3 to 150 cm^3.)
(d) Remove the Bunsen burner from under the beaker. Add spatula measures of one of the given solids, stirring after each addition, until no more solid will dissolve and there is some excess solid at the bottom of the beaker.
(e) Stir the mixture again for one minute, and make sure that there is still some solid left undissolved. The solution is now saturated at this particular temperature.
(f) Place the thermometer in the solution, wait until the mercury thread has stopped moving, and note the exact temperature of the solution. Record this temperature at the top of your first column of results.
(g) Quickly but carefully pour approximately 20 cm^3 of the *clear solution* (not excess solid) into the evaporating basin. Weigh the basin and solution, and record the mass in the first column of your results table. Keep the solution left in the beaker for later stages.
(h) Carefully evaporate the liquid in the evaporating basin so as to leave the pure solid which was dissolved in it. This is the stage where you are most likely to make mistakes. Wear goggles during the evaporation, because sometimes solid 'spits out', especially if you evaporate the solution too quickly. Evaporate the last few cm^3 particularly carefully using the smallest possible flame and make sure that the solid is really dry but do not overheat. (This last stage can be achieved by heating on a water bath if there is sufficient time.)
(i) When the evaporating basin is cool, reweigh it and record the mass in the first column of your table.
(j) Wash and dry the evaporating basin. Repeat steps (d) to (i) at least once, so as to obtain values at different temperatures. It will probably be necessary to heat the solution in the beaker, stirring constantly, until its temperature is about 10–15° below the temperature used in the first set of results. Record the various results in different columns in your table.
(k) Work out the solubility of your solid at each of the temperatures used in the experiment, using the worked example on page 171 as a guide. Note that your results may include those obtained by other groups, and that all your working should be shown in your notebook.

Results

(i) Work out the various solubilities, as shown, and plot the results on a graph, using solubility values on the *y* (vertical) axis, and temperature on the *x* (horizontal) axis.
(ii) Label the graph fully (include the name of the solute) and fasten it in your notebook.
(iii) If other groups have used different solids, compare the various graphs.

Notes about the experiment

Whenever you record the solubility of a solid, it is essential to state the temperature at which the solubility is measured, because (as the experiment shows) the solubility of a solid varies with temperature.

The results of experiments of this type can be plotted on a graph, known as a *solubility curve*. Sometimes it is useful to plot several curves on one graph, as shown in Figure 11.7. Some of the information which can be obtained from solubility curves is discussed after the experiment.

Conclusion

How does the solubility of a solid generally vary with temperature? (Base your answer on all the class results.)

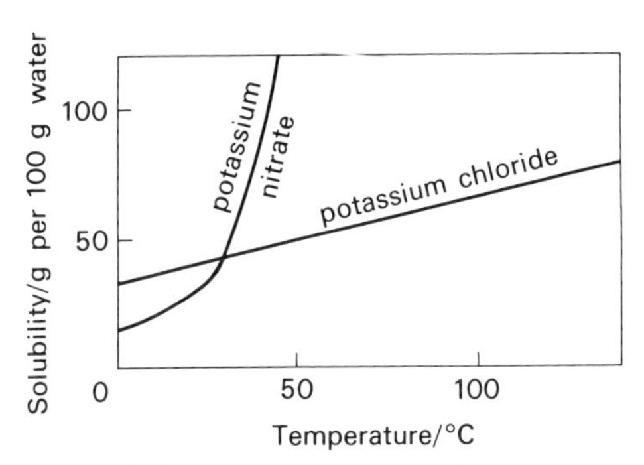

Figure 11.7 The solubility curves for potassium nitrate and potassium chloride in water

Points for discussion

1. If you start with a *hydrated* solid (page 175) in an Experiment like 11.1(b), it is particularly difficult to obtain accurate results during procedure (h). Why is this so?

2. A set of instructions for an experiment on solubility included the following statement. 'When taking the temperature of the solution, it does not matter what the exact temperature is, as long as you know the exact temperature.' Can you explain what this means?

Some uses of solubility curves and solubility data

1. It is possible to find the solubility of a solid at temperatures not actually used in the experiment. For example the readings in the experiment might be taken at 18 °C, 38 °C, 55 °C, 73 °C, and 90 °C, but from the solubility curve it is possible to deduce the solubility of the solid at *any* temperature between, say, 10 °C and 100 °C.

2. It is possible to calculate the mass of crystals which would be obtained by cooling a volume of the hot, saturated solution from one known temperature to another.

For example, the solubility curve may show that the solubility of a solid X is 23 g per 100 g of water at 100 °C, and 12 g at 50 °C. From this we can conclude that if 100 g of water saturated with X at 100 °C were cooled to 50 °C, $(23-12)\,g = 11$ g of solid X would crystallize out.

Remember that solubilities are normally measured in g per 100 g of solvent, so if 50 g of water were cooled as described above, only 5.5 g of solid X would crystallize out.

3. If the solubilities of two solids, A and B, in a mixture are known and are quite different, it is possible to separate B from A by a process called *fractional crystallization*.

Suppose, for example, that A is more soluble than B in water. It is possible to make a hot solution of the two solids so that the solution is saturated or nearly saturated with respect to B, but is not saturated with respect to A at either this temperature *or* when it will be later cooled to room temperature. When the solution is then cooled, B crystallizes out because it can no longer completely dissolve at lower temperatures, but A remains in solution even at the lower temperature because A is more soluble than B and the solution is not saturated with respect to A. B can be filtered off, washed with a little water, and dried. The remaining solution will still contain both A and B so it is not possible to obtain the more soluble salt (A) in a pure state by this simple technique. The technique is frequently used to separate one solid from a mixture or to purify a solid from soluble impurities.

Check your understanding

1. Use your own solubility curve to calculate solubility values at two temperatures different from any temperatures used in the experiment.

2. Calculate the mass of solid which would be formed if 50 g of water saturated with dissolved solid was cooled from 80 °C to 40 °C in your experiment.

3. Obtain solubility data for another solid, and plot a second solubility curve on the same axes as your first one. If you had a *mixture* of the same two solids, which one could you obtain in a pure state from the mixture by the process of fractional crystallization? Explain your answer.

The solubility of gases in water

Experiment 11.2
Is air soluble in water?***

Apparatus
Set up the apparatus as in Figure 11.8. The flask and the delivery tube must be completely filled with water and the end of the delivery tube must be level with the bottom of the bung.

Procedure
Heat the flask until the water boils. Continue the heating until there is no further change. Record any changes which take place.

Results
Describe what you saw happen, and record the volume of any gas collected.

Notes about the experiment
(i) No gases can enter the apparatus, so any gas collected must have come from the water. When water is boiled it does not decompose chemically to form gases.
(ii) You have found that the solubility of most *solids* increases with a rise in temperature. What happens to the solubility of a *gas* in water if the temperature rises? (Think carefully about this; many students get it wrong. It may help to think of what happens inside a bottle of lemonade on a hot day.)

Conclusion
Write your own conclusion, based on the title of the experiment, and also on the question above.

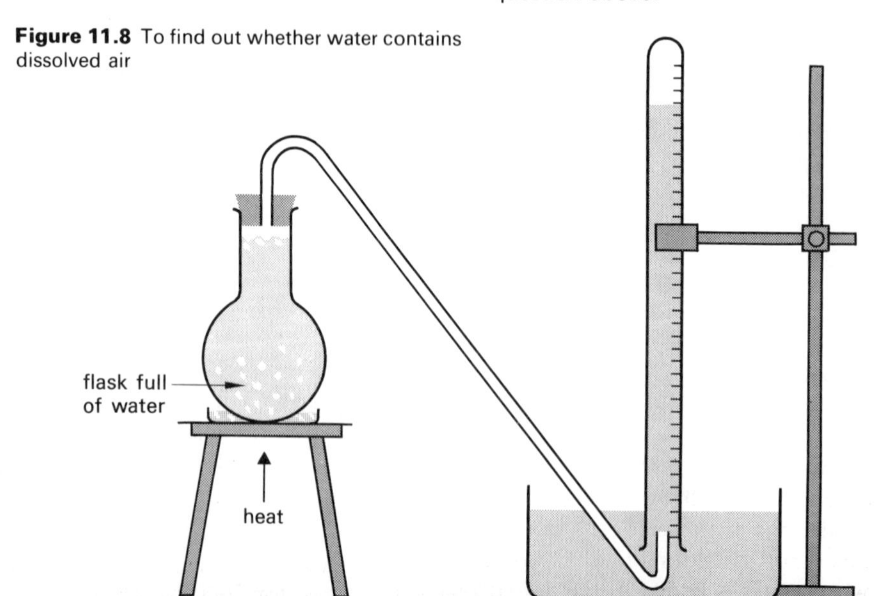

Figure 11.8 To find out whether water contains dissolved air

Some other points about the solubilities of gases in water

Some gases appear to be extremely soluble in water, even at low pressures, for example hydrogen chloride and ammonia. However, these gases actually react with the water and so are special cases. Over 80 000 cm^3 of ammonia can 'dissolve' in 100 cm^3 of water at normal atmospheric pressure, for example. Such gases can be used for the fountain experiment (p. 376) and their solutions are normally made using an inverted funnel as described on page 378.

Air is soluble in water to some extent, and this dissolved air enables fish to live in water and also helps to give water a 'taste' (p. 167). However, air is a mixture and not a compound, so dissolved air is also a mixture. The different components of air (e.g. oxygen, nitrogen, carbon dioxide) dissolve to different extents in water, so that air boiled out of water does not have the same percentage composition as ordinary air.

Pressure also affects the solubility of a gas in water. Carbon dioxide gas is not very soluble in water at ordinary pressure, but its solubility increases if the pressure is increased. Fizzy drinks contain dissolved carbon dioxide under pressure, so when the pressure is released (e.g. when the stopper is removed, or when a can is opened) some of the gas can no longer dissolve, and it 'bubbles out'.

Points for discussion

1. If 100 cm^3 of the air boiled out of water is transferred to a syringe and analysed as in Experiment 9.1 (p. 134), it is found that about 33 per cent of the gas is removed by the copper, compared with about 20 per cent of normal air. What conclusion can you make from this information?

2. What do you see happen inside a beaker of cold water if it is left in a warm room?

3. Have you heard about deep sea divers suffering from a painful and dangerous condition often called 'the bends'? How is it caused?

4. If a gas which is moderately soluble in water is being collected over water, why is warm water sometimes used in the trough?

11.4 WATER OF CRYSTALLIZATION. DELIQUESCENT, HYGROSCOPIC, AND EFFLORESCENT SUBSTANCES

Water of crystallization

Crystals of many salts contain water molecules which are chemically combined with the salt. Such salts are called *hydrated* salts, and this combined water is called *water of crystallization*. These salts can usually also exist in the *anhydrous* form, which is when they are no longer chemically combined with water molecules.

An anhydrous salt and its corresponding hydrated salt are often quite different in appearance, e.g. in colour, and size and shape of crystals. This is because the water of crystallization is an essential part of the structure of the crystals, and when this water is removed the structure of the solid (and often its colour) changes. For example, hydrated copper(II) sulphate consists of blue crystals, whereas the anhydrous salt appears as a white powder. (It is still crystalline, but the crystals are now too small to be seen with the naked eye.) Usually the hydrated form is changed to the anhydrous form by heating, and the reverse process occurs when water is added to the anhydrous salt.

'Are you sure it was only water you gave him. He's gone blue!'

$$CuSO_4.5H_2O(s) \xrightarrow{+\text{heat}} CuSO_4(s) + 5H_2O(g)$$

hydrated crystals, blue — anhydrous salt, white powder

(Note that heat energy is needed to remove water of crystallization, and heat energy is released when an anhydrous salt is hydrated.) Similarly, sodium carbonate decahydrate has large, colourless crystals whereas the monohydrate appears as a white powder.

A hydrated salt with five molecules of water of crystallization per formula unit is called a pentahydrate. Similarly there are monohydrates ($X.H_2O$), dihydrates ($X.2H_2O$), hexahydrates ($X.6H_2O$), heptahydrates ($X.7H_2O$), and decahydrates ($X.10H_2O$). It is possible for a salt to form several hydrates. It is of interest that some metals form chlorides which are hexahydrates and sulphates which are heptahydrates, e.g. $MgCl_2.6H_2O$, $MgSO_4.7H_2O$, and $ZnCl_2.6H_2O$, $ZnSO_4.7H_2O$.

Some common examples of hydrated salts, together with their anhydrous forms, are given in Table 11.2.

Note that in hydration the water is *chemically combined* with the substance as water of crystallization. A hydrated solid may be perfectly dry even though it contains water. For this reason, *dehydration* is a *chemical* reaction which removes *combined* water and is not the same as the physical process called drying (see Figure 11.9, p. 179).

Table 11.2 Some common hydrated and anhydrous salts

Name, formula, and appearance of hydrated form	*Formula and appearance of anhydrous form*
Copper(II) sulphate pentahydrate, $CuSO_4.5H_2O$, blue crystals	$CuSO_4$, white powder
Iron(II) sulphate heptahydrate, $FeSO_4.7H_2O$, green crystals	$FeSO_4$, white powder
Calcium chloride hexahydrate, $CaCl_2.6H_2O$, white crystals	$CaCl_2$, white powder
Sodium carbonate decahydrate $Na_2CO_3.10H_2O$, colourless crystals	Na_2CO_3 (rarely encountered) or $Na_2CO_3.H_2O$, white powder
Cobalt chloride hexahydrate, $CoCl_2.6H_2O$, pink crystals	$CoCl_2$, blue powder

Experiment 11.3
To find the percentage by mass of water in a hydrated salt*

Apparatus

Crucible, pipe clay triangle, Bunsen burner, tripod, glass rod, tongs, spatula, access to balance.

Hydrated magnesium sulphate.

Procedure

(a) Make out a table for your results as shown in Table 11.3.

(b) Weigh the crucible empty and again when it contains between 1 g and 2 g of hydrated magnesium sulphate. (Best results are obtained by using a fairly small mass of solid; it is difficult to heat large quantities evenly.) Record your results in the table.

(c) Heat the crucible *gently* on the pipe clay triangle. At frequent intervals, hold the crucible with tongs and stir the solid with a glass rod, taking care not to bring any of

Table 11.3 Setting out the results for Experiment 11.3

	First weighing	*Second weighing*	*Third weighing, if necessary*
Mass of crucible + hydrated salt (g)			
Mass of crucible empty (g)			
(1) Mass of hydrated salt (g) =			
Mass of crucible + anhydrous salt (g)			
Mass of crucible empty (g)			
(2) Mass of anhydrous salt (g) =			

the solid out of the crucible. If a black colour appears in the solid, you are overheating.

(d) When the solid looks the same all over and no further change appears to be taking place, remove the Bunsen and allow the crucible to cool. Reweigh the crucible and contents when cool, and enter the result in your table.

(e) Again gently heat the crucible and contents, with stirring, for a few minutes. After cooling, reweigh the crucible and contents and enter the result in your table. If the last two weighings are the same, the experimental stages are complete. If not, repeat this step until a constant mass is obtained.

Results

Complete the table, and calculate the mass of the water of crystallization in your sample, as follows.

Mass of water of crystallization = (mass of hydrated salt) − (mass of anhydrous salt) = reading 1 − reading 2.

Conclusion

The mass of the water of crystallization in a compound is often calculated as a percentage of the total mass of the hydrated salt. Per cent water of crystallization

$$= \frac{\text{mass of water of crystallization}}{\text{mass of hydrated salt}} \times 100 \text{ per cent}$$

Calculate the percentage by mass of the water of crystallization in your salt. Name the salt, and set out your calculation so that each stage is easy to follow.

More to do

1. It is possible to take the experiment a stage further by using the results to calculate a simple formula for the hydrated salt, as shown on page 311. If you understand this type of calculation, work out the formula of the salt from your results.

2. You might like to make a check on the accuracy of your conclusion to Experiment 11.3, as follows:

(i) Find out the formula of the hydrated magnesium sulphate.

(ii) Work out its formula mass (as shown on page 301), including the water of crystallization. For example, for $XSO_4 \cdot 6H_2O$, include (6 × 18) for the six molecules of water present in the formula unit.

(iii) Work out the formula mass of the water molecules *only* in one formula unit of the salt (6 × 18 in the above example).

(iv) Work out the percentage of water in a formula unit of the compound, i.e.

$$\frac{\text{formula mass of water in formula unit}}{\text{formula mass of hydrated salt}} \times 100 \text{ per cent}$$

(v) This should be the same answer as you found in your experiment. How accurate was your experimental result? Can you suggest any sources of error?

Points for discussion

1. Why is it important to heat the crystals gently in Experiment 11.3?

2. Why is it important to stir the crystals during the heating?

3. What was the purpose of procedure (e)?

Deliquescent, hygroscopic, and efflorescent substances

Some substances gain or lose water of crystallization naturally, e.g. without being heated. The next experiment provides some examples of this kind, and introduces some new terms.

Experiment 11.4
To illustrate deliquescent, efflorescent, and hygroscopic substances*

Apparatus
Spatula, three watch-glasses, balance sensitive to at least 0.01 g. Anhydrous copper(II) sulphate, sodium hydroxide flakes or pellets, clear crystals of sodium carbonate decahydrate (washing soda).

Procedure
(a) Make out a table for your results, as in Table 11.4.
(b) Place two spatula measures of anhydrous copper(II) sulphate on a watch-glass and weigh the glass and contents. Record the mass in your results table. Leave the watch-glass in the open laboratory for a few days.
(c) Repeat (b) using one spatula measure of sodium hydroxide. Use this chemical with great care, and preferably wear gloves.
(d) Repeat (b) using two spatula measures of sodium carbonate decahydrate.
(e) Reweigh all three samples after a few days and record the masses in your table.

Results
Complete your table of results. For each mass change, say whether it is an increase or a decrease.

Conclusion
Copy out the following definitions and match each one with an example from the experiment.

A deliquescent substance absorbs water vapour from the air and dissolves in the water to form a saturated solution, e.g. ______.

A hygroscopic substance absorbs moisture from the air but does not dissolve in it, e.g. ______.

An efflorescent substance loses some or all of its water of crystallization when exposed to the air, e.g. ______.

Note that Figure 11.9 summarizes the main differences between some of the terms used in this chapter.

Points for discussion
1. You may be shown some examples of deliquescent substances. Some are laboratory chemicals which gradually change to solutions if left in bottles which are opened and closed frequently over a period of time.
2. Fresh crystals of sodium carbonate decahydrate should be clear (ice-like). They have the formula $Na_2CO_3 . 10H_2O$. Crystals which have been exposed to the air

'I wasn't as heavy as this before; I must be hygroscopic'

Table 11.4 Setting out the results for Experiment 11.4

	Anhydrous copper(II) sulphate	*Sodium hydroxide*	*Hydrated sodium carbonate*
Initial appearance			
Appearance after several days			
Mass of watch-glass + solid after several days (g) =			
Mass of watch-glass + solid initially (g) =			
Change in mass (g) =			

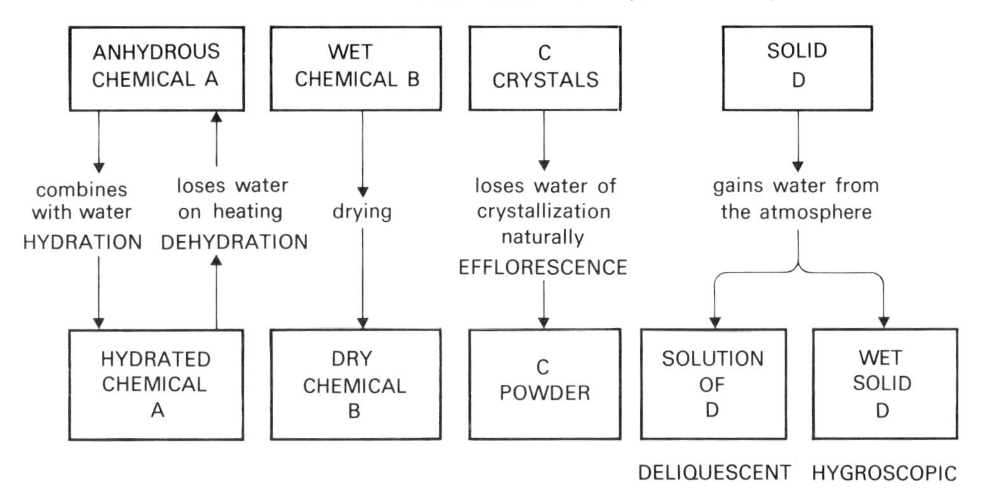

Figure 11.9 A summary of some of the terms used in the chemistry of water

become covered with a *white powder* of formula $Na_2CO_3.H_2O$. Do you think that the 'missing' water of crystallization is necessary for the crystalline appearance of sodium carbonate decahydrate? Explain your answer.

3. The hygroscopic nature of anhydrous copper(II) sulphate is used as a test for water. If anhydrous copper(II) sulphate turns blue when added to a liquid X, is it correct to conclude that liquid X is water? Explain your answer.

4. Cooking salt sometimes absorbs moisture from the atmosphere, because something in it is hygroscopic. If this happens, the salt does not flow freely. What is present in cooking salt which causes this effect? What can be added to counteract it?

5. A hygroscopic substance
A is too small to be seen with the naked eye
B dissolves in moisture from the air
C absorbs moisture from the air
D absorbs hydrogen very rapidly
E reacts rapidly with hydrogen

6. Which of the following is not true about the hydrated and anhydrous forms of copper(II) sulphate?
A One is blue and the other is white
B One is dry and the other is wet
C One is crystalline and the other looks powdered
D They have different formula masses
E Energy changes are involved in their interconversions.

11.5 A SUMMARY OF CHAPTER 11

The following 'check list' should help you to organize the work for revision.

1. Definitions

(a) A saturated solution (p. 170)
(b) The solubility of a solid (p. 170)
(c) A deliquescent substance (p. 178)
(d) A hygroscopic substance (p. 178)
(e) An efflorescent substance (p. 178)

2. Other ideas

(a) The important physical properties of water.
(b) Colour changes with anhydrous copper(II) sulphate (white → blue) or anhydrous cobalt(II) chloride (blue → pink) can be used to detect the *presence* of water, but a boiling point or freezing point test must be used to confirm that a liquid is pure water.

(c) The water cycle (page 163 and Figure 11.1).
(d) How water is processed at a water treatment works (p. 164), especially chlorination.
(e) Tap water is safe to drink, but is not pure water. Pure water is obtained by distillation and deionization. The principle of deionization (p. 167) and how the process compares with distillation.
(f) Water is a good solvent for many chemicals, be they ionically or covalently bonded. There are no rules in deciding solubility at this level, only guide lines (p. 169).
(g) What supersaturated solution is (p. 170).
(h) How to calculate the solubility of a substance from experimental results (p. 170).
(i) In general, the solubility of a *solid* in a solvent increases with temperature. For some solids, the solubility only changes slightly with temperature (e.g. potassium chloride) but for others there is a considerable change in solubility as the temperature changes (e.g. potassium nitrate).
(j) As a complete contrast, the solubility of a gas in water *decreases* with a rise in temperature, so that gas then bubbles out of solution. An increase in pressure increases the solubility of a gas in water, but has no noticeable effect on the solubility of a solid.
(k) How solubility curves can be used to calculate (i) the mass of solid which would crystallize out of a saturated solution when it is cooled to a certain temperature (or how much more would dissolve if the temperature was increased), and (ii) the solubilities at temperatures other than those used in an experiment, page 173.
(l) The principle of fractional crystallization (p. 173).
(m) Air boiled out of water is richer in oxygen than ordinary air, because oxygen is more soluble in water than is nitrogen. Similarly, air boiled out of water contains more carbon dioxide than ordinary air.
(n) The typical differences between hydrated and anhydrous salts (p. 175 and Table 11.2) and between dehydration and drying (p. 176 and Figure 11.9).
(o) How to calculate the percentage (by mass) of water of crystallization in a hydrated compound from experimental results *and* from its formula (p. 176 and p. 311).
(p) Deliquescent substances include sodium hydroxide and calcium chloride, hygroscopic substances include most anhydrous salts and copper(II) oxide, and the only common efflorescent substance is sodium carbonate decahydrate.

3. *Important experiments*

(a) An experiment to determine the solubility (and/or solubility curve) of a solid in a solvent such as water, e.g. Experiment 11.1, page 171.
(b) An experiment to show that water contains dissolved air, e.g. Experiment 11.2, page 174.
(c) An experiment to find the percentage of water of crystallization in a hydrated salt, e.g. Experiment 11.3, page 176.

QUESTIONS

Some questions relevant to this chapter are included at the end of Chapter 12.

12 Water (2): detergents, hardness of water, and water pollution

12.1 SOAPS AND SOAPLESS DETERGENTS

Soaps

Soap was invented about 2000 years ago, and until the 17th century most people made their own soap at home. They did this by boiling *animal* fats (like lard) with an alkali such as potassium carbonate obtained from wood ash. In Experiment 12.1 you will make a soap by using a *vegetable* fat such as castor oil, but the principle is the same. The home-made soap was usually an unpleasant smelling, yellow solid which was only suitable for washing clothes. Even as recently as 120 years ago, only the rich could afford to buy 'nice' soap to wash themselves. It is said that Queen Elizabeth I was 'rich enough' to be able to afford to take a bath 'every month, whether she needed it or no'. Modern soaps are much more refined. Perfume, colouring, and other substances may be added, and the product may be sold as soft soaps, hard soaps, soap flakes, or toilet soaps.

In the early part of this century most soap was made by using whale fat (blubber) as the animal fat, but many people objected to the killing of whales just for the blubber. Chemists turned to vegetable fats (oils) instead of animal fats for soap making. Typical vegetable oils include olive oil, peanut oil, maize oil, etc.

Experiment 12.1
The laboratory preparation of a soap**

Apparatus
Test-tubes, beaker, tripod, gauze, Bunsen burner, protective mat, filter funnel, filter paper, glass rod, spatula, 10 cm^3 measuring cylinder.
Castor oil, sodium hydroxide solution of concentration 5 mol dm^{-3}, sodium chloride, deionized water, perfume and colouring matter (if desired).

Procedure
(a) Mix approximately 2 cm^3 of castor oil (a compound derived from organic acids and alcohols) and about 10 cm^3 of the sodium hydroxide solution in the beaker.
Warning: the alkaline solution is much more concentrated than that which you normally use, so take great care not to spill any. Report any accident immediately.
(b) Slowly bring the contents of the beaker to the boil, stirring constantly. After the mixture has boiled gently for five minutes add about 10 cm^3 of deionized water and five or six spatula measures of sodium chloride. Stir constantly and boil mixture for a further two or three minutes.
(c) Add the perfume and colouring matter as the mixture cools, and when it is cool break up the solid pieces with a spatula. Filter or pour off the liquid and wash the solid two or three times with a *little* deionized water.
(d) When the solid is dry test its detergent qualities by scraping a little into a test-tube and shaking with a little deionized water. Are you convinced that you have made a soap? *Note*: The sample is not sufficiently pure to be applied to the skin.

Notes about the experiment

Most soaps are the sodium or potassium salts of an organic acid. To be effective as a soap, the salt must be formed from an organic acid which contains a fairly long carbon chain, e.g. stearic acid, $C_{17}H_{35}COOH$ (see below).

A typical soap formed from this acid would be sodium stearate. Vegetable oils and animal fats are often compounds (esters, see page 421) formed from a combination of long-chain organic acids like stearic acid and an alcohol called glycerol. When the fat or oil is boiled with alkali it breaks down, releasing the free organic acids and the glycerol. The acids are immediately neutralized by the alkali to form salts, which are in this case soaps.

Stearic acid

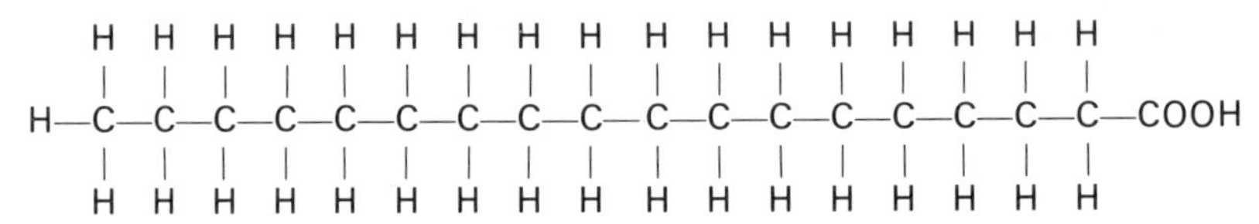

Formation of a soap

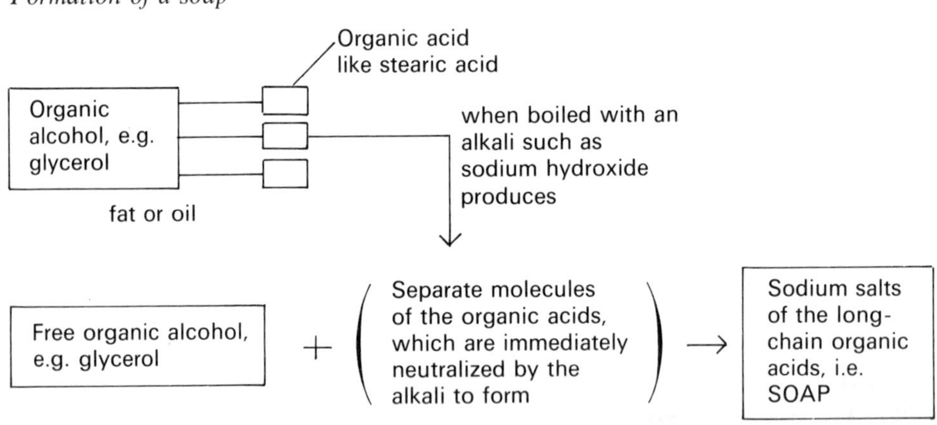

Soapless detergents

Soaps are sodium or potassium salts of organic acids such as stearic acid, palmitic acid, etc. Soaps are just one type of a group of cleaning agents called detergents. Ordinary soaps are now outnumbered in use by the *soapless* or *synthetic detergents*. Synthetic detergents were first made in the 1930s because vegetable and animal fats are potential food substances (e.g. vegetable oils are used to make margarine, see page 159) and there was not enough to make soap as well. Scientists invented a new kind of detergent which could be made from crude oil instead of using animal fat or vegetable oil.

'I want my money back, there is no soap in this'
'No Madam, it's a soapless detergent'

The invention of synthetic detergents is a very good example of the way chemists learn about the structure and properties of natural substances and then try to modify them to produce new materials even more useful to mankind. Synthetic detergents had to have the same kind of structure as traditional soaps, because it was realized that this structure was essential for them to act as cleaning agents. Synthetic detergents thus have a 'backbone' of carbon and hydrogen atoms to which is attached an ionic group, in exactly the same way that a typical soap such as sodium stearate has a long backbone of carbon and hydrogen atoms, $C_{17}H_{35}$, and an ionic group, the $-COO^-Na^+$ group.

Synthetic detergents have important advantages over ordinary soaps because, unlike soaps, they are not affected by hard water, nor are they affected by acids and alkalis. The number of different soaps which can be made is limited by the fact that only a few organic acids are found in the vegetable oils which are available in commercial quantities. On the other hand it is possible to make a very large variety of different synthetic detergents because the carbon–hydrogen skeleton can be varied at will, and also the ionic group can be one of several different kinds. Another important advantage is therefore that synthetic detergents can be 'tailor-made' for a particular purpose, e.g. dispersing an oil slick at sea. Most washing up liquids, washing powders, and shampoos are synthetic detergents.

This revolution in detergents (like that in plastics, page 424) came about at a time when crude oil was readily available in apparently unlimited quantities, and was a comparatively cheap starting material for many organic-based processes. This was a short-sighted view, and of course crude oil is now limited in supply and very expensive. In the future chemists will again have to use their skills and ingenuity to develop new starting materials for the manufacture of detergents.

How detergents work

When a detergent is added to water it dissolves as two separate ions, e.g. from an ordinary soap such as sodium stearate,

$$C_{17}H_{35}COONa(s) \rightarrow C_{17}H_{35}COO^-(aq) + Na^+(aq).$$

The large stearate ion is the actual cleansing agent, and similar large organic ions are formed when any detergent dissolves in water. The important feature of these ions is that they have a long covalently bonded 'tail' of carbon and hydrogen atoms, and also an ionic group (the 'head'). The covalent tail is 'water-hating' (*hydrophobic*) but will dissolve in grease and oil, which are covalently bonded. The ionic head will join with (or dissolve in) water molecules, and it is said to be *hydrophilic* ('water-loving') (Figure 12.1). Thus in one ion there is the ability to mix with both grease and water, and it is this type of effect which allows a detergent to clean greasy dishes etc. and to allow the grease to be 'washed away' with the water. Dirt is normally removed at the same time because it usually adheres to the grease (Figure 12.2).

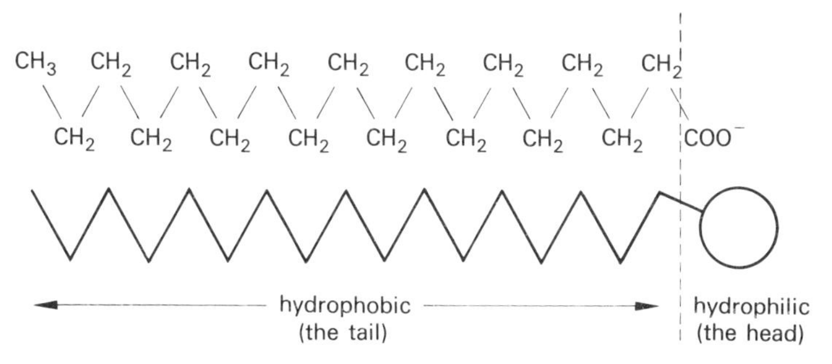

Figure 12.1 The essential features of a detergent

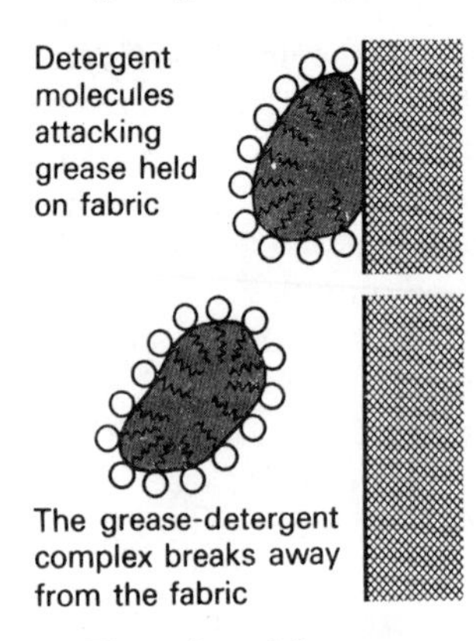

Figure 12.2 The action of detergents on grease (simplified). The ionic heads are carried away with the water and take the grease with them

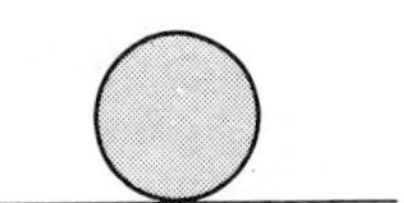

(a) Droplet of water on a surface. The surface is not 'wetted' efficiently.

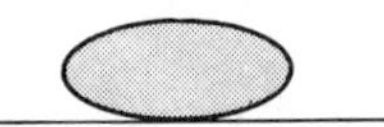

(b) The droplet spreads when a detergent is added because the water becomes a better wetting agent. Molecules of water in the droplet are dispersed because the hydrophilic head of each detergent molecule is attracted to a water molecule, and the hydrophobic tail repels other water molecules.

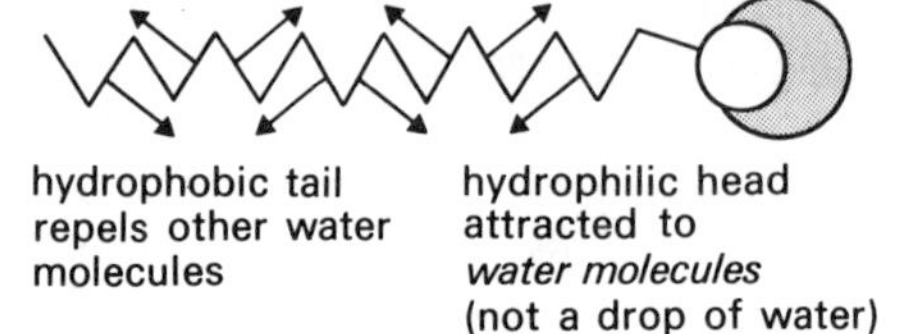

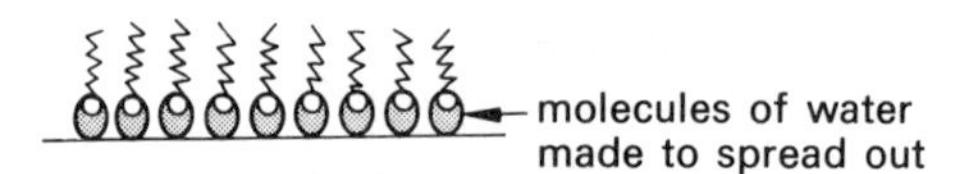

(c) This effect lowers the surface tension of water and makes it a better wetting agent.

Figure 12.3 How detergents lower the surface tension of water

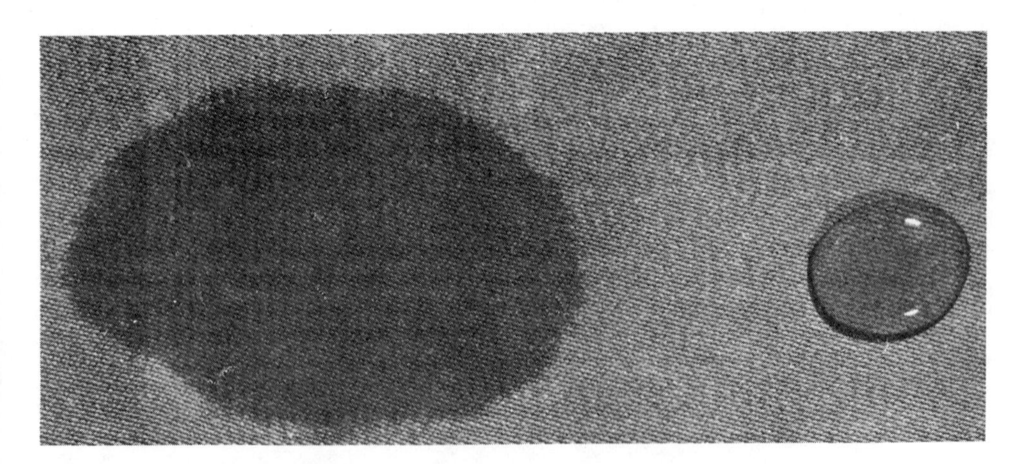

Figure 12.4 Detergents make water a better wetting agent

A detergent also cleans because of another effect. Water on its own is not a good 'wetting' agent because it does not readily spread completely over a surface or penetrate the fibres of fabrics. This is mainly due to the relatively large inter-molecular forces between water molecules which give it a high surface tension and

Figure 12.5 Surface tension of water (*Unilever*)

tend to prevent it spreading easily. A detergent lowers the surface tension of water by reducing the intermolecular forces, and makes water a better wetting agent. This happens because when a detergent is added to water, the hydrophilic groups attract water molecules but the hydrophobic groups repel them. The effect is to reduce the intermolecular forces and thus to make water molecules spread out, as shown in Figure 12.3. The effect is also shown in Figure 12.4 where a drop of water (on the right-hand side) does not spread, or penetrate the fabric, until detergent is added (on the left-hand side of the photograph). Similarly, Figure 12.5 shows a coin suspended on a water surface because of the surface tension of the water. If a drop of detergent were to be added to the water, the coin would sink because the surface tension of the water would be lowered.

12.2 HARDNESS OF WATER

Hard water

Although drinking water is normally free from harmful bacteria and dangerous dissolved chemicals, it still contains some dissolved substances. Usually the dissolved substances which remain are harmless to health and indeed some are beneficial, but certain dissolved materials can be inconvenient for other reasons.

Experiment 12.2

Do dissolved ions react with soap? *

Apparatus

Burette, 10 cm^3 pipette, funnel, conical flask with bung or cork.
Standard soap solution, deionized water, solutions of sodium chloride, potassium chloride, magnesium chloride, and calcium chloride of concentrations 0.005 mol dm^{-3} (0.005 M).

Procedure

(a) Make a table for your results as follows.

	Deionized water	Sodium chloride solution	Potassium chloride solution	Calcium chloride solution	Magnesium chloride solution
Volume of soap solution needed to make a lather (cm^3)					
Does a precipitate appear?					

(b) Pipette 10 cm^3 of deionized water into the conical flask. Pour the soap solution into the burette. Add 0.5 cm^3 of soap solution to the water, cork the flask and shake vigorously. Continue adding the soap solution 0.5 cm^3 at a time, and shake after each addition, until a lather is formed which lasts for half a minute. Record the total volume of soap solution used.

(c) Wash out the flask thoroughly and repeat (b) with 10 cm^3 of (i) sodium chloride solution, (ii) potassium chloride solution, (iii) calcium chloride solution, and (iv) magnesium chloride solution, instead of the deionized water.

Notes about the experiment

Water which does not readily form a lather with soap is called hard water.

The experiment should tell you not only which chemicals cause water to be hard, but also whether it is the metal part (cation) or the non-metal part (anion) which is responsible for the hardness.

Conclusion

When you are quite clear about what the experiment shows, make your own conclusion, including a definition of hard water, saying which ions appear to cause it, and whether hard water causes a precipitate to form when mixed with soap.

How the ions responsible for hardness get into water

In Britain calcium salts are mainly responsible for hardness of water. There are two main ways in which calcium salts get into natural water.

A calcium compound which simply dissolves

Gypsum ($CaSO_4.2H_2O$) and anhydrite ($CaSO_4$) are two forms of calcium sulphate which occur naturally in Britain. These minerals are only slightly soluble in water, but streams flowing over them dissolve enough of the mineral to make the water hard. You will learn later that dissolved calcium sulphate causes so-called 'permanently hard water', whereas dissolved calcium hydrogencarbonate causes so-called 'temporarily hard water'.

A calcium compound which reacts and then dissolves

The following experiment illustrates how this occurs. There is no need to write up the experiment; it simply illustrates a fact.

Experiment 12.3
The action of water containing carbon dioxide on calcium carbonate*

Apparatus

Small conical flask, spatula, glass rod. Carbon dioxide generator with delivery tube, finely divided calcium carbonate.

Procedure

Pour about 30 cm^3 of water into the conical flask. Add a *very small* amount of calcium carbonate and stir to obtain a fine suspension. (The amount of solid added should be just enough to make a *faint* suspension.) Pass carbon dioxide through the suspension until no further change takes place.

Notes about the experiment

You should have seen how the normally insoluble calcium carbonate gradually 'disappeared' when carbon dioxide was passed into the suspension. This is because a chemical reaction takes place between the calcium carbonate, water, and carbon dioxide:

$$CaCO_3(s) + H_2O(l) + CO_2(aq) \rightarrow Ca(HCO_3)_2(aq).$$

insoluble *soluble*

The calcium hydrogencarbonate is *soluble* and so it dissolves. The reaction occurs in nature whenever rain water (which contains carbon dioxide dissolved from the air) passes over limestone or chalk rock (calcium carbonate). The dissolved calcium hydrogencarbonate is carried away in the streams and rivers, and causes the water to be hard.

The ions which cause water to be hard

Iron and magnesium ions from soluble or sparingly soluble iron and magnesium salts also make hard water, and they are also found dissolved in water. Calcium ions, however, are normally found in higher concentrations than those of magnesium and iron. Note that only *dissolved* calcium (and magnesium and iron) compounds can make water hard. The factors responsible for the hardness are the calcium (and magnesium and iron) *ions*. The anion (e.g. sulphate or hydrogencarbonate) is irrelevant as far as hardness is concerned.

Figure 12.6 Stalactites and stalagmites in limestone caves

The reaction of water containing dissolved carbon dioxide on calcium carbonate has other important effects. Caves are often to be found in limestone areas because of the 'dissolving' of the mineral by the reaction with rain water containing dissolved carbon dioxide. These caves often contain stalagmites and stalactites (Figure 12.6), which are formed as another consequence of the same reaction.

The slow reaction between rain water and calcium carbonate produces a dilute solution of calcium hydrogencarbonate. This reaction is reversible, i.e. it can proceed 'the other way round' and thus form calcium carbonate, water, and carbon dioxide again.

$$Ca(HCO_3)_2(aq) \rightarrow CaCO_3(s) + CO_2(aq) + H_2O(l).$$

This reverse reaction readily occurs if a solution of calcium hydrogencarbonate is allowed to evaporate. As drops of the solution hang from the roof of a cave, water evaporates and the original reaction is reversed, resulting in the formation of a deposit of calcium carbonate on the roof of the cave. During the course of time this builds up and grows down from the roof of the cave so that stalactites are produced. A similar reaction occurs on the floor of the cave after water has dropped from the roof, so that a stalagmite often grows up from the floor of a cave directly below a stalactite, and they may eventually join.

Why hard water is inconvenient

Hard water wastes soap

You should have noticed in Experiment 12.2 that when soap was added to solutions of magnesium or calcium salts, a precipitate or 'scum' appeared before the solution could form a lather with the soap. You should also remember that the solutions of calcium and magnesium salts needed more soap than the other solutions in order to form a lather. These two facts (the appearance of a precipitate and the reluctance to form a lather) are connected.

The formula of a typical soap was given on page 182, and it was explained that when such a soap dissolves in water it forms large negatively charged ions (e.g. stearate ions) and small, positively charged ions, e.g. sodium ions. The stearate ion is the factor responsible for the cleaning action of the soap.

Although sodium stearate and potassium stearate are soluble in water, the stearates of calcium, magnesium, and iron are insoluble. If a soap dissolves in hard water, the stearate (or similar) ions from the soap join with the (e.g.) calcium ions from the hard water and form a precipitate (the scum) of calcium stearate. This is a very similar type of reaction to those discussed on pp. 89–90, and in Experiment 6.6. It is like saying that sodium carbonate is soluble in water and calcium carbonate is insoluble, so that if sodium carbonate solution is added to calcium chloride solution (or any solution containing calcium ions), there will be a precipitate of calcium carbonate.

$$\underset{\left(\text{calcium ions from hard water}\right)}{Ca^{2+}(aq)} + \underset{\left(\text{stearate ions from dissolved soap}\right)}{\text{stearate}^{-}(aq)} \rightarrow \underset{\left(\text{precipitate of insoluble calcium stearate}\right)}{\text{Ca stearate(s)}}$$

This means, of course, that soap cannot at first function as a cleaning agent in hard water; free stearate ions are needed for it to be active as a cleaning agent. Only when *all* of the dissolved calcium ions in the water have *reacted* with the soap to form a scum will the remaining soap be able to form free stearate ions and behave as a cleaning agent. Hard water thus causes an expensive waste of soap. (Soapless detergents, e.g. washing up liquids, are not affected by hard water because the calcium salts of their negative ions are soluble in water.)

Hard water causes an unpleasant scum to appear

The precipitate formed when soap is used in hard water is seen as an unpleasant scum which is difficult to remove from sinks, baths, and fabrics.

Hard water can cause pipes to be blocked

As explained on page 188, the natural reaction which produces a solution of calcium hydrogencarbonate can be reversed, and this is particularly obvious if such a solution is boiled. When calcium hydrogencarbonate solution (one type of hard water) is heated, a reaction therefore takes place which produces a deposit of insoluble calcium carbonate. A much slower natural version of this is responsible for the formation of stalactites and stalagmites, but when hard water is boiled the formation of calcium carbonate occurs comparatively rapidly. This builds up inside kettles as 'kettle fur' in hard water areas, and is inconvenient, but the problem is far worse on an industrial scale where millions of litres of water may be used for heating purposes. Pipes and boilers which contain hot, hard water thus become coated with increasingly thick layers of calcium carbonate. At best their efficiency is reduced, and at worst they could become completely blocked with consequent risk of explosion.

Points for discussion

1. Why does soap not form a lather when first added to a sample of hard water? Why does it form a lather eventually?

2. Why is it better to use soap in a fixed volume of hard water, say in a sink or bowl, rather than in running hard water?

3. What is an enzyme detergent, and what advantages do they have over ordinary detergents?

Softening water

Temporarily hard and permanently hard water

It is obviously important to remove calcium and magnesium ions from water for a variety of reasons. This process is called *softening*, and water which contains only a small proportion of calcium and magnesium ions is called soft water. Before softening methods are examined in detail, it is important to understand the difference between water which is described as 'temporarily' hard and water which is described as 'permanently' hard. These terms are in fact very misleading, for permanent hardness is far from permanent, and is quite easily softened. The next experiment illustrates the meaning of the terms; there is no need to write up the experiment.

Experiment 12.4
To distinguish between temporarily hard water and permanently hard water*

Apparatus

10 cm^3 pipette, conical flask, burette, Bunsen burner, protective mat, three beakers.

Soap solution, calcium sulphate solution (saturated solution diluted by water in the ratio 1:2), calcium hydrogencarbonate solution (made as in Experiment 12.3 but diluted by an equal volume of water), a mixture of the calcium sulphate and calcium hydrogencarbonate solutions (1:1).

Procedure

(a) Make out a table for your results as follows.

	Calcium sulphate solution	*Calcium hydrogencarbonate solution*	*Mixture of the two solutions*
Volume of soap solution needed to form a lather with untreated sample (cm^3)			
Volume of soap solution needed to form a lather with boiled sample (cm^3)			

(b) Arrange for bulk samples of the calcium salt solutions to be boiled for ten minutes and then allow them to cool. This can be shared between groups.
(c) Pipette 10.0 cm^3 of the original (unboiled) solution of calcium sulphate into a conical flask. Pour the soap solution into the burette. Run the soap solution slowly into the conical flask, with vigorous shaking, until a lather is formed which lasts for half a minute. Record the volume of soap solution used.
(d) Repeat (b) using 10.0 cm^3 of each of the other two unboiled solutions and then with 10.0 cm^3 of each of the cooled, boiled samples.

Notes about the experiment

(i) *Temporarily hard water is hard water which is softened by boiling.*
Permanently hard water is not softened by boiling, but it can be softened by other methods.
(ii) Your experimental results should convince you that one of the two chemicals normally responsible for hardness can be changed by boiling (i.e. it makes water temporarily hard), whereas the other is not (and makes the water permanently hard). On page 188 it was explained that if calcium hydrogencarbonate solution is boiled or allowed to evaporate, the natural reaction which formed the solution in the first place is reversed and the soluble calcium hydrogencarbonate is converted back to insoluble calcium carbonate. Temporarily hard water (calcium hydrogencarbonate solution) is thus softened by boiling because the $Ca^{2+}(aq)$ ions which cause the hardness are removed from the solution to form an insoluble compound, calcium carbonate. Permanently hard water (calcium sulphate solution) is unaffected by boiling; there is no chemical change when it is boiled.
(iii) Note that the mixture used in the experiment was a mixture of temporarily hard water and permanently hard water. It is possible for such a mixture to occur naturally and, as you should have found from the experiment, a mixture like this is *partially* softened by boiling (i.e. the permanent hardness within it is still there after boiling.)
(iv) Make sure that your notes include definitions of temporarily and permanently hard water, the names of the chemicals which cause them, and how temporarily hard water is softened by boiling (with an equation).

'Permanently' hard water	*'Temporarily' hard water*
Cannot be softened by boiling, although it can be softened by other methods. Usually contains dissolved calcium or magnesium sulphate only.	Can be softened by boiling. It usually contains dissolved calcium hydrogencarbonate. On boiling, insoluble calcium carbonate is formed and thus $Ca^{2+}(aq)$ ions are removed from solution in the water, so that it is softened.

Check your understanding

1. The table at the top of p. 191 gives information concerning tests done on samples of water from the same tap.
(a) What can you deduce concerning the tap water?
(b) Why is a scum formed with the unboiled tap water?

Sample	*Volume of sample* (cm^3)	*Effect of adding soap flakes and shaking the sample*
Tap water	10.0	After adding a few flakes, a white scum forms; more scum forms as more flakes are added until after 28 flakes a permanent lather is formed.
Tap water which has been boiled for five minutes	10.0	Only 7 flakes required to get a permanent lather

(c) What would have been observed if a soapless detergent had been added to each sample instead of soap flakes?
(d) What are the health benefits from drinking this tap water?
(e) Give the equation for the reaction which occurs when this water is used in a boiler, and mention one problem that is caused by this. (A.E.B. 1979)

2. Samples of water were taken from three towns, P, Q, and L. 10 cm^3 portions of each were titrated with soap solution in the usual way. This was repeated with 10 cm^3 of boiled water from each town. 10 cm^3 of deionized water was used as a reference test. The results are shown in the table.

Type of water	*Volume of soap solution needed to produce a lather with 10* cm^3 *of untreated water* (cm^3)	*Volume of soap solution needed to produce a lather with 10* cm^3 *of boiled water* (cm^3)
Deionized	1	1
P	10	10
Q	15	10
L	20	5

What can you conclude about the water from each of the towns? Over which kind of rocks do you think the water flowed in the catchment area for each town? Explain your answers.

3. Why is sea water not suitable for washing in? Is it hard?

Methods of softening water

As only *dissolved* $Ca^{2+}(aq)$ and/or $Mg^{2+}(aq)$ ions normally cause hardness, water can be softened by any method which either *removes* these ions completely from the water, or which *converts* the ions into insoluble compounds. The main methods are summarized in Table 12.1. Note that methods 1, 2, and 3 involve chemical reactions which convert dissolved $Ca^{2+}(aq)$ ions into *insoluble* calcium compounds. Methods 4 and 5 (in effect) *remove* $Ca^{2+}(aq)$ ions completely from the water. The principle of an ion exchange resin has been explained on page 167. To remove hardness only, a cation exchange resin may be used (i.e. one which 'swaps' only positive ions such as $Ca^{2+}(aq)$), because the anions do not cause hardness. Compare this carefully with the principle used to deionize water completely (p. 168).

Experiment 12.5
To illustrate some softening methods other than boiling*

Apparatus

Test-tubes and rack, 10 cm^3 pipette, burette, 25 cm^3 measuring cylinder, conical flask, 100 cm^3 beaker, stirring rod.

Soap solution, suitable ion exchange resin (e.g. in a burette), temporarily hard water (dilute solution of calcium hydrogencarbonate prepared as in Experiment 12.3), permanently hard water (saturated solution of calcium sulphate diluted by water in the ratio 1:2), concentrated sodium carbonate solution.

Table 12.1 Methods of softening water

Method	*Type of hardness removed*	*How hardness is removed*	*Any other points*
1. Boiling	Temporary hardness only	Ca^{2+}(aq) converted into insoluble calcium carbonate (learn the equation)	Totally unsuitable on a large scale, as a great deal of energy is required, and thus uneconomic to use. Solid formed can be inconvenient
2. Addition of calcium hydroxide	Temporary hardness only	$Ca(OH)_2 + Ca(HCO_3)_2 \downarrow 2CaCO_3 + 2H_2O$ Dissolved Ca^{2+}(aq) converted into insoluble calcium carbonate	Known as Clark's method. Cheap, used on large scale at water treatment plants. Only a calculated amount must be added; excess only makes the water hard again by addition of Ca^{2+}(aq) ions
3. Addition of sodium carbonate	Both types	Dissolved ions converted into insoluble carbonates and precipitated, e.g. $Ca^{2+}(aq) + CO_3^{2-}(aq) \downarrow CaCO_3(s)$	Very convenient. On a large scale it is more economic to use (2) to remove temporary hardness and then this method to complete the softening
4. Ion exchange	Both types	Water flows through cation exchange resin. Sodium, potassium, or hydrogen ions on resin exchange with calcium or magnesium ions in hard water; ions responsible for hardness entirely removed	Can be used on a small scale (e.g. as Permutit) in appliances which are fitted to taps. Also used to 'deionize' water, in which case anions are also removed from the water
5. Phosphate treatment	Both types	Complex polyphosphates (e.g. Calgon) 'lock up' calcium and/or magnesium ions by forming a complex structure with them	Has additional advantage of preventing corrosion of pipes carrying water. Calgon is a component of most washing up liquids

Procedure

(a) Note that you will be using only either the permanently hard water or the temporarily hard water in your own experiments, but you will need the results for the other types of hardness from other groups. Organize your work accordingly.

(b) Make out a table for your results as shown in Table 12.2.

(c) Pipette 10.0 cm³ of the type of hard water which you are using into a clean conical flask. Determine the volume of soap solution needed to form a permanent lather by adding soap solution from a burette, a little at a time, and shaking after each addition. Record the result in the table.

(d) Pour about 15 cm³ of the type of hard water you are using through an ion exchange column, and collect the liquid in a beaker. Pipette 10.0 cm³ of this sample into a clean conical flask and determine the

Table 12.2 Setting out the results for Experiment 12.5

	Permanently hard water	*Temporarily hard water*
Volume of soap solution needed to form a lather with 10.0 cm³ solution, after		
(a) no treatment		
(b) treating with sodium carbonate solution		
(c) passing through an ion exchange resin		

volume of soap solution now needed to form a permanent lather. Record your result in the table.

(e) Pour about 15 cm^3 of your hard water solution into a clean beaker. Add about 2 cm^3 of sodium carbonate solution, and stir. Allow any precipitate to settle, and then pipette 10.0 cm^3 of the mixture in the beaker into a clean conical flask, determine the volume of soap solution needed to form a permanent lather, and again enter the result in your table.

(f) Fill in the results of procedures (c) to (e) for the other kind of hard water, as obtained by another group.

Notes about the experiment

If you compare the results with the information given in Table 12.1, you should be able to explain your results. Note that sodium carbonate is commonly called washing soda; you should now know why. Procedure (e) involves a slight error, but this should not hide the point of the experiment.

Points for discussion

Consider the experimental results below.

(a) Can you explain these results (see Table 12.1)?

(b) Was the hard water temporarily hard or permanently hard?

Sample	*Volume of soap solution needed to form a lather*
10.0 cm^3 of hard water	7.5 cm^3
10.0 cm^3 of the same hard water + 0.5 cm^3 of calcium hydroxide solution	4.0 cm^3
10.0 cm^3 of the same hard water + 1.0 cm^3 of calcium hydroxide solution	6.0 cm^3
10.0 cm^3 of the same hard water + 2.0 cm^3 of calcium hydroxide solution	11.0 cm^3

12.3 WATER POLLUTION

The nature of the problem

Water is often called the universal solvent because it dissolves most common substances, even if only to a small extent. We have already seen how this property can be inconvenient in the formation of hard water but a much more serious aspect is the way that water dissolves many toxic materials which are in the waste products of modern society. In one sense it is convenient that we can use water to remove much of the waste from industrial reactions, and that we also remove some of the waste products from our bodies in aqueous solution. However, the consequences of using water in this way are potentially very serious. It was realized many years ago that human waste should be treated before it is returned to rivers etc., and Britain has a very well developed system of sewage works, particularly in inland areas. (As explained later, however, a sewage works does not solve all the problems connected with domestic waste.) We also have a reasonably good record of disposing of sewage at sea, although beaches in some parts of Europe are bathed by heavily polluted water.

Unfortunately, in the past we have taken a rather short-sighted view of the consequences of the removal of industrial waste. Water has always seemed cheap and plentiful, and to wash waste into rivers seemed a simple and convenient way of removing it; it became somebody else's problem! To illustrate the scale of the problem, it has been estimated that at one time 3 million kg of chemical waste entered just one part of Lake Erie (on the border between Canada and the USA) in a single day. The removal of industrial waste has resulted in heavy pollution of some of our major rivers, and one or two 'scares' about the quality of drinking water. In addition, an evil-smelling open sewer (which is how many rivers could be described in the not too distant past) is socially undesirable.

'We have often taken a short-sighted view of the consequences of the removal of industrial waste'

We deliberately use water to remove waste from industrial reactions, and we must be prepared to deal with (and pay for) any problems caused by such actions. Just as important, however, are the *indirect* ways by which pollutants enter water courses because of our way of life. For example, chemicals which are essential in modern agriculture, such as herbicides (weed killers), insecticides, and fertilizers, can be washed into rivers and streams. Similarly, accidental leaks of pollutants occur in industry, sometimes because of the enormous volumes of water used for cooling and heating. In the past, accidental pollution of this kind was difficult to detect until it caused a problem, but now Water Authorities are very experienced in looking for a wide range of possible pollutants, and rivers are constantly monitored for pollution. Figure 12.7 illustrates another example of an accidental type of pollution!

The main aspects of water pollution are considered under the following headings.

Direct contamination of drinking water

When water evaporates and forms clouds, the rain which may be produced from the clouds is in effect like distilled water, and any metal ions etc. which may be present in sea water will not be found in rain when it is first formed. If pollutants do enter drinking water, they must therefore do so between the formation of rain and its arrival as drinking water at a tap.

Most drinking water is obtained from reservoirs which are situated in areas away from residential and industrial developments, and so there is little risk of pollution in these cases. However, the demand for pure water is now so great that some of the larger rivers in Britain are used as sources of drinking water. Although the water is treated as on page 164, *dissolved* substances in the water are not normally removed at a water treatment works. It is therefore important that water which is to be treated for drinking should not be taken from a river which contains toxic chemicals.

If certain chemicals enter drinking water supplies they can cause serious medical problems. This is particularly important if the substances are *cumulative*, (i.e. can be excreted only very slowly) such as the ions of heavy metals like lead, cadmium, and mercury. Remember that it is also possible for toxic materials dumped on *land* to be washed into rivers. (In 1964 people living in the Californian city of Montebello noticed that their drinking water had a strange taste and smell. This city obtained its water supply from underground wells. It was found that a small factory had dumped a batch of herbicide into a ditch, and that this had drained underground to a well. The water still smelled of the herbicide 5 years later.) Approved tipping sites are owned by Local Authorities in Britain, and they may grant licences to factories for the disposal of solid waste (e.g. in sealed drums). Such sites are carefully chosen to reduce the risk of water draining from the site entering rivers before it has been sufficiently diluted and before it has been given the chance to be affected by

Figure 12.7 Open-cast butter mining! (*Severn Trent Water Authority*) The aftermath of a fire in a butter warehouse. Large quantities of butter melted and entered water courses. The photographs show (a) testing the effect of the butter on the water and (b) removing some of the solidified butter from a channel

biological and chemical processes occurring in the soil, many of which reduce toxicity. Generally, people are taking a much more responsible attitude to the disposal of waste than they were a few years ago.

Drinking water in Britain is monitored very carefully. The World Health Organization has recommended 'limits of tolerance' for many chemicals which could be found in drinking water, such as 0.1 mg dm^{-3} for lead, 0.05 mg dm^{-3} for copper, and 0.01 mg dm^{-3} for cadmium. Even better standards are required in some countries, and the standards are likely to be further improved. The only problem so far has been due to dissolved nitrates.

Nitrates can enter rivers by run-off (drainage) from agricultural land on which fertilizers have been applied, but more important sources are treated sewage and some industrial effluents. Nitrates are serious pollutants for two reasons. They increase the biological oxygen demand of the water (see later), and they are changed into nitrite ions, NO_2^-, in the body. This latter change is particularly likely to happen in young children. Nitrite ions are toxic; they interfere with the ability of the haemoglobin in the blood to carry oxygen around the body. Babies suffering from an excess intake of nitrates have been known to go 'blue' due to lack of oxygen in their blood. The maximum concentration of nitrate ions in drinking water, as recommended by the World Health Organization, is 10 ppm (parts per million), and in certain parts of Britain this has been exceeded.

The accumulation of toxic substances in food

In the past people imagined that the volume of water in the oceans was so great that they would provide a virtually limitless 'sink' for our waste material. We now realize that this is simply not the case. Although it is true that the water which *evaporates* from the sea is pure, there are other important considerations. For example, the careless dumping of human sewage at sea has made bathing unpleasant and dangerous in certain areas. Perhaps more serious is the fact that we tend to forget that the seas are an important source of food.

Fish, shellfish, etc., take an extremely large volume of sea water into their bodies every day. If this water is polluted there is a risk that these fish etc. will trap some of the pollutants in their bodies. It is thus possible to have a low and apparently safe concentration of ions such as $Hg^+(aq)$ in sea water (due to the discharge of industrial waste) but for fish to concentrate such chemicals in their bodies. If we then eat the fish, we also absorb the cumulative poisons.

This type of natural concentration of toxic material has produced several tragedies and several 'scares'. It is particularly likely to happen in estuaries where polluted river water reaches the sea, and where sea life may be exposed to comparatively high concentrations of pollutants before they are diluted by the ocean. Serious medical problems occurred in the fishing community of Minamata in Japan where mercury levels in the water were quite high because of a permitted discharge from a factory. Similarly, levels of up to 40 parts per million of arsenic have been found in shellfish caught off Britain, and cadmium is found in salmon and in fish pastes, although at present this is well within the required standards.

Similar problems occur with other substances which are concentrated in a *food chain*, i.e. which gradually accumulate in different animals and plants as they use 'contaminated' food. In this way, very low concentrations of toxic substances in sea water are gradually concentrated in the cells of animals and plants at each stage in the chain to such an extent that levels can be potentially dangerous in the food eaten by man. Another important example of this kind of pollution was caused by the widespread use of DDT as an insecticide. This particular substance is *persistent*, i.e. it is not decomposed naturally in living cells nor by reactions occurring in soil, air, or water. Concentrations of the chemical built up, first in plants and then in the animals and humans which ate them. Very strict controls are now in operation to check the level of toxic substances in food and to control the output of toxic industrial waste, and insecticides like DDT have been replaced by less persistent varieties.

The effect of pollution on the dissolved oxygen in water

The dissolved oxygen in the water of rivers, lakes, and seas is essential for the life of the animals and plants which live in the water. Water with a relatively high level of dissolved oxygen can support many forms of life, including active forms with a comparatively high consumption of oxygen, such as fish. Water which contains very little dissolved oxygen can only support the life of leeches, worms, and so on. The type of life which can be found in a particular stretch of water is therefore a rough guide to the level of dissolved oxygen within it.

'No wonder he brings his own oxygen!'

The dissolved oxygen in water is also important in another way. River water normally has a natural way of destroying many of the toxic chemicals which enter it. This happens because there is usually enough oxygen dissolved in the water to allow relatively harmless bacteria (most of which are aerobic, i.e. they need oxygen to live) to oxidize many of the harmful chemicals into harmless waste. We make use of this kind of process in a controlled way at a sewage works, as explained on page 201. These bacteria are also important in oxidizing (i.e., in this case, causing the decay of) the large quantities of dead plant material which are produced as plants which live in water die naturally.

Some waste material has a particularly high *biological oxygen demand* (BOD), i.e. it requires a comparatively large amount of oxygen for its natural oxidation in river water etc. Dead plant material, for example, has a high BOD and some rivers give up nearly all of their dissolved oxygen in order to allow the natural decay (oxidation) of the dead plant material to take place. Other organic-based substances also have a high BOD, e.g. the soluble carbohydrates discharged in the effluent from sugar refining, breweries, paper making, and milk processing.

It is obvious that any substance which has a high BOD, or which results in the formation of something with a high BOD, is a potential pollutant of water. Such substances cause a decrease in the concentration of dissolved oxygen in the water, and this causes serious problems. Nitrogen compounds (especially nitrates) are particularly important pollutants because they give rise to a high BOD in two different ways.

Some nitrogen compounds are converted by natural processes into other nitrogen-containing compounds (e.g. as in the nitrogen cycle, page 383) and many of these reactions also require oxygen. Thus the sequence $NH_4^+ \rightarrow NO_2^- \rightarrow NO_3^-$ involves an oxidation at each step, and ammonium ions therefore also have a high BOD. In addition, nitrogen compounds are important plant foods (p. 390) and provide a 'soup of plant nutrients' in which aquatic plants grow and reproduce rapidly. This in turn leads to increased amounts of dead plant material and an increased BOD for its decay. A river choked with weeds is also an unpleasant sight.

If any water is 'overloaded' with waste materials, the dissolved oxygen is quickly used up as the bacteria oxidize the waste, and this results in important changes in the river. The lack of dissolved oxygen means that certain forms of life (e.g. fish) cannot be supported, and the harmless aerobic bacteria can no longer 'feed' on the waste so that the river soon becomes heavily polluted. The aerobic bacteria disappear and are replaced by anaerobic bacteria which are capable of living in water which contains hardly any dissolved oxygen. Many of these anaerobic bacteria are harmful to man (some are responsible for serious diseases such as cholera and typhoid) and they do

Table 12.3 How dissolved oxygen is removed from river water

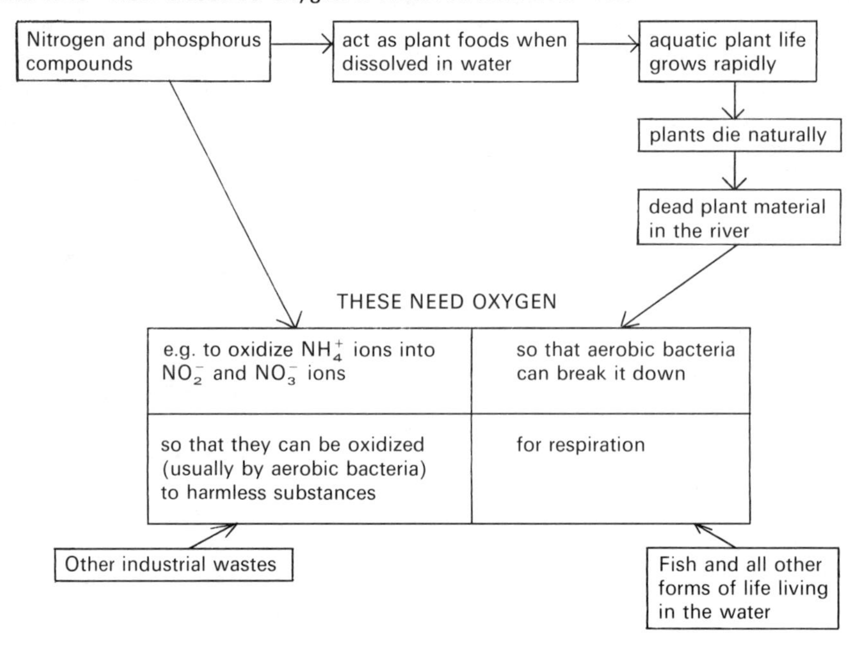

What happens when the dissolved oxygen in the water is used up

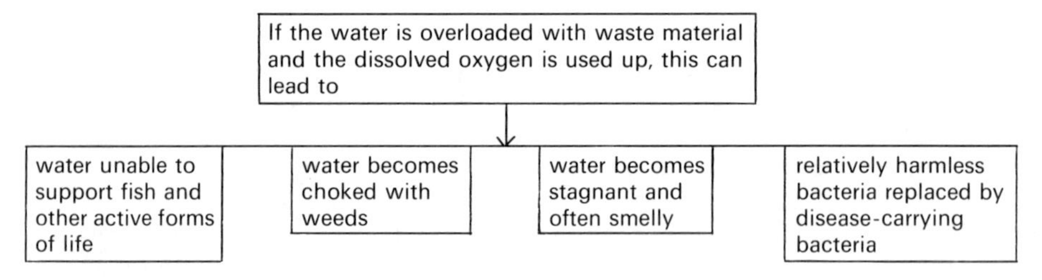

not help to destroy waste. In Britain and in most parts of the world it is not possible to contract diseases such as cholera by drinking water supplied through a tap, because such water is chlorinated. It is possible, however, to catch the disease by bathing in polluted water. These ideas are summarized in Table 12.3.

Great improvements have been made with regard to the pollution of rivers, and many stretches of water which could not support fish a few years ago (e.g. the lower reaches of the Thames) can now do so in ever increasing numbers. The conditions of water in the Norfolk Broads is, however, gradually worsening, largely due to plant nutrients encouraging the growth of algae and the resultant increase in BOD.

Detergents

Some of the early soapless detergents caused their own pollution problems. When a detergent is used to clean anything, it does not react chemically; its cleaning action is a purely physical process. This means that when detergents are washed into the drains (e.g. from washing up water, or bath water), their cleaning properties remain unchanged. If the detergent remains in the rivers, the agitation of the water can cause a blanket of foam to form over the water surface. In some cases large areas of lakes were covered in foam, sewage works were similarly affected, and foam from the drains was reported as reaching sinks on the seventh floor of one block of flats! Apart from looking unpleasant, the foam reduced the amount of oxygen dissolving in river water, and severely restricted the amount of light entering the water.

This particular feature had never been a problem when ordinary soaps were the only detergents used, and chemists soon realized what was responsible. Ordinary soaps are made from natural potential foods (vegetable oils or animal fats), and bacteria living in water can rapidly 'feed' on them and break them down. Ordinary soaps do not therefore persist in water; they are said to be *biodegradable* (i.e. capable of being broken down by bacteria). Synthetic detergents, on the other hand, were often not biodegradable, then 'covalent tail' (p. 183) did not have the same kind of carbon chain as those found in typical food substances such as oils and fats. The main difference was that the carbon chains in synthetic detergents often had 'side branches' which bacteria could not act upon. Chemists quickly learned to make new synthetic detergents which could be broken down by bacteria, and these are now readily available.

Detergents have two other undesirable effects on water. Many detergents contain phosphates (see Calgon, page 192). Phosphates, like nitrogen compounds, are important plant nutrients and so they increase the BOD of river water in the same way that nitrogen compounds do. (Note that the use of phosphate fertilizers does not cause much pollution because phosphates are not readily washed out of soil, as explained on page 392.) In addition, detergents which persist in river water are toxic to fish.

Steps are being taken to replace phosphates in detergents by more suitable substances, and to establish special treatment systems which remove phosphates from effluent.

Oil spillages

Crude oil is essential for the modern world, and enormous volumes have to be transported by sea. When oil is being loaded or unloaded there is always a risk of spillage, but the oil companies should have such problems well under control. Figure 12.8 shows an inflatable oil boom as used by Esso at their refinery in Milford Haven.

Figure 12.8 An inflatable oil boom (*Esso*)

The boom is used to enclose any spillage of oil which may occur during loading or unloading, so that the spillage can be recovered or dispersed. Any oil refinery processes a very large volume of crude oil and also uses a great deal of water during the processing, and so there is always a risk of oil contaminating the water which leaves the refinery. Such water is subjected to very careful controls before it is returned to the sea. Figure 12.9 shows part of the control system used by Esso at Milford Haven to ensure that oil is completely removed from water leaving the site.

A more likely source of an oil spillage is a ship at sea. There have been occasional examples of the deliberate discharge of oil at sea during, for example, a tank-washing operation. Such occasions are now rare, usually detected, and the culprits prosecuted. Unfortunately, serious accidents have also occurred from time to time. Considering the volume of oil carried at sea, the number of accidents has been small, but such is the size of a modern oil tanker that an accident involving one that is fully laden can have very serious consequences. In the recent past the coasts of Britain and France have been seriously affected by the wrecks of the oil tankers Torrey Canyon in 1967 and the Amoco Cadiz in 1978.

A spillage of a large volume of crude oil at sea is always a serious problem. The oil is immiscible with water, and floats on top of it. If it is washed ashore it causes the most appalling devastation of beaches, and is extremely difficult to remove. Sea birds which are covered in it have little chance of survival. It kills a wide variety of aquatic life, and in addition is thought to contain carcinogenic chemicals which may accumulate in fish.

Figure 12.9 Oil separators (*An Esso photograph*)

Great advances have been made in combating pollution of this kind, although it is ironical that the problem will inevitably decline as do the world's stocks of crude oil. Large spillages are sometimes dispersed by the use of detergents, which break up the oil into smaller droplets and make it miscible with water. At first the detergents used were more toxic to marine life than the oil itself! Fortunately it is possible to make many varieties of soapless detergents (p. 183), and some very effective and apparently non-toxic detergents have been formulated. Another common method of dispersal is the sand-sink method, in which spilt oil is sprayed with a mixture of chemically treated sand and water. The oil clings to the sand and sinks to the bottom of the sea, where it remains until it is broken up naturally.

Nuclear waste

This is more of a potential problem than an actual one. We are faced with an energy crisis and several countries are committed to an expansion in nuclear power. This inevitably means that more and more radioactive waste will be formed, much of it in liquid form. There are various ways of disposing of this. For example, liquid which is likely to remain radioactive for a very long time is sealed in appropriate containers and either stored on site (theoretically until it is completely safe to be discharged or until it can be treated in some other way), buried down deep unused mine shafts, or buried at sea. Modern advances include mixing the radioactive material with special glass-making substances so that it becomes part of the glass when processed. Such special glass is then buried in some way. Glass is not subject to corrosion and such methods should provide safe storage for thousands of years.

Pollution problems caused by the disposal of radioactive waste have so far been very few. This is largely because the potential seriousness of the problem was appreciated from the beginning, and technology has been developed to deal with it. We have been a little slow to recognize the dangers of other forms of pollution, and often did very little to prevent it. Now that we have a positive attitude towards the issue, there is every possibility that we will develop new processes and eliminate most of the major pollution problems.

APPENDIX

How used water is treated

In Britain, waste water can be discharged into a public sewer (from which it goes to a sewage works and is treated before being allowed to enter a river), or directly into a river. There are clearly defined controls which apply to the materials which can be discharged in either of these ways, and the first recorded statute of this kind in Britain was in 1338.

What happens at a sewage works

In Britain, numerous Public Health Acts and Control of Pollution Acts (etc.) have been passed in recent years. An industry can only discharge effluent into a sewer with the approval of the Local Authority, who have considerable powers under the various Acts. Before permission is granted, detailed consideration is given to the nature and composition of the effluent, the maximum quantity which it is expected will be discharged in one day, the highest rate of discharge, and the actual sewer into which the effluent will be discharged.

In Chapter 11, the role of the Water Authorities in the treatment of drinking water was explained. Other important jobs of the local Water Authority are to treat *used water* at a sewage works (so that harmful pollutants are removed) and also to check on pollution in all the rivers within their area. Remember that sewage is not just waste from toilets; it also includes rainwater from the streets of a town, waste from factories, and waste from sinks and baths. All the water in the sewers should go to a sewage works or to an outlet in the sea.

Figure 12.10 Sludge tanks at a sewage works (*Severn Trent Water Authority*)

The essential process at a sewage works is a biological one, in which pollutants are oxidized to harmless materials by bacteria. The waste is brought into contact with appropriate types of bacteria under controlled conditions of aeration (i.e. supply of air), temperature, nutrients, pH, etc. Bacteria are remarkably adaptable to a wide variety of conditions, and some can detoxify (make harmless) wastes which are highly poisonous to man. The process produces a 'sludge' which consists of (i) living and dead bacteria, (ii) solids which passed through the sieves and sedimentation tanks (see Figures 12.10 and 12.11), and (iii) the products of bacterial action. This sludge has to be disposed of, but fortunately a large proportion of it is safe to be used in agriculture as a valuable fertilizer (see Figure 12.12) and this helps to maintain the nitrogen balance in the soil. Sludge which contains a significant proportion of heavy metal compounds is not suitable for use in this way, and most of it is dumped (e.g. at sea), incinerated, or used in land reclamation.

The liquid effluent which has passed through the process is discharged into rivers, and it is highly purified compared to its condition on arrival at the works. Purification is continued naturally in the river (as described on page 197). If the sewage was discharged directly into the river, before treatment, the river would be 'overloaded' with pollutant material, and the natural process of purification would only take place to a very limited extent. In some cases river water containing discharged sewage effluent is suitable for further purification at a water treatment works so that it can then be drunk. However, it is important to realize that effluent from a sewage works still presents problems, no matter how effective the treatment has been. For example, bacteria only *change* nitrogen compounds into other (less toxic) nitrogen compounds, they do not *remove* nitrogen from the material. The effluent therefore still contains nitrogen compounds, and still has an oxygen demand. Treated sewage is in fact a more important source of nitrates etc. in river water than either run-off from agricultural land or industrial waste.

The discharge of water directly into a river

As a sewage works is so efficient, you may wonder why *all* liquid waste is not poured into the sewers. There are two main reasons why this is not done. There are not enough sewage works to deal with the large quantities of liquid waste produced by industry. In addition, some chemicals present in liquid waste will also be toxic to bacteria, and if these chemicals arrive at a sewage works they could kill the very bacteria on which the treatment depends. Some industrial effluent must therefore be dealt with differently.

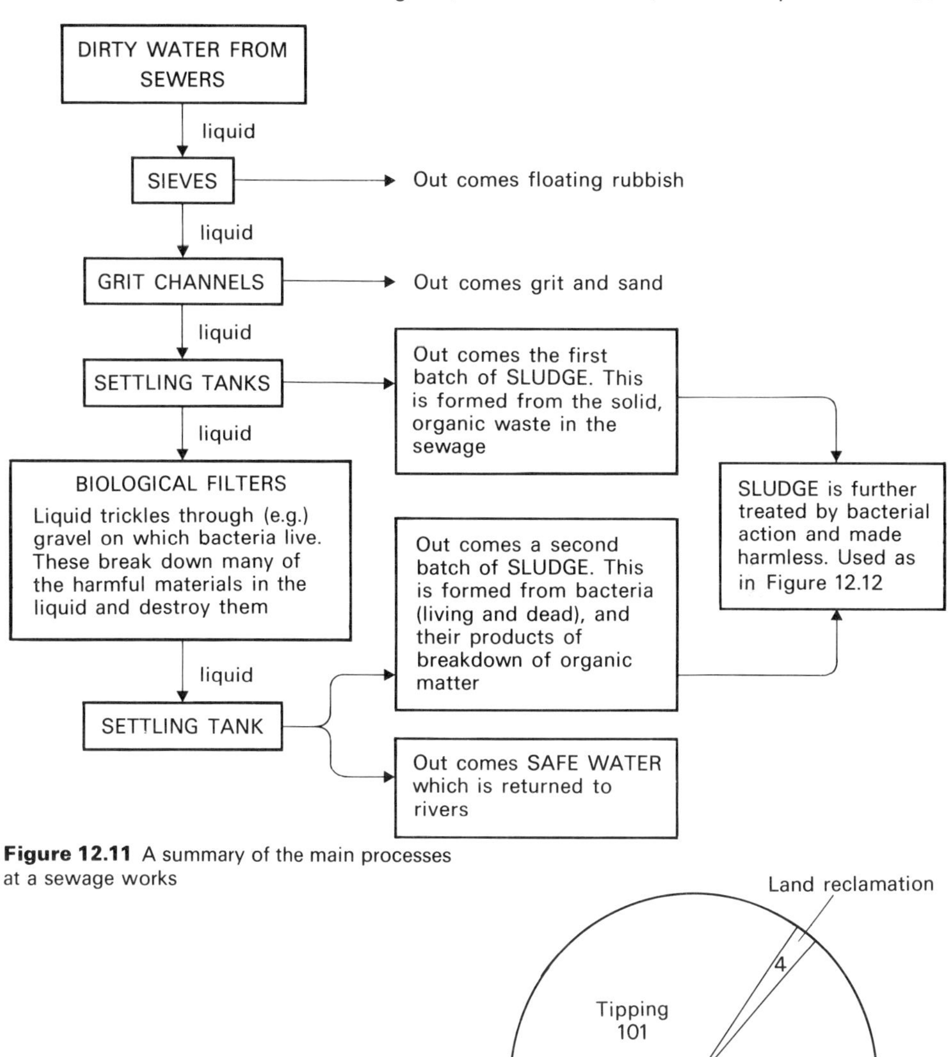

Figure 12.11 A summary of the main processes at a sewage works

Figure 12.12 How sewage sludge is disposed of by one Water Authority (*Information provided by the Severn Trent Water Authority for the year 1978*)

In Britain, various Acts such as the Rivers (Prevention of Pollution) Act and the Dumping at Sea Act control the discharge of effluent into the rivers and seas, and this is normally the concern of the Water Board or Ministry of Agriculture, Fisheries and Food. Apart from normal considerations such as the chemical nature of the effluent and the rate of discharge, the maximum temperature of the effluent at the time of discharge is also considered. The discharge of warm water into a river (e.g. water which has been used for cooling at a power station) can have an important effect on the temperature of the water in localized regions of the river, and therefore on the balance of life within it. Also, energy is now so expensive that the discharge of warm water into a river is a serious waste of heat energy. Schemes are now in operation to use such water to boost central heating systems in the locality, and to help provide the warmth to grow such crops as tomatoes, etc.

Some waste is discharged into a river without any preliminary treatment, but if the waste is likely to cause a hazard of some kind it is normally passed through some purification procedure at the works. Wastes which are very acidic or alkaline (e.g. the strongly alkaline waste from a paper mill) are first neutralized. Many chemical methods are used to remove heavy metals and other toxic substances which may be present in liquid waste. For example, sulphides and cyanides can be precipitated out as insoluble compounds. Such treatment, however, may produce further problems of a similar kind because the solid sludge from such systems is often toxic and still has to be disposed of.

The cost of waste disposal is high and ever increasing as new and higher standards are demanded by Local Authorities. Clearly, these costs have to be borne by the people who buy the products of industrial processes, i.e. as increased prices for the products, and by the rate increases of the Local Authority and Water Board. More sewage works are being constructed, and some factories have their own biological purification systems (like those used at a sewage works), in which organisms are used to purify a wide variety of toxic substances.

Point for discussion

The activities pursued by man on land cause pollutants to get into rivers. Complete the table on the right, which relates to such pollutants.

Explain why these pollutants stay in the sea and are not carried any further around the 'water cycle'. (A.E.B. 1976)

Pollutants in the rivers and seas	*Cause of this pollution*	*Effect of this pollution on sea or river*
Nitrate ions		
Excess hydrogen ions		
DDT		

12.4 A SUMMARY OF CHAPTER 12

The following 'check list' should help you to organize the work for revision.

1. Definitions

(a) Hard water (p. 186)
(b) Temporarily hard and permanently hard water (p. 190)

2. Other ideas

(a) Detergent molecules all have the same basic structure, i.e. a relatively long (covalent) carbon–hydrogen chain containing a small ionic group. The former is hydrophobic ('water-hating'), the latter hydrophilic ('water-loving').

(b) Why synthetic detergents were produced (to avoid using a potential food), and the advantages they have over traditional soaps (not affected by pH or hard water, and can be 'tailor made' for a particular purpose).

(c) How the hydrophobic and hydrophilic parts of a detergent molecule help to bring oil and water together. Detergents also clean by lowering the surface tension of water, thus making it a better 'wetting' agent.

(d) Dissolved calcium (and to a lesser extent, magnesium) ions are mainly responsible for hardness of water. The calcium ions enter the water either by simple solution (from calcium sulphate, which produces permanently hard water) or from insoluble calcium carbonate by a chemical reaction which produces soluble calcium hydrogencarbonate (which is responsible for temporarily hard water).

$$CaCO_3(s) + H_2O(l) + CO_2(aq) \rightleftharpoons Ca(HCO_3)_2(aq)$$

(e) Why soaps form a scum in hard water (i.e. a precipitate of, for example, calcium stearate) and how this wastes soap and is inconvenient.

(f) How hard water can cause blockages in pipes, and fur in kettles, by a reversal of the reaction which forms calcium hydrogencarbonate from calcium carbonate.
(g) Hard water can be softened by either removing the ions responsible, or converting them into insoluble compounds. The main methods used for softening water should be known (Table 12.1), and you should be able to explain how each of them works.
(h) How to recognize hard water, and the two different kinds, from titration values with soap solution, e.g. using boiled and unboiled samples of water and comparing them with a sample of deionized water.
(i) The problem of water pollution, and what is being done to overcome it. It is best to consider this section as something which is not to be learned 'by heart', but rather as material which requires a common sense appreciation of the overall picture. Note the meaning of specific terms such as 'biodegradable detergents'.

3. *Important experiments*

(a) A laboratory preparation of a soap from a natural fat or oil and sodium hydroxide, e.g. Experiment 12.1. (Include a *simple* understanding of the chemistry of the preparation.)
(b) The softening of water, and deciding the relative amounts of temporary hardness and permanent hardness in a sample.

QUESTIONS

1. Explain the terms 'efflorescent', 'deliquescent', and 'hygroscopic' and give examples to illustrate your answer.

2. (a) You are given an aqueous solution of potassium chloride which is saturated at room temperature.
(i) On what evidence would you confirm that the solution is saturated?
(ii) Give a careful account of the way you would use this saturated solution to measure the solubility of potassium chloride at room temperature. Mention the precautions you would take and show how the result is calculated.
(b) Outline how you would obtain a sample of pure water from a solution of potassium chloride. Describe *two* tests by which you could prove that the sample was free from dissolved potassium chloride. (A.E.B. 1977)

3. (a) It is required to separate a mixture of *three* solid dyes; one red, one yellow, and one blue. The following facts are known about the dyes. The blue and yellow dyes are soluble in cold water, while the red dye is insoluble. When an excess of aluminium oxide is added to a stirred, green aqueous solution of the mixed blue and yellow dyes and the aluminium oxide is filtered off and washed with water, it is found that the solid residue is yellow and the filtrate blue. When the yellow solid is stirred with ethanol and the mixture filtered, the solid residue is white and the filtrate yellow.
Describe how you would obtain dry samples of the three dyes.
(b) A green dye is known to be a hydrate. Describe how you would determine the percentage of water of crystallization in the hydrate. (J.M.B.)

4. Give three reasons in each case why (a) air is considered to be a *mixture* of nitrogen and oxygen, (b) water is considered to be a *compound* of hydrogen and oxygen.
Draw a diagram of the apparatus you would use to obtain a sample of the air dissolved in tap water. How would you determine the proportion of oxygen in the air so obtained? How and why would your result differ from the proportion of oxygen in ordinary air? (J.M.B.)

5. Below is a simple representation of a detergent molecule.

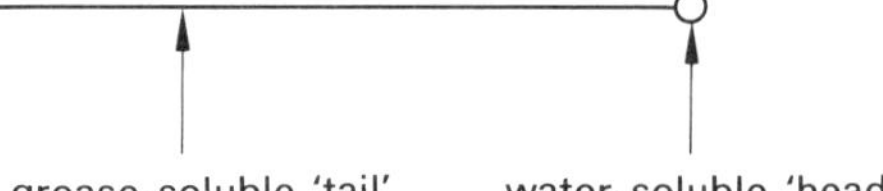

Using representations like the one above, draw a diagram to show how you would expect detergent molecules to interact with a globule of grease.
Give *two* factors that are essential for a detergent to remove grease efficiently from fibres and surfaces.
Explain how large amounts of detergents get into rivers and lakes. Most detergents produced today are 'biodegradable'. Explain what this term means. (A.E.B. 1976)

6. A sample of water from a limestone area needed a large quantity of soap in order to provide a lather. After the water was boiled and allowed to cool, a lather was easily formed with a very small quantity of soap.
(a) Name a compound which may have been causing the hardness in this water.
(b) Explain why it is difficult to form a lather by adding soap to water containing this compound.
(c) Write an equation to show the effect of heat on this compound.
(d) Explain the chemical reactions by which this compound gets into the water. (J.M.B.)

7. (a) Describe carefully what happens when carbon dioxide is passed through calcium hydroxide solution until no further change occurs. Give equations.
(b) Explain how stalactites and stalagmites are formed. Why does one usually form opposite the other?

8. (a) Describe what happens when a soapy detergent is first added to a sample of hard water.
(b) Why does a lather eventually form?
(c) What similarities and differences are there between soapy and soapless detergents?

9. Approximately 80 per cent of the surface of the earth is covered by water or ice and a significant proportion of the atmosphere is water vapour. Discuss:
(a) the importance of the sea as a source of chemicals and food.
(b) the pollution of water by man due to industry, transport, agriculture, and waste disposal.
(c) the ways in which water is purified naturally. (A.E.B. 1978)

10. (a) Describe how you would prepare a solution of potassium nitrate that is saturated at room temperature. How would you use this solution to determine the solubility of potassium nitrate (in g/100 g water) at room temperature? Mention the precautions you would take to ensure accuracy, and show how you would calculate the result.
(b) Explain the following experiments which were carried out with a single sample of *hard water*.
(i) When boiled, a small amount of a white solid was deposited.
(ii) Some hardness remained even after the water had been boiled for a considerable time.
(iii) The water could be softened completely by adding sodium carbonate.
(A.E.B. 1975)

11. (a) Explain what is meant by *hardness of water*. What chemical compounds cause (i) temporary hardness, (ii) permanent hardness? Describe the processes which cause these compounds to be present in the water in rivers or lakes.
(b) Describe and explain the most suitable method by which the hardness of water is removed from (i) the water in a bath, (ii) the water supply to a house.
(c) Explain why (i) hard water forms an insoluble scum with soap, (ii) an insoluble deposit (fur) may form in a kettle in which water is boiled. How may the insoluble deposit be removed from a kettle by chemical means? (A.E.B. 1979)

12. (a) Calcium carbonate, present in rocks or soil, is one of the causes of hardness of water. Explain why this is so.
(b) Explain the use of (i) calcium hydroxide, (ii) sodium carbonate, in the softening of hard water.
(c) Why does the presence of dissolved sodium carbonate not make water hard?
(d) A copper boiler used in the preparation of distilled water is encrusted with a layer of white scale caused by the hardness in the water used. Explain how this scale was formed from the hard water.
(e) If supplies of dilute sulphuric, hydrochloric, and nitric acids were available, which of these acids would you use to remove the scale from the boiler? Give reasons for your choice. (C.)

13 Electrolysis

13.1 WHICH SUBSTANCES CONDUCT ELECTRICITY?

Simple conduction in solids

Your earlier experiences in chemistry, and your own common sense, should tell you that the only pure substances which conduct electricity in the *solid* state are the metallic elements and the non-metallic element carbon (in the form of graphite). Those solids which do allow a current to pass through them are called *conductors*. All other pure *solids* (i.e. the remainder of the non-metallic *elements* and all the thousands of solid *compounds*) will not allow a current to pass through them, i.e. they are *non-conductors*. If any of these non-conductors are used to protect something from electricity, they are also known as *insulators*. An insulator is not quite the same as a non-conductor; there are many non-conductors of electricity, but they are not all used as insulators.

Can pure liquids or solutions conduct electricity?

It is important to understand the difference between a pure liquid (which contains no water) and a solution, which is a substance dissolved in a solvent, e.g. sodium chloride solution. A pure liquid is *one* substance only, but a solution contains at least two substances. You may have seen some solid substances both in solution and also as pure liquids (when melted). For example, sodium chloride will dissolve in water to form a solution, but it can also be melted to form liquid (molten) sodium chloride which contains no water.

In order to make a solid change to a pure liquid, we need to melt it. Some substances have 'already melted' at room temperature and are already in the liquid state, e.g. paraffin oil and cooking oil. (Each of these substances is actually a mixture of liquids.) In the next experiment you will try to pass electricity through compounds in the liquid state. Note that a metal or carbon rod which is used to conduct electricity into (or out of) a liquid is called an *electrode*. You will use carbon electrodes in the next experiment. The *anode* (+) is the electrode from which electrons leave the liquid and go back to the d.c. supply. The *cathode* (−) is the electrode at which electrons enter the liquid from the d.c. supply.

Experiment 13.1
Trying to pass electricity through substances in the liquid state**

Apparatus

Six volt battery or similar source of d.c. supply, six volt lamp and holder, connecting wires and crocodile clips, carbon electrodes in holder, crucibles, pipe clay triangle, tripod, Bunsen burner, tongs, safety screen, access to fume cupboard.

Substance	*Type of bonding*	*Does it conduct when in liquid form?*	*Observations (e.g. chemical changes) at the electrodes* ANODE	CATHODE

Paraffin wax, sugar, naphthalene, lithium chloride, sodium hydroxide, lead(II) bromide, cooking oil, liquid paraffin, water.

Procedure
Note that a *low* voltage *d.c.* supply is used in these and similar experiments; this is completely different from the *mains a.c.* supply, which must not be used on any account in these experiments. Some of the chemicals used in this experiment are potentially dangerous, especially when heated. Lithium chloride, sodium hydroxide, lead(II) bromide, and naphthalene will be demonstrated only. Even then, the first two should be heated behind a safety screen, and the other two in the fume cupboard.
(a) Make out a table for your results as shown at the top of the page.
(b) Half fill a crucible with one of the substances you are allowed to use. If the substance is a solid, *gently* heat the solid, using the tripod and pipe clay triangle, until it just melts.
(c) Set up the circuit as shown in Figure 13.1, warming the crucible from time to time if necessary to keep the chemical molten. Record your observations in the table.
(d) If you have melted a solid, remove the electrodes, place them on a suitable mat, and allow the liquid to cool again. Write the name of the chemical you have used on a piece of paper next to your apparatus, and then move to another set of apparatus which has been used for another chemical. (This avoids having to clean the electrodes after each experiment.)
(e) Repeat steps (c) and (d) if necessary with the new chemical, making sure that you allow time for any 'crust' on the electrodes to melt before you connect the d.c. supply.
(f) Continue until you have used liquid paraffin, paraffin wax, sugar, cooking oil, and water.
(g) Add further results to your table from the demonstrations. Note that information about column 2 in the table is given under the next heading.

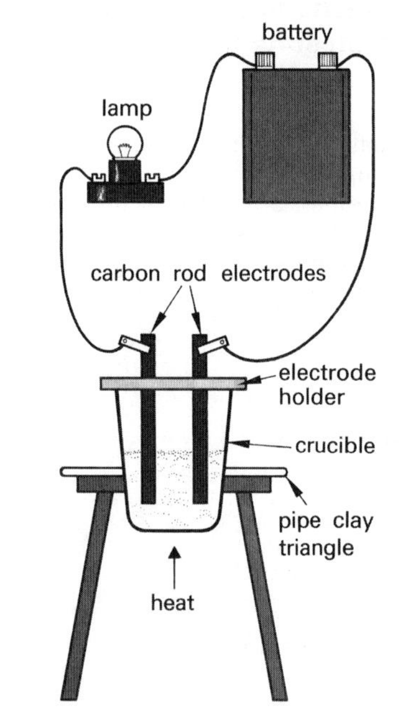

Figure 13.1 Investigating the effect of electricity on substances in the liquid state.

Notes about the experiment
Some of the substances used in the experiment are ionically bonded and some are covalently bonded and contain only molecules. You should be able to pick out three of the compounds as being ionically bonded because of the nature of the elements and/or radicals within them, and also because they have higher melting points than the others. All of the other compounds used in the experiment are covalent and molecular; they have either 'already melted' or are easy to melt, i.e. they have low melting points. Your results should enable you to make a conclusion as follows. (Note that cooking oil, liquid paraffin, and paraffin wax are actually mixtures of compounds, but the bonding in all of the compounds in each mixture is the same.)

Conclusion
Some compounds conduct electricity when in the ______ state. The compounds which conduct in this way are ______ bonded, and they change chemically when conducting. Such liquids are called electrolytes. Compounds which contain only ______ do not conduct electricity when molten; they are non-electrolytes.

Experiment 13.2
To determine whether some substances conduct electricity when dissolved in water*

Apparatus
Six volt battery or alternative d.c. supply, six volt lamp and holder, two carbon electrodes in holder, 50 cm^3 beaker, connecting wire.
Deionized water, aqueous solutions of sodium chloride, lithium chloride, potassium iodide, sodium hydroxide, sulphuric acid, cane sugar, glucose, and starch.

Procedure
(a) Make out a table as follows:

(b) Set up the circuit as shown in Figure 13.2. Half fill the beaker with deionized water, put in the electrodes and record your results.

(c) Thoroughly rinse out the beaker, and repeat (b) with the other solutions in turn. Make sure that the beaker is thoroughly rinsed after each solution has been used. Information for the second column in the table is given under the next heading.

Solution	*Bonding type (contains ions, or molecules?)*	*Does it conduct?*	*Observations (e.g. chemical changes) at the electrodes ANODE*	*CATHODE*

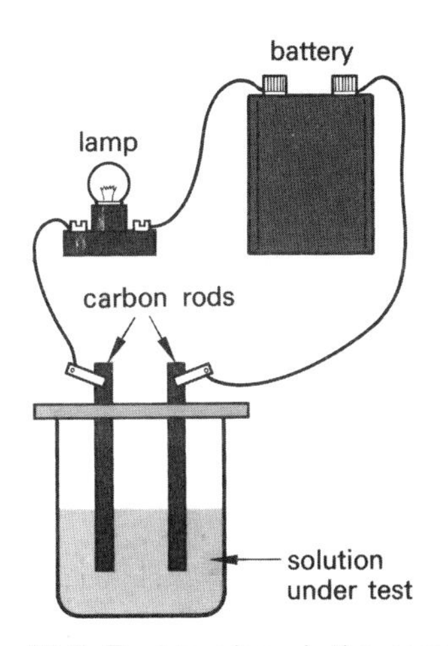

Figure 13.2 To determine whether some substances will conduct electricity in aqueous solution

Notes about the experiment
You should be able to recognize three of the compounds as being ionically bonded because of the nature of the elements and/or radicals within them; dilute sulphuric acid also contains ions. The substances in the other solutions are covalently bonded and consist of molecules. Complete the second column in the table of results by writing either 'contains ions' or 'covalent; contains molecules'. Your results should enable you to complete the following conclusion, and we can now also define what we mean by electrolysis.

Conclusion
Compounds do not conduct electricity in the ______ state, but when they are either ______ or ______, some of them do so. The compounds which conduct in this way all contain ______, but solutions or pure liquids which contain only ______ never conduct electricity. The solutions or liquids which conduct have one other thing in common; as they conduct, they also change chemically.

An electrolyte is a substance which, when molten or dissolved in water, conducts an electric current and is decomposed by it.
Electrolysis is the chemical change which takes place when an electric current is passed through an electrolyte.

A comparison between simple conduction in solids and the conduction of electricity by electrolytes

The conduction of electricity in a solid is very different from that which occurs during electrolysis. A solid conductor is completely unchanged *chemically* when electricity passes through it (although it may get hot, e.g. in electric light filaments), but this is not the case with an electrolyte. Chemical changes occur at both anode and cathode during electrolysis. Another difference between a solid conductor and an electrolyte is the nature of the current passing through the chemical. An electric current is a flow of charged particles; in a metallic conductor it is a flow of electrons (p. 271), but in an electrolyte it is a flow of ions, as is explained in the next section. An electrolyte must be melted or in solution before it can conduct, because the ions are not able to move from place to place in the solid state.

13.2 WHAT HAPPENS DURING ELECTROLYSIS?

In order to understand what happens during electrolysis it is best to consider a pure liquid (molten) electrolyte, as there is only one substance present and no water to complicate matters. When lithium chloride was melted and electrolysed in Experiment 13.1, you probably recognized at least one of the products. You may see the experiment repeated so that the products can be confirmed; chlorine gas is formed at the anode, and lithium metal at the cathode.

Molten lithium chloride contains only two kinds of ion; Li^+ and Cl^-. When the circuit is completed, electrons from the d.c. supply go to the cathode and make it negatively charged. This attracts the positive ions (cations) to the cathode. Cations are atoms which have lost one or more electrons, i.e. they have a 'vacancy' for electrons. When free electrons are available, as at a cathode, each cation picks up the number of electrons that it requires to turn it back into a neutral atom. In molten lithium chloride, the cations are Li^+ and on reaching the cathode these each receive one electron and become lithium atoms. We say that lithium ions are *discharged* at the cathode. Electrons are thus constantly removed from the cathode, and the neutral atoms formed join together (i.e. with others of their own kind) to form stable structures. Remember that individual atoms only occur naturally in the noble gases; in this example, lithium atoms join to form a giant metallic structure, i.e. metallic lithium. A chemical change thus occurs at the cathode (see figure on p. 211).

Negative ions (anions) are atoms which have gained extra electrons, and are attracted to the anode (positively charged). At the anode, the anions give up the 'extra' electrons to the anode and become neutral atoms. In the example, the chloride, Cl^-, ions each lose an electron to form chlorine atoms. The atoms formed join together in pairs to form chlorine molecules, Cl_2. A chemical change thus occurs at the anode. We say that chloride ions are discharged at the anode (see figure on p. 211).

Are they attracted to one electrode or both?

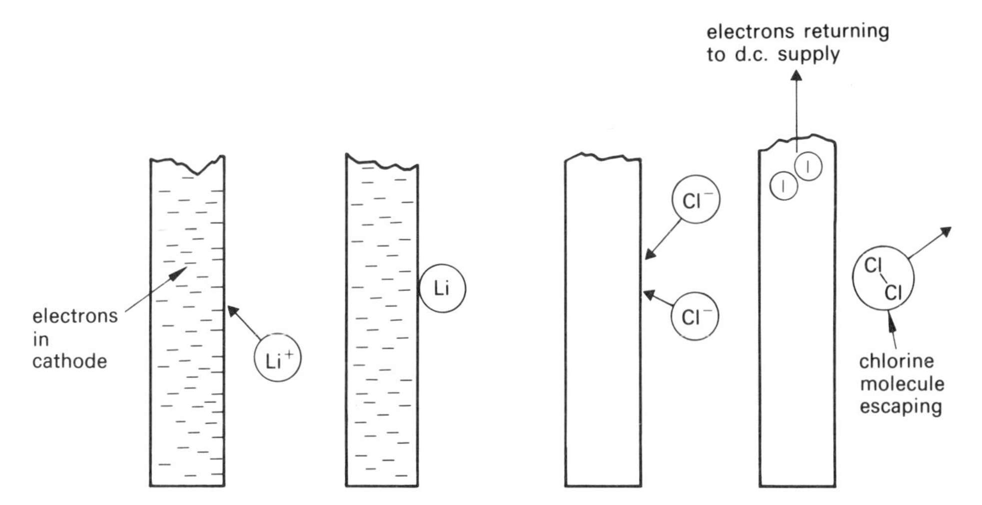

Lithium ions receive electrons from the cathode and are discharged as lithium atoms

Chloride ions give electrons to the anode and are discharged as chlorine atoms which form chlorine molecules

The overall effect is that electrons are continually removed from the cathode and an *equal number* are fed in at the anode, from which they return to the d.c. source and complete the circuit. Although electrons do not actually pass through the liquid, they *appear* to do so and a complete circuit is formed. The ions act as electron carriers, and are chemically changed in the process. These changes are always summarized by ionic equations, e.g.

$$\text{at the cathode } 2Li^{+}(l) + 2e^{-} \rightarrow 2Li(l)$$

$$\text{at the anode } 2Cl^{-}(l) - 2e^{-} \rightarrow Cl_2(g)$$

(You may be puzzled about the appearance of the number 2 in the equations. It is needed to balance the anode equation because the product at the anode is chlorine molecules, each of which is formed from two chloride ions. It is included in the cathode equation so that the same number of electrons are then involved in each of the equations. Remember that the number of electrons which leaves the cathode must be the same as the number which enter the anode. If the equations which summarize a particular electrolysis are to be comparable, they must therefore involve the same number of electrons.)

A simple analogy may help you to understand more clearly the difference between conduction in a metal and in an electrolyte. Imagine that a group of people are using buckets of water to put out a fire. They could form a human chain, and pass buckets filled with water along the chain. This is how electrons are 'passed along' a metal when it conducts. Alternatively, some individuals could take full buckets from the tap to the fire and another group of individuals could take empty buckets back to the tap. This kind of process occurs in electrolysis, where anions (full buckets) are 'full of electrons', and cations (empty buckets) can take up more electrons. This is summarized in Figure 13.3 (p. 212).

The part water plays in electrolysis

You may think that the water (in an electrolyte which is in solution) is just there to allow the ions freedom of movement, and that it takes no part in the actual electrolysis. However, it is not quite as simple as this. If water took no part in the

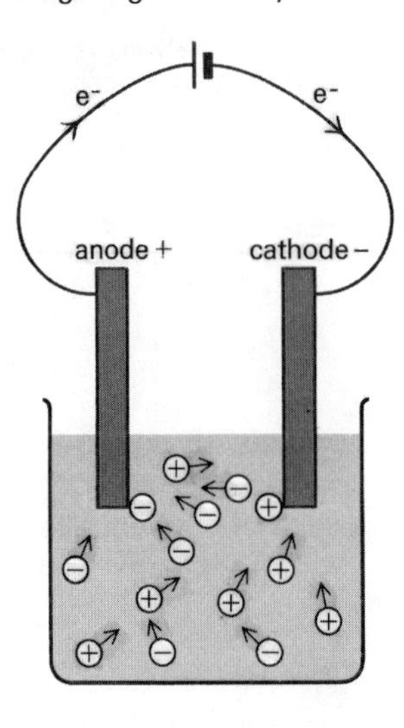

⊕ positive ions (empty buckets)

⊖ negative ions (full buckets)

Figure 13.3 To show the electron flow during electrolysis

electrolysis of lithium chloride solution the products would be lithium and chlorine, but if you refer back to Experiment 13.2, your results should show that when a solution of lithium chloride is electrolysed, gases are formed at both electrodes. You probably recognized one of them by its smell, because chlorine is formed at the anode. You may repeat the experiment in such a way that the products can be collected and then identified, e.g. using an apparatus such as that shown in Figure 13.7. The gas formed at the cathode can be shown to be hydrogen.

The solute, lithium chloride, LiCl, does not contain any hydrogen so the hydrogen formed at the cathode must have come from the water, which is the only other substance present. Water therefore *takes part* in the electrolysis. Note that when pure lithium chloride is melted, the products are lithium metal at the cathode (not hydrogen) and chlorine at the anode. Water thus provides freedom of movement of the ions and may also affect what products are formed in an electrolysis.

Pure liquid electrolyte	*Electrolyte solution*
Only two types of ion present, products easy to predict	Two types of ion present from solute, *and* also water molecules. Products less easy to predict; they may include hydrogen or oxygen from the water

This may seem rather odd, because water itself is not an electrolyte. However, water molecules do *ionize* to a very small extent:

$$H_2O(l) \rightleftharpoons H^+(aq) + OH^-(aq)$$

Also, you may imagine that, because the concentrations of $H^+(aq)$ and $OH^-(aq)$ ions in water are low, the electrolysis of aqueous solutions can produce only very small volumes of hydrogen or oxygen. This is not the case. As $H^+(aq)$ or $OH^-(aq)$ ions are discharged (removing these ions from the electrolyte), more water molecules ionize, adding both $H^+(aq)$ and $OH^-(aq)$ ions to the electrolyte. Water molecules are thus a *potential* source of large numbers of $H^+(aq)$ and $OH^-(aq)$ ions, although the number of ions present at any one time is very small.

The discharge of $OH^-(aq)$ ions in electrolysis

You may be puzzled how oxygen can be formed from water molecules which ionize to form hydroxide ions, $OH^-(aq)$, and not oxide ions, $O^{2-}(aq)$. It is a very common error to think that oxygen and hydrogen are formed from hydroxide ions and so it is important to make quite sure of the following. Whenever $OH^-(aq)$ ions (e.g. from water, or from an alkali such as sodium hydroxide) are discharged in electrolysis, they *always* form water and oxygen:

$$4OH^-(aq) - 4e^- \rightarrow 2H_2O(l) + O_2(g)$$

Learn this ionic equation; the discharge of $OH^-(aq)$ ions is very common.

Selective discharge

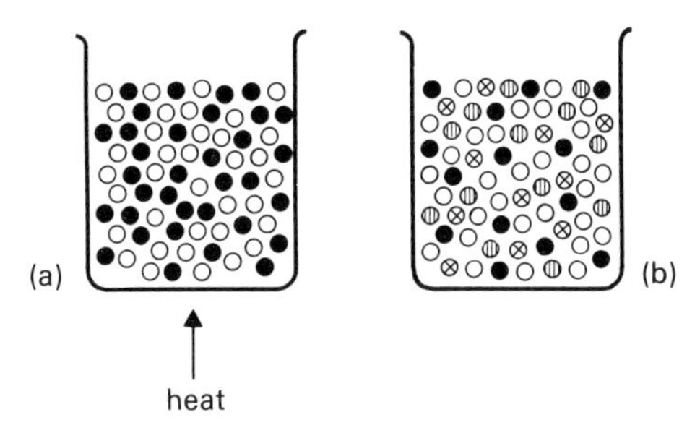

Figure 13.4 The ions present in (a) molten lithium chloride and (b) its aqueous solution. Note that most of the particles in (b) are water *molecules*, and these are not shown

We can now see that when an ionic solid is dissolved in water to form an electrolyte, there are usually four different kinds of ions present, i.e. two from the ionic solute and two from the water. Figure 13.4 shows that in a solution of lithium chloride the ions present are $Li^+(aq)$, $Cl^-(aq)$, $H^+(aq)$ and $OH^-(aq)$, whereas molten lithium chloride contains only Li^+ and Cl^- ions. The solution conducts in the same way as described earlier for pure liquids (p. 210), and usually only *one* kind of ion is discharged at each of the electrodes, even though there may be two different types of cation and two different types of anion present in the solution. We say that at each electrode one of the ions is *selectively discharged*. (Note that 'selection' is not always perfect, see page 215.)

In the electrolysis of lithium chloride solution, $H^+(aq)$ ions are selectively discharged at the cathode (in preference to $Li^+(aq)$). At the anode the $Cl^-(aq)$ are selectively discharged in preference to $OH^-(aq)$. In molten lithium chloride we say that Li^+ ions are discharged at the cathode and that Cl^- ions are discharged at the anode. In this case we do not say that they are *selectively* discharged, as they are the only ions present; there is no alternative.

	At the cathode	*At the anode*
Electrolysis of a solution of lithium chloride	Hydrogen ions selectively discharged, hydrogen gas formed	Chloride ions selectively discharged, chlorine gas formed
Electrolysis of molten lithium chloride	Lithium ions discharged, lithium metal formed	Chloride ions discharged, chlorine gas formed

Where more than one kind of anion or cation is present, which particular type of ion is selected depends upon the concentrations of the various ions, how easy it is for them to gain or lose electrons, and the nature of the electrode. Some of these ideas are discussed in more detail under the next heading, but at this stage you may find it difficult to use them to work out which particular ion will be selectively discharged in a given situation, and it is perhaps better simply to *learn* what happens in the detailed examples of electrolysis which follow, and which occur elsewhere in the book. For the moment it is enough if you understand that where there is a 'choice' of ion for discharge, it is usually easier to discharge one of them, and this type of ion is the type which is selectively discharged. Even if one of the very small number of $H^+(aq)$ or $OH^-(aq)$ ions formed from water molecules is/are discharged, the process can continue because 'fresh supplies' are formed by the ionization of more water molecules.

Some of the factors which govern selective discharge

If we consider an electrolyte containing two metal ions, each present in equal concentration, only one of the two types of metal ion will be discharged at the cathode and it will be the one which most readily *gains* electrons. If you have studied the activity series (p. 249) you will understand that the metals in the series are placed in order according to their ability to *lose* electrons and so form positive ions. The ease with which these various positive ions can gain electrons to reform the corresponding metal atoms (as happens when they are discharged in electrolysis) therefore follows a sequence opposite to that of the metals in the activity series. For example, sodium metal is near the top of the activity series because sodium atoms easily lose electrons. Sodium ions, however, are near the bottom of a series which measures the ability of ions to gain electrons, because this reverse process is comparatively difficult.

An 'order of discharge' of cations in electrolysis can therefore be shown as below, and it is the reverse of the sequence in the activity series.

Na^+
Mg^{2+}
Al^{3+}
Zn^{2+}
Fe^{2+}
Pb^{2+} — increasing ease of discharge at a cathode (↓)
H^+
Cu^{2+}
Hg^+
Ag^+
Au^+

It is normally possible to use this series to predict which cation, from a mixture of two, will be discharged in the electrolysis examples encountered in elementary chemistry, but remember that the *concentrations* of the various ions are also important. For example, silver ions and mercury ions are very similar in their ability to gain electrons (see their electrode potentials, p. 258). If these two ions are present in the same electrolyte, silver ions will be selectively discharged in preference to mercury ions providing that the two ions are present in equal concentrations or the silver ions have a greater concentration than the mercury ions. If the mercury ions are present in a greater concentration than the silver ions, it is *possible* that the mercury ions could then be selectively discharged, depending on the relative concentrations of the two ions. The nature of the electrode (and some other, more complicated factors) can also affect the discharge of ions, but the only examples of this kind that you are likely to encounter are the electrolysis of copper(II) sulphate solution using *copper* electrodes (p. 220), which affects the discharge at the anode, and the Mercury Cell (p. 538) which affects the discharge at the cathode.

It is also possible to list the common anions in order of 'ease of losing electrons', i.e. in increasing order of ease of discharge at an anode.

SO_4^{2-}
NO_3^-
OH^-
Cl^-
↓ increasing ease of discharge at an anode

Note that aqueous solutions containing $Cl^-(aq)$ ions (e.g. sodium chloride solution or dilute hydrochloric acid) sometimes produce a mixture of chlorine and oxygen at the anode when they are electrolysed. This happens because both $Cl^-(aq)$ and $OH^-(aq)$ ions may be discharged at the same time, depending on the concentration of the $Cl^-(aq)$ ions.

Point for discussion

If you have studied the activity series, you will remember that aluminium metal rarely behaves as it should according to its true position in the series. The discharge of its ions in electrolysis, however, follows the pattern we would expect from its position in the series. Can you explain this apparent discrepancy?

13.3 SOME DETAILED EXAMPLES OF ELECTROLYSIS

General

Before you begin this section, make sure that you fully understand the following points about ions. Reactions during electrolysis are *always* shown by ionic equations, and so there is absolutely no point in continuing until you understand how to use them and what they mean. Note in particular whether electrons are being *added* or *removed*, and that the *number* of electrons in the equation depends on the charge on the ion and also on how many atoms are in a molecule of the product.

1. Hydrogen and metals form positively charged ions.
2. Non-metals and radicals form negatively charged ions.
3. Positive ions (cations) are attracted to the cathode, and the ones which are discharged *gain* electrons.
4. Negatively charged ions (anions) are attracted to the anode, and the ones which are discharged *lose* electrons.
5. The charge on an ion is equal to its normal valency or combining power.

The following ionic equations include all the examples normally needed. Note that the products must be capable of *existing* as you describe them, e.g. H_2 *not* H, and O_2 but *not* O.

l = liquid g = gas s = Solid aq: aqueus

At a cathode

$2H^+(aq) + 2e^- \rightarrow H_2(g)$

$Na^+(l) + e^- \rightarrow Na(l)$—*Only in molten compounds*

$Cu^{2+}(aq) + 2e^- \rightarrow Cu(s)$

$Al^{3+}(l) + 3e^- \rightarrow Al(l)$—*Only in molten compounds*

At an anode

$2Cl^-(aq) - 2e^- \rightarrow Cl_2(g)$

$4OH^-(aq) - 4e^- \rightarrow 2H_2O(l) + O_2(g)$

$Cu(s) - 2e^- \rightarrow Cu^{2+}(aq)$—*Special case—see page 220*

Note that sometimes these equations need to be 'multiplied up', e.g.

$$4H^+(aq) + 4e^- \rightarrow 2H_2(g)$$

in order that the equations for the anode and cathode reactions in a particular example both involve the same number of electrons.

Some types of voltameter in common use

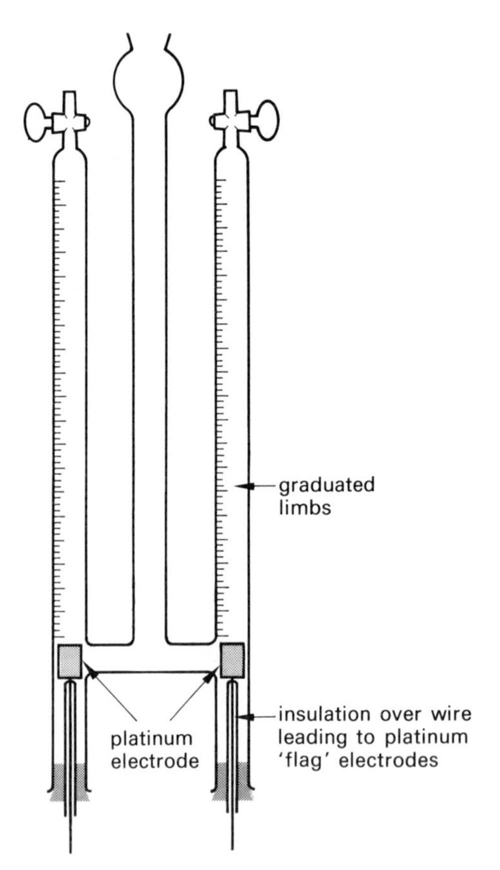

Figure 13.5 Hofmann voltameter

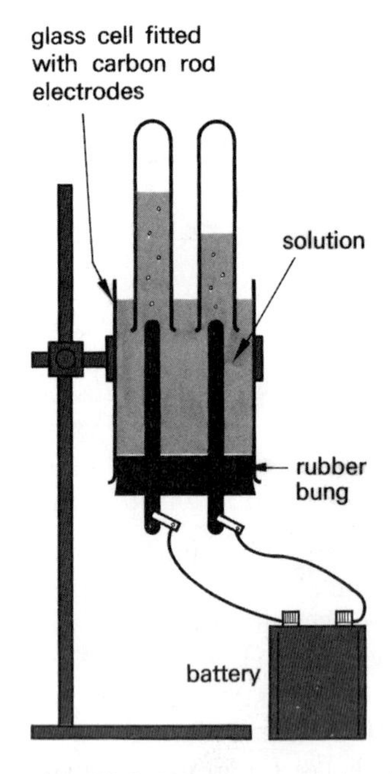

Figure 13.6 Simple electrolysis cell

Any piece of apparatus used to hold an electrolyte during electrolysis can be called a *voltameter*. The Hofmann voltameter, Figure 13.5, is just one such apparatus. This voltameter is particularly useful where gases are to be collected, and where it is necessary to measure the volumes of the gases. It is expensive to use on a class scale, however.

You may use other, simple voltameters such as those shown in Figures 13.6 and 13.7. The former is sometimes simply called an electrolysis cell. Either is perfectly adequate for collecting small samples of gases, but in 13.6 the test-tubes need to be raised slightly above the bottom of the cell so that ions can flow towards the electrodes and the solution can be displaced from the tubes.

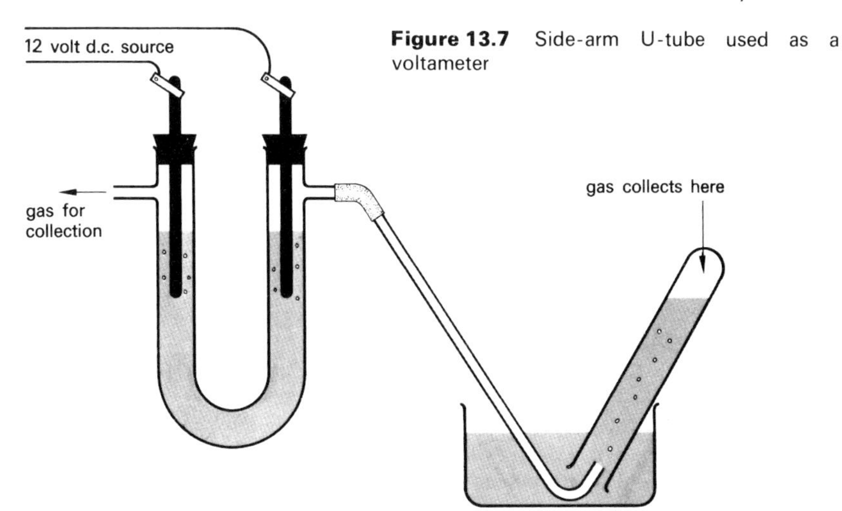

Figure 13.7 Side-arm U-tube used as a voltameter

Writing up an electrolysis experiment

You may be instructed to write up an electrolysis experiment in a way different from other experiments. The following are typical subheadings used in describing a full electrolysis experiment:

(*a*) *Apparatus*, for which a fully labelled diagram is normally adequate. Make sure that the chosen voltameter is suitable for the particular experiment, name the materials used for the electrodes, and label anode and cathode, adding (+) and (−) signs.

(*b*) *Ions present*, under which you show how the chemicals present dissociate to form separate ions in the electrolyte. For copper(II) sulphate solution this would be shown:

from copper(II) sulphate $CuSO_4(aq) \rightarrow Cu^{2+}(aq) + SO_4^{2-}(aq)$

from water $H_2O(l) \rightleftharpoons H^+(aq) + OH^-(aq)$

(*c*) *At the cathode*, under which you state which ions are attracted to the cathode, which of them (if there is more than one type) is selectively (or preferentially) discharged, and show by means of an ionic equation what happens at the electrode. If there are any colour changes etc., state them.

(*d*) *At the anode*, under which you repeat (*c*) but with respect to the anode. Check at this stage that the products are capable of independent existence, and that the two ionic equations are balanced in terms of the number of electrons involved.

(*e*) *Changes in the electrolyte*, under which you state how the electrolyte changes as the result of losing some ions. These changes are referred to under the individual experiments which follow.

Some specific examples of electrolysis

Experiment 13.3
The electrolysis of sodium chloride solution (brine)*

Apparatus
A suitable voltameter with carbon electrodes, connecting wire, and crocodile clips, source of low voltage d.c., test-tubes and bungs, Bunsen burner, splint.
Sodium chloride solution, litmus paper.

Procedure

(a) Set up the apparatus in the usual way, prepare for collection of gases at anode and cathode, and pass a current until enough gas has been collected at each electrode to perform a test on it.

(b) Test the gas from the anode with moist, blue litmus paper.

(c) Test the gas from the cathode with a lighted splint.

An honourable discharge

Notes about the experiment

(i) From the results of procedures (b) and (c) you should be able to write up the experiment in the way instructed, explaining which ions are selectively discharged, etc.

(ii) Note that the electrolyte slowly becomes a solution of sodium hydroxide as the other ions are discharged.

$$\begin{array}{ccc} & Na^+ & \text{(}Cl^-\text{)} \rightarrow \text{discharged} \\ \text{discharged} \leftarrow \text{(}H^+\text{)} & OH^- & \end{array}$$

(iii) If the sodium chloride solution is not previously saturated with chlorine, most of the chlorine formed initially will simply dissolve in the electrolyte, and its presence will not at first be obvious. A carbon electrode is used wherever chlorine is evolved, because chlorine attacks platinum and other materials that are used as electrodes.

Experiment 13.4
The electrolysis of water containing dilute sulphuric acid***

Apparatus

Hofmann voltameter with either carbon or platinum electrodes, low voltage d.c. supply, connecting wires and crocodile clips, two test-tubes with bungs, Bunsen burner, splint.

Water containing dilute sulphuric acid.

Procedure

(a) With the taps open, pour the electrolyte into the voltameter until the liquid level reaches the taps; close the taps. If necessary, open the taps slightly until there is no air space above the liquid within them.

(b) Connect the electrodes to the low voltage d.c. supply, noting which is the anode and which is the cathode.

(c) Allow the current to pass until enough gas has been collected for testing, and it is obvious what the relative volumes are.

(d) Note the volumes of gas produced at the two electrodes.

(e) Place an inverted test-tube over the tap above the cathode, open the tap to allow gas into the test-tube, but close the tap before liquid emerges from it. Quickly place a bung in the tube before it is turned back over.

(f) Repeat (e) with another test-tube, for the gas at the anode.

(g) Test the gas from the cathode with a lighted splint, and the gas from the anode with a glowing splint.

Notes about the experiment

You might have been puzzled by the title of the experiment, which suggests that water is the electrolyte. If you work out which ions are present, and what is happening at the electrodes, you will see that there is only one kind of cation. The cations discharged therefore include some which come from water molecules, and some which come from the acid. Each time a cation is discharged, another one is formed by the ionization of water molecules. The total number of cations present in the solution is more or less constant, because water molecules are continuously ionizing to replace those which are discharged. The sulphuric acid, on the other hand, is completely dissociated into ions from the beginning.

Things may be clearer if you think of the anode reaction. The anion which is *selectively* discharged only comes from water molecules, and so as each anion is discharged, more water molecules ionize. It therefore *appears* as if water is acting as the electrolyte because its molecules are continuously ionizing and being discharged, and it can be shown that there is exactly the same amount of sulphuric acid at the end of the experiment as there is at the beginning. The acid is more concentrated at the end, however, for in effect water has been used up during electrolysis.

The reactions taking place appear to be:

$$H_2O(l) \rightarrow H^+(aq) + OH^-(aq)$$
(continuous ionization of water)

at the cathode
$$4H^+(aq) + 4e^- \rightarrow 2H_2(g)$$

at the anode
$$4OH^-(aq) - 4e^- \rightarrow 2H_2O(l) + O_2(g)$$

Note that when the two ionic equations are balanced in terms of the number of electrons lost and gained (i.e. 4 in each case), they show that two molecules of hydrogen are produced for every molecule of oxygen, and this is why (see Avogadro's Law, page 437) the *volume* of hydrogen produced is twice the volume of oxygen produced. These results thus confirm that it is the water which is effectively electrolysed, and that water is produced by the combination of hydrogen and oxygen in the volume (molecular) ratio 2:1.

$$2H_2(g) + O_2(g) \rightarrow 2H_2O(l)$$

For this reason, the experiment is sometimes called the electrolysis of water.

Write up the experiment in the usual way, drawing attention to the volumes of gas produced and the conclusions which can be drawn from them.

Experiment 13.5
The electrolysis of copper(II) sulphate solution using carbon or platinum electrodes *

Apparatus
Suitable voltameter (gas collection needed) with carbon or platinum electrodes, low voltage d.c. supply, connecting wires and crocodile clips, test-tube, splint.
Copper(II) sulphate solution.

Procedure
(a) Set up the circuit in the usual way. Look for signs of activity at each electrode, and collect any gases formed.
(b) When enough gas has been collected (you may need to increase the voltage to achieve this), disconnect the d.c. supply.
(c) Test the gas formed at the anode with a glowing splint.

Notes about the experiment
The results are straightforward, and you should be able to explain what is happening and to write up the experiment in the usual way. Explain what changes occur in the solution, and what you would see if the current was passed for some time. Note that the reaction at the cathode is the one which *always* occurs when copper(II) sulphate solution is electrolysed, but the reaction at the anode depends upon the choice of material used for the anode, as the next experiment shows.

Experiment 13.6
The electrolysis of copper(II) sulphate using a copper anode *

Apparatus
Small beaker with copper strips for anode and cathode (as in Figure 13.8), low voltage d.c. supply, connecting wires and crocodile clips, access to sensitive balance. (Note that the *cathode* could be carbon or platinum instead of copper, but copper is used here for convenience.) Best results are obtained if an ammeter and variable resistance are included in the circuit, and the current is kept constant at about 150 milliamps.
Copper(II) sulphate solution.

Procedure

(a) Weigh the dry anode and cathode, and record the weighings. (Make sure that you can identify which is the cathode, later.)

(b) Electrolyse the solution in the usual way. Look for signs of chemical activity at each electrode, and compare your observations with those from the previous experiment.

(c) After about 45 minutes, stop the current. Remove the anode from the solution, examine it, dry it, and reweigh it. Record the weighing and compare it with the initial one.

(d) With great care, remove the cathode from the solution. Examine it, and allow it to dry naturally, preferably by suspending it in the air so that air circulates around it. Reweigh when dry, record the weighing, and compare it with the initial one.

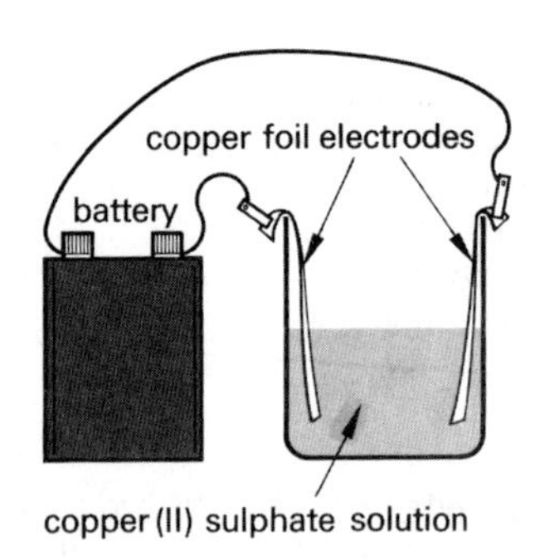

Figure 13.8 The electrolysis of copper(II) sulphate solution using copper electrodes

Notes about the experiment

(i) You should be able to show which ions are present in the electrolyte, and explain the cathode reaction in the usual way. Your weighings should help to confirm what happened at the cathode.

(ii) The anode reaction is a special case. When electrodes are made of platinum or carbon they cannot themselves take part in any chemical changes as they are inert under these conditions. When a copper *anode* is used with copper(II) sulphate solution (or a solution of any other copper(II) salt), there is, however, a third way in which the anode can receive electrons (i.e. in addition to the possibility of gaining electrons from the two different anions in the solution). The atoms in the anode release electrons and form copper(II) ions which pass into solution, and the electrons pass back to the d.c. supply. The anode thus gradually dissolves and loses mass, which your weighings should confirm.

$$Cu(s) - 2e^- \rightarrow Cu^{2+}(aq)$$

(iii) There appears to be no change in the intensity of the blue colour of the electrolyte, because for every $Cu^{2+}(aq)$ ion discharged at the cathode, another $Cu^{2+}(aq)$ ion is formed from the anode. The process can thus continue until all the copper anode has dissolved. The decrease in mass of the anode is equal to the increase in mass of the cathode. Your results should have confirmed this, although in practice the weighings are affected by the fact that some particles of copper are often 'washed off' the cathode, and the finely divided copper which is plated on the cathode is sometimes oxidized to copper(II) oxide.

Points for discussion

1. Explain *why*, in the last experiment, the loss in mass of the anode should equal the increase in mass of the cathode.

2. In Experiment 13.5, the blue colour of the solution gradually fades and the solution becomes acidic if the electrolysis is allowed to continue. Can you explain how this happens?

3. Suppose, in Experiment 13.5, the electrolysis had been allowed to continue for ten minutes and then the connections had been reversed so that the initial cathode became the anode. The electrolysis was then continued for another fifteen minutes, using the same current as before. What would you see happening at the electrodes during the twenty-five minutes?

4. In Experiment 13.3, (i) equal volumes of chlorine and hydrogen are produced, but (ii) the volume of hydrogen collected at the cathode is often greater than the volume of chlorine collected at the anode. Can you explain each of these statements?

Check your understanding

1. An electric current was passed through an unknown solution. The gases which were evolved were collected and tested. The gas from the anode bleached damp litmus paper and the gas from the cathode burned with a squeaky pop. The solution was probably

A sodium hydroxide
B hydrochloric acid
C nitric acid
D sulphuric acid
E copper(II) sulphate

2. Copy out and complete the following paragraph, by adding *one* word in each space.

Solid sodium chloride will not conduct electricity. This is because the ______ in its crystals are not free to move. When the crystals are ______ strongly, they ______ and a current will then flow when a voltage is applied. If carbon electrodes are used, ______ is released at the cathode and ______ at the anode. The process is called ______. (J.M.B.)

13.4 SOME USES OF ELECTROLYSIS

The industrial extraction of reactive metals

Atoms of *reactive* metals easily lose electrons to form positive ions and they occur in nature as these ions. The reverse process (i.e. the conversion back to atoms of the metal by making the ions gain electrons) is quite difficult, and can usually only be achieved by the electrolysis of molten compounds of the metal.

The use of this principle in extracting sodium is described on page 533, and for extracting aluminium on page 534.

Electroplating

Electroplating is the process of coating an object (usually metallic) with a thin layer of another metal by electrolysis. This is a useful way of protecting a metal from corrosion (e.g. by plating with nickel, chromium, or gold, as in Figure 13.9) or of making a metal object more attractive.

The principle is to clean thoroughly the object to be plated, and to make it the *cathode* of a cell. The electrolyte is a solution of a salt of the metal which is to be plated on to the cathode. The anode is usually made of the same metal as that which is present as ions in the solution. You may be allowed to plate an object with copper or nickel in the laboratory.

Figure 13.9 Bumpers emerging from an electroplating bath (*Wilmot-Breeden*)

The purification of copper

This is basically a large scale version of the laboratory electrolysis of copper(II) sulphate solution using copper electrodes (Experiment 13.6). The anodes are impure copper and pieces of pure copper form the cathodes. Only pure copper atoms dissolve from the anodes and are eventually deposited on the cathodes, which increase in mass. This is an important process, for large amounts of copper are used for electrical wiring, and even slight traces of impurities make it a less efficient conductor.

The electrolysis of sodium chloride solution (brine)

Experiment 13.3 is modified on a large scale in order to prepare hydrogen, chlorine, and sodium hydroxide (three important industrial chemicals) all by the one reaction, from a relatively cheap and readily available material. The industrial details are given on page 536.

13.5 FARADAY'S LAWS

Michael Faraday (1791–1867) was a brilliant scientist. He invented the dynamo and did a great deal of pioneer work on the chemical effects of electricity. To him we owe the terms anode, cathode, and ion (Greek: *ana*, 'up'; *kata*, 'down'; *odos*, 'path'; *ion*, 'a traveller'). His most important contribution in the field of electrolysis was his discovery of the relationship between the amount of electricity used and the quantity of chemical change that it produced.

Some electrical units

Before you can understand Faraday's work and the results he achieved, you must know the meaning of several of the units used. You are probably already familiar with the ampere as the unit of current strength and the volt which is a measure of potential difference. Quantity of electricity is measured in *coulombs* and this is found by multiplying the current in amperes by the time in seconds that the current flows. Thus if a current of two amperes flows for five minutes, the quantity of electricity used is $(2 \times 5 \times 60)$ coulombs, i.e. 600 coulombs.

Number of coulombs used = current in amps × time of current flow in seconds.

Finding the relationship between the quantity of electricity used and the mass of the product

A solution of copper(II) sulphate is electrolysed using copper electrodes. The copper cathode is weighed before and after the electrolysis. The quantity of electricity is found from the current strength and the time. The experiment is repeated many times using different quantities of electricity. The experiment is too lengthy and exacting to be carried out in a school laboratory.

The table shows a typical set of results:

Current (A)	*Time (s)*	*Quantity of electricity (C)*	*Mass of copper (g)*
0.5	600	300	0.099
1.0	600	600	0.198
1.5	600	900	0.297
2.0	600	1200	0.396
2.5	600	1500	0.495
2.0	900	1800	0.594
2.0	1000	2000	0.660
2.0	1200	2400	0.792

Points for discussion

1. Draw a graph of these results, plotting quantity of electricity along the x axis and mass of product along the y axis.

2. Faraday carried out many similar experiments in which he measured the masses of the products of electrolysis and the quantity of electricity used. From his results Faraday formulated his *First Law of Electrolysis*: 'The mass of a substance dissolved off, or produced at, an electrode during electrolysis, is proportional to the quantity of electricity which passes through the electrolyte.'
Does your graph support this law?

The relationship between the masses of different substances released by the same quantity of electricity

The discovery of this relationship was another of Faraday's brilliant pieces of work and is embodied in *Faraday's Second Law*: 'The masses of different elements released by the same quantity of electricity, form simple whole number ratios when divided by their relative atomic masses'.

We can illustrate this more clearly by giving you some results of experiments with a number of different voltameters. For example, when the same current is passed for the same length of time through a number of different voltameters (Figure 13.10), the masses of the elements released are as follows:

Silver	*Magnesium*	*Hydrogen*	*Lead*	*Aluminium*
0.535 g	0.060 g	0.005 g	0.515 g	0.045 g

Dividing the mass of each metal by its relative atomic mass:

$\frac{0.535}{108}$	$\frac{0.060}{24}$	$\frac{0.005}{1}$	$\frac{0.515}{207}$	$\frac{0.045}{27}$
$=0.005$	$=0.0025$	$=0.005$	$=0.0025$	$=0.0017$

An interesting pattern has now emerged from this division, for the number obtained in the case of silver is the same as for hydrogen, twice that obtained for magnesium and lead, and three times that obtained for aluminium.

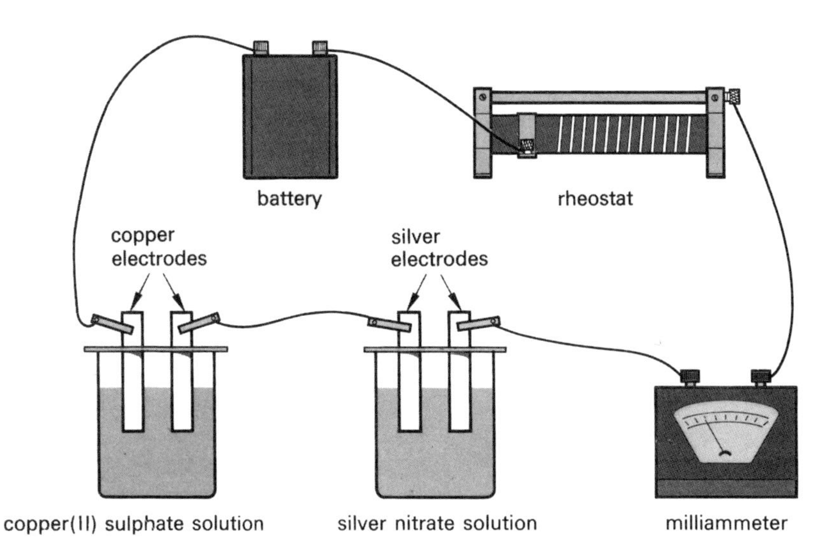

Figure 13.10 Comparing the masses of silver and copper produced by the same quantity of electricity

Faraday's laws and the electronic structure of the atom

The *smallest* quantity of electricity that will release one mole of atoms (discharge one mole of ions) of any element is 96 500 coulombs. This quantity of electricity is termed the *faraday*. One faraday will discharge one mole of silver ions, one mole of sodium ions, one mole of hydrogen ions, but two faradays are needed to discharge one mole of magnesium or calcium ions, and three faradays are needed for one mole of aluminium ions.

As a faraday will discharge a mole of silver ions and as each electron discharges one silver ion it follows that a mole of silver ions will be discharged by a mole of electrons. A faraday can thus be considered as a *mole of electrons.*

Each silver ion carries one positive charge (Ag^+) and one faraday will discharge a mole of silver ions. As two faradays are needed to discharge a mole of magnesium ions (i.e. the same number of ions), each magnesium ion must need two electrons for its discharge and so the magnesium ion can be written Mg^{2+}.

One faraday of electricity (96 500 C) consists of a mole of electrons.
One faraday is needed to discharge a mole of a monovalent ion during electrolysis, two faradays for a mole of a divalent ion, and three faradays for a mole of a trivalent ion.

These facts, derived from Faraday's Laws, can be used to calculate the masses of substances taking part in, and produced by, electrochemical reactions.

Two typical examples are given.

1. The same current is passed for the same time through a solution of silver nitrate and through a separate voltameter containing dilute sulphuric acid. If 0.108 g is deposited on the silver cathode, what mass of hydrogen gas will be liberated?

Ag^+ and H^+ are both univalent, so the same quantity of electricity will liberate an equal number of moles of atoms of each.

$$0.108 \text{ g of silver is } \frac{0.108}{108} \text{ moles of silver} = 0.001 \text{ mol}$$

$\therefore$ 0.001 moles of hydrogen atoms (*not* molecules) will also be liberated.

$$0.001 \text{ moles of hydrogen atoms} = \frac{0.001}{1} \text{ g} = 0.001 \text{ g}$$

2. Calculate the mass of magnesium produced by the electrolysis of fused magnesium chloride, if a current of 1.93 A is passed for 16 minutes and 40 seconds.

$$16 \text{ minutes and } 40 \text{ seconds} = 1000 \text{ seconds.}$$

Quantity of electricity used $= 1000 \times 1.93$ coulombs $= 1930$ C

$$1930 \text{ C} = \frac{1930}{96\,500} \text{ faradays}$$

Two faradays will discharge one mole of magnesium ions, Mg^{2+}, i.e. 24 g

$$\therefore \frac{1930}{96\,500} \text{ faradays will discharge } \frac{1930}{96\,500} \times 12 \text{ g magnesium.}$$

$\therefore$ Mass of magnesium liberated $= 0.24$ g

Check your understanding

1. In the manufacture of aluminium (p. 534), a current of 30 000 A was passed through the cell for 24 hours. What mass of aluminium was produced? (Take the faraday to be 96 500 C.)

2. In an electrolysis experiment the same current was passed for the same time through two cells in series. In the first voltameter 0.081 g of copper were produced and in the second, 0.162 g of copper were formed. If the symbol for the copper ion in the first case is Cu^{2+} what is the symbol for the ion of copper in the second case?

3. A current of 0.36 A produces 0.23 g of lead in ten minutes. Calculate (a) the quantity of electricity used, (b) the number of faradays needed to produce one mole of lead atoms.

4. If an electric current of 0.1 A is passed through an electrolyte for 1.25 hours, how much electricity (in coulombs) is used?

5. How many faradays are needed to produce:
(a) 2.70 g of aluminium
(b) 6.0 g of magnesium
(c) 10 g of hydrogen
(d) 71 g of chlorine

13.6 A SUMMARY OF CHAPTER 13

The following 'check list' should help you to organize the work for revision.

1. Definitions

(a) Electrode, anode and cathode (p. 207)
(b) Electrolyte and electrolysis (p. 210)
(c) Coulomb (p. 222) and the faraday (p. 224)
(d) Faraday's First and Second Laws of Electrolysis (p. 223)

2. Other ideas

(a) Conductors, non-conductors, and insulators.
(b) Electrical conductors can conduct a flow of electrons, and are unchanged chemically when they do so. Electrolytes effectively conduct a flow of electrons, but the current in the liquid is transferred by moving ions; electrolytes are decomposed chemically when they conduct.
(c) Electrolytes must contain ions, and they only conduct electricity when they are molten or dissolved in water.
(d) What happens during electrolysis, e.g. how the flow of electrons is maintained and why there are chemical changes at the electrodes (p. 210).
(e) Cations are positively charged ions which are attracted to the cathode during electrolysis; they are either hydrogen ions or ions of a metal, and if they are discharged they *gain* electrons from the cathode and become neutral atoms and/or molecules.
(f) Anions are negatively charged ions which are attracted to the anode during electrolysis; they are ions of non-metals or radicals, and if they are discharged they *give* electrons to the anode and become neutral atoms and/or molecules.
(g) In aqueous electrolytes, some water molecules ionize and provide additional ions, $H^+(aq)$ and $OH^-(aq)$. Either or both of these may be discharged during electrolysis, i.e. the $H^+(aq)$ ions becoming H_2 molecules and the $OH^-(aq)$ ions becoming water and O_2 molecules.
(h) What is meant by selective discharge (p. 213).
(i) The ionic equations normally used in electrolysis (p. 215).
(j) Common voltameters and their uses (p. 216).
(k) How to write up an electrolysis experiment (e.g. p. 217).
(l) The choice of the material used for the anode in the electrolysis of copper(II) sulphate solution can affect the reaction which occurs at the anode. If a copper anode is used, this 'dissolves':

$$Cu - 2e^- \rightarrow Cu^{2+}(aq)$$

(m) Applications of electrolysis, as described on pages 221–2.
(n) Calculations involving Faraday's Laws of Electrolysis.

3. *Important experiments*

The experimental details, ionic equations and explanations involved in

(a) the electrolysis of sodium chloride solution (e.g. Experiment 13.3);
(b) the electrolysis of dilute sulphuric acid (electrolysis of water), e.g. Experiment 13.4;
(c) the electrolysis of copper(II) sulphate solution using carbon or platinum electrodes (e.g. Experiment 13.5);
(d) the electrolysis of copper(II) sulphate solution using a copper anode (e.g. Experiment 13.6).

Note that some syllabuses also include the electrolysis of sodium hydroxide solution. The important details of all these electrolysis examples are summarized in Table 13.1.

QUESTIONS

1. A constant current of 2.0 A (amperes) was passed for nearly three hours through a concentrated solution of copper(II) chloride, using inert electrodes. The following observations were recorded.

Stage I—after 10 minutes
A pink deposit was visible on one electrode and a gas which bleached moist litmus paper was evolved at the other electrode.

Stage II—after 1 hour
The same pink deposit and the same gas were formed as at stage I, but the solution was much paler in colour.

Stage III—after 2 hours
The pink deposit was no longer being formed: instead a colourless gas was evolved. At the other electrode, there were still bubbles of gas but moist litmus paper was no longer bleached by the gas.

Stage IV—after 3 hours
All bubbling had ceased and the liquid would no longer conduct electricity.

(a) Give the formulae of all the ions present in the copper(II) chloride solution.
(b) (i) At which electrode (positive or negative) was the pink deposit formed?
(ii) Name the gas responsible for bleaching the moist litmus paper.
(c) Why had the colour of the copper(II) chloride solution faded by stage II?
(d) Name the two gases being evolved at stage III. Give the equation for each electrode reaction and describe the tests you would carry out to confirm the identity of each gas.
(e) Why had the electrolysis ceased at stage IV?
(f) How many coulombs of electricity had been used in 2 hours? (C.)

2. Explain clearly why, when copper(II) sulphate solution is electrolysed using platinum electrodes, oxygen is the product at the anode, but a different result is obtained when copper electrodes are used.

How is electrolysis used for the purification of crude copper? Describe two other commercial uses of electrolysis.

3. (a) Give the detailed description of the preparation of a small sample of lead by electrolysis of a fused compound.
(b) Give an outline of the *industrial* preparation of a metal by electrolysis.
(c) A current of 0.5 A, flowing for six minutes twenty-six seconds, through two cells in series, was found to deposit 0.216 g of silver on the cathode of the first cell and 0.059 g of nickel on the cathode of the second. The relative atomic masses of silver and nickel are 108 and 59 respectively. Calculate (i) the quantity of electricity passed through the two cells, (ii) the quantity of electricity needed to deposit one mole of silver atoms, (iii) the quantity of electricity needed to deposit one mole of nickel atoms. Comment on the results of your calculations in (ii) and (iii). (C.)

Table 13.1 Some examples of electrolysis

Ions present	*At the cathode*	*At the anode*	*Ions left in solution*
Copper(II) sulphate solution with platinum or carbon electrodes			
$CuSO_4 \rightarrow Cu^{2+}(aq) + SO_4^{2-}(aq)$ $H_2O \rightleftharpoons H^+(aq) + OH^-(aq)$	Copper ions selectively discharged $2Cu^{2+}(aq) + 4e^- \rightarrow 2Cu(s)$	Hydroxide ions selectively discharged $4OH^-(aq) - 4e^- \rightarrow 2H_2O(l) + O_2(g)$	$H^+(aq)$ and $SO_4^{2-}(aq)$. Solution gradually becomes colourless dilute sulphuric ac d
Sodium chloride solution with carbon electrodes			
$NaCl \rightarrow Na^+(aq) + Cl^-(aq)$ $H_2O \rightleftharpoons H^+(aq) + OH^-(aq)$	Hydrogen ions selectively discharged $2H^+(aq) + 2e^- \rightarrow H_2(g)$	Chloride ions selectively discharged $2Cl^-(aq) - 2e^- \rightarrow Cl_2(g)$	$Na^+(aq)$ and $OH^-(aq)$. Solution gradually becomes dilute sodium hydroxide solution
Dilute sulphuric acid with platinum or carbon electrodes			
$H_2SO_4 \rightarrow 2H^+(aq) + SO_4^{2-}(aq)$ $H_2O \rightleftharpoons H^+(aq) + OH^-(aq)$	Hydrogen ions discharged $4H^+(aq) + 4e^- \rightarrow 2H_2(g)$	Hydroxide ions selectively discharged $4OH^-(aq) - 4e^- \rightarrow 2H_2O(l) + O_2(g)$	$H^+(aq)$ and $SO_4^{2-}(aq)$. The sulphuric acid gradually becomes more concentrated
Sodium hydroxide solution with platinum or carbon electrodes			
$NaOH \rightarrow Na^+(aq) + OH^-(aq)$ $H_2O \rightleftharpoons H^+(aq) + OH^-(aq)$	Hydrogen ions selectively discharged $4H^+(aq) + 4e^- \rightarrow 2H_2(g)$	Hydroxide ions discharged $4OH^-(aq) - 4e^- \rightarrow 2H_2O(l) + O_2(g)$	$Na^+(aq)$ and $OH^-(aq)$. The sodium hydrox de solution slowly becomes more concentrated
Copper (II) sulphate solution with copper electrodes			
$CuSO_4 \rightarrow Cu^{2+}(aq) + SO_4^{2-}(aq)$ $H_2O \rightleftharpoons H^+(aq) + OH^-(aq)$	Copper ions selectively discharged $Cu^{2+}(aq) + 2e^- \rightarrow Cu(s)$	Copper atoms from the anode dissolve as ions $Cu(s) - 2e^- \rightarrow Cu^{2+}(aq)$	Colour of solution remains unchanged, because a $Cu^{2+}(aq)$ ion is formed for every $Cu^{2+}(aq)$ discharged The anode loses mass and the cathode gains in mass by the same amount

14 Using the Periodic Table

14.1 THE PERIODIC TABLE

The structure of the table

Imagine what it would be like to have to learn the chemical and physical properties of all the elements and compounds now known. This would be a frightening task, and fortunately we do not have to do it as the elements fall naturally into groups showing similar properties. Chemists have classified the elements in such a way that if we study just one member of a group of similar elements, it is possible to predict the likely behaviour of other elements (and their compounds) in the same group.

One of the important classification systems used in chemistry is the Periodic Table, in which the elements are arranged in order of increasing atomic number, i.e. the number of protons (which is the same as the number of electrons) in their atoms. As we have already seen, the number of electrons in the outer shell of an atom determines the chemical properties of the atom. Each row of elements in the Periodic Table ends with an element which has atoms containing a full outer shell of electrons. It will be helpful to think of the first part of the Periodic Table (i.e. the first twenty elements) as shown in Figure 14.1, which illustrates the points made so far. A full Periodic Table is reproduced at the front of the book.

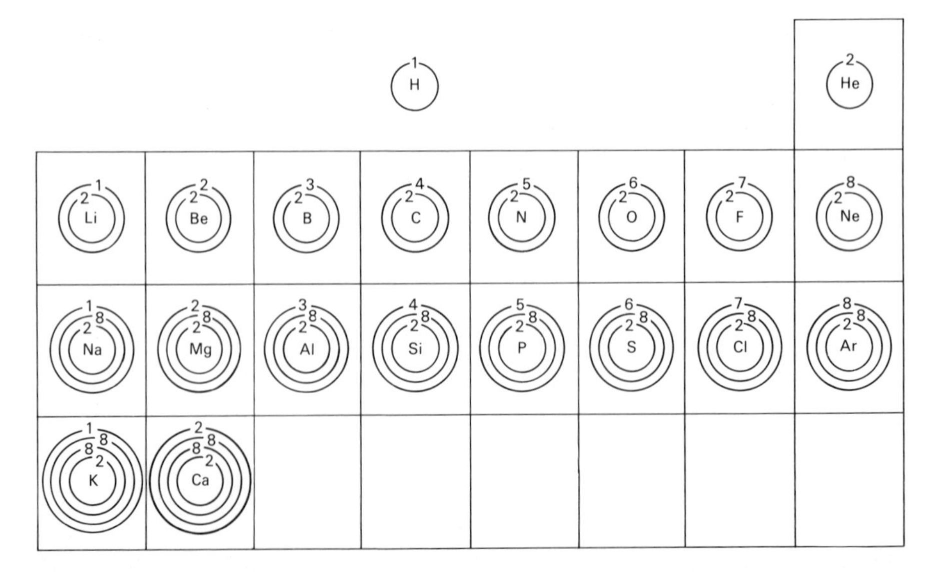

Figure 14.1 The electronic structure of the first twenty elements in the Periodic Table

Vertical columns of elements in the Periodic Table are called *groups*. Horizontal rows of elements in the table are called *periods*.

Certain parts of the table have special names or common names, as shown in Figure 14.2.

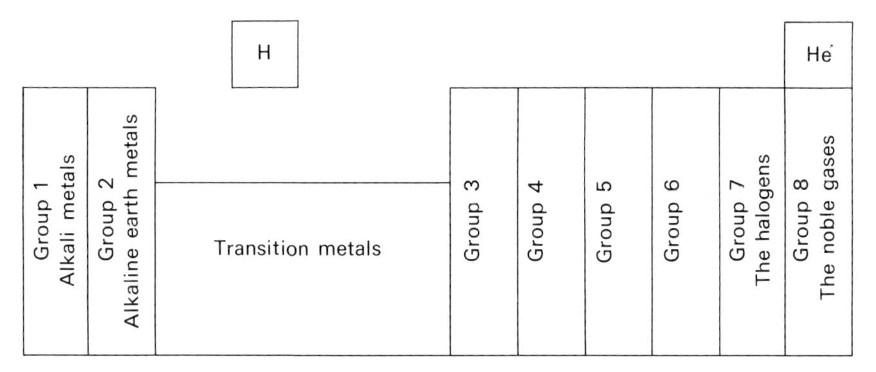

Figure 14.2 Regions of the Periodic Table

The elements in any one group of the table have similar properties

The elements in any one group all have the same number of electrons in the outer shells of their atoms (Figure 14.1). As the arrangement of electrons decides chemical properties, it follows that all the members of a group will have similar chemical properties.

You will probably study some of the properties of the elements in Groups 1, 2, and 7, and you will then be able to give examples showing how the members of a group have similar chemical properties.

Note that hydrogen is not placed in the same group as other elements; it is unique, and does not really belong to any group. You might imagine that hydrogen should be placed in Group 1 as its atoms have one electron in their outer shell. We could equally well argue that it should be placed in Group 7 because, like the halogens, its atoms are one electron short of a fully filled shell! Clearly, hydrogen is difficult to place in the table, and it is best left on its own (see also page 126).

Hydrogen was placed on its own

The elements in any one group all show the same main valency

Atoms of the elements in Group 1 all need to *lose* one electron to gain the electron structure of a noble gas, and they therefore all have a valency of one. Similarly, elements of Groups 2 and 3 have valencies of 2 and 3 respectively. Group 4 elements have a valency of four, although they nearly always *share* electrons in order to form stable structures rather than lose or gain four electrons, which is very difficult.

The elements in Groups 5, 6, and 7 have valencies of three, two, and one respectively, NOT five, six, and seven. This is because their atoms usually *gain* electrons in order to obtain stable electron structures. For example, as the elements of Group 7 need to gain one electron to achieve stable electron structures, they all have a valency of one.

Group number	1	2	3	4	5	6	7	8
Usual valency of elements	1	2	3	4	3	2	1	0

(Note that we do not include the transition elements in these eight normal groups.)

An important consequence of this is that if you know the formula of a compound of an element in, say, Group 3 (e.g. Al_2O_3), then it is possible to predict the formula of a similar compound of another Group 3 element, even if you have never heard of it, e.g. Ga_2O_3.

14.2 CHANGES ACROSS A TYPICAL PERIOD

Metallic and non-metallic properties

The elements in Groups 1, 2, and 3 (i.e. those on the left-hand side of the Periodic Table) tend to be metals because their atoms have 1, 2, or 3 electrons in their outer shells. Elements on the right-hand side (Groups 5, 6, 7, and 8) are non-metals as their atoms have 5, 6, 7, or 8 electrons in their outer shells. Elements therefore become more non-metallic in character from left to right across a typical period. Note that Group 4, in the middle of a typical period, contains both metals and non-metals, but the metals are found at the bottom of the group.

Chemical reactivity

An element of Group 1 is more reactive than the Group 2 element in the same period because it is easier to lose one electron (in forming stable electronic structures) than it is to lose two. This trend continues across the table so that chemical reactivity is high at the beginning of a period and falls to a fairly low point in Group 4. From this point, elements need to *gain* electrons and as it is easier to gain two electrons (Group 6 elements) than three electrons (Group 5 elements), reactivity increases again and reaches another peak at Group 7. Group 8 elements have little or no chemical reactivity.

You may have done experiments which show this trend, by comparing the chemical behaviour of each of the elements in a period with the same substance, e.g. water or steam. The trend in chemical reactivity is summarized in Figure 14.3.

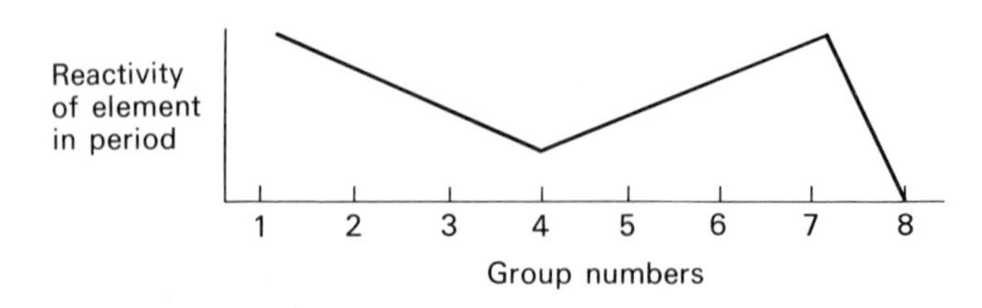

Figure 14.3 Reactivity trends across a period

The type of oxide formed

There is a change in the type of oxide formed by the elements in a period, because of the increase in non-metallic character from left to right across a period. As metal oxides are generally basic, oxides of elements on the left-hand side of a period tend to be basic. As non-metallic oxides are usually acidic (or occasionally neutral), the oxides of elements on the right-hand side of a period tend to be acidic (or neutral). Sometimes, this gradual change across a period produces an 'intermediate' oxide which is amphoteric, e.g. aluminium oxide in Group 3 of Period 3 (see the different types of oxide, page 153).

Structural changes

The structures of both the elements and their compounds follow a general trend across a period, and this is discussed in Chapter 16.

The special case of the transition elements

These blocks of elements (e.g. the block of 10 containing iron, chromium, manganese, and copper) in the 'centre' of the table are very similar when considered as a *horizontal* row. They (or their compounds) usually have the following properties.

1. The elements are always hard, dense metals with high melting and boiling points.
2. The elements often have more than one valency, e.g. iron, 2 and 3.
3. Their compounds are often coloured.
4. The elements and their compounds often show catalytic properties, and are used as catalysts in many major industrial processes, examples of which are given in Chapter 27.

Check your understanding

1. Name (a) three alkali metals, (b) two halogens, (c) two noble gases, (d) one insoluble non-metallic oxide.

2. Three elements, with atomic numbers between 10 and 19, have symbols X, Y, and Z. An X atom has two electrons more than a noble gas, a Y atom has one electron less than a noble gas, and a Z atom three electrons less than a noble gas. Using the above information and the letters X, Y, and Z:

(a) Write the simplest formula for the compound which would be formed between (i) elements X and Y, (ii) elements X and Z, (iii) elements Y and Z.

(b) In which of the compounds, if any, would there be ionic bonding? Give an explanation for your answer.

(c) State, giving an explanation for your answer, which one of the elements, if any, would conduct electricity at room temperature and pressure. (J.M.B.)

14.3 A STUDY OF THE GROUP 1 ELEMENTS (THE ALKALI METALS)

Atomic and ionic sizes

Table 14.1 gives some information about the elements of Group 1. Simply by looking at the data about the radii of the atoms, or by drawing scale diagrams to compare the diameters of the atoms, you will see that the atomic radii increase as the atomic number increases down the group. An atom of lithium (atomic number 3) is smaller than an atom of sodium (atomic number 11) because it contains fewer shells of electrons.

Note that the elements of Group 1 all form ions with a charge of 1+, e.g. Na^+. The ionic radii also increase as the atomic numbers increase because each ion has one more electron shell than an ion of the element above it in the group. The radius of each ion is, however, less than that of the 'parent' atom from which it was formed because an electron shell has been lost. For example, a sodium ion is smaller than a sodium atom. Again, scale diagrams (from the data in Table 14.1) make this point more obvious.

Table 14.1 Some physical properties of the Group 1 metals

	Symbol	*Atomic number*	*Mass number*	*Melting point °C*	*Radius of atom (nm)*	*Radius of ion (nm)*	*Density ($g\ cm^{-3}$)*
Lithium	Li	3	7	79	0.133	0.078	0.535
Sodium	Na	11	23	98	0.157	0.098	0.971
Potassium	K	19	39	64	0.203	0.133	0.862
Rubidium	Rb	37	85	39	0.216	0.149	1.53
Caesium	Cs	55	133	29	0.235	0.165	1.90
Francium	Fr	← Unstable radioactive element →					

A metallic ion (i.e. a positive ion) is always smaller than the atom from which it came. (In most of the examples encountered at an elementary level this is because the ion has one shell of electrons less than the atom (see Figure 14.4).)

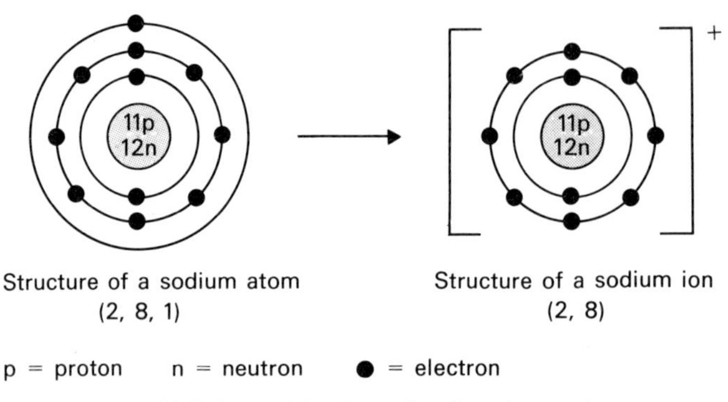

Figure 14.4 A metal ion is smaller than the metal atom

The reactions of the elements with cold water

The next experiment investigates how the elements of Group 1 react with water, and it also shows some of their physical properties.

Experiment 14.1
The reactions of sodium, potassium, and lithium with cold water**

Apparatus

Large trough, 450 cm^3 beaker, sharp knife, tongs, filter paper, white tile, safety screen. Sodium, lithium, and potassium metals (the lithium should be already cut into small pieces), universal indicator solution.

Procedure

Notes: (i) The metals sodium and potassium must be used only by the teacher.
(ii) This experiment is very much a test of your ability to make observations. Watch carefully, and see how many facts you can note about the metals being used.

(a) Cut off a small piece of potassium about the size of a rice grain on a dry white tile (away from any sink or water supply), making sure that no pieces fly off. Return the unwanted metal to the bottle, and replace the cap.
(b) Carefully drop the piece of potassium on to the centre of a large trough of water behind a safety screen.
(c) When the reaction has finished, add a few drops of universal indicator to the 'water' in the trough. Empty the trough and refill with clean water.

(d) Repeat steps (a) to (c) but using sodium instead of potassium.
(e) Repeat steps (a) to (c) but using lithium instead of potassium.
This reaction with lithium can be done on a smaller scale, using a large beaker rather than the trough.

Results
Set out your results in a table, using headings such as 'chemical reactivity' and 'does it melt easily?'. It is possible to make suitable headings for eight or more observations.

Notes about the experiment

If you saw a flame when using potassium, it was not just the metal which was burning. What else do you think was burning? (The heat of the reaction vaporizes some of the metal, and this burns together with one of the products. This can be demonstrated by holding a Bunsen burner flame horizontally about 10 cm above any of the reacting metals during the experiment. The vaporized metals then produce their characteristic flame tests, as on page 451).

Conclusion
Make your own conclusion to the experiment (it should be longer than usual) based on answers to the following questions; do not just answer the questions.
(a) Is it reasonable to place these metals in the same group? (Give evidence to support your answer.)
(b) Are these metals different from other metals, i.e. metals in other groups? (Again give evidence to support your answer.)
(c) *Why* are these elements so similar?
(d) Although these metals are similar (i.e. they show the same kinds of properties), do they differ in their reactivity (i.e. the vigour with which they give their chemical reactions)? If the answer is yes, *how* do they differ in reactivity, e.g. does their reactivity increase as their atomic number increases, or what?
(e) The equation for the reaction between potassium and water is as shown below. What will be the equations for the other two reactions?

$$2K(s) + 2H_2O(l) \rightarrow 2KOH(aq) + H_2(g)$$

Points for discussion

1. Why are these metals called the alkali metals? (Care: the metals are not alkalis!)
2. Which of the Group 1 elements would sink in water?
3. Why do you think that rubidium was not used in the experiment?

The reactions of the elements with air

Remember that the air is a mixture, and the Group 1 metals react with the oxygen, the water vapour, and (indirectly) the carbon dioxide in the air, so the reactions can be quite complex. However, we are normally concerned only with the reactions between the metals and the oxygen in the air. A summary of a more complicated *series* of reactions (using sodium as an example) is given on page 338.

In the cold
Lithium, sodium, and potassium are normally stored under paraffin oil, to prevent them coming into contact with the air. When a piece of an alkali metal is removed from the paraffin oil, it does not look like a metal because in spite of its protection by the oil, it has still reacted slightly with oxygen in the air to form a dull coating of the metallic oxide. This is usually a thin layer only, and if the metal is cut open, the freshly cut surface is very shiny and looks metallic. The shiny surface does not last long, however, for it rapidly reacts in the cold with oxygen in the air and forms a new coating of the oxide, e.g.

$$4Na(s) + O_2(g) \rightarrow 2Na_2O(s).$$

The reactions of the elements when heated in the air
It is always potentially dangerous to heat the alkali metals, and these reactions will probably be demonstrated for you, behind a safety screen. Typical observations are given below. This time the oxidation is complete (compared with the surface coating formed in the cold) and the metals *burn*.

Metal	Observations when heated in air
Sodium	Melts to a silvery liquid. Burns with a yellow flame to form a white solid. $$2Na(s)+O_2(g) \rightarrow Na_2O_2(s)$$ Note that sodium forms two oxides, Na_2O and Na_2O_2. The latter, sodium peroxide, is formed when the metal burns, but the former is the product when an oxide coating forms in the cold
Lithium	Melts to a silvery liquid. Burns with a scarlet flame to form a white, solid oxide
Potassium	Melts to a silvery liquid. Burns with a lilac flame to form a white, solid oxide

There is a reactivity trend (e.g. potassium is more reactive than sodium and lithium) but it is not quite so obvious in this experiment. You may have noticed that lithium can be melted without catching fire, but potassium ignites as soon as it melts.

The reactivity trend in Group 1

You should have noticed in Experiment 14.1 that the reactivity of the elements in Group 1 with water increases as the atomic number increases, i.e. as we 'go down the group'. The same trend is true if their reactivity is compared in other experiments.

The reason for this trend is that these elements (like all metals) tend to *lose* electrons to form stable electron structures. (The elements of Group 1 all lose one electron when they react.) To remove an electron requires energy, and the further away the electron is from the nucleus, the less energy is required to remove it. Look at the atomic structures of lithium, sodium, and potassium. It is much easier for an atom of potassium to lose its outer electron than it is for an atom of lithium to do so, because the outer electron is further from the nucleus in a potassium atom. Potassium is therefore more reactive than sodium, which is more reactive than lithium.

Another way of looking at this is to imagine that the outer electron of potassium has fewer orbits to 'jump' into before it escapes from the atom. Similarly, lithium is less reactive because its outer electron has to move through more orbits before it escapes, and this needs more energy.

Check your understanding

1. Element X is in Group 1 of the Periodic Table. It is likely to be
A a very reactive non-metal
B an element which readily forms X^- ions
C a dense, hard metal with a high melting point
D a light, soft metal with a low melting point
E a dense, soft metal with a high melting point

2. Which of the following statements about the Periodic Table is *not* correct?
A Elements with similar properties occupy the same vertical columns
B There are many more metals than non-metals
C The most reactive metals are in Group 1
D The non-metals are found in the bottom right-hand corner of the table
E The elements are listed in order of atomic number

3. The elements of Group 1, going down the column,
A become less reactive
B have smaller ions
C become less electronegative
D have decreasing atomic numbers
E show an increasing tendency to gain an electron

4. The element with atomic number 10 is likely to have similar properties to the element with atomic number
A 9; B 11; C 16; D 18; E 28

14.4 A STUDY OF THE GROUP 2 ELEMENTS

Atomic and ionic radii

Table 14.2 Some physical properties of the Group 2 elements

	Symbol	*Atomic number*	*Mass number*	*Melting point °C*	*Radius of atom (nm)*	*Radius of ion (nm)*	*Density ($g\ cm^{-3}$)*
Beryllium	Be	4	9	1283	0.112	0.031	1.85
Magnesium	Mg	12	24	650	0.160	0.065	1.74
Calcium	Ca	20	40	850	0.197	0.099	1.55
Strontium	Sr	38	88	770	0.215	0.113	2.58
Barium	Ba	56	137	710	0.221	0.135	3.50
Radium	Ra	← Unstable radioactive element →					

Information about the Group 2 elements is given in Table 14.2. Note (and if possible draw scale diagrams to represent) the trends in atomic and ionic radii of the first few elements. These trends are the same as for Group 1 (p. 232), and for the same reasons.

Note that once again the ions are smaller than the 'parent' atoms, because each atom completely loses its outer shell of electrons in forming an ion. The ions of Group 2 elements have a charge of +2, e.g. Mg^{2+}.

Point for discussion

Compare the radii of the Group 2 ions with those of the corresponding members of Group 1. Can you suggest why they are smaller?

The reactions of the elements of Group 2 with water

Although these elements do react with water, they tend to do so less vigorously than the corresponding elements of Group 1. This is because the Group 2 elements all have atoms which need to lose two electrons (to become stable) but the atoms of Group 1 elements only need to lose one electron. It requires more energy to remove two electrons than to remove one. It is therefore important to use more drastic conditions, if necessary, to bring about a reaction, and this is why magnesium is reacted with both cold water and steam in the next experiment.

Experiment 14.2
The reactions of calcium and magnesium with cold water or steam**

Apparatus
Apparatus as in Figure 14.5(a), 250 cm^3 beaker, spatula, test-tube, splint, steam can, safety screen.
Magnesium ribbon (cleaned with emery paper), calcium turnings.

Procedure
(a) Make out a table for your results as follows:

Metal	*Action on cold water*	*Reaction with steam*

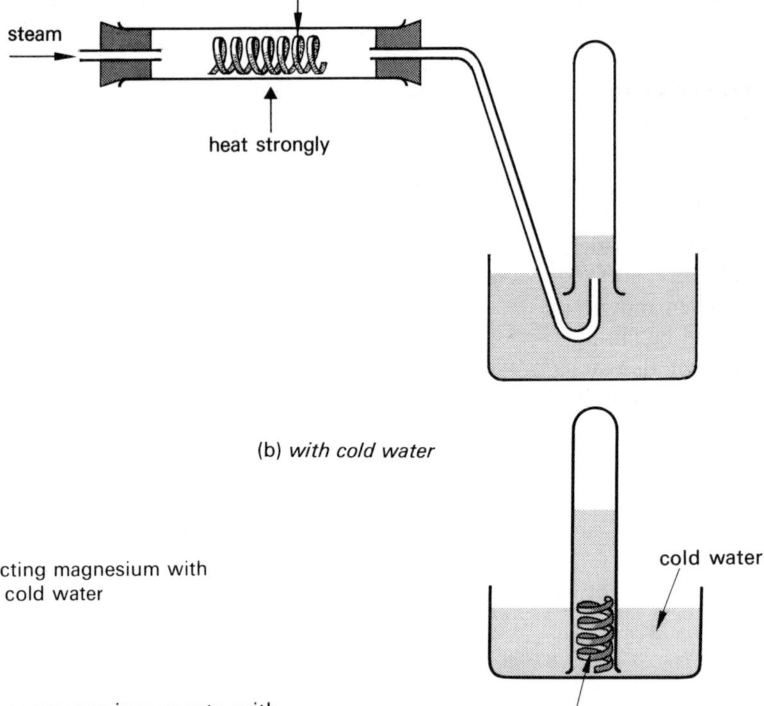

Figure 14.5 Reacting magnesium with (a) steam and (b) cold water

(b) Find out how magnesium reacts with cold water by setting up the apparatus shown in Figure 14.5(b). Look for signs of immediate reaction, and leave the apparatus for a week. Test any gas formed for hydrogen.
(c) You will find out how magnesium reacts with steam by watching a demonstration using the apparatus shown in Figure 14.5(a), behind a safety screen. The steam can is heated until the water boils and steam passes over the magnesium. The magnesium is then heated strongly so that steam is passing over the hot metal. Any gas formed is collected in the test-tube, and tested when the reaction is over. Care is needed to prevent 'sucking back' of the water into the hot combustion tube. Under no circumstances do you stop heating the tube while the end of the delivery tube is under water. Remove the test-tube and lift the apparatus clear of the water before heating is stopped.
(d) Half fill a beaker with cold water, and carefully add a few small pieces of calcium. Try to collect some of the gas given off by holding a test-tube (filled with water) over one of the reacting pieces but do not trap a piece of metal inside the test-tube. Test the gas for hydrogen. Record all of your observations in the table.

Notes about the experiment

(i) The equations for the reactions are:

$$Ca(s) + 2H_2O(l) \rightarrow Ca(OH)_2(aq) + H_2(g)$$
$$Mg(s) + H_2O(g) \rightarrow MgO(s) + H_2(g)$$

(ii) Note that the products are not quite comparable. In fact magnesium also first forms the hydroxide, $Mg(OH)_2$, but this then decomposes to the oxide because of the high temperature inside the combustion tube.
(iii) Your observations should have included the appearance of a white solid or cloudiness with the calcium reaction. This is calcium hydroxide, one of the products, which is only slightly soluble in water.

Conclusion
Give equations for the reactions, and make your own conclusion based on answers to the following questions.

1. Is there a difference in reactivity between these two metals? Give evidence for your answer.
2. Can you explain why there is (or is not) a reactivity trend like that shown by the Group 1 elements?

Points for discussion

1. What is meant by 'sucking back', and how is it prevented in the last experiment?

2. Why can you be certain that any gas collected in the test-tube does not contain steam?

The reactions of the elements of Group 2 with the air

As with the Group 1 elements, magnesium and calcium also react with oxygen in the air (without heating), although they are less likely to react with water vapour. They are not as reactive as the elements of Group 1, however, and so magnesium is not kept in paraffin oil, although calcium sometimes is. The metals are always covered in a layer of oxide (and also nitride in the case of calcium), which makes them look dull.

When magnesium is *heated* in air, the oxidation is complete and it burns; you should be familiar with this reaction. Calcium also burns when heated, but then the product also includes a significant proportion of the nitride, as well as the oxide. You may see a demonstration of the combustion of calcium; remember that both elements burn even more vigorously in pure oxygen (p. 150).

The main reactions occurring on combustion are:

$$2Mg(s) + O_2(g) \xrightarrow[\text{dazzling white flame}]{} \underset{\text{white powder}}{2MgO(s)}$$

$$2Ca(s) + O_2(g) \xrightarrow[\text{brick-red flame}]{} \underset{\text{white powder}}{2CaO(s)}$$

Calcium is again more reactive than magnesium, but the effect is less obvious because the calcium is covered with a thicker protective layer of oxide and nitride (it has reacted with cold air more than magnesium has) and is slightly more difficult to ignite.

The reactions of magnesium and calcium with dilute acids

You may have studied these reactions previously (e.g. Experiment 5.3, page 60), but the emphasis was probably different from that which is needed here. It is not safe to add the Group 1 metals to acids, but in the next experiment you will investigate how calcium and magnesium react with dilute solutions of the mineral acids.

Experiment 14.3
The reactions of calcium and magnesium with dilute acids*

Apparatus

Rack of test-tubes, spatula, splint.
Samples of magnesium and calcium metals, dilute solutions of nitric, hydrochloric, and sulphuric acids.

Procedure

(a) Make out a table for your results as follows:

Dilute acid	*Calcium*	*Magnesium*

(b) Pour about 5 cm³ of one of the dilute acids into a test-tube, and add a small amount of one of the two metals, e.g. a 2 cm length of (cleaned) magnesium ribbon or two/three granules of calcium.

(c) Observe carefully what happens, and test any gas evolved for hydrogen. Record your results in the table.

(d) Repeat (b) and (c), using a fresh sample of the same acid but with the other metal. When you record your results, say which of the two metals seemed to react more vigorously with this acid.

(e) Repeat these procedures using the other two acids, and compare the reactivity of the two metals with *one* acid at a time.

Notes about the experiment

(i) The reaction between calcium and dilute sulphuric acid is slightly unusual. You probably saw a fairly rapid reaction to start with, but the reaction then slowed down as a white solid was formed. The white solid is the salt calcium sulphate, which is one of the relatively few insoluble sulphates. This insoluble salt coats some of the metal, preventing the acid from reacting with all of the metal surface.

$$Ca(s) + H_2SO_4(aq) \rightarrow CaSO_4(s) + H_2(g)$$

(ii) The reactions of metals with nitric acid are not like those with other acids (page 251). You may have detected hydrogen when *these* metals reacted with nitric acid, but this is unusual. Most metals react with nitric acid to produce oxides of nitrogen, not hydrogen, and the gases given off in your experiment with nitric acid were probably a mixture of hydrogen and oxides of nitrogen. (You may have seen, for example, a brown gas, called nitrogen dioxide, NO_2.)

Conclusion

Make your own conclusion to the experiment, including (i) equations for all the reactions except those involving nitric acid, and (ii) a comment about whether the experiment has convinced you, once again, that calcium is more reactive than magnesium.

The reactivity trend in Group 2

The experiments should have convinced you that calcium is generally more reactive than magnesium, even though the two elements react in the same way. Their reactions with water or steam show this trend particularly clearly.

As with the Group 1 elements, the elements of Group 2 become more reactive as their atomic numbers increase. The reasons for this are similar to those given for Group 1 metals on page 234, but of course atoms of the Group 2 metals need to lose two electrons instead of one.

Check your understanding

1. Make sure that you can use chemical and physical properties to support the fact that the Group 1 metals are similar to each other but different from other metals, and that the same applies to the Group 2 metals.

2. How do you think that (a) beryllium and (b) strontium would react with water or steam?

3. Why are the Group 2 elements less reactive than the Group 1 elements?

4. Elements W, X, Y, and Z have atomic numbers 9, 14, 18, and 19 respectively.

(a) Write the electronic configuration of the atoms of each element.

(b) State which, if any, of the elements will (i) form positive ions, and give the symbol(s) of the ion(s) formed, (ii) form negative ions and give the symbol(s) of the ion(s) formed.

(c) Which one of the elements is most likely to be least chemically reactive?

(d) State which two of the elements are most likely to react with each other to form a covalent compound. Give the formula for the compound formed and draw a diagram indicating the formation of the covalent bond or bonds in a molecule of the compound. (J.M.B.)

5. A metal, M, atomic number 19, reacts with cold water to form hydrogen. The metal also reacts with chlorine to form a chloride which dissolves in water giving a solution which conducts electricity.

(a) Write the electronic structure of (i) an atom of M, (ii) an ion of M.

(b) Write the formula for the chloride of M showing the charges on the ions.

(c) What substance, other than hydrogen, would be formed when the metal reacted with water? Write an equation for the reaction of the metal M with water.

(d) A concentrated solution of the metal chloride was electrolysed using carbon electrodes. (i) Name the product formed at the cathode. (ii) Write an ionic equation for the reaction at the anode. (iii) Describe one chemical test, and its expected result, which could be used to confirm the presence of the substance formed at the anode. (J.M.B.)

6. A metal, A, reacts vigorously with cold, dilute acids but has little reaction with water. When A is heated in steam, a colourless gas, B, is evolved and a white solid, C, remains. Universal indicator paper (or red litmus paper), when shaken with a mixture of C and water, was *slowly* turned blue. What do you think the substances A, B, and C might be? Explain why the indicator was turned blue *slowly*. (J.M.B.)

14.5 A STUDY OF THE GROUP 7 ELEMENTS (THE HALOGENS)

The properties of chlorine, a member of this family, are studied in more detail in Chapter 24. In this chapter we are concerned only with some general relationships between the members of this group. The elements exist as simple, covalent molecules, Cl_2, Br_2, etc. (which are gases or volatile liquids or solids), contrasting with the giant metallic lattices of the Group 1 and Group 2 elements, which are all solids.

Atomic and ionic radii

Table 14.3 Some physical properties of the Group 7 elements

Element	*Symbol*	*Atomic number*	*Mass number*	*Melting point (°C)*	*Boiling point (°C)*	*Radius of atom (nm)*	*Radius of ion (nm)*
Fluorine	F	9	19	−218	−188	0.064	0.136
Chlorine	Cl	17	35	−101	−34	0.099	0.181
Bromine	Br	35	80	−7	59	0.114	0.195
Iodine	I	53	127	114	185	0.133	0.216
Astatine	At	← Unstable radioactive element →					

Table 14.3 gives some information about the elements of Group 7. The atomic radii in this group increase as the atomic numbers increase for the same reasons as with the elements of Groups 1 and 2 (or indeed the members of *any* group in the table).

The atoms of elements of Group 7 have seven electrons in their outer shells and so all gain one electron when forming an ion. The ions therefore have a charge of −1, e.g. Cl^-, and, as expected, the ionic radii increase as the atomic radii increase (i.e. down the group). These elements also frequently form *covalent* bonds, by sharing a pair of electrons. The metals of Groups 1 and 2 do not do this.

An important difference from the trends observed in Groups 1 and 2 is that the ions of Group 7 elements have *larger* radii than those of the parent atoms. A simple explanation of this is that the nucleus of the ion now has to 'hold' an extra electron. As the total number of negative charges is now greater than the number of positive charges in the nucleus, the electrons are not attracted by the nucleus to the same extent as in the neutral atom, and the ion 'expands'. Also, electrons *repel* one another and so one extra electron causes extra repulsion. You may wish to draw scale diagrams to illustrate these trends.

A non-metallic ion (i.e. a negatively charged ion) is always larger than the atom from which it was formed.

The physical states of the elements

Table 14.3 gives the melting points of the halogens, and these increase as the atomic number of the element increases. This is also true of the boiling points, so that fluorine, the member with the lowest atomic number in the group, has the lowest melting point and boiling point of the group. In fact, fluorine has both melted and boiled at room temperature, so it is normally seen as a gas. The same is true of chlorine, but the boiling point of bromine is just high enough to keep it as a liquid at room temperature (although it is so near to its boiling point that bromine vapour is always evaporating rapidly from the liquid). Iodine and astatine have much higher boiling points and are solids at room temperature, although iodine has a considerable vapour pressure at room temperature. (The vapour is irritating and solid iodine should never be left in open vessels in the laboratory.) This trend in boiling points and melting points is opposite to that in Groups 1 and 2, all of which are solids at room temperature.

Note that the four common halogens all form coloured vapours, which is unusual. This emphasizes that they are like each other but different from other non-metals. (Other common gaseous elements, such as oxygen, nitrogen, and hydrogen, are colourless.)

Halogen	*Appearance at room temperature*	*Colour of vapour*
Fluorine	Yellow gas	Yellow
Chlorine	Yellowish-green gas	Yellowish-green
Bromine	Red-brown liquid	Red-brown
Iodine	Grey-black solid	Purple

Displacement reactions of the halogens

Experiment 14.4
Halogen displacement reactions*

Apparatus
Chlorine generator (or supply of chlorine water, which contains chlorine molecules, Cl_2), rack of test-tubes, spatula, access to fume cupboard.
Bromine water (this contains bromine molecules, Br_2, dissolved in water but is safer to use than pure bromine), solutions of potassium chloride, potassium bromide and potassium iodide, solid iodine.

Procedure
(a) Make out a table as shown below, and record the results in the table at each stage.
(b) Bubble chlorine into (or add some chlorine water to) a sample of water in the fume cupboard.
(c) Bubble chlorine into (or add some chlorine water to) a sample of (i) potassium bromide solution, (ii) potassium iodide solution.
(d) Add a little bromine water to a sample of water and mix the solution.
(e) Add a little bromine water to a sample of potassium chloride solution, and mix the two solutions.
(f) Add a little bromine water to a sample of potassium iodide solution and mix the two solutions.
(g) Add a small crystal of iodine to (i) a sample of water, (ii) a solution of potassium chloride, and (iii) a solution of potassium bromide.
(h) Put a cross in the table in the 'squares' where chlorine 'meets' a chloride, bromine 'meets' a bromide, and iodine 'meets' an iodide as there is no reaction in these cases.

Halogen element	*Water (used as a control)*	*Potassium chloride solution*	*Potassium bromide solution*	*Potassium iodide solution*

Notes about the experiment

(i) The point of the experiment is to decide which of the mixtures produced a *chemical* change. In these experiments, a chemical change is obvious because there is a colour change when it happens, and this is different from the effect the halogen has on water alone. The colour of the final solution should give you a clue as to what the products include, wherever there is a change. Remember that bromine *molecules* (the element) are coloured, but bromide *ions* are not. This applies to the other halogens, too.

(ii) Having decided in which mixtures chemical changes have taken place, the next thing is to decide which of the three halogen elements used is the most reactive. In this particular experiment, your answer will depend not upon the *vigour* of the reactions (they appear to react at about the same rate in this experiment), but rather upon which halogen has produced the most reactions. Is your answer what you would have expected from your knowledge of reactivity trends in Groups 1 and 2?

Conclusion

1. Say which of the three halogens appears to be the most reactive, and support your answer by using the experimental results. (Statements such as 'bromine displaced iodine from an iodide, but did not displace chlorine from a chloride' will be useful.)
2. Compare the reactivity trend in Group 7 with that in Groups 1 and 2.
3. Where reaction has occurred, there has been a transfer of electrons from *ions* of a halogen (halide ions) to *atoms* of a more reactive halogen. Such transfers are best shown by ionic equations, e.g.

$$Cl_2(g) + 2Br^-(aq) \rightarrow 2Cl^-(aq) + Br_2(aq)$$

Use ionic equations to summarize all the reactions which you observed in the experiment.

Points for discussion

1. Would it have been possible to use sodium chloride, bromide, and iodide solutions (instead of the potassium salts) in the experiment? Explain your answer.

2. Can you suggest why fluorine was not used in the experiment?
(There is more than one answer.)

Reactions of the halogens with hydrogen

All the halogens react with hydrogen, but some of the reactions are not safe to use in a laboratory. The conditions under which reactions occur are as follows.

'Don't open the curtains, I've got chlorine and hydrogen in here'

Fluorine explodes when mixed with hydrogen, even in the dark and at a very low temperature.

$$F_2(g) + H_2(g) \rightarrow 2HF(g)$$

Chlorine reacts with hydrogen (without heating) in diffused daylight, and in bright sunlight the mixture explodes.

$$Cl_2(g) + H_2(g) \rightarrow 2HCl(g)$$

Sunlight does not provide enough energy to start the reaction between bromine and hydrogen; the mixture must be heated to about 200 °C before any reaction takes place.

$$Br_2(g) + H_2(g) \rightarrow 2HBr(g)$$

Iodine and hydrogen, even when heated to 500 °C and in the presence of a catalyst, do not react completely.

$$I_2(g) + H_2(g) \rightleftharpoons 2HI(g)$$

This evidence should support one of your conclusions to the last experiment, i.e. that in Group 7 the elements become *less* reactive as their atomic numbers increase.

Reactions of the halogens with metals

The same trend of reactivity is shown when the halogens react with metals. You may see some demonstrations of this, e.g. the great reactivity of chlorine with sodium or magnesium (p. 464). Bromine is less reactive, and iodine only reluctantly reacts with metals.

An effective way of comparing the reactivity of the halogens towards metals is to try reacting each of them in turn with heated iron wool. (These reactions must be demonstrations.) A typical apparatus is shown in Figure 14.6. (This must be used in the fume cupboard, with the usual safety precautions.) It is found that chlorine reacts vigorously with hot iron to form a black, shiny solid called iron(III) chloride, $FeCl_3$, which sublimes as a brown vapour.

$$2Fe(s) + 3Cl_2(g) \rightarrow 2FeCl_3(s)$$

When bromine vapour passes over hot iron, the reaction is less vigorous and a mixture of products is formed, iron(II) bromide (yellow crystals) and iron(III) bromide (black crystals).

$$2Fe(s) + 3Br_2(g) \rightarrow 2FeBr_3(s)$$

When iodine vapour is passed through very hot iron wool, there is little sign of chemical reactivity, although some iron(II) iodide is formed (a red solid), e.g.

$$Fe(s) + I_2(g) \rightarrow FeI_2(s)$$

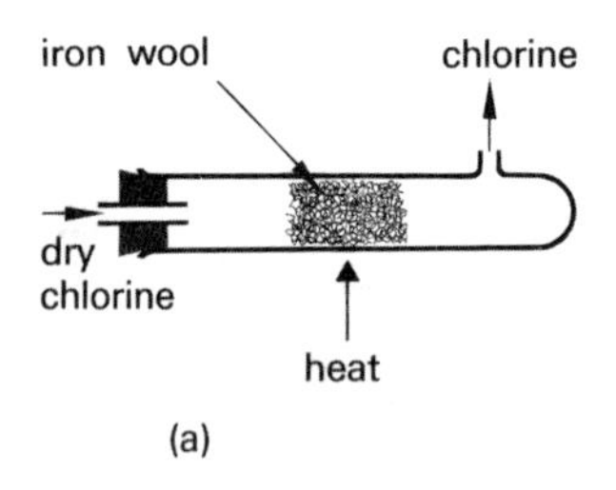

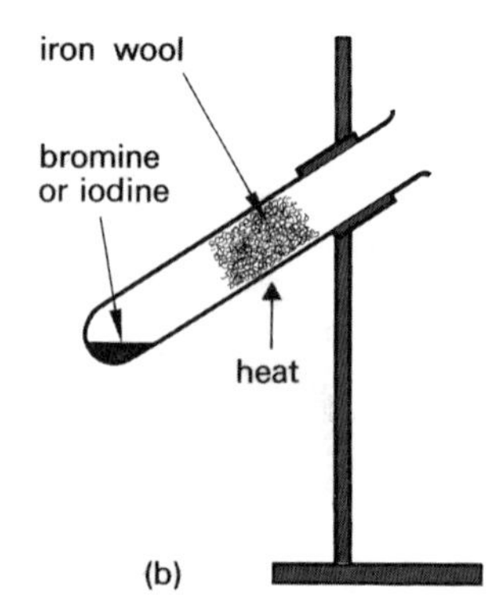

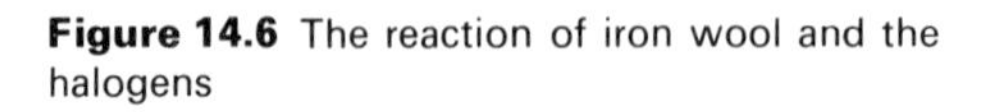
Figure 14.6 The reaction of iron wool and the halogens

Oxidizing power of the halogens

One of the definitions of oxidation (p. 290) is the loss of one or more electrons. A substance which removes electrons from something else (i.e. oxidizes it) is therefore acting as an *oxidizing agent*. Atoms of the halogen elements need to gain an electron in order to achieve a stable electron structure. When they do this they are therefore acting as oxidizing agents, because they are oxidizing something else by taking electrons from it.

In this section you have learned that fluorine is the most reactive of the halogens, i.e. it is the best 'taker' of electrons. Fluorine is therefore a powerful oxidizing agent, and the oxidizing power of the halogens follows the same order as their chemical reactivity. In Experiment 14.4, chlorine displaced bromine from bromide ions because chlorine is a 'better grabber of electrons' than bromine

$$2Br^-(aq) + Cl_2(g) \rightarrow Br_2(aq) + 2Cl^-(aq)$$

and so the chlorine ended up with the 'extra' electrons. The reactions in Experiment 14.4 are also oxidations; chlorine oxidizes bromide ions to bromine and iodide ions to iodine (by loss of electrons in each case), and bromine similarly oxidizes iodide ions to iodine. Iodine, the weakest oxidizing agent of the halogens studied, cannot oxidize (displace) either bromides or chlorides.

The reactivity trend in Group 7

You should now understand that the reactivity of the halogens follows a trend opposite to that shown by the elements of Groups 1 and 2, i.e. the halogens become *less* reactive as their atomic numbers increase. This should not be surprising, as the elements of Group 7 are 'opposite' to the elements of Groups 1 and 2 in their 'needs'. The non-metals (e.g. the halogens) need to *gain* electrons to form stable structures, whereas the metals need to *lose* electrons to form stable structures.

Fluorine is more reactive than chlorine because the outer shell of fluorine atoms is fairly close to the nucleus, and the attraction between nucleus and electrons helps to 'pull in' the extra electron the atom needs in order to have a fully filled shell. The outer shell of an atom of iodine, on the other hand, is a long way from the nucleus and it is more difficult for iodine to gain an electron.

Check your understanding

1. In any one group of the Periodic Table, the elements

A become more reactive as their atomic numbers increase

B become less reactive as their atomic numbers increase

C lose electrons more easily as their atomic numbers increase

D lose electrons less readily as their atomic numbers increase

E are all equally reactive

2. This question refers to the elements of the Periodic Table with atomic numbers from 3 to 18. Some of the elements are shown by letters but the letters are not the symbols of the elements.

3 A	4	5	6	7	8 E	9	10 G
11 B	12 C	13	14 D	15	16	17 F	18

(a) Which of the elements lettered A to G,
 (i) is a noble gas
 (ii) is a halogen
 (iii) would react most readily with chlorine

(b) Give
 (i) the formula of the hydride of D
 (ii) the formula of the oxide of C

(c) Indicate whether the bonding in the oxide of C will be ionic or covalent.

(J.M.B.)

3. Use the Periodic Table to answer the following.

(a) Is astatine (At) likely to be a solid, a liquid or a gas at room temperature?

(b) Is rubidium (Rb) a metal or a non-metal?

(c) Are the compounds of vanadium (V) likely to be colourless or coloured?

(d) What is the formula of gallium oxide? (The atomic number of gallium (Ga) is 31.) Would you expect this oxide to be basic, acidic, or amphoteric?

4. The following are ionic and atomic radii (in nm) of members of the same group of the Periodic Table.

	Atomic radius	*Ionic radius*
A	0.133	0.078
B	0.157	0.098
C	0.203	0.133
D	0.216	0.149
E	0.235	0.165

(a) Is this a metallic group or a non-metallic group?
(b) Which element would have the lowest atomic number?
(c) Which element would be the most reactive?

5. The following data applies to five elements in the Periodic Table.
Element A valency 1, m.p. 97 °C, atomic number 11.
Element B valency 4, m.p. 1410 °C, electronic structure, 2, 8, 4.
Element C valency 1, m.p. −101 °C.
Element D valency 1, electronic structure 2, 8, 8, 1.
Element E valency 0, electronic structure 2, 8.
(a) Which elements are in the same group?
(b) Which element is in the same group as carbon?
(c) Name two elements which would react very vigorously together.
(d) Which element has a giant structure?
(e) Which element is most likely to form only covalent bonds?
(f) Which element would resemble helium?
(g) Which element could most easily form a negative ion?

6.

1	2	3	4	5	6	7	0
Li	Be	B	C	N	O	F	Ne
Na	Mg	Al	Si	P	S	Cl	Ar

The above two rows of the Periodic Table may help you to answer the questions below. One of the numbers, symbols, or words given at the end of each of the following sentences completes the sentence correctly.
(a) The atomic mass of $^{14}_{7}N$ is
(i) 7, (ii) 14, (iii) 98, (iv) 21, (v) 2.
(b) The number of neutrons in one atom of $^{23}_{11}Na$ is
(i) 0, (ii) 11, (iii) 12, (iv) 23, (v) 253.
(c) A fluoride anion has the same number of electrons as
(i) Be^{2+}, (ii) O, (iii) Cl^-, (iv) Ne.
(d) A sodium cation has a different number of electrons from
(i) O^{2-}, (ii) F^-, (iii) Li^+, (iv) Al^{3+}.
(e) $^{35}_{17}Cl$ and $^{37}_{17}Cl$ are
(i) allotropes, (ii) isomers, (iii) isotopes, (iv) molecules. (J.M.B.)

14.6 A SUMMARY OF CHAPTER 14

The following 'check list' should help you to organize the work for revision.

1. General points

Table 14.4 Some changes across a short period (i.e. one which does not contain transition elements)

	Group 1	*Group 2*	*Group 3*	*Group 4*	*Group 5*	*Group 6*	*Group 7*	*Group 8*
Reactivity (comparative, within the period)	High	Lower	Lower still	Low	Higher	Higher still	High	No reactions
Type of oxide	Basic	Basic	May be amphoteric	Acidic	Acidic	Acidic	Acidic	No oxide formed
Element	Metal	Metal	Metal	Metal or non-metal	Non-metal	Non-metal	Non-metal	Non-metal
Charge on ion (if formed)	1+	2+	3+	4+ (rare, only occurs in Pb^{4+})	3−	2−	1−	—
Ionic radius compared to atomic radius	← Smaller than atomic radius →				← Larger than atomic radius →			

Table 14.5 Some group relationships in Groups 1, 2, and 7 of the Periodic Table

	Group 1	*Group 2*	*Group 7*
Similarity	All members are chemically similar to each other but different from other groups	All members are chemically similar to each other but different from other groups	All members are chemically similar to each other but different from other groups
Reactivity	Increases down the group	Increases down the group	Decreases down the group
Atomic and ionic radii	Increase down the group	Increase down the group	Increase down the group
Similarities in physical properties	Low density (some float on water), soft, kept in paraffin oil. Low melting points. Metals	Hard metals, all more dense than water	Molecular. Gases or volatile liquids or solids. Non-metals. Form coloured vapours (p. 240). Toxic
Melting points and boiling points	Decrease down the group (in general), elements all solids at room temperature	Decrease down the group (in general), elements all solids at room temperature	Increase down the group hence change from gas, through liquid to solid down the group
Similarities in chemical properties	(a) React vigorously with cold water to form the hydroxide (alkaline solution) and hydrogen (p. 252)	React with cold water (Mg slow) to form solid hydroxides (which dissolve slightly to form alkaline solutions) and hydrogen. React vigorously with steam	A halogen will displace a halogen lower in the group from a solution of one of its salts (Table 14.6)
	(b) Form surface coating of oxide in cold, burn when heated, to form the oxide (or peroxide), (p. 237)	Form surface coating of oxide in cold, burn when heated, to form the oxide (p. 234)	Reactivity trend is also shown by the reactions with hydrogen (p. 241) and metals (p. 242)
	(c) Too reactive to add to acids	React rapidly with dilute acids (p. 237)	Oxidizing power (i.e. ability to remove electrons from other chemicals) decreases down the group. Fluorine and chlorine are powerful oxidizing agents

(a) In the Periodic Table, elements are arranged in order of increasing atomic number (Figure 14.1).
(b) Horizontal rows in the table are called periods, and vertical columns are called groups; regions of the table have special names (Figure 14.2).
(c) Elements in any one group have the same number of electrons in the outer shells of their atoms, and they therefore have similar chemical properties (you should be able to quote examples for Groups 1, 2, and 7) and the same main valency.
(d) Hydrogen is unique; it does not belong to any group in the table.
(e) Transition elements have special properties of their own (p. 231).
(f) The main changes which occur across a period are summarized in Table 14.4.
(g) The main changes which occur down Groups 1, 2, and 7 are summarized in Tables 14.5 and 14.6.

2. *Important experiments*

You could be asked to describe in detail an experiment which shows how a metal reacts with steam, e.g. Experiment 14.2. The *results* of the other experiments should be known, as outlined below.

(a) Details of the reactions of lithium, sodium, and potassium on cold water, including equations.
(b) Colours of flames, nature of products, and equations for the combustion of lithium, sodium, and potassium in air (p. 233).
(c) Details and equations for the reactions of calcium on cold water and magnesium on steam (p. 252).
(d) Colours of flames, nature of products, and equations for the combustion of calcium and magnesium in air (pp. 237 and 252).
(e) Details of the reactions between calcium and magnesium and dilute acids; equations, the unusual reaction between calcium and dilute sulphuric acid (which slows down as insoluble calcium sulphate is formed), and the production of oxides of nitrogen (and perhaps *some* hydrogen) with nitric acid.
(f) Colour changes and ionic equations for the displacement reactions of the halogens (Table 14.6).
(g) Details of the ways halogens react with hydrogen (conditions important, p. 241) and with metals (p. 242), to provide further evidence for the reactivity trend within the group.

Table 14.6 The halogen displacement reactions

Halogen element	*How added*	*Potassium chloride solution*	*Potassium bromide solution*	*Potassium iodide solution*
Fluorine	NOT USED because (a) it is very toxic and corrosive, (b) it is so reactive that it reacts with the water in the solutions	—	—	—
Chlorine	Gas bubbled through the solution under test, or chlorine water mixed with solution	NO REACTION	Bromine displaced. Orange-brown colour appears $Cl_2(aq) + 2Br^-(aq) \rightarrow Br_2(aq) + 2Cl^-(aq)$	Iodine displaced. Dark colour appears in solution. Pieces of solid iodine visible $Cl_2(aq) + 2I^-(aq) \rightarrow I_2(aq) + 2Cl^-(aq)$
Bromine	Bromine water mixed with solution under test	NO REACTION	NO REACTION	Observations as above $Br_2(aq) + 2I^-(aq) \rightarrow I_2(aq) + 2Br^-(aq)$
Iodine	A few small crystals are added to the solution	NO REACTION	NO REACTION	NO REACTION

QUESTIONS

Multiple choice questions

1. X, Y, and Z are in the same period of the Periodic Table. X is a non-metal, Y is a metal, and Z shows the properties of both metals and non-metals. Which one of the following represents the order of these three elements in the Periodic Table?
A XYZ; B YXZ; C YZX; D ZXY; E ZYX
(C.)

2. The metal rubidium is three places below lithium in Group I of the Periodic Table. If the reaction of lithium with water is described as moderately slow, the reaction of rubidium with water is likely to be
A very fast
B moderately fast
C as slow as that of lithium
D rather slower than that of lithium
E so slow that no reaction is noticeable
(C.)

3. The elements with the code letters W, X, Y, and Z are placed in the Periodic Table as shown.

W															X		
																Y	Z

Which one of the following is a correct statement regarding W, X, Y, and Z?
A The atomic numbers increase in the order W, X, Y, Z.
B W and X are solids but Y and Z are gases under room conditions.
C The valency of W and of Y is one, but the valency of X and Z is two.
D W, X, Y, and Z are in the same period of the Periodic Table.
E W is in Group 1, X in Group 5, Y in Group 6, and Z in Group 7. (C.)

4. In the Periodic Table, the metal rubidium (Rb) is placed below potassium. The formula of the chloride of rubidium is
A Rb_2Cl; B $RbCl$; C $RbCl_2$;
D $RbCl_3$; E $RbCl_4$ (C.)

5.

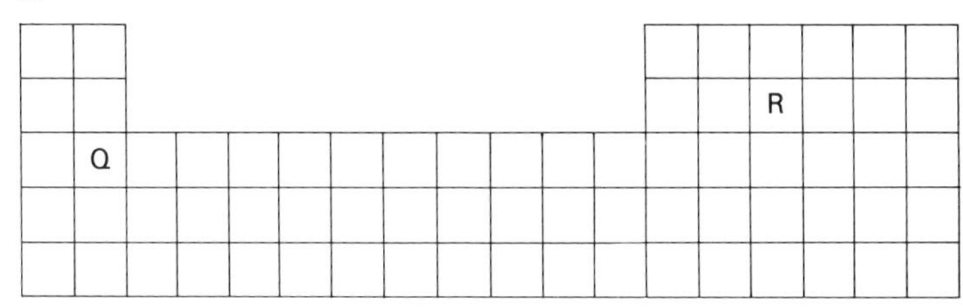

Elements with the code letters Q and R occupy the positions shown in the outline Periodic Table. The most likely formula of a compound formed between them is

A QR; B QR_2; C Q_2R; D Q_2R_3;
E Q_3R_2 (C.)

6. Sodium and potassium are members of a family of elements called
A the alkaline earth metals
B the transition metals
C metalloids
D alkali metals
E the unreactive metals

Structured questions and questions requiring longer answers

7. (a) When dry chlorine is passed through hot iron wool in a combustion tube, the iron glows strongly and black crystals are deposited in the cooler parts of the tube.

(i) Name the product of, and write an equation for, this reaction.

(ii) Name three members of the halogen family other than chlorine.

(iii) State whether each of the halogens mentioned in (i) would react with iron more vigorously or less vigorously than would chlorine.

(b) (i) Describe the reaction of chlorine with hydrogen under any one stated set of conditions.

(ii) Compare the vigour of the reaction of hydrogen with chlorine with that of the reactions of hydrogen with the other halogens you have named.

(c) Name two halogens which react with potassium bromide solution. Write an equation for one of the reactions which occur. (J.M.B.)

8. Lithium, sodium, and potassium are in Group 1 of the Periodic Table. These three metals are known collectively as 'alkali metals'.

Chlorine, bromine, and iodine are in Group 7 of the Periodic Table. These three non-metals are known collectively as 'halogens'. Halogen means 'salt producer'.

Helium, neon, and argon are in Group O of the Periodic Table. These three gases are known collectively as 'noble (inert) gases'.

(a) What is *an alkali*? Give the name and formula of *one* alkali.

(b) What is a *salt*? Give the name and formula of a salt formed between an alkali metal and a halogen and describe how a pure sample of this salt could be prepared by a titration method.

(c) Why are helium, neon, and argon unreactive?

(d) Write down, in each case, the formula of (i) the alkali metal ion, (ii) the halogen ion, that has the same arrangement of electrons as the neon atom.

(e) 142 g of chlorine gas and 168 g of krypton gas (Kr) occupy equal volumes under the same conditions of temperature and pressure. What can you deduce about the gas krypton from this information?

(Relative atomic masses: Cl 35.5; Kr 84)

(C.)

9. (a) Sodium and aluminium have atomic numbers of 11 and 13 respectively. They are separated by one element in the Periodic Table, and have valencies of one and three respectively. Chlorine and potassium are also separated by one element in the Periodic Table (they have atomic numbers of 17 and 19 respectively) and yet they both have a valency of one. Explain the difference.

(b) The halogens, fluorine, chlorine, bromine, and iodine, show a gradation in properties. Illustrate this by reference to their ease of combination with hydrogen, and the ease of replacement of one halogen by another. (J.M.B.)

15 The activity series: cells and corrosion

15.1 THE ACTIVITY SERIES

Classification systems are very useful in chemistry. You have already met the classification of substances as elements, compounds, and mixtures, and also the classification of elements in the Periodic Table. The activity series is another important method of classification.

The reactions of the metals with air

When metals are heated in air, if they react at all they usually react with the oxygen to form oxides. The reactions are complicated by two factors.

1. Many metals react with *other* components of the air, e.g. burning calcium also reacts with nitrogen, and sodium reacts with water vapour. For the purposes of comparison in this section, only the reactions with oxygen will be considered.
2. Some metals burn *completely* to form a high yield of their oxides, but others react briefly to become coated with a layer of oxide which then stops further reaction with the metal.

The reactions of the common metals with oxygen have already been discussed, e.g. in the chapter on oxygen (p. 150) and in the Periodic Table trends in Groups 1 and 2 (pp. 233 and 237). The only other common metals which 'burn' in air are zinc and aluminium (when used as the *powdered* metals), and you may see demonstrations of these reactions. The reactions between metals and oxygen are summarized in Table 15.1.

The order of the elements in Table 15.1 follows the order of reactivity, i.e. sodium is the most reactive of the common metals (in its action with oxygen), and silver and gold are the least reactive. When elements are placed in order of their chemical reactivity like this, the sequence is known as an *activity series*. It is normally used only for metals, together with the non-metals carbon and hydrogen. You will see where these non-metals fit into the series later in the chapter.

You may be puzzled as to why aluminium appears fairly high in the series, when it does not 'burn' in oxygen unless finely divided. Aluminium, given a chance, *is* a reactive metal but it rarely appears to be so because its surface layer of oxide is very strong and hard to penetrate and it prevents the metal underneath from showing its true chemical reactivity. You may be shown a simple demonstration of how reactive aluminium can be when the surface layer is destroyed. If a piece of aluminium foil is dipped into mercury(II)

'Oxide coats are always tight on me'

Table 15.1 The reactions of common metals with oxygen in the air

Increasing chemical reactivity	*Metal*	*Reaction*	*Effect of water on oxide*
↑	Sodium	Metal instantly tarnishes, i.e. forms oxide layer in cold. Burns readily when heated, with yellow flame. Off-white solid formed $2Na(s) + O_2(g) \rightarrow Na_2O_2$	Oxide reacts with water to form strongly alkaline solution
	Calcium	Rapidly covered with oxide layer in the cold. Sometimes difficult to ignite because of this. Fresh specimens burn readily when heated, with brick red flame. White solid formed $2Ca(s) + O_2(g) \rightarrow 2CaO(s)$	Oxide reacts with water to form strongly alkaline solution
	Magnesium	Oxide layer formed slowly when cold. Burns easily when heated, with dazzling white flame. White solid formed $2Mg(s) + O_2(g) \rightarrow 2MgO(s)$	Oxide is basic and reacts slightly to form weakly alkaline solution
	Aluminium	Oxide layer formed instantly in cold. Layer difficult to penetrate—protects metal beneath, which does not burn (unless in powder form) even when heated strongly $4Al(s) + 3O_2(g) \rightarrow 2Al_2O_3(s)$	Oxide insoluble in water. Amphoteric. May affect indicators due to impurities
	Zinc	Oxide layer formed in cold. Turnings and pieces will not burn when heated, but the powder burns when heated, with blue-white flame. Oxide yellow when hot, white when cold $2Zn(s) + O_2(g) \rightarrow 2ZnO(s)$	Oxide insoluble in water. Amphoteric. May affect indicators due to impurities
	Iron	Reacts in cold with water *and* oxygen (i.e. rusts) so not really comparable with the others. Heated *powder* sparkles as it oxidizes but does not really burn $3Fe(s) + 2O_2(g) \rightarrow Fe_3O_4(s)$	Basic oxide, insoluble in water, so no effect on indicators
	Lead	Oxide layer formed in the cold, but does not burn when heated $2Pb(s) + O_2(g) \rightarrow 2PbO(s)$	Amphoteric oxide, insoluble in water, so no effect on indicators
	Copper	Does not form any appreciable oxide layer in cold, although may react with gases in the air to form a green layer. Forms black oxide coating when heated, but this protects the metal below, which does not burn $2Cu(s) + O_2(g) \rightarrow 2CuO(s)$	Basic oxide, insoluble in water, so no effect on indicators
	Silver Gold	No oxide layer, even when heated	

chloride solution (care: this is poisonous) and then left exposed to the air, feathery growths of a white powder rapidly form on the metal, which also gets hot. This powder is aluminium oxide, and it forms because the mercury(II) chloride deposits mercury on the foil, which breaks down the oxide layer protecting the metal, and allows the metal to show its true chemical reactivity. Thus given the right circumstances, the metal reacts (and continues to react) rapidly with oxygen even in the cold. (The spilling of mercury from broken instruments, followed by the rapid oxidation and subsequent failure of aluminium components, was thought to be responsible for several aeroplane crashes before the cause was found and prevented.)

Aluminium is usually placed fairly high in the activity series because of its *potential* chemical reactivity. *In practice*, because of the oxide layer, it is not so reactive.

The reactions of the common metals with dilute acids

You have already studied the reactions of the common metals with typical dilute acids, e.g. in the chapter on acids (p. 60) and in studying Group 2 of the Periodic Table (p. 237). These reactions are summarized in Table 15.2. Once again, the order of chemical reactivity (with acids) follows the same trend, and the metals appear in the same order in the table. Aluminium is again apparently misplaced, but its potential reactivity is clearly shown by the fact that it reacts rapidly with *hot* dilute hydrochloric acid.

Note that nitric acid is not included in Table 15.2 because it is not 'typical' in its reactions with metals. Reactive metals such as calcium and magnesium may react with nitric acid to give some hydrogen, but the main product of the reaction between a metal and nitric acid is likely to be one or more oxides of nitrogen, e.g. NO_2. Remember that the unreactive metal copper *does* react with warm dilute nitric acid, and this is a convenient way of obtaining copper(II) ions from the metal. You will understand the reference to lead being 'only just above hydrogen in the activity series' later in the chapter.

Table 15.2 The reactions of metals with typical acids

Increasing chemical reactivity	*Metal*	*Reaction with dilute hydrochloric acid*	*Reaction with dilute sulphuric acid*
↑	Sodium Potassium	Too reactive to be used with any acids	
	Calcium	Rapid effervescence in the cold to form hydrogen	Rapid effervescence initially in the cold, slowing down as insoluble calcium sulphate coats the metal
	Magnesium	Rapid effervescence in the cold to form hydrogen	As with hydrochloric
	Aluminium	Slow reaction in cold until oxide layer is penetrated; rapid if warmed, hydrogen formed	Very little reaction, oxide layer unbroken
	Zinc	Steady effervescence to liberate hydrogen in the cold	Usually slow effervescence to form hydrogen unless the metal is impure
	Iron	Fairly slow effervescence in the cold to give hydrogen	As with hydrochloric
	Lead	No reaction—it is only just above hydrogen in the activity series	As with hydrochloric
	Copper Silver Gold	No reaction with these dilute acids	

Typical equations:

$Mg(s) + 2HCl(aq) \rightarrow MgCl_2(aq) + H_2(g)$

$Mg(s) + H_2SO_4(aq) \rightarrow MgSO_4(aq) + H_2(g)$

$Zn(s) + 2HCl(aq) \rightarrow ZnCl_2(aq) + H_2(g)$

$Fe(s) + H_2SO_4(aq) \rightarrow FeSO_4(aq) + H_2(g)$

The reactions of the metals with water or steam

You have studied some of these reactions already, e.g. the reactions of sodium and potassium on cold water (p. 232), and the reactions of calcium and magnesium on water or steam (p. 235). None of the other metals in the activity series reacts with *cold* water, as they are less reactive.

Experiment 15.1
To determine the reactions, if any, when zinc, iron, aluminium, and copper are heated in steam***

Apparatus
As for Experiment 14.2 and Figure 14.5, page 236.

Procedure
As in (c) for Experiment 14.2, but using zinc, iron, aluminium, and copper in turn instead of the magnesium. Make sure that the metals are heated *strongly* in the current of steam, and take care to avoid 'sucking back'.

Results
Record your observations, and the relative vigour of any reactions, in a suitable table, adding the results of earlier experiments you have done with metals and cold water or steam.

Use your table of results to make an appropriate conclusion, linking your observations with the activity series (page 250), and including all relevant equations. You will need help to explain what happens when iron reacts with steam (see Table 15.3).

Points for discussion

1. Gas sometimes collects in hot water radiators made of cast iron. The gas is a mixture, and includes air which had dissolved in the water. Explain why the air comes out of solution, and name another gas likely to be present in the mixture, explaining how it is formed.

2. Suggest reasons why copper is a good choice of metal for plumbing.

Table 15.3 The reactions of the metals with water or steam

Metal	*Reaction*
Sodium	Reacts vigorously with *cold* water $2Na(s) + 2H_2O(l) \rightarrow 2NaOH(aq) + H_2(g)$
Calcium	Fairly vigorous reaction in *cold* water $Ca(s) + 2H_2O(l) \rightarrow Ca(OH)_2(aq) + H_2(g)$
Magnesium	*Very* slow in cold water, rapid when heated in steam $Mg(s) + H_2O(l) \rightarrow MgO(s) + H_2(g)$
Aluminium	No reaction (water or steam) due to protective oxide layer
Zinc	No reaction in the cold, fairly rapid when heated in steam $Zn(s) + H_2O(g) \rightarrow ZnO(s) + H_2(g)$
Iron	Reacts when heated in steam, reaction reversible $3Fe(s) + 4H_2O(g) \rightleftharpoons Fe_3O_4(s) + 4H_2(g)$
Lead Copper Silver Gold	No reaction in water or steam

Displacement reactions between metals and ions

In Experiment 14.4 you learned that a reactive halogen, like chlorine, will displace a less reactive halogen, like iodine, from a solution of one of its salts, e.g. an iodide. If this is a general pattern of behaviour, then a reactive metal (i.e. one that is high in the activity series) should displace a less reactive metal from a solution of one of its salts. The next experiment investigates this.

Experiment 15.2
Investigating the ability of the metals to displace each other from solution*

Apparatus
Rack of test-tubes.
Clean strips of copper, aluminium, zinc, and magnesium, clean iron nails, separate solutions of copper(II), lead(II), silver, mercury(I), magnesium, zinc, and aluminium nitrates, and a solution of iron(II) ammonium sulphate.

Procedure
Note: (i) Mercury salts, lead salts, copper salts, silver salts, and zinc salts are poisonous.
(ii) It is helpful to boil the copper solutions before they are used so as to remove dissolved air, which can oxidize some of the products. If this is done, the solutions must be allowed to cool slightly before metals are added.
(iii) In order to save on time (and metal strips!) each group can use just a few of the metals, and the results can then be compared on a class basis.
(a) Prepare a table for your results as shown below.

Metal added	*Solution of* $Cu^{2+}(aq)$	*Solution of* $Pb^{2+}(aq)$	*Solution of* $Ag^{+}(aq)$	*etc. as appropriate*

(b) Pour copper(II) nitrate solution into a test-tube to a depth of about 2 cm, and repeat with the other solutions in the other test-tubes. Place a strip of zinc in each tube. Record your observations in the table of results, looking for colour changes and the formation of any precipitate or deposit on the metal. Note that sometimes there will be no reaction.
(c) Repeat the experiment with strips of the other metals and fresh samples of the solutions, until each metal has been tested with each solution.

Notes about the experiment
(i) The results are sometimes a little difficult to interpret because when some metals are precipitated in a finely divided form, they are rapidly oxidized by oxygen dissolved in the water, and they may then not look like metals. If the solutions are boiled before use, much of the dissolved oxygen is boiled out, and the chances of oxidation are reduced.
(ii) Whenever a reaction occurred, it was between the *atoms* of the metal being added and the metal *ions* from the dissolved salt. When there is a reaction, these 'change places', i.e. the added metal displaces the other metal from its ions, and dissolves as ions to take its place, e.g.

$$Zn(s) + Cu^{2+}(aq) \rightarrow Cu(s) + Zn^{2+}(aq)$$

(iii) We use ionic equations to show what is happening in these metal/metal ion reactions because the other ion present in the salt solution (e.g. nitrate) takes no part in the reaction; it is a spectator ion. (Compare this with the similar use of ionic equations for the halogen displacement reactions on page 241.)
(iv) You will understand from the next section why these reactions occur. Note that sodium, potassium, and calcium metals are not used in displacement reactions because they also react with the water present, and so the reactions are complicated. Sometimes bubbles of gas are seen when magnesium is used in displacement reactions, for a similar reason.

Conclusion
1. Make a list of the metals in order of reactivity, putting the most reactive one (i.e. the one which displaces most other metals) first.
2. How does your list compare with the activity series (page 250)? Does the same pattern of reactivity still show?
3. Give ionic equations for all of the reactions which took place, making sure that they are properly balanced.

Points for discussion

1. You may have noticed that the blue colour of the copper(II) salt solution faded when zinc metal was added. Explain why this occurs, and state what would be formed if excess zinc was added, and the solution was filtered and then heated to dryness.

2. If a piece of aluminium foil is cleaned with emery paper, dipped into mercury(II) chloride solution, rinsed in water, and then dipped into boiled copper(II) sulphate solution, it is rapidly covered with a red-brown coating. Explain why this occurs, with an equation.

The action of heat on metal compounds

Generally, one of three things happens when a metal compound is heated.

1. The compound may not change chemically, although it might melt.
2. The compound may react with oxygen in the air, i.e. it is oxidized.
3. Many metallic compounds, particularly those containing oxygen, *decompose* when heated, i.e. they split up into simpler substances. Such decompositions can take place in two main ways.

Thermal decomposition occurs when a compound is decomposed on heating, and the products do not recombine together on cooling.

e.g.
$$CuCO_3(s) \xrightarrow{\text{heat}} CuO(s) + CO_2(g)$$

Thermal dissociation occurs when a compound is decomposed on heating, but the products, on cooling, reform the original compound.

e.g.
$$NH_4Cl(s) \underset{\text{cooling}}{\overset{\text{heating}}{\rightleftharpoons}} NH_3(g) + HCl(g)$$

Other common examples of thermal dissociation include the dehydration of many hydrated salts when they are heated. The anhydrous salts often start to recombine with water from the air when they are cool.

In Experiment 15.3, you will investigate what happens when some compounds of the metals in the activity series are heated. Notice that the emphasis here is on the *compounds* and not on the metals themselves. You will be investigating the *thermal stability* of these compounds (i.e. whether or not they decompose easily when heated), and the products of such reactions. Some students will predict that a reactive metal will form reactive compounds, and therefore that compounds of reactive metals will easily decompose, i.e. they are thermally unstable. Others may think that a reactive metal will form very stable compounds, which are therefore difficult to decompose and are thermally stable. You may like to make your own forecast, and then check it by doing Experiment 15.3.

Experiment 15.3
Investigating the thermal stability of the nitrates and carbonates of metals in the activity series*

Apparatus

Rack of test-tubes, retort stand and clamp, test-tube holder, Bunsen burner, teat pipette, spatula, clock, access to fume cupboard. Solid samples of the nitrates and carbonates of sodium, calcium, magnesium, zinc, and copper. Calcium hydroxide solution.

Procedure

Note: (i) The nitrates must be heated in a fume cupboard, although the carbonates may be heated in the open laboratory.

(ii) Work out which gases *could* be formed from each chemical before you heat it, and prepare in advance for any gas tests which may be necessary. Gases may be given off for only a short time during the heating, and you may not have time to organize the appropriate chemical tests etc. during the actual heating. Remember that hydrated salts will give off steam when heated, and other changes may then follow.

(iii) When testing for oxygen it is important that pieces of carbon from a burning splint do NOT fall into a molten compound. To avoid this possibility the tubes should be heated in an almost horizontal position, and the sodium nitrate should be used only by the teacher.

(iv) One of the compounds to be heated *appears* to melt when heated, but in fact it dissolves in its own water of crystallization which is given off, and with continued heating it decomposes further. Make sure that you recognize this particular substance, and that you describe what happens to it in your table of results.

(a) Make out a table for your results as shown below.

(b) Place about 2 spatula measures of sodium nitrate into a clean, dry test-tube. Clamp the tube so that it is in a suitable (almost horizontal) position for heating with a Bunsen. Do not readjust the position of the clamp, or the setting of the Bunsen, so that all of the compounds are heated with the same flame and at the same height above it.

(c) Place the burner in position below the tube and note the time. Heat the compound, looking for signs of chemical change, and noting the time taken to produce such a change. Most compounds of sodium, calcium, magnesium, and zinc are white, so that there may be little visible indication of chemical change, and you may have to rely on testing for gases.

(d) Repeat (b) and (c) but using, in turn, separate samples of the other nitrates instead of sodium nitrate. Allow the product from the reaction of heat on zinc nitrate to cool before you record your final observation.

(e) Repeat (b) and (c), but this time using the carbonates, in turn, instead of sodium nitrate, and work in the open laboratory. Again allow the product from the reaction with zinc carbonate to cool before making your final observation.

Notes about the experiment

(i) It is not easy to identify some of the gases produced in the experiment, because they are often given off for only a short time. Another difficulty, already referred to, is that some decompositions are not accompanied by colour changes, in which case it is difficult to decide when changes occur. However, zinc oxide has an unusual property which may help you to explain what happened when zinc carbonate and nitrate were heated. Zinc oxide is yellow when hot and white when cold.

(ii) If you find it difficult to interpret your results, you can find the necessary information in various tables etc. in Chapter 19.

(iii) The experiment should have convinced you that the reactive metals, near the top of the activity series, generally form very stable compounds. Compounds of reactive metals (e.g. sodium compounds) are therefore difficult to decompose on heating. They are thermally stable, although they may become anhydrous or melt when they are heated. Most of them need to be heated to a high temperature before they melt, because they usually have giant ionic lattices with high melting points. If they do decompose, they usually do so in a way different from those of the similar compounds of other metals (e.g. note what happens to sodium nitrate when heated).

Metal compound	*Initial colour and state*	*Time taken for decomposition to start*	*Observations during/after heating*	*Name of any gas(es) evolved*

As a complete contrast, compounds of the less reactive metals (e.g. copper compounds) are less stable, and they often decompose when heated. There is once again a trend of reactivity down the activity series, but this time it is applied to the compounds of the metals rather than the metals themselves. The way in which the compounds react when heated follows a trend opposite to that of the reactivity of the metals themselves; the compounds of metals near the top of the series are very unreactive when heated.

Make sure that you understand how the effect of heat on the nitrates and carbonates of the metals varies according to the position of the metal in the activity series. Remember that although some compounds react in the same *kind* of way, the ease with which they react will vary.

Conclusion
Make your own conclusion, linking your results with the activity series, and writing equations for the reactions which took place.

Point for discussion

Are the reactions which occurred in the last experiment thermal dissociations or thermal decompositions, or were there examples of both types? Explain your answer(s).

15.2 WHY METALS FORM AN ACTIVITY SERIES

What does 'reactive' mean?

The chemical reactivity of any element depends largely upon the arrangement of electrons within its atoms, and particularly on the number of electrons in their outer shells. Sodium is a very reactive metal because its atoms only need to lose one electron each in order to form a stable structure. Potassium is like sodium, but is even more reactive because it is easier for potassium to lose its outer electron than it is for sodium to do so (see page 234). It is not surprising, therefore, to see sodium and potassium at the top of an activity series of metals. Similarly, magnesium and calcium will be fairly high in the series, because their atoms need to lose only two electrons to become stable, and calcium will be higher than magnesium because it can lose its two outer electrons more easily than magnesium can. These ideas have been encountered in other parts of the book, such as the discussions on Periodic Table trends (p. 234) and on bonding (p. 116).

All metal atoms tend to *lose* electrons when they react; in doing so they become *ions* which usually have a noble gas electron structure. The reactivity of a metal depends upon how easily it loses its outer electron(s) to form ions; sodium is reactive and near the top of the activity series because it loses an electron very easily.

The reactivity of a metal depends upon how easily it can lose its outer electron(s) to become a positively charged ion.

Many students make the mistake of thinking that all reactivity is measured by the ability to *lose* electrons, but this is only true for metals. Non-metals *gain* electrons when they form ions, and so their reactivity can be measured by their ability to gain electrons. We must also remember that non-metals often *share* electrons to form covalent bonds.

Battles for electrons

Many chemical reactions can be imagined as a 'battle for electrons', in which the better 'giver' of electrons loses electrons to the other element or substance. All metals are fairly good at losing electrons, but some are better than others. We can show this, and help to confirm the ideas just given, by a simple experiment which you may see demonstrated.

Strips of two different metals are placed in an electrolyte such as dilute sulphuric acid, and connected externally through a suitable voltmeter as shown in Figure 15.1. As long as two *different* metals are used, a current flows between the metals and the voltmeter registers a potential difference. What is happening is that one of the two metals is a better 'giver' of electrons than the other, and is passing electrons to the other metal. This flow of electrons is an electric current, and the electrolyte completes the circuit by enabling ions to move between the electrodes. The reading on the voltmeter depends upon the metals being used; the larger the potential difference, the greater is the driving force which pumps electrons from one metal to the other.

Some metals are good at losing electrons

The ability of metals to lose electrons can be compared by making one of the electrodes a standard (e.g. making the anode a carbon rod) and using each metal in turn as the cathode, noting the voltage reading each time. Even sodium and potassium can be included, but they must not be dipped into the acid; a modified apparatus as shown in Figure 15.2 is used instead. A pea-sized piece of sodium is held in a crocodile clip and carefully touched on the filter paper; the voltmeter is read as quickly as possible, and the sodium withdrawn.

The results of experiments like this fully support the ideas given earlier. The biggest voltages are recorded when sodium or potassium are connected to carbon, because these two metals are the best 'losers' of electrons, and therefore the most reactive metals. Other metals follow in the same order as they appear in the activity series, which supports the idea that chemical reactivity (for metals) depends on their ability to lose electrons. A particularly interesting result is obtained with aluminium.

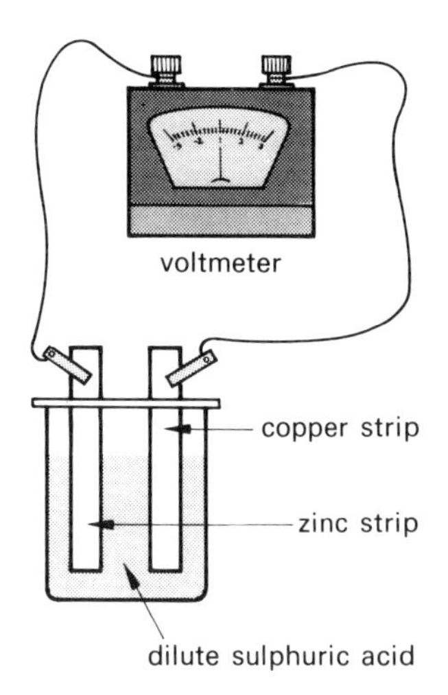

Figure 15.1 A chemical reaction produces an electric current

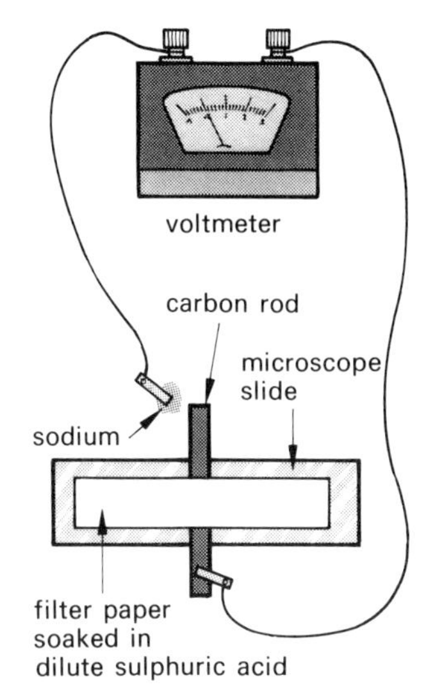

Figure 15.2 A simple cell using sodium and carbon

If the element is used as normally encountered, i.e. with its oxide coating, it produces a voltage which suggests that it should be placed quite low in the series. If the oxide layer is first removed, by dipping in mercury(II) chloride solution, the voltage produced is much higher and aluminium shows its 'true potential'.

You might like to think again about the displacement reactions in Experiment 15.2, which can also be considered as 'battles for electrons'. The metal which is the better loser of electrons (i.e. the more reactive metal) finishes as ions (having lost electrons), and the less reactive metal is changed from ions to metal atoms because it is forced to gain electrons.

Electrode potentials

Carbon is the only non-metallic element which conducts, and it is useful as a 'reference electrode' simply because metals are better losers of electrons than carbon. Another non-metallic element, hydrogen, can also be given a place in the activity series.

Table 15.4 Standard electrode potentials of the more common metals

Element	*Standard electrode potential* (*V*)
Sodium	−2.71
Magnesium	−2.36
Aluminium	−1.67
Zinc	−0.76
Iron	−0.44
Lead	−0.12
Hydrogen	0.00
Copper	+0.34
Mercury	+0.78
Silver	+0.79
Gold	+1.50

When chemists compare the abilities of different metals to lose electrons, they use a similar experiment to that just described, but they use carefully controlled standard conditions of temperature, concentration of electrolyte, etc. The reference electrode used in these exact measurements is one known as the *hydrogen electrode*, the theory of which is too complicated to explain at this level. The voltage recorded when a metal is connected to a standard hydrogen electrode under standard conditions is known as the *standard electrode potential* of the metal. Some typical values are given in Table 15.4. Thus when a standard zinc electrode is connected to a standard hydrogen electrode and the circuit is completed by the movement of ions in solution, a voltage of 0.76 volts is recorded between the two electrodes. This voltage is said to be the standard electrode potential of zinc. It is given a negative sign for reasons explained below.

Note the following points.

1. The standard hydrogen electrode is assumed to have a standard electrode potential of 0.
2. Metals which normally lose electrons to the standard hydrogen electrode are given negative electrode potentials. These are the more reactive metals. The more negative the potential, the greater is the ability to lose electrons.
3. Metals which normally gain electrons from a standard hydrogen electrode are given positive electrode potentials. These are the less reactive metals.
4. The list of elements in order of their ability to lose electrons (i.e. in order of their

standard electrode potentials) is sometimes called the electrochemical series. It is virtually the same as the activity series, except that it can be extended to include many other substances rather than just the metals and carbon. The hydrogen electrode is just one example of the way non-metals which are neither solids nor conductors can be included in the electrochemical series.
5. Calcium is included in the activity series but not in Table 15.1. This is because the electrode potential of calcium is not quite as expected from its chemical reactivity, for reasons which are not normally explained at this level.
6. The value given for aluminium in Table 15.1 is that when its oxide layer has been removed. In practice it is often less reactive than zinc and iron.

15.3 CARBON AND HYDROGEN IN THE ACTIVITY SERIES; SOME FURTHER APPLICATIONS OF THE SERIES

Hydrogen in the activity series

Metal plus acid reactions

Experiment 15.2 showed that a metal high in the activity series will displace another (lower in the series) from a solution of its ions. In exactly the same way, a metal above hydrogen in the series will displace hydrogen from a solution of its ions, i.e. H_2 molecules will be formed from $H^+(aq)$ ions. As all acidic solutions contain $H^+(aq)$ ions, they are solutions of hydrogen ions, and this is why some metals (the more reactive ones) react with dilute acids to form hydrogen (Table 15.2, page 251) but other less reactive metals will not. It might help for you to think of 'acid–metal' reactions as further examples of displacement reactions, and to use ionic equations for these, too, e.g.

$$Mg(s) + 2H^+(aq) \rightarrow Mg^{2+}(aq) + H_2(g)$$

You might expect lead, which is above hydrogen in the electrochemical series, to react with dilute acids and produce hydrogen. The electrode potential of lead is, however, very close to that of the hydrogen electrode (Table 15.4) and no reaction appears to take place.

Reduction of metal oxides by hydrogen

Hydrogen gas will reduce the heated oxides of lead (and the oxides of all the metals below lead in the series) to the metal, itself being oxidized to steam (see page 158).

$$PbO(s) + H_2(g) \rightarrow Pb(l) + H_2O(g)$$

(Note that hydrogen is below lead in the electrochemical series, and might be expected *not* to reduce oxides of lead. The position of hydrogen in the electrochemical series is decided by reactions of hydrogen *ions* in solution, which behave differently from the hydrogen gas considered in the above example.)

Carbon in the activity series

Although it is not usual to include carbon in the activity series, it is useful to think of some of its properties in connection with the series. As with hydrogen molecules in the previous example, carbon is also a reducing agent and will remove the oxygen from the heated oxides of less reactive metals, i.e. those in the mid or low region of the activity series. You may have seen an experiment in which a sample of an oxide of lead is placed on a carbon block in a fume cupboard and then heated strongly, e.g. using a blowpipe and Bunsen burner. Silvery beads of molten lead form as the lead oxide is reduced to metallic lead (which melts at the temperature used) by the carbon:

$$PbO(s) + C(s) \rightarrow Pb(l) + CO(g)$$

The carbon monoxide formed burns immediately to form carbon dioxide, so the overall reaction is

$$2PbO(s) + C(s) \rightarrow 2Pb(l) + CO_2(g)$$

If it helps, you can think of this kind of reaction as a 'battle for oxygen'. Carbon and hydrogen only win the battle (i.e. reduce a metal oxide) if they are competing with a metal in the mid or low region of the activity series. Carbon and carbon monoxide are used in industry for reducing metallic oxides (e.g. the oxides of iron, page 527, and lead) to the corresponding metal.

The extraction of metals

The activity series is almost the reverse of the order in which metals have been discovered and used by mankind. Gold and silver are so unreactive that they are found *native*, i.e. as the uncombined metal mixed with rock, and so they are *comparatively* easy to separate from any ore which contains them. Gold and silver were used by the earliest civilizations.

Most other metals occur *combined* with other elements to form a compound, often called a mineral. This introduces an extra step into the extraction. The mineral still has to be separated from rock and other impurities in the ore, but it also has to be chemically changed so as to form the metal itself, free from the other elements with which it is combined in the compound. Many metallic minerals are oxides, and so the extra step needed to produce the metal is often a reduction of the oxide. This extra step is fairly easy if the metal is an unreactive one, because compounds of unreactive metals are often easy to decompose (as you found in Experiment 15.3), and their oxides are easy to reduce (e.g. by carbon or hydrogen). Even when a metal does not occur naturally as an oxide, its ore is often roasted and changed into an oxide, after which it is reduced. Men of the Bronze Age were able to make the alloy bronze by heating ores that contained both of the fairly unreactive metals copper and tin with charcoal (carbon).

The Iron Age came after the Bronze Age because iron can only be obtained from iron ore (usually Fe_2O_3) when a high temperature (produced by burning carbon in a draught of air) is used to help the reduction of the ore by carbon and carbon monoxide. Iron ore is more difficult to reduce because iron is a more reactive metal than copper or tin, and it is more difficult to obtain it from its compounds.

By the time of the Roman occupation of Britain, gold, silver, mercury, bronze, iron, lead, and brass (an alloy of copper and zinc) were known and used, but it was not until comparatively recently that the very reactive metals such as sodium, potassium, and aluminium were discovered. Sodium, for example, was first seen in 1807 when Humphrey Davy electrolysed molten sodium hydroxide. These reactive metals occur in nature as ions and it is very difficult to make the ions gain electrons and form atoms of the metals. Electrolysis of molten salts is normally used to prepare these metals, the cathode providing the necessary electrons, e.g. aluminium from the 'melted' oxide (p. 534) and sodium from the melted chloride (p. 533). On the other hand, lead, iron, and copper are obtained by reducing the heated oxides with carbon or carbon monoxide. The method of extraction of a metal therefore depends upon its position in the activity series, the extraction becoming easier the lower the position of the metal in the series.

Check your understanding

Carbon at red heat will remove oxygen from the oxides of the metals A, B, and C, but not from the oxide of metal D. Metal C will remove oxygen from the oxide of A, but not from the oxide of B. List the metals A, B, C, and D in order of decreasing reactivity. If metal A is divalent and metal B is trivalent, write (a) the formula for the nitrate of A, (b) the formula for the sulphate of B.

(J.M.B.)

15.4 ELECTRICITY FROM CHEMICAL REACTIONS; SIMPLE CELLS

Electricity from chemicals

As explained on page 257, when two different metals (or a metal and carbon) are used as electrodes in an electrolyte, as in Figure 15.1, a current flows if the electrodes are joined externally by a wire. There is no battery or conventional source of electricity in this simple experiment; the substances themselves are producing the electricity. This is like electrolysis in reverse. In electrolysis, electricity is used to bring about chemical changes. In this 'opposite' sense, the chemicals are reacting and producing electricity.

Simple cells

A simple arrangement of the kind shown in Figure 15.1, using metals (or a metal and carbon) and an electrolyte to provide electricity, is called a *simple cell.* In a situation of this type, electrons leave the more reactive metal of the two (zinc in the example in Figure 15.1) and travel through the wire to the other metal. In providing electrons, atoms of the more reactive metal 'dissolve' as ions:

$$Zn(s) - \underset{\hookrightarrow \text{(to external circuit)}}{2e^-} \rightarrow Zn^{2+}(aq)$$

The electrolyte completes the circuit by allowing ions to move between the electrodes. There is at least one other chemical change, involving the electrolyte, but at this level attention is normally focused only on the metals themselves.

In a simple cell, therefore, the more reactive metal slowly dissolves, decreases in mass, and a current flows in the external circuit. This current should continue until the reactive metal has completely dissolved.

The zinc/copper cell used in the example can be modified by using a single piece of apparatus with two electrolytes, in which the electrolytes are kept apart by a porous pot but can make electrical contact through the pot. Such an arrangement is called a Daniell cell (Figure 15.3).

Many other combinations of metal/metal ion systems are possible, and the voltage between the electrodes depends upon the degree of separation of the two metals in the activity series. As zinc and copper are relatively far apart in the series, the Daniell cell will produce a reasonably good voltage, approximately equal to the difference between their standard electrode potentials.

$$\left.\begin{array}{ll} \text{Zn} & -0.76 \\ & 0.00 \\ \text{Cu} & +0.34 \end{array}\right] \text{Difference} = 0.76 + 0.34 \text{ volts} = 1.10 \text{ volts}$$

A magnesium/magnesium salt solution plus copper/copper salt solution cell would produce a higher voltage.

Simple cells like these are rarely used today as sources of electricity, because they are not really portable, they are large, and they soon 'run out' of chemicals of one kind or another. Nevertheless, an understanding of such cells is important in trying

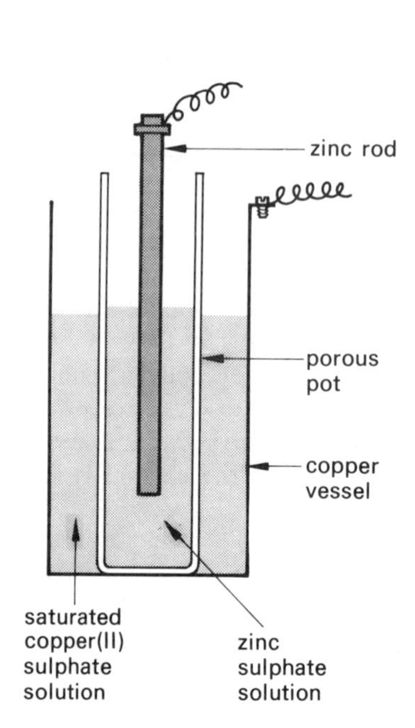

Figure 15.3 A Daniell cell

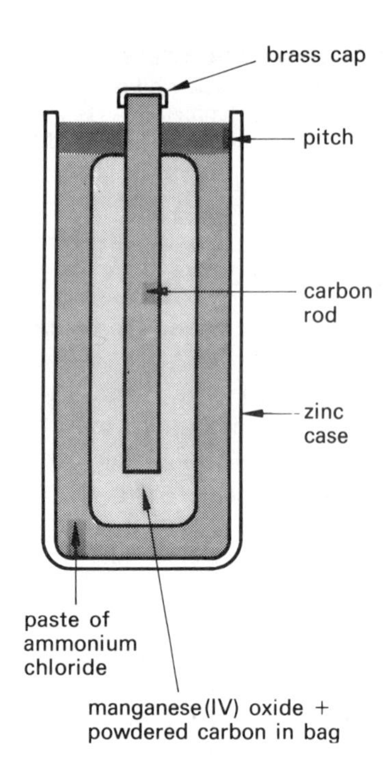

Figure 15.4 A dry cell

to explain many chemical reactions, and it has led to more useful forms of 'battery' such as the common 'dry' cell (Figure 15.4). In the dry cell, the electrodes are zinc (the case) and a carbon rod, and the electrolyte is a paste.

Points for discussion

If a cell was set up in the same way as a Daniell cell, but using a magnesium rod and magnesium sulphate solution instead of the copper rod and copper(II) sulphate solution, what would be the maximum voltage produced by the cell under standard conditions? Explain your answer (refer to Table 15.4).

15.5 CORROSION

Corrosion and rusting

The word corrosion is normally applied only to metals. It describes a reaction between a metal and the atmosphere (or with other substances with which it comes into contact during normal use). Corrosion is a serious problem when it causes the metal to lose its normal structural properties. Copper metal reacts with both oxygen and carbon dioxide in the air and is eventually covered with a green coating. You may have seen this on bronze statues and copper lightning conductors. This particular example is not a serious form of corrosion because the green powder (which is a kind of carbonate) forms a thin layer only; it is quite attractive, and, more important, the metal underneath is as strong as before. Similarly, the coating of aluminium by an oxide layer is not an inconvenient example of corrosion; the metal is protected in this way.

Unfortunately, the corrosion of iron and its steel alloys continues right through the metal, unlike the surface coatings of copper and aluminium referred to above. In time, the strength of the iron or steel is greatly reduced as the metal slowly crumbles to a red-brown powder during the corrosion process. This particular type of corrosion is called rusting; the term rusting should not be used for any other metal. When iron is exposed to both oxygen and moisture, it rusts, and this is one of the biggest disadvantages of iron and most forms of steel. Rust is mainly hydrated iron(III) oxide, formed as the result of a series of reactions. Iron and (particularly) steel are used on a large scale in spite of the fact that they rust so easily, because they are relatively cheap to produce, they are strong, easy to work with, and different types can be made to satisfy demanding requirements. Nevertheless, the cost of combating corrosion in Britain alone runs into millions of pounds each year.

Experiment 15.4
To demonstrate the factors which cause iron to rust*

Apparatus
Four dry test-tubes (two with rubber bungs), labels, test-tube rack, test-tube holder, Bunsen burner, emery paper, beaker, tripod, and gauze.
Iron nails, anhydrous calcium chloride, cotton wool, paraffin wax.

Procedure
(a) Boil some deionized water in a beaker, and leave it boiling until you are ready for step (g).
(b) Thoroughly rub four clean iron nails with emery paper, so that they are completely clean.
(c) Label the four dry test-tubes A to D.
(d) Place a nail in tube A and leave it open to the air.
(e) Place a nail in tube B and add a little water. Leave uncovered.
(f) Place a nail in tube C, put a plug of cotton wool above the nail, and then add anhydrous calcium chloride on top of the plug to a depth of about 2 cm. Close the tube with a bung.
(g) Place a nail in tube D, and add sufficient freshly boiled (and still hot) water to cover the nail. Cover the surface of the water with melted paraffin wax and close the tube with a bung.
(h) Leave the tubes in a rack, undisturbed for about a week, then remove the nails and examine them. If there is no change in the nail in tube A, leave it for another week.

Results
Set out the results in a suitable table. Make sure that it is obvious what each tube contains (or does not contain).

Conclusion
Explain how the experiment demonstrates that *both* air and water are needed to make iron rust. Make sure that you can explain what was present (or missing) in each of the tubes, and why the water was boiled for (g) and then covered with wax. Note that the experiment does not prove that it is the *oxygen* in the air which is important, but the next experiment helps to confirm this.

Experiment 15.5
A more detailed examination of the factors which make iron rust***

Apparatus
Five glass tubes approximately 60 cm long with bungs to fit (the diameter of the tubes is not critical, but the bungs should fit tightly), beakers (in which the tubes will be clamped), five retort stands and clamps, 4 small beakers.
Steel wool, propanone (acetone), sodium chloride solution, oil (e.g. engine oil), magnesium ribbon, strip of tin, emery paper.

Procedure

(a) Five glass tubes are set up as shown in Figure 15.5, but with the following variations. Note that degreasing can be done by rinsing in propanone and allowing to dry. Bungs should only be inserted in the tubes after they have been prepared and placed in the beakers.

Tube 1. The steel wool is first degreased, then soaked in water and drained.

Tube 2. The steel wool is first degreased, then soaked in sodium chloride solution and drained.

Tube 3. The steel wool is soaked in oil and drained.

Tube 4. The steel wool is degreased and then moistened with water. A piece of *cleaned* magnesium ribbon is pushed right through the ball of steel wool so that a length of the ribbon sticks out at each side of the ball.

Tube 5. As for tube 4, but using a strip of tin (or copper) instead of the magnesium.

(b) Leave the tubes assembled (with bungs inserted) for a week or so, and examine them regularly.

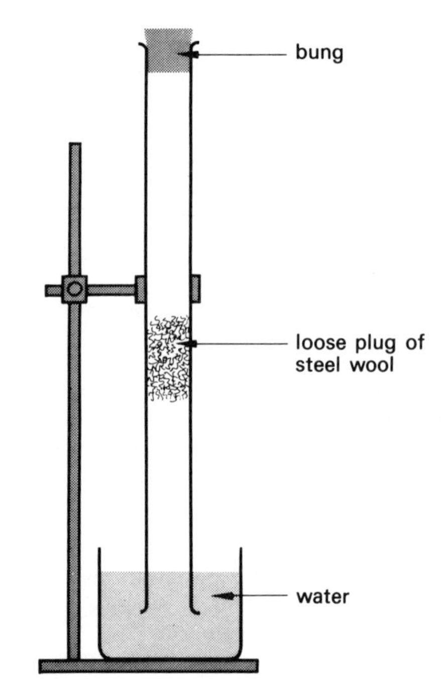

Figure 15.5 A more detailed examination of rusting

Results

Record all your observations in a table. Note that the water levels inside and outside each tube should be the same at the beginning of the experiment. Look for any changes in level. If water rises in any of the tubes, wait until it stops rising and then measure the fraction of the air space in the tube which has been taken up by the water. This fraction will only be approximate, as the air pressure may be different from what it was at the beginning of the experiment.

Notes about the experiment

(i) When iron rusts, a kind of electrical cell is set up of the type explained in the previous unit. Some iron atoms lose electrons to another part of the structure, and form iron ions:

$$Fe(s) - 2e^- \rightarrow Fe^{2+}(aq).$$

(The reasons why some iron atoms lose electrons to other parts of the structure are complicated, and need not concern us at this stage.) The 'circuit' is completed by ions in an electrolyte, as before. Often the electrolyte is ordinary rain water. Rain water or tap water normally contains dissolved carbon dioxide and other substances, which form ions in solution and thus make it an electrolyte.

(ii) The iron(II) ions formed by these cells are oxidized to iron(III) ions, $Fe^{3+}(aq)$, and they then form hydrated iron(III) oxide, which is the red-brown powder we call rust. The important thing is that this process (i.e. the gradual disintegration of iron metal into ions) is continuous. The corrosion gradually 'eats away' the metal, and the product (rust) has no structural strength whatsoever.

(iii) If you understand the principle of rusting and of simple cells, you should be able to explain the effects the magnesium and tin metals had on the rusting of iron. You should also be able to explain in more detail why it is that both air and water are needed for rusting, although rusting is rather more complicated than this simple picture suggests.

Conclusion

Make your own conclusion, explaining all the things which happened in as much chemical detail as possible. The explanations should include:

1. why the volume of gas changed inside some tubes;
2. how magnesium and tin affect the rusting process; and

3. at least one suggestion as to why a solution of common salt causes rusting to occur at a rate different from that with tap water.

Point for discussion

1. Explain why car bodies corrode more rapidly in winter than in summer. There should be more than one reason.

2. If a car body is made of stainless steel, it is very resistant to corrosion. Suggest reasons why this is not done on a commercial scale.

Methods used to combat corrosion

The methods used to prevent iron and steel from corroding usually work by ensuring that both oxygen and water are unable to come into contact with the metal, e.g. painting, tin plating, greasing, galvanizing, and coating with plastics.

A food can is usually made of steel (because steel is relatively cheap) which is then plated with a thin layer of tin. The tin itself resists corrosion and protects the steel from both oxygen and water. It would be too expensive to make a food can out of tin alone. In galvanizing, iron or steel is similarly protected by a thin layer of zinc. These two apparently similar methods have an important difference. If the coating of tin *is* broken (e.g. this may happen if the food can is dropped or hit), the steel then rusts more quickly than it would have done on its own. On the other hand, if the layer of zinc on galvanized iron is broken, the corrosion of the iron is still prevented.

This last point can be understood by considering a large scale method used to combat corrosion. In this, a smaller piece of a metal more reactive than iron (e.g. magnesium) is bolted to the iron or steel (e.g. the hull of a ship). This more reactive metal sets up a kind of cell with the iron and an electrolyte (e.g. salt water or even moist air). The more reactive metal (magnesium) donates electrons to the iron and slowly dissolves (corrodes) away, but the iron does not corrode at all. Remember that in cells of this type only one metal dissolves, and this is the more reactive of the two. A more reactive metal used in this way is called a *sacrificial anode* because it forms the anode of the cell (from which electrons are donated) and is 'sacrificed' (i.e. it corrodes) to save the iron or steel. This simple system is very effective in reducing corrosion on a large scale, e.g. on the hull of a ship, because the block of more reactive metal does not form part of the structure of the ship (and therefore the fact that it corrodes is of no importance) and it can be renewed quickly and regularly, at far less cost than taking the ship out of service and repainting or repairing the hull.

You can now understand the role of the strip of magnesium in the last experiment. As magnesium is more reactive than the iron, it was acting as a sacrificial anode, and it corroded but prevented some of the steel from rusting (not all of the steel, perhaps, because the magnesium did not make electrical contact with all of it).

We can now explain why galvanized iron does not rust to a serious extent when damaged, but a damaged tin can will rust very quickly indeed. With galvanized iron, even if the surface is scratched and water and oxygen reach the iron, there is a second line of defence: the zinc acts as a sacrificial anode, because it is a more reactive metal than iron. The zinc itself corrodes (i.e. starts to dissolve as $Zn^{2+}(aq)$ ions) but in doing so it protects the iron. When this happens, the zinc should eventually be used up, and the iron would then be unprotected. Usually this does not happen, because there is also a third line of defence. The zinc ions formed by the corrosion of zinc metal react with other ions (e.g. from the electrolyte) to form a precipitate of zinc hydroxide, which tends to seal up the crack which exposed iron to the atmosphere. The corrosion slows down or stops, because air and water can no longer reach the iron. The formation of zinc hydroxide is very similar to the formation of rust, but with the important difference that it provides a better seal against the atmosphere than rust does; water and air can easily penetrate through rust.

On the other hand, iron is more reactive than tin, so that when a tin can is scratched the iron acts as the anode of the cell and it dissolves (corrodes) rapidly, i.e. the iron in this case is the anode. Iron atoms give electrons to tin and form $Fe^{2+}(aq)$ ions which soon become rust. When this occurs inside a damaged food can, gases formed by the reactions in the electrolyte of the cell can build up pressure inside, so that when the can is opened the contents may spurt out under pressure.

Points for discussion

1. Explain why corrosion often takes place in water pipes where a steel pipe is connected to a copper pipe.

2. Ornamental gates are often made of wrought iron which is a relatively pure form of iron. Why is rusting not such a problem in this case as it would be if the gates were made of steel?

3. Instead of using a sacrificial anode, a modern method of preventing rusting of large steel structures is to pump electrons into the steel by means of a low voltage direct current. Explain how this will prevent corrosion.

15.6 A SUMMARY OF CHAPTER 15

The following 'check list' should help you to organize the work for revision.

1. Definitions

(a) Thermal decomposition (p. 254)
(b) Thermal dissociation (p. 254)

2. Other ideas

(a) Common metals can be placed in order of chemical reactivity. You should learn the order, and be able to support the idea by describing the ways in which they react with (i) oxygen (Table 15.1, page 250), (ii) dilute acids (Table 15.2, page 251), (iii) water or steam (Table 15.3, page 252), and (iv) the way they displace each other from solutions of their ions,

e.g. $$Zn(s) + Cu^{2+}(aq) \rightarrow Zn^{2+}(aq) + Cu(s)$$

(b) Aluminium metal should appear fairly high in the series, but it rarely shows its true chemical reactivity because it is covered with a protective layer of oxide.
(c) Compounds of reactive metals are difficult to decompose on heating, e.g. their carbonates (Tables 19.1 and 19.4) and their nitrates (Tables 19.1 and 19.5). Reactive metals are difficult to extract from their ores for the same reason; electrolysis is generally used. Compounds of unreactive metals tend to decompose more easily when heated. Oxides of such metals can be reduced by carbon, carbon monoxide, or (much more rarely) hydrogen to form the metal.
(d) Metals lose electrons when they react. They differ in reactivity because some metals lose electrons more easily than others. Metals near the top of the activity series lose electrons easily. Many reactions involving metals and their ions can be thought of as 'battles for electrons'; the more reactive metals always lose electrons to the others.
(e) A way of measuring the ability of a metal to lose electrons is to measure its standard electrode potential. Reactive metals are given negative electrode potentials; the more negative the electrode potential, the more reactive is the metal.
(f) Metals above hydrogen in the activity series will displace hydrogen from *typical* dilute acids (i.e. not necessarily from nitric). This is similar to the idea that a metal A will displace a metal B from a solution containing ions of B, provided that A is above B in the activity series. Remember that dilute acids contain hydrogen ions, $H^+(aq)$.
(g) Gaseous hydrogen and solid carbon will remove oxygen from (i.e. reduce) the heated oxides of metals low in the activity series.
(h) When two different metals (or a metal and carbon) are placed in an electrolyte, a current of electrons flows when the two electrodes are joined externally by a wire. This reaction forms the basis of the dry cell (the common battery), and other forms of battery. In these cases, the more reactive metal gives electrons to the other through the wire, and 'dissolves' as ions in the process.

(i) Rusting is one type of corrosion, and it is the result of a cell reaction like that in (h). If iron atoms lose electrons, they dissolve as $Fe^{2+}(aq)$ ions which are oxidized to $Fe^{3+}(aq)$ and eventually form hydrated iron(III) oxide, $Fe_2O_3.xH_2O$, rust. This can be prevented by stopping air and water from coming into contact with the iron, e.g. by painting, galvanizing, tin plating, greasing, coating with plastics, etc. Alternatively, a metal more reactive than iron can be placed in contact with the iron, when any corrosion which does occur will involve the more reactive metal rather than iron. Metals used in this way are called sacrificial anodes.
(j) Galvanized iron is protected from corrosion by a layer of zinc. If the zinc layer is broken, the zinc acts as a sacrificial anode, but does not continue to be sacrificed in this way because a deposit of insoluble zinc hydroxide soon seals up the break or scratch in the zinc coating.

3. Important experiments

(a) One to demonstrate the factors which cause iron to rust, e.g. Experiment 15.4.
(b) An experiment which demonstrates the effect of a sacrificial anode (e.g. Experiment 15.5), and explains how it works.
(c) You could be asked to describe any of several experiments which show that metals form an activity series, e.g. their reactions with acids, with water, with oxygen, and in displacements. Make sure that you learn the detailed results of experiments such as these, and that you can use them to justify an activity series and write equations for the reactions involved.
(d) It is important to learn the *results* of experiments such as Experiment 15.3, including exceptions to general behaviour, equations, and colour changes.

QUESTIONS

Short answer and structured questions

1. Say whether the following are true or false.
(a) Copper will replace zinc from a solution of zinc nitrate.
(b) Iron will replace copper from a solution of copper(II) sulphate.
(c) Aluminium will replace magnesium from a solution of magnesium chloride.
(d) Iron will liberate hydrogen from dilute hydrochloric acid.
(e) Copper will liberate hydrogen from dilute sulphuric acid.

2. Explain why an electric current passes along a wire joining a zinc rod to a copper rod with both metals dipping into a solution of sulphuric acid.

3. The following is a list of standard electrode potentials:

Magnesium	−2.36
Zinc	−0.76
Copper	+0.34
Silver	+0.79

(a) Which two metals, if used together in a cell, would produce the largest e.m.f.?
(b) Explain the meaning of the positive and negative signs.

Questions requiring longer answers

4. (a) For each of the metals iron, copper, and magnesium, describe the reaction (if any) when the metal is (i) heated in air, (ii) added to dilute hydrochloric acid. On the basis of these reactions, place the metals in order of chemical reactivity, with the most reactive first.
(b) The atomic number of caesium, Cs, is 55.
(i) How would you expect caesium to react with water? What would be the nature of the aqueous product?
(c) Write down the electronic structure of the fluorine, F, atom.
Would you expect this element to form anions or cations? Give reasons for your answer and state the charge carried by these ions.

5. Describe how you would set up a cell to produce electricity from a chemical reaction and explain the working of the cell.

6. Write ionic equations for the following:
(a) the reaction between zinc and copper(II) sulphate solution;
(b) the reaction between magnesium and dilute hydrochloric acid;

(c) the reaction at the cathode when copper is deposited;
(d) the reaction at the anode when chlorine ions are discharged.

7. Describe in detail, three experiments you could carry out to place lead, magnesium, and zinc in their correct order in the electrochemical series.

Explain why the tests you choose might not place aluminium correctly in the series.

8. (a) Design and describe a quantitative experiment to find out whether a new alloy is oxidized by the atmosphere.

What results would you expect to observe if the new alloy were (i) easily oxidized, (ii) rust resistant?
(b) If the alloy were attacked, what further experiments would you set up to discover which parts of the air caused the reaction?
(c) How is air prevented from attacking iron (i) on the blades of a lawn mower; (ii) on a dust bin; (iii) on cutlery? (J.M.B.)

9. The following is a list of symbols of some of the elements in order of an 'activity series': K, Mg, Al, Zn, Fe, H, Cu, Ag.
(a) Which of these elements will not displace hydrogen from a dilute acid?
(b) Which of these elements has the most stable hydroxide?
(c) A piece of zinc is placed in iron(II) sulphate solution and a piece of iron is placed in zinc sulphate solution. In which solution would there be a reaction and why? Give the equation for the reaction.
(d) From these elements name (i) a metal which reacts with cold water, (ii) a different metal which reacts with hot water but only very slowly with cold, (iii) any other metals which will react when heated in a current of steam.
(e) Name any metals in the list whose heated oxides can be reduced by hydrogen. For one of these metals give an equation for the reaction.
(f) If mixtures of aluminium oxide and iron, and of iron oxide and aluminium, are heated, in which mixture is there a reaction and why? Give the equation for the reaction.

Outline any one experiment by which you could prepare dry crystals of zinc sulphate from a different zinc compound. (A.E.B.)

10. Draw a diagram of an apparatus suitable for the electrolysis of copper(II) sulphate solution using platinum electrodes and for the collection of the products. Give the names and polarities of the electrodes, the names of the products, and the equations for the electrode reactions.

After passing the current for, say, ten minutes, what would be the effect of reversing the current and passing it in the opposite direction for about twenty minutes?

In the example given above, electricity is used to bring about a chemical change. Describe a way in which a chemical change can be used to release electrical energy. Carefully indicate the reactions which occur and where they take place. (J.M.B.)

16 Structure

16.1 WHAT DO WE MEAN BY STRUCTURE?

You know that all substances are made up from atoms, molecules, or ions, but as yet we have not explained how these particles are *arranged* in different substances. For example, they could be packed together *tightly* or *loosely*, in a *regular* pattern or in a *random* way. It should come as no surprise to learn that the packing of the particles affects the properties of a substance. An understanding of packing arrangements has helped scientists to produce synthetic (man-made) materials that sometimes have more useful properties than natural substances. The way in which the particles which make up a substance are arranged is known as its structure. The structure of a substance depends upon the bonding within it, the size and shape of the particles within it, and whether it is in the gas, liquid, or solid state.

From your knowledge of the kinetic theory (p. 14) you will remember that gases have no shape and no fixed arrangement of particles, and therefore have no structure. The gas particles have overcome the attractive forces between them and so are free to move at random and will fill whatever space is available. This explains why most gases have some very similar physical properties, e.g. they all diffuse quickly, they have no shape of their own, they always fill their container, and they are poor conductors of heat and electricity.

The particles of a liquid are free to move within the liquid, but there are still fairly strong forces between the particles. Liquids therefore have a surface and definite shapes, i.e. those of the containers in which they are placed. This does not mean that liquids have a definite structure, because the relative positions of the particles are constantly changing. In a solid the particles do not have enough energy to overcome the forces between them and so remain in the same fixed position. This particular arrangement of the particles of a solid is known as its structure.

You might imagine that all solids have a similar structure, because the particles of solids are packed fairly tightly together and are not free to move from place to place. This is true in a very general sense, but in this chapter you will learn that there are different ways in which the particles of solids can be arranged, even though they are always packed fairly close together. Some solids are very different indeed from each other, e.g. rubber and steel. These differences arise because the particles which occur in solids can vary in their shape, size, bonding, and perhaps most important of all, the ways, in which they can be packed together. You may be puzzled by the statement that particles

They are always packed closely together

can vary in shape, because you may think of all particles as being like atoms, i.e. spherical. If the particles in a substance are molecules, not atoms, they are almost certainly not spherical (see the shape of a sulphur molecule, S_8, Figure 16.7). Different shapes can be packed together in different ways, and they can also be *held* together in different ways according to how they are bonded. There are many possible structural arrangements in the solid state, and in this chapter we will be concerned with a few of them.

A great deal of evidence supports the idea that the particles in different solids are arranged in different ways. Diamond and graphite are both forms of the same element, carbon, and they therefore consist only of carbon atoms. Their physical properties, however, are very different (p. 273), and this difference can only be explained by the fact that their particles are packed differently and/or because they are bonded differently. Further support for this idea comes if we consider the volumes taken up by the same number of atoms of different forms of the same element. For example, 12 g of diamond and 12 g of graphite will each contain the same number of carbon atoms, but 12 g of diamond have a volume of 3.4 cm^3 whereas 12 g of graphite have a volume of 5.4 cm^3. This suggests that the atoms are packed more closely in diamond than they are in graphite.

The packing structure in a solid is a regular pattern and not a random arrangement. If you have ever watched crystals grow from a solution, you may have wondered why crystals of the same substance usually have the same shape, no matter how big or small they are. We can imagine that some particles in the solution join together to form a three-dimensional unit which is characteristic of the substance. Layer after layer of particles then 'grow' on top of this unit so that the shape is kept constant and eventually the crystal is big enough to see. This suggests that many solids have a highly ordered arrangement of particles, and the shape of a crystal is often a clue to the arrangement of particles within it.

Some of the packing arrangements in solids are very complicated, but modern methods of determining structure (e.g. X-ray diffraction) and the use of computers have enabled scientists to reveal the secrets of the internal packing of many solids. An understanding of structure has helped to explain how DNA and other complicated biological substances do their job, has led to the production of a wide range of plastics, and to the use of graphite in re-entry cones for space modules. Methods of determining structure such as X-ray diffraction cannot be used in a school laboratory, but it is possible for you to consider some of the structural arrangements that have been discovered and to see how these structures are related to the properties of the solid. All of the ideas in this chapter will be easier to understand if suitable models are available for you to look at.

16.2 THE STRUCTURES OF SOLID ELEMENTS

Every metallic element has a special structure called a *metallic lattice*. Some non-metallic elements (in the solid state) have structures formed by joining *atoms* together by covalent bonds, and the others consist of *molecules* joined together by intermolecular forces.

Metallic lattices

About 70 per cent of the elements are metallic, and their properties are so different from those of other elements and also from those of compounds that it seems reasonable to suppose that a different kind of bonding is present. As all metals are good conductors of heat and electricity, it is probable that they all have the same kind of structure.

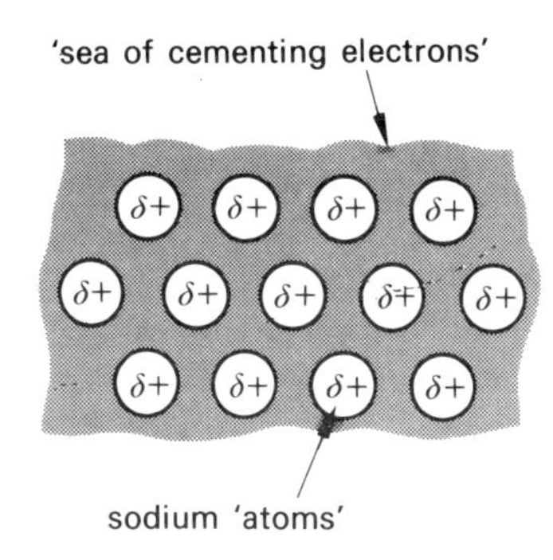

Figure 16.1 Illustration of bonding in sodium metal

In very simple terms we can imagine that the metal atoms in a metallic lattice have all 'donated' (given up) the electrons in their outer energy level to a general 'pool' of electrons in the lattice. These electrons are free to move throughout the structure, and help to bind the metal particles together. Metallic lattices are often likened to being 'islands of positive ions surrounded by a mobile sea of electrons'. For example, in the metallic lattice of sodium, each atom in the lattice gives up its outer electron to an 'electron pool', which holds the whole metallic crystal together. This is rather like the way in which bricks of a wall (sodium 'atoms') are bound together by cement (the pool of electrons). This is shown in simple form in Figure 16.1. Metallic bonding is one of the "stronger' forms of bonding (see Table 16.6, page 285).

Metals are good conductors of electricity because of the mobile 'sea of electrons' within them. If electrons (i.e. electric current) enter a metal at one side, a similar number of electrons are 'pushed out' at the other side, and the metal conducts (Figure 16.2). Similarly, metals are good conductors of heat because the 'mobile electrons' can spread heat energy quickly. Solids with other forms of bonding do not conduct electricity (except graphite) because they do not have these 'mobile electrons', or other charged particles which are free to move. (Ionic *solids* do not conduct, but if they are melted free ions are produced which can carry a current, page 210).

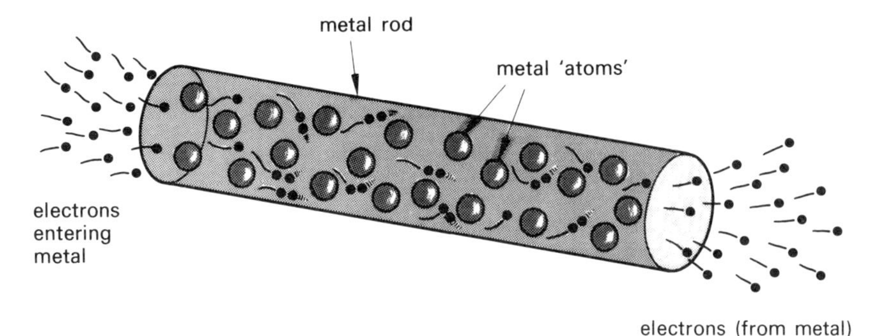

Figure 16.2 Electrical conduction by a metal

This simple description of metallic lattices has emphasized the special bonding which holds metal 'atoms' together. It is also possible to *pack* metal 'atoms' together in different ways, and so the detailed structure of a metallic lattice involves both bonding and packing. The packing aspects of metals are not normally studied at an elementary level. However, you might guess that sodium and potassium, for example, have packing arrangements different from that in copper, for it is possible to cut sodium and potassium with an ordinary knife but this cannot be done with copper. Similarly, sodium and potassium float on water whereas copper sinks. These facts can be partly explained by understanding that the particles in sodium (or potassium) are further apart from each other in the lattice than are those of copper.

Metals with more than one structure

The ways in which the particles of metals are packed together are now understood in great detail. You may not have realized that some metals have more than one structure (i.e. allotropes, page 273). This happens because their particles can pack in more than one arrangement. For example, normal tin is the so-called 'white' allotrope. If white tin is kept at very low temperatures for some time, it begins to change to 'grey' tin, which is a powdery form. This change occurs because of a different packing arrangement, and articles made of tin can disintegrate in very cold climates because of this change. Paraffin for Scott's ill-fated polar expedition was stored in metal containers, many of which disintegrated in the sub-zero temperatures. This was because of a structural change in the tin-based solder used on the cans. It has been suggested that the loss of this fuel was a contributory factor in the failure to complete the expedition.

Alloys

Pure metals are not used very often, because it is possible to produce thousands of alloys (i.e. combinations of two or more metals or a metal and a non-metal), with a wide range of properties, from less than 80 pure metals. An alloy composed of metals A and B may have properties entirely different from those of either A or B. Even before World War II, over 5000 alloys were in use and the science of metallurgy has advanced considerably since then.

The fact that it is possible to 'improve' the basic properties of a metal (particularly its strength and resistance to wear and corrosion) by converting it into an alloy is of tremendous practical importance. The properties of an alloy depend to a large extent on the way in which the different particles within it pack together. Scientists can often prepare an alloy 'tailor made' for a particular task. For example, by far the greater part of a modern car is metal, but only a very small proportion is *pure* metal. Most of the metallic components are alloys, such as the mild steel used for the body, the many types of alloy steels used for moving components, tin alloys for bearings, brass radiators, nickel alloys in spark plugs, etc.

More to do

Find out which metals are used to make each of the following alloys, and the proportions in which they are mixed.
(a) stainless steel, (b) duralumin, (c) brass, (d) bronze, (e) solder, (f) the alloy used to make 'silver' coins.

Elements with giant atomic (macromolecular) structures

These consist of atoms joined together by strong covalent bonds to form a three-dimensional network of atoms (a giant structure). Very few elements have this kind of structure, and the examples normally studied at an elementary level are diamond and graphite, the two allotropes of carbon.

Allotropes (or polymorphs) are different structural forms of the same element in the same physical state.

Diamond and graphite are both forms of the same element, carbon, and they each consist only of carbon atoms. The data in Table 16.1 show that these two allotropes have very different physical properties, and for centuries scientists made the mistake of thinking that they were two different elements. Remember, however, that although allotropes of the same element may have very different physical properties (because of the different ways in which their particles are packed), they usually have the same *chemical* properties because they have the same atoms within them. For example, if a sample of each carbon allotrope is burned completely in an excess of oxygen, carbon dioxide is the only product in each case.

Table 16.1 Some properties of diamond and graphite

Property	*Diamond*	*Graphite*
Internuclear distance (nm)	0.154	0.142
Appearance	Colourless, transparent, crystals	Black, shiny, solid
Hardness	Hardest natural substance known	Very soft, flakes easily
Density (g cm^{-3})	3.5	2.2
Electrical conductivity	Non-conductor	Conducts in the direction parallel to the hexagonal planes
Molar volume at room temperature (cm^3)	3.4	5.4

The structure of graphite

If you rub a few flakes of graphite between your fingers and then examine the skin, you will notice that the graphite has 'rubbed off' and that it feels slippery. These properties are explained by the structure of graphite.

The atoms in graphite are bonded together in layers, in which a hexagonal pattern of atoms can be seen (see Figure 16.3). In the layers each carbon atom is strongly bonded to three other atoms. The bonds between the layers, however, are more like weak intermolecular forces than strong covalent ones. The layers can thus slide over each other, and this explains the slippery feel of graphite and why the layers flake off. These properties are utilized by suspending graphite in oil to act as a lubricant, and by mixing graphite with clay for use in pencil leads, where the graphite 'rubs off' onto paper. Another substance which forms plate-like layers is mica (Figure 16.4), and this can be split horizontally (along the layers) by using a finger nail.

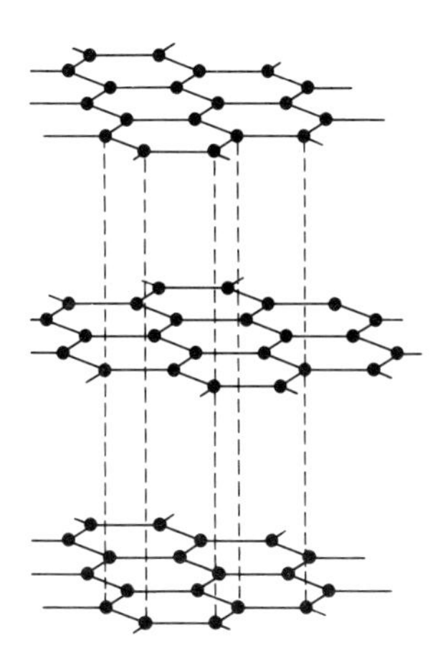

Figure 16.3 The structure of graphite. The dotted lines represent weak bonds between the layers

Figure 16.4 Mica, showing its flat-sheet crystal structure

The structure of diamond

As a complete contrast, each atom in diamond forms *four* strong bonds and no weak ones (Figure 16.5), each atom being joined to four others arranged in a tetrahedron around it. (A model will make this easier to understand.) There are also no flat layers which can easily slide over each other, as there are in graphite. The whole structure is a strong, rigid mass of atoms, and we can easily understand why diamond is so different from graphite. It is the hardest natural substance, and is thus used to make cutting instruments and drilling equipment, as well as for jewellery. Silicon carbide, SiC, is almost as hard as diamond, and it is interesting to note that it has the same type of structure, although it contains two different kinds of atom. The two elements within it are, however, members of the same group in the Periodic Table, and it represents the chemist's way of creating a substance with the same kind of structure as diamond, but more cheaply. Silicon carbide is used as an abrasive.

When the structures of the two carbon allotropes were known, attempts were made to convert graphite into diamond by repeating the conditions which produce diamonds in nature, i.e. enormous pressures and temperatures. We can imagine the atoms in graphite being forced into the tighter, diamond arrangement by these drastic conditions. Artificial diamonds can be made in this way, and they are much used as cutting-stones on large drill bits for boring oil wells etc., but are not yet of high enough quality to be used as gem-stones. It is not yet possible to make large artificial diamonds.

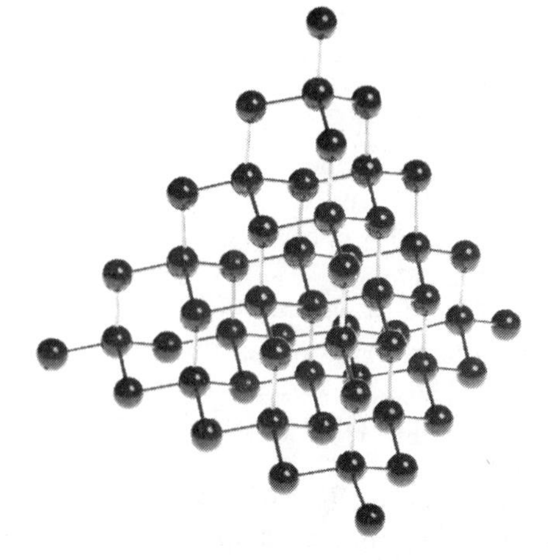

Figure 16.5 The structure of diamond

Elements with simple molecular structures

Most of the non-metals have these kinds of structure, and typical common examples include iodine and sulphur. In each of these elements, simple separate molecules are held together in a solid structure by intermolecular bonds. The *intra*molecular bonds are strong covalent bonds and cannot easily be broken, but the *inter*molecular bonds are much weaker and can be broken by heating, when the solid usually melts.

The structure of iodine

The molecules in iodine are simple I_2 molecules. The intramolecular (I–I) bond is a strong, covalent bond but the intermolecular bonds (Figure 16.6) between I_2 molecules are easily broken when the solid is heated. In this case, however, the iodine does not usually melt when these bonds are broken; it sublimes, i.e. changes straight from a (black) solid to a (purple) vapour.

The intermolecular forces between iodine molecules are of a special kind called *van der Waals forces*. You do not need to understand what this term means, but it is important to understand that these forces are *weak* (Table 16.6, page 285). (A simple explanation of van der Waals and other intermolecular forces is given on page 285.) You will remember from your work on the Periodic Table that iodine is the only common halogen element which is a solid at room temperature. Fluorine, chlorine, and bromine also exist as diatomic molecules (F_2, etc.), but the van der Waals forces between their molecules are even smaller than those between iodine molecules. This means that even at room temperature, the van der Waals forces between fluorine (and chlorine) molecules are broken and the elements are already gases. At lower temperatures, fluorine, chlorine, and bromine also become solids and exist as molecular crystals, held together by the special type of intermolecular forces called van der Waals forces.

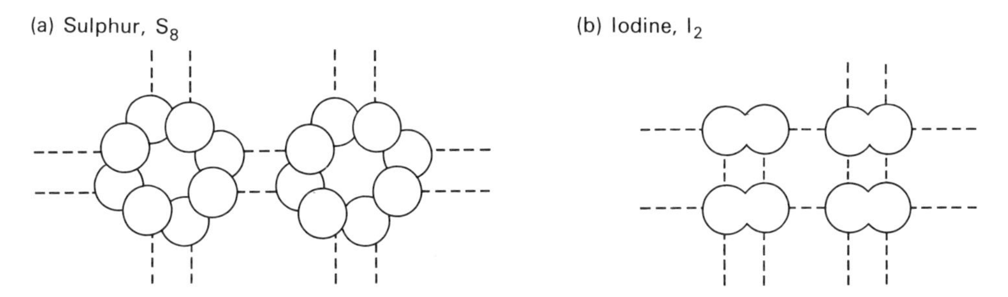

Figure 16.6 Simple representations of intermolecular forces between (a) sulphur molecules and (b) iodine molecules in the solid state. The dotted lines represent intermolecular forces which in these examples are van der Waals forces

The structure of the allotropes of sulphur

Sulphur molecules are much bigger than the molecules of iodine, because they consist of eight atoms of sulphur joined together by strong covalent bonds (intramolecular bonds). The molecule has the shape of a 'puckered ring' (Figure 16.7). These molecules can be packed together in more than one way, and so solid sulphur has allotropic forms. The allotropes of sulphur are studied in the next section.

The intermolecular forces between sulphur molecules are again van der Waals forces, and as before these are easily broken by heating. Sulphur melts at about 119 °C and then the yellow solid becomes an amber coloured liquid. The melting

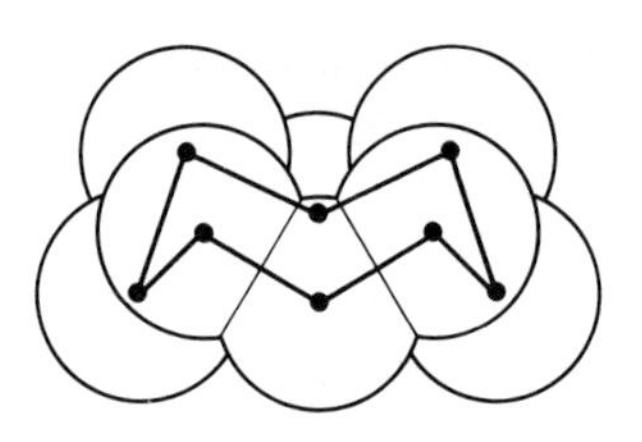

Figure 16.7 A sulphur molecule, S_8

point is slightly higher than that of iodine because there are more intermolecular forces between the larger sulphur molecules than between the smaller iodine molecules, but the melting point is still low compared with those typical of ionic solids (Figure 16.6).

The allotropes of sulphur

The physical changes which occur when sulphur melts When sulphur is *gently* heated until it melts, and then slowly heated to its boiling point, the following changes take place.

1. The yellow powder melts to form an amber coloured liquid (119 °C).
2. At first the liquid sulphur is very mobile (i.e. runny), but it gradually darkens and becomes more viscous (i.e. 'treacly').
3. The liquid changes through a red-brown colour to a very dark, almost black liquid, which is very viscous.
4. Near the boiling point, the dark liquid suddenly becomes mobile again.
5. At 444 °C the liquid boils.

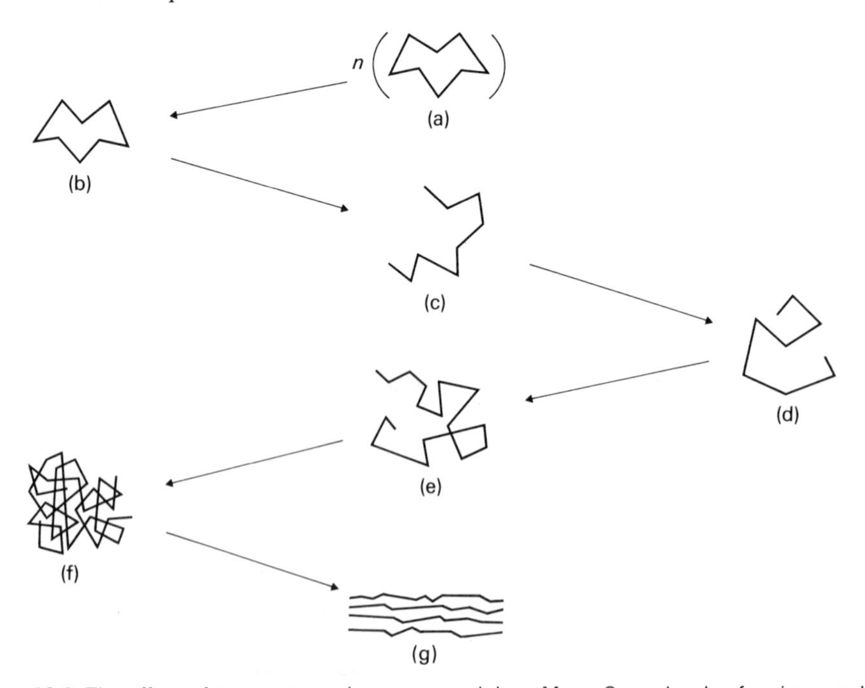

Figure 16.8 The effect of temperature changes on sulphur. Many S_8 molecules forming a solid allotrope (a), become individual S_8 molecules (b) at the melting point. The molecules open up (c and d) and join together to form long chains (e), which become entangled in each other so that the liquid becomes viscous (f). Long free chains of sulphur atoms are found in boiling sulphur and plastic sulphur (when stretched) (g)

The various changes in the liquid are due to structural changes which take place at different temperatures. If you do an experiment to show these changes, you may also be given instructions on how to pour the almost-boiling liquid sulphur into a beaker of cold water so as to produce a strange solid form called plastic sulphur. (This experiment is not to be attempted without proper instructions and supervision.) These changes are summarized in Figure 16.8 and should be discussed in detail. Note that plastic sulphur is *not* an allotrope of sulphur. Sulphur has only two true allotropes, rhombic sulphur and monoclinic sulphur, which are produced in the next two experiments.

In Experiment 16.2 you will again see some of these physical changes, but the emphasis is on the solid which crystallizes from the molten sulphur rather than on the liquid itself.

Experiment 16.1
To prepare rhombic (α) sulphur**

Apparatus
Watch-glass, filter paper, elastic band, hand lens or microscope, teat pipette.
A saturated solution of powdered roll sulphur in carbon disulphide (kept in a fume cupboard).

Procedure
(a) Using a teat pipette, place a few drops of the solution of sulphur in carbon disulphide on to a watch-glass (in the fume cupboard). *Carbon disulphide is poisonous, easily absorbed through the skin, and has an unpleasant smell. Make sure that the chemical is used only in the fume cupboard, and that only a few drops of the solution are used.*

Leave the watch-glass in the fume cupboard until the solvent has evaporated.
(b) Examine the crystals of rhombic (α) sulphur under a hand lens or microscope. Draw the shape of a typical crystal.

A general point about rhombic sulphur This is the stable allotrope of sulphur at room temperature. Monoclinic (β) sulphur, which you will make in the next experiment, is only stable at higher temperatures, and gradually changes to rhombic sulphur if left at room temperature. However, as structural changes in the solid state are very slow (because the molecules do not readily move from place to place), this conversion takes a long time.

Experiment 16.2
To prepare monoclinic (β) sulphur**

Apparatus
Filter paper, hand lens or microscope, test-tube and holder, Bunsen burner, paper clip, tongs, spatula, protective mat.
Powdered roll sulphur.

Procedure
(a) Fold together two thicknesses of filter paper as for filtering, and then clip them together with a paper clip so that a cone is formed.
(b) Place powdered roll sulphur in a test-tube to a depth of about 5 cm. You are going to melt this sulphur and pour it into the paper cone to solidify. Heat the sulphur slowly, with a low flame (preferably in a fume cupboard) and rotate the tube constantly until all the sulphur has *just* melted. *If this stage of the experiment is done slowly, the sulphur should not catch fire, but before you begin the heating make sure that you know what to do if the sulphur does catch fire. The exact instructions which you are given will depend upon the circumstances in which you do the experiment.*

(c) Hold the paper cone over the mat by means of a pair of tongs, and pour the liquid sulphur into it, *quickly* but carefully. *Make sure that the liquid is not poured near the hand.* Place the cone on the mat and leave until a thin crust forms over the liquid.
(d) Remove the paper clip, split open the cone, and allow any remaining liquid to drain away. *Remember that any such liquid will be hot.* Crystals of monoclinic sulphur will be seen growing from under the crust. Examine them with a hand lens or microscope, and draw the shape of a typical crystal.

Notes about the experiment

(i) Remember that both allotropes of sulphur contain the same units (S_8 molecules), but the molecules are packed together in different ways. The rhombic form has a density of 2.07 g cm^{-3}, and the monoclinic form has a density of 1.96 g cm^{-3}, so the molecules are packed more tightly in the rhombic form. Unlike diamond and graphite, the allotropes of sulphur have no obvious uses or properties which can be explained by their structures. (The detailed structure of monoclinic sulphur was not known until 1965.)
(ii) Monoclinic sulphur is only stable between 96 °C and the melting point of sulphur, and at lower temperatures it slowly changes to rhombic sulphur. Note that it is relatively easy to change rhombic sulphur to monoclinic sulphur, and the other way round, but the interconversion of diamond and graphite is very difficult indeed.

The structures of elements in a typical period in the Periodic Table
The way in which the structure of elements varies within a short period is summarized in Table 16.2. The period from sodium to argon is used in the table, but similar trends occur in other periods. Note that most of the elements are either metallic lattices or molecular; the giant atomic lattices are relatively rare, and tend to occur in the first few elements of Group 4, e.g. carbon and silicon.

Table 16.2 Structures of some elements and their compounds in a short period

	Sodium	*Magnesium*	*Aluminium*	*Silicon*	*Phosphorus*	*Sulphur*	*Chlorine*	*Argon*
Elements	←	Metallic	→	←		Non-metallic		→
Structure of elements	←	Giant metallic lattice	→	Giant atomic lattice	←	Molecular	→	Free atoms
Oxides	←	Basic →	Amphoteric	←	Acidic		→	
Structure of oxides	←	Giant ionic lattice →	Giant (ionic-covalent) lattice	Giant atomic lattice	←	Molecular	→	
Structure of chlorides	←	Giant ionic lattice →	←		Molecular	→		
Structure of hydrides	←	Giant ionic lattice →		←		Molecular	→	

Check your understanding

1. In which substances would you expect to find free atoms at room temperature?
2. Give three ways in which solids differ from gases or liquids, and explain the differences.
3. Different solid forms of the same element are called
A isobars; B isotopes; C allotropes; D isomers; E monotropes

Questions 4 to 8 are concerned with the substances listed below.

A sulphur; B hydrogen; C diamond; D sodium; E graphite
Choose from this list the substance which you think best fits each of the following descriptions.

4. At room temperature most of the volume occupied by this element is space.
5. This substance has a giant structure, but some bonds are weaker than others.
6. This substance has a molecular structure, each molecule containing eight atoms.
7. This substance has a giant structure but it has a relatively low melting point and is quite soft.
8. All bond angles and bond lengths in this substance are the same. It does not conduct electricity and has a high melting point.
9. The data in Table 16.3 refer to five *elements* lettered V, W, X, Y, and Z.
(a) Which pair of elements are metals?
(b) Which pair are allotropic forms of the same element?
(c) Which element could be a noble gas?
(W.J.E.C.)

Table 16.3 Some structural information (W.J.E.C.)

Element	*Atomic mass*	*Melting point °C*	*Electrical conductivity of solid element*	*Reaction with oxygen*
V	31	590	Non-conductor	Burns readily
W	40	851	Good	Burns readily
X	20	−249	Non-conductor	No reaction
Y	207	327	Good	Oxidizes slowly
Z	31	44	Non-conductor	Burns readily

16.3 THE STRUCTURES OF COMPOUNDS

The structure of a solid compound depends to a large extent on the bonding within it. If the compound is ionic, the structure will always be a giant ionic crystal lattice, consisting of ions held together in a giant network. If the bonding is covalent, the structure could consist of simple molecules held together by intermolecular forces. (This is like the structure of the *elements* iodine and sulphur, the only difference being that there will be at least two *different* kinds of atom within a molecule of a compound.) Alternatively, it could consist of atoms held together by covalent bonds in a giant network to form a macromolecule. Silicon carbide, SiC, has a structure of the latter type (p. 274) but *compounds* with such structures are not studied in detail at this level.

Note: Giant *ionic* structures and metallic lattices are not called macromolecules (unlike the giant *atomic* lattices) because they do not contain covalent bonds.

The structure in any one compound then depends upon the way in which the atoms, ions, or molecules *pack* together. This in turn depends upon the shape of the particles, their size, the ratio in which they occur in the compound, etc. The ionic lattices thus include several different packing systems, as do the other main types. It is usual to study just one or two examples of these main types of structure at this level.

Giant ionic structures (lattices)

These are giant structures in which a large number of ions (of at least two different types) are held together in a three-dimensional arrangement. The ions can differ in size, charge, and relative numbers, so there are many possible arrangements. The whole structure must be electrically neutral, however, i.e. the total number of negative charges must equal the total number of positive charges. At an elementary level, sodium chloride is normally used as an example of an ionic lattice (Figure 16.9).

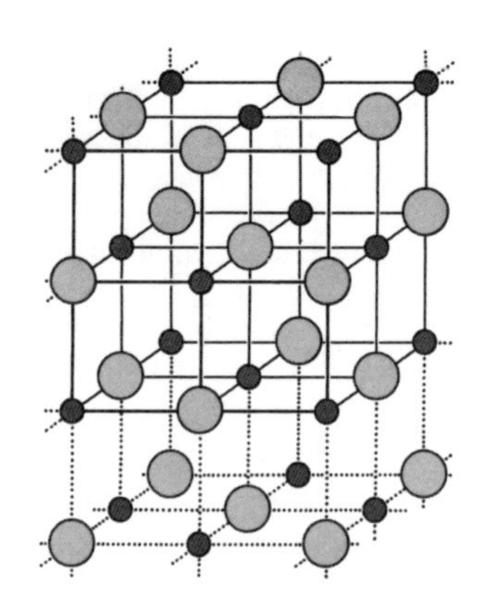

Figure 16.9 The structure of sodium chloride, NaCl. The larger spheres represent the chloride ions; the solid lines indicate the unit cell

Note the following.

1. In all ionic lattices the ions are packed together (so that they 'touch' each other) in a regular fashion.
2. The arrangement or pattern is the one which gives the structure with the greatest stability (i.e. the minimum potential energy).
3. An ionic lattice is held together because the attractive forces between oppositely charged ions (+ and −) are greater than the repulsive forces between ions of the same charge. This is because the distances between the *oppositely* charged ions (arranged in horizontal planes) are less than the distances between the *similarly* charged ions (which are positioned diagonally with respect to the planes) (see Figure 16.9).
4. The formula of sodium chloride, NaCl, does not mean that ions are grouped in the structure in pairs, or that pairs of ions are 'free' in the structure. It just means that there are equal numbers of sodium and chloride ions in the structure. Similarly, the lattice of magnesium chloride, $MgCl_2$, does not contain groups or molecules of $MgCl_2$ which are separate from each other, but it does contain twice as many chloride ions as magnesium ions. The actual numbers involved in the lattice may be millions, and they are all locked together in a three-dimensional pattern. In sodium chloride, each sodium ion does not have just one chloride ion 'partner'; it is surrounded equally by six chloride ions, and similarly each chloride ion is surrounded by six sodium ions. It is not possible to recognize a particular pair of ions as forming a separate unit.

More advanced terms used in considering ionic lattices

For most requirements, the information given above and in Figure 16.9 is adequate in considering a typical ionic lattice such as that of sodium chloride. Some courses may require an understanding of the following terms.

The co-ordination number of a structure is the number of particles which surround and 'touch' any given particle. If more than one type of particle is present, the co-ordination number for one of them is the number of particles of the opposite type which surround and 'touch' it.

A unit cell is the simplest three-dimensional arrangement which, by moving a distance equal to its own dimensions in various directions, can form the complete lattice.

Figure 16.9 includes the unit cell of sodium chloride. The co-ordination number around a chloride ion is 6, and that around a sodium ion is 6. We refer to the co-ordination in sodium chloride as being 6:6. Notice that whenever the charge on the anion is equal to the charge on the cation, the co-ordination numbers must be the same (as they are in sodium chloride), but they do not have to be 6.

The sodium ions (considered alone) form a *face-centred cube*, and so do the chloride ions. The structure of sodium chloride is often described as *two interpenetrating face-centred cubic lattices, one of sodium ions and one of chloride ions*. The unit cell of sodium chloride is a face-centred cube consisting of both sodium and chloride ions. By convention this unit cell is considered to have four sodium ions at the corners of each face (as in Figure 16.9) but it would not be incorrect to use a unit cell with chloride ions at the corners.

As a crystal grows by repeated layering on top of a basic shape (the unit cell), you might imagine that crystals of sodium chloride should have the shape of a face-centred cube. In fact there is no such shape; to the eye, it is simply a cube. You should know that sodium chloride usually crystallizes as cube-shaped crystals (Figure 16.10). Its crystals can only be cleaved easily (i.e. split) along lines which follow the lines of ions within them, e.g. at right angles to each other. You may see a demonstration of this, or be allowed to examine the cracks in a large crystal.

Note that if the layers of ions are displaced by just one interionic distance, similar charges come opposite each other, and there is repulsion between the layers rather than attraction. This is the main reason why ionic structures are so brittle when subjected to shock or strain.

Figure 16.10 A crystal of sodium chloride showing its regular shape. The front edge of the crystal is about 1 cm long

Simple molecular compounds

The solid form of water, ice, has a simple molecular structure. The individual molecules have the formula H_2O, and the molecules are held together to form a structure which takes up more volume than the liquid water which produced it; water expands on freezing, unlike most other liquids.

The main intermolecular bonds in ice belong to a special class of bonding called *hydrogen bonding*. (A simple description of this and other forms of intermolecular bonding is given on page 285.) Compare, however, the fact that the intermolecular forces in iodine and sulphur (Figure 16.6) are called van der Waals forces, whereas those between water molecules are called hydrogen bonds. These bonds are broken when ice is heated, and then the H_2O molecules separate from each other and the ice melts (Figure 16.11). Hydrogen bonds are stronger than van der Waals forces, but they are weaker than ionic bonds and covalent bonds (see Table 16.6, page 285).

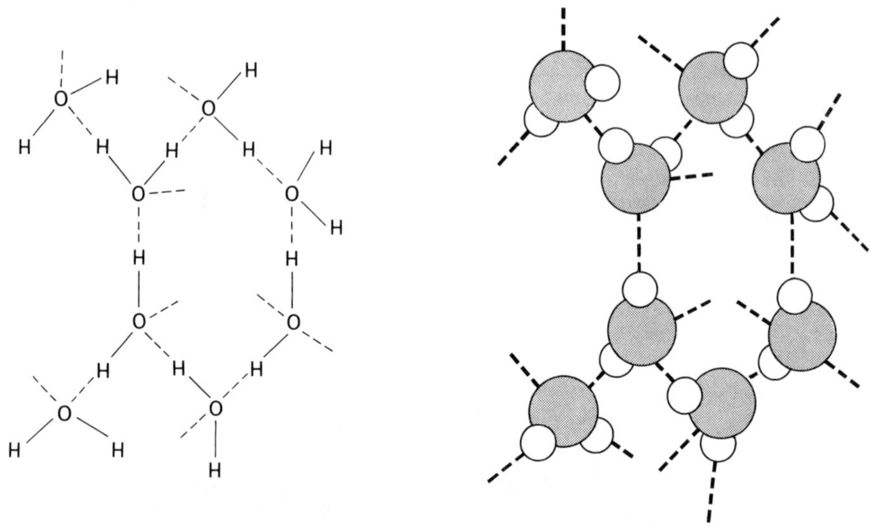

Figure 16.11 The structure of ice. The dotted lines represent intermolecular bonds (hydrogen bonds in this case). The other bonds, represented by solid lines in the left-hand diagram, are intramolecular bonds

Trends in the structure of compounds across a typical short period

The way in which the structure of compounds varies within a short period is summarized in Table 16.2 (p. 278). The period from sodium to argon is used in the table, but similar trends occur in compounds of elements within other short periods.

16.4 A COMPARISON BETWEEN SIMPLE MOLECULAR STRUCTURES AND GIANT STRUCTURES

The examples of structure discussed in this chapter fall into two main types. Both elements and compounds can exist in the solid state as simple molecular structures, (e.g. S_8, I_2, and H_2O), in which separate units (molecules) are held together by intermolecular forces. Other elements and compounds can exist in the solid state as giant structures, in which there are no simple units but rather a massive three-dimensional network of particles, e.g. metals, diamond, graphite, and the ionic lattices. It is absolutely essential that you fully understand the differences between these two types of structure.

The particles in a giant structure are joined together by *strong* bonds (although there are also some weaker ones in graphite) to form a large, three-dimensional arrangement in which no free, individual units exist. In a typical structure of this type, there may be millions of ions or atoms, all joined together, and there are no weak forces of the type which correspond to the intermolecular forces in simple molecular structures. There are three main types of giant structure: (i) metallic lattices, (ii) ionic lattices, and (iii) atomic lattices (macromolecules). As all of these contain strong bonds they are very difficult to 'pull apart' and so have high melting points and boiling points, and they are nearly all solids at room temperature.

Simple molecular structures, on the other hand, contain 'separate' units (molecules). The atoms *within* each molecule are held together by strong, covalent, intramolecular bonds. The molecules themselves are held together by weaker, intermolecular bonds. Read again the points made on pages 125–6, e.g. why molecular structures tend to have low melting points. One of the most misunderstood points in elementary chemistry is the difference between *inter*molecular and *intra*molecular bonding; 'inter' means 'between', and 'intra' means 'within'.

Simple molecular substances usually have low melting points and boiling points because their molecules are easily separated from each other, and the structure then 'comes apart', e.g. the substance melts or evaporates. Many molecular substances are already liquids or gases at room temperature. If they are solids at room temperature, they are usually easily melted. These ideas are summarized in Tables 16.4 and 16.5.

Note that many students imagine that all covalent substances have simple molecular structures, and therefore low melting points. This may be true of many covalently bonded substances, but there are some very common examples where covalently bonded atoms form a macromolecule (a giant covalent structure), e.g. in diamond and graphite.

Apart from the melting points and boiling points, these various types of structure can also be distinguished by electrical effects, as summarized in Table 16.4.

Table 16.4 Some differences between giant and molecular structures

Giant substances			*Simple molecular substances*
Metallic lattices	*Ionic lattices*	*Atomic (macromolecular) lattices*	
Usually high melting point and boiling point	High melting point and boiling point	High melting point and boiling point	Low melting and boiling points
Conduct electricity in solid state	Will not conduct in solid state, but will when molten or dissolved, and are then decomposed at the same time	Will not conduct in solid or molten state (except graphite)	Do not conduct electricity when solid or molten
Insoluble in most solvents, although may react and then appear to dissolve	More soluble in water than in organic solvents such as hexane	Rarely dissolve or react with any solvent	May dissolve in water, but more soluble in molecular solvents such as hexane

Check your understanding

1. Say what structures you think the following substances will have.

(a) Solid A has a melting point of 1500 °C and conducts electricity when in the solid state.

(b) Solid B has a melting point of 27 °C and does not conduct electricity even when molten.

(c) Solid C has a melting point of 880 °C and does not conduct electricity when in the solid state, but does so when molten.

(d) Solid D has a very high melting point and conducts electricity when in the solid state, but it is not a metal.

2. Elements or compounds may exist in the following forms:

A free ions

B ionic crystals (giant ionic structures)

C macromolecular structures (giant covalent structures)

Table 16.5 The particles from which matter is made

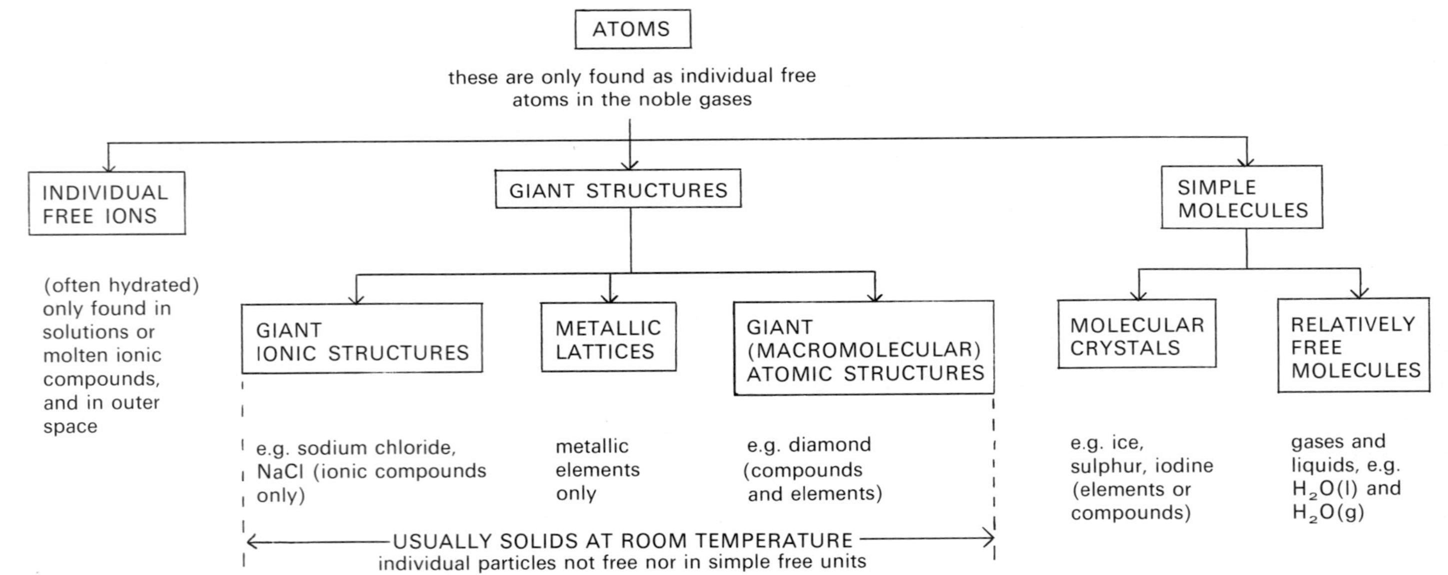

D simple atoms
E simple molecules
Which best describes the structures of each substance given below?
(a) argon, (b) gaseous carbon dioxide, (c) diamond, (d) iodine crystals, (e) molten sodium chloride, (f) solid sodium chloride.
(J.M.B.)

3. Explain clearly, in your own words, the differences between giant structures and molecular structures. Give examples to illustrate your answer.

4. Which of the following pairs are allotropes?
A Carbon (graphite) and sulphur (monoclinic)
B Carbon (diamond) and sulphur (monoclinic)
C Sodium chloride and caesium chloride
D Phosphorus (white) and phosphorus (red)
E ^{31}P and ^{30}P

APPENDIX A MORE DETAILED CONSIDERATION OF INTERMOLECULAR FORCES

The energies of various kinds of bonds

A true ionic bond (in which one or more electrons are *completely transferred* from one atom to another) or a true covalent bond (in which electron pairs are *equally shared* between the combining atoms) is only found in rare cases. In many cases the bonds between atoms are intermediate between these two extremes and display both ionic and covalent character. This situation also gives rise to other types of bonding such as polar bonding, and hydrogen bonding. In addition, there are other more specialized forms of bonding called metallic bonding and van der Waals forces.

There is a wide variation of bond energies associated with these different types of bonds. The approximate values are given in Table 16.6. The bond energies are found by calculating the heat energy required (in kilojoules) to break the bonds between a mole of molecules. The first three kinds of bond shown in the table are strong bonds, and the latter three are weak bonds. Ionic bonds and covalent bonds are usually *intramolecular* bonds, i.e. bonds within the molecule itself, whereas hydrogen bonds, polar bonds and van der Waals forces are usually *intermolecular* bonds, i.e. bonds between the molecules. It is obviously much easier to split up a structure into 'molecules' than it is to split up the molecules themselves; intermolecular bond strengths are small compared to intramolecular bond strengths. The bonding in metals is completely different from all other types of bonding (p. 270).

Table 16.6 Energies of various types of bond

Type of bond	*Strength of bond*	*Bond energy ($kJ\ mol^{-1}$)*
Metallic bond	strong	approx. 80 to 580
Ionic bond	strong	120 to 450
Covalent bond	strong	120 to 550
Hydrogen bond	fairly weak	13 to 30
Dipolar bond	weak	<13
van der Waals forces	very weak	<4

Polar molecules

When the atoms in a molecule are all the same kind (e.g. in Cl_2, H_2, P_4, etc.), the bonding within the molecule is 'true' covalent bonding. The shared electrons which form each bond can be considered to be equidistant from each atom,

As the atoms on each 'side' of the bond are identical, neither will obtain more than a half share of the shared electrons.

However, if the two combining atoms are not the same, for example in a molecule of hydrogen chloride (HCl), then one of the atoms will attract the shared pair of electrons more strongly than the other atom. This will result in the atom with the greatest electron attracting power becoming slightly negatively charged with respect to the other atom which will, relatively, become slightly positively charged. In the case of hydrogen chloride, the chlorine atom has a greater electron attracting power than the hydrogen atom and thus the former becomes slightly negatively charged ($\delta-$) and the latter becomes slightly positively charged ($\delta+$). This can be represented as follows:

H ×× Cl or
$\delta+$ $\delta-$

The covalently bonded molecule is thus showing a slight ionic character. Molecules in which unequal sharing of electrons takes place are called *polar molecules* and are often referred to as *dipoles*.

The power of an atom to attract electrons to itself is called its *electronegativity*. A chlorine atom is more electronegative than a hydrogen atom. The larger the difference in the electronegativities of the combining atoms, the greater is the polarization of the covalent bond between them, i.e. the greater is the *ionic character* of the covalent bond. If the electronegativity difference is very large, then complete electron transfer takes place and a totally ionic bond is formed.

Intermolecular forces between dipoles

Molecules possessing dipoles are attracted to each other in much the same way that positive and negative ions attract each other. However, as the charges involved are small compared to ionic charges, the bonds formed are much weaker. This effect can be illustrated by considering hydrogen chloride molecules, i.e.

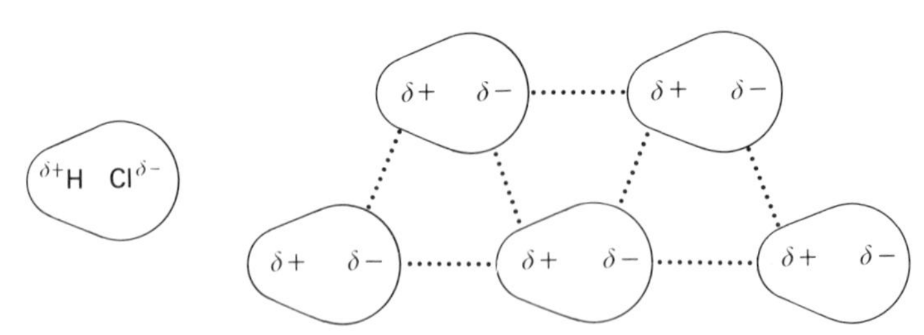

(The dipolar bonds are shown by dotted lines.)

This ordered structure only exists at low temperatures, (up to -114 °C in the case of hydrogen chloride), as only a small amount of heat energy is required to give the molecules sufficient kinetic energy to overcome the intermolecular dipolar attractive forces. Compounds whose molecules only possess small dipoles thus have low melting points and boiling points and exist as gases at room temperature. However, compounds composed of molecules with larger dipoles (i.e. having a greater ionic character) will have higher melting points and boiling points.

Hydrogen bonding

Hydrogen bonding is a special form of dipole-dipole bonding; hydrogen bonds are much stronger than ordinary dipole-dipole bonds.

In order for a hydrogen bond to form, some molecules must contain a hydrogen atom which has a positive charge greater than that normally formed by polarization. This usually happens only when hydrogen atoms are joined directly to an atom of one of the three most electronegative elements, i.e. nitrogen, fluorine, or oxygen.

Thus in a molecule of hydrogen fluoride, HF, the difference in electronegativity between the hydrogen atom and the very electronegative fluorine atom is so great that the covalent bond between them is very polarized, and the hydrogen atom has a distinct δ^+ charge:

δ^+H F$^{\delta-}$

$\overset{\delta^+}{H}$ ×× $\overset{\delta^-}{F}$

A similar situation arises in the N—H bonds in ammonia, NH_3, and in the O—H bonds in water, H_2O.

The effectiveness of the δ^+ charge on the hydrogen atom in these cases is further increased by the fact that the hydrogen atom is the smallest kind of atom, and therefore the positive charge is 'concentrated' in a small volume. When a polar hydrogen atom *of this type* is attracted to a very electronegative atom (F, N, or O) in either a different kind of molecule or another molecule of the same kind, a hydrogen bond is formed. The δ^+ and δ^- charges which are attracted to each other in such a situation are larger than those normally encountered in dipole-dipole bonds. Hydrogen bonds normally occur as intermolecular bonds, although examples of intramolecular hydrogen bonding are known in organic chemistry.

The energy associated with a hydrogen bond is fairly small compared to the strength of ionic and covalent bonds (Table 16.6), but the presence of such bonding between molecules can produce a very marked effect on some of their properties. For example, the melting and boiling points of water are much higher than would be expected by comparison with the hydrides of the other elements in Group 6 (i.e. H_2S, H_2Se, and H_2Te) all of which are gases at room temperature. This is due to the formation of hydrogen bonds between water molecules (Figure 16.11). Only when a certain quantity of thermal energy has been supplied to the ice molecules do they have sufficient kinetic energy to break enough of the hydrogen bonds and cause the crystalline framework to collapse. The water molecules can then pack more closely together and so water also shows the unusual property of contracting when melting, or expanding on freezing.

Van der Waals forces

If there were no intermolecular forces between certain substances, then these substances would exist only as gases, because there would be no force to attract their molecules together. However, *every* substance can be liquefied if it is cooled to a low enough temperature, i.e. a temperature at which the kinetic energy possessed by its molecules is insufficient to overcome the intermolecular forces between them. Thus every substance must possess some form of intermolecular force, even when molecules consist of like atoms (e.g. Cl_2) and do not possess a permanent dipole. Van der Waals forces are the weakest kind of intermolecular forces and are named after the

Dutch scientist who first studied them, and they exist between molecules in *any* substance.

You will remember that it is sometimes convenient to imagine that the electrons of an atom circulate the nucleus in given orbits. At any moment in time it is unlikely that these orbiting electrons will be evenly distributed. This will result in a minute dipole ($\delta\delta+$ and $\delta\delta-$) being set up in the atoms, i.e.

$\delta\delta+$ $\delta\delta-$

This minute dipole can then *induce* a dipole in a neighbouring atom (in the same way as a magnet can induce an opposite pole in a piece of soft iron). The two atoms then attract each other, i.e.

$\delta\delta+$ $\delta\delta-$ $\delta\delta+$ $\delta\delta-$

The weak forces of attraction between these atoms are known as *van der Waals forces*. It must be stressed that any particular dipole formed is transient, i.e. it only exists for an instant of time, because in our simplified picture the electrons are constantly orbiting the nuclei at tremendous speeds. The distribution of the electrons thus changes almost instantaneously, another momentary dipole is set up, and so on.

The magnitude of the van der Waals forces in a given situation depend to a large extent upon the size of the molecules. The larger the molecules, the more electrons are present, and the greater the van der Waals forces. This is reflected in the melting and boiling points of the halogens. Fluorine, which has relatively small F_2 molecules, has low melting and boiling points and is a gas at room temperature. The same is true of chlorine. The van der Waals forces between the larger Br_2 molecules are greater, and bromine is a liquid at room temperature. The forces between iodine molecules are even greater, and iodine is a solid at room temperature. Nevertheless, the *relatively* weak nature of its van der Waals forces is illustrated by the fact that iodine sublimes, even at low temperatures.

16.5 A SUMMARY OF CHAPTER 16

The following 'check list' should help you to organize the work for revision.

1. *Definitions*

(a) Allotropes (polymorphs), page 273
(b) Co-ordination number (if needed), page 280
(c) Unit cell (if needed), page 280

2. *Other ideas*

(a) The main features of a metallic lattice, and why such a lattice conducts electricity.
(b) What an alloy is (a mixture of a metal and one or more other elements), and a few common examples, e.g. solder (lead and tin), brass (copper and zinc), steels (iron, carbon, and perhaps other elements), and bronze (copper and tin).
(c) The structures of graphite and diamond. You should be able to use the differences in structure to explain why the two allotropes of carbon have such different properties.
(d) Sulphur, phosphorus, and carbon are common elements which form allotropes.
(e) The difference between intermolecular (between molecules) and intramolecular (inside molecules) bonds.

(f) The elements iodine and sulphur as examples of molecular solids, with the molecules held together by van der Waals intermolecular forces. The molecules in iodine are I_2 and in sulphur S_8.
(g) The crystalline shapes of the two allotropes of sulphur.
(h) The physical changes which occur when solid sulphur is heated (p. 276).
(i) Rhombic (α) sulphur is the stable allotrope of sulphur at room temperature, and monoclinic (β) sulphur slowly changes to the rhombic form if left at room temperature.
(j) The way in which structures of both elements and compounds vary across a short period in the Periodic Table (Table 16.2).
(k) The main characteristics of an ionic lattice, e.g. that of sodium chloride. Sodium chloride forms crystals which are cube-shaped.
(l) Ice has a molecular structure, the intermolecular forces in this case being of a type called hydrogen bonding.
(m) The main differences between molecular structures and giant structures, as on pages 282–283 and in Table 16.4.
(n) The main divisions of structure are also summarized in Table 16.5.
(o) More details about intermolecular forces, if appropriate, as in the Appendix.

3. *Important experiments*

(a) An experiment to prepare rhombic (α) sulphur, e.g. Experiment 16.1.
(b) An experiment to prepare monoclinic (β) sulphur, e.g. Experiment 16.2.

QUESTIONS

1. Changes from one liquid allotrope of sulphur to another are very rapid but it takes much longer for both 'plastic' and β sulphur to form the α variety. Explain these facts.

2. (a) Name the particles present in the crystal lattices of (i) sodium, (ii) sodium chloride, (iii) diamond, (iv) carbon dioxide.
(b) Explain the following in terms of the structures of the substances involved:
(i) in the solid state sodium is a good conductor of electricity and is deformable whereas diamond is a non-conductor and is extremely hard;
(ii) sodium chloride has a high melting point whereas carbon dioxide is a gas at room temperature and pressure.
(c) What do you understand by *allotropy*? Name *one* of the crystalline allotropes of sulphur. Describe its appearance and how you would prepare a sample of it in the laboratory. (A.E.B. 1973)

3. Compare the properties of diamond and graphite and explain how the differences are related to their structures.

4. Describe and explain the changes which take place when sulphur is slowly heated from room temperature to 444 °C, rapidly cooled, and then left at room temperature for some time.

17 Oxidation and reduction

17.1 INTRODUCTION

Chemists have used the words oxidation and reduction for many years. Oxidation was first defined as a process in which an element or compound gains oxygen or loses hydrogen, and reduction as the opposite. These definitions are still used where a simple transfer of oxygen or hydrogen takes place, but the idea has been extended to include *most* chemical reactions.

More chemical reactions can be regarded as 'battles for electrons' rather than as 'battles for oxygen or hydrogen', and definitions of oxidation and reduction now also refer to electron transfer. In these cases we are using one of the most important principles in chemistry, that atoms or molecules exchange or share electrons in order to become more stable. We have seen that some atoms prefer to gain electrons and others prefer to lose electrons. Definitions of oxidation and reduction now include these ideas because it is useful to be able to summarize some reactions by saying which chemical has gained electrons and which has lost them, i.e. by saying which has been reduced and which has been oxidized. Similarly, if we need to make a chemical lose electrons, we might choose to react it with another chemical which easily gains them, i.e. which causes the first chemical to lose electrons. It is useful to know which chemicals are good at doing this, i.e. which are good oxidizing agents. These ideas will be easier to understand when you have learned some definitions and had practice in using them.

17.2 DEFINITIONS AND OTHER USEFUL POINTS

A substance is **oxidized** *if it gains oxygen, loses hydrogen, or loses electrons. Such a reaction is an oxidation.*

A substance is **reduced** *if it loses oxygen, gains hydrogen, or gains electrons. Such a reaction is a reduction.*

A substance which brings about an oxidation is an **oxidizing agent**, *and it is itself reduced during the reaction.*

A substance which brings about a reduction is a **reducing agent**, *and it is itself oxidized during the reaction.*

Note: An oxidizing agent must be itself reduced in performing its task, for if it (for example) removes electrons from something (i.e. oxidizes it), it must itself *receive* those electrons, i.e. be reduced. Think about this carefully; it is frequently misunderstood. As **red**uction normally occurs with **ox**idation, these reactions are often called **redox** reactions.

It is useful to think of electrolysis and the activity series in connection with oxidation and reduction.

All electrolytic changes at a cathode must be reductions as ions gain electrons.
All electrolytic changes at an anode must be oxidations as ions or atoms lose electrons.
Hydrogen is more likely to reduce oxides of metals low in the activity series (i.e. below hydrogen itself) rather than oxides of metals high in the series.
A metal will reduce an oxide of a metal lower than itself in the activity series.

Note: Think of these last two points as 'battles for oxygen'; the more reactive element wins.

There is no point in learning countless examples of oxidations and reductions, for a large proportion of all chemical reactions involve oxidation and reduction. It is more important that you should be able to recognize what is happening in a *given* reaction in terms of oxidation and reduction. This point will become more obvious if you look at typical questions on oxidation and reduction. You should be able to recognize what is being oxidized and what is being reduced in a given reaction, and you should be able to suggest the kinds of oxidation and reduction which various oxidizing and reducing agents can do. You will meet appropriate detailed examples throughout your course, and the emphasis in this chapter is therefore on the general principles.

17.3 EXAMPLES OF OXIDATION AND REDUCTION INVOLVING HYDROGEN AND OXYGEN

Some general points in answering redox questions

Remember the following points when answering questions on redox reactions.

1. If something is oxidized or reduced in a full chemical reaction, represented by a complete equation, there is also a chemical which is reduced or oxidized. *Some* full equations do not appear to indicate both oxidation and reduction, e.g.

$$H_2(g) + Cl_2(g) \rightarrow 2HCl(g)$$

We can say that the chlorine is reduced to hydrogen chloride by gaining hydrogen, but it is not easy to see why the hydrogen is oxidized. Whenever possible, however, look for both oxidation and reduction in a full equation.

A reaction such as the one represented by the equation

$$Zn(s) \rightarrow Zn^{2+}(aq) + 2e^-$$

is *not* a complete reaction, as it involves only one substance. It is a half reaction, and the equation used as an example is an ion–electron half equation. In cases like these, only *one* of the processes oxidation or reduction occur; the other half reaction, which *must* happen at the same time, is not shown.
2. In describing a reaction in terms of oxidation and reduction, always use expressions such as 'X is oxidized to Y by gaining oxygen, and W is reduced to Z by losing oxygen' rather than the temptingly simple 'X is oxidized and W is reduced'.
3. Describe what you would *see* as evidence of oxidation or reduction if this is relevant to a question, e.g. 'colourless X is oxidized to yellow Y by________'.

The following examples should make these points clear; colour changes have been ignored so as to emphasize the oxidation/reduction.

Examples involving hydrogen and oxygen

Note: Do not be put off by complicated-looking examples, or even by chemicals which may not be studied in your course. The actual principle is very simple.

$$CuO(s)+H_2(g) \rightarrow Cu(s)+H_2O(g)$$

The copper(II) oxide is reduced to copper (by loss of oxygen) and the hydrogen is oxidized to steam (by gaining oxygen).

$$PbO(s)+C(s) \rightarrow Pb(l)+CO(g)$$

The lead(II) oxide is reduced to molten lead (by loss of oxygen), and the carbon is oxidized to carbon monoxide (by gaining oxygen).

$$3Fe(s)+4H_2O(g) \rightarrow Fe_3O_4(s)+4H_2(g)$$

The iron is oxidized to iron(II) di-iron(III) oxide (tri-iron tetraoxide) by gaining oxygen, and steam is reduced to hydrogen by losing oxygen.

$$SO_2(g)+2Mg(s) \rightarrow 2MgO(s)+S(s)$$

Sulphur dioxide is reduced to sulphur (by losing oxygen), and the magnesium is oxidized to magnesium oxide by gaining oxygen.

$$H_2S(g)+Cl_2(g) \rightarrow 2HCl(g)+S(s)$$

Hydrogen sulphide is oxidized to sulphur (by losing hydrogen), and the chlorine is reduced to hydrogen chloride (by gaining hydrogen).

17.4 EXAMPLES OF OXIDATION AND REDUCTION USING ELECTRON TRANSFER

The general points referred to under section 17.3 still apply to electron transfer reactions, but you are more likely to meet equations for half reactions as well as full equations. For example, changes at a cathode during electrolysis, e.g.

$$Cu^{2+}(aq)+2e^- \rightarrow Cu(s)$$

are given by half equations. This particular example is a reduction only. Only in full equations can there be both oxidation and reduction.

Note also that electron transfer reactions are sometimes difficult to detect because electrons are not always shown in the equations. Thus in the reaction

$$2FeCl_2(aq)+Cl_2(g) \rightarrow 2FeCl_3(aq)$$

there appears to be no transfer of electrons, oxygen, or hydrogen. Nevertheless, oxidation and reduction are occurring because of electron transfer. Always look for atoms becoming ions (or ions becoming atoms), and for an ion changing 'valency'. If either of these happens there must be a transfer of electrons. In the example above, iron(II) ions are changing to iron(III) ions and are thus *oxidized* by loss of electrons. At the same time some chlorine atoms are becoming chloride ions, and they are being reduced by gaining electrons. This is more obvious if the equation is broken into two half equations, in which case electrons are shown in the equations. The electrons cancel out, however, if the two half equations are combined.

$$2Fe^{2+}(aq)-2e^- \rightarrow 2Fe^{3+}(aq)$$

$$Cl_2(g)+2e^- \rightarrow 2Cl^-(aq)$$

Metals near the top of the activity series are powerful reducing agents because they easily lose electrons to something else (i.e. reduce something). Similarly, the halogens are powerful oxidizing agents because they easily *take* electrons from something else (i.e. oxidize something). However, chlorine is a better oxidizing agent than bromine, which in turn is better than iodine. This is shown by the displacement reactions referred to on page 240. Note that reactive metals and reactive non-metals are opposite to each other as reducing agents and oxidizing agents, and they are also to be found at opposite sides of the Periodic Table.

Metals near the top of the activity series are powerful reducing agents

Examples

$$Fe^{2+}(aq) + 2e^- \rightarrow Fe(s)$$

The iron(II) ions are reduced to iron metal by each gaining two electrons.

$$Fe^{3+}(aq) + e^- \rightarrow Fe^{2+}(aq)$$

The iron(III) ions are reduced to iron(II) ions by each gaining an electron.

$$Zn(s) - 2e^- \rightarrow Zn^{2+}(aq)$$

Zinc atoms are oxidized to zinc ions by each losing two electrons.

$$Zn(s) + Cu^{2+}(aq) \rightarrow Zn^{2+}(aq) + Cu(s)$$

Zinc atoms are oxidized to zinc ions by each losing two electrons, and copper(II) ions are reduced to copper atoms by each gaining two electrons.

Experiment 17.1
To show that redox reactions involve electron transfers*

Apparatus
Petri dish containing an agar gel to which potassium chloride has been added, number 6 cork borer, two carbon electrodes joined by wires and crocodile clips. Solutions of potassium iodide, starch, iron(III) ions, acidified potassium manganate(VII), acidified potassium dichromate(VI).

Procedure
(a) Punch out 6 holes in the gel using the cork borer.
(b) Pour a mixture of potassium iodide solution and starch solution into three of the holes.
(c) Pour a sample of one of the other reagents into one of the other holes.
(d) Place one carbon rod into an iodide/starch hole and the second one in the hole containing the other reagent. Record carefully what happens.
(e) Repeat (d) using the other solutions, but clean the electrodes and use a fresh hole of iodide/starch mixture each time.

Notes about the experiment
(i) Acidified potassium manganate(VII) and acidified potassium dichromate(VI) solutions are both powerful oxidizing agents (for formulae and colour changes see Table 17.2). This is because they easily add oxygen to, or remove electrons from, other substances. You are not expected to write equations which show how they oxidize. It is sufficient to understand *how* they work, i.e. by adding oxygen or removing electrons, and also the colour changes they undergo.
(ii) Iodide ions have no effect on starch, but iodine molecules make starch turn blue-black.

Conclusion

Try to write your own conclusion, based on answers to the following questions.

1. What was happening to the iodide ions in the three experiments? Were the iodide ions being oxidized or reduced? (Explain your answer.)
2. Write a half equation to show what happens to the iodide ions.
3. What was the purpose of the carbon electrodes and the connecting wire?
4. Why was potassium chloride added to the gel?
5. What do iron(III) ions, acidified potassium manganate(VII), and acidified potassium dichromate(VI) have in common when they are in contact with iodide ions? What are they acting as in these reactions?
6. Write a half equation to show what happens to the iron(III) ions when in contact with iodide ions, and then combine the two half equations together to make a full equation for the reaction.

Check your understanding

Study the following equations carefully and decide which substances are oxidized and which are reduced, giving reasons for your answers. In some of the examples you will need to look for hidden electron transfers or valency changes.

1. $Cu(s) - 2e^- \rightarrow Cu^{2+}(aq)$
2. $Cl_2(g) + H_2(g) \rightarrow 2HCl(g)$
3. $2Ag^+(aq) + Mg(s) \rightarrow 2Ag(s) + Mg^{2+}(aq)$
4. $2Ca(s) + O_2(g) \rightarrow 2CaO(s)$
5. $Fe_2O_3(s) + 3CO(g) \rightarrow 2Fe(l) + 3CO_2(g)$
6. $N_2(g) + 3H_2(g) \rightarrow 2NH_3(g)$
7. $C(s) + H_2O(g) \rightarrow CO(g) + H_2(g)$
8. $2FeSO_4(s) \rightarrow Fe_2O_3(s) + SO_2(g) + SO_3(g)$
9. $Fe(s) + CuSO_4(aq) \rightarrow FeSO_4(aq) + Cu(s)$
10. $Cl_2(g) + 2KBr(aq) \rightarrow 2KCl(aq) + Br_2(aq)$
11. $2Na(s) + Cl_2(g) \rightarrow 2NaCl(s)$

17.5 OXIDIZING AGENTS AND REDUCING AGENTS

Students often get confused about oxidation and oxidizing agent, etc. They forget that when an oxidizing agent oxidizes something, it is itself reduced. The opposite is true for reducing agents. For example, in the reaction

$$CuO(s) + H_2(g) \rightarrow Cu(s) + H_2O(g)$$

hydrogen is the reducing agent, and reduces copper(II) oxide to copper. In doing this, however, the reducing agent is itself oxidized to steam by gaining oxygen.

It is useful to be familiar with the common reducing and oxidizing agents, what they are capable of doing, and the colour changes or other signs which indicate that they have acted in this way. The common agents are described in Tables 17.1 and 17.2. Typical uses are described in the appropriate chapters, e.g. the conversion (oxidation) of iron(II) salts to iron(III) salts on pages 342–5.

Suppose that you are asked to suggest how a certain substance A can be converted to a substance B, and that you work out that such a reaction would be an oxidation (e.g. A could be Fe^{2+} and B could be Fe^{3+}). You will obviously need an oxidizing agent to bring about the reaction, and you might be tempted to name *any* of the chemicals listed in Table 17.2. In this particular example, there would be no problem because it is comparatively easy to oxidize iron(II) ions to iron(III) ions, and any of the reagents listed in the table would be successful. However, it is important to realize that the substances listed in Tables 17.1 and 17.2 are referred to as reducing agents and oxidizing agents not because they always react in that way, but because they usually do so. A reducing agent can be made to act as an oxidizing agent if it reacts with a reducing agent stronger than itself, and the opposite is true of an oxidizing agent. You should therefore use these terms with care. Detailed examples of their use are given in appropriate parts of the book. To be absolutely certain that you are using the correct term, always state the reaction involved. For example, it is better to say 'hydrogen is a reducing agent for copper(II) oxide' rather than 'hydrogen is a reducing agent'.

Table 17.1 Common reducing agents

Reagent	*Formula*	*Usually changes to*	*External signs of its action*
Hydrogen	H_2	Water, by removing oxygen	Depends on the other reagent
Moist sulphur dioxide or its solution	SO_2 or H_2SO_3	Sulphate ions	Depends on the other reagent
Reactive metals	—	Metal ions (by losing electrons) or metal oxide (by gaining oxygen)	Depends on the other reagent
Carbon	C	Carbon monoxide, which often burns to form carbon dioxide	Depends on the other reagent
Carbon monoxide	CO	Carbon dioxide by removing oxygen	Depends on the other reagent
Potassium iodide	KI	Iodine	Brown colour (due to the iodine formed dissolving in potassium iodide solution)

Table 17.2 Common oxidizing agents

Reagent	*Formula*	*Usually changes to*	*External signs of its action*
Acidified (H_2SO_4) potassium manganate(VII)	$KMnO_4$	Colourless Mn^{2+}(aq)	Colour change, purple to colourless
Acidified (H_2SO_4) potassium dichromate(VI)	$K_2Cr_2O_7$	Green Cr^{3+}(aq)	Colour change, orange to green
Oxygen	O_2	An oxide	Depends on the other reagent
Chlorine	Cl_2	A chloride or chloride ions	Depends on the other reagent
Concentrated sulphuric acid (sometimes needs to be hot)	H_2SO_4	Sulphur dioxide	Gas produced (sulphur dioxide) with characteristic smell
Concentrated nitric acid	HNO_3	Nitrogen dioxide	Brown gas evolved
Acidified hydrogen peroxide	H_2O_2	Water	Depends on the other reagent
Manganese(IV) oxide	MnO_2	Colourless Mn^{2+}(aq)	Depends on the other reagent
Sodium chlorate(I) (sodium hypochlorite)	NaOCl	Cl^-(aq)	Depends on the other reagent

Remember the following points.

1. Full equations or complete half equations are not normally required for reactions which involve potassium manganate(VII), potassium dichromate(VI), concentrated sulphuric acid, or concentrated nitric acid when they are used as oxidizing agents. This is because the equations are fairly complex, and in these cases it is sufficient to show their part in the reaction by writing a half equation summarizing what happens to the *chemical being oxidized only*. For example, the oxidation of sodium sulphite to sodium sulphate by concentrated sulphuric acid could be shown by

$$Na_2SO_3(aq) + \underset{\text{(from oxidizing agent, i.e. concentrated sulphuric acid)}}{[O]} \rightarrow Na_2SO_4(aq)$$

Similarly, the oxidation of an iron(II) salt by acidified potassium dichromate(VI) could be shown by

$$Fe^{2+}(aq) \quad - \quad \underset{\text{(to oxidizing agent, i.e. acidified potassium dichromate(VI))}}{e^-} \quad \rightarrow \quad Fe^{3+}(aq)$$

2. Remember that potassium manganate(VII) and potassium dichromate(VI) are normally used in acid solution, and if so they *must* be acidified with dilute *sulphuric* acid. It would be an important mistake to refer to such an oxidizing agent as, e.g. just 'potassium manganate(VII)'; it behaves rather differently when in a neutral or in an alkaline solution.

3. Acidified potassium manganate(VII) and potassium dichromate(VI) solutions are particularly useful as oxidizing agents because they show very obvious colour changes when they act in this way (see Table 17.2). For example, we know when potassium dichromate(VI) has oxidized something because it goes green in doing so. Typical examples of uses of these colour changes include the following.

(a) Distinguishing between alkanes and alkenes in organic chemistry (p. 412). Alkenes are oxidized by these acidified oxidizing agents, but alkanes are not.

(b) Detecting the presence of a gas which is a reducing agent, e.g. sulphur dioxide, page 459.

(c) In a police 'breathalyser test', in which the orange crystals are potassium dichromate(VI) moistened with dilute sulphuric acid. The crystals turn green when they have oxidized a certain mass of ethanol (alcohol) vapour contained in expired air.

Check your understanding

1. For each of the following equations write RO if the reaction involves both reduction and oxidation,
O if the reaction involves oxidation only,
R if the reaction involves reduction only,
X if the reaction involves neither oxidation nor reduction.

(a) $2CuO(s) + C(s) \rightarrow 2Cu(s) + CO_2(g)$
(b) $Na^+ + e^- \rightarrow Na$
(c) $H_2S(g) + Cl_2(g) \rightarrow 2HCl(g) + S(s)$
(d) $Fe^{2+} - e^- \rightarrow Fe^{3+}$
(e) $2Cl^- - 2e^- \rightarrow Cl_2$
(f) $NaOH(aq) + HCl(aq) \rightarrow NaCl(aq) + H_2O(l)$

(J.M.B.)

2. Which of the following statements describes oxidation?
A Addition of hydrogen to a compound
B A gain of one or more electrons
C An increase in the valency of a metal
D A decrease in the number of negatively charged ions present in the formula of a compound
E Removal of oxygen

More to do

1. The formula of the manganate(VII) ion is MnO_4^-. Work out a fully balanced half equation which shows how it oxidizes (by gaining electrons from another substance) in acid solution. You will need to know that $H^+(aq)$ ions (from the acid) and electrons (from the substance being oxidized) are also reactants, and that the products are $Mn^{2+}(aq)$ ions and water molecules. Note that charges must be balanced as well as atoms.

2. Repeat 1, but for acidified dichromate(VI) instead of manganate(VII). The reactants are $Cr_2O_7^{2-}(aq)$, $H^+(aq)$ and electrons, and the products are $Cr^{3+}(aq)$ ions and water molecules.

3. If you would like to try balancing some fairly complex equations, use your answers to questions 1 and 2 to work out fully balanced complete equations for the reactions between (a) acidified potassium manganate(VII) solution and iron(II) sulphate solution, and (b) acidified potassium dichromate(VI) solution and iron(II) sulphate solution.

4. Convert your equations from question 3 into fully balanced ionic equations (not half equations).

17.6 HYDROGEN PEROXIDE, H_2O_2

Hydrogen peroxide is an oxidizing agent commonly used in chemistry. It does not conveniently fit into other main chapters, and so it is mentioned here in a little more detail.

Hydrogen peroxide, H_2O_2, is a colourless liquid when pure, but this liquid is very unstable and it is normally used in dilute aqueous solution. The concentration of such a solution is normally measured by its 'volume strength'. When hydrogen peroxide decomposes, oxygen is formed

$$2H_2O_2(aq) \rightarrow 2H_2O(l) + O_2(g)$$

and the volume strength is the number of volumes of oxygen which can be obtained (in theory) from one volume of solution. Thus 10 volume hydrogen peroxide is of such a concentration that 1 cm^3 of it can be decomposed to yield 10 cm^3 of oxygen.

The equation given above shows that hydrogen peroxide solution readily provides oxygen, and so it is obviously an oxidizing agent. Normally, this is the only chemical property of hydrogen peroxide which is studied at an elementary level. You have used this reaction in the laboratory preparation of oxygen, when it was speeded up by the use of a catalyst (p. 149).

It is sometimes more convenient to regard the solution as an oxidizing agent because of its ability (in the presence of dilute acids) to accept electrons, rather than to provide oxygen. In these cases it oxidizes something else by taking electrons from it:

$$H_2O_2(aq) + \underset{\text{from dilute acid}}{2H^+(aq)} + \underset{\text{from substance being oxidized}}{2e^-} \rightarrow 2H_2O(l)$$

For example, hydrogen peroxide in acid solution will oxidize an iron(II) salt to an iron(III) salt. We can summarize such a reaction by considering the oxidation and reduction steps separately, and writing half equations for each of them. The oxidation of the iron(II) salt:

$$Fe^{2+}(aq) - e^- \rightarrow Fe^{3+}(aq)$$

The reduction of the hydrogen peroxide in acid solution:

$$H_2O_2(aq) + 2H^+(aq) + 2e^- \rightarrow 2H_2O(l)$$

To compare these two half equations properly, we ought to balance them so that the same number of electrons is shown in each of them. In this example we 'double' the iron reaction ($2Fe^{2+}(aq) - 2e^- \rightarrow 2Fe^{3+}(aq)$) and leave the other half equation unchanged. The full equation, if required, can then be obtained by combining the two balanced half equations:

$$H_2O_2(aq) + 2H^+(aq) + 2Fe^{2+}(aq) \rightarrow 2H_2O(l) + 2Fe^{3+}(aq)$$

Note that electrons do not appear in the full equation, as they cancel out when the two half equations are added together.

'Painting' with hydrogen peroxide

White pigments used in paints often contained lead compounds, although the use of lead in paints is now restricted because its compounds are poisonous. Such pigments (e.g. in old oil paintings) often become discoloured over a period of time because the lead ions react with the gas hydrogen sulphide. This gas is found, to a very small extent, as a pollutant in the atmosphere and it reacts with lead ions to form black lead sulphide. The original white colour can be restored by carefully treating the appropriate areas with a dilute solution of hydrogen peroxide. This oxidizes the black lead sulphide to white lead sulphate:

$$Pb(s) + 4H_2O_2(aq) \rightarrow PbSO_4(s) + 4H_2O(l)$$

17.7 A SUMMARY OF CHAPTER 17

The following 'check list' will help you to organize the work for revision.

1. *Definitions*

(a) Oxidation (p. 290)
(b) Reduction (p. 290)
(c) Oxidizing agent and reducing agent (p. 290)

2. *Other points*

(a) Electrolytic changes at a cathode are reductions, and those at an anode are oxidations.
(b) Hydrogen and carbon will reduce oxides of metals fairly low in the activity series.
(c) In a full equation, oxidation is accompanied by reduction. In a half equation, oxidation or reduction may be occurring, but not both.
(d) Give full accounts (and colour changes if relevant) in describing redox reactions, e.g. points (1) and (2) on page 291.
(e) It is sometimes difficult to decide which atoms or ions are losing or gaining electrons in the course of a reaction; always look for atoms becoming ions (or the other way round), or a change in valency of an atom or ion.
(f) In oxidizing something, an oxidizing agent is itself reduced (or the other way round for a reducing agent).
(g) The names and formulae of the common reducing agents, and any visible signs of their activity when they react in this way (Table 17.1).
(h) The names and formulae of the common oxidizing agents, and any visible signs of their activity when they react in this way (Table 17.2).
(i) Full equations and ionic equations are not normally expected for reactions showing concentrated sulphuric and nitric acids, acidified potassium manganate(VII) and acidified potassium dichromate(VI); their part in a redox reaction can be shown by '$\ldots + [O]$ (from oxidizing agent)', or by '$\ldots - e^-$ (to oxidizing agent)', as appropriate.
(j) Hydrogen peroxide is a useful oxidizing agent. Its action can be shown by electron transfer in acid solution or by addition of oxygen, as appropriate (see equations page 297).

QUESTIONS

1. In the reaction between acidified potassium dichromate(VI) and iron(II) ions, which may be shown by the equation

$$Cr_2O_7^{2-}(aq) + 6Fe^{2+}(aq) + 14H^+(aq) \rightarrow 2Cr^{3+}(aq) + 7H_2O(l) + 6Fe^{3+}(aq)$$

A there is both oxidation and reduction
B the colour of the solution changes from green to yellow
C the iron(II) ions are reduced
D the dichromate(VI) ions are oxidized
E hydrogen ions are reduced

2. (a) Define oxidation and reduction in terms of electron transfer.
(b) Write an *ionic* equation for the following reaction. State, with reasons, which is the oxidizing agent and which is the reducing agent.
Magnesium is added to sulphuric acid.
(c) (i) State what is observed at each electrode when copper(II) sulphate solution is electrolysed using copper electrodes.
(ii) Write equations for the reactions occurring at each electrode.
(iii) Classify the reactions as oxidation or reduction. (W.J.E.C.)

3. What type of reaction is represented by the following equation?

$$Fe^{2+} \rightarrow Fe^{3+} + e^-$$

Name *two* reagents which could be used separately to bring about this reaction. What reagent would you use to test that this reaction had taken place? What would you expect to observe during this test? (J.M.B.)

4. (a) Name an oxidizing agent, and write an equation for a reaction in which it is used for oxidation. (b) Name a reducing agent, and write an equation for a reaction in which it is used for reduction. (J.M.B.)

5. Name the product of each of the following reactions, and in each case state whether the reactant named in italics undergoes oxidation or reduction: (a) *hydrogen* burning in oxygen, (b) *chlorine* reacting with iron(II) chloride, (c) *hydrogen* reacting with nitrogen in the presence of a catalyst. (J.M.B.)

6. Describe how you would bring about the following conversions.

(a) $Cu^{2+}(aq) \rightarrow Cu(s)$
(b) $CuO(s) \rightarrow Cu(s)$
(c) $PbS(s) \rightarrow PbSO_4(s)$
(d) $MnO_4^-(aq) \rightarrow Mn^{2+}(aq)$
(e) $Fe^{2+}(aq) \rightarrow Fe^{3+}(aq)$
(f) $2I^-(aq) \rightarrow I_2(aq)$

7. Why is it that oxidation and reduction reactions occur together? Select *two* reducing agents and for *each* describe a different reaction illustrating the above statement.

Name an oxidizing gas (not oxygen nor ozone), an oxidizing liquid (not a solution) and an oxidizing solid, and in *each* case give a different substance that can be oxidized by them, writing an appropriate equation. (C.)

8. Describe briefly and write equations for the reactions by which you could convert (a) oxygen gas into a compound containing oxide ions, (b) hydrogen gas into a solution containing hydrogen ions, (c) a solution containing calcium ions into an insoluble compound of calcium, (d) metallic iron into a solution containing Fe^{3+} ions, (e) sulphur into a solution containing sulphate ions, (f) carbon dioxide into a solution containing carbonate ions. (C.)

18 The mole The laws of chemical combination: formulae and equations

18.1 THE MOLE

When chemists compare the chemical properties of substances, they prefer to compare a fixed *number of particles* of the substances rather than to compare a certain mass of each, e.g. 1 g. This is because chemical reactions take place between the particles of the reacting substances, and 1 g of hydrogen, for example, contains a different number of atoms from 1 g of helium. It may seem very difficult to 'count the particles in a substance', especially as we cannot even see them. This can be done, however, and chemists use a certain standard number of particles as a basis for comparing different substances. This standard number is contained in an amount of substance known as *a mole*.

Suppose that we were to weigh out 1 g of hydrogen, and that the number of hydrogen atoms in this 1 g was x. If we then weighed out 1 g of helium, this would contain fewer atoms (i.e. less than x) because each atom of helium weighs more than an atom of hydrogen. This is like buying a kilogram of apples and a kilogram of grapes; there would be fewer apples than grapes because an apple weighs more than a grape.

We can take this a step further. We know that an atom of helium is four time as heavy as an atom of hydrogen, because their relative atomic masses are $A_r(\text{H}) = 1$ and $A_r(\text{He}) = 4$. The number of atoms of helium in our 1 g of helium will thus be $\frac{1}{4}x$. This is like saying that if there are x grapes to the kilogram and an apple is four times as heavy as a grape, there will be $\frac{1}{4}x$ apples in a kilogram. Similarly, there will be $\frac{1}{16}x$ atoms of oxygen in 1 g of oxygen, and $\frac{1}{12}x$ atoms of carbon in 1 g of carbon, because their relative atomic masses are 16 and 12 respectively.

Therefore

1 g of hydrogen contains x atoms
1 g of helium contains $\frac{1}{4}x$ atoms
1 g of carbon contains $\frac{1}{12}x$ atoms
and 1 g of oxygen contains $\frac{1}{16}x$ atoms.

If we now calculate the mass of each of these substances we would need in order to have x atoms of it, these masses become

1 g of hydrogen contains x atoms
4 g of helium contains x atoms
12 g of carbon contains x atoms
16 g of oxygen contains x atoms.

These masses are the relative atomic masses of the elements *expressed in grams*. This means that if we measure out elements by taking a mass of each equal to its relative atomic mass expressed in grams, we will always be working with the same number of atoms (x in the above example). The value of x is now known to be 6.02×10^{23}, i.e. 6.02 multiplied by 10 twenty-three times. This is a very large number!

An amount of a substance which contains 6.02×10^{23} *particles is called a mole of the substance.* 12 g of carbon is a mole of carbon atoms.

Note that the relative atomic mass of an element has no units; it is simply a number. A mole of an element is an amount of material, however, and must have units. Thus the relative atomic mass of carbon is 12, but a mole of carbon is 12 g.

The information which you need to learn about moles is summarized below.

1. If different elements are weighed out in the masses obtained by expressing their relative atomic masses (A_r) in grams (e.g. 1 g of hydrogen, 16 g of oxygen, etc.), it can be shown that the quantities so obtained contain the same number of *atoms*. Similarly, 0.1 mol of any element will also always contain the same number of atoms, but the number will only be $\frac{1}{10}$ of the number in a mole, i.e. $\frac{1}{10}$ of 6.02×10^{23}, or 6.02×10^{22}.
2. The number of atoms in a mole of an element is known to be 6.02×10^{23} and this number is called the *Avogadro number* (named after an Italian scientist, Amadeo Avogadro).
3. Similarly, the relative molecular mass (M_r) of any molecular element or compound (expressed in grams) contains 6.02×10^{23} *molecules* of the substance.
4. We sometimes use the term *relative formula mass* (rather than relative molecular mass) when we are referring to ionic compounds, because they do not contain molecules.* The relative formula mass is obtained in just the same way as a relative molecular mass (i.e. by adding together the individual relative atomic masses in the formula of the compound) and is the same number, but by avoiding any mention of molecules we avoid confusion about the structure of the substance. The relative formula mass (or molar mass) of sodium chloride, NaCl, is thus $23 + 35.5 = 58.5$.

Avogadro's number!

The distinction between the two terms is not very important, because we rarely need to *write* either of them. No matter whether we think of relative molecular mass or relative formula mass, we obtain the same number, e.g. 58.5 for sodium chloride. From this we can say that 58.5 g is a mole of sodium chloride, and this is the statement we often need to write down. For most students, the difficulties arise when there is a need to state exactly what we have *in* a mole of an ionic compound. It would be quite wrong, for example, to say that 58.5 g of sodium chloride contains a mole of *molecules*. This point is fully explained under 'moles of ionic compounds' on page 302.
5. *The amount of substance containing* 6.02×10^{23} *particles is a basic scientific unit called the mole*. 'Amount of substance' can be a mass (e.g. 12 g of carbon), a volume of gas (e.g. 24 dm^3 of hydrogen), or a volume of solution (e.g. 1 dm^3 of a solution containing 1 mol dm^{-3} of a dissolved substance), but it must always have units.

In the same way that we buy eggs in packs of six, chemists count particles in packs of 6.02×10^{23}. Each unit pack of particles (i.e. 6.02×10^{23}) is called a mole, and we

* The term relative molecular mass *is acceptable* when applied to ionic compounds, even although they do not contain molecules. The use of the term in this way can be confusing, and the term relative formula mass avoids the problem.

can have 0.1 mol, 0.5 mol, 3 mol, etc. The unit pack has to contain a large number of particles or otherwise we would not be able to work with them conveniently, e.g. by weighing. Remember that 6.02×10^{23} atoms of hydrogen still only weigh 1 g!

Note that when we weigh out chemicals in this way, we are 'counting the atoms' (or molecules, etc.) by weighing, even though we cannot see them. This is a like a bank clerk weighing out a bag of 10p coins and being able to state the number of coins in the bag without actually seeing them. Similarly, gas volumes (p. 441) and the volumes of standard solutions (p. 445) are measured in order to 'count the particles'.

Some simple examples of using moles

Moles of atoms

23 g of sodium contains 6.02×10^{23} atoms of sodium (a mole of sodium atoms).
2.3 g of sodium contains 6.02×10^{22} atoms of sodium (0.1 mol of sodium atoms).
128 g of copper (A_r(Cu) = 64) contains $2 \times 6.02 \times 10^{23}$ atoms of copper, $= 12.04 \times 10^{23}$ atoms of copper, i.e. 2 mol of copper atoms.

Moles of molecules

Many substances, both elements and compounds, consist of molecules and so when we are working with them it is usual to refer to moles of molecules rather than moles of atoms. A mole of molecules is an amount of substance containing 6.02×10^{23} molecules, and it will have a mass equal to the relative molecular mass of the substance expressed in grams. For example, the relative molecular mass of water $M_r = 1+1+16 = 18$. Therefore 1 mole of water weighs 18 g, and contains 6.02×10^{23} molecules of water. Similarly the relative molecular mass of carbon dioxide $= 12+16+16 = 44$. A mole of carbon dioxide weighs 44 g and contains 6.02×10^{23} molecules of carbon dioxide.

Note: The statement '16 g of oxygen' could mean 16 g of oxygen atoms (O), 16 g of oxygen molecules (O_2) or 16 g of oxygen ions (O^{2-}). Always state the type of particle being considered and make sure that you understand the difference between, for example, I_2 (an iodine molecule), 2I (two atoms of iodine), and $2I^-$ (two iodide ions). In the absence of any clear statement about the type of particle being considered, you must assume that the substance is in the form in which you would normally find it. Thus '16 g of oxygen' would be taken to mean 16 g of oxygen molecules (O_2), as we normally encounter oxygen as O_2 molecules. It is far better to state '16 g of oxygen molecules', so that the statement is absolutely clear.

Moles of ionic compounds

As explained earlier, ionic compounds can be given a relative molecular mass or a relative *formula* mass (calculated by adding together the individual relative atomic masses of the 'atoms' in its formula), and if this mass is expressed in grams we have a mole of the compound. We must be very careful, however, in describing the particles within an ionic substance. We must never refer to molecules being within ionic compounds, but we can refer to moles of ions or *moles of formula units*. For example, the relative formula mass of sodium chloride is 58.5. Therefore a mole of sodium chloride is 58.5 g. If we wish to refer to the particles present *in* a mole of sodium chloride, we can say that 58.5 g of sodium chloride contains a mole of *formula units* of sodium chloride. This means a mole of 'NaCl' units, where each unit is a pair of ions (one sodium ion and one chloride ion). This does not mean that the 'NaCl' units are separate units (like molecules). It is simply a convenient way of saying that we have a *total* number of sodium and chloride ions equal to the number in a mole's worth of 'NaCl' units, i.e. a mole's worth of ion pairs.

Similarly, the relative molecular mass (or relative formula mass) of the ionic solid sodium hydroxide, NaOH, is $23+16+1 = 40$. Therefore a mole of sodium hydroxide is 40 g, and this will contain a mole of formula units of sodium hydroxide, i.e. the *equivalent* of 6.02×10^{23} 'NaOH' units or pairs of ions (Na^+ and OH^-). We cannot say 'a mole of sodium chloride ions' or 'a mole of sodium hydroxide ions' because there are no such ions.

The term mole of ions can be used, but it refers to 6.02×10^{23} ions of the *same type*, e.g. a mole of sodium ions, Na^+. It cannot, therefore, be used to describe a *solid* ionic compound, which must be made up of at least two types of ion. The term is useful when considering solutions of ionic compounds, in which ions separate from each other. In the solid ionic compound we refer to a mole of sodium chloride formula units (NaCl), and in solution these ions separate from each other to produce a mole of sodium ions *and* a mole of chloride ions. It would also be correct to say that a mole of solid sodium chloride contains a mole of sodium ions and a mole of chloride ions, joined together.

Similarly, a mole of magnesium chloride, $MgCl_2$, will contain a mole of formula units of $MgCl_2$. If this quantity is dissolved in water, the solution will contain the equivalent of a mole of $MgCl_2$ units, i.e. a mole of magnesium ions and *two* moles of chloride ions.

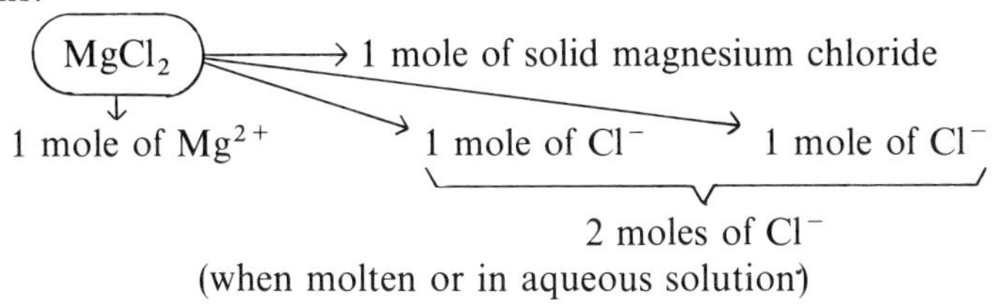

Note: When the formula of a compound is written down, it often means a mole of that substance. Formulae should *not*, therefore, be used as abbreviations for names. If you are making a written description, write the *name* of a chemical in full and use formulae only for equations and for mole calculations.

e.g. H often means 1 mole of hydrogen atoms.

H_2 often means a mole of hydrogen molecules, each molecule containing two atoms.

$2H_2$ often means two moles of hydrogen molecules, etc.

Moles and solutions

When substances are dissolved in a solvent, we usually measure the concentration of the solution in moles of dissolved substance per dm^3 (litre) of solution. For example, as a mole of sodium chloride is 58.5 g, a solution which contains 58.5 g of sodium chloride in 1 dm^3 of solution will have a concentration of 1 mol dm^{-3} with respect to sodium chloride.

Notes

1. 58.5 g of sodium chloride dissolved in 1 dm^3 of *water* is not quite the same as 58.5 g of sodium chloride dissolved in 1 dm^3 of *solution*. If 58.5 g of sodium chloride is added to 1 dm^3 of water, the solution formed would have a volume slightly different from 1 dm^3. We always calculate concentrations in mol dm^{-3} of *solution*, not solvent. For example, in order to produce a solution which is 1 mol dm^{-3} in sodium chloride, 58.5 g of sodium chloride is dissolved in water and the volume is *made up* to 1 dm^3 by adding more water. This is done because our apparatus is designed to measure volumes of solution and so, if we measure out for example

100 cm^3 of a solution containing 1 mol dm^{-3}, we know that we have 'counted out' $\frac{1}{10}$ mol of solute particles.

2. We must always state what particles the concentration refers to, e.g. 1 mol dm^{-3} *with respect to sodium chloride*. This means that 1 dm^3 of such a solution contains the equivalent of one mole (6.02×10^{23}) formula units ('NaCl') of sodium chloride. It would be equally correct to say that the solution contains 1 mol dm^{-3} with respect to sodium ions, Na^+(aq), and 1 mol dm^{-3} with respect to chloride ions, Cl^-(aq). Similarly, a mole of magnesium chloride ($MgCl_2$) is $24 + 35.5 + 35.5$ g $= 95$ g. A solution containing 95 g of magnesium chloride in 1 dm^3 of solution could be labelled as 1 mol dm^{-3} with respect to magnesium chloride, 1 mol dm^{-3} with respect to magnesium ions, Mg^{2+}(aq), or 2 mol dm^{-3} with respect to chloride ions, Cl^-(aq). Think about this last point carefully.

3. A concentration of 1 mol dm^{-3} is sometimes referred to as a *molar solution*, and given the symbol M (see page 445). Some students confuse molarities with moles. Molarity is a *concentration*, a mole is an *amount* of substance. All concentrations in this book are referred to as, e.g., 1 mol dm^{-3} and not 1 M etc. If you will not be using the term molarity in your course, ignore this point completely.

Some basic calculations involving concentrations

An understanding of the following types of calculation is needed for work in volumetric analysis, which is considered in Chapter 23.

(a) What mass of sodium hydroxide must be dissolved in 250 cm^3 of solution to produce a solution which has a concentration of 0.10 mol dm^{-3} with respect to sodium hydroxide?

Answer

A mole of sodium hydroxide (NaOH) is $23 + 16 + 1$ g $= 40$ g. 1 dm^3 of a solution which is 1.0 mol dm^{-3} with respect to sodium hydroxide will contain 40 g of sodium hydroxide.

1 dm^3 of a solution which is 0.1 mol dm^{-3} with respect to sodium hydroxide will contain 4.0 g of sodium hydroxide.

$\therefore$ 250 cm^3 ($\frac{1}{4}$ dm^3) of such a solution will contain $\frac{1}{4} \times 4.0$ g $= 1.0$ g.

(b) How many moles of formula units of hydrochloric acid are contained in 25.0 cm^3 of a solution which has a concentration of 0.1 mol dm^{-3} with respect to the acid?

Answer

1.0 dm^3 of the acid contains 0.1 moles of formula units.

$\therefore$ 1.0 cm^3 of the acid contains $\frac{0.1}{1000}$ moles of formula units.

$\therefore$ 25.0 cm^3 of the acid solution contains $25 \times \frac{0.1}{1000}$ moles of formula units $= 0.0025$ moles of formula units.

(c) How many moles of chloride ions are contained in 250 cm^3 of a solution which has a concentration of 0.500 mol dm^{-3} with respect to magnesium chloride, $MgCl_2$?

Answer

1 mole of magnesium chloride contains 2 moles of chloride ions.

$\therefore$ 0.5 mole of magnesium chloride contains 1.0 mole of chloride ions.

$\therefore$ a solution which contains 0.5 mol dm^{-3} of magnesium chloride contains 1.0 mol dm^{-3} of chloride ions, Cl^-(aq).

$\therefore$ 250 cm^3 ($\frac{1}{4}$ dm^3) of such a solution contains $\frac{1}{4} \times 1.0$ moles of Cl^-(aq) $= 0.25$ moles of Cl^-(aq) ions.

(d) What is the concentration (in mol dm^{-3}) of a solution containing 2.00 g of sodium hydroxide in 500 cm^3 of solution?

Answer
2 g in 500 cm^3 = 4 g per dm^3
One mole of sodium hydroxide, NaOH, is 23+16+1 g = 40 g.
4 g of sodium hydroxide is 0.1 mole of sodium hydroxide
∴ 0.1 mol are dissolved per dm^3, and the solution has a concentration of 0.1 mol dm^{-3}.

Check your understanding

(You will need to refer to the relative atomic masses in the Periodic Table at the front of the book.)

1. How many grams of (a) magnesium, (b) copper, and (c) zinc would you need to measure out in order to have a mole of atoms of each?

2. How much would you need to weigh out in order to have (a) 0.5 mole of sulphur atoms, (b) 0.1 mole of iron atoms, and (c) 0.25 mole of magnesium atoms?

3. How much would you need to weigh out in order to have (a) 3.01×10^{23} atoms of zinc, (b) 12.04×10^{23} atoms of carbon, and (c) 6.02×10^{23} atoms of sulphur?

4. In each of the following pairs, which mass contains the greater number of atoms?
(a) 18 g of carbon or 6.5 g of zinc
(b) 7 g of oxygen or 7 g of nitrogen
(c) 16 g of sulphur or 8 g of oxygen

5. What mass would you need in order to have a mole of *molecules* of each of the following?
(a) hydrogen; (b) chlorine; (c) nitrogen; (d) sulphuric acid; (e) ammonia; (f) nitric acid

6. What mass would you need in order to have a mole of formula units of each of the following?
(a) sodium chloride; (b) calcium chloride; (c) lithium fluoride; (d) potassium oxide

7. Why is the term formula unit used in question 6?

8. Which of the following substances should be referred to in moles of formula units rather than moles of molecules?
(a) sodium hydroxide; (b) ice; (c) iodine; (d) sulphur; (e) sodium chloride; (f) calcium oxide; (g) hydrogen chloride

9. 28 g of nitrogen
A contains 6×10^{23} atoms of nitrogen
B contains 6×10^{23} molecules of nitrogen
C contains two moles of nitrogen
D is heavier than 28 g of hydrogen
E contains the same number of atoms of nitrogen as 35.5 g of chlorine (N = 14, Cl = 35.5)

10. A new illuminated sign erected at Heathrow Airport contains 2 g of a rare monatomic gas (X) of relative atomic mass 20. How many atoms of X are present in this sign?
A 6×10^{22}
B 6×10^{23}
C 6×10^{24}
D 12×10^{23}
E 10×10^{23}

11. One mole of carbon dioxide
A contains the Avogadro Number of carbon dioxide atoms
B contains the same number of molecules as 1 g of hydrogen
C contains the same number of molecules as 32 g of oxygen
D has a mass of 28 g
E contains 602×10^{23} molecules of carbon dioxide (C = 12, O = 16)

12. (a) Complete the following sentence in your notebook. The Avogadro Number 6.02×10^{23} is the number of

(b) A packet of sugar contains crystals of average mass 3.42×10^{-3} g. How many sugar molecules will an average crystal contain? (Formula of sugar: $C_{12}H_{22}O_{11}$.) (J.M.B.)

13. (a) How many moles of anhydrous sodium carbonate are present in 200 cm^3 of a solution which is of concentration 2.0 mol dm^{-3} with respect to sodium carbonate?
(b) How many moles of sodium chloride are there in 500 cm^3 of a solution which is of concentration 0.1 mol dm^{-3} with respect to sodium chloride?

14. (a) What mass of magnesium sulphate (anhydrous) would be required to make 1.5 dm^3 of a solution of concentration 1.0 mol dm^{-3}?
(b) What mass of sodium hydroxide would have to be dissolved in 100 cm^3 of solution in order to produce a solution of concentration 0.50 mol dm^{-3}?

15. (a) What is the molar mass of (i) calcium nitrate, (ii) zinc sulphate?
(b) What are the molar masses of (i) propanol, C_3H_7OH, (ii) tetrachloromethane, CCl_4, and (iii) benzene, C_6H_6?
(c) How many moles are present in (i) 100 g of calcium, (ii) 13 g of zinc, (iii) 28 g of nitrogen, and (iv) 8 g of sulphur? (Note that the question does not refer to moles of atoms, molecules, etc. Make sure that your answers are quite clear and unambiguous.)

18.2 USING MOLES TO FIND FORMULAE

Introduction

A chemist uses moles as 'measurable packages of particles' in order to understand how *individual particles* take part in chemical reactions. In this unit you will learn how this idea is used to calculate formulae. In section 18.3 you will learn of its application to chemical equations, and in Chapter 23 you will learn how the idea is applied to reactions taking place between standard solutions.

Thus if it can be shown that 12 g of carbon react exactly with 32 g of oxygen, both of which are easily measurable quantities, we can say

1 mole of carbon atoms (12 g) react with 1 mole (32 g) of oxygen molecules.
$\therefore$ 6.02×10^{23} atoms of carbon react with 6.02×10^{23} molecules of oxygen.

From this we can see that each carbon atom reacts with one molecule of oxygen, and we can write

$$C(s) + O_2(g) \rightarrow CO_2(g)$$

We have constructed a chemical equation to show how *individual* atoms and molecules react (even though we cannot see them or weigh them) by working in moles, which are manageable and measurable quantities. As you will see, the main point in most of the calculations which follow in this chapter is to *work in moles*. The mole is a very important part of a chemistry course, so make sure that you understand the principle fully.

The Law of Constant Composition

The Law of Constant Composition (also called the Law of Definite Proportions) states that all pure samples of the same chemical compound contain the same elements combined together in the same proportions by mass.

The results of a typical experiment used to illustrate this law are summarized below. You may participate in an experiment of this kind, e.g. by performing reactions (a) and (b) below, whilst the teacher does reaction (c) and then demonstrates the reduction of all three samples.

1. Samples of copper(II) oxide are made from three completely different chemicals, which are separately heated to constant mass in a crucible.
(a) From copper(II) hydroxide,

$$Cu(OH)_2(s) \rightarrow CuO(s) + H_2O(g)$$

(b) From copper(II) carbonate,

$$CuCO_3(s) \rightarrow CuO(s) + CO_2(g)$$

(c) From copper(II) nitrate (in a fume cupboard),

$$2Cu(NO_3)_2(s) \rightarrow 2CuO(s) + 4NO_2(g) + O_2(g)$$

2. A small portion of one of the three samples of copper(II) oxide is placed in a previously weighed porcelain boat which is then reweighed.
3. The boat is placed in a combustion tube as in Figure 10.4 on page 158 and heated in a stream of hydrogen gas until the mass of the boat is constant. Note the precautions to be used in experiments with hydrogen (p. 156). The hydrogen reduces the hot copper(II) oxide to copper,

$$CuO(s) + H_2(g) \rightarrow Cu(s) + H_2O(g)$$

4. The previous two steps are repeated twice, using the other two samples of copper(II) oxide. Typical results from such an experiment might be as follows.

	Sample 1	*Sample 2*	*Sample 3*
Weight of boat + copper(II) oxide (g)	10.92	10.10	10.23
Weight of empty boat (g)	8.42	7.21	8.12
Mass of copper(II) oxide (g)	2.50	2.89	2.11
Weight of boat + copper metal (g)	10.42	9.52	9.81
Weight of empty boat (g)	8.42	7.21	8.12
Mass of copper metal (g)	2.00	2.31	1.69
Mass of oxygen in compound (g)	(2.50 − 2.00) = 0.50	(2.89 − 2.31) = 0.58	(2.11 − 1.69) = 0.42

In sample 1, 0.5 g of oxygen combine with 2.0 g of copper to produce 2.50 g of copper(II) oxide.

The percentage of copper in the sample is $\frac{2.00 \times 100}{2.50}$ per cent = 80 per cent.

In sample 2, 0.58 g of oxygen combine with 2.31 g of copper to form 2.89 g of copper(II) oxide.

The percentage of copper in the sample is $\frac{2.31 \times 100}{2.89}$ per cent = 80 per cent.

In sample 3, 0.42 g of oxygen combine with 1.69 g of copper to form 2.11 g of copper(II) oxide.

The percentage of copper in the sample is $\frac{1.69 \times 100}{2.11}$ per cent = 80 per cent.

Such results confirm the Law of Constant Composition, because although the samples of copper(II) oxide were made by three completely different methods, their composition was the same, i.e. they each contained 80 per cent copper and 20 per cent oxygen by mass.

The significance of the Law of Constant Composition

Experiments such as the one just discussed also show that other compounds have a fixed composition. This must mean that every compound has a fixed proportion of each of the elements within it, and these proportions cannot be changed without making it into a different compound. It must also be true, therefore, that each compound has a *formula* which tells us the proportions in which the atoms combine together in the compound. For example, copper(II) oxide is known to have the formula CuO. As this compound always contains copper particles and oxygen particles in the ratio 1:1, the percentage of copper and the percentage of oxygen will not vary from one sample to another, as the experimental results confirm. (Note that although the particles in copper(II) oxide occur in the ratio 1:1 *by number*, their mass ratios are different because the particles have different relative atomic masses.) There is also another oxide of copper called copper(I) oxide, Cu_2O, which has a fixed percentage by mass of copper, but this percentage is different from that in copper(II) oxide. There are no oxides of copper with any other formula, such as Cu_3O or Cu_2O_3.

Always take care in writing chemical formulae. You should know how to 'work out' a formula by using the symbols and valencies of the common elements and radicals (p. 32). However, you must realize that the 'rules' used to work out a formula in this way only help us to understand *why* the formula is so written. It is not possible to use these 'rules' in more complex situations. The chemical formulae of *all* compounds are determined by carefully conducted analysis experiments of the type just described. The *experimental* method is the only way of determining a formula. Using valencies etc. to work out a formula is really only a check on what has already been shown by experiment.

The results of experiments such as those with the copper(II) oxide samples on page 307 enable us to work out an actual formula for a compound, as well as its composition by mass. The way in which this can be done is described under the next heading.

Finding an empirical formula by experiment

The experiment discussed on page 307 illustrates how a compound can be 'broken apart' and analysed to find its percentage composition by mass. It is also possible in some cases to 'build up' a compound and find its composition by mass, although this is only possible for binary (two-element) compounds and even then only in a few cases where direct combination of the elements is possible. Such experiments are useful, however, because in addition to verifying the Law of Constant Composition they can also be used to calculate the simplest formula for the compound being analysed.

Before you actually do an experiment from which a formula can be calculated, it is important that you understand the method of calculation. In order to determine the empirical (simplest ratio) formula from experimental data, proceed as follows. (The experimental results could be given as the masses of the different chemicals which combine together, or as a percentage by mass of the different chemicals in the compound.)

1. Use the data to write a statement about the masses of the elements which combine together, e.g.
8.32 g of lead combine with 1.28 g of sulphur and 2.56 g oxygen.
(Percentage compositions are used in the same way. Thus if a compound contains 40 per cent Cu, 20 per cent S, and 40 per cent O, we would write: 40 g of copper combine with 20 g of sulphur and 40 g of oxygen.)

Note: The data which will enable you to write a statement like this may be given in an indirect way, and will first have to be 'unravelled'. Examples of this are given after the basic procedure so as not to confuse the calculation.

2. Rewrite the statement by dividing the mass of each element by the mass of one mole of atoms of that element, so that we are now working in moles of atoms.
$\frac{8.32}{207}$ *moles of lead atoms combine with* $\frac{1.28}{32}$ *moles of sulphur atoms and* $\frac{2.56}{16}$ *moles of oxygen atoms.*

∴ 0.04 moles Pb atoms combine with 0.04 moles S atoms and 0.16 moles O atoms.
3. Convert these mole ratios into whole numbers by dividing each of them by the smallest (0.04 in this case).
1 mole of lead atoms combine with 1 mole of sulphur atoms and 4 moles of oxygen atoms.
1 mole of lead atoms (6×10^{23} atoms) contains the same number of atoms as 1 mole of sulphur atoms (6×10^{23} atoms of sulphur), but there are four times as many atoms in 4 moles of oxygen atoms (i.e. $4 \times 6 \times 10^{23}$ atoms of oxygen.)
In a simple ratio, therefore,
1 atom of Pb combines with 1 atom of S and 4 atoms of O and the empirical formula is $PbSO_4$.

It is important to set out all calculations clearly, using a new line for each step. It is possible to gain good marks in a chemical calculation even if your final answer is incorrect, *providing that you show your working clearly and show that you understand the chemistry*. In the above example, the statements you would be expected to write down in your answer are printed in italics. Calculations of this type are also conveniently set out in table form:

Element	*Mass (or per cent mass)*	*Molar mass*	$\frac{\textit{Mass}}{\textit{Molar mass}}$	*Ratio of moles*	*Simplest ratio*
Lead	8.32 g	207 g	8.32/207	0.04	1
Sulphur	1.28 g	32 g	1.28/32	0.04	1
Oxygen	2.56 g	16 g	2.56/16	0.16	4

Ratio of atoms 1:1:4
Empirical formula $PbSO_4$

Note that the formula obtained in this way is not necessarily the actual formula (i.e. the molecular formula); it is only the simplest ratio in which the atoms combine. You must always make this clear by referring to it as the empirical formula, i.e. the simplest formula.

Suppose that the empirical formula of a compound has been found by experiment to be HO. The true formula is $(HO)_n$ where n is an integer (i.e. a whole number). For example, if $n = 1$ then the true formula would be HO, if $n = 2$ the true formula would be H_2O_2, etc. Further experimental work is needed to find the value of n, and this is explained in the section on molecular formulae (pp. 311–13).

Sometimes the data needed to produce the first statement in a calculation to determine an empirical formula are 'disguised' as experimental readings. These are often weighings obtained by combining chemicals together (e.g. burning magnesium in oxygen to form magnesium oxide). Alternatively, the weighing could be obtained by breaking down a chemical (e.g. by reducing a metal oxide to the metal in a stream of hydrogen). The important thing is to use the weighings from the experiment to find the masses of the chemicals which actually *react* together, and then to continue as shown previously.

Experiment 18.1

Determining the empirical formula of magnesium oxide by experiment*

Apparatus
Crucible and lid, tripod, Bunsen burner, pipe clay triangle, protective mat, tongs, balance.
Magnesium ribbon.

Procedure

(a) Weigh a crucible and lid on the balance and record this mass.

(b) Scrape the ribbon, if necessary, to remove any oxide film. Coil about 15 cm of the ribbon around a pencil and place the ribbon in the crucible.

(c) Replace the lid on the crucible and reweigh.

(d) Place the crucible containing the magnesium ribbon on a pipe-clay triangle supported by a tripod, and heat using a Bunsen burner as in Figure 18.1. Make sure the flame is low at first and then gradually increase it. Heat strongly for a few minutes. Remove the Bunsen burner and slightly lift the crucible lid by means of the tongs. Quickly replace the lid taking care to lose as little magnesium oxide 'smoke' as possible. Repeat this process until the magnesium ceases to flare up. When this stage has been reached remove the crucible lid and heat strongly to make sure that the combustion is complete.

(e) Remove the flame and allow the crucible to cool. When cool, replace the lid and reweigh the whole.

(f) Reheat the crucible and contents, cool and reweigh. If the mass is not the same as at the end of part (e), repeat the heatings until consecutive final mass measurements agree (i.e. the mass remains constant).

Results

Record all your weighings in a table.

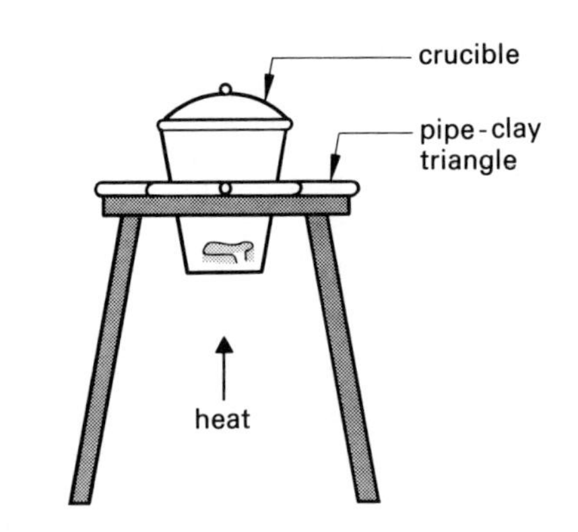

Figure 18.1 Heating magnesium in a crucible

Notes about the experiment

(i) In your conclusion, explain why it is important to keep the crucible lid in place at first, and why it is necessary to lift it *slightly* from time to time during the experiment.

(ii) Use your experimental results to work out the empirical formula of magnesium oxide, using the steps given in the previous example. Remember that the first step in the procedure is to convert your readings into a statement of the type 'X g of magnesium combine with Y g of oxygen', but you must also show the working which produces this statement.

Remember also that your final answer is only an *empirical* formula. It so happens that in this case the empirical formula is the same as the actual formula, but in order to prove this an extra step is needed in the calculation, as shown on page 311.

Check your understanding

1. If the experimental results had shown that 6 g of magnesium combined with 4 g of oxygen, show that these figures could also be used to calculate the same empirical formula for magnesium oxide.

2. Hydrogen was passed over a sample of copper oxide as shown in Figure 10.4 on page 158. The following results were obtained.

Mass of porcelain boat empty = 14.4 g
Mass of porcelain boat and copper oxide = 18.4 g
Mass of porcelain boat and copper = 17.6 g

Give an equation for the reaction taking place, and then use the results to calculate an empirical formula for the copper oxide.

3. Why would it not be possible to use magnesium oxide in the way described in question 2?

4. If another experiment showed that 2 g of calcium combined with 3.55 g of chlorine, what would be the empirical formula of the calcium chloride formed?

5. Determine the empirical formula of each of the following compounds for which the composition by mass is given:

(a) magnesium 9.5 g, chlorine 28.4 g
(b) copper 40 per cent, sulphur 20 per cent, oxygen 40 per cent
(c) nitrogen 1.40 g, hydrogen 0.40 g, carbon 0.60 g, oxygen 2.40 g
(d) carbon 75 per cent, hydrogen 25 per cent
(e) carbon 40 per cent, hydrogen 6.67 per cent, oxygen 53.3 per cent

6. Why, in Experiment 18.1, was it necessary to heat the crucible to constant mass?

7. An oxide of lead was weighed in a porcelain boat and it was then reduced to lead by heating it in a stream of hydrogen. The boat with the lead in it was then allowed to cool with the hydrogen still passing over it, and it was then weighed. It was reheated in hydrogen, recooled and reweighed until a constant weight was attained for the boat and the lead.
The following weightings were obtained:

Weight of boat = 10.20 g
Weight of boat + lead oxide = 17.37 g
Final weight of boat + lead = 16.41 g
(Relative atomic masses, Pb = 207, O = 16)

(a) Name a drying agent which could be used to dry the hydrogen used in the experiment.
(b) Why was (i) the boat cooled with the hydrogen still passing over it, (ii) the experiment repeated until a constant weight was attained?
(c) (i) What weight of lead was produced in the experiment? (ii) What weight of oxygen was originally combined with this weight of lead? (iii) Calculate the weight of oxygen which combines with 1 mole of lead atoms. (iv) How many moles of oxygen atoms combine with 1 mole of lead atoms? (v) Write the formula and the name of the oxide of lead used in the experiment.
(d) In a second experiment 4.14 g of lead was obtained from 4.46 g of another oxide of lead. (i) Calculate how many moles of oxygen atoms combine with 1 mole of lead atoms to form this oxide. (ii) Give the name and formula of this oxide. (J.M.B.)

Finding the empirical formula of a hydrated salt

In the case of hydrated salts, the mass of water of crystallization is taken as one unit. It is converted into a number of moles of water *molecules* by dividing it by the relative molecular mass of water, 18. The other elements are converted into moles of atoms, as before.

Suppose that a hydrate has the following percentage composition: iron 20.15 per cent, sulphur 11.51 per cent, oxygen 23.02 per cent, and water 45.32 per cent. We are asked to determine the empirical formula of the salt. The method of calculation is summarized in the table below. If you prefer, each stage can be set out as shown on pages 308–9.

Element or group	*Per cent mass*	*Molar mass*	*Per cent mass / Molar mass*	*Ratio of moles*	*Simplest ratio*
Fe	20.15	56 g	20.15/56	0.36	1
S	11.51	32 g	11.51/32	0.36	1
O	23.02	16 g	23.02/16	1.44	4
H_2O	45.32	18 g	45.32/18	2.52	7

Ratio of atoms or groups Fe:S:O:H_2O is 1:1:4:7
Formula is $FeSO_4\ 7H_2O$

Check your understanding

Calculate the empirical formulae of hydrated salts with the following percentage compositions:
(a) Cu = 25.6 per cent, S = 12.8 per cent, O = 25.6 per cent, H_2O = 36.0 per cent
(b) Na = 16.09 per cent, C = 4.20 per cent, O = 16.78 per cent, H_2O = 62.93 per cent
(c) Mg = 11.82 per cent, Cl = 34.98 per cent, H_2O = 53.20 per cent

Calculating the actual (molecular) formula

The molecular formula gives the actual number (and not just the ratio of the numbers) of each type of atom in one molecule of the compound. A molecular formula can be used only for molecular compounds; for ionic substances, the empirical formula is the only one that can be written, as ionic substances have giant structures in which there are no free units such as molecules. Molecular substances can have both empirical formulae and molecular formulae, e.g. the empirical formula for ethane is CH_3 and its molecular formula is C_2H_6. Sometimes the empirical formula is the same as the molecular formula, even for molecular compounds. For example, CH_4 is both the empirical and the molecular formula for methane.

In order to obtain a molecular formula from an empirical formula, the relative molecular mass of the compound is needed. This may be given directly as a statement, or indirectly in the form of some other data. For example, the mass of 24 dm^3 of a gas at room temperature and pressure may be given, which is the same as a mole of the gas (the relative molecular mass in grams), as explained on page 441.

Suppose that an empirical formula of HO has been calculated as in the previous section, and the molecular mass is 34. The molecular formula is always a simple multiple of the empirical formula, or the same as the empirical formula.

If the empirical formula = HO, the molecular formula = $(HO)_x$, $= H_xO_x$ *(where x is a small whole number)*
$\therefore$ *Molecular mass of* $H_xO_x = x + 16x = 17x = 34$
$\therefore$ $x = 2$ *and the molecular formula is* H_2O_2

Once again, the statements you would be expected to write down in your answer are shown in italics. Remember that sometimes the empirical formula is the same as the molecular formula, but you always need the molecular mass in order to find out.

Check your understanding

1. Determine the molecular formulae of compounds having the following percentage composition by mass:
(a) Carbon = 80 per cent, hydrogen = 20 per cent, relative molecular mass = 30
(b) Hydrogen 5.9 per cent, oxygen 94.1 per cent, relative molecular mass = 34
(c) Carbon 38.75 per cent, hydrogen 16.1 per cent, nitrogen 45.2 per cent, relative molecular mass = 31

2. Calculate the empirical formula of a hydrocarbon which contains 92.3 per cent of carbon (relative atomic masses, H = 1, C = 12).
(b) The relative molecular mass of the hydrocarbon is 78. What is its molecular formula? (J.M.B.)

3. Calculate the empirical formula of a compound that has the composition: 52.0 per cent zinc, 9.6 per cent carbon, 38.4 per cent oxygen. (A.E.B.)

4. Calculate the formula of a hydrocarbon containing 82.8 per cent by mass of carbon and having a relative molecular mass of 58. (C.)

Percentage composition from a formula

If we *know* the formula of a compound, we can calculate its percentage composition by mass without doing an experiment. The procedure is as follows (using ammonium nitrate as an example).

1. Write down the formula of the compound: NH_4NO_3.
2. Find the relative formula (molar) mass: $14 + (4 \times 1) + 14 + (3 \times 16) = 80$.
3. Express each relative atomic mass as a percentage of the relative formula (molar) mass. (If more than one atom of a particular element is present, then the 'total mass' of these atoms must be used, e.g. as two nitrogen atoms appear in the formula, the mass used is 28 and not 14.)

$$\text{per cent nitrogen} = \frac{28}{80} \times 100\% = 35\%$$

$$\text{per cent hydrogen} = \frac{4}{80} \times 100\% = 5\%$$

per cent oxygen $= \frac{48}{80} \times 100\% = 60\%$

4. Check that the percentages add up to 100.

Note: If a hydrated compound is being considered, the percentage of water should be calculated as a complete separate unit rather than considering the water as the individual elements hydrogen and oxygen. Thus in hydrated magnesium chloride, $MgCl_2.6H_2O$, molar mass 203, the percentage composition by mass is

per cent magnesium $= \frac{24}{203} \times 100\% = 11.82\%$

per cent chlorine $= \frac{71}{203} \times 100\% = 34.98\%$

per cent water $= \frac{108}{203} \times 100\% = 53.2\%$

Experiment 11.3 (p. 176) describes how to find the percentage of water of crystallization in a compound by experiment.

Check your understanding

1. Calculate the percentage composition by mass of the following compounds.
(a) calcium carbonate, (b) anhydrous copper(II) sulphate, (c) methane, (d) potassium hydrogen carbonate.

2. Determine the percentage composition by mass of each of the following hydrates.
(a) iron(II) sulphate heptahydrate,
(b) sodium carbonate decahydrate,
(c) copper(II) sulphate pentahydrate.

A summary of section 18.2

Some of the inter-relationships in section 18.2 are summarized in Table 18.1.

Table 18.1 Types of calculation involving the mole and percentage composition

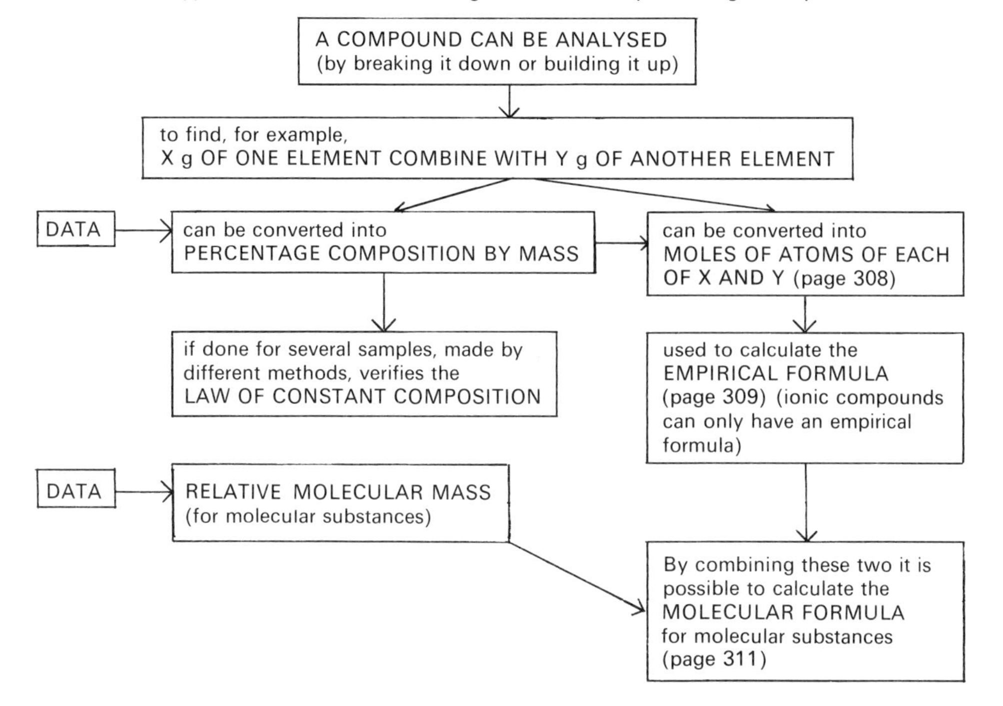

18.3 WRITING EQUATIONS

The Law of Conservation of Mass

The Law of Conservation of Mass states that matter is neither created nor destroyed in a chemical reaction.

Experiment 18.2
To illustrate the Law of Conservation of Mass*

Apparatus
250 cm^3 conical flask and bung, small test-tube with rim, piece of cotton, access to balance.
Solutions of silver nitrate and sodium chloride.

Procedure
(a) Pour some silver nitrate solution into the flask, using a volume approximately equal to that which would almost fill the test-tube.
(b) Tie a piece of cotton tightly round the neck of the test-tube so that the tube can (later) be held in the flask as shown in Figure 18.2.
(c) Pour sodium chloride solution into the test-tube until it is nearly full. Wipe the outside of the tube.
(d) Suspend the tube inside the flask by wedging the cotton with the bung as in Figure 18.2.
(e) Weigh the complete apparatus without mixing the liquids, and record the result.
(f) Gently tip the flask so that the two solutions mix, making sure that liquid does not leak from the flask.

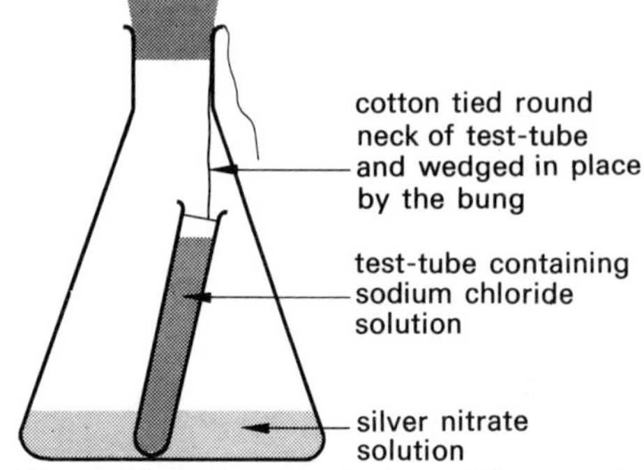

Figure 18.2 Verifying the Law of Conservation of Mass

(g) Note what happens inside the flask, and reweigh the apparatus.

Results
Record your observations and the weighings.

Conclusion
Write an equation for the reaction, and use the results to verify the Law of Conservation of Mass.

The significance of the Law of Conservation of Mass
The Law may seem a matter of common sense, but it is very important. It shows that when chemicals react, the same 'atoms' that are there at the beginning are still there at the end, and therefore there is no change in the total mass of the chemicals. All that happens is that the 'atoms' rearrange themselves or 'swap partners'. Note that in some reactions there *appears* to be a loss in mass because one or more of the products escape (e.g. as a gas), but if the masses of *all* of the products and reactants are considered, all chemical reactions 'obey' the Law.

The importance of the Law is in its application to equations. The work on the mole has shown that it is possible to work with large numbers of atoms and molecules and then to scale this down to what is happening between individual atoms, molecules, or ions. We then summarize the reaction by means of an equation. The Law of Conservation of Mass tells us that when we are writing an equation we must end up with the same total number of atoms as we had at the beginning; we can rearrange them, but we cannot 'gain' any or 'lose' any during the reaction. In other words, an equation must be *balanced*, i.e. show the same total number of atoms on both sides. Many students find difficulty in balancing equations.

Balancing equations

Note: (i) It is a waste of time trying to write chemical equations if you do not know the common symbols and valencies.

(ii) Experimental evidence has enabled us to write an equation for every chemical reaction, (e.g. by finding the masses of chemicals which react together, and the masses of the products, and by converting them into moles, or by experiments such as Experiment 18.3). You will be expected either to *learn* all of the equations you have used, or to be able to *work them out* from basic principles. It must be remembered that a fully balanced equation is a very important piece of chemistry. As a general rule, every chemical reaction you describe in a written answer should be given an equation, and credit will often be given for each equation that is relevant. Beware, however: it is easy to write balanced equations for reactions which *do not occur*! Always look on equations as an important summary of (and source of information about) reactions which have *actually been shown to happen*, even though they may not have been performed by you personally. The next experiment shows how an equation can be obtained by studying a precipitation reaction. You will not be able to use the results to produce an equation unless you have a thorough understanding of the mole concept.

Experiment 18.3
Using a precipitation reaction to determine an equation*

Apparatus
Rack of test-tubes, suitably placed communal burettes for the two solutions, glass rod, labels.
Solutions of lead(II) nitrate and of a metallic chloride, both of concentration 1.0 mol dm^{-3} (with respect to formula units).
Note: Solutions of lead salts are POISONOUS and must be used with care.

Procedure
(a) Make out a table for your results as follows:

Tube number	*Volume of lead(II) nitrate solution (cm^3)*	*Volume of metal chloride solution (cm^3)*	*Height of precipitate (mm)*
1	2.0	2.0	
2	2.0	4.0	
3	2.0	6.0	
4	2.0	8.0	
5	2.0	10.0	

(b) Make up test-tube 1 as shown, using 2.0 cm^3 of the lead(II) nitrate solution from one of the burettes and 2.0 cm^3 of the metal chloride solution from another.
(c) Carefully mix the contents of the tube with a glass rod, and then leave the tube to settle in the rack for 24 hours.
(d) Repeat (b) and (c) four more times, using volumes as shown in the table.

Results
Complete the table, making sure that the tubes have been left to settle for some time before noting the heights of the precipitates.

Notes about the experiment
Read the following points, and then use your results to work out the formula of your chloride and then a full chemical equation for the reaction between lead(II) nitrate solution and the metal chloride solution. Use the symbol X for the cation in your chloride. Remember that your chloride could be XCl, XCl_2, XCl_3, or XCl_4.

(i) Write an ionic equation for the reaction, and note the ratio in which the lead(II) ions and chloride ions react together.
(ii) If your metal chloride had the formula XCl, a solution of concentration 1.0 mol dm^{-3} of XCl would contain 1.0 mol dm^{-3} of chloride ions. If the chloride had the formula XCl_3, a solution of concentration 1.0 mol dm^{-3} with respect to XCl_3 would contain 3.0 mol dm^{-3} of chloride ions.
(iii) If 2.0 cm^3 of a solution A react completely with 2.0 cm^3 of a solution B, then 1000 cm^3 (1 dm^3) of A will react completely with 1000 cm^3 (1 dm^3) of B.

Point for discussion

Solutions of a metal sulphate and of barium chloride, each of concentration 1.0 mol dm^{-3}, were used in a reaction. Five tubes were taken, containing the following mixtures.

Tube	*Volume of metal sulphate solution (cm^3)*	*Volume of barium chloride solution (cm^3)*	*Volume of water (cm^3)*	*Total volume of mixture (cm^3)*
1	2	2	8	12
2	2	4	6	12
3	2	6	4	12
4	2	8	2	12
5	2	10	0	12

A precipitate (of barium sulphate, $BaSO_4$) appeared in each tube. The tubes were centrifuged and the heights of the precipitate were recorded as follows.

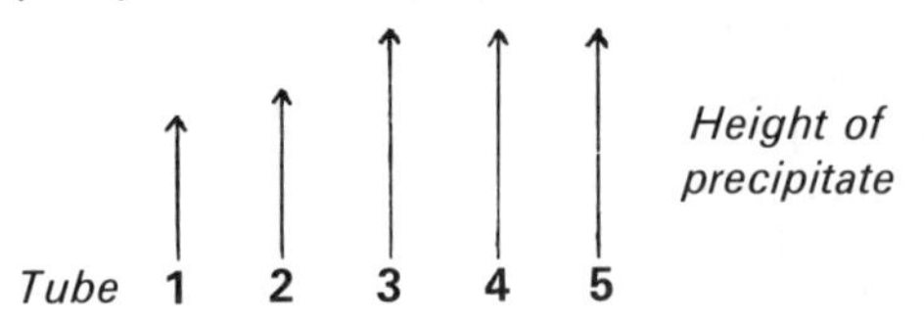

Use the results to explain the reaction which has taken place, to work out the formula of the metal sulphate, and the equation for the reaction. Use the symbol X for the cation in the metal sulphate solution.

Working out equations from reacting masses

Experiments which find the ratio of the masses in which chemicals combine together can be used to find an equation for the reaction. In each case, it is necessary to convert the reacting masses into moles of atoms (or formula units, or molecules) as in the following example.

2.4 g of magnesium react with 7.1 g of chlorine to form magnesium chloride.

$\therefore \frac{2.4}{24}$ moles of magnesium atoms react with $\frac{7.1}{35.5}$ moles of chlorine atoms to form magnesium chloride.

$\therefore$ 0.1 moles of magnesium atoms react with 0.2 moles of chlorine atoms to form magnesium chloride.

$\therefore$ 1 mole of magnesium atoms react with 2 moles of chlorine atoms to form magnesium chloride.

The equation is $Mg(s) + Cl_2 \rightarrow MgCl_2$.
(Note Cl_2 and not 2Cl, because chlorine gas exists as molecules.)

Check your understanding

1. Use the following experimental results to produce chemical equations for each reaction. Set out your working clearly.
(a) 0.90 g of aluminium react with 3.55 g of chlorine to form aluminium chloride.
(b) 1.52 g of iron(II) sulphate when heated produce 0.64 g of sulphur dioxide, 1.60 g of iron(III) oxide, and 0.80 g of sulphur(VI) oxide (sulphur trioxide).
(c) 4.00 g of iron(III) sulphate reacts with 6.24 g of barium chloride to form 3.25 g of iron(III) chloride and 6.99 g of barium sulphate.
2. Explain how and why each of the examples (b) and (c) above also confirm the Law of Conservation of Mass.
3. The following information is not complete. What is missing, and how do you know the mass of the 'missing chemical'?
1.60 g of copper(II) sulphate reacts with 0.80 g of sodium hydroxide to make 0.98 g of copper(II) hydroxide.

Rules for writing equations

Obviously you do not have time to check every equation used in your course by doing an experiment, but scientists *have* done this for every reaction, and the equations you use summarize known facts. You must learn the equations you have used and you must also be able to 'work out' equations if necessary. For example, you may have prepared only four salts by experiments such as those used in Chapter 6, but you could be asked to write an equation for the preparation of almost *any* salt which can be made by similar methods. If you have to *construct* an equation, proceed as follows.

1. Be sure that you know the names of *all* the reactants and products.
2. Write the equation in words. (If you know your formulae, this step can be omitted.)
3. Write down a 'skeleton' equation, using the exact formulae for all the substances taking part and leaving a space in front of each formula for adding numbers.
4. Balance the equation.
5. Add the state symbols behind each reagent.

Example 1
Write the equation for the reaction between sodium and water.

1. Sodium reacts with water to produce a solution of sodium hydroxide; hydrogen is evolved.
2. Sodium + water → sodium hydroxide + hydrogen
3. $Na + H_2O \rightarrow NaOH + H_2$
(Remember that most of the gaseous elements are molecular, usually diatomic.)
4. There are two atoms of hydrogen on the L.H.S. (left-hand side) and three atoms on the R.H.S. (right-hand side). We must first make the number of hydrogen atoms on the R.H.S. an even number, because we are dealing in 'units of two' in the molecule H_2O. This can be achieved by having two NaOH units.

$$Na + H_2O \rightarrow 2NaOH + H_2$$

On the R.H.S. we now have 'in excess' one Na, two H, and one O and the equation can be balanced by adding these units to the L.H.S.

$$2Na + 2H_2O \rightarrow 2NaOH + H_2$$

5. Finally we add the state symbols.

$$2Na(s) + 2H_2O(l) \rightarrow 2NaOH(aq) + H_2(g)$$

The whole operation is, of course, carried out in one line.

Example 2
Write the equation for the combustion of butane in oxygen.

1. Butane (C_4H_{10}) will react with oxygen to form carbon dioxide and steam.
2. Butane + oxygen → carbon dioxide and steam
3. $C_4H_{10} + O_2 \rightarrow CO_2 + H_2O$
4. To balance an equation in which rather large numbers of atoms are involved, consider first the substance which has the largest numbers of atoms and 'balance' the equation with respect to that substance, e.g. C_4H_{10}.

$$C_4H_{10} + O_2 \rightarrow 4CO_2 + 5H_2O$$

We can now balance the equation by putting $6\frac{1}{2}$ in front of the oxygen molecule. This would be *acceptable*,

$$C_4H_{10} + 6\tfrac{1}{2}O_2 \rightarrow 4CO_2 + 5H_2O$$

but as we cannot have half molecules it is neater to complete the equation by multiplying throughout by two. The state symbols are also added.

$$2C_4H_{10}(g) + 13O_2(g) + 8CO_2(g) + 10H_2O(g)$$

Check your understanding
Construct fully balanced equations (e.g. 10 in total) which summarize the formation of named salts of your own choice. The examples should be chosen so as to illustrate all of the methods of salt preparation discussed in Chapter 6, and should *not* include reactions you have actually used yourself. Be careful in making your choice; you will need to think about which salts are insoluble in water, which metals react with which acids, etc.

Ionic equations

These have been introduced wherever appropriate throughout the book, and their use is explained on page 91. It is possible to construct a set of 'rules' for writing ionic equations, but in practice they are best learned off by heart.

Check your understanding
Test your understanding of ionic equations by writing them for the following reactions, which cover all of the main *situations* where such equations are used.

1. Aqueous solutions of barium chloride and magnesium sulphate.
2. Aqueous solutions of calcium nitrate and sodium carbonate.
3. Aqueous solutions of sodium hydroxide and nitric acid.
4. Zinc and an aqueous solution of copper(II) sulphate.
5. Aqueous solutions of sodium carbonate and hydrochloric acid.

Points for discussion
Explain what is wrong with each of the following equations. (They *may* be correctly balanced already.)

(a) $Mg(s) + Ag^{+}(aq) \rightarrow Mg^{2+}(aq) + Ag(s)$
(b) $2Na(s) + Cu^{2+}(aq) \rightarrow 2Na^{+}(aq) + Cu(s)$
(c) $Cu(s) + Fe^{2+}(aq) \rightarrow Cu^{2+}(aq) + Fe(s)$

18.4 REACTING MASSES FROM EQUATIONS

A properly balanced equation provides a great deal of information. It shows the mole ratios in which the chemicals react, and from these we can find the actual *masses* which would react together in a certain situation, and calculate the masses (or volume, if gases) of products which would be formed. This sort of information is particularly important when calculating how efficient an industrial reaction is.

The state symbols also tell us about any changes of state which might occur during a reaction, and the ΔH sign, if present (see page 477) tells us whether the reaction is endothermic or exothermic. The ⇌ sign, if present, tells us when a reaction is reversible (can go in both directions) and that an equilibrium might result.

Example

What is meant by the equation:

$$CaO(s) + H_2O(l) \rightarrow Ca(OH)_2(s) \qquad \Delta H = -67.2\,kJ?$$

Solid calcium oxide and liquid water react together in the ratio one unit: one unit (formula units and molecules respectively) to form one unit of solid calcium hydroxide (formula unit). This is in itself not particularly useful in the practical sense; we wish to deal with *measurable* quantities of the substances. It is more useful to interpret the equation in terms of units X times as large as a formula unit or a molecule, where X is Avogadro's number, i.e. 6.02×10^{23}. These are measurable quantities.

State symbols?

∴ X formula units of solid calcium oxide react with X molecules of water to form X formula units of solid calcium hydroxide.

∴ 1 mole of calcium oxide reacts with 1 mole of water to form 1 mole of calcium hydroxide.

This is the practical interpretation of a typical equation.

The equation also shows the energy change involved and its nature and this is normally quoted for the quantities shown in the equation measured in moles. Thus when 1 mole of calcium oxide reacts with 1 mole of water, 67.2 kJ are liberated.

Check your understanding

Balance, and then interpret, the following equations.

(a) $NH_3 + H_2O + X_2SO_4 \rightarrow (NH_4)_2SO_4 + XOH$

(b) $XOH + Cl_2 \rightarrow XCl + XClO_3 + H_2O$

(c) $X_2O_5 + H_2O \rightarrow H_3XO_4$

(d) $H_2XO_4 + C \rightarrow H_2O + XO_2 + CO_2$

Finding reacting masses from equations

This is really the reverse of the process discussed on page 308, where we found reacting masses by experiment and then converted them into moles, from which equations can be worked out. Here we start with the equation, and work back to find the masses of substances which react in a given situation.

Suppose that we are asked to calculate the mass of magnesium oxide produced by the complete combustion of 9.6 g of magnesium in pure oxygen. (The actual statements that you would be expected to show in an examination answer are shown in italics.)

1. Write down a fully balanced equation for the reaction. (This is vital; you cannot perform these calculations without an equation.)

$$2Mg(s)+O_2(g) \rightarrow 2MgO(s)$$

2. From the equation, write down the numbers of moles involved with each of the substances mentioned in the problem. (We are not asked to calculate the mass of oxygen used.)

2 moles of magnesium atoms → 2 moles magnesium oxide

3. Convert moles into masses.

48 g (2 × 24) magnesium → 80 g (2 × 40) magnesium oxide

and then proceed as follows.

4. $1\ g\ magnesium \rightarrow \frac{1}{48} \times 80\ g\ magnesium\ oxide$

$9.6\ g\ magnesium \rightarrow 9.6 \times \frac{1}{48} \times 80\ g\ magnesium\ oxide$

9.6 g magnesium → 16 g magnesium oxide

There are a few variations of this type of question, but the basic idea is the same. Suppose that we are asked to calculate the mass of calcium hydroxide that will react with 21.4 g of ammonium chloride, and also the mass of the ammonia produced. Using the steps as in the previous example,

$$2NH_4Cl(s)+Ca(OH)_2(s) \rightarrow 2NH_3(g)+CaCl_2(s)+2H_2O(g)$$

2 moles ammonium chloride react with 1 mole calcium hydroxide → 2 moles ammonia.

107 g NH_4Cl react with 74 g $Ca(OH)_2$ → 2 moles NH_3

1 g NH_4Cl react with $\frac{1}{107} \times 74$ g $Ca(OH)_2$ → $\frac{1}{107} \times 34$ g NH_3

21.4 g NH_4Cl react with $21.4 \times \frac{1}{107} \times 74$ g $Ca(OH)_2$ → $21.4 \times \frac{1}{107} \times 34$ g NH_3

21.4 g NH_4Cl react with 14.8 g $Ca(OH)_2$ to produce 6.8 g NH_3

Check your understanding

1. Calculate the mass of calcium oxide produced when 20.0 g of calcium carbonate is decomposed by heat.

2. A solution containing 8.0 g of sodium hydroxide is neutralized by hydrochloric acid. What mass of sodium chloride will be produced if the solution is evaporated to dryness?

3. What mass of copper will remain if 3.2 g of copper(II) oxide are completely reduced to the metal?

4. (a) (i) Write the equation for the reaction which occurs when 0.1 moles of sodium hydroxide and 0.1 moles of sulphuric acid are mixed in solution. (ii) Name the salt formed in this reaction.

(b) (i) How many moles of the salt would be formed? (ii) Calculate the mass of the salt formed.

(c) What would be seen if a strip of clean magnesium ribbon were added to a solution of the salt in water? Explain your answer.

(d) (i) Name another salt which can be made from sodium hydroxide and sulphuric acid. (ii) Write the equation for the preparation of the second salt. (iii) State how many moles of sodium hydroxide must be added to 0.1 moles of sulphuric acid to prepare the second salt. (J.M.B.)

5. (a) When copper(II) sulphate crystals are heated, the water of crystallization is evolved.

$$CuSO_4.5H_2O(s) \rightarrow CuSO_4(s) + 5H_2O(g)$$

Calculate the mass of anhydrous copper(II) sulphate which can be obtained from 25.0 g of crystals.

(b) When 16.0 g of anhydrous copper(II) sulphate are heated strongly for about 30 minutes weight is lost until only 8.0 g of solid remain. Name the remaining solid. Write a possible equation for the reaction. (J.M.B.)

Reactions between solids and standard solutions

The procedure is as just described, except that this time one of the reactants is in solution. We need to know the concentration of the solution, but as these are normally given in mol dm^{-3}, we can still work in moles and the basic idea is the same.

Suppose that we are asked to calculate the concentration of a hydrochloric acid solution, 100 cm^3 of which dissolves 3.0 g of magnesium ribbon. (Remember, the first stage *must* be to write a balanced equation for the reaction.)

$$Mg(s) + 2HCl(aq) \rightarrow MgCl_2(aq) + H_2(g)$$

1 mole of magnesium atoms react with 2 moles of hydrochloric acid

$\therefore$ 24 g of magnesium react with 2 moles of hydrochloric acid

$\therefore$ 1 g of magnesium reacts with $\frac{1}{24} \times 2$ moles of hydrochloric acid

$\therefore$ 3 g of magnesium reacts with $3 \times \frac{1}{24} \times 2$ moles $= 0.25$ moles of hydrochloric acid

$\therefore$ 0.25 moles of hydrochloric acid must be present in 100 cm^3 of solution

$\therefore$ 2.5 moles of the acid must be present in 1000 cm^3 (1 dm^3) of solution

The concentration of the acid is 2.5 mol dm^{-3}.

Suppose that we are asked to calculate the mass of pure iron that would be dissolved by 500 cm^3 of a solution of sulphuric acid of concentration 0.1 mol dm^{-3}.

$$Fe(s) + H_2SO_4(aq) \rightarrow FeSO_4(aq) + H_2(g)$$

1 mole of iron atoms reacts with 1 mole of sulphuric acid

$\therefore$ 56 g of iron atoms react with 1 mole of sulphuric acid

1 dm^3 (1000 cm^3) of the acid solution contain 0.1 moles of the acid

$\therefore$ 500 cm^3 of the acid solution contain $\frac{1}{2} \times 0.1$ moles $= 0.05$ moles of the acid

$\therefore$ 56×0.05 g of iron react with 0.05 moles of the acid

$\therefore$ 2.8 g of iron react with 0.05 moles of the acid

Check your understanding

A solution of sodium carbonate, containing 5.30 g of the dissolved anhydrous salt, will exactly neutralize 200 cm^3 of hydrochloric acid solution. Calculate the concentration of the acid solution.

Finding reacting volumes from equations

These calculations, though similar to those involving masses, may involve the extra step of converting a number of moles of gas into the *volume* it would occupy under certain conditions of temperature and pressure. These calculations are considered in Chapter 23, where you will also gain more experience in interpreting equations.

18.5 A SUMMARY OF CHAPTER 18

The following 'check list' should help you to organize the work for revision.

1. Definitions

(a) A mole (p. 301)
(b) The Law of Constant Composition (p. 306)
(c) The Law of Conservation of Mass (p. 314)

2. Other points

(a) The Avogadro number is the number of particles in a mole of substance, and is equal to 6.02×10^{23}.
(b) The importance of recognizing and stating the type of particle being referred to in mole calculations, and understanding the differences between them, e.g. atoms, molecules, formula units, and ions.
(c) How to convert a mass of a substance into a number of moles of the substance (naming the type of particles being referred to), and the other way round. For example, a mass is converted into moles of atoms by dividing the mass by the relative atomic mass of the substance.
(d) When we work in moles, we are really working with a certain *number* of the stated particles. If the relative atomic mass of any element is converted into a mass in grams, that mass will always contain the same number of *atoms*, no matter which element is used.
(e) Similarly, we can express the relative molecular mass in grams (for molecular substances) and the relative formula mass in grams (for formula units of ionic compounds); these masses will contain the same number of molecules or formula units.
(f) How to calculate the concentration of a solution in mol dm^{-3} (given the mass dissolved, e.g. as on page 304). How to express these concentrations in terms of the individual ions which may be present, e.g. a solution which has a concentration of 1.0 mol dm^{-3} with respect to $MgCl_2$ has a concentration of 2.0 mol dm^{-3} with respect to $Cl^-(aq)$.
(g) How to use experimental results to verify the Law of Constant Composition (p. 306). The significance of the law (p. 307).
(h) What is meant by an empirical (simplest) formula, and how to calculate one given experimental data about a compound or its percentage composition by mass (p. 308), or data for a hydrated salt (p. 311).
(i) How to calculate a molecular formula from an empirical formula (for molecular substances), as on page 311. For this calculation, it is necessary to know (or calculate) the relative molecular mass of the substance.
(j) How to calculate the percentage composition by mass of a compound from its formula (p. 312).
(k) The inter-relationships between these points are summarized in Table 18.1 (p. 313); you may find this helpful in understanding the chapter as a whole.
(l) The significance of the Law of Conservation of Mass (p. 314) in balancing equations (pp. 315, 317).
(m) How to calculate reacting masses from equations (p. 319) and when standard solutions are involved (p. 321).

3. Important experiments

(a) An experiment which verifies the Law of Constant Composition (e.g. like the one summarized on page 306).
(b) An experiment to determine the empirical formula of a compound (e.g. Experiment 18.1 on page 309).
(c) An experiment to verify the Law of Conservation of Mass, (e.g. Experiment 18.2 on page 314).
(d) An experiment to explain how a precipitation reaction can be used to find an equation, (e.g. Experiment 18.3 on page 315).

QUESTIONS

Multiple choice and short answer questions

1. Ammonia reacts with hot copper(II) oxide to form metallic copper. In this reaction, the number of moles of copper(II) oxide which react with one mole of ammonia is
A 1; B $1\frac{1}{2}$; C 2; D $2\frac{1}{2}$; E 3 (C.)

2. The mass of one mole of a chloride formed by a metal Y is 74.5 g. The formula of the chloride could be
A Y_3Cl; B Y_2Cl; C YCl; D YCl_2; E YCl_3
(Relative atomic mass: Cl, 35.5) (C.)

3. You are required to make 100 cm^3 of a solution which is 0.1 mol dm^{-3} with respect to Cu^{2+} ions, using copper(II) sulphate pentahydrate, $CuSO_4.5H_2O$. How much of the salt should you weigh out?
A 0.64 g; B 1.60 g; C 1.78 g; D 2.50 g; E 6.40 g (C.)

4. One mole of magnesium was heated in a closed vessel in the presence of one mole of chlorine (Cl_2) and one mole of bromine (Br_2). If the solid remaining was soluble in water and had a mass of 139.5 g, it could be deduced that
A all of the chlorine had reacted
B all of the bromine had reacted
C some magnesium was left
D the mass of halogen left was 115.5 g
E the solid consisted only of the compound MgBrCl.
(Relative atomic masses: Mg 24; Cl 35.5; Br 80) (C.)

5. One mole of carbon-12 has a mass of
A 0.012 kg; B 0.0224 kg; C 0.024 kg; D 1 kg; E 12 kg
(The relative isotopic mass of ^{12}C is 12) (C.)

6. The equation for the burning of hydrogen in oxygen is

$$2H_2(g)+O_2(g) \rightarrow 2H_2O(g)$$

This equation indicates that
A 2 atoms of hydrogen combine with 2 atoms of oxygen
B 2 moles of steam can be obtained from 0.5 moles of oxygen
C 2 moles of steam can be obtained from 1 mole of oxygen
D 2 g of hydrogen combine with 1 g of oxygen
E 1 mole of steam can be obtained from 1 mole of oxygen (C.)

7. In a compound of magnesium (A_r(Mg) = 24) and nitrogen (A_r(N) = 14), 36 g of magnesium combine with 14 g of nitrogen. The simplest formula for the compound is
A MgN; B Mg_2N; C Mg_3N; D Mg_2N_2; E Mg_3N_2

8. (a) 3.5 g of nitrogen combine with 2 g of oxygen to form an oxide of nitrogen. What is the empirical formula of this oxide?
(b) On analysis, a compound was found to contain 2.3 g of sodium and 0.8 g of oxygen. What is the empirical formula of this oxide?

9. (a) Write *balanced* equations for the reaction of potassium hydroxide solution with dilute sulphuric acid to form (i) the acid salt, (ii) the normal salt.
(b) What volume of potassium hydroxide solution of concentration 0.1 mol dm^{-3} will need to be added to 50 cm^3 of dilute sulphuric acid of concentration 0.05 mol dm^{-3} to form (i) the acid salt, (ii) the normal salt? (W.J.E.C.)

10. (a) 27.40 g of an oxide of lead (A) is found on reduction to give 24.84 g of lead. Calculate the simplest (or empirical) formula of oxide (A).
(A_r(O) = 16; A_r(Pb) = 207)
(b) on reaction with nitric acid, oxide (A) gives lead(II) nitrate, water, and a second oxide of lead (B). The oxide (B) is found to contain 32 g of oxygen combined with the molar mass of lead. Calculate the simplest formula for oxide (B). (W.J.E.C.)

Structured questions and questions which require longer answers

11. The corrosion of iron is estimated to cost several hundred million pounds per year in this country. Explain why this corrosion, or rusting, costs so much.
Give *two* factors which help rusting to occur.
Give *two* different methods of preventing rusting.
With very large ships several kilograms of rust can form on a ship in a few days. Calculate how many kg of oxygen and how

many kg of iron would be used up if 8 kg of rust accumulated on a ship in one day. Assume the formula of rust is Fe_2O_3.

(A.E.B. 1976)

12. The hydrogencarbonate of a monovalent metal M reacts with hydrochloric acid in accordance with the following equation:

$$MHCO_3(s) + HCl(aq) \rightarrow MCl(aq) + H_2O(l) + CO_2(g)$$

In an experiment to find the relative atomic mass of M, it was found that 4.2 g of the hydrogencarbonate ($MHCO_3$) reacted with 100 cm^3 of hydrochloric acid of concentration 0.5 mol dm^{-3}.
(a) What fraction of a mole of hydrochloric acid is contained in 100 cm^3 of the acid?
(b) From the answer to (a) and the given equation, deduce the fraction of a mole of the hydrogencarbonate contained in 4.2 g of $MHCO_3$.
(c) From the answer to (b), calculate the *relative molecular* mass of $MHCO_3$.

(W.J.E.C.)

13. (a) (i) Describe how you would prepare dry crystals of sodium sulphate ($Na_2SO_4.10H_2O$), starting from dilute sulphuric acid and dilute aqueous sodium hydroxide. (ii) What mass (in grams) of the *hydrated* crystals would you expect to obtain from 0.1 mole of sodium hydroxide?
(b) Excess aqueous barium chloride was added to a solution of 0.025 mole of sodium sulphate dissolved in water. Write down the ionic equation for the reaction and calculate the mass of the dry product. (O. & C.)

14. (a) Describe how you would prepare dry crystals of hydrated copper(II) sulphate ($CuSO_4.5H_2O$) using copper(II) oxide and dilute sulphuric acid as the starting materials. Calculate the mass (in grams) of the *hydrated* crystals which could be obtained from 0.01 moles of copper(II) oxide.
(b) Describe what happens when the hydrated crystals are gently heated until no further change in mass occurs. Calculate the percentage change in mass of the hydrated crystals. (O. & C.)

15. When aqueous solutions of silver nitrate and calcium chloride are mixed, solid silver chloride is precipitated, the equation for the reaction being

$$2AgNO_3(aq) + CaCl_2(aq) \rightarrow 2AgCl(s) + Ca(NO_3)_2(aq)$$

(a) What volume of 0.5 molar calcium chloride would you add to 100 cm^3 of molar silver nitrate just to complete the above reaction?
(b) What mass of solid silver chloride (formula mass 143.5) would be formed?

(W.J.E.C.)

16. 4.76 g of the hydrated chloride of a divalent metal M, of relative atomic mass 59, contain 2.60 g of the anhydrous salt. Find the number of molecules of water of crystallization in each molecule of the hydrated salt.

(W.J.E.C.)

19 The metals and their compounds

Introduction

The metals and their compounds are among the most important substances that you are likely to meet. Metals were vital even to people of early civilizations. This is why whole eras of history have been named after those metals which were important at the time, for example the Bronze Age and the Iron Age.

Modern civilization depends greatly on metals and their alloys. Steel has many different forms and thousands of uses. More specialized metals such as tantalum, titanium, and zirconium have been developed for use in space missiles, jet aircraft, and nuclear reactors. Our use of metals and their compounds is so great, that at the present rate of consumption a number of metals will soon be completely used up, and recycling (p. 513) is extremely important.

We tend to forget that many *compounds* of metals are also used in daily life. Ordinary substances like soap and common salt are compounds of metals, and most of the bottles on the shelves of any chemistry laboratory contain substances which are compounds of metals. In this chapter we shall see how some important compounds of metals are used and made. The extraction of some metals from their ores is explained in Chapter 27.

You may have already studied some properties of the common metals when investigating the reactions of acids, and while studying the Periodic Table and the activity series. This chapter contains more facts to add to your knowledge, and it is important to organize these facts.

The work in this chapter has been divided into sections where the emphasis is on the non-metal part of a compound (e.g. sulphates, chlorides) and other, smaller, sections where the emphasis is on the metal part of the compound (e.g. properties of calcium compounds). If you learn these facts conscientiously, you should be able to remember the essential chemistry of all the common compounds of metals without learning each one individually. The tables in this chapter list metals in the same order as that of the activity series (except Table 19.3) because the properties of their compounds often follow a similar pattern of reactivity changes.

A revision hint

When you need to remember something about a particular metal compound (e.g. calcium carbonate) try to think in the following way.

1. Ask yourself what *type* of compound it is (in this case a metal carbonate) and then try to remember the *general* properties of this type and whether the named example is exceptional in any way. It is far easier to do this than to learn the properties of every metal compound you have studied; in a typical course this might involve the study of 70 or more compounds of metals.
2. Then look at it from 'the other way round', e.g. calcium carbonate is also a

compound of *calcium* as well as a carbonate, and if you remember the general properties of calcium compounds you should be able to add further to your knowledge of this particular substance. For example, as all common calcium compounds are white and ionic, then so is calcium carbonate.

If you do not think of *types* of chemicals in this way, there is a real danger that you could fail to answer a question which is in fact quite easy. Suppose that you are asked to describe the preparation of a pure sample of zinc sulphate. Many students would rack their brains trying to think of how they have prepared the substance in the laboratory. This may be the wrong approach, for it is quite possible that they may never have prepared this particular substance. They are making the mistake of thinking of it as an individual substance rather than as a general type. If we ask ourselves what *type* of chemical we are referring to, it becomes a problem of how to prepare a metal sulphate, i.e. as we learn from the general properties of sulphates, a salt. Our general knowledge of sulphates tells us that zinc sulphate is a soluble salt, and so we can pick any one of several methods of preparation, such as sulphuric acid + zinc oxide (acid + base), sulphuric acid + zinc hydroxide (acid + base), or sulphuric acid + zinc carbonate (acid + carbonate). The *details* of the method we choose will be the same as when we used the method in the laboratory, but perhaps with different chemicals.

19.1 OXIDES OF METALS

An oxide is a compound of oxygen with one other element, i.e. element + oxygen → a compound called an oxide. Many naturally occurring substances are oxides, e.g. iron ore and aluminium ore (bauxite).

Methods of preparation in the laboratory

Direct combination with oxygen

This is achieved by heating the metal in the air or oxygen. This is not a good nor convenient method, for some metals only form a thin layer of oxide on the metal (see individual details on page 250).

Heating a metal carbonate

This is a useful method, particularly for oxides of less reactive metals, the carbonates of which decompose easily, e.g.

$$CuCO_3(s) \rightarrow CuO(s) + CO_2(g)$$

Carbonates of more reactive metals either do not decompose when heated (those of sodium and potassium) or decompose only when heated to a high temperature (e.g. calcium carbonate). The carbonates of aluminium and iron(III) do not exist. (See Experiment 15.3, p. 254.)

The carbonates of aluminium and iron(III) do not exist

Heating a metal nitrate

This is a useful method, particularly for the oxides of metals below calcium in the activity series, the nitrates of which readily decompose, e.g.

$$2Cu(NO_3)_2(s) \rightarrow 2CuO(s) + 4NO_2(g) + O_2(g)$$

As nitrogen dioxide is toxic, these reactions should be carried out in a fume cupboard.

Nitrates of elements at the top of the activity series behave differently when heated. The nitrates of sodium and potassium produce the nitrite and oxygen, e.g.

$$2NaNO_3(s) \rightarrow 2NaNO_2(s)+O_2(g)$$

Heating a metal hydroxide

This is not as useful as heating a metal carbonate or nitrate because the hydroxides are not as readily available. Again, this is suitable for the hydroxides of less reactive metals, e.g.

$$Cu(OH)_2(s) \rightarrow CuO(s)+H_2O(g)$$

The hydroxides of sodium and potassium do not decompose on heating, and that of calcium requires a very high temperature.

A summary of methods used for making oxides is given in Table 19.1.

Table 19.1 A summary of oxide preparations

	Reaction with oxygen	*Heat on carbonate*	*Heat on nitrate*	*Heat on hydroxide*
K	↑ Not Recommended ↓	×	×	×
Na		×	×	×
Ca		✓	✓	✓
Mg		✓	✓	✓
Al		Carbonate does not exist	✓	✓
Zn		✓	✓	✓
Fe(II)		×*	Rarely used	×*
Fe(III)		Carbonate does not exist	✓	✓
Pb		✓	✓	✓
Cu		✓	✓	✓

* If these compounds are heated, iron(III) oxide is formed, not iron(II) oxide.

Points for discussion

1. Why were copper compounds used as examples for the above reactions, rather than, say, calcium compounds?

2. You may remember that when copper(II) nitrate is heated (Experiment 15.3, p. 254), it *appears* to melt to form a turquoise coloured liquid, then solidifies again, and then goes black. Can you explain what is happening during these changes?

Properties of metal oxides

1. All oxides of metals can act as bases; in other words, they react with acids to form salts. A few (e.g. aluminium oxide and zinc oxide) can also act as acids and are therefore amphoteric (see page 155).

2. Calcium oxide reacts violently with water. If water is added to a small sample of the freshly prepared oxide (often formed by roasting a piece of limestone in the air, as in Experiment 19.7), an exothermic reaction occurs (i.e. heat is given out). Steam is formed, and also a white powder which takes up a greater volume than the original solid. This is the 'slaking of lime' and the product is called calcium hydroxide (slaked lime),

$$CaO(s)+H_2O(l) \rightarrow Ca(OH)_2(s)$$

3. Information about a number of oxides of metals is given in Table 19.2.

Table 19.2 The common metal oxides

Formula	*Name*	*Colour*	*Other points to note*
Na_2O	Sodium monoxide	White	Reacts with water to form sodium hydroxide
Na_2O_2	Sodium peroxide	Pale yellow	A typical peroxide
CaO	Calcium oxide	White	Reacts with water (p. 327). Good drying agent for ammonia. Commonly called quicklime
MgO	Magnesium oxide	White	———————————————
ZnO	Zinc oxide	White	Yellow when hot. Amphoteric
Al_2O_3	Aluminium oxide	White	Amphoteric
FeO	Iron(II) oxide	Black	———————————————
Fe_2O_3	Iron(III) oxide	Red	———————————————
Fe_3O_4	Iron(II) di-iron(III) oxide (tri-iron tetraoxide)	Black	Magnetic
PbO	Lead(II) oxide	Yellow	Amphoteric, but usually only encountered as a typical base. The most stable oxide of lead
PbO_2	Lead(IV) oxide	Brown	An oxidizing agent, like MnO_2; e.g. oxidizes concentrated hydrochloric acid to chlorine
Pb_3O_4	Dilead(II) lead(IV) oxide (tri-lead tetraoxide)	Red	Behaves as if a mixture of $2PbO + PbO_2$; when dilute nitric acid is added the PbO reacts and dissolves to form soluble lead(II) nitrate, leaving the insoluble PbO_2. Also an oxidizing agent as it appears to contain PbO_2. When heated strongly, forms PbO + oxygen
CuO	Copper(II) oxide	Black	———————————————

19.2 HYDROXIDES OF METALS

These are ionic compounds made up of metal ions and hydroxide ions.

Methods of preparation

Precipitation

As most metal hydroxides are insoluble, this is an excellent method of preparation. The hydroxides of sodium and potassium, which are soluble, are the only exceptions. Mix together two solutions, one containing OH^- ions (i.e. sodium hydroxide solution or ammonia solution) and the other containing the appropriate metal ion (use a solution of the nitrate if in doubt, because all nitrates are soluble), e.g.

$$Cu^{2+}(aq) + 2OH^-(aq) \rightarrow Cu(OH)_2(s)$$

Unfortunately there are complications in making the hydroxides in this way, except for those of magnesium and iron. The hydroxides of the other metals must be prepared by adding the solution of OH^- ions (alkali) slowly, drop by drop, to the metal ion solution until precipitation is just complete. If excess alkali is added, the precipitate first formed may then dissolve. The actual effect depends on both the individual alkali and the metal ion, and these and other factors are summarized in Table 19.3.

Table 19.3 The common metal hydroxides

Formula	*Name*	*Colour*	*Solubility in water*	*Other points*
NaOH	Sodium hydroxide	White	Soluble	Solid is deliquescent
KOH	Potassium hydroxide	White	Soluble	Solid is deliquescent
$Ca(OH)_2$	Calcium hydroxide	White	Slightly soluble	Solid often called slaked lime, solution lime water
$Mg(OH)_2$	Magnesium hydroxide	White	Insoluble	Precipitated by any alkali even if excess is used
$Fe(OH)_2$	Iron(II) hydroxide	Dirty green	Insoluble	
$Fe(OH)_3$	Iron(III) hydroxide	Brown	Insoluble	
$Al(OH)_3$	Aluminium hydroxide	White	Insoluble	Precipitated by using any alkali, but redissolves if the alkali is NaOH or KOH and excess is used, because the solids are amphoteric
$Pb(OH)_2$	Lead(II) hydroxide	White	Insoluble	
$Cu(OH)_2$	Copper(II) hydroxide	Blue	Insoluble	Precipitated by any alkali but redissolves if the alkali is ammonia and excess is used because a soluble complex salt is formed
$Zn(OH)_2$	Zinc hydroxide	White	Insoluble	Precipitated by any alkali but redissolves if an excess of *any* alkali is used as the solid is amphoteric and also forms soluble complexes with ammonia

The action of water on a metal oxide

This is not a general method, and it is restricted to the oxides of metals near the top of the activity series, i.e. potassium, sodium, and calcium, e.g.

$$CaO(s) + H_2O(l) \rightarrow Ca(OH)_2(s)$$

Experiment 19.1
Making some insoluble metal hydroxides*

Apparatus

Rack of test-tubes, teat pipette, test-tube holder, Bunsen burner, stirring rod.

Dilute solutions of any soluble salts of calcium, magnesium, aluminium, iron(II), iron(III), zinc, lead(II), and copper(II). Dilute sodium hydroxide and ammonia solutions.

Remember that compounds of zinc, lead, and copper are poisonous, and that sodium hydroxide solution must be used with great care.

Procedure

(a) Make a table for your results as follows:

Solution	*Appearance of precipitate (if any) with sodium hydroxide*	*Effect of excess sodium hydroxide*	*Appearance of precipitate (if any) with ammonia solution*	*Effect of excess ammonia solution*

(b) Pour about 1 cm^3 of each salt solution into separate test-tubes.
(c) Add sodium hydroxide solution *drop-wise* from a teat pipette to one of the salt solutions. Stir. Repeat until a precipitate can be seen.
(d) Add a further 2 cm^3 of sodium hydroxide solution and stir the mixture. Warm the contents of the tube *gently*.
(e) Repeat (c) and (d) with each of the other salt solutions in turn.
(f) Repeat (c) and (d) with each of the salt solutions in turn, but using ammonia solution each time in place of the sodium hydroxide solution.

Results
Complete the table.

Conclusion
Make sure that you can use Table 19.3 to explain your results. Write equations (ionic if possible) for the reactions in which the hydroxides are formed. There is no need to construct equations for the reactions in which metal hydroxides dissolve in excess alkali.

Points for discussion

1. The precipitation of metal hydroxides can be a useful way of identifying the metal ion(s) in a solution, for the colour of the precipitate and/or its behaviour when excess alkali is added varies from one metal ion to another. Some of the tests given on page 451 depend upon these reactions.
2. Using the knowledge gained from this experiment, work out tests for distinguishing between solutions of (a) sodium chloride and magnesium chloride, (b) iron(II) nitrate and iron(III) nitrate, (c) aluminium chloride and zinc chloride.

Properties of metal hydroxides

1. All metal hydroxides can act as bases. Those which dissolve in water are called alkalis.
2. Some metal hydroxides decompose to the oxide when heated (see preparation of oxides, page 327).
3. Information about the common metal hydroxides is given in Table 19.3.

Check your understanding

1. Which one of the following metals forms a hydroxide which only melts (i.e. does not decompose) when heated: aluminium, magnesium, sodium, calcium, copper? Explain your answer.
2. Name another metal hydroxide which would probably only melt when heated.
3. Write equations for the reactions which would take place when the other metal hydroxides mentioned in **1** are heated.
4. Which would be the most unstable (to heat) of the hydroxides listed in **1**? Explain your answer.
5. If a metal hydroxide decomposes when heated in an open tube, is the reaction a thermal decomposition or is it a thermal dissociation?

19.3 CARBONATES OF METALS

Methods of preparation

All metal carbonates are insoluble except for those of sodium and potassium, so they can be made by precipitation. Solutions of sodium carbonate or potassium carbonate are used to provide the $CO_3^{2-}(aq)$ ions needed for the precipitation of the others, e.g.

$$CO_3^{2-}(aq) + Mg^{2+}(aq) \rightarrow MgCO_3(s)$$

As with all precipitations, remember that the metal ion must be provided in a *solution*; use the nitrate if you are not sure about the solubilities of any other compounds. In the example above, the precipitate could be made by mixing sodium carbonate solution and magnesium nitrate solution.

The only (minor) complication is that some of the precipitates will not be pure carbonates, for they precipitate as a mixture of the carbonate and the hydroxide; such solids are sometimes called *basic carbonates*.

Remember that the carbonates of iron(III) and aluminium do not exist. See Table 19.4 for a general summary of carbonates.

Table 19.4 The common metal carbonates

Formula	*Name*	*Solubility in water*	*Usual appearance*	*Other points*
Na_2CO_3	Sodium carbonate	Soluble	Colourless crystals or white powder	The crystals, $Na_2CO_3.10H_2O$, effloresce (p. 178) to a white powder, the monohydrate
$CaCO_3$	Calcium carbonate	Insoluble	White powder	Occurs in nature as chalk, limestone, marble
$MgCO_3$	Magnesium carbonate	Insoluble	White powder	—
$ZnCO_3$	Zinc carbonate	Insoluble	White powder	Commonly called calamine
—	Aluminium carbonate	—		Does not exist
$FeCO_3$	Iron(II) carbonate	Insoluble	Green powder	—
—	Iron(III) carbonate	—		Does not exist
$PbCO_3$	Lead(II) carbonate	Insoluble	White powder	—
$CuCO_3$	Copper(II) carbonate	Insoluble	Green powder	—

Experiment 19.2
Preparing insoluble metal carbonates*

Apparatus
Rack of test-tubes.
Solutions of soluble salts of sodium, calcium, magnesium, zinc, iron(II), lead(II), and copper(II). Sodium carbonate solution.
Remember that compounds of zinc, lead, and copper are poisonous.

Procedure
(a) Make out a table for your results as shown below.
(b) Pour about 1 cm^3 of each salt solution into separate test-tubes.
(c) Add about 2 cm^3 of sodium carbonate solution to each tube.

Results
Complete the table.

Solution	*Appearance*	*Appearance after addition of sodium carbonate solution*

Conclusion
1. Make sure that you can explain what happened in each of the test-tubes (including any cases where there was *no* reaction).
2. Write equations (ionic if possible) for each reaction.
3. Note that solutions of iron(III) and aluminium were not included in the experiment, because they do not form carbonates. Sodium carbonate *does* give a precipitate when added to solutions containing iron(III) or aluminium ions, but the precipitates are iron(III) and aluminium *hydroxides*, not carbonates. This happens because sodium carbonate solution always contains some hydroxide ions as well as carbonate and sodium ions.

Properties of metal carbonates

1. Carbonates of *some* metals decompose to the metal oxide and carbon dioxide when heated (Table 19.1). The effect of heat on metal carbonates depends upon the position of the metal in the activity series (p. 326).
2. Carbonates (and hydrogencarbonates) cause effervescence when added to dilute acids, forming a salt, water, and carbon dioxide. This reaction is used as a standard salt preparation (p. 83), as well as providing a test for a carbonate (or hydrogencarbonate) and a laboratory preparation of carbon dioxide (p. 370).
3. The only hydrogencarbonate of any importance is sodium hydrogencarbonate, $NaHCO_3$, and as its properties are very similar to those of sodium carbonate Na_2CO_3, it is useful to be able to distinguish between them:
(a) Sodium hydrogencarbonate (in solid form or in solution) will liberate carbon dioxide on heating, but sodium carbonate will not do so.

$$2NaHCO_3(s) \rightarrow Na_2CO_3(s) + CO_2(g) + H_2O(g)$$

(b) Sodium carbonate solution will give a white precipitate (of magnesium carbonate) when magnesium sulphate solution is added to it. Sodium hydrogencarbonate solution will not give this test.

$$Mg^{2+}(aq) + CO_3^{2-}(aq) \rightarrow MgCO_3(s)$$

Experiment 19.3
Distinguishing between some metal compounds*

The experiment consists of several parts. In each part, you will be given some 'unknown' substances and a little information, and then asked to identify the chemicals by USING ONLY THE INFORMATION YOU HAVE LEARNED SO FAR IN THIS CHAPTER. All the normal laboratory chemicals you may need will be available for you. For each chemical you identify, explain how you know what it is.

Procedure
(a) You have three white solids labelled A, B, and C. One of them is sodium hydrogencarbonate, one is sodium carbonate and the other is magnesium carbonate. Using *only* information you have learned in this chapter, work out tests to identify each of the solids A, B, and C.
(b) You have available four solutions; they are labelled D, E, F, and G. The solutions are (but not in the correct order!) sodium carbonate, magnesium chloride, zinc chloride, and iron(II) sulphate. Using dilute sodium hydroxide solution as the *only* other reagent, identify each solution.
(c) Two white solids are provided, labelled J and K. One of them is magnesium hydroxide and the other is magnesium carbonate. Find out which is which.

More to do
There may be more than one way of solving the problems in procedures (a), (b), and (c). When you have finished your practical work, write notes about other ways of solving each of the problems.

19.4 NITRATES OF METALS

Methods of preparation

All nitrates are soluble, and all nitrates are salts, so they are prepared by one of the methods used for making salts (Chapter 6), i.e. by the action of nitric acid on the appropriate metal or compound. The only thing to remember is that nitric acid is also a powerful oxidizing agent, and so if a compound of a metal with more than one valency is used, the solution formed will contain the nitrate of the metal in its higher valency. Thus the action of nitric acid on iron(II) hydroxide will produce some iron(III) nitrate. For similar reasons, dilute nitric acid will not attack aluminium metal, for the protective oxide layer on the metal is made even more effective by the oxidizing acid.

Properties of metal nitrates

Information about some of the common metal nitrates is given in Table 19.5. It is very important to remember that all nitrates are soluble in water, and this has been referred to on several occasions when solutions are mixed to form precipitates. For example, lead(II) nitrate is one of the few compounds which can provide lead(II) ions in solution since nearly all other lead(II) compounds are insoluble in water. The only important chemical properties of metal nitrates are their reactions when heated and the reaction used as a test for a nitrate.

The action of heat on metal nitrates is summarized on pages 326–7, and you may remember that you heated some nitrates in Experiment 15.3, p. 254.

Table 19.5 Some common metal nitrates

Formula	*Name*	*Solubility in water*	*Usual appearance*	*Other points*
$NaNO_3$	Sodium nitrate	All soluble	White crystals	When heated, decomposes to the nitrite and oxygen
$Ca(NO_3)_2$	Calcium nitrate		White crystals	
$Mg(NO_3)_2$	Magnesium nitrate		White crystals	
$Zn(NO_3)_2$	Zinc nitrate		White crystals	
$Al(NO_3)_3$	Aluminium nitrate		White crystals	
$Fe(NO_3)_2$	Iron(II) nitrate		Usually only encountered in solution	Not prepared by usual salt preparations, which produce iron(III) salts. See top of page
$Fe(NO_3)_3$	Iron(III) nitrate			
$Pb(NO_3)_2$	Lead(II) nitrate		White crystals	No water of crystallization
$Cu(NO_3)_2$	Copper(II) nitrate		Blue crystals	

The test for a nitrate

1. Place a small volume of a solution of the suspected nitrate in a boiling tube.
2. Add an equal volume of sodium hydroxide solution.
3. Warm the tube gently until the contents boil. If ammonia gas is given off (this gas turns damp red litmus paper blue), the solution contains an ammonium compound

(p. 63). It may or may not contain a nitrate (see step 4). If the ammonium test is positive, boil gently until ammonia is no longer given off.
4. Allow the solution to cool slightly then add about 0.5 g of Devarda's alloy. Warm gently to boiling and again test for ammonia gas. If the result is positive in this stage, the solution contains a nitrate. The Devarda's alloy has reduced the nitrate to ammonia.

Check your understanding

Write equations for the action of heat on (a) copper(II) nitrate, (b) sodium nitrate, and (c) lead(II) nitrate.

Experiment 19.4
Testing for nitrates in solution*

You will be supplied with five liquids labelled A, B, C, D, and E. Two of them are solutions of metal nitrates in water, one is a solution of an ammonium compound, and the other two are ordinary water. Use the Devarda's alloy test to decide which of the labelled chemicals are nitrates, and which is the ammonium compound.

Point for discussion

Name a compound which would give a positive test for ammonia in *both* steps 3 and 4 in the test for a nitrate (page 333).

19.5 SULPHATES OF METALS

Methods of preparation

1. Sulphates are salts of sulphuric acid and so the soluble ones can be prepared by the action of dilute sulphuric acid on a suitable metal or compound, as described in Chapter 6. Of the common sulphates, only barium sulphate and lead(II) sulphate are insoluble in water, and calcium sulphate is only slightly soluble in water. These three sulphates are prepared by precipitation.
2. The insoluble sulphates referred to above are prepared by adding a *solution* containing the appropriate metal ion to a *solution* containing sulphate ions (e.g. sodium sulphate solution). The resulting precipitate is filtered, washed, and dried, e.g.

$$SO_4^{2-}(aq) + Pb^{2+}(aq) \rightarrow PbSO_4(s)$$

3. Sulphuric acid is a *dibasic* acid (p. 79) and can be used to prepare hydrogen-sulphates as well as the normal sulphates. A typical method has been described on page 88.

Properties of metal sulphates

Information about the common sulphates is given in Table 19.6.

The test for a soluble sulphate

To a small volume of a solution of the suspected sulphate, add a little dilute hydro*chloric* acid followed by some barium *chloride* solution. A white precipitate (of barium sulphate) confirms the presence of sulphate ions,

$$Ba^{2+}(aq) + SO_4^{2-}(aq) \rightarrow BaSO_4(s)$$

Table 19.6 Some common metal sulphates

Formula	*Name*	*Solubility in water*	*Usual (often hydrated) appearance*	*Other points*
Na_2SO_4	Sodium sulphate	Soluble	White crystals	
$CaSO_4$	Calcium sulphate	Insoluble	White powder	Occurs naturally as gypsum and anhydrite
$MgSO_4$	Magnesium sulphate	Soluble	Colourless crystals	Commonly called Epsom salts
$ZnSO_4$	Zinc sulphate	Soluble	White crystals	
$FeSO_4$	Iron(II) sulphate	Soluble	Green crystals	Decomposes in an unusual way when heated, forming red iron(III) oxide (change of valency) and two different oxides of sulphur $2FeSO_4(s) \rightarrow Fe_2O_3(s) + SO_2(g) + SO_3(g)$
$PbSO_4$	Lead(II) sulphate	Insoluble	White powder	
$CuSO_4$	Copper(II) sulphate	Soluble	Blue crystals	When heated loses water of crystallization to form white powder. The reverse colour change sometimes used to detect the presence of (not necessarily pure) water
$BaSO_4$	Barium sulphate	Insoluble	White powder	Normally only seen as the precipitate in a positive sulphate test

Note: Many students confuse the chemicals used in the sulphate test with those used in the chloride test, particularly as the result, if positive, is a white precipitate in each case. To avoid this problem, remember that the reagents to be added go in 'matching pairs' (i.e. hydro*chloric* acid and barium *chloride*, or *nitric* acid and silver *nitrate*) and that you must never add extra chloride ions (i.e. barium chloride and hydrochloric acid) when doing a chloride test.

Points for discussion

A student has suggested that some salts can be made by the following reactions. In actual fact, there is something wrong with each of the suggestions. For each salt, explain why the suggested reaction would not work *and* suggest one which would. Give equations for the reactions you suggest.

1. To make aluminium sulphate, react aluminium carbonate with dilute sulphuric acid.

2. To make copper(II) sulphate, mix a solution containing $Cu^{2+}(aq)$ ions with a solution containing $SO_4^{2-}(aq)$ ions, and filter off the precipitate of copper(II) sulphate.

3. To make zinc sulphate, mix solutions of zinc chloride and dilute sulphuric acid, and crystallize the salt formed in the usual way.

4. To prepare calcium sulphate, mix solutions containing sodium sulphate and calcium carbonate.

19.6 CHLORIDES OF METALS

Methods of preparation

1. All metal chlorides are salts, and most of them are soluble in water. They are made by the typical salt preparation methods discussed in Chapter 6, using dilute hydrochloric acid on an appropriate metal or compound. Only two common chlorides are insoluble, those of lead and silver.

2. Lead and silver chlorides are made by precipitation. A *solution* containing the appropriate metal ion is added to a *solution* containing chloride ions (e.g. sodium chloride). The resulting precipitate is filtered, washed, and dried, e.g.

$$Pb^{2+}(aq) + 2Cl^{-}(aq) \rightarrow PbCl_2(s)$$

Properties of metal chlorides

Information about some common metal chlorides is given in Table 19.7.

The test for a soluble chloride

To a small volume of a solution of the suspected chloride, add a little dilute *nitric* acid followed by some silver *nitrate* solution. A white precipitate (of silver chloride) confirms the presence of chloride ions:

$$Ag^{+}(aq) + Cl^{-}(aq) \rightarrow AgCl(s)$$

Note: A dissolved bromide will give a near-white precipitate in this test. It can be distinguished from a chloride by the fact that a precipitate of silver chloride will dissolve easily when dilute ammonia solution is added, but a precipitate of silver bromide will not dissolve in dilute ammonia solution. An iodide will give a bright yellow precipitate with silver nitrate solution.

Experiment 19.5
The reaction between a solid chloride and concentrated sulphuric acid**

Apparatus

Rack of test-tubes, test-tube holder, Bunsen burner, spatula.

Sodium chloride, manganese(IV) oxide, concentrated sulphuric acid, litmus papers, ammonia solution, access to fume cupboard.

Procedure

(a) Make a table for your results as follows:

Mixture	*Observations*	*Gas evolved*

(b) Put a spatula measure of sodium chloride into a test-tube. CAREFULLY add about 0.5 cm³ concentrated sulphuric acid. Blow across the top of the tube and test the gas with moist litmus paper. Hold the stopper of the ammonia solution near the top of the tube.

(c) Put half a spatula measure of sodium chloride in a test-tube. Add half a spatula measure of manganese(IV) oxide and then 0.5 cm³ concentrated sulphuric acid (CARE). Warm the tube and contents in a fume cupboard, and test any evolved gas with moist blue litmus and moist red litmus papers. (See note (ii) below.)

Notes about the experiment

(i) Your observations should enable you to decide the name of the gas which was formed by the reaction in (b), and to write an equation for the reaction. (The *solid* product of the reaction is sodium hydrogensulphate). Note that *concentrated* sulphuric acid gives this reaction with other solid metal chlorides, as well as with sodium chloride. *Dilute* sulphuric acid does not give these reactions.

(ii) In explaining what happened in (c), note that the gas you detected in (b) is also *first* formed in (c). This gas then reacts with manganese(IV) oxide to form a different gas. Manganese(IV) oxide acts as an oxidizing reagent in this reaction, and you should be able to write a *simple* equation for its action. *Note*: Other oxidizing agents must *not* be used with concentrated sulphuric acid, unless you are given special instructions to do so.

(iii) The reaction in (b) is useful for preparing a common laboratory gas (page 467) and in testing for an *insoluble* chloride.

Conclusion
Explain what happened in both reactions, and explain how the reaction in (b) is useful in the laboratory.

Table 19.7 Some common metal chlorides

Formula	*Name*	*Solubility in water*	*Usual appearance (i.e. hydrated)*	*Other points*
$NaCl$	Sodium chloride	Soluble	White crystals	
$CaCl_2$	Calcium chloride	Soluble	White crystals	Anhydrous salt often used as a drying agent, but not for ammonia
$MgCl_2$	Magnesium chloride	Soluble	White crystals	Deliquescent
$ZnCl_2$	Zinc chloride	Soluble	White crystals	Deliquescent
$AlCl_3$	Aluminium chloride	Soluble	White crystals	Anhydrous salt must be made by dry methods
$FeCl_2$	Iron(II) chloride	Soluble	Colourless crystals	Anhydrous salt made as above
$FeCl_3$	Iron(III) chloride	Soluble	Brown solid	Anhydrous salt made as for aluminium
$PbCl_2$	Lead(II) chloride	Insoluble	White crystals	Soluble in hot water
$CuCl_2$	Copper(II) chloride	Soluble	Blue-green crystals	
$AgCl$	Silver chloride	Insoluble	White solid	Normally only seen as the precipitate in a positive chloride test

Experiment 19.6
Testing solutions for chloride and sulphate ions*

You will be provided with six labelled liquids, two of which are ordinary water. The others are sodium chloride solution, sodium sulphate solution, dilute hydrochloric acid, and dilute sulphuric acid. Do *simple* tests to find out what each of the labelled liquids is, and explain your results.

19.7 SODIUM AND POTASSIUM COMPOUNDS

The metals

1. They are very reactive (e.g. at the top of the activity series) because their atoms need to lose only one electron to attain a stable structure.
2. They are very soft and have low densities (they float on water).

3. They react with oxygen in the air, and the compounds formed then react with other compounds in the air. A typical series of reactions is:

$$Na(s) \xrightarrow{O_2(g)} Na_2O(s) \xrightarrow[\text{with water vapour}]{\text{deliquesces (p. 178), reacts}} NaOH(aq)$$

$$NaOH(aq) \xrightarrow{CO_2(g)} Na_2CO_3.10H_2O(s)$$

$$\text{surface coating of } Na_2CO_3.H_2O \xleftarrow{\text{effloresces (p. 178)}} Na_2CO_3.10H_2O(s)$$

To avoid these changes the metals are stored under a liquid hydrocarbon.
4. They react with cold water to form an alkaline solution and hydrogen gas, e.g.

$$2Na(s)+2H_2O(l) \rightarrow 2NaOH(aq)+H_2(g)$$

5. Other reactions of the metals (e.g. combustion in oxygen, reaction with chlorine) are considered under appropriate headings elsewhere in the book.

The compounds of sodium and potassium

1. All sodium and potassium compounds are soluble in water. Whenever you need to provide anions in solution (e.g. SO_4^{2-}), the corresponding sodium or potassium salts will *always* be satisfactory.
2. All sodium and potassium compounds are ionic.
3. All sodium and potassium compounds are white unless the compound also contains a transition element such as chromium or manganese. The only common coloured compounds of sodium and potassium are their orange dichromate(VI) compounds and the purple potassium manganate(VII).
4. Tests for sodium and potassium ions are given on page 451.

Some uses of sodium and its compounds

The metal: in street lamps, and as a coolant in some nuclear reactors.
Sodium hydroxide: manufacture of paper, artificial silk, and soap.
Sodium carbonate: glass and paper making.
Sodium hydrogencarbonate: baking powder.
Sodium chloride: food industry, glazing pottery, and the manufacture of sodium, chlorine, and sodium hydroxide.

19.8 CALCIUM, MAGNESIUM, AND ZINC COMPOUNDS

Calcium metal

1. This is a reactive metal (near the top of the activity series) as it easily loses two electrons to form a stable ion.
2. It is sometimes stored under a hydrocarbon oil, for it reacts with oxygen in the air and then with other substances, e.g.

$$Ca(s) \xrightarrow{O_2(g)} CaO(s) \xrightarrow{H_2O \text{ vapour}} Ca(OH)_2(s) \xrightarrow{CO_2(g)} CaCO_3(s)$$

3. It reacts fairly vigorously with *cold* water to form an alkaline solution of calcium hydroxide (lime water),

$$Ca(s)+2H_2O(l) \rightarrow Ca(OH)_2(aq)+H_2(g)$$

In contrast to sodium and potassium, the calcium sinks. In addition, the water soon turns milky because calcium hydroxide is not very soluble and forms a suspension.

Magnesium and zinc metals

It is likely that you have studied the properties of these fairly reactive metals in detail, e.g. on steam (p. 252), dilute acids (p. 251), oxygen (p. 250), and displacement reactions with salts of less reactive metals (p. 253).

The compounds of calcium, magnesium, and zinc

These metals always have a valency of two, so that if you know the formulae of the compounds of one of them, the formulae of the corresponding compounds of the other metals are easily determined.

1. All the common compounds of these metals are white and ionic.
2. The tests for these metals are given on page 451.
3. Note the rather unusual behaviour of calcium oxide when water is added to it (p. 327). The following experiment illustrates this reaction.

Experiment 19.7
(i) Making a sample of calcium oxide from calcium carbonate and (ii) reacting the product with water**

Apparatus
Tripod, gauze, Bunsen burner, protective mat, mouth blowpipe, piece of aluminium foil 6 cm × 6 cm, universal indicator paper, teat pipette, tongs.
Marble chips.

Procedure
(a) Place a piece of marble chip on a clean gauze supported on the tripod, and heat it strongly for about five minutes. The Bunsen burner should be held above the chip, and the blowpipe used to direct a flame on to the chip.
(b) Allow the chip to cool on the mat.
(c) Fold up the edges of the aluminium foil to form a shallow tray. Place the tray on the protective mat and, using tongs, place the cooled chip in the centre of the tray.
(d) Very slowly and carefully add de-ionized water, drop by drop, to the chip.
(e) Watch and listen for any signs of reaction as the water meets the solid. Gently touch the edge of the foil to see if its temperature has changed BUT DO NOT ALLOW THE CHEMICAL IN THE TRAY TO TOUCH THE SKIN.
(f) Touch the wet surface of the chip with a piece of universal indicator paper.

A marble chip?

Results
Record all your observations.

Conclusion
Explain the chemical reactions which took place during the experiment, giving an equation for each of them, and explain all of your observations. Why was it necessary to heat the calcium carbonate *strongly* in order to make it decompose?

Point for discussion

When you heated the marble chip with the blowpipe, you may have noticed that the portion of chip being heated gave off a bright white light (incandescence). This was the basis of an old form of stage lighting in which a container of calcium oxide (quicklime) was heated by means of an oxyhydrogen blowpipe. The intense white light produced was called 'limelight', which explains how the term 'being in the limelight' came to be used.

Some uses of magnesium, calcium, zinc, and their compounds

The metals: magnesium is present in a range of important low density alloys; zinc is used in alloys (e.g. brass, which also contains copper), for protecting iron from corrosion by galvanizing, and in dry cells.

Calcium carbonate: in the blast furnace; to prepare the oxide and hydroxide, and as marble in building.

Calcium hydroxide: in the manufacture of mortar and cement, as a cheap industrial base, for neutralizing acid soils, and for softening temporarily hard water.

Calcium sulphate: gypsum is used to prepare Plaster of Paris, $(CaSO_4)_2H_2O$, plaster and plasterboard.

Magnesium compounds: sometimes used in medicines, e.g. a suspension of the hydroxide is the active ingredient of 'milk of magnesia' (to neutralize acidity) and the sulphate is a laxative (Epsom salts).

Zinc compounds: are used in paints (e.g. 'rust curing' paint) and ointments, e.g. the carbonate in calamine lotion.

19.9 ALUMINIUM AND ITS COMPOUNDS

The metal

1. Aluminium does not normally show its reactive nature because it is protected by a thin but very strong film of oxide.
2. If a piece of aluminium is made the anode of a cell in which oxygen is being formed by electrolysis, the oxygen reacts with the aluminium anode to thicken the thin oxide film already present. This process is called *anodizing*. Figure 19.1 illustrates a typical experiment of this kind, and you may anodize some aluminium in the laboratory. The thicker layer of oxide produced by this process protects the metal even more than usual, and readily adsorbs a dye. Materials made of anodized aluminium (e.g. saucepans) therefore resist corrosion and can be attractively coloured.

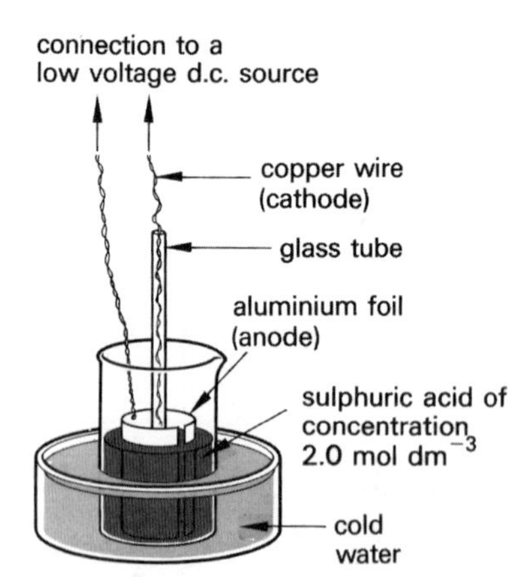

Figure 19.1 Anodizing aluminium foil

3. Although the oxide film normally stops the metal from being reactive, aluminium *is* reactive in certain conditions (see page 249). Its reactivity is also shown by the 'thermite' reaction which you may see demonstrated. (This experiment must only be attempted by a trained scientist under appropriate conditions.) In the reaction, aluminium metal reacts with iron(III) oxide (reduces it), the reaction being so exothermic (p. 477) that the iron formed is molten.

$$2Al(s) + Fe_2O_3(s) \rightarrow Al_2O_3(s) + 2Fe(l)$$

As aluminium 'wins' this battle for oxygen, it is clearly more reactive than iron, and the violence of the reaction suggests that it should be placed *well above* iron in the activity series.

Aluminium compounds

1. All the common aluminium compounds are white.
2. It is quite difficult for an aluminium atom to lose three electrons in order to form a stable ion, and so you cannot assume that its compounds are always ionic.
3. The test for aluminium compounds is given on page 451.

Some uses of aluminium

The metal is particularly useful because it has a low density, is non-poisonous, and resists corrosion. For one or more of these reasons it is used to make products such as saucepans and storage tanks, and also as a foil for packaging. The metal is also often alloyed and used in aircraft bodies and 'aluminium' framed greenhouses, and in other materials where a low density and resistance to corrosion are required. Note that many examples of so-called aluminium in daily life are in fact not the element but rather aluminium alloys.

19.10 IRON AND ITS COMPOUNDS

The metal

Iron is the second most common metal in the Earth's crust, and it occurs as various oxides (haematite, Fe_2O_3, and magnetite, Fe_3O_4) and also as sulphides, e.g. FeS_2, iron pyrites. The extraction of the metal in industry is described in Chapter 27. This process produces cast iron which can be converted into wrought iron and steel. The pure element is difficult to obtain, but the iron filings you have used to investigate many of the properties of the metal (e.g. its reaction with acids, p. 251, and with steam, p. 252) are reasonably pure.

The alloys of iron are very important. An alloy is usually made of two or more metals, although it may be made up of a metallic element and elements such as carbon, silicon, nitrogen, and phosphorus. Alloys are usually prepared by melting metals together and allowing the melt to cool. The atoms in a metal are closely packed and the metal can be deformed by causing one layer of atoms to slide over another. However, the introduction of some larger metal atoms into the structure makes it more difficult to dislocate the layers in this way. On the other hand the presence of very small atoms, such as carbon or nitrogen which can fit into the small spaces between the metal atoms, will also affect the physical properties of the metal (Figure 19.2). An alloy is usually less malleable and ductile, with a lower melting point and electrical conductivity than the pure metal.

Cast iron	*Wrought iron*	*Steel*
Contains about 4 per cent carbon with a small percentage of silicon, phosphorus, and sulphur. This alloy of iron expands when it changes from liquid to solid. It is brittle and is used where tensile strength is not important.	Much purer form obtained, for example, by heating cast iron with iron(III) oxide. This is a softer form and can be worked, e.g. made into chains etc. It is purer because it contains less sulphur, phosphorus, and silicon but it still contains some carbon.	Steels are alloys of iron containing between 0.15 and 1.5 per cent carbon. Their properties are determined by: (a) the percentage of carbon, (b) heat treatment, (c) adding other metals, such as chromium, manganese, etc.

Iron compounds

1. All the iron compounds you will study are coloured, and so it is important to learn the colours of the individual compounds. Remember that there are iron(II) and iron(III) compounds.

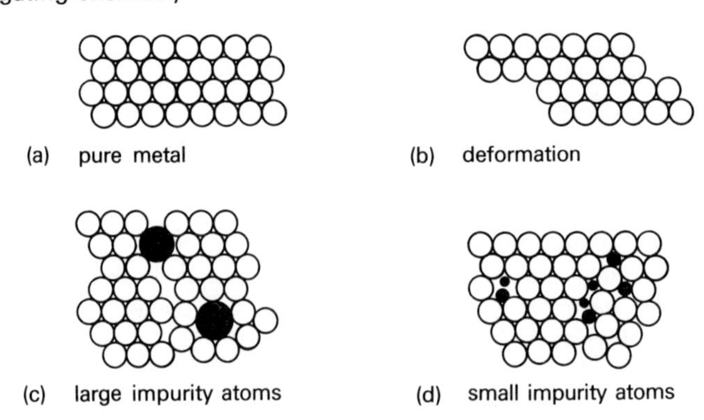

Figure 19.2 Illustration of the effect of the presence of foreign atoms on the properties of a pure metal. (a) Uniform arrangement of atoms in the pure metal, (b) deformation of the metal by dislocation, (c) the effect on the atomic arrangement of large impurity atoms, and (d) the effect of small impurity atoms

2. The oxidation of an iron(II) compound to an iron(III) compound
Note: Oxygen in the air slowly oxidizes iron(II) compounds to iron(III) compounds, so samples from opened bottles are rarely pure.

This reaction is an oxidation because the iron(II) ions lose electrons,

$$Fe^{2+}(aq) - e^{-} \rightarrow Fe^{3+}(aq)$$

Many oxidizing agents will bring about this change, e.g. hydrogen peroxide, concentrated nitric acid, acidified potassium manganate(VII), acidified potassium dichromate(VI), concentrated sulphuric acid, and chlorine. Wherever possible, try not to add any 'new' anions during the oxidation, e.g. chlorine would be used to oxidize iron(II) chloride to iron(III) chloride. Concentrated sulphuric acid would also do this, but would also provide sulphate ions so that if the final solution were crystallized, a mixture of iron(III) sulphate and iron(III) chloride could be produced.

3. The reduction of an iron(III) compound to an iron(II) compound
The best reducing agent to use is a combination of iron metal and a dilute acid containing the same anion as that in the compound. The mixture must be warmed. For example, to reduce iron(III) chloride to iron(II) chloride, add iron filings and warm dilute hydrochloric acid. To reduce iron(III) sulphate to iron(II) sulphate, add iron filings and warm dilute sulphuric acid. In each case, the metal reacts with the acid, liberating electrons

$$Fe(s) \rightarrow Fe^{2+}(aq) + 2e^{-}$$

Some of these electrons combine with hydrogen ions from the acid to liberate hydrogen gas,

$$2H^{+}(aq) + 2e^{-} \rightarrow H_2(g)$$

and others reduce the iron(III) ions already present,

$$2Fe^{3+}(aq) + 2e^{-} \rightarrow 2Fe^{2+}(aq)$$

4. The tests which distinguish between iron(II) and iron(II) compounds are given on page 451.

Experiment 19.8
The oxidation of iron(II) compounds**

Apparatus
Rack of test-tubes and a boiling tube, Bunsen burner, stirring rod.
Freshly prepared solution of iron(II) sulphate in boiled water, twenty-volume strength solution of hydrogen peroxide, sodium hydroxide solution, chlorine generator, freshly prepared iron(II) chloride solution, access to fume cupboard.

Procedure
(a) Pour about 1 cm^3 of iron(II) sulphate solution into a test-tube. Add a few drops of sodium hydroxide solution.
(b) Pour about 1 cm^3 of iron(II) sulphate solution into a boiling tube. Add about 0.5 cm^3 of hydrogen peroxide solution and about 1 cm^3 of dilute sulphuric acid. Boil the contents of the tube. Cool and add sodium hydroxide solution slowly, with stirring, until a precipitate forms. Compare the colour of this precipitate with that formed in (a). Are they the same?
(c) Pour about 1 cm^3 of iron(II) chloride solution into a test-tube and add about 1 cm^3 of sodium hydroxide solution.
(d) Pour about 2 cm^3 of iron(II) chloride solution into a test-tube, and pass chlorine gas through the solution (in the fume cupboard) for half a minute or so. Add sodium hydroxide solution to the contents of the test-tube until a precipitate forms. Compare the colour of this precipitate with that formed in (c). Are they the same?

Results
Describe all of your observations.

Notes about the experiment
Explain your results and make a conclusion by answering as many of the following questions as possible.
(i) Did the acidified hydrogen peroxide solution oxidize the iron(II) sulphate solution? How do you know?
(ii) What was the point of procedures (a) and (c)?
(iii) Did the chlorine oxidize the iron(II) chloride? How do you know?
(iv) Give suitable equations for all of the reactions in the experiment. (The work on page 294 may be helpful.)

Points for discussion
1. Can you suggest why the hydrogen peroxide was acidified with dilute *sulphuric* acid in the experiment?
2. Other oxidizing agents could also oxidize iron(II) compounds, e.g. concentrated nitric acid, potassium manganate(VII), etc. If any of these other oxidizing agents had been used in the last experiment they would have caused complications if the iron(III) salt formed in the reaction was crystallized. Why, for example, should concentrated nitric acid not be used in order to prepare the salt iron(III) sulphate from iron(II) sulphate?

Experiment 19.9
The reduction of iron(III) sulphate to iron(II) sulphate**

Apparatus
Rack of test-tubes, Bunsen burner, stirring rod, spatula.
Iron filings, dilute sulphuric acid solution, iron(III) sulphate solution, sodium hydroxide solution.

Procedure
(a) Pour about 1 cm^3 of iron(III) sulphate solution into a test-tube. Add sodium hydroxide solution slowly, with stirring, until a precipitate appears.
(b) Pour about 1 cm^3 of iron(III) sulphate solution into a test-tube. Add 1 cm^3 of dilute sulphuric acid and some iron filings. Warm the mixture. Allow the hydrogen produced to bubble through the solution until there is no further change in colour. Carefully add sodium hydroxide solution until a precipitate forms. Compare the colour of this precipitate with that formed in (a).

Results
Record all of your observations.

Conclusion
Make your own conclusion, explaining all of the changes which took place in the experiment, and giving appropriate equations.

Points for discussion
1. To obtain a pure sample of iron(II) sulphate from iron(III) sulphate why is it necessary to use (a) sulphuric acid rather than hydrochloric acid, and (b) iron instead of zinc?
2. What reagents would you use to reduce iron(III) chloride to iron(II) chloride?
3. Cast iron was traditionally used for a variety of purposes, many of which have been replaced by modern plastics. Can you name some of these uses? Why was cast iron preferred to other forms of iron or steel in these cases?
4. Wrought iron can be worked (wrought). What is it used for?

19.11 LEAD, COPPER, AND THEIR COMPOUNDS

Lead metal
Lead is a relatively unreactive metal. The freshly cut metal is shiny, but it rapidly tarnishes in air due to the formation of the basic carbonate, which then prevents further corrosion. Lead is very dense but also very soft.

Lead compounds
The important compounds are the lead(II) compounds and the various oxides. The oxides are a little unusual (see Table 19.2). The only common soluble lead(II) salt is lead(II) nitrate. The test for lead ions is given on page 451.

Copper metal
This is also relatively unreactive. The only chemical reactions of any importance are (i) with dilute nitric acid (this is the only convenient way of turning the metal into a solution of its ions) and (ii) with concentrated sulphuric acid (the laboratory preparation of sulphur dioxide, page 458).

Copper compounds
The metal forms copper(I) compounds and copper(II) compounds, but only the latter are normally considered at an elementary level.

All copper(II) compounds are coloured. They show no unusual behaviour apart from their reactions with ammonia (p. 329). The test for copper(II) ions is given on page 451.

Some uses of copper and its compounds
The metal is much used in plumbing, for it is resistant to corrosion and fairly easy to work with. It is also used to make electrical wiring as it resists corrosion and is an excellent conductor of electricity. Common alloys containing the metal include the bronzes (copper and tin), the brasses (copper and zinc), and the coinage alloys (copper, tin and zinc for pennies, etc. and copper and nickel for 'silver' coins). The attractive appearance of the pure metal is also utilized in decorative and ornamental work.

Copper(I) oxide is used in making red glass, and copper(II) sulphate is used in garden fungicides and in electroplating.

Note: It is useful in the cases of aluminium and copper, which have very many common uses, to ask yourself why they are chosen for a particular purpose rather than *any other* metal.

19.12 A SUMMARY OF CHAPTER 19

The important facts are summarized in the tables. Note also:

1. Whenever possible, link the reactivity of the metals and their compounds with the activity series.
2. When concentrated sulphuric acid is added to a solid metallic chloride, hydrogen chloride gas is formed, e.g.

$$NaCl(s) + H_2SO_4(l) \rightarrow HCl(g) + NaHSO_4(s)$$

(The mixture may need warming.) If manganese(IV) oxide is added as an oxidizing agent to such a reaction mixture, the hydrogen chloride first formed is then oxidized to chlorine,

$$2HCl(g) + \underset{\text{from oxidizing agent}}{[O]} \rightarrow H_2O(l) + Cl_2(g)$$

(see Experiment 19.5, p. 336).

QUESTIONS

Multiple choice questions

1. The oxide of a metal was found to react both with hydrochloric acid and with sodium hydroxide solution. Which one of the following is the best description of the oxide?
A acidic; B alkaline; C amphoteric;
D basic; E neutral (C.)

2. Which one of the following hydroxides does *not* give a good yield of a salt with dilute sulphuric acid?
A iron(II) hydroxide
B magnesium hydroxide
C zinc hydroxide
D calcium hydroxide
E copper(II) hydroxide. (C.)

3. Sodium metal is
A hard
B denser than water
C white in colour
D a metal melting below 100 °C
E a metal melting above 100 °C

4 Sodium metal is kept under
A water; B ethanol; C nitric acid;
D paraffin oil; E mercury

5. Iron is galvanized by coating it with
A copper; B tin; C zinc;
D aluminium; E lead

6. A white crystalline compound exposed to air changed into a white powder. The crystalline compound may have been
A potassium hydroxide
B calcium oxide
C sodium nitrate
D sodium carbonate
E potassium sulphate

7. Zinc oxide is
A unaffected by heat
B decomposed to the metal by heat
C turned permanently yellow by heat
D turned temporarily yellow by heat
E blackened by heat

8. When sodium hydroxide is added to a solution of zinc sulphate it produces
A no visible change
B a white precipitate insoluble in excess alkali solution
C a white precipitate soluble in excess alkali solution
D a greenish precipitate insoluble in excess alkali solution
E a green precipitate soluble in excess alkali solution

9. The protective film of oxide on the surface of aluminium metal may be strengthened by
A cathodizing
B anodizing
C galvanizing
D hydrolysing
E sherardizing

Structured questions and questions requiring longer answers

10. (a) Iron wire is placed in aqueous copper(II) sulphate and left for a few hours. (i) Describe what is seen. (ii) Give the ionic equation for this reaction. (iii) Explain, in terms of electron transfer, why this is classified as a redox reaction.

(b) Write down the equation for the reaction which takes place when (i) steam is passed over heated iron filings, (ii) chlorine is passed over heated iron filings.
(c) (i) Name a reducing agent which will convert iron(III) oxide to the metal.
(ii) Give the equation for this reaction.
(O. & C.)

11. (a) Describe how crude copper is purified electrolytically, indicating the reaction which takes place at each electrode. State what happens to the impurities during the purification process.
(b) Describe, with essential experimental details, how you would prepare from copper(II) oxide a sample of copper(II) sulphate crystals.
(c) State and explain what you would see happen, giving reasons, when the following are added separately to samples of copper(II) sulphate solution until no further change takes place: (i) aqueous ammonia solution, (ii) zinc powder. (A.E.B. 1979)

12. (a) Name *one* metal in each case (i) which reacts with cold water, (ii) which reacts with steam but not with cold water, (iii) the oxide of which can be reduced by hydrogen, (iv) which will not react with dilute hydrochloric acid, (v) the carbonate of which will not decompose on heating, (vi) the oxide of which decomposes on heating, leaving the metal. Give equations where appropriate.
(b) Give reasons for the following: (i) Zinc is often used as a protective coating for iron. (ii) Copper is frequently used for water pipes. (iii) Aluminium is much more resistant to atmospheric corrosion than iron. (iv) Sodium is usually kept in liquid paraffin.
(A.E.B. 1979)

13. (a) Describe the differences between metals and non-metals by referring to (i) their physical appearances, (ii) their electrical properties, (iii) the chemical properties of their oxides, (iv) the type of bond present in their chlorides. (You may find it useful to give your answers in the form of a table.)
(b) Place the metals calcium, copper, and iron in order of decreasing activity (most reactive first) and justify your order by comparing their reactions, if any, with (i) water or steam, (ii) dilute hydrochloric acid. Write equations for the reactions that you describe.
(C.)

14. (a) Describe *three* different *chemical* properties of metals or their compounds, using a different metal or its compound to illustrate *each* property.
(b) State what would be observed (if anything) in *each* of the following experiments: (i) small samples of sodium and potassium are added separately to cold water, (ii) samples of sodium carbonate and zinc carbonate are strongly heated, (iii) zinc and copper are added separately to dilute hydrochloric acid.

From the results of these experiments, place the elements *hydrogen*, *sodium*, *potassium*, *copper*, and *zinc* in their correct order of reactivity, placing the most reactive first and giving reasons for your order.
(c) State *one* important property of each of the following metals, that makes them important industrial materials. Give a different property in each case. (i) aluminium, (ii) iron, (iii) copper. (W.J.E.C.)

20 Rates of reaction

20.1 WHAT DO WE MEAN BY RATE OF REACTION?

Some chemical reactions proceed very slowly (e.g. the rusting of iron). Others are extremely fast (e.g. explosions). In industry, it is more economical to use a reaction which produces a product quickly, rather than to use one which takes several days, so scientists look for ways of making some important reactions proceed at a faster rate. On the other hand some reactions proceed at a faster rate than is desirable, e.g. certain decompositions. In such cases, scientists seek ways of slowing down the process. If we can understand how to change the rate of a chemical reaction, this knowledge could be very useful indeed.

In investigating the rate of a reaction, a *rough guide* can be obtained by noting the time taken for the reaction to proceed from start to finish. The reaction can then be repeated using the same quantities but under a different set of conditions. If it is found to take less time from start to finish than before, we can assume that the rate of reaction has been increased. This approach does not measure the true *rate of reaction*, however, because it only measures the *average* speed. All reactions proceed more quickly at some stages than at others, and a more scientific method measures the *actual* rate of reaction at a given moment in time.

Scientists usually measure rates of reaction by finding how the concentration of a product increases with time. Less often, they measure how the concentration of a reactant decreases with time. The units of rate of reaction, when measured like this are moles per dm^3 per second, i.e. $mol\ dm^{-3}\ s^{-1}$.

When we take the measurements which enable us to calculate the rate of reaction, it is important that we do not interfere with the reaction itself. For example, if we take a sample away from the reaction vessel, or if we add other chemicals to find out how much of a certain substance is present, we will probably change the course of the main reaction. If possible we should use physical methods such as colour changes, weight loss, or the volume of a gas liberated, as a measure of reaction rate. Measurements such as these can show how a reaction is proceeding without disturbing the course of the reaction.

Factors which affect reaction rates

Obvious methods for speeding up a chemical reaction are raising the temperature, increasing the concentration of the reacting chemicals, and using a catalyst. Other ways include dividing a solid reagent into smaller pieces (i.e. increasing its surface area), increasing the pressure (if possible), and using light as a source of energy. All these are investigated in this chapter. You will need to know not just *how* reaction rates can be changed, but also *why* they are changed by these effects.

20.2 HOW CONCENTRATION CHANGES AFFECT REACTION RATES

This section includes two different ways of investigating the effect of concentration changes on reaction rates. Two experiments are used because they are quite different in experimental technique. In the first experiment you will measure the *average* speed of a chemical reaction over a period of time; in the second you will take measurements which will allow you to determine the rate of the reaction at a particular instant in time. It is important that you should understand both experiments, because variations of them can be used for other investigations. You might also see a demonstration of a third technique, where average rates can be compared by measuring the time taken for a colour change to occur. A typical reaction of this type is shown in the so-called 'iodine clock' experiment.

It is important to remember in these experiments, *and in all other rate investigations*, that every variable except the one being investigated must be kept constant throughout the experiment. For example, in Experiment 20.2 each mini-experiment must use the *same volume* of liquid, the *same mass* of calcium carbonate, and must be conducted at the *same temperature*. Only the *concentration* of the acid solution is changed because we are trying to find how concentration changes affect the reaction rate.

The principle of Experiment 20.1

When solutions of sodium thiosulphate ($Na_2S_2O_3$) and dilute hydrochloric acid are mixed, a chemical reaction takes place which produces a precipitate of sulphur. This precipitate is not immediately visible as it is formed quite slowly. The solution becomes increasingly 'milky' as the precipitate becomes 'thicker'. The rate of reaction can be estimated by comparing how quickly the precipitate reaches a certain 'milkiness' under different conditions.

A convenient way of estimating when a certain amount of precipitate has been formed is to place the flask containing the reaction mixture on a piece of paper which has been marked with a cross. At first the cross can be seen when looked at from above, through the liquid, but as the liquid becomes milky it becomes more and more difficult to see the cross. If we measure the time it takes for the cross to 'disappear', we are measuring the time taken for the reaction mixture to produce a certain amount of sulphur. If we repeat this using different reaction mixtures, we can see how the altered conditions are affecting the reaction rate.

It is important that you understand that this technique does *not* measure the rate at any given moment in time. It only gives an indication of *average* rate (compare with Experiment 20.2).

Experiment 20.1
The reaction between sodium thiosulphate solution and dilute hydrochloric acid *

Apparatus
100 cm³ conical flask, 10 cm³ measuring cylinder, 100 cm³ measuring cylinder, stop-clock, white paper or tile.
Hydrochloric acid solution of concentration approximately 2 mol dm^{-3}, sodium thiosulphate solution of concentration approximately 40 g dm^{-3}.

Procedure
(a) Make out a table for your results as follows:

Volume of sodium thiosulphate solution (cm³)	*Volume of acid (cm³)*	*Volume of water (cm³)*	*Total volume (cm³)*	*Time taken to obscure cross (s)*

(b) Mark a pencil cross on the paper or tile so that when a flask stands on it the cross can be seen through the bottom of the flask.

(c) Measure out 30 cm^3 of the sodium thiosulphate solution and mix it with 20 cm^3 of water in the flask. Add these details to the *third* line of your table, as you will later be using two lower concentrations and two higher concentrations.

(d) Pour 5 cm^3 of the hydrochloric acid solution into the flask and, at the same time, start the stop-clock. Swirl the contents of the flask and then place the flask over the cross.

(e) Look down at the cross through the flask. As soon as the reaction has produced enough precipitate to hide the cross, stop the clock and note the time. Complete line 3 of your table. (If you are working in pairs or groups, the same person should observe the disappearance of the cross each time. Why?)

(f) Predict how a higher concentration and a lower concentration of sodium thiosulphate solution will affect the rate of the reaction. Then repeat procedures (b) to (e), but using in turn 10 cm^3, 20 cm^3, 40 cm^3, and 50 cm^3 of sodium thiosulphate solution, together with the appropriate volume of water needed to keep the *total* volume constant (50 cm^3) before adding the acid.

Results

Complete the table.

Notes about the experiment

Before chemical reactions can take place, collisions must occur between the particles of the reacting substances. The particles which collide must also have enough energy (e.g. kinetic energy) to break chemical bonds, so that the atoms and ions may recombine to form different chemicals. Any factor which increases the rate of a reaction must make the particles collide more frequently and/or must give the particles more kinetic energy. An increase in kinetic energy may cause the particles to 'hit each other' with enough energy to break bonds and to rearrange themselves, and it also increases the possibility of the molecules 'shaking themselves apart' before collision. Only *one* of these ideas is involved in this investigation.

Conclusion

Make a conclusion based upon the following two points.

(a) Do concentration changes affect the rate of a reaction taking place between solutions? If so, how (not why) do they do so?

(b) Explain *why* you think this happens.

Points for discussion

1. Do you think that the concentration of a *solid* can be increased? Explain your answer.

2. One way of increasing the concentration of a gas is to increase the pressure of the gas. Explain what this means. State what you think an increase of pressure would do to the rate of a reaction between gases, explaining your answer.

Increasing the concentration?

The principle of Experiment 20.2

When any carbonate is mixed with a dilute acid, a chemical reaction produces carbon dioxide gas, more water, and a salt,

e.g. $$CaCO_3(s) + 2HCl(aq) \rightarrow CaCl_2(aq) + H_2O(l) + CO_2(g)$$

According to the Law of Conservation of Mass, there should be no total change of mass during the reaction, but if the reaction takes place in an open container such as a conical flask, the carbon dioxide gas escapes and the total mass of the *flask and liquid* gradually decreases. If the mass of flask and liquid decreases *rapidly*, then the reaction is taking place rapidly, and if the rate of the loss in mass slows down this is because the chemical reaction is slowing down.

This is a different principle to that used in the previous experiment. This time the readings can be taken regularly so that a graph can be plotted. From the graph we can see what the rate of the reaction is at any time during the reaction. We could not do this in Experiment 20.1, which only compared average rates.

Experiment 20.2
The reaction between calcium carbonate and dilute hydrochloric acid***

Apparatus

Direct reading balance, 100 cm^3 measuring cylinder, graph paper, 100 cm^3 conical flask, stop-clock, cotton wool.
Marble chips, hydrochloric acid solution of concentration 2 mol dm^{-3}.

Procedure

(a) Make out a table for your results as shown below.
(b) Place 6.0 g of marble chips in the conical flask. Dilute 20 cm^3 of the hydrochloric acid solution with 20 cm^3 of water so as to produce an acid solution of concentration 1 mol dm^{-3}. Obtain a plug of cotton wool to fit loosely into the neck of the flask.
(c) Place the flask containing marble chips on the balance. Reset the balance to take into account an increase in mass of about 40 g, which will happen when you add the 40 cm^3 of acid solution.
(d) Remove the flask from the balance, quickly pour the 40 cm^3 of acid of concentration 1 mol dm^{-3} into the flask, and loosely insert the plug of cotton wool. Replace the flask on the balance and immediately note the weight of flask and contents. Start the stop-clock.
(e) Record the mass of flask and contents every 30 seconds for about fifteen minutes.
(f) Repeat procedures (b) to (e) but use 40 cm^3 of acid of concentration 2 mol dm^{-3} (instead of the acid of concentration 1 mol dm^{-3}) and make out a fresh table.

Results

Complete the tables of results. Plot graphs of mass of carbon dioxide formed (*y* axis) against time (*x* axis). Use the same set of axes for the two graphs, but plot the two lines in different colours. Make sure that each line is labelled in some way.

Notes about the experiment

(i) The plug of cotton wool is used to prevent any acid spray escaping from the apparatus. If this did happen, the weighings would be misleading as the masses recorded would be less than they should be.
(ii) If each reaction had proceeded at the same rate throughout, each graph would have been a straight line. Think carefully about the fact that the graphs are curves. Why are they steeper at some points than others? You will need to explain these facts in your conclusion.

Conclusion

Make your own conclusion, based on the following points.

1. Does the experiment show that increasing the concentration of the acid also increases the rate of the reaction? Explain your answer. (Note that to make a fair comparison you must compare the graphs at the same point in time, e.g. after thirty seconds.)

Time (s)	*Initial mass of flask and contents (M_1) (g)*	*Mass at given time of flask and contents (M_2) (g)*	*Mass of carbon dioxide formed = $M_1 - M_2$ (g)*

2. At which stage of each reaction was the rate the fastest? How did you decide your answer? Why do you think this is true of most reactions?

3. Why does each graph eventually become a horizontal straight line?

Check your understanding

1. Another variation of Experiment 20.2 is shown in Figure 20.1. In this experiment, magnesium metal reacts with dilute hydrochloric acid as follows:

$$Mg(s) + 2HCl(aq) \rightarrow MgCl_2(aq) + H_2(g)$$

The acid is in excess.

(a) What readings would you take to investigate how the rate of the reaction depends upon the concentration of the acid? (b) How would you vary the experiment to obtain the required results? (c) What could be the shape of the graph(s)?

2. Each of the following mixtures contains the same number of 'molecules' of both hydrochloric acid and sodium thiosulphate. Which combination would you expect to produce a precipitate of sulphur most quickly?

A 400 cm^3 of HCl (concentration 1 mol dm^{-3}) and 400 cm^3 of $Na_2S_2O_3$ (concentration 1 mol dm^{-3})

B 200 cm^3 of HCl (concentration 2 mol dm^{-3}) and 200 cm^3 of $Na_2S_2O_3$ (concentration 2 mol dm^{-3})

C 100 cm^3 of HCl (concentration 4 mol dm^{-3}) and 100 cm^3 of $Na_2S_2O_3$ (concentration 4 mol dm^{-3})

D 200 cm^3 of HCl (concentration 2 mol dm^{-3}) and 100 cm^3 of $Na_2S_2O_3$ (concentration 4 mol dm^{-3})

E 400 cm^3 of HCl (concentration 1 mol dm^{-3}) and 100 cm^3 of $Na_2S_2O_3$ (concentration 4 mol dm^{-3})

3. Which of the following statements is *not* true about the reaction between dilute hydrochloric acid and marble chips (calcium carbonate)?

A It is faster after three seconds than it is after ten seconds.

B It slows down with time.

C It eventually stops.

D It proceeds at a constant rate.

E It causes a loss of mass in the reaction vessel contents.

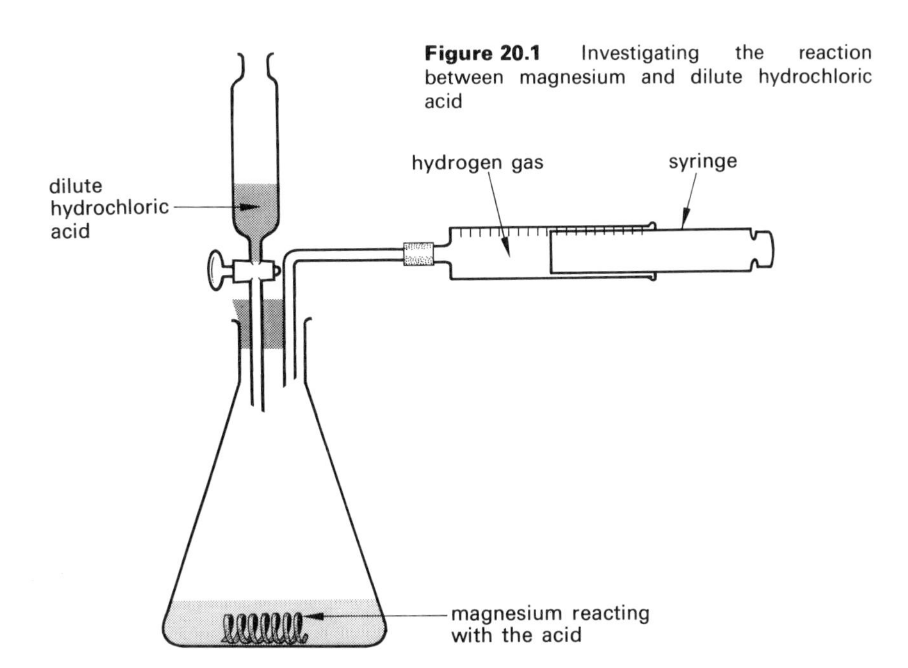

Figure 20.1 Investigating the reaction between magnesium and dilute hydrochloric acid

4. The element strontium lies immediately below calcium in Group 2 of the Periodic Table. The symbol for the element is Sr.
(a) How many electrons are there in the outer energy level of the strontium atom?
(b) Write the formula for strontium nitrate.
(c) Write the equation for the reaction between strontium and water.
(d) 0.1 g of strontium is reacted with cold water. How would you expect (i) the rate of reaction and (ii) the volume of gas liberated to compare when similar observations are made upon reacting 0.1 g of *calcium* with cold water? Give a reason for your answer in each case.

20.3 HOW CHANGES IN TEMPERATURE AFFECT REACTION RATES

To make sure that you really do understand the need to fix all the conditions except the factor being investigated in these experiments, you could plan your own experiment to find out how temperature affects reaction rates.

A suitable method would be the one used for the reaction between sodium thiosulphate solution and dilute hydrochloric acid, as in Experiment 20.1. Think about the experiment carefully before you begin. Remember the following points in planning your procedure. (a) It is possible to cool solutions, as well as to heat them. (b) There is no point in trying to start each experiment at an exact temperature. For example, there is no point in deciding to heat a sample of sodium thiosulphate solution to exactly 50 °C, because as soon as you add the acid, the temperature will no longer be 50 °C (unless you also take the trouble to warm the acid to exactly 50 °C). It is far simpler to heat a sample of sodium thiosulphate solution to *approximately* the temperature you need, then to add the acid, start timing, stir with a thermometer, and note the temperature of the mixture.

Writing up the experiment

When you write up the experiment, record all of your results in a table, and emphasize which factors you kept constant. Before you make a conclusion, you will need to know that although raising the temperature *does* increase the frequency of collisions between particles, this plays only a very small part in changing the rate of the reaction. The most important effect is that an increase in temperature increases the kinetic energy of the particles so that the collisions are 'more energetic'. This means that the collisions take place with more force, and therefore more bonds in the reactants will be broken.

20.4 HOW SURFACE AREA AFFECTS REACTION RATES

It is possible to increase the frequency with which particles collide in a solution by making the solution more concentrated. You can increase the frequency of collisions in a gas by increasing the pressure, which is like making the gas 'more concentrated'. It is not possible to increase the concentration of a solid because its particles are already tightly packed together. If a solid is divided up more finely, however, there will be more contact between the solid and any surrounding liquid, or gas, or even another finely divided solid. A finely divided solid has a greater *surface area* than the same amount of solid in 'lump' form. This affects the rate of a reaction in which the solid is involved. For example, when a cube of sugar is put into a cup of tea, the liquid only makes contact with the outer surfaces of the cube. All of the sugar *inside* the cube (i.e. most of the sugar) does not at first touch the liquid. If the same mass of sugar is used as a powder, far more of the sugar makes contact with the liquid. Which dissolves faster, the cube or the powder? Do you think that the same kind of thing will happen in *chemical* reactions? The next experiment will check your idea.

Experiment 20.3
To investigate how a change in surface area affects the rate of a reaction***

Apparatus and procedure
Refer to Experiment 20.2, but repeat the operations using first 6 g of *partly crushed* calcium carbonate, and then 6 g of *powdered* calcium carbonate. The acid solution should be of concentration 1 mol dm^{-3} in each case.

Results
Record your results in a table. The results should then be plotted in the same way as those for Experiment 20.2. You can either add two new plots to the graph used earlier (in different colours), or you can transfer the line for acid of concentration 1 mol dm^{-3} from your first graph on to a new sheet of graph paper, and then add two new lines from Experiment 20.3. Whichever of these methods you use, make sure that each line is clearly labelled.

Conclusion
Make your own conclusion, explaining your results.

Point for discussion
Your three lines (two from Experiment 20.3 and one from Experiment 20.2) should all become horizontal at the same point on the *y* axis. Why does this happen?

20.5 CATALYSTS

You probably already know something about catalysts, but the following experiments should help you to understand their properties more fully. Note that the details of Experiment 20.5 need not be learned; the experiment is used only to produce a full definition of what a catalyst does.

Experiment 20.4
The catalytic thermal decomposition of potassium chlorate(V)***

Necessary information
When potassium chlorate(V), $KClO_3$, is heated it decomposes and produces potassium chloride and oxygen:

$$2KClO_3(s) \rightarrow 2KCl(s) + 3O_2(g)$$

Apparatus
Tripod, Bunsen burner, stop-clock, glass rod, spatula, three boiling tubes (hard glass), splints, clamp stand(s), access to balance, filtration apparatus, oven, safety screen.
Potassium chlorate(V), copper(II) oxide.

Procedure
NOTE THAT THIS REACTION MUST BE DEMONSTRATED, AND A SAFETY SCREEN MUST BE USED.
(a) Add five spatula measures of potassium chlorate(V) to one of the tubes.
(b) Weigh accurately about one spatula measure of copper(II) oxide and record the mass. Mix the copper(II) oxide with five spatula measures of potassium chlorate(V) and place this mixture in the second tube.
(c) Place five spatula measures of copper(II) oxide in the third tube.
(d) Clamp the tube containing potassium chlorate(V) horizontally, place the Bunsen below the tube and immediately start the clock. Continuously test for oxygen with a glowing splint. As soon as oxygen is given off, stop the clock and note the time. [*Note*: Take care not to allow any carbon from the end of a splint to fall into the tubes.]
(e) Repeat (d) with the other two tubes in turn. Oxygen may not be evolved from both of them.
(f) After cooling add water to the tube which originally contained the mixture. Potassium chlorate(V) and potassium chloride are both soluble in water but copper(II) oxide is not. Stir and if necessary add more water until all the white solid has dissolved. Warm if necessary.
(g) Filter the suspension from (f) to obtain the insoluble copper(II) oxide. Wash the

residue and paper thoroughly with de-ionized water. Dry the paper and residue in an oven, along with another piece of wet filter paper, until the mass is constant. Note the mass of the 'blank' piece of filter paper and deduct it from the mass of 'paper +copper(II) oxide' to obtain the mass of the copper(II) oxide.

Results

Set out the times in a table, and record your weighings. Compare the mass of copper(II) oxide at the end of the experiment with that at the beginning.

Conclusion

Write your own conclusion. Include an equation for the thermal decomposition, and answer the following points.

1. Does the addition of copper(II) oxide alter the rate of the reaction?
2. Is any of the copper(II) oxide used up as a result of the reaction?

Note: You cannot quite describe how a catalyst works at this stage. You do not know whether the catalyst actually *takes part* in the reaction, or whether it somehow affects the rate of a reaction just by 'being there'.

Experiment 20.5
The reaction between sodium potassium tartrate (Rochelle salt) and hydrogen peroxide solution**

Necessary information

When hydrogen peroxide solution and sodium potassium tartrate solution are heated together, a reaction takes place in which carbon dioxide gas is evolved. The details of the reaction (e.g. an equation) are not relevant to this discussion.

The rates of reaction with and without catalyst can be compared by noting the rate at which carbon dioxide bubbles are formed in each reaction mixture, and also the temperature to which the mixture must be raised to produce this effervescence.

Apparatus

Spatula, boiling tube, measuring cylinder, glass rod, test-tube holder, Bunsen burner, thermometer (−10 °C to 110 °C).
'Twenty volume' hydrogen peroxide solution, sodium potassium tartrate, cobalt(II) chloride.

Procedure

(a) Pour 10 cm^3 of the hydrogen peroxide solution into the boiling tube and add two spatula measures of sodium potassium tartrate. Stir until dissolved.
(b) Heat the contents of the tube until vigorous effervescence occurs. It may be necessary to heat the liquid almost to boiling. Note the temperature at which the reaction occurs. (Do not confuse 'boiling' with the effervescence caused by the chemical reaction.)
(c) Rinse out the tube, add a fresh mixture as in (a), and add a few crystals of cobalt(II) chloride. Stir to dissolve. The solution should be pale pink.
(d) Warm the tube *gently* and *slowly*, stirring the solution with the thermometer. As soon as vigorous effervescence occurs, remove the tube from the burner and hold it over a sink. Take the temperature of the solution immediately, MAKING SURE THAT YOU POINT THE TUBE AWAY FROM YOUR FACE. Note any changes which you see in the tube over the next few minutes.

Results

Record the temperatures at which reaction took place in each case, the comparative rate of effervescence in the two mixtures, and any other changes you observed, particularly colour changes.

Conclusion

This experiment provides some extra information (i.e. in addition to that from Experiment 20.4). Make your own conclusion after thinking about the following questions. End your conclusion by choosing a definition of a catalyst from question 5 below.

1. Does cobalt(II) chloride catalyse this reaction?
2. Is the cobalt(II) chloride used up as a result of the reaction?
3. Does it *take part* in the reaction?
4. Does it reappear after taking part in the reaction?
5. A catalyst is a substance which

A changes the speed of a chemical reaction. It is unchanged chemically and in

mass at the end of the reaction, even though it may have taken part in the reaction.
B changes the speed of a chemical reaction. It does not take part in the reaction, and is unchanged at the end of the reaction.

More information about catalysts

Catalysts can be divided into inhibitors (sometimes called *negative catalysts*) and *positive catalysts*. A positive catalyst changes the rate of a chemical reaction by speeding it up. An inhibitor slows a reaction down. In both cases, the catalyst does actually take part in the reaction but it is then reformed so that at the end of the reaction it appears unchanged.

The decomposition of hydrogen peroxide solution provides an example of the use of both an inhibitor and a positive catalyst. The decomposition of hydrogen peroxide solution (into oxygen and water) takes place slowly even in the absence of a catalyst, but an inhibitor (e.g. glycerine) is often added to the solution by the manufacturer to keep it stable. On the other hand, if we wish to use the solution to prepare oxygen (p. 149), we accelerate the natural rate of decomposition by adding a positive catalyst, such as manganese(IV) oxide, MnO_2.

A more familiar example is the use of tetraethyl lead(IV), $Pb(C_2H_5)_4$ in petrol. Tetraethyl lead(IV) is an inhibitor which helps to stop pre-ignition ('knocking'). This occurs when the petrol/air mixture in the cylinders ignites before it should do so. (A pollution effect due to the use of this inhibitor is referred to on page 146.) The use of rust inhibitors in anti-freeze solutions, of chemicals to retard atmospheric oxidation in plastics, and of antidotes in certain cases of poisoning are further familiar examples of inhibitors. You should be able to quote several examples of the use of positive catalysts, e.g. in some of the industrial processes mentioned in Chapter 27.

Enzymes

Enzymes are a special subgroup of catalysts. They are special proteins produced by living cells, and are often called 'biological catalysts'. All living cells are like complicated chemical factories, with hundreds of reactions taking place to provide the substances and the energy needed for life to continue. These reactions are linked together so that they proceed at just the right rates and involve just the right quantities to sustain the life of a cell, no matter what is happening to it. This remarkable control of many interconnected reactions is due largely to the presence of enzymes in the cells.

Digestion provides some simple examples of the working of enzymes. The protein we eat is made up of very large molecules, each of which contains many much smaller molecules joined together (i.e. it is a natural macromolecule, see page 427). When we digest this protein, enzymes catalyse the breakdown of the large molecules into the much smaller units within them, so that the small units can then enter the blood stream and be used by our bodies. If we try to break down proteins in a chemical laboratory, we need to boil them up with hydrochloric acid for several hours. Fortunately, enzymes in our bodies do the same job more efficiently, more quickly, and without the need to subject our cells to such drastic treatment!

Enzymes have all the properties of the catalysts used in the laboratory, but there are two important differences. Many ordinary catalysts such as platinum can catalyse a wide variety of different reactions, but enzymes are *highly specific*, i.e. a certain enzyme will catalyse only one particular reaction. Secondly, the structure of an enzyme changes when it is heated or when it is in the presence of acids or alkalis. It can then no longer function as a catalyst. Ordinary catalysts are not 'deactivated' by such treatment.

Points for discussion

1. What are 'enzyme detergents', and what advantages are they claimed to have over ordinary detergents?

2. You might like to find out what the following terms mean, and find some examples of their use in industry: promoter, catalytic poison.

20.6 LIGHT AS A FORM OF ENERGY

Light is a form of energy; it can speed up some chemical reactions and alter the course of others. For example, methane and chlorine react together only slowly in the dark, but in diffused daylight the reaction proceeds more rapidly (p. 407). In bright sunlight, however, a *different* reaction occurs; a mixture of methane and chlorine then explodes, producing carbon and hydrogen chloride instead of the substitution products obtained in diffused daylight (p. 407). Those of you studying biology will be familiar with experiments which show that plants are unable to make starch when kept in the dark, and on page 241 information is given about the way in which light affects the reactions between hydrogen and the halogens.

Light is the least important of the factors which can affect reaction rates. This is because the other factors apply in nearly all cases, but light only noticeably affects a few reactions. Nevertheless, one of these, photosynthesis (p. 139), is probably the most important reaction in the world.

Another example of the way in which chemicals can be affected by light is the 'reaction' used in many aspects of photography. You may do an experiment in which you make precipitates of silver chloride, silver bromide, and silver iodide (the silver halides), and leave some samples in the dark and others in the light. Those left in the light show colour changes due to the formation of metallic silver. Photographic films, plates, and papers are coated with silver halides mixed with gelatine. Areas of the material which are exposed to light produce metallic silver. Development increases the formation of silver in these areas, which therefore appear as dark areas on a negative. 'Fixing' dissolves away the unused silver halides, so that the negative can then be exposed to light without being affected any further. In these examples, light is not affecting the *rate* of the reaction, but the examples do serve to remind us that light can influence a chemical reaction in several ways.

More to do

1. Make a list of common laboratory chemicals which are normally stored in brown bottles, and suggest why they are stored in this way.

2. Explain the process by which a negative is used to make a normal photographic print.

20.7 REVERSIBLE REACTIONS AND EQUILIBRIA

Reversible reactions

Reversible reactions are those which can proceed both from left to right (as shown by an equation) and also from right to left. They are indicated by using the sign $\rightleftharpoons$ (rather than $\rightarrow$) in the equation. You may have come across reactions of this type, some of which are listed below. Remember that *everything* is reversible in such reactions, including energy changes; if heat is given out (ΔH – ve, p. 477) when the reaction goes from left to right, then the reverse reaction will take in heat (ΔH + ve). By convention, the ΔH symbol given with any equation for a reversible reaction refers to the forward reaction only, i.e. the reaction from left to right as shown in the equation. The back reaction will have a ΔH of the same value but opposite in sign:

$$5H_2O(l) + CuSO_4(s) \rightarrow CuSO_4.5H_2O(s), \qquad \Delta H \ -ve$$

$$CuSO_4.5H_2O(s) \rightarrow CuSO_4(s) + 5H_2O(l), \qquad \Delta H \ +ve$$

These can be summarized as:

$$5H_2O(l) + CuSO_4(s) \underset{\text{endothermic}}{\overset{\text{exothermic}}{\rightleftharpoons}} CuSO_4.5H_2O(s)$$

Similarly,

$$CaCO_3(s) + H_2O(l) + CO_2(g) \underset{\text{softening temporary hard water, formation of kettle fur, etc.}}{\overset{\text{formation of hard water}}{\rightleftharpoons}} Ca(HCO_3)_2(aq)$$

Experiment 20.6
A simple illustration of a reaction which 'goes both ways'*

Apparatus
Glass rod, 100 cm³ beaker, white tile or paper, two teat pipettes, small measuring cylinder.
Dilute bromine water, dilute sulphuric acid, dilute sodium hydroxide solution.

Procedure
(a) Pour approximately 10 cm³ of the bromine water into the beaker and stand it on the tile or paper.
(b) Using one of the teat pipettes add dilute sodium hydroxide solution slowly, with stirring, until a colour change takes place.
(c) Using the other teat pipette, add dilute sulphuric acid, with stirring, to the solution obtained in (b) until a colour change takes place.
(d) Repeat (b) and (c) alternatively several times, using the same 'bromine' solution.

Can both reactions in a reversible reaction take place at the same time?

The previous experiment may suggest to you that a reversible reaction goes completely in one direction or completely in the other, and that it is not possible to have both reactions going on at the same time. If both reactions *can* go on at the same time, such systems will always contain a mixture of both 'reactants' (substances appearing on the left-hand side of the equation) and 'products' (substances appearing on the right-hand side of the equation). This will happen because, although some substances are being used up by the forward reaction, they are also being continually re-formed by the reverse reaction. The next experiment investigates whether this actually happens.

Experiment 20.7
A simple experiment with litmus solution*

Apparatus
Three Petri dishes, white paper or tile, glass rod, small measuring cylinder, teat pipette.
Neutral (purple) litmus, dilute sulphuric acid, dilute sodium hydroxide solution, deionized water.

Procedure
(a) Pour 10 cm³ of dilute sulphuric acid into Petri dish A and add approximately 1 cm³ of neutral litmus solution. Stir.
(b) Wash the glass rod in deionized water. Pour 10 cm³ of dilute sodium hydroxide

solution into Petri dish B. Add approximately 1 cm^3 of neutral litmus solution and stir.
(c) Repeat (b) using Petri dish C but use 10 cm^3 of deionized water instead of the alkali.
(d) Stand Petri dish A on the paper or tile and place dish B on top. View the two from above. Compare the 'combined' colour of A+B with that in dish C.

Notes about the experiment

(i) $H^+(aq)$ ions (which cause acidity) and $OH^-(aq)$ ions (which cause alkalinity) are still present in a neutral solution, but they are present in equal concentrations. A solution is described as being acidic when it contains a higher concentration of $H^+(aq)$ ions than $OH^-(aq)$ ions, NOT when it contains *only* $H^+(aq)$ ions.
(ii) In the presence of excess $H^+(aq)$ ions, the red variety of litmus is formed. In the presence of excess $OH^-(aq)$ ions, the blue variety of litmus is formed. This colour change is reversible because the blue colour in alkaline solutions can be made to go red by adding acid, and back to blue again by adding alkali.

$$\text{Acid form of litmus (red)} \underset{+\text{acid}}{\overset{+\text{alkali}}{\rightleftharpoons}} \text{Alkaline form of litmus (blue)}$$

Note that there are only two forms of litmus, red and blue. Purple litmus is not a third variety. It is the colour produced when the red and blue forms are present together and in equal concentrations, as procedure (d) should indicate.

Conclusion
Explain why litmus appears to be purple in deionized water, and how this helps to explain whether it is possible to have a mixture of reactants and products in a reversible reaction.

Equilibrium

Imagine a reversible reaction $A+B \rightleftharpoons C+D$, starting with only A and B, and allowed to continue without inteference. When A and B are mixed, the forward reaction is at first rapid, because the concentrations of A and B are at their highest. At the same time, the back reaction is very slow because there are virtually no molecules of C and D. As time goes on the rate of the forward reaction decreases (as A and B are used up) and the rate of the back reaction increases (as more of C and D are formed) until the point is reached when the rate of both reactions is the same. This situation is then described as an equilibrium (balance point), and there will be no further change in the concentrations of the various chemicals present unless the conditions are changed, e.g. unless more reagents are added, or something is removed, or the pressure is changed, etc. When a system reaches equilibrium, it therefore contains a mixture of both reactants and products. These ideas can be emphasized by following a simple analogy, as used in the next experiment.

Experiment 20.8
Using a simple analogy to illustrate equilibrium*

Apparatus
2 × 50 cm^3 measuring cylinders, two pieces of glass tubing a few centimetres longer than the height of the measuring cylinders. (There should be several types of bore so that different groups can investigate different 'reactions'. Each group should have two dissimilar pieces of tubing.)

Procedure
(a) Pour 50 cm^3 of water into one of the measuring cylinders.
(b) Take one of the pieces of tubing (A_1) and place it in measuring cylinder A so that it touches the bottom. Place a thumb or finger tightly over the other end of the tube A_1, lift the tube out of the water and hold it over the other measuring cylinder B. Release your thumb or finger so that the water 'trapped' in the tube falls into B (Figure 20.2).

This procedure represents a chemical reaction; some of the contents of the first

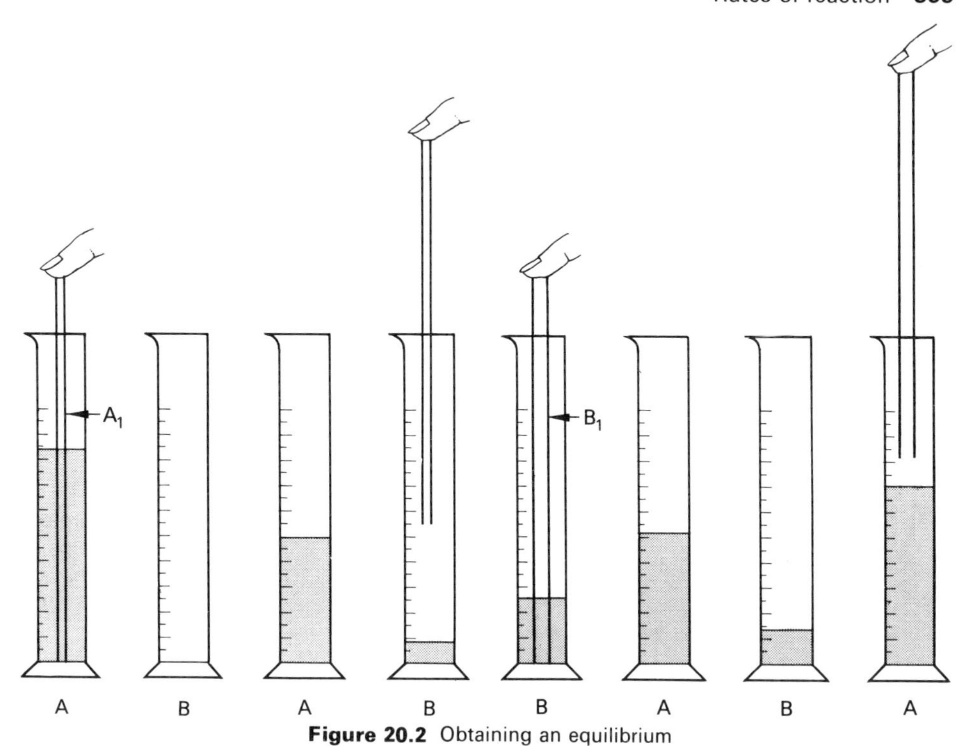

Figure 20.2 Obtaining an equilibrium

tube have 'reacted' (been used up) and become products in the second tube. The volume of water transferred represents the rate of the 'reaction', A → B.

(c) Using the second piece of tubing, B_1, transfer some of the water now in measuring cylinder B back into A by the same procedure as before.

This represents the 'reverse reaction', B → A. The volume of water transferred is proportional to the rate of the 'reaction' B → A. Note the volumes of water in each of the measuring cylinders.

(d) Repeat procedures (b) and (c) alternately and always in pairs (i.e. one forward reaction and one backward reaction = a pair) taking volume readings after each pair until no further change takes place in the volumes of the liquids in the tubes. This is when 'equilibrium' has been established; just as much water is being transferred from A to B as is being transferred from B to A. The rates of the two reactions are equal.

(e) Mark the level of the liquid in the two cylinders. Compare your volumes with those of other groups who have established equilibrium.

(f) Plot 'volume in measuring cylinder' against 'number of transfers' for each tube, on the same axes.

Notes about the experiment

This experiment is used only as an illustration, so there is no need to write it up in the usual way. Make sure that you can use your observations and the graphs to answer the following questions.

1. Why did the reaction from A to B slow down during the experiment, and the rate of reaction from B to A increase?
2. Does an equilibrium position always involve equal quantities of reactants and products? Do you think that this is ever likely to happen?
3. What is equal at equilibrium, every time?

Points for discussion

1. Do you think that the same equilibrium position would have been produced in Experiment 20.8 if you had started with the same volume of water but in test-tube B? (This corresponds to the same 'reaction' but written the other way round.) Check your answer by doing the 'reverse' experiment, if in doubt.

2. What do you think would happen if the bore of both tubes was the same? What would your graph look like?

Other points about equilibria

Chemical equilibria are *dynamic*, i.e. both reactions still continue (at equal rates) at equilibrium although they do not appear to be continuing as there is no further change in the concentrations. This can be shown by the fact that a saturated solution of lead(II) chloride becomes radioactive when placed in contact with a sample of radioactive (solid) lead(II) chloride. The saturated solution cannot dissolve any more lead(II) chloride, but it 'swaps' dissolved lead ions in the solution for radioactive lead ions in the solid, thus showing that the equilibrium $Pb^{2+}(aq) \rightleftharpoons Pb^{2+}(s)$ is dynamic. An equilibrium would be *static* if both reactions stopped at equilibrium.

Experiment 20.8 illustrates this last point. When equilibrium was reached, you could have continued the 'reaction' but it would have made no difference to the volume of water in each cylinder, because both forward and backward reactions would have reached the same rate; i.e. a forward reaction transfers the same volume of water as a backward reaction. If you had continued the reactions after they had reached equilibrium, you would have had a dynamic equilibrium. Your graphs also show this point.

Students frequently imagine that equilibrium is reached when equal concentrations of reactants and products are obtained. This is very rarely the case, and equilibrium positions vary considerably, e.g. it is possible to have a situation consisting of 99 per cent products and 1 per cent reactants, but another equilibrium may contain 1 per cent products and 99 per cent reactants. Again, Experiment 20.8 illustrates this point, because groups will have produced different equilibria, according to the bores of the tubes used in the experiment.

How equilibrium positions can be changed

If a chemical is being manufactured by a reversible reaction, an equilibrium mixture may contain only a small proportion of the desired product. It would be very useful if the equilibrium position could be changed so as to produce a new equilibrium position richer in the desired chemical. Remember that the only way in which this can be achieved is by making a change in the *conditions* of the process.

Once equilibrium has been established, a new equilibrium can only be achieved by altering the conditions

Earlier in this chapter, five factors were considered, each of which can influence the rate of a reaction. These factors still apply to reversible reactions, and they apply to *both* reactions in a reversible reaction. However, a change in one of these factors can influence one of the two reactions in a reversible reaction more than the other, and this is how an equilibrium position can be changed. The changes (explained more fully in the examples which follow) can be predicted by using the *Principle of Le Chatelier*: 'If a change is made to a system in equilibrium, the system alters so as to oppose the change, and a new equilibrium is formed.' This can be rephrased in simple terms as 'whatever is done to a system in equilibrium, the system does the opposite'.

How concentration changes affect a system in equilibrium

Consider the equilibrium $A + B \rightleftharpoons C + D$. If more A (or B) is added at equilibrium, the rate of the forward reaction will be increased (because there will be more collisions between A and B) but the back reaction will not be affected as it does not involve collisions between molecules of A and B. A new equilibrium position will be reached, richer in products. Similarly, if C (or D) is removed, the rate of the back reaction decreases and more products are made. Each of these can be predicted by the Principle of Le Chatelier, for whatever change is made (e.g. adding A) the system does the opposite (e.g. removes A).

If reactants are added or products are removed from a system at equilibrium, a new equilibrium is produced which contains a higher proportion of the products.

You may have done an experiment to illustrate this idea, e.g. the reaction between bismuth(III) chloride and water:

$$BiCl_3(aq) + H_2O(l) \rightleftharpoons BiOCl(s) + 2HCl(aq)$$

Alternative additions of water and concentrated hydrochloric acid force the equilibrium first one way and then the other, and the changes can be observed because of the appearance and disappearance of the precipitate.

Reversible changes occur naturally. The reaction

$$CaCO_3(s) + H_2O(l) + CO_2(g) \rightleftharpoons Ca(HCO_3)_2(aq)$$

does not reach equilibrium in a cave system where the solution is constantly removed in the streams which leave the cave; the reaction in such circumstances is constantly driven from left to right and the calcium carbonate is continually 'dissolved'. On the other hand, if temporary hard water (e.g. $Ca(HCO_3)_2$ solution) is boiled in a kettle or allowed to evaporate, the same reaction is forced the other way because the carbon dioxide and steam are constantly escaping. This produces kettle fur in kettles or stalactites and stalagmites in caves (p. 188).

How temperature changes affect a system in equilibrium

An increase in temperature in a reversible reaction will increase the rates of both forward and backward reactions, and so equilibrium is reached more quickly. There is also an additional effect, which changes the equilibrium position, because one of the reactions is speeded up more than the other.

When the temperature of a system in equilibrium is raised, a new equilibrium is formed which contains a higher proportion of the material(s) made by the endothermic reaction (ΔH +ve). A similar change in favour of the exothermic process is produced by a lowering of the temperature.

The above statements can be predicted by the Principle of Le Chatelier. If the temperature is *increased* the system tries to *decrease* it. This is achieved by increasing the rate of the endothermic process, which 'absorbs' some of the extra heat energy.

How catalysts affect a system in equilibrium

If a positive catalyst is used in a reversible reaction, the rates of *both* forward and backward reactions are increased *equally*.

A catalyst does not *alter the equilibrium position (i.e. the yield) but it ensures that equilibrium is reached more quickly than without it; this is an important economic consideration.*

How pressure changes affect a system in equilibrium

Pressure changes can only affect reactions between *gases*, for although gases can be compressed, solids and liquids cannot. An increase in pressure has the same effect as increasing the concentration of a gas, so that both forward and backward reactions involving gases are speeded up and equilibrium is reached more quickly. There can also be a second effect, which is usually more important, when one of the reactions is speeded up more than the other and a new equilibrium is formed.

If the equation for the reaction shows a different number of gas *molecules on the left-hand side than there are* gas *molecules on the right-hand side, then a change in pressure will produce a different equilibrium position.*

If the pressure is increased, the system tries to decrease the pressure (Principle of Le Chatelier) by increasing the rate of the reaction which produces fewer molecules, e.g.

$$2A(g) + B(g) \rightleftharpoons 2C(g)$$

(3 molecules of gas) (2 molecules of gas)

Reaction → produces fewer molecules, so increasing the pressure increases the rate of → more than that of ←, and a new equilibrium is formed with a higher proportion of C.

A summary of the factors which affect equilibria

Table 20.1 The factors which affect chemical equilibria

Factor	*Type of reversible reaction*	*Attainment of equilibrium*	*Effect on equilibrium position*
Increase in the concentration of a substance X	Any	Faster	New equilibrium position containing a lower concentration of X
Increase in temperature	Most	Both reactions speeded up so equilibrium attained more quickly	New equilibrium with a higher proportion of the substance(s) made by the endothermic reaction
Decrease in temperature	Most	Opposite to above, therefore slower	New equilibrium with a higher proportion of the substance(s) made by the exothermic reaction
Increase in pressure	Gas reactions where the equation shows an equal number of molecules of reactants and products	Faster (concentration effect)	No change
Decrease in pressure	Gas reactions as above	Slower (concentration effect)	No change
Increase in pressure	Gas reactions where the equation shows an unequal number of reactant and product molecules	Faster (concentration effect)	New equilibrium with a higher proportion of the substance(s) produced by the reaction causing a reduction in the number of molecules (and hence a reduction in pressure)
Decrease in pressure	Gas reactions as above	Slower (concentration effect)	New equilibrium with a higher proportion of the substance(s) produced by the reaction causing an increase in the number of molecules (and hence the pressure)
Positive catalyst	Most	Faster	No change
Negative catalyst	Most	Slower	No change

The factors which affect equilibria are summarized in Table 20.1. The following examples illustrate the various ideas in the table.

1. For a reaction

$$A_2(g)+B_2(g) \rightleftharpoons 2AB(g);\ \Delta H = -210\ kJ\ mol^{-1},$$

the following changes would produce a new equilibrium with a higher yield of product:
(a) lowering the temperature,
(b) adding more of A and/or B, and also removing AB. Pressure changes will not affect the yield. The use of a positive catalyst will not increase the yield, but will reduce the time taken to reach equilibrium.

2. In the reaction

$$D_2(g)+3E_2(g) \rightleftharpoons 2DE_3(g);\ \Delta H = +810\ kJ\ mol^{-1},$$

the following changes would produce a new equilibrium with a higher yield of product:
(a) raising the temperature (particularly useful because it also reduces the time taken to reach equilibrium),
(b) adding more of D_2 and/or E_2, and also removing DE_3,
(c) raising the pressure.

A positive catalyst will not increase the yield, but it is important because it reduces the time taken to reach equilibrium.

The conditions used for industrial processes are often decided by a consideration of the above factors. However, the theoretical conditions may have to be modified for economic reasons. For example, a lowering of the temperature may in theory produce a higher yield, but this will also result in the slowing down of both reactions. A longer time will be needed to reach equilibrium, and a compromise temperature may have to be used.

Check your understanding

1. (a) The formation of methanol (methyl alcohol) from hydrogen and carbon monoxide can be represented by

$$CO(g)+2H_2(g) \rightleftharpoons CH_3OH(l) \qquad \Delta H = +91\ kJ\ mol^{-1}$$

What mass of hydrogen would react to cause a heat change of 91 kJ?
(b) What would be the effect on the equilibrium concentration of methanol in this endothermic reaction if (i) the temperature was increased, (ii) the pressure was increased, (iii) the hydrogen concentration was increased? (J.M.B.)

2. The equation for the reaction by which ammonia is manufactured is:

$$N_2(g)+3H_2(g) \rightleftharpoons 2NH_3(g)$$

(a) What would be the effect on the equilibrium concentration of ammonia of (i) increasing the pressure, (ii) increasing the nitrogen concentration?
(b) The equilibrium concentration of ammonia increases as the temperature is lowered. Is heat evolved or absorbed when ammonia is formed?
(c) Why is a catalyst used in this reaction? (J.M.B.)

3. $2SO_2(g)+O_2(g) \rightleftharpoons 2SO_3(s)$ + heat evolved $\quad \Delta H = -189\ kJ\ mol^{-1}$

The equation represents a system in equilibrium. State the changes in the equilibrium concentration of sulphur trioxide which would be caused by (a) adding oxygen, (b) heating the mixture, (c) increasing the pressure. (J.M.B.)

4. In the reaction

$$2Y+W \rightleftharpoons Y_2W,\ \Delta H = -8400\ kJ\ mol^{-1}$$

which of the following would result in a higher yield of Y_2W?
A The use of a suitable positive catalyst
B Lowering the temperature
C Removal of W
D Reducing the pressure
E Increasing the surface area of W

5. When calcium hydrogencarbonate solution decomposes in nature, an equilibrium is not established because
A the reaction is not reversible
B one or more of the products escapes from the system
C the reaction is too slow
D a negative catalyst is present
E the optimum temperature is too high

20.8 A SUMMARY OF CHAPTER 20

The following 'check list' should help you to organize the work for revision.

1. General points

(a) Reaction rates are measured by considering how quickly products are formed (or reactants used up) in a chemical reaction.
(b) The units of rate of reaction (at a given moment in time) are mol dm^{-3} s^{-1}. Sometimes we *compare* rates simply by measuring the time taken to complete a reaction, or by measuring the time taken for the reaction to reach a certain stage. In these cases we are *not* finding the rate at a particular point in time.
(c) In experiments used to investigate rates of reaction, it is important to fix all of the variables (concentration, temperature, surface area, etc.) throughout the experiment *except* the one which is being investigated.
(d) A reaction rate increases if the concentration of one or more of the reactants is increased. The concentration of a solid cannot be increased, but the concentration of a solution can be increased by adding more solute, and the concentration of a gas can be increased by increasing the pressure.
(e) The statements in (d) are explained because increased concentrations cause a greater frequency of collisions between particles, and this increases the rate of reaction.
(f) The rate of a typical chemical reaction is fastest at the beginning (because the reagents are then at their highest concentrations) and gradually decreases (as substances are used up and their concentrations decrease). It stops when one or more of the reagents is used up.
(g) In experiments such as that on page 353, for a given amount of reagent the reaction will always produce the same loss in mass (or volume of gas), i.e. the graphs will always 'tail off' at the same level in a particular experiment. The *time* taken to reach this point may vary, however, according to the concentration of a solution or the surface area of a solid. Thus, samples of 0.2 g of magnesium ribbon may be added separately to 20 cm^3 of a hydrochloric acid solution of concentration 1 mol dm^{-3}, and then 20 cm^3 of hydrochloric acid solution of concentration 2 mol dm^{-3}. If the acid is in excess, the same *volume* of hydrogen will be formed in each case, but it will be formed *more quickly* with the more concentrated acid.
(h) The rate of a chemical reaction is increased if the temperature is increased. An increase in temperature increases the rate of collision between particles, but the main reasons for the faster rate are that molecules start to 'shake themselves apart' and collisions between them become 'more energetic'.
(i) If the surface area of a solid reactant is increased (i.e. it is made more finely divided), the rate of reaction increases. This is because there is more contact between the reacting particles when the surface area of a solid is increased.
(j) A catalyst is a substance which alters the speed of a chemical reaction. It is unchanged chemically and in mass at the end of the reaction and so does not always appear to take part in the reaction, although it actually does so and is re-formed.
(k) A positive catalyst speeds up a reaction, an inhibitor slows down a reaction.
(l) Enzymes are 'biological catalysts'. They are 'deactivated' by heating, or by exposure to high concentrations of acid or alkali. They are highly specific in their action.
(m) Light can affect a small number of chemical reactions.
(n) If you are asked to describe an experiment which shows that a certain chemical is a catalyst, remember that you should use the results to indicate that the catalyst (i) is unchanged chemically and in mass at the end of the reaction, and (ii) alters the rate of the reaction. Many students forget part (i) of the argument.
(o) Reversible reactions are reactions which can go 'both ways'.
(p) Equilibrium is the situation reached in a reaction when the rate of the forward reaction is the same as the rate of the back reaction. There is a mixture of 'reactants' and 'products' at

equilibrium. Once equilibrium has been reached, the concentrations of the chemicals present cannot be changed unless the conditions of the experiment are changed (e.g. altering the pressure).

(q) The factors which can change equilibrium positions are summarized in Table 20.1. Examples are given on page 363.

2. Important experiments

Make sure that you can describe typical experiments to show how each of these variables can affect the rate of a reaction: temperature, concentration, surface area, or the use of a catalyst.

QUESTIONS

Multiple choice questions

1. The graph represents the progress of reaction between calcium carbonate and an excess of hydrochloric acid. The curve shows how the total volume of carbon dioxide liberated (at s.t.p.) varied with time.

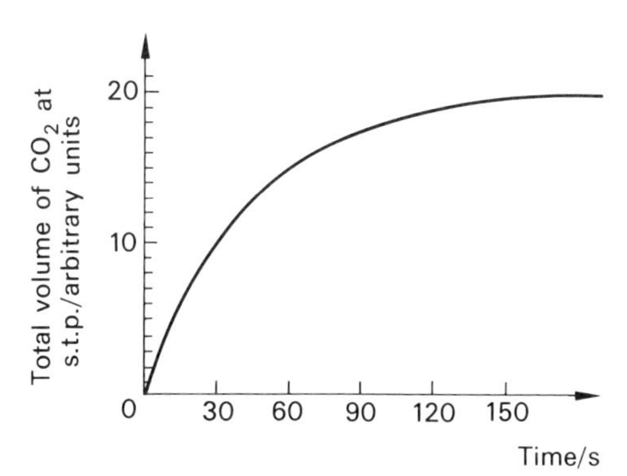

How long did it take for half of the calcium carbonate to react?
A 30 s; B 60 s; C 90 s; D 120 s;
E 150 s (C.)

2. Calcium carbonate was placed in a flask on a top-pan balance and dilute hydrochloric acid was added. The total mass of the flask and its contents was recorded every five seconds. The following figure shows a plot of the results.

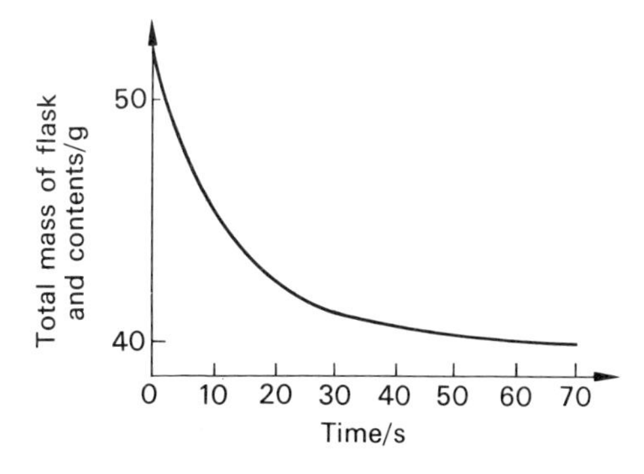

At which one of the following times was the reaction fastest?
A 10 s; B 20 s; C 30 s; D 40 s;
E 50 s (C.)

3. A conical flask containing excess dilute nitric acid is weighed after a lump of magnesium carbonate has been added. A loose plug of glass wool is placed in the mouth of the flask to prevent the loss of acid spray and the mass of apparatus is recorded at regular intervals. The graph of the 'mass of the flask + contents' against 'time' is plotted. Which one of the following curves, A, B, C, D, or E is most likely to be the graph obtained? (C.)

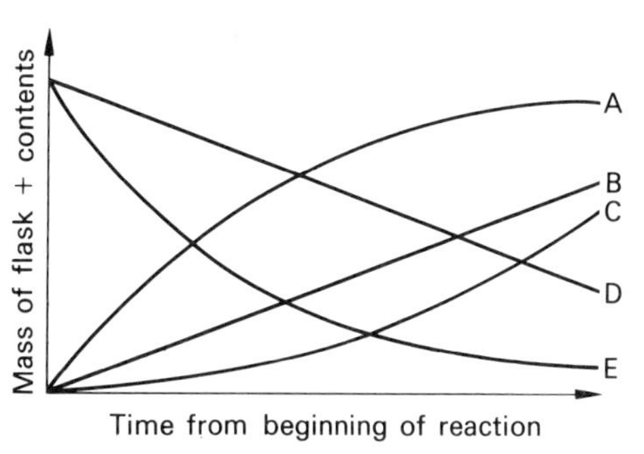

4. Which one of the curves (p. 366), A, B, C, D, or E, corresponds to the decomposition of 1 mole of hydrogen peroxide using manganese(IV) oxide as a catalyst? (C.)

5. Explain why an increase in the concentration of one or more of the reactants increases the rate of a chemical reaction.

6. An increase in pressure does not noticeably affect the rate of a reaction between solids, but increases the rate of a reaction between gases. Explain.

7. Briefly describe three different ways in which the rate of a chemical reaction can be measured.

Graphs for question 4 on page 365.

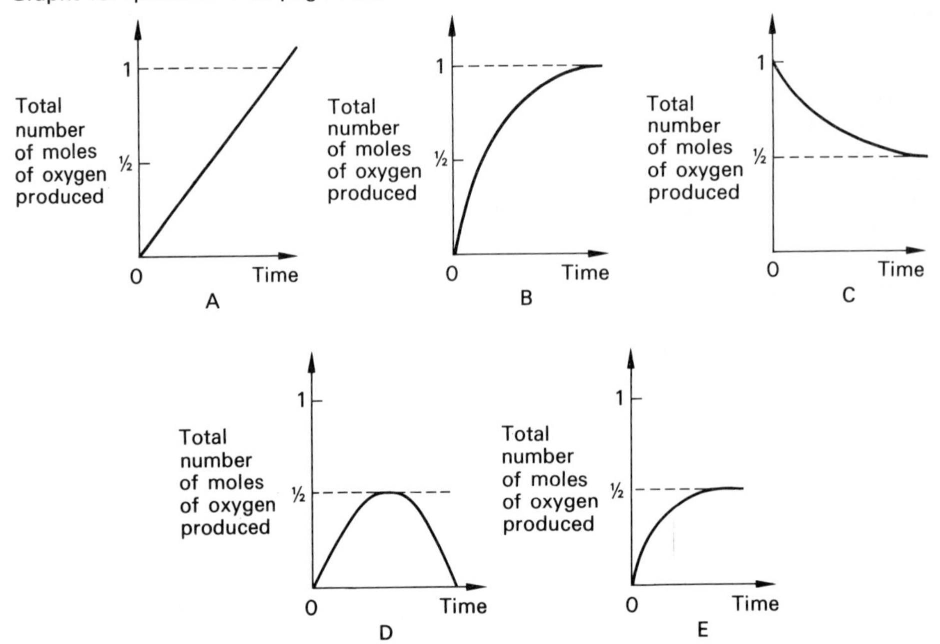

Questions requiring longer answers

8. (a) An aqueous solution of hydrogen peroxide decomposes to water and oxygen. This reaction is rapid at room temperature when a catalyst is added to a moderately concentrated solution. (i) Name a suitable catalyst. (ii) Write the equation for the reaction. (iii) You are asked to investigate the rate at which oxygen is evolved at room temperature when one gram of catalyst is added to 50 cm^3 of a suitable hydrogen peroxide solution. Draw the apparatus you would use and outline the experiment you would do. (iv) Sketch a graph showing how you would expect the volume of oxygen to vary with time. (It is not necessary to use graph paper.) Label this graph P. (v) On the same axes, sketch another graph to indicate how the results would be changed if the experiment were repeated with warm hydrogen peroxide solution. Label this graph Q. (vi) Apart from temperature, what other factors might affect the rate of this reaction?
(b) State *two* reactions of industrial importance in each of which a named catalyst is used. Give the conditions under which *one* of these reactions is carried out. (You are not required to describe how the products are isolated.) (A.E.B. 1977)

9. The progress of the reaction

$$Zn + H_2SO_4 \rightarrow ZnSO_4 + H_2$$

may be followed by recording the total volume of hydrogen evolved after known times.
(a) Give a diagram of an apparatus suitable for carrying out the experiment. Sketch a graph showing the type of results you would expect to obtain. Label the axes of your graph as 'total volume of hydrogen' (vertical axis) and 'time from start of reaction' (horizontal axis).
(b) In a second experiment, a more concentrated solution of sulphuric acid was used, and the reaction was found to proceed faster. What explanation can you give for this observation? Name two other changes in the experimental conditions that you would also expect to lead to a faster reaction.
(c) What would be the maximum volume of hydrogen evolved if 6.5 g of zinc and 100 cm^3 of 0.10 mol dm^{-3} sulphuric acid were used? (Relative atomic mass: Zn, 65. Assume that one mole of hydrogen occupies 24 dm^3 under the conditions of the experiment.) (C.)

10. Explain what is meant by the term catalyst. Describe a simple experiment or experiments to demonstrate the use of a catalyst. Give two industrial reactions in which a catalyst is used, naming the initial substances, the product(s), and the catalyst used as well as any essential conditions. (Long ac-

counts are not wanted, nor are diagrams. The manufacture of nitric acid is excluded.) (C.)

11. The following statements are made in a textbook. 'The rates of most chemical reactions are approximately doubled by raising the temperature at which the reactions are carried out by 10 °C.'

'The rate at which a chemical substance reacts is directly proportional to its concentration.'

(a) Describe the experiments you would carry out to test the truth of these two statements when applied to either the reaction between a metal and a dilute acid or the decomposition of hydrogen peroxide catalysed by manganese dioxide (manganese(IV) oxide).

(b) Explain simply, in terms of the ions or molecules present, why the rate of a reaction is increased both by raising the temperature and also by increasing the concentration of the reagents. (C.)

21 Carbon and nitrogen: two important non-metals

Some general chemistry of the non-metallic elements

Out of the hundred or so elements, only about twenty are non-metals. Two of these, however, make up about 75 per cent of the mass of the Earth's crust (oxygen, 46.4 per cent and silicon, 27.8 per cent).

The physical differences between metal elements and non-metal elements are discussed on page 25. If you have studied some elements in detail, you will understand the main chemical differences between metal and non-metal elements which are summarized in Table 21.1.

Several very important principles, such as the activity series and periodicity, are used in the study of the metals and their compounds. These principles are not normally used for the study of non-metals at an elementary level, except for the halogen group. A study of non-metals is based more upon an understanding of their redox reactions, acid–base reactions, and the preparations of gases. These aspects are emphasized in this chapter.

Table 21.1 Some chemical differences between metallic and non-metallic elements

Non-metallic elements	*Metallic elements*
Atoms usually *share* electrons or *gain* 1, 2, or 3 electrons, in forming stable electron structures	Atoms usually *lose* 1, 2, or 3 electrons to form stable electron structures
When ions are formed, they are always negatively charged except for hydrogen, which can form both H^+ and H^- ions	Ions formed are always positively charged
Their compounds may be ionic (e.g. when they consist of a metal combined with a non-metal) or covalent (e.g. when a non-metal combines with another non-metal)	Their compounds are nearly always ionic
Never liberate hydrogen when added to dilute acids	*May* liberate hydrogen when added to dilute acids (depends upon both the acid used and the reactivity of the metal)
Their oxides are either acidic or neutral, never basic	Their oxides are usually basic or amphoteric. Common ones are never acidic
Liberated at the anode during electrolysis (except for hydrogen when present as H^+ ions)	Liberated at the cathode in electrolysis
Are often oxidizing agents (because they tend to gain electrons and often also combine with hydrogen) except for hydrogen itself, which is a reducing agent	Are reducing agents because they tend to lose electrons when reacting

21.1 CARBON AND ITS COMPOUNDS

Carbon the element

Physical properties

Carbon occurs in the pure state as the two allotropic (polymorphic) forms diamond and graphite (p. 273), and in many impure forms such as coal, coke, soot, and charcoal (which are also made up of the graphite lattice).

Diamond is the hardest known natural substance, and this is because of its structure (p. 274). Over 90 per cent of the world's diamonds come from South Africa, where they are found in the craters of extinct volcanoes. They were formed during the Earth's cooling when carbon was exposed to enormous pressures and to very high temperatures. Synthetic diamonds can now be made by treating carbon in a similar way, using pressures in excess of 100 000 atmospheres and temperatures in the region of 2000 °C, but only small diamonds have so far been made in this way. Nevertheless, synthetic diamonds have important uses; they are used in the manufacture of cutting tools and drill bits.

The other allotrope of carbon, graphite, occurs to a small extent in many countries, including Britain; it was mined in Borrowdale in Cumbria. Nowadays, graphite is usually prepared synthetically by the Acheson process, in which impure carbon is heated with sand in an electric furnace. In contrast to diamond, graphite is one of the softest solids known and is used as a lubricant, in electrodes (note that diamond does *not* conduct electricity), as a 'moderator' in atomic reactors, and (mixed with clay) as 'lead' in pencils. The very different properties of diamond and graphite can be explained by their structures (p. 273).

Powdered forms of carbon have good adsorbing power and are used to purify substances. For example, solutions which are contaminated with a coloured impurity can often be purified by boiling with 'activated charcoal' and then filtering off the suspended charcoal. The coloured impurity is adsorbed on the surface of the charcoal. This principle is made use of in many forms of gas mask. The granular carbon they contain has the ability to adsorb many gases, particularly some of the common toxic ones.

Both forms of carbon have giant atomic structures (i.e. they are macromolecular), and thus have high melting points around 3700 °C.

Chemical properties

Carbon atoms need to lose, gain, or share 4 electrons in order to gain a stable electronic structure, and so carbon is not very reactive and always forms covalent bonds.

1. When heated to a high temperature in a good supply of air or oxygen, it glows and produces carbon dioxide,

$$C(s) + O_2(g) \rightarrow CO_2(g)$$

2. When carbon is heated strongly in carbon dioxide, the gas is reduced to carbon monoxide. This happens in the blast furnace (p. 527).

$$C(s) + CO_2(g) \rightarrow 2CO(g)$$

3. It will reduce the oxides of less reactive metals to the metal. A high temperature is needed. You may have done this on a small scale using a blowpipe and charcoal block with, for example, lead(II) oxide,

$$PbO(s) + C(s) \rightarrow Pb(l) + CO(g)$$

4. When carbon is heated to white heat it reacts with steam,

$$C(s) + H_2O(g) \rightarrow CO(g) + H_2(g)$$

The gases produced are both flammable, and the mixture was traditionally used as an industrial fuel ('water gas') and to manufacture hydrogen, e.g. for the synthesis of ammonia.

Carbon dioxide

Sources of the gas

Carbon dioxide is produced whenever carbon or any of its compounds (including many fuels) are completely burned in a good supply of air or oxygen. In industry the gas is made by heating limestone in a lime kiln (Figure 21.1) and it is also obtained as a by-product of fermentation reactions. Large quantities also come from the steam-reforming of natural gas in the production of hydrogen for ammonia synthesis (p. 549).

Figure 21.1 A vertical lime kiln (*ICI Ltd*)

Laboratory preparation

The usual method of making carbon dioxide is to add dilute hydrochloric acid to marble chips (a pure form of calcium carbonate) as in Figure 21.2. The gas is only slightly soluble in water, and so it can be collected over water. It may also be collected by downward delivery.

$$CaCO_3(s) + 2HCl(aq) \rightarrow CaCl_2(aq) + CO_2(g) + H_2O(l)$$

Physical properties of carbon dioxide

Solubility in water	*Colour*	*Odour*	*Density relative to air*	*Toxicity*
Slight. More soluble under pressure—used in 'fizzy' drinks	Colourless	Virtually none	More dense	Only toxic at relatively high concentrations

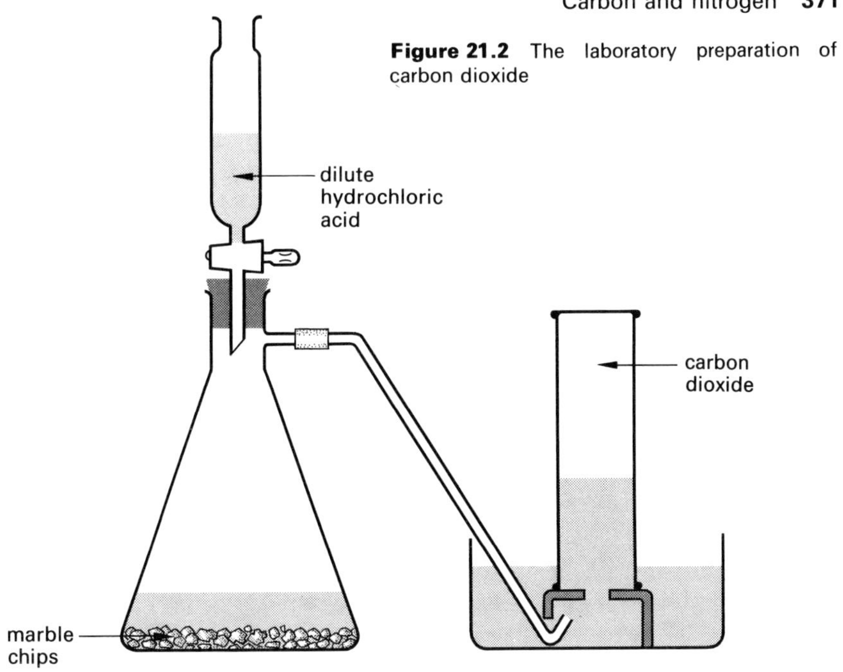

Figure 21.2 The laboratory preparation of carbon dioxide

Experiment 21.1
Some properties of carbon dioxide **

Apparatus
Carbon dioxide generator (e.g. as in Figure 21.2, or a Kipps apparatus as in Figure 21.3), three gas jars and covers, Bunsen burner, tongs, three boiling tubes (with bungs) in rack, tapers.
Magnesium ribbon, calcium hydroxide solution, blue litmus paper, universal indicator solution.

Procedure
(a) Collect three gas jars and three boiling tubes of the gas.
(b) Wrap a piece of magnesium ribbon around the tongs so that a piece about 6 cm long hangs from the tongs. Ignite the magnesium ribbon, remove the cover from one of the gas jars of carbon dioxide, and quickly plunge the burning magnesium into the gas. DO NOT look directly at the burning magnesium. Does carbon dioxide allow magnesium to burn in it? If so, what products are formed?
(c) Moisten a piece of blue litmus paper with water, and drop it into a boiling tube filled with carbon dioxide, immediately replacing the bung. Does carbon dioxide dissolve in water to form an acidic solution? If so, does the colour change tell you whether the solution is very acidic or only slightly acidic?
(d) Add a few drops of universal indicator to another tube of the gas, replace the bung, and shake the tube. Does any colour change help to support your conclusion from (c)?

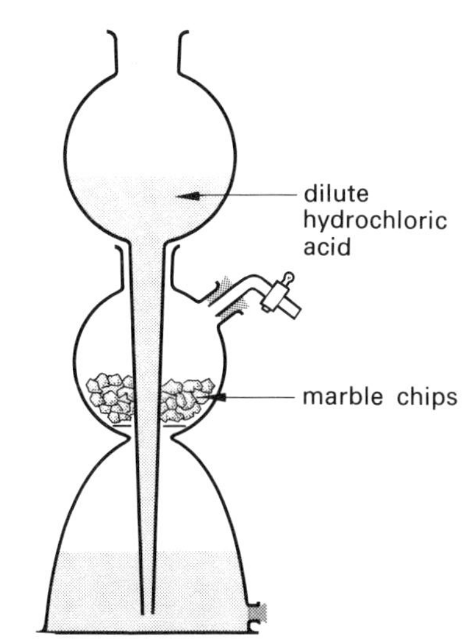

Figure 21.3 The Kipp's apparatus (as used for the preparation of carbon dioxide)

(e) Add a few cm^3 of calcium hydroxide solution to a third tube of gas, replace the bung, and shake the tube.
(f) Light a taper and plunge it into a gas jar of the gas. Compare your observations with what happened in (b).
(g) Light a taper, remove the cover from another gas jar of carbon dioxide, and 'pour' the gas over the lighted taper from just above the flame.

Notes about the experiment

Make sure that you can explain all your observations, giving equations where possible. If you need help, the reactions are summarized on page 397. Note that the extinguishing of a lighted taper is a much misused test. If a lighted taper is plunged into an ordinary test-tube (i.e. a relatively narrow one) containing *air*, it will be extinguished! This is why the test was done with a gas jar of gas in the experiment.

Points for discussion

1. Can you explain how a Kipps apparatus works (Figure 21.3), and why it is so useful?
2. To remove carbon dioxide from a mixture of gases, the mixture may be bubbled through sodium hydroxide solution. Why do you think that sodium hydroxide solution is used to *absorb* the gas, but calcium hydroxide solution is used to detect the presence of the gas?

Uses of carbon dioxide

1. It is dissolved in water under pressure to make 'fizzy' drinks.
2. The solid form (dry ice) is used as a refrigerant.
3. It is used in fire extinguishers; being a dense gas, it blankets the fire and prevents oxygen from reaching it.
4. It is produce *in situ* by baking powder, and also in health salts. Baking powder consists of a dry mixture of sodium hydrogencarbonate and a solid acid such as tartaric or citric acid. Reaction only takes place when water is added, when the acid reacts with the hydrogencarbonate to form carbon dioxide. A similar principle is used in health salts.
5. It is used to transfer heat energy from certain types of atomic reactors.

Carbon monoxide, CO

Sources of the gas

Carbon monoxide gas is formed when carbon or its compounds (e.g. many fuels) are burned in a *limited* supply of oxygen. Large volumes of the gas enter the atmosphere in this way, from car exhaust fumes, power stations, aircraft, etc.

Physical properties of carbon monoxide

Solubility in water	*Colour*	*Odour*	*Density relative to air*	*Toxicity*
Insoluble	None	None	Slightly less dense	Toxic; particularly dangerous as has no smell

Carbon monoxide is poisonous because it combines with the haemoglobin in the red blood corpuscles in preference to oxygen. This prevents the supply of oxygen from reaching the cells of the body.

Chemical properties of carbon monoxide

1. Burns in air or oxygen with a blue flame to form carbon dioxide,

$$CO(g)+\tfrac{1}{2}O_2(g) \rightarrow CO_2(g)$$

This reaction can be used to distinguish it from carbon dioxide, which does not burn.

2. Good reducing agent. Used in industry to reduce oxides of some metals to the metal, e.g. in the blast furnace (p. 527). This is an important use of the gas.
3. Unlike carbon dioxide, it is a neutral oxide and has no reaction with acids, alkalis, or indicators under normal laboratory conditions.

The carbon cycle

Carbon is an essential component of all living things, and carbon dioxide is the 'transfer agent' for the carbon atoms which are continually circulating in Nature. Remember that nearly all of the carbon atoms on Earth have been present since the Earth began, and that these same carbon atoms are constantly circulating through various carbon compounds (see Figure 21.4). For example, it is possible that some of the carbon atoms in your body once formed part of a tree in a primeval forest, then were part of a dinosaur, then part of the wood of a Viking ship, then part of Shakespeare's body, etc. Note that carbon dioxide is the 'transfer agent' for this circulation. Carbon atoms in living things generally enter the atmosphere as carbon dioxide when the organism respires or dies. Plants then use the carbon dioxide to build up carbon compounds by photosynthesis.

Until recently, there was a balance between the amount of carbon dioxide liberated into the atmosphere and that used up from the atmosphere. The main sources of carbon dioxide are the combustion of carbon compounds (e.g. fuels), industrial processes, and respiration. The main processes which remove the gas from the atmosphere are photosynthesis and the absorption of the gas by sea water. The processes which add carbon dioxide to the atmosphere have gradually increased in scale during the last fifty years or so, especially steps A and B in Figure 21.4. This is because we now burn more fuel than ever before and we live in a society dependent upon industrial development. The processes which absorb carbon dioxide from the

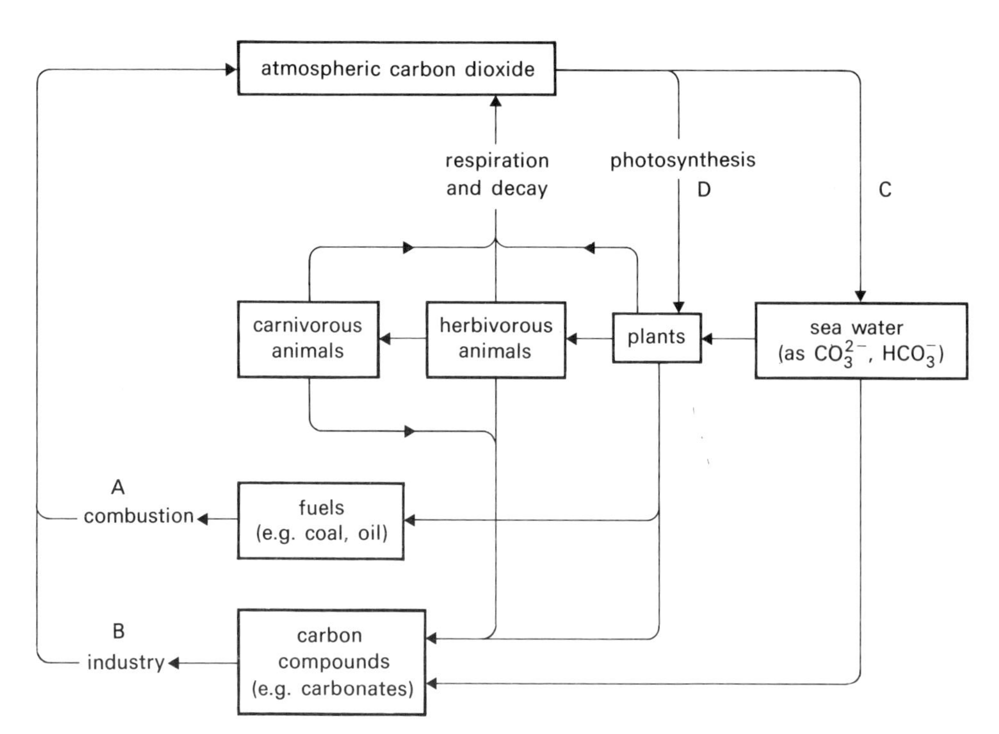

Figure 21.4 The carbon cycle

Most reactions take place more quickly when the temperature rises

atmosphere, however, cannot take place at the same increased rate. For example, the absorption of carbon dioxide by step D is virtually constant because the amount of vegetation on the Earth is practically unchanged. (If anything, there is less vegetation than before because of the rate we fell trees to provide timber and to clear land for roads and cities!) Similarly, the rate of absorption by step C is almost constant because the surface area of the oceans is constant, and the absorption of the gas by sea water is a slow process.

Carbon atoms still cycle through Nature, but the small proportion of carbon dioxide in the atmosphere is slowly but gradually increasing at the expense of other carbon compounds. This may have serious long-term consequences, as explained on page 142. Eventually a balance will be restored, because we are using up our carbon-based fuels and at some future date this source of carbon dioxide may no longer be so important. The rate at which the gas enters the atmosphere will therefore fall, giving the oceans the chance to 'catch up' by absorbing comparatively more carbon dioxide until the absorption and formation of carbon dioxide are in balance. In the short term the carbon dioxide concentration in the atmosphere will continue to rise (by about 0.2 per cent per year). It is not toxic at these levels, but nobody is quite sure what its effect on our climate may be before the natural balance is restored.

21.2 NITROGEN AND ITS COMPOUNDS

Nitrogen the element

Nitrogen is in Group 5 of the Periodic Table as its atoms have five electrons in their outer shells. Atoms of nitrogen do not form N^{5+} ions by losing five electrons, but they can gain three electrons and form ions of N^{3-}, e.g. in magnesium nitride, Mg_3N_2. These ions are comparatively rare, however, and the great majority of compounds formed by the element are covalently bonded.

Nitrogen was first isolated (from air) in 1772 by the Swedish chemist Carl Wilhelm Scheele. It forms about 78 per cent by volume of air, and is also present in a number of compounds found naturally, particularly nitrates and proteins. Impure nitrogen can be prepared from the air in the laboratory (p. 136), and pure nitrogen is obtained industrially from liquid air (p. 540).

Physical properties of nitrogen

Solubility in water	*Colour*	*Odour*	*Density relative to air*	*Toxicity*
Virtually insoluble	None	None	Same	Non-toxic

Chemical properties of nitrogen

The chief chemical feature of nitrogen is its inertness. The only reaction you are likely to study is its combination with hydrogen in the Haber process (p. 548). There is no positive test for the gas.

Uses of nitrogen

1. It is converted by the Haber process into ammonia, and subsequently into fertilizers and nitric acid.
2. Liquid nitrogen is used as a refrigerant.
3. It provides an inert atmosphere for certain chemical reactions.

Oxides of nitrogen

Nitrogen forms several oxides, but the only one you are likely to encounter in the laboratory is the dioxide, NO_2. (Nitrogen monoxide, NO, occurs as an intermediate in the industrial manufacture of nitric acid, which is explained on page 552.) Nitrogen dioxide is seen as a product of the action of heat on most metal nitrates (p. 326) and of the reaction of metals with nitric acid (p. 251).

Physical properties of nitrogen dioxide, NO_2

Solubility in water	*Colour*	*Odour*	*Density relative to air*	*Toxicity*
Reacts and dissolves to form acidic solution	Dark brown gas or pale yellow-brown liquid	Sharp, unpleasant	More dense	Very toxic

Chemical properties of nitrogen dioxide

1. Nitrogen dioxide reacts with water to form a mixture of nitric and nitrous acids,

$$2NO_2(g) + H_2O(l) \rightarrow HNO_2(aq) + HNO_3(aq)$$

2. The pale yellow-brown liquid form (N_2O_4) dissociates when heated into the dark brown gas, NO_2. (Pure dinitrogen tetraoxide, N_2O_4, is colourless but the liquid is usually coloured because it contains dissolved nitrogen dioxide.)

$$N_2O_4(l) \rightleftharpoons 2NO_2(g)$$

Ammonia, NH_3

Laboratory preparation

Ammonia is formed when any ammonium salt is heated with an alkali. The usual laboratory method is to heat a finely ground solid mixture of ammonium chloride and dry calcium hydroxide, as in Figure 21.5. (If a little water is added and the mixture made into a paste, this helps to avoid cracking the flask.) The gas must be dried by using calcium oxide; it reacts with other common drying agents such as concentrated sulphuric acid and anhydrous calcium chloride.

$$Ca(OH)_2(s) + 2NH_4Cl(s) \rightarrow CaCl_2(s) + 2NH_3(g) + 2H_2O(l)$$

or, ionically,

$$NH_4^+(aq) + OH^-(aq) \rightarrow NH_3(g) + H_2O(l)$$

Physical properties of ammonia gas

Solubility in water	*Colour*	*Odour*	*Density relative to air*	*Toxicity*
Extremely soluble	None	Unpleasant, choking	Less dense	Toxic

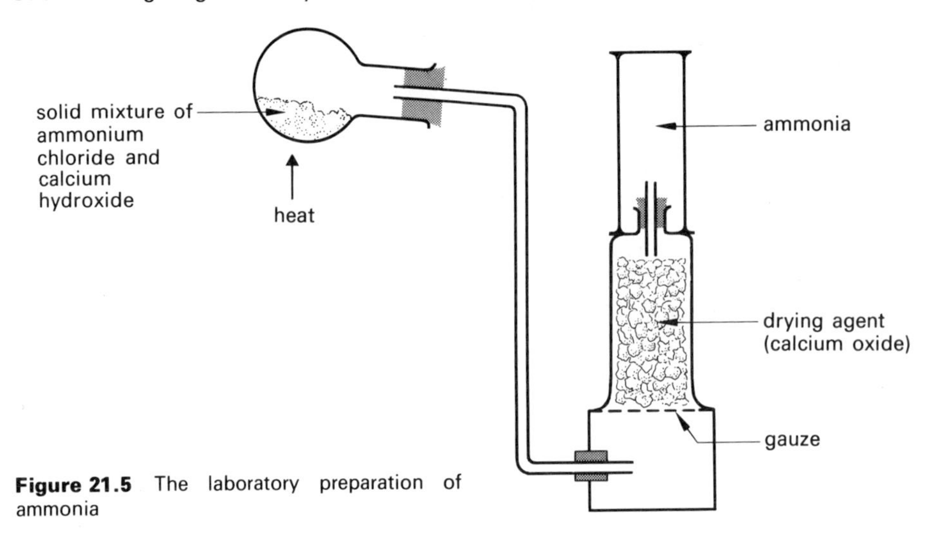

Figure 21.5 The laboratory preparation of ammonia

Note: (i) The fact that ammonia is *very* soluble in water is normally shown by the 'fountain experiment' as in Experiment 21.2.
(ii) Ammonia gas is easily liquefied under pressure.

Experiment 21.2
The fountain experiment***

Apparatus
Ammonia generator, *dry* round bottom flask and other apparatus as shown in Figure 21.6, access to fume cupboard.
Universal indicator solution.

Procedure
(a) Working in the fume cupboard, and with both screw clips open, pass dry ammonia gas through the flask for several minutes. Stop the flow of gas and immediately tighten both clips so as to prevent the gas escaping from the flask.
(b) Set up the apparatus as shown in Figure 21.6; this part of the experiment may be conducted in the open laboratory.
(c) Colour the water green by adding a few drops of universal indicator solution.
(d) With the open end of the rubber tubing well below the surface of the water in the trough, unscrew its clip and observe what happens. If nothing happens after a few minutes, cup the hands around the flask for a while so as to warm it, and then remove the hands.

Results
Record all of your observations.

Notes about the experiment
Ammonia gas is *very* soluble in water, and when the screw clip is opened some of the gas can dissolve in water in the tube. This *immediately* reduces the pressure of the gas in the flask, and so the atmospheric

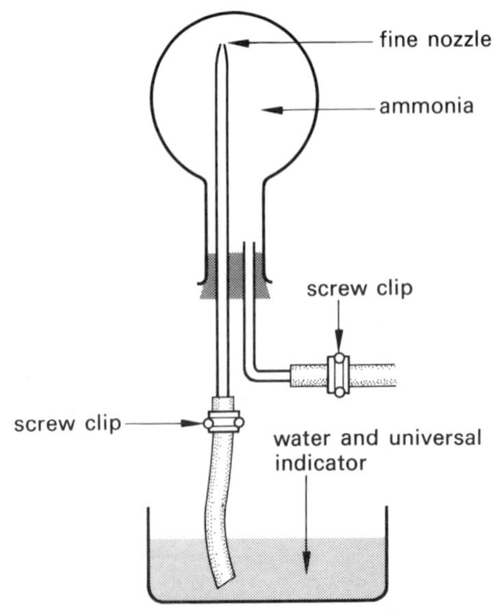

Figure 21.6 The fountain experiment

pressure is suddenly greater than the pressure in the flask. The atmospheric pressure acting on the surface of the water in the trough forces water up the tube, where more gas dissolves and the process continues. The water is thus forced rapidly up the tube and emerges from a fine nozzle in the form of a fountain. The fountain continues until all the gas has dissolved.

Conclusion
Make sure that you can explain clearly why the fountain was formed, and any other observations you made.

Points for discussion
1. Was the flask filled completely with ammonia when the clip was opened? Explain your answer.
2. Other very soluble gases will also give the fountain experiment, e.g. hydrogen chloride and sulphur dioxide. Thêse two gases would produce one different effect (compared with ammonia) if they were used instead of ammonia in the last experiment. What is this difference?

Chemical properties of ammonia gas
1. Ammonia is the only common gas which is alkaline, and this property is used as a test for the gas (p. 450). It neutralizes acid solutions and acid gases to form ammonium salts, e.g.

$$NH_3(g) + HCl(g) \rightarrow NH_4Cl(s)$$

2. It will reduce the heated oxides of metals below iron in the activity series to the metal, itself being oxidized to steam and nitrogen, e.g.

$$2NH_3(g) + 3CuO(s) \rightarrow 3Cu(s) + 3H_2O(g) + N_2(g)$$

Experiment 21.3
Some properties of ammonia**

Apparatus
Boiling tubes (with bungs) in rack, teat pipettes, access to fume cupboard. Concentrated ammonia solution, concentrated hydrochloric acid, red litmus paper.

Procedure
Note: Only open the bottle of concentrated ammonia in the fume cupboard. Keep the bottles of concentrated ammonia and concentrated hydrochloric acid well apart from each other. Keep both liquids well away from your face throughout the experiment, and do not breathe *in* near a sample of either liquid. Read the note about using concentrated ammonia solution on page 378.

(a) Using a teat pipette, place three drops of concentrated ammonia solution into each of two boiling tubes (in the fume cupboard), and fit each with a bung as soon as the liquid is inside.
(b) Using a DIFFERENT teat pipette, place three drops of concentrated hydrochloric acid into a third boiling tube, and seal it with a bung.
(c) Dampen a piece of red litmus paper and drop it into one of the tubes of ammonia solution, replacing the bung immediately.
(d) Working in the fume cupboard, remove the bungs from a tube of ammonia solution and from the tube of concentrated hydrochloric acid, and hold the tubes next to each other.

Conclusion
Make sure that you can explain all of your observations, and that you can write equations for the reactions. Note that the ammonia solution is used simply as a convenient source of ammonia gas, which escapes rapidly from the solution. The reactions were not those of concentrated ammonia *solution*, nor of hydrochloric *acid*; make sure that you name the chemicals correctly.

Uses of ammonia

Most of the ammonia which is manufactured (by the Haber process) is converted into nitric acid and into fertilizers (see page 390). It is also used to prepare urea (for fertilizers and plastics) and, when liquefied, as a refrigerant.

Ammonia solution, $NH_3(aq)$, and ammonium salts

The preparation of ammonia solution

Ammonia gas is prepared in the normal way and, instead of being dried and collected, is passed into water as shown in Figure 21.7. A solution of any other very soluble gas (e.g. hydrogen chloride dissolving to produce hydrochloric acid) is made in the same way. If the gas was passed straight into water 'sucking back' would occur. (This is a misleading term, for atmospheric pressure actually *pushes* the water up the tube when the pressure inside drops, as in the fountain experiment.)

Points for discussion

1. What is meant by 'sucking back'? Can you explain why it might happen when ammonia gas is passed into water?
2. Can you explain why the use of the filter funnel in Figure 21.7 prevents 'sucking back'?

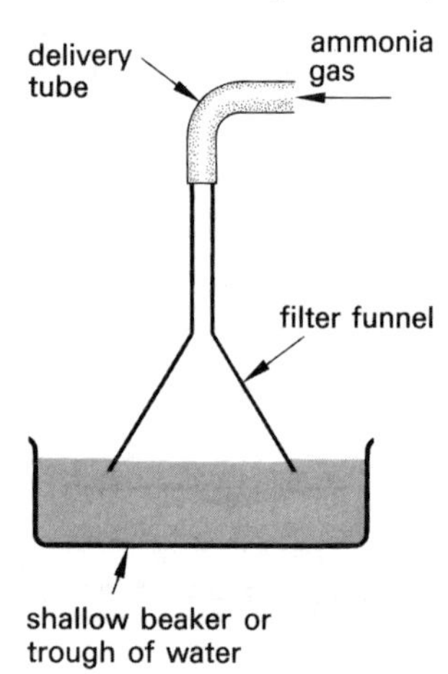

Figure 21.7 Preparation of a solution of ammonia

Properties of ammonia solution

Physical properties When ammonia gas dissolves in water, the solution formed is an alkali because *some* of the ammonia molecules which dissolve also *react* with water molecules to form hydroxide ions:

$$NH_3(g)+H_2O(l) \rightleftharpoons NH_4^+(aq)+OH^-(aq).$$

Most of the molecules of ammonia which dissolve in the water do not react, however, and so the concentration of $OH^-(aq)$ ions is very low and the solution is described as *ammonia solution* rather than ammonium hydroxide solution. Ammonia solution is only a weak alkali, even when concentrated, because it only contains a low concentration of hydroxide ions.

The solution smells strongly of ammonia, particularly the saturated (concentrated) solution which is often called '880 ammonia' because it has a density of 0.880 g cm^{-3}. Such a solution must be used with great care. Ammonia gas rapidly escapes from it as soon as it is exposed to the air. The gas can cause the eyes to stream, and the sudden shock of inhaling the pungent gas can cause people to panic and to drop the bottle.

Chemical properties Ammonia solution is an important (although weak) alkali. It shows the usual properties of alkalis, neutralizing acids (to form ammonium salts) and precipitating metal hydroxides from solutions of their salts (but not without 'complications', as explained on page 329).

Point for discussion

The density of pure water is 1.0 g cm^{-3}. A saturated solution of ammonia in water contains a high concentration of dissolved ammonia molecules, and yet its density is significantly *less* than that of pure water, being only 0.88 g cm^{-3}. Can you offer an explanation for this?

Properties of ammonium salts

1. They are all soluble in water.
2. They all give off ammonia gas when heated with an alkali; this property is used as a test for the ammonium ion,

$$NH_4^+(aq) + OH^-(aq) \rightarrow NH_3(g) + H_2O(l)$$

The ammonia given off is detected in the usual way.
3. Some ammonium salts show *thermal dissociation* (p. 254) when heated. For example, ammonium chloride changes into ammonia gas and hydrogen chloride gas when heated, but these gases recombine on cooling to form the original solid.

$$NH_4Cl(s) \rightleftharpoons NH_3(g) + HCl(g)$$

As this particular thermal dissociation also involves a direct change from solid to vapour, it is also called a sublimation.
4. Some ammonium salts show *thermal decomposition* when heated (p. 254). For example, the carbonate decomposes when heated but the products do not recombine on cooling.

$$(NH_4)_2CO_3(s) \rightarrow 2NH_3(g) + CO_2(g) + H_2O(l)$$

5. Ammonium salts rarely form *neutral* solutions when they dissolve in water; their solutions are usually acidic, and this property is useful in providing both soil acidity and a nitrogenous fertilizer at the same time (p. 395).

Experiment 21.4
The preparation and properties of ammonium compounds*

Apparatus

Burette, 10 cm^3 pipette and pipette filler, conical flask, evaporating basin, Bunsen burner, teat pipette, tripod, gauze, spatula, rack of test-tubes.

Dilute ammonia solution, dilute sulphuric acid, phenolphthalein, calcium hydroxide solution, concentrated hydrochloric acid, sodium hydroxide solution, barium chloride solution, silver nitrate solution, dilute nitric acid, dilute hydrochloric acid, solid samples of ammonium chloride, ammonium carbonate, and ammonium sulphate, five solutions labelled A to E for analysis (see procedure (f)).

Procedure

(a) Pipette 10 cm^3 of dilute sulphuric acid into a conical flask, and add a few drops of phenolphthalein indicator. Slowly add dilute ammonia solution from a burette, swirling constantly, until the indicator just changes colour. Note the volume of ammonia solution used.
(b) Repeat (a) but without the indicator, and add the correct volume of ammonia solution to neutralize the acid.
(c) Evaporate the solution from (b) in the usual way so as to prepare crystals of the salt.
(d) Dissolve a spatula measure of ammonium chloride in a few cm^3 of deionized water in a test-tube, and find the pH of the solution.
(e) Repeat (d) using ammonium sulphate instead of ammonium chloride.
(f) The five 'solutions' labelled A to E include two samples of water, ammonium chloride solution, ammonium sulphate solution, and ammonium carbonate solution. Find out which of them are ammonium compounds, and then do further tests so that each solution can be fully identified.

(g) Pour five drops of concentrated hydrochloric acid into a test-tube, and stand it in a rack. Pour about 2 cm^3 of calcium hydroxide solution into another test-tube in the rack but do not place this tube near the first one. Add two spatula measures of ammonium carbonate to a third, dry test-tube, and heat it. During the heating, take samples of gas from the tube by using a teat pipette, and bubble the samples into the calcium hydroxide solution. When this has produced a positive test, hold the test-tube containing the concentrated acid alongside the tube being heated, and gently blow fumes from the acid over the mouth of the other tube.

Results
Record all of your observations in the usual way.

Conclusion
Make sure that you can explain all of your observations and conclusions, and give equations for all of the reactions. Choose your equation for the salt preparation from the following, and explain why you choose one of them rather than the other.

$$2NH_3(aq) + H_2SO_4(aq) \rightarrow (NH_4)_2SO_4(aq)$$

$$2NH_4OH(aq) + H_2SO_4(aq) \rightarrow (NH_4)_2SO_4(aq) + 2H_2O(l)$$

Points for discussion

1. You were not invited to perform 'rough' and then 'accurate' titrations in the last experiment before the final, 'exact' neutralization to make the salt. In this case such accuracy is not essential, because the concentration of the ammonia solution is changing slightly throughout the experiment, and the crystals obtained from the solution can *only* be the salt ammonium sulphate. Can you explain these facts?

2. The volume of ammonia solution used to neutralize the 10 cm^3 of dilute sulphuric acid may have been considerably more than 10 cm^3. You should be able to suggest *three* different reasons, any or all of which could account for the greater volume.

3. Note that you were not asked to detect the ammonia given off in procedure (g) by using damp litmus paper in the usual way. Such a test might indeed work in this reaction, but in theory it could be confused by another effect. Can you explain how this confusion could arise?

Some uses of ammonium compounds

1. Ammonium chloride, NH_4Cl, is used in dry batteries (electrolyte) and in soldering.
2. Ammonia solution is used in household cleaning agents.
3. Ammonium carbonate is used in smelling salts.
4. Ammonium nitrate is used in large quantities as a fertilizer, and also in explosives.
5. Ammonium sulphate is widely used as a fertilizer.

Nitric acid, HNO_3

Laboratory preparation of nitric acid

A typical method of preparation is shown in Figure 21.8. Excess heating is avoided, to prevent thermal decomposition of the acid. An all-glass apparatus must be used as the concentrated acid rapidly attacks cork and rubber.

$$H_2SO_4(l) + KNO_3(s) \rightarrow HNO_3(g) + KHSO_4(s).$$

(Compare this reaction with that between concentrated sulphuric acid and a chloride, on page 345).

Physical properties of nitric acid

The acid prepared as above contains more than 90 per cent nitric acid, and is called fuming nitric acid. Concentrated nitric acid contains about 68 per cent nitric acid. The pure acid is a dense, oily, colourless liquid, but it and its concentrated solution

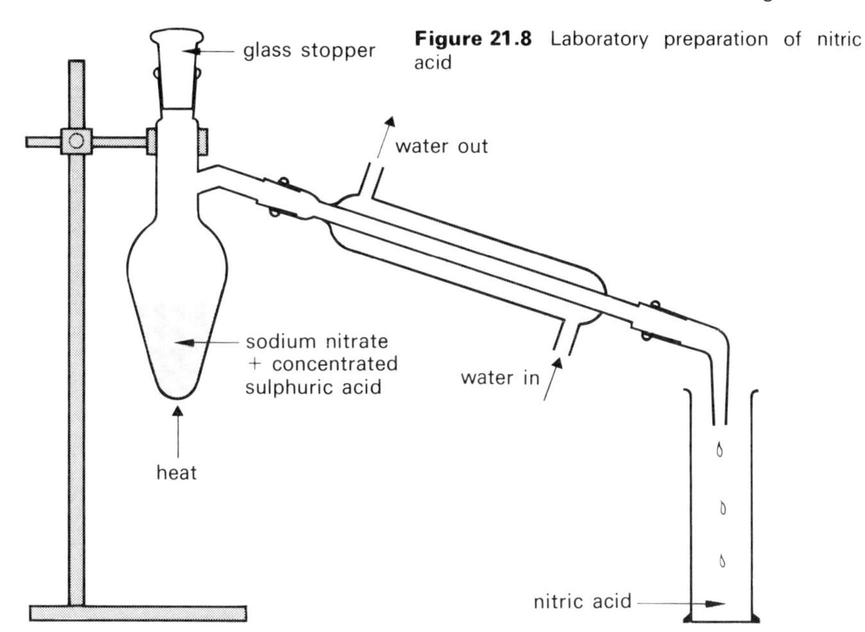

Figure 21.8 Laboratory preparation of nitric acid

are often yellow because of dissolved oxides of nitrogen formed by the slow decomposition of the acid in daylight. This is why the concentrated acid is often stored in brown bottles.

Chemical properties of dilute nitric acid

The properties of dilute nitric acid are considered separately from the chemical properties of the concentrated acid, which has some *very different* properties. (See also Table 5.5, page 75.)

The dilute acid is a typical strong acid except in its reactions with metals (p. 251). For example, it reacts with metal carbonates to form a salt, carbon dioxide, and water, e.g.

$$MgCO_3(s) + 2HNO_3(aq) \rightarrow Mg(NO_3)_2(aq) + CO_2(g) + H_2O(l)$$

It will neutralize bases and alkalis to form a salt and water only, e.g.

$$CuO(s) + 2HNO_3(aq) \rightarrow Cu(NO_3)_2(aq) + H_2O(l)$$

or

$$NaOH(aq) + HNO_3(aq) \rightarrow NaNO_3(aq) + H_2O(l)$$

and it gives the usual colour changes with indicators.

The chemical properties of concentrated nitric acid

Thermal decomposition When heated, concentrated nitric acid decomposes rapidly, forming brown fumes of nitrogen dioxide:

$$4HNO_3(aq) \rightarrow 4NO_2(g) + 2H_2O(l) + O_2(g)$$

This can be shown experimentally by using the apparatus shown in Figure 21.9. Fairly strong heating is required, and rubber and cork cannot be used, as explained earlier. The nitrogen dioxide produced dissolves in the water and the presence of oxygen in the collection tube can be confirmed in the usual way.

Oxidizing properties of concentrated nitric acid The concentrated acid is a

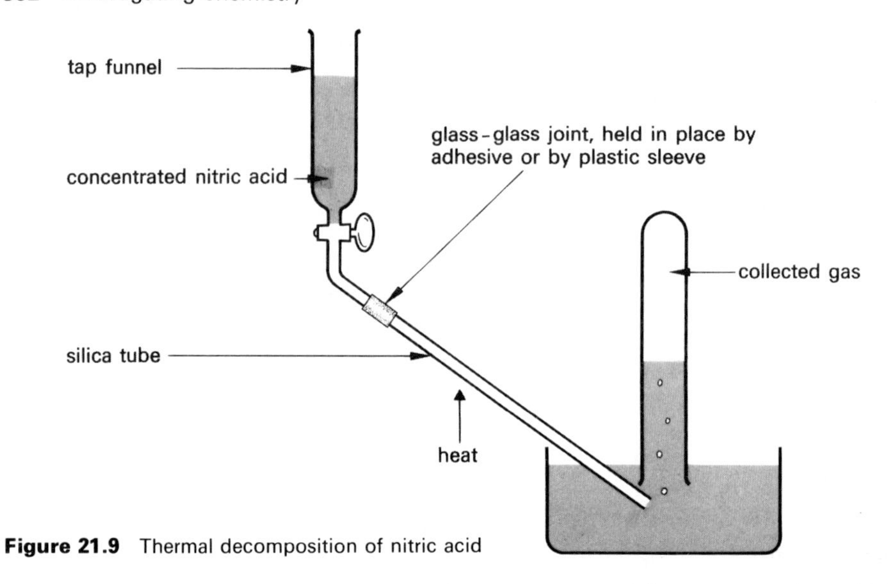

Figure 21.9 Thermal decomposition of nitric acid

powerful oxidizing agent. (The dilute acid also shows oxidizing properties, which is why it does not liberate hydrogen when added to metals; the hydrogen first formed is oxidized to water, and the acid is reduced to oxides of nitrogen in the process.)

It is sometimes convenient to assume that concentrated nitric acid is a powerful oxidizing agent because it can easily provide *oxygen* (as in its thermal decomposition above). In other cases it is more convenient to explain its oxidizing power by its ability to remove *electrons* from the chemical being oxidized. It is unlikely that you will need to write full, balanced equations when nitric acid acts as an oxidizing agent, and equations of the following type are perfectly acceptable.

$$Fe^{2+}(aq) \quad - e^- \quad \rightarrow Fe^{3+}(aq)$$

(from solution of iron(II) salt) (to concentrated nitric acid)

$$2I^-(aq) \quad - \quad 2e^- \quad \rightarrow I_2(s)$$

(e.g. from potassium iodide solution) (to concentrated nitric acid)

$$H_2S(g) + [O] \rightarrow H_2O(l) + S(s)$$

(from concentrated nitric acid)

Note that when nitric acid oxidizes something, it is always reduced to brown fumes of nitrogen dioxide, and water. This is an important factor in describing what is *seen* during these reactions.

Experiment 21.5
Some oxidizing reactions of concentrated nitric acid**

Apparatus
Rack of test-tubes, teat pipette, stirring rod. Concentrated nitric acid, sodium hydroxide solution, freshly prepared iron(II) sulphate solution, potassium iodide solution.

Procedure
(a) Pour a little potassium iodide solution into a test-tube and add a few *drops* of concentrated nitric acid (CARE!). Note any changes taking place.

(b) Pour a few cm^3 of the iron(II) sulphate solution into a test-tube, and add sodium hydroxide solution slowly, with stirring, until a precipitate appears. Note the colour of the precipitate *carefully*.

(c) Pour a few cm^3 of iron(II) sulphate solution into a third test-tube. Add a few *drops* of concentrated nitric acid, slowly and carefully. Stir the mixture. Add sodium hydroxide solution slowly, with stirring, until a precipitate forms. Compare the colour of the precipitate with that *first* formed (it may have changed slightly by now) in (b).

Results

Record all of your observations in the usual way.

Notes about the experiment

You should know the significance of the colours of the precipitates produced when working with solutions of iron salts. The colour of the precipitate obtained in (b) may not have been quite what you expected, but there should have been an obvious difference between it and the colour obtained in (c). You should be able to explain why the colours appeared as they did, and why one of the precipitates may have changed colour.

Conclusion

Explain all of your observations, and their significance. Give equations for the reactions, and describe clearly what is being reduced or oxidized in each case.

Points for discussion

1. The procedure in (c) could not have been used in order to prepare a pure sample of the solid salt iron(III) sulphate, starting from iron(II) sulphate. Why not? How could the procedure have been modified so as to produce a pure sample of iron(III) sulphate?

2. You may have noticed that it took a greater volume of sodium hydroxide solution to produce a precipitate in (c) than it did to produce one in (b). Can you explain why?

Uses of nitric acid

Large quantities of the acid are used in the manufacture of fertilizers, such as ammonium nitrate. It is also a very important reagent in many organic reactions, being used industrially to prepare organic-based chemicals such as explosives and dyes.

21.3 THE NITROGEN CYCLE; FERTILIZERS AND OTHER AGRICULTURAL CHEMICALS

The nitrogen cycle

Nitrogen is essential to all living organisms. It is present in their tissues in compounds called proteins. The soil contains nitrate ions and ammonium ions, both of which contain nitrogen atoms. Nitrogen compounds enter the soil in animal manure, and by the natural decay of plant and animal tissue. They are converted by bacterial and chemical action into nitrates.

Nitrogen compounds also enter the soil in other ways. Lightning causes some of the nitrogen and oxygen in the air to combine to form oxides of nitrogen. These then dissolve in water to form very dilute nitric acid which is washed into the soil. Oxides of nitrogen also enter the air from other sources (p. 145). The amount of nitrogen added to the soil from rain water is greater than one might think; on average, between 10 and 15 kg of nitrogen fall in rain on each hectare of land each year in Britain. (One hectare = 10 000 m^2 or 2.47 acres.)

Other sources of 'soil nitrogen' include the roots of leguminous plants, e.g. peas, beans, and clover. Such plants have nodules on their roots which contain colonies of bacteria capable of converting atmospheric nitrogen directly into soluble nitrogen compounds. All of these various nitrogen compounds either enter the soil directly as nitrates, or are converted into nitrates by natural reactions in the soil.

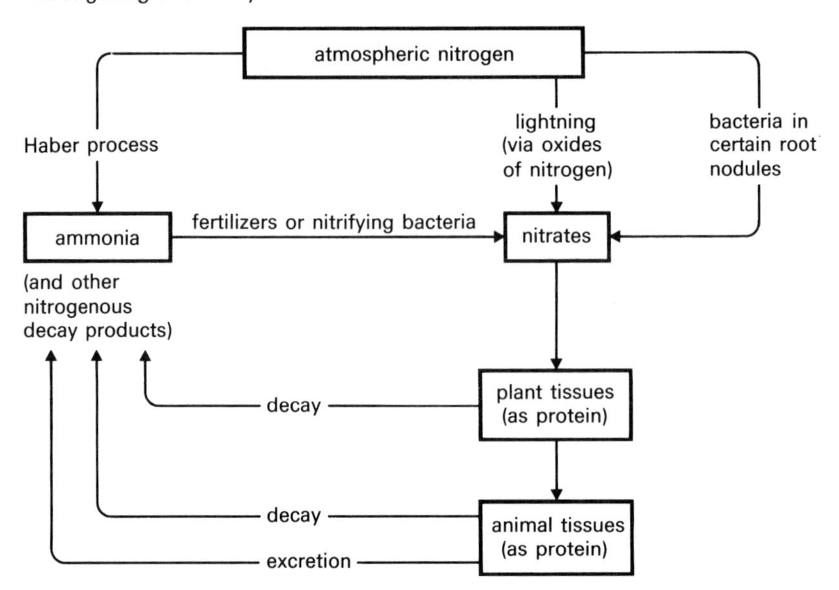

Figure 21.10 The nitrogen cycle

Plants obtain their 'nitrogen' by absorbing these nitrates in solution through their root systems, and then use them to make proteins. Herbivorous animals eat plants and so obtain their necessary nitrogen; they in turn are eaten by carnivorous animals, who thus obtain the proteins they need. These processes are part of the nitrogen cycle, which is illustrated in Figure 21.10. The world contains an almost constant number of nitrogen atoms, which are circulating continually in different compounds. The use of nitrogen by living things is a cyclic process like the carbon cycle (p. 373), and in a similar way nitrogen atoms present in your body tissue will also have had an interesting 'history'!

Most plants can only absorb nitrogen in the form of nitrates because these are all *soluble*. Although the air contains about 80 per cent of nitrogen by volume, this cannot be used directly by most plants and so it has to be converted into a soluble form, i.e. it has to be *fixed*. This fixed nitrogen needed by plants was at one time supplied naturally from the nitrogen cycle, but the natural cycle has been disturbed by man's need to produce ever-increasing supplies of food. If crops are constantly harvested (i.e. the vegetation is not left to decay on the spot) the soil becomes low in nitrogen compounds. It then becomes important to fix nitrogen artificially (usually via Haber process → ammonia → fertilizers) and to feed it to the soil to supplement the natural cycle. In other words, we are forcing some of the nitrogen atoms which at the moment are unhelpful to us (e.g. those in the air) to become part of useful, soluble nitrogen compounds. These soluble nitrogen compounds may vary (e.g. ammonium salts, urea, nitrates, or ammonia itself), but natural reactions in the soil convert the nitrogen atoms in the compounds into nitrate ions, and these are then absorbed by plants.

Why else do we need to add chemicals to the soil?

A variety of chemicals are available for use in agriculture. Some of these are very simple, e.g. fertilizers and the chemicals used to control the pH of the soil. Others are quite complicated organic compounds, such as the many pesticides, fungicides, and herbicides now available.

Chemists have to find answers to many problems facing modern civilization, especially increasing demands for energy and for food. The world population continues to increase at an alarming rate, and it is becoming more and more difficult to provide anything like enough food. At least half of the people in the world are suffering from some kind of food shortage. This is particularly serious in the developing countries, where natural resources are often limited or unexploited and where technology is improving only slowly. Each year there are some 80 million extra mouths in the world to be fed, i.e. a world increase much greater than the total population of the United Kingdom!

One solution to the problem would be to use more land for agricultural purposes, so that more food crops could be grown. This is not a long-term answer, however, for most of the undeveloped land which remains is not suitable for agricultural use. The problem is that the number of mouths to be fed is increasing but soon no more land can be made available to grow food.

The problem can only be solved (or partially solved) by improving the amount of food which can be obtained from a given area of land and by taking steps to reduce the birth rate. Biologists are involved with these problems, e.g. in helping to control the birth rate and in developing new strains of wheat, maize, etc. which produce better yields of food. Chemists are concerned with the development of fertilizers, pesticides, herbicides, fungicides, and various minerals and vitamins which are added to the food given to farm animals. Pesticides kill insects and other pests which attack food crops and reduce their yield (Figure 21.11). Fungicides kill fungi which cause plant disease. In Britain alone, over £50 million of cereal crops are sometimes lost through disease in one year, and this figure would be far greater if fungicides and pesticides were not available. Herbicides kill weeds which would otherwise prevent some soil nutrients, water, air, and light reaching the crop. Similarly, farm animals produce food more efficiently if they are given supplements in their diet, and are kept free from disease by the use of sprays, dips, and injections.

Figure 21.11 Complete destruction of foliage of potato plant by potato blight (*Shell*)

Steps have to be taken to ensure that agricultural chemicals are not wrongly used. Some fertilizers, if used in excess and in the wrong situation, can be washed into water courses and lead to pollution (p. 196). Careless use of pesticides can also kill *helpful* insects (e.g. bees and ladybirds) as well as the harmful pests. If a pesticide is *persistent* (i.e. is not broken down quickly to harmless substances by animal cells, plant cells, bacteria, or ultraviolet light from the sun) it may become concentrated in a food chain (see page 196). Chemists are constantly improving chemicals used in agriculture, and people are being trained more thoroughly in the correct use of these chemicals.

The use of fertilizers and other chemicals has greatly improved the yields obtained from food crops. The correct use of suitable pesticides etc. is just as important in increasing these yields as the use of fertilizers. The technological advances in farm machinery, and in improving irrigation also play a part. In this chapter we will consider only one of these factors in detail: the use of fertilizers.

Fertilizers

What chemicals do plants need?

In photosynthesis, plants produce carbohydrate foods (sugars and starches) by using carbon dioxide from the air, water taken up by the roots, and energy from sunlight. In addition plants need many other substances for growth and for the chemical reactions which take place in their cells. These other substances are usually absorbed in solution through the roots. All plants probably need at least traces of the great majority of the elements, although it is difficult to be sure about some of them. The need for some elements is fairly obvious, however, and we will concentrate on these.

The chemicals needed in the soil can be divided into three main groups. There are some elements which are needed in very small amounts, but which nevertheless are essential for plant life. These elements are called *trace elements*; they include boron. copper, zinc, and manganese. Soils normally contain enough of these trace elements. They are rarely added, except where a particular crop has a special need or where the soil lacks one or more trace elements.

Another group of elements is needed in fairly small amounts, but in greater quantities than the trace elements. These are often called the *secondary nutrients*. Examples include magnesium, sodium, sulphur, and calcium (see Figure 21.12). The main secondary nutrient, calcium, is normally provided by adding lime to the soil. In the past it was sometimes necessary to add sulphur compounds to soil to replace the sulphur taken up by plants. We now put so much sulphur dioxide into the air (p. 143) that between 15 and 50 kg of sulphur each year fall (in rain) on each hectare of land in the UK and sulphur is no longer applied artificially to the soil.

Three elements are needed in the soil in much greater quantities. These are nitrogen, phosphorus, and potassium, and they are known as *major nutrients*. The 'big three' are often referred to in agriculture by their chemical symbols, i.e. N, P, and K. To put all of this in perspective, a particular vegetable crop could remove 200 kg of nitrogen, 18 kg of phosphorus, and 144 kg of potassium from a hectare of soil, but only 30 kg in total of trace elements and secondary nutrients.

The functions of the major nutrients in plants, and the effects of excess and deficiency are summarized in Table 21.2.

The big three

Figure 21.12 The effect of magnesium deficiency on the leaf of a sugar beet plant. Note the pale leaf colour due to lack of chlorophyll (*Fisons*)

Table 21.2 Why nitrogen, phosphorus, and potassium are needed by plants

	Nitrogen (N)	*Phosphorus (P)*	*Potassium (K)*
Why plants need the mineral	Increases growth of stem and leaf. Helps build up chlorophyll	For root growth (∴ especially important for root crops). Accelerates ripening of the crop and also seed formation. Strengthens stems. Essential for seed germination	Not fully understood, but seems particularly important in the development of fruit and seeds, and in maintaining the general chemical reactions occurring in the plant. Promotes resistance to disease
Outward signs of deficiency	Undersized leaves, poor growth rate. Leaves pale green or often yellow	Poor root system. Stunted growth, small leaves. Leaves fall prematurely, less blossom than normal, low yield of grain or fruit. Dull, blue-green leaves	Edges of leaves turn yellow and eventually brown. Leaves may die early
Outward signs of excess	Growth too rapid—plant becomes too soft and is less resistant to disease and bad weather	Crop ripens too early, ∴ yields low	—

Experiment 21.6
To illustrate the effect of nutrients on plant growth**

(This experiment is a modification of one published by ICI.)
Note: Unlike many ordinary chemistry experiments, this one will take some time. Regard it as a long-term experiment, and one not to be used in winter unless appropriate facilities are available. Each group can be responsible for one of the plant pots.

Apparatus
Thirteen 8 cm plant pots, each with 'saucer', grass seed, horticultural sand (thoroughly washed), stones or broken pot

Table 21.3 The nutrient solutions for Experiment 21.6

Chemical formula	*Concentration in mol dm^{-3}*	*Concentration of main nutrient(s) in mol dm^{-3}*	*Concentration of solute in g dm^{-3}*
K_2SO_4	0.03	K 0.015	1.30
KNO_3	0.015	K 0.015 and N 0.015	1.50
$CaCl_2.6H_2O$	0.005	Ca 0.005	1.10
$NaH_2PO_4.2H_2O$	0.0015	P 0.0015 and Na 0.0015	0.23
$FeSO_4.7H_2O$	0.0001	Fe 0.0001	0.02
NaCl	0.0001	Na 0.0001	0.006
H_3BO_3	0.000 015	B 0.000 015	0.001
$CuSO_4.5H_2O$	0.000 01	Cu 0.000 01	0.0025
$ZnSO_4.7H_2O$	0.000 001	Zn 0.000 001	0.0003

(thoroughly washed) for the bottom of each plant pot, measuring cylinder.
Solutions of plant nutrients as shown in Table 21.3. (If possible, the chemicals should be 'Analar' grade, and the solutions must be made up in deionized water.)

Procedure
(a) Make out a table for your results as shown in the lower part of Table 21.4.
(b) Place a layer of stones or broken pot at the bottom of each plant pot, and then fill each pot with washed sand.
(c) Decide upon a suitable mass of grass seed which will thickly cover the surface of one of the pots. Weigh out this amount for each of the pots, and sow the seed evenly over the surface of the sand in each pot. Stand each pot in its saucer.
(d) Each pot must be 'watered' regularly with exactly the same volume of the appropriate solution. Study Table 21.4 carefully, and make up labelled solutions (i.e. mixtures) for each pot. Each of your solutions must be made up to the same total volume, i.e. that for the complete set of nutrients, by adding deionized water. Make sure that the pots are labelled too. Note that some elements appear in more than one solution, and the potassium sulphate solution is used *only* to provide potassium ions (but not nitrogen) when the potassium nitrate is omitted. The measuring cylinder must be *thoroughly* washed out with deionized water before a new solution is used.
(e) 'Water' each pot with the appropriate liquid at regular intervals. Use the same volume of each liquid each time. The exact volume you use, and the frequency with which it should be applied, depend upon the circumstances in which you do the experiment. (The volumes in Table 21.4 are for guidance only.) Be consistent, and do not allow the sand to dry out.
(f) When the grass in any of the pots is more than 2.5 cm high, cut the grass down to 2.0 cm and keep all of the cuttings. Weigh the cuttings immediately and enter the mass in the appropriate column of the results table. The first time you make a cutting counts as 'week 1', and from this point cut the grass at the same time each week, always cutting down to 2.0 cm height and weighing the cuttings from each pot. Repeat this for several weeks until you have sufficient data.

Results
Complete the table of results, and plot the results on a suitable bar graph.

Conclusion
Make your own conclusion.

Points for discussion

1. Why do you think that sand was used in Experiment 21.6, rather than soil or a seed compost?
2. Can you explain why the concentrations of copper and zinc compounds used in the experiment were kept very low?
3. Why was it important to make up the *total* volume of each of your mixtures to the same volume, by using deionized water?

Table 21.4 Typical mixtures for Experiment 21.6

	Volume of nutrient solution used (cm^3)												
Pot number →	1	2	3	4	5	6	7	8	9	10	11	12	13
Nutrient ↓													
K_2SO_4	—	—	—	5	—	—	—	—	—	—	—	—	—
KNO_3	—	5	—	—	5	5	5	5	5	5	5	—	5
$CaCl_2.6H_2O$	—	5	5	5	—	5	5	—	5	5	5	5	—
$NaH_2PO_4.2H_2O$	—	5	5	5	5	—	5	5	5	5	5	—	5
$FeSO_4.7H_2O$	—	5	5	5	5	5	—	5	5	5	5	5	—
NaCl	—	5	5	5	5	5	5	—	5	5	5	5	5
H_3BO_3	—	5	5	5	5	5	5	5	—	5	5	5	—
$CuSO_4.5H_2O$	—	5	5	5	5	5	5	5	5	—	5	5	—
$ZnSO_4.7H_2O$	—	5	5	5	5	5	5	5	5	5	—	5	—
Deionized water	50	10	15	10	15	15	15	20	15	15	15	20	35
Total volume	50	50	50	50	50	50	50	50	50	50	50	50	50
Results													
Components:	Water only	Complete set	K and N missing	N missing	Ca missing	P missing	Fe missing	Cl missing	B missing	Cu missing	Zn missing	N, P, K, missing	Fe, Cu, Zn, Ca, B missing
Mass of cut grass week 1													
Mass of cut grass week 2													
etc. as required													

The chemicals used as fertilizers

It is important to remember that the 'elements' which plants need must be provided in *soluble* compound form so that they can be absorbed in solution by the roots. As explained earlier, plants are surrounded by nitrogen (the element) in the air, but they cannot use this nitrogen directly; it has to be *fixed*, i.e. converted into a soluble compound containing nitrogen. The nitrogen cycle provides a continuing supply of nitrogen in the soil, but this no longer replaces the nitrogen taken out by the intensive cropping programmes we now need to use. Similarly, phosphorus and potassium are removed at faster rates than they are replaced naturally.

Animal manure and rotted vegetable waste (compost) are added to soils, and these contain nitrogen, phosphorus, potassium, and secondary and trace elements. The percentages of nitrogen, phosphorus, and potassium in manures and composts are very small (see Table 21.5) and variable. They cannot replace the amounts removed by most crops. However, manure and compost are important in other ways, for they do add trace elements to the soil and they provide *organic* material (humus) which greatly improves the condition of the soil.

In the past, a shortage of plant nutrients in the soil was not common because we did not need to grow as much food. Farmers used crop rotation, in which each field would have a turn to be left fallow for a year. Only small quantities of plant foods were removed from the field during this year, and a dressing of manure and the action of natural processes (e.g. the nitrogen cycle) would put the soil in 'good heart' for a future crop, especially if the crop was to be of a different type to that grown previously. Better still, farmers would sow a field with clover (or a similar type of plant) for a year. Clover is a *leguminous* plant (p. 383), and bacteria living in the root nodules can fix nitrogen directly from the air and convert it into soluble nitrogen compounds. The clover would be cut and fed to animals, usually in the form of hay, whilst the roots with their valuable nitrogen-containing nodules would be ploughed back into the soil, thus adding more nitrogen to it during its 'recovery' programme. These practices are still carried out, but to a much lesser degree than before. Nowadays, farmers in some areas simply cannot afford to have a plot of land which is not producing crops for a year, and the only way of replacing the plant foods in agricultural land is to add suitable soluble compounds containing the essential elements.

Nitrogenous fertilizers Ammonium nitrate is made from ammonia (produced by the Haber process) and nitric acid, which is itself made from ammonia. It is perhaps the most important source of nitrogen for agriculture in the world. Ammonium sulphate is usually obtained from ammonia produced as a byproduct of the steel and fibre industries, but is now less widely used as a source of nitrogen. Other sources of nitrogen include liquid ammonia (which is injected into the soil), metal nitrates such as sodium nitrate, and, of growing importance, urea $(NH_2)_2CO$. Urea is now used all over the world, but particularly in the Tropics. It is made by reacting ammonia from the Haber process with carbon dioxide.

Fertilizers containing phosphorus Most fertilizers containing phosphorus are phosphates, and these are made from phosphate rock, which is mainly $Ca_3(PO_4)_2$. Most of the phosphate rock imported by Britain comes from Morocco, Senegal, Florida, and Tunisia. 1.8 million tonnes were imported in 1977, of which 1.3 million tonnes were converted into fertilizers (Figure 21.13). The rock is normally ground to a powder in the country of origin, and then transported. Rock phosphate itself is almost insoluble in water and is therefore not very useful as a fertilizer, so it is converted into more soluble forms. Other sources of phosphate used as fertilizers

Figure 21.13 Ground rock phosphate in the bulk raw materials store at Fisons' Immingham factory

include bone meal and the basic slag produced as a byproduct of the steel industry, but only a very small amount (in percentage terms) of bone meal is used for fertilizers nowadays, and supplies of basic slag are diminishing as new steel-making processes are adopted.

Potassium fertilizers Until recently, most of the potassium compounds used as fertilizers in the United Kingdom (see Figure 21.14) were imported as potassium chloride from Germany, France, and Israel. Britain has large deposits of sylvinite, which contains a mixture of potassium chloride and sodium chloride, near Whitby in north-east Yorkshire. In the 1970s the Boulby mine was developed to utilize this resource. The sylvinite occurs in layers over 10 m thick at a depth of 1000 m, and the Boulby mine now supplies an increasing proportion of our potassium needs.

In practice, many commercial fertilizers contain mixtures of several individual fertilizers, as is explained in the next section.

Figure 21.14 Mining a mineral containing potassium compounds, for use as a fertilizer (*ICI, Agricultural Division*)

Points for discussion

What are the advantages of leaving the *roots* of cereal and other crops in the ground after harvesting? Why do the roots of beans, peas, and clover bring special advantages?

Types of fertilizer

Most fertilizers are used in solid form, although there is a gradually increasing demand for liquid fertilizers. Fertilizers can be *straight fertilizers*, in which only one of the elements nitrogen, phosphorus, and potassium is supplied, or *compound fertilizers* in which two or all three of the elements are supplied. Compound fertilizers may also include trace elements. In general the compound fertilizers tend to be in greater demand than the straight fertilizers.

NPK values

The proportion of a particular element in a fertilizer is expressed as its per cent by total mass, and the symbol of the element is used for identification. We therefore refer to NPK values, i.e. the per cent by mass of nitrogen, phosphorus, and potassium in a given fertilizer. Unfortunately, old names for chemicals are still commonly used in agriculture, and it is important to understand how these per cent values are measured and to what they refer.

People often refer to potassium compounds as 'potash' compounds, e.g. sulphate of potash is potassium sulphate. The proportion of potassium in a fertilizer was traditionally measured by quoting the per cent by mass of potassium in the *oxide form*, K_2O. Similarly, the per cent of phosphorus was traditionally measured as the per cent by mass of phosphorus as the *oxide form*, P_2O_5. (The use of the oxide as the standard was largely determined by the methods of analysis then available, which depended on combustion of the materials to the oxide.) The element we are interested in (phosphorus or potassium) is only *part* of the percentage thus quoted (i.e. the per cent by mass of potassium is only part of the value quoted for K_2O, which includes oxygen as well). All bags of fertilizer must, by law, show a clear statement of the nutrient composition in terms of the element as well as the oxide form, and so if you *look* at the declaration on the container there should be no ambiguity (see Figure 21.16). On the other hand, the *oral* statement '10 per cent potash' may mean 10 per cent of K_2O or 10 per cent of the element potassium, which are very different things. Nitrogen, fortunately, is only quoted in elemental form.

Why are many different kinds of fertilizer made?

Each of the elements nitrogen, phosphorus, and potassium is essential for plant growth, but the ratios in which they are required (by mass) vary from one plant species to another. For example, a farmer may apply 160 kg of nitrogen, 300 kg of phosphorus (as P_2O_5) and 300 kg of potassium (as K_2O) per hectare in order to grow potatoes. For the same soil conditions, the applications for a hectare of wheat would be of the order of 75 kg of nitrogen, 60 kg of phosphorus (P_2O_5), and 60 kg of potassium (K_2O).

The fertilizer requirements for a particular crop also depend upon the soil conditions, e.g. the type of soil, the way in which it has been treated for previous crops, and rainfall in the area. Phosphates, which provide most of the phosphorus which plants need, are firmly held in the soil (especially in clays) and therefore only a very small proportion of phosphate fertilizer, if any, is washed away by rain water (i.e. *leached* from the soil). This is also true of potassium compounds, but nitrogen

Figure 21.15 This farmer is about to apply a special 'grass' fertilizer to land which is used for grazing. He is checking that the mechanism is at the correct height above the foliage to ensure even application. The compound fertilizer analysis is shown clearly on the bag: 29% nitrogen 5% phosphate (as P_2O_5) and 5% potassium (as K_2O). Note the high 'N' factor to ensure rapid growth of the grass (*Fisons*)

compounds (e.g. nitrates) are more readily leached out of soil. (Nutrients are leached more readily from sandy soils than from clay soils. The fertilizer requirements for a potato crop given earlier were for a sandy soil; much smaller quantities would be applied to clay soils and those previously treated with fertilizers.)

Thus nitrogen compounds are usually added more frequently and/or in greater proportions than phosphates and potassium compounds, partly to replace natural losses (e.g. by leaching) but mainly because plants respond very quickly to nitrogenous fertilizers (see Figure 21.15). On the other hand, the temptation to add *excess* nitrogen must be avoided. An excess of nitrogen can provide problems both for the plant (see Table 21.2) and also by the excess being washed into rivers (p. 196). It would also be a wasteful use of an expensive product.

To account for variations in soil conditions and crop needs, a fertilizer manufacturer may produce 9 different NPK fertilizers, 1 PK fertilizer, 1 NK fertilizer, and several straight nitrogen fertilizers (see Figure 21.16).

Figure 21.16 Some of the range of fertilizers produced by ICI Agricultural Division

The main chemicals used to provide nitrogen, phosphorus, and potassium are summarized in Table 21.5, together with any particular advantages each may have to offer.

Table 21.5 Some common examples of straight and compound fertilizers

Fertilizer	*Formula*	*Nutrients supplied, with per cent by mass of nutrient in elemental form*	*Other features*
		(Assuming that the compound is supplied in the pure, anhydrous form, which is not always the case)	
Manure	—	N (e.g.) 0.5% P (e.g.) 0.15% K (e.g.) 0.4%	Provides humus and trace elements
Urea	$(NH_2)_2CO$	N 46%	Very economic way of transporting N in 'solid' form. Can also be used in solution to spray foliage, when it is absorbed by the leaves
Liquid ammonia	$NH_3(l)$	N 82.5%	Must be injected into the soil
Ammonium nitrate	NH_4NO_3	N } See question **2** below	Very soluble—rapid action. Slightly acidic
Ammonium sulphate	$(NH_4)_2SO_4$	N } See question **2** below	As above
Potassium sulphate	K_2SO_4	K 44.8%	Very soluble
Potassium chloride	KCl	K 52.3%	Very soluble
Triple superphosphate		P 20.0%	Dissolves slowly
Diammonium phosphate		N 18%, P 20%	Dissolve slowly
Monoammonium phosphate		N 12%, P 23%	Dissolve slowly
Bone meal	—	P (% varies)	Dissolve slowly
Basic slag	—	P (% varies)	Dissolve slowly
Compound fertilizers (e.g. Growmore) Varies according to type	—	e.g. N 7% P 7% K 7%	Contains all three main elements in a balanced form, and sometimes trace elements

Check your understanding

1. In Table 21.5, which of the compound fertilizers (a), (b), or (c) might be described as a high potash fertilizer?
2. In Table 21.5, the percentage composition by mass of some of the fertilizers has not been included. Calculate the N values in (i) ammonium sulphate and (ii) ammonium nitrate.
3. Why is it more efficient to use a tonne of ammonium nitrate rather than a tonne of ammonium sulphate as a nitrogenous fertilizer?

Points for discussion

1. An application of soot is beneficial for soil, but soot is now more difficult to acquire. (a) Why is soot more difficult to obtain now than it used to be? (b) Can you suggest why soot is beneficial to the soil? (If in doubt, see the destructive distillation of coal, page 501.)
2. Although only small amounts of 'nitrogen' enter river water because of fertilizer applications, they increase the 'Biological Oxygen Demand' (BOD) of the water. Can you explain what this means, and why nitrogen compounds have this effect? (See page 197 if in doubt.)

3. Here is a typical example of some information available to a farmer.
To convert per cent by mass of P_2O_5 to per cent by mass of phosphorus (the element), multiply by 0.4366. To convert per cent by mass of K_2O to per cent by mass of potassium (as element), multiply by 0.83.
Can you explain how the numbers 0.4366 and 0.83 are derived?

4. Why do you think that of all the solid nitrogenous fertilizers, urea, $(NH_2)_2CO$, is the most economical to transport over long distances?

5. Another old name still used generally by farmers and fertilizer manufacturers is muriate. Can you find out the modern name for muriate of potash?

6. Suppose that you had to produce a 50 kg bag of mixed compound fertilizer of the following specification (all figures are for the *elemental* form): N 25 per cent, P 12 per cent, and K 9 per cent. You can make up the fertilizer by using ammonium nitrate, ammonium phosphate, $(NH_4)_3PO_4$, and potassium chloride. How much of each substance should you weigh out? (Care: this is not an easy calculation!) Show your working clearly.

7. Suppose that you bought a 25 kg bag of fertilizer which you thought had a P value of 15 per cent, but in reality the 15 per cent referred to phosphorus as P_2O_5. (i) What mass of phosphorus (as 'element') would you have expected to have in the bag? (ii) What was the actual mass of 'elemental phosphorus' in the bag?

8. Liquid ammonia, b.p. $-33.4\ ^\circ C$, is used as a relatively cheap nitrogenous fertilizer in the USA and on the Continent, and is injected about 12 cm below the soil surface when conditions are suitable. Aqueous ammonia (containing about 25 per cent N), is also commonly used as a straight N fertilizer.
(a) Why is the liquid ammonia not sprayed directly on to the soil?
(b) What are the suitable conditions of the soil needed before injecting the liquid?
(c) Why has this fertilizer not found widespread application in Britain?
(d) Which form of 'liquid' ammonia is sometimes referred to as *anhydrous ammonia*?

9. Why is it unwise to apply lime to soil which has been recently manured?

Soil pH

If soils are left untreated they tend to become acidic because calcium compounds (which are used to neutralize acidity) are gradually leached out, and 'acids' are continuously added in rain water (e.g. as oxides of sulphur and oxides of nitrogen). Although an acid soil is useful for growing certain plants, many vegetables and farm crops give their best yields when grown in soils which are neutral or only slightly acidic (see Table 21.6). Farmers and gardeners therefore add either calcium hydroxide ('hydrated lime') or powdered calcium carbonate (ground limestone) to ensure that soil pH is kept fairly close to the neutral point. Both of these react with acids, but the former must be used more carefully than the latter for it is a weak alkali and an excess of it could make areas of the soil alkaline. An excess of limestone causes no problems because it is almost insoluble.

The application of a calcium compound also supplies the calcium ions which plants need, and helps to prevent diseases connected with 'lime deficiency' or acid soils, such as attacks by the club root fungus which affects members of the brassica family (cabbages, sprouts, etc.).

Some kinds of fruit and other plants are best grown on an acid soil, e.g. blueberries and heathers. In cases like these, gardeners often use ammonium salts to 'feed' nitrogen to the plants (instead of other nitrogenous fertilizers) because most ammonium salts are acidic in solution.

The effect of fertilizer on crop production

Now that you have a simple understanding of the kind of chemical principles involved in agriculture, it is appropriate to quote some facts to illustrate the actual gains which have been achieved by developing agricultural chemistry.

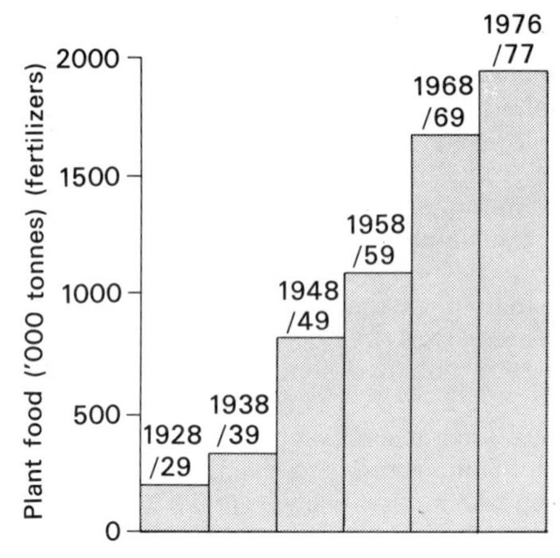

Figure 21.17 The growth in fertilizer useage in the UK (*Fertilizer Manufacturers Association*)

In 1910, some 10 per cent of the crop production in Western Europe could be attributed to fertilizer usage. By 1930 this had grown to 20 per cent, and in 1980 more than 35 per cent of crop production was due to the application of fertilizers. Figure 21.17 shows the increased use of fertilizers in the United Kingdom between 1928 and 1977, and the improved yields of wheat, barley, and potatoes produced by these applications of fertilizer are summarized in Table 21.7.

It has been estimated that, in order to keep pace with the growth in population, about £5000 million will have to be invested every year for the remainder of this century in order to produce new fertilizer factories (including Haber process etc.) and to improve distribution methods. It is also important to realize that meat-producing animals and dairy cattle feed on grass and that grassland is also given fertilizers. It has been shown that a hectare of grazing land can easily support twice the normal number of cows if nitrogenous fertilizers are applied to it (Figure 21.15).

Table 21.6 Optimum pH range for growing some common crops

Crop	*Optimum pH*
Barley	6.5–7.8
Beetroot	6.0–7.5
Broccoli	6.0–7.5
Brussels sprouts	6.0–7.5
Cabbages	6.0–7.5
Carrots and parsnips	5.5–7.0
Cauliflower	6.0–7.5
Lettuce	6.0–7.0
Leeks	6.0–8.0
Onions	6.0–7.0
Peas and beans	6.0–7.5
Potatoes	4.8–6.5
Rhubarb	5.5–7.0
Tomatoes	5.5–7.0
Turnips and swedes	5.5–6.8

Table 21.7 Improved yields of wheat, barley, and potatoes due to application of fertilizers

	Crop yield in tonnes per hectare		
Year	*Wheat*	*Barley*	*Potatoes*
1934–8	2.31	2.09	16.92
1948	2.60	2.44	19.20
1958	3.08	2.89	17.00
1968	3.55	3.45	24.60
1977	4.92	4.45	28.60

(Information reproduced by courtesy of the Fertiliser Manufacturers Association)

Point for discussion

It would be wrong to assume that the improved yields of wheat, barley, and potatoes quoted in Table 21.7 are due to the action of fertilizers alone. Make a list of other factors which will have helped to improve the yields.

21.4 A SUMMARY OF CHAPTER 21

Chapter 21 contains many facts. This summary is mainly a list of 'revision headings' which should help you to organize the work when you need to revise it. Make sure that you learn the following.

(a) The general chemical properties of the metallic and non-metallic elements (Table 21.1, p. 368).

(b) The physical properties of diamond and graphite (p. 369), and the chemical properties of elemental carbon.

(c) The laboratory preparation of carbon dioxide, its physical properties, and some of its uses.

(d) The chemical properties of carbon dioxide.

(i) Normally carbon dioxide does not support combustion (e.g. it puts out a lighted taper) but burning magnesium metal continues to burn in it.

$$2Mg(s) + CO_2(g) \rightarrow 2MgO(s) + C(s)$$

A crackling is heard, and white magnesium oxide is formed, together with black specks of carbon. *Note*: The putting out of a lighted taper is *not* a test for the gas; other gases do this, e.g. nitrogen.

(ii) It is weakly acidic, and turns damp blue litmus paper a wine-red colour. Its solution *behaves* as if it were carbonic acid,

$$H_2O(l) + CO_2(g) \rightarrow H_2CO_3(aq)$$

but the acid is weak, unstable, and cannot be isolated from solution.

(iii) The normal test for the gas is to pass it into a solution of calcium hydroxide (lime water). A teat pipette is used to collect a sample of the gas and the sample is then pumped into calcium hydroxide solution. A positive test is shown by the appearance of a milkiness in the solution. The milkiness is caused by a very fine precipitate of calcium carbonate.

$$CO_2(g) + Ca(OH)_2(aq) \rightarrow CaCO_3(s) + H_2O(l)$$

Note that if carbon dioxide continues to pass into the solution, the cloudiness eventually clears again as the soluble hydrogencarbonate is formed:

$$CaCO_3(s) + CO_2(g) + H_2O(l) \rightarrow Ca(HCO_3)_2(aq)$$

This reaction is responsible for the formation of temporarily hard water (p. 187), and again shows the acidic nature of the gas. In solution it acts as though it is carbonic acid which, being a dibasic acid, reacts with the base calcium hydroxide to form *two* types of salt, a carbonate and a hydrogencarbonate (cf. sulphuric acid, page 79).

(iv) In a way similar to that described in (iii), the gas will react with the alkali sodium hydroxide, again acting as if it were carbonic acid. It forms two salts, the first one being the carbonate (which in this case is *soluble*, so that no visible change is observed).

$$2NaOH(aq) + CO_2(g) \rightarrow Na_2CO_3(aq) + H_2O(l)$$

If the gas continues to pass into the solution, a white precipitate is eventually seen, because sodium hydrogencarbonate is much less soluble than sodium carbonate.

$$Na_2CO_3(aq) + CO_2(g) + H_2O(l) \rightarrow 2NaHCO_3(aq+s)$$

Compare (iii) and (iv) very carefully.

(e) The physical and chemical properties of carbon monoxide.

(f) The carbon cycle.

(g) The physical properties of nitrogen, its lack of chemical reactivity, and some properties of nitrogen dioxide.

(h) The laboratory preparation of ammonia gas, and its physical properties including the fountain experiment.

(i) The chemical properties of ammonia gas and some of its uses.

(j) How to prepare a solution of a very soluble gas by using an inverted funnel.

(k) The physical and chemical properties of ammonia solution, and properties of ammonium compounds.

(l) The physical properties of nitric acid.

(m) Dilute nitric acid as a typical acid in all of its reactions except those with metals.

(n) The properties of concentrated nitric acid, especially its reactions as an oxidizing agent.
(o) The nitrogen cycle, its importance in agriculture and how it is related to the world food problem. The meaning of fixed nitrogen, i.e. 'nitrogen' which is in soluble form.
(p) A *general* understanding of our attempts to combat the world food problem. (Most of the facts quoted in Section 21.3 are used to illustrate the problem; they need not be learned.)
(q) Pollution which could be caused by careless use of agricultural chemicals.
(r) Why plants need nitrogen, phosphorus, and potassium.
(s) Some of the *main* fertilizers.
(t) NPK values and how they are calculated.
(u) How soil acidity can be neutralized, and how soils can be made more acidic if necessary.

Important experiment

The fountain experiment, e.g. Experiment 21.2.

QUESTIONS

1. (a) Give the names and percentages by volume of the two major elemental gases of the air.
(b) Which of these two gases cannot be used directly by green plants?
(c) What group of plants are able to utilize this gas with the aid of bacteria?
(d) Plant growth may also be encouraged by the use of artificial fertilizers. Give the name and formula of one chemical compound suitable for such use which contains the element named in (b) and calculate the percentage mass of that element in the compound.

2. Describe carefully how you would separate a pure sample of *each* substance in the following mixtures:
(a) sodium chloride solution,
(b) magnesium and sulphur,
(c) iodine and carbon (graphite). (J.M.B.)

3. (a) Draw a labelled diagram to show how gas jars of dry ammonia can be prepared and collected. Write an equation for the reaction.
(b) Describe a method of showing that ammonia is very soluble in water.
(c) Under what conditions does ammonia react with copper(II) oxide? State what would be seen during the reaction and *either* write an equation for the reaction *or* state the products.
(d) State briefly how crystals of ammonium sulphate can be made in the laboratory starting from ammonia solution. (J.M.B.)

4. (a) Explain how pesticides and artificial fertilizers could help to solve the world food problem.
(b) What environmental and social effects are likely to result from the production and use of pesticides and fertilizers on such a large scale?
(c) What part can leguminous crops (peas, beans, and clover) play in the world food problem? (A.E.B. 1976)

5. Ammonia gas can be prepared in the laboratory using the reaction represented by the equation:

$$(NH_4)_2SO_4 + Ca(OH)_2 \rightarrow CaSO_4 + 2NH_3 + 2H_2O$$

(a) State whether it is necessary to heat the mixture.
(b) Ammonia may be dried using calcium oxide. (i) How does calcium oxide dry the gas? (ii) Why is concentrated sulphuric acid an unsuitable drying agent for this purpose?
(c) State how the dry ammonia may be collected.
(d) State two simple tests by which you could identify ammonia.
(e) Calculate the volume of ammonia, measured at room temperature and pressure, obtained when 0.1 mol of ammonium sulphate reacts with calcium hydroxide according to the above equation.
(f) Explain the following experiments: (i) Ammonia is passed over heated copper(II) oxide in a combustion tube. The residue is reddish-brown, whilst a colourless gas and a colourless liquid are also formed. (ii) Aqueous ammonia can be used to distinguish a solution containing iron(II) ions from one containing iron(III) ions. (A.E.B. 1977)

6. (a) Name all the products formed and write equations for the reactions taking place at the electrodes during the electrolysis of a concentrated aqueous solution of sodium chloride between carbon electrodes.
(b) Explain the following statements and write equations for the reactions involved.
(i) A solution of sodium hydroxide may be

used to distinguish between iron(II) chloride and iron(III) chloride.

(ii) An aqueous solution of carbon dioxide can react with sodium hydroxide solution to form two different salts.

(iii) Bottles of sodium hydroxide solution on laboratory shelves frequently have a white deposit around their stoppers.

(c) Name a white solid which, when warmed with an aqueous solution of sodium hydroxide, gives an alkaline gas. Write the equation for the reaction involved. (C.)

7. Ethanedioic acid has the molecular formula $H_2C_2O_4$. The sodium salt, $Na_2C_2O_4$, is an ionic solid which is soluble in water. The barium salt, BaC_2O_4, is not soluble in water but forms a solution with dilute hydrochloric acid. When solid ethanedioic acid is warmed with concentrated sulphuric acid, a mixture containing equal volumes of carbon monoxide and carbon dioxide is given off.

(a) Draw a labelled diagram of the apparatus you would use to prepare and collect a sample of carbon monoxide free from carbon dioxide, starting from ethanedioic acid.

(b) Explain why equal volumes of carbon monoxide and carbon dioxide are formed when ethanedioic acid reacts with concentrated sulphuric acid. State clearly the function of the sulphuric acid in this reaction.

(c) Briefly explain the poisonous nature of carbon monoxide.

(d) Suggest reactions by which you could distinguish between aqueous solutions of sodium carbonate, sodium sulphate, and the sodium salt of ethanedioic acid.

(ethanedioic acid = oxalic acid) (C.)

8. Identify a colourless liquid which has the properties given below. Explain the significance of *each* test and make a general conclusion about the nature of the liquid. Give equations where appropriate.

(a) When the liquid is warmed with potassium hydroxide solution a gas is evolved which turns moist red litmus paper blue.

(b) When the liquid is added to anhydrous copper(II) sulphate a blue colour results.

(c) When barium chloride solution is added to the liquid, followed by dilute hydrochloric acid, a white precipitate is formed.

9. Describe an experiment that you could perform to find, as accurately as you can, the percentage by volume of oxygen in the air. Nitrogen as normally obtained from the air has a slightly greater density than nitrogen prepared from a compound. Give the reason for this greater density.

Give two natural ways by which nitrogen is returned to the soil. Describe what you would observe and say what is formed when (a) solid sodium hydroxide is left in a dish and exposed to the air, (b) copper foil is heated in the air. (C.)

10. (a) Describe the process by which ammonia is manufactured in industry. Give the chemical equation and the reaction conditions.

(b) Ammonia is one of the reagents used to make artificial fertilizers. Give the name of a reagent which would be reacted with ammonia to produce a fertilizer. Name two other compounds which might be mixed with the product in order to produce a general fertilizer.

(c) A typical artificial fertilizer might have the following analysis: 15 per cent N, 5 per cent P_2O_5, 11 per cent K_2O.

What does this mean and why is this sort of information often used in agriculture?

22 Organic chemistry

22.1 THE DIFFERENT WORLD OF ORGANIC CHEMISTRY

What is organic chemistry?

Organic chemistry developed as a separate branch of chemistry at the beginning of the nineteenth century. At that time, interest in chemistry was increasing rapidly and many chemists began to study compounds such as sugars and alcohols which occurred naturally in plant and animal 'organisms'. In 1808 the Swedish chemist Jöns Berzelius gave the name *organic* to such compounds. This distinguished them from substances such as salt and chalk which came from mineral sources, and which he called *inorganic* compounds. (Many of the substances which Berzelius called organic were known well before his time, and indeed alcohol was prepared as long ago as 7000 BC.)

Scientists used to think that organic chemicals were 'special' because they could be made only by living cells. This was proved to be incorrect when it was shown that organic chemicals could be made in a laboratory from inorganic chemicals. However, Berzelius' classification into organic and inorganic chemicals was not abandoned, because organic chemicals did have something in common. When analysed they were all found to contain the element carbon. Consequently, in 1848, the German chemist Leopold Gmelin defined organic chemistry as the chemistry of the compounds of carbon. Gmelin's definition still stands today. The only exceptions are the oxides of carbon and the carbonates, which by his definition are organic compounds. These are always studied in the inorganic section of a chemistry course.

Almost all organic compounds contain hydrogen in addition to carbon, and oxygen also occurs fairly frequently. Some organic compounds also contain nitrogen, or sulphur, or a halogen, and in rare cases phosphorus or metal atoms may be present.

In other words, all organic compounds are made up from only a handful of elements, and most contain just two or three of these elements joined together. Even so, nearly three million organic compounds are known today, and many more are being discovered each year. Compare this with the number of inorganic compounds known, about 50 000, and remember that inorganic compounds are formed using the hundred or so elements available.

Organic compounds play an important part in our lives. All plants and animals are composed largely of organic compounds, and many of the organic compounds produced by the coal and petroleum industries are formed (directly or indirectly) from decayed living matter. More recently, synthetic manufacture of organic compounds has expanded to produce materials which we now take for granted in everyday life, such as plastics, detergents, synthetic fibres, drugs, and insecticides. Work is still expanding rapidly in this field, and new materials are being produced every day.

Organic chemistry is the study of the chemical compounds of carbon, excluding compounds such as the oxides of carbon and the carbonates.

The structure of organic compounds. Isomerism

Carbon has a valency of four, and normally forms four covalent bonds. Other atoms such as those of nitrogen and silicon can also form several covalent bonds, but carbon is unique in that its atoms have a remarkable ability to join up with each other in an apparently unlimited way, to form chains (which may be 'straight' or branched) or closed rings (cyclic structures).

For example, when one carbon atom joins up with hydrogen atoms, the simplest molecule that can be formed is

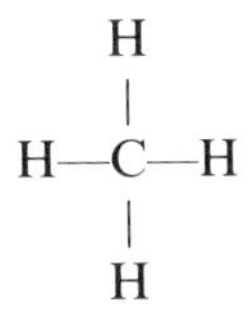

This is the methane molecule (CH_4). The formula above tells us three things: (a) the number of atoms in the molecule, (b) what kinds of atom they are, and (c) the way in which they are linked together. This is known as a *graphical* formula. A graphical formula shows the individual atoms in a molecule and how they are arranged. Molecules are not really flat as graphical formulae suggest; they have a three-dimensional shape. Graphical formulae are very useful for showing what happens to individual atoms during a chemical reaction, but they are cumbersome to write for large molecules. The *structural* formula is an abbreviated formula, which shows the arrangement of the atoms in the molecule by means of groups of atoms (see Table 22.1). A third and still simpler type of formula is the *molecular* formula, e.g. CH_4, which shows only the number of atoms of the different elements in one molecule of the compound, and not how they are arranged (Table 22.1).

When *two* carbon atoms join with hydrogen atoms various arrangements are possible. They could form two methane molecules, or the two carbon atoms could link together covalently and 'fill up' their remaining valencies with hydrogen atoms, in which case a molecule of ethane (C_2H_6) would be formed:

```
   H  H
   |  |
H—C—C—H
   |  |
   H  H
```

(*Note*: Do not think that carbon atoms can 'choose' how they join together! What happens in a particular situation depends upon the compounds which are reacting, and the conditions under which they react.)

Similarly, if three carbon atoms joined with hydrogen atoms a molecule of propane, C_3H_8, could be formed:

```
   H  H  H
   |  |  |
H—C—C—C—H
   |  |  |
   H  H  H
```

or, in terms of the carbon skeleton,

```
  |  |  |
 —C—C—C—
  |  |  |
```

When four carbon atoms join together the carbon skeleton could be

```
  |  |  |  |
 —C—C—C—C—
  |  |  |  |
```

This is a 'straight' chain of four carbon atoms, but as mentioned earlier carbon also has the ability to form branched chains and cyclic structures. Thus four carbon atoms could also join to form

```
     |
    —C—
     |
 —C—C—C—
  |  |  |
```

a branched chain structure, or

```
  |  |
 —C—C—
  |  |
 —C—C—
  |  |
```

a cyclic structure. The four carbon atoms have been arranged in three different ways in this example, and if hydrogen atoms are added to the 'free' valencies of the carbon skeletons, the completed graphical formulae are obtained as shown in Table 22.1.

Table 22.1 Structures of some C_4 saturated hydrocarbons

Name of compound	*Graphical formula*	*Structural formula*	*Molecular formula*
Butane	H H H H H—C—C—C—C—H H H H H	$CH_3—CH_2—CH_2—CH_3$	C_4H_{10}
2-Methylpropane	H H—C—H H	H H—C—C—C—H H H H	CH_3 (on central C) $CH_3—CH—CH_3$	C_4H_{10}
Cyclobutane	H H H—C—C—H H—C—C—H H H	$CH_2—CH_2$ $CH_2—CH_2$	C_4H_8

Isomerism

The two compounds with formula C_4H_{10} shown in Table 22.1 illustrate an important aspect of organic chemistry. They both have the same molecular formula, C_4H_{10} (see Table 22.1), but different structures within the molecule; they are *isomers*. Isomers are not interchangeable; they are different compounds. The fact that they have the same molecular formula is a coincidence.

When two or more structures exist which have the same molecular formula (same type and number of atoms) but different structural formulae (different arrangements of the atoms), the individual forms are called isomers.

Points for discussion

1. Draw as many structures with formula C_4H_{10} as you can, and then build a model of each of your structures. How many isomers are there of formula C_4H_{10}?
Note: Some students imagine that a 'branch' could form at the end of a three carbon chain:

```
  |   |   |
 —C——C——C—
  |   |   |
         —C—
          |
```

However, this molecule is exactly the same as the first isomer shown in Table 22.1; its structural formula is still $CH_3.CH_2.CH_2.CH_3$. This will be more obvious if you see models of the isomers of butane.
2. The isomers with molecular formula C_4H_{10} provide a simple example of isomerism. The larger the molecule, the greater the number of possible structures. As carbon atoms can join together by either single, double, or triple covalent bonds, as well as forming cyclic structures and branched chains, you can understand that as the number of carbon atoms increases, the number of different arrangements (i.e. isomers) increases. The number of possible branched chain isomers of molecular formula $C_{25}H_{52}$, for example, is more than 36 million, although this does not mean that they actually exist in nature or have been made.
3. There are three isomers of formula C_5H_{12}. Can you draw their structures?
4. There is a fourth way of arranging the carbon skeleton in a four-carbon hydrocarbon (i.e. in addition to the three shown in Table 22.1). Draw the graphical formula for the other structure.
5. Is your structure drawn in answer to question 4 an isomer of butane? Explain your answer.
6. Any reference to a straight carbon chain is slightly misleading, which is why such statements have been qualified by referring to 'straight' chains. Can you explain the reason for this? (Hint: try building a model of pentane, C_5H_{12}.)

Classifying organic chemicals

Since so many organic compounds are known to exist, it is obviously important to have some way of classifying them, i.e. of dividing them up into groups with something in common, in the same way that the Periodic Table is used in inorganic chemistry. In the early days of organic chemistry, each compound was given a name which was completely unconnected with any similar compounds, e.g. carbolic acid, fire damp, oil of wintergreen, etc. To try to remember such a large number of 'trivial' names would be almost impossible, and so a systematic (i.e. logical) way of naming organic compounds has been introduced. Before you can use the system, you must understand a little more about organic compounds.

Functional groups

Our attention so far has been concentrated on organic *hydrocarbons*, i.e. compounds composed of carbon and hydrogen only. However, many organic compounds contain other elements as well as carbon and hydrogen.

If one of the hydrogen atoms in ethane:

$$\begin{array}{ccccccc} & & H & & H & & \\ & & | & & | & & \\ H & — & C & — & C & — & H \\ & & | & & | & & \\ & & H & & H & & \end{array}$$

is replaced by a chlorine atom, then a new compound, monochloroethane, is formed:

$$\begin{array}{ccccccc} & & H & & H & & \\ & & | & & | & & \\ H & — & C & — & C & — & Cl \\ & & | & & | & & \\ & & H & & H & & \end{array}$$

This compound has properties different from those of ethane because of the presence of the chlorine atom. Similarly, if a hydrogen atom of ethane is replaced by a hydroxyl radical, ethanol,

$$\begin{array}{ccccccc} & & H & & H & & \\ & & | & & | & & \\ H & — & C & — & C & — & OH \\ & & | & & | & & \\ & & H & & H & & \end{array}$$

is formed, which again differs in its properties from ethane or monochloroethane.

Most organic molecules are similar to these last two examples. They consist of a fairly unreactive hydrocarbon portion (i.e. carbon and hydrogen atoms, which form a kind of backbone in the molecule) joined to another atom or group of atoms (e.g. a chlorine atom or a hydroxyl group) which is known as the *functional group*, e.g.

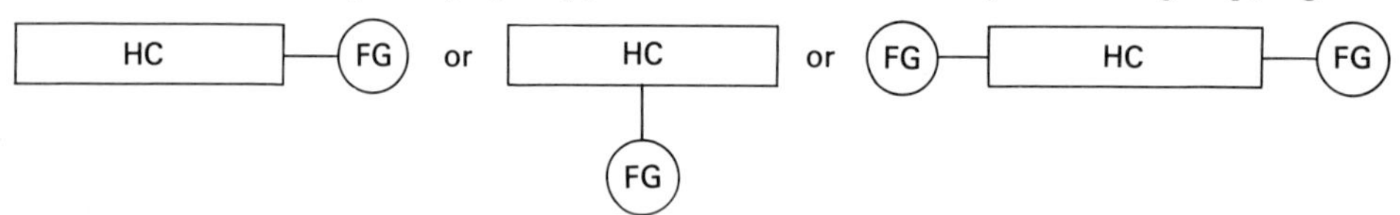

(where HC = the hydrocarbon 'backbone' and FG = a functional group).

Some molecules have more than one kind of functional group, or they may have the same functional group repeated several times. The functional group is mainly responsible for the *chemical* reactions of the compound. All those organic chemicals which contain the same functional group will show similar chemical properties. This is the start of organic classification. Just as all the elements in one particular group in the Periodic Table show similar properties, and we have families of elements such as the halogens or the alkali metals, so we have families of organic compounds. For example, there is a family of compounds which all contain the hydroxyl group, OH, and these are called the alcohols. At an elementary level you are likely to study only a few types of organic compounds, but you may have heard of families of organic chemicals called the ketones, the aldehydes, the amines, etc. Each member of one of these families contains the same functional group, which is different from the functional group present in molecules of a different family.

Homologous series

Just as a 'group' in the Periodic Table refers to a family of elements with similar properties, an *homologous series* is the equivalent in the organic world. It refers to a series of organic chemicals which have very similar properties, because the members all contain the same functional group.

An homologous series is a group of organic compounds, all of which have the same general formula and similar chemical properties.

(The general formula for the alkane family is C_nH_{2n+2}. Thus when $n = 1$, the alkane is methane, CH_4, and when $n = 2$ ethane, C_2H_6, is formed.) You will probably be studying three or four homologous series in a very simple way, and it is useful to remember the following points which apply to any homologous series.

1. As all members of a series have similar chemical properties, it is only necessary to study in detail the behaviour of *one* typical member, because then the properties of all other members of the series can be predicted with reasonable accuracy. It should be remembered, however, that the activity of the functional group becomes less obvious in the later members of a series.
2. Members of a series show a gradual change in physical properties. As the lengths of the hydrocarbon chains are increased the number of intermolecular attractions increases, and in turn the melting points and boiling points increase. You will remember that sulphur has a higher melting point than iodine (p. 276) and this is partly explained by the fact that there are more intermolecular attractions between sulphur molecules (which contain eight atoms) than there are between the smaller iodine molecules (which contain only two atoms). The intermolecular forces between hydrocarbons are of the same kind as those between iodine molecules and between sulphur molecules, i.e. van der Waals forces.

Check your understanding

Write down the molecular formulae for (a) the sixth member of the homologous series of alkanes, and (b) the ninth member of the same series.

Naming organic chemicals

The homologous series of alkanes

The alkanes are a series of hydrocarbons in which each member contains only single covalent bonds between carbon atoms. You will be studying some of the properties of this series in section 22.2. There are 70 known 'straight' chain members of this series, and their names form the basis of the organic naming system.

Some members of the alkane series are shown in Table 22.2. Notice that all the names end in *ane*, which tells us that all bonds between carbon atoms in the molecules are single covalent. The first part of each name tells us how many carbon atoms are in the molecule. Thus *meth* in a name tells us that the molecules of the compound in question contain only one carbon atom. *Hex* always means that there are six carbon atoms per molecule, etc. It is not necessary to learn the more complicated examples; in elementary work you will need to go no further than the first six members of the alkane series.

Naming branch chains

When an alkane like methane reacts, it usually breaks one or more of its C—H bonds so that the hydrogen atom is 'released' and then replaced by another atom or group. If one of the C—H bonds in a molecule of methane is broken, for example, the group CH_3— would be left from the original CH_4 molecule. This CH_3— radical is called a methyl radical. Similarly, ethane, C_2H_6, forms the ethyl radical, C_2H_5—. When a branched chain forms, radicals of this type normally form the branches. You have seen one example so far, one of the isomers of butane (Table 22.1). You should

Table 22.2 Physical properties of some alkanes

Name of alkane	*Formula*	*Melting point (°C)*	*Boiling point (°C)*	*Density (g cm^{-3})*	*Physical state at room temperature*
Methane	CH_4	−182	−164	—	Gas
Ethane	C_2H_6	−183	− 89	—	Gas
Propane	C_3H_8	−188	− 42	—	Gas
Butane	C_4H_{10}	−138	− 1	—	Gas
Pentane	C_5H_{12}	−129	36	0.63	Liquid
Hexane	C_6H_{14}	− 95	69	0.66	Liquid
Heptane	C_7H_{16}	− 91	98	0.68	Liquid
Pentadecane	$C_{15}H_{32}$	10	270	0.77	'Liquid'
Hexadecane	$C_{16}H_{34}$	18	287	0.77	Solid
Heptadecane	$C_{17}H_{36}$	22	302	0.78	Solid

now know why it is called 2-methylpropane. It consists of a propane chain with a methyl radical joined to carbon atom number 2 to form a branch.

It is unlikely that you will meet examples more complicated than this at an elementary level. If you later study organic chemistry at a higher level you will find that there is a set of rules for naming more complicated examples, but the basic idea is as above.

Naming compounds with other functional groups

Each functional group has its own 'code' which is used in a name. The only homologous series included in most elementary syllabuses are the alkanes (which do not have a functional group), the alkenes (hydrocarbons with a C=C double covalent bond), and the alcohols (which contain the OH functional group). An alkene has the code *ene* in its name, and an alcohol has the code *ol*. For example, the compound of formula

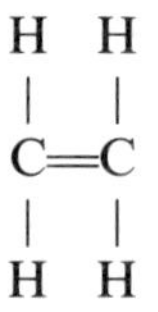

is named ethene; 'eth' because it contains two carbon atoms, and 'ene' because it contains a C=C double bond. Similarly, the compound with formula CH_3OH is named methanol; 'meth' because it contains one carbon atom, and 'ol' because it contains the OH functional group.

Points for discussion

1. Draw two different structural formulae for the molecular formula C_2H_6O. You should be able to name one of the isomers you draw. (Remember that the atoms must all keep their normal valencies in your structures.)

2. Draw two different structural formulae for the molecular formula $C_2H_4Br_2$. Suggest names for your structures.

22.2 HYDROCARBONS

The homologous series of alkanes

The alkanes are a series of saturated hydrocarbons of general formula C_nH_{2n+2}. (A *saturated compound* contains only single carbon–carbon covalent bonds.) The alkanes with small molecules are gases at room temperature and pressure, but those with larger molecules are either liquids or solids. Some alkanes are shown in Table 22.2. Methane is normally studied as a typical member of the series.

Sources of methane and other alkanes

Alkanes are formed in nature by the decomposition of vegetable and animal matter in the absence of air. Over millions of years nature has built up vast stocks of coal and oil from such decompositions and large quantities of alkanes are among the products obtained when coal and oil are distilled in the absence of air. Methane is always found associated with coal and oil deposits, and it is indeed the main constituent of natural gas (>78 per cent) which is used widely as a fuel. Methane is also obtained as a byproduct of the distillation of coal and oil and at sewage works. It can also be observed bubbling in ponds and marshy regions where vegetable matter is decomposing and for this reason the gas was originally named 'marsh gas'. Animals also break down vegetable matter during digestion and so produce methane, and it has been estimated that a cow produces over 500 000 cm^3 per day. Methane is also the main constituent of the atmospheres of the planets Jupiter and Saturn.

A cow produces half a million cm^3 of methane per day!

Physical properties of methane

Solubility in water	*Colour*	*Odour*	*Density relative to air*	*Toxicity*
Insoluble	None	None	Less dense	Not toxic

Note: (i) The liquid alkanes are less dense than water and, being immiscible with it, will float on its surface.

(ii) Methane has no smell, but natural gas has an odour because a volatile substance is deliberately added to it during processing to make any escape of gas easily detectable.

Chemical reactions of methane

There are only two reactions of any importance.

Substitution reactions with halogen elements Alkanes are saturated compounds, and saturated compounds usually take part in substitution reactions.

A substitution reaction occurs when an organic molecule reacts with an element or compound of the type X—Y so that X enters the organic molecule, replacing an atom (usually hydrogen) which then combines with Y.

Methane and the other alkanes readily take part in substitution reactions with chlorine and bromine. Bromine reacts slowly in diffused daylight, and chlorine more quickly; the reactions with chlorine can be explosive in sunlight. In each case a series of steps occurs as a hydrogen atom is substituted by a halogen atom, e.g.

$$CH_4(g) + Cl_2(g) \rightarrow \underset{\text{monochloromethane}}{CH_3Cl(g)} + HCl(g)$$

and then

$$CH_3Cl(l) + Cl_2(g) \rightarrow \underset{\text{dichloromethane}}{CH_2Cl_2(l)} + HCl(g)$$

(sequence continued overleaf)

and then

$$CH_2Cl_2(l) + Cl_2(g) \rightarrow \underset{\text{trichloromethane}}{CHCl_3(l)} + HCl(g)$$

and then

$$CHCl_3(l) + Cl_2(g) \rightarrow \underset{\text{tetrachloromethane}}{CCl_4(l)} + HCl(g)$$

Similar reactions occur with bromine (but not 'bromine water'), and if other alkanes are used the principle is the same but the number of steps is increased.

Note: It is often useful to show a graphical formula in an equation rather than a molecular formula, e.g.

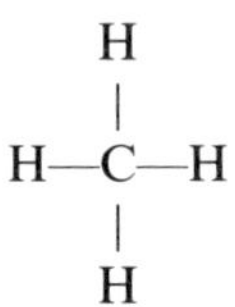

rather than CH_4, for it is easier to see what is happening to individual bonds during the reaction.

Combustion In a plentiful supply of oxygen, methane burns completely with a clean blue flame to form carbon dioxide and water vapour. Other alkanes react in exactly the same way, e.g.

$$CH_4(g) + 2O_2(g) \rightarrow CO_2(g) + 2H_2O(g)$$
$$C_4H_{10}(g) + 6\tfrac{1}{2}O_2(g) \rightarrow 4CO_2(g) + 5H_2O(g)$$

If insufficient oxygen is present for complete combustion (as is the case in underground mine workings) then poisonous carbon monoxide is produced instead of carbon dioxide. This is the deadly 'after damp' that remains in coal mines after explosions. A similar situation also exists in the cylinders of petrol engines (petrol is a mixture containing some alkanes), and when incomplete combustion occurs the exhaust fumes contain carbon monoxide and occasionally even carbon. This is why it is dangerous to run a car engine inside a garage and why petrol engines occasionally need a 'de-coke'.

Experiment 22.1
Some properties of methane*

These simple tests will help to illustrate some of the physical and chemical properties of methane. Note that there is no reaction in some of the tests; you will see the point of this later in the section.

Apparatus

Apparatus for collecting supply of natural gas (over water) from a gas tap, Bunsen burner, boiling tubes in rack with bungs to fit, spills.

Red and blue litmus papers, solution of bromine in water ('bromine water'), acidified solution of potassium manganate(VII).

Procedure

(a) Collect three boiling tubes full of natural gas (which is almost pure methane) over water in the usual way. Stopper each tube immediately. (The tubes will each need to be refilled later in the experiment.)

(b) Cautiously sniff a small amount of the gas in one of your boiling tubes. Pure methane has no smell (page 407). Can you explain your observation?

(c) Test another sample of gas with both red and blue litmus papers, dampening the papers with water before the test. Does methane dissolve in water to form either an acidic or an alkaline solution?

(d) Ignite another sample of gas using a lighted spill (CARE!). Does methane burn with a 'clean' flame or a smoky flame? What colour is the flame?
(e) Add a small volume of bromine water to a tube of methane, quickly replace the stopper, and shake the tube so that the contents are well mixed. Does methane react with the bromine water?
(f) Repeat (e) but using a small volume of acidified potassium manganate(VII) solution instead of the bromine water. Does methane react with the reagent?

Results and conclusion
Record your results in the usual way.
1. Use them to explain the general chemistry of methane. Do not confuse bromine water (a dilute solution of bromine molecules in water) with the pure element bromine.
2. Draw attention also to 'reactions' which methane does *not* give.

Uses of the alkanes

The combustion reactions of alkanes are very exothermic, and this is why many of the alkanes are used as fuels, e.g. natural gas (methane), petrol (a mixture of several alkanes), paraffin (another mixture, of less volatile alkanes), Calor gas and Camping Gas (mainly butane, liquefied under pressure), etc.

Higher alkanes are also used as solvents and in the manufacture of other chemicals. Solid alkanes have a variety of uses, e.g. Vaseline is a mixture of paraffin wax (a solid alkane) and oil.

The homologous series of alkenes

The alkenes are a series of unsaturated hydrocarbons, each containing a carbon–carbon double covalent bond and having the general formula C_nH_{2n}, e.g. ethene, C_2H_4 or

```
H       H
 \     /
  C═C
 /     \
H       H
```

The alkenes are not found to any great extent in nature. Coal gas and natural gas contain small amounts of ethene, but most of the alkenes used in industry (and they are used in large quantities) are obtained by cracking petroleum (p. 413). The alkenes are named by replacing the ending *ane* of the corresponding alkane by *ene* (careless writers beware!). The series shows the usual gradual increase in melting points and boiling points as the number of carbon atoms in the molecules increases; the first four members are gases, but pent-1-ene is a liquid at room temperature and pressure. Ethene is normally studied as being typical of the series.

Physical properties of ethene

Solubility in water	*Colour*	*Odour*	*Density relative to air*	*Toxicity*
Insoluble	None	Virtually odourless	About the same	Non-toxic

Chemical properties of ethene

As alkenes are unsaturated (unsaturated compounds contain at least one double or triple covalent bond), their reactions are mainly *addition* reactions.

An addition reaction occurs when two or more molecules combine to form just one, larger molecule. The term is usually confined to organic chemistry.

An unsaturated compound can be thought of as a compound having 'room' for more atoms, which it gains by addition. The C=C bond is more reactive than C—C or C—H bonds, and so unsaturated compounds usually take part in addition reactions rather than substitution reactions. Compare this with the alkanes, which *cannot* add and usually substitute.

Addition with halogen elements If ethene is mixed with bromine vapour, a rapid addition reaction takes place. The colour of the bromine disappears, and sometimes a few drops of the product (a liquid) can be seen coating the inside of the gas jar.

$$C_2H_4(g) + Br_2(g) \rightarrow C_2H_4Br_2(l) \text{ (1,2-dibromoethane)}$$

or

$$\mathrm{H_2C{=}CH_2} + Br_2(g) \rightarrow \mathrm{Br{-}CH_2{-}CH_2{-}Br}$$

Chlorine reacts similarly, but far more rapidly.

Addition with aqueous solutions of the halogen elements If a small volume of a *solution* of bromine in water (bromine water) or in trichloromethane is added to an alkene, the brown colour of the halogen disappears almost instantly as it adds to the alkene (see previous point).

Note that in both the above examples, the *same* reaction is taking place but the 'source' of the bromine molecules is different. The important thing is the *visual* effect, which is the same in each of these reactions. There is a *rapid decolorization* due to an *addition* reaction.

Note: Students are often confused by the fact that both the alkanes and the alkenes react with the halogens, but by different reactions. Remember that only bromine itself (and not bromine water or any other solution of bromine) reacts with alkanes, and then it does so in a *series* of *slow*, multi-step *substitution* reactions. This should be contrasted with the *rapid*, *addition* reaction of alkenes which takes place with *either* bromine *or* a solution of bromine.

Other addition reactions of ethene Ethene will react with hydrogen by addition to form ethane,

$$C_2H_4(g) + H_2(g) \xrightarrow[\text{high pressure, temperature}]{\text{nickel or platinum catalyst}} C_2H_6(g)$$

with hydrogen halides,

$$C_2H_4(g) + HCl(g) \rightarrow C_2H_5Cl(l) \text{ (monochloroethane)}$$

and with 'water' to form an alcohol, ethanol,

$$C_2H_4(g) + H_2O(g) \xrightarrow[H_3PO_4 \text{ catalyst}]{300\,^\circ C,\ 60 \text{ atmospheres}} C_2H_5OH(g)$$

The polymerization of ethene and other alkenes is a special type of addition reaction, discussed in section 22.5.

Reaction of ethene with acidified potassium manganate(VII) This reaction is not given by alkanes. If this reagent is added to an alkene, and the two substances shaken together, the purple colour of the manganate(VII) ion rapidly changes to a colourless solution. The reagent is a very powerful oxidizing agent, and oxidizes the alkene by an addition reaction, itself being reduced to the colourless Mn^{2+} ion.

Combustion of ethene Ethene and other alkenes burn in air, but as their molecules contain a higher proportion of carbon than do the alkanes, they do not burn completely with a clean blue flame. The flame is yellow, and some smoke and soot are formed although most of the carbon does become carbon dioxide, e.g.

$$C_2H_4(g)+3O_2(g) \rightarrow 2CO_2(g)+2H_2O(g)$$

Experiment 22.2
Some properties of alkenes**

Apparatus
Bunsen burner, teat pipette, small gas jar, access to fume cupboard.
Rack of stoppered test-tubes of ethene and small gas jar of ethene OR supply of a liquid alkene such as cyclohexene, red and blue litmus papers, bromine, dilute solution of bromine (in either water or trichloromethane), very dilute solution of potassium manganate(VII) acidified with dilute sulphuric acid.

Procedure
(a) Cautiously sniff a *small* sample of the gas or liquid.
(b) Test the gas or liquid with (i) damp blue litmus paper, (ii) damp red litmus paper. Does the alkene dissolve in water to form either an acidic or an alkaline solution?
(c) Ignite a test-tube of the gas using a lighted splint (CARE; ethene/air mixtures are explosive) or ignite a small sample of the liquid in an evaporating basin. Does the alkene burn with a clean flame? Are soot and smoke produced? What is the colour of the flame? What do your answers tell you about molecules of an alkene?
(d) Add a small volume of a solution of bromine to a test-tube of ethene (or to a test-tube containing a few drops of the liquid alkene), quickly replace the stopper, and shake the tube so that the contents are well mixed. Does the alkene react with the bromine solution? If so, does it react rapidly or slowly, and is the reaction an addition or a substitution?
(e) Repeat procedure (d) using acidified potassium manganate(VII) solution instead of bromine solution. Does the alkene react with this reagent? If so, does it react rapidly or slowly, and is the reaction an addition or a substitution?
(f) **This part of the procedure must be a demonstration, using ethene and not the liquid alkene.**
Working in a fume cupboard, place two drops of bromine into a small gas jar by means of a teat pipette. Allow *all* the bromine to vaporize and then invert this gas jar over a gas jar of ethene. Are there any signs of a reaction? If so, is the reaction slow, or fast? If there is a reaction, is it an addition or a substitution?

Results and conclusion
Record all your observations in the usual way. Make sure that you can explain all of the results, using the correct scientific terms such as addition, substitution, unsaturated etc. wherever possible. Write equations for all of the reactions except for the one with potassium manganate(VII).

Uses of alkenes

Many alkenes are made by cracking petroleum (p. 413), and their readiness to take part in addition reactions makes them very important starting materials for the manufacture of ethanol (see other addition reactions of ethene, page 410), other alcohols, anti-freezes, coolants, detergents, and (above all) polymers.

How to distinguish between methane and ethene

Test	*Methane*	*Ethene*
Combustion	Clean, blue, flame	Yellow flame, some soot
Bromine water	No reaction	Rapid decolorization
Acidified $KMnO_4$ solution	No reaction	Rapid decolorization

Points for discussion

1. What problem(s) could be caused by the fact that 'liquid alkanes are less dense than water, and, being immiscible with it, float on its surface'?

2. Why is a substance added to natural gas so as to give it an odour?

3. Why do you think that one of the alkenes is called pent-1-ene rather than just pentene? Draw the structure of as many of its isomers as you can.

4. Why is there no alkene called methene?

Check your understanding

1. Write balanced equations for the combustion of (a) ethane, (b) pentane and (c) but-1-ene (all in a plentiful supply of air).

2. Ethane, C_2H_6, butane C_4H_{10}, and octane, C_8H_n (where n is a whole number) are all members of the same hydrocarbon series.

(a) (i) Explain what is meant by the term 'hydrocarbon'. (ii) State the name of the above series. (iii) Deduce the value of n in the formula for octane.

(b) Give the name of another member of the same hydrocarbon series and state one of its uses.

(c) Butane has two isomers. Give their structural formulae.

(d) (i) Write an equation for the combustion of ethane in an excess of oxygen. (ii) If 20 cm^3 of ethane are burned in an excess of oxygen, state the minimum volume of oxygen required for complete combustion and the volume of carbon dioxide formed. Assume that all volumes are measured at the same temperature and pressure. (J.M.B.)

3. There are large numbers of organic compounds because

A carbon atoms can join together to form long chains
B carbon reacts vigorously with many elements
C carbon has a variable valency
D carbon forms covalent bonds easily
E there are millions of living organisms

4. A sample of oil was vaporized and passed over a heated catalyst; a gas formed which was found to decolorize bromine water. The gas

A is ethane
B contains ethane
C contains a saturated hydrocarbon
D contains an unsaturated hydrocarbon
E contains a mixture of alkanes

22.3 THE OIL INDUSTRY

Some of the basic processes which occur at a refinery

Separation into fractions

The crude oil is desulphurized (see page 144) before it is fractionally distilled. This preliminary treatment provides a valuable supply of sulphur and makes sure that the combustion products of any fuels which may be made from the oil are relatively free from sulphur dioxide. The sulphur content of crude oil varies according to its origin; oil from the Middle East has a higher sulphur content than that extracted in the USA, for example.

A simplified description of the fractional distillation of crude oil at a refinery is given on page 43, and Experiment 4.3 was used to illustrate the principle. The main fractions into which crude oil is separated are shown in Figure 4.9, page 45.

Cracking

There are many processes carried out at a refinery after fractional distillation, but of these, cracking is probably the most important.

There is an enormous demand for the gaseous and volatile liquid hydrocarbons from crude oil. These low molecular mass and low boiling point fractions are much used as liquid fuels (petrol, paraffin) and as raw materials for the manufacture of plastics and other organic chemicals, and so it is important to obtain them in high yields. However, large quantities of the heavier, less volatile fractions are also produced by the fractional distillation of crude oil. Some of these have important uses (lubricating oils, etc.) but they are not used on anything like the same scale as the simpler hydrocarbons.

The excess heavy fractions are not wasted, but reprocessed by cracking. This is essentially a brief exposure to a high temperature (e.g. by mixing with steam and heating to 900 °C for one second) which splits up the larger molecules into smaller molecules by breaking C—C bonds. This produces further supplies of the simpler hydrocarbons, e.g. for petrol. In addition, whenever a hydrocarbon molecule is cracked into two or more smaller molecules, at least one of the products is unsaturated. These small, unsaturated molecules such as ethene and propene are very reactive, and readily take part in polymerizations and other addition reactions to produce important chemicals. Cracking thus ensures much higher yields of important and reactive chemicals by reprocessing excess quantities of less important fractions (see Figure 22.1).

The following example, using pentane, is a simple illustration of what happens in cracking. In practice, of course, a pentane molecule would not need to be cracked; it is already a small and useful molecule. The situation is more complicated, however, with larger molecules.

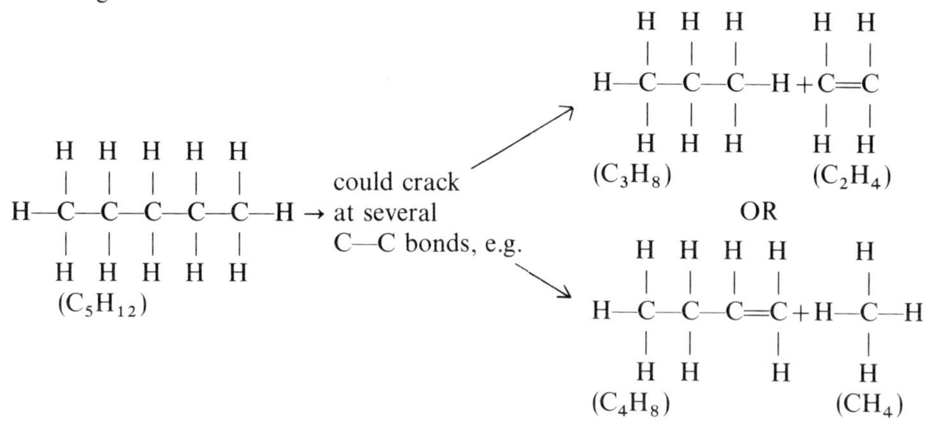

The *total* number of carbon atoms in each pair of product molecules is 5, and the total number of hydrogen atoms is 12, but each pair of product molecules can only use the 12 hydrogen atoms 'available' if one molecule of each pair contains a double bond.

The importance of the oil industry; future problems

Crude oil is one of the most important of the world's raw materials, and the demand for it is still increasing in spite of the fact that reserves of crude oil are rapidly being used up. More than half of the total production of organic chemicals, and over one-third of the world's power, are obtained from oil. It is vital that we should make every effort to ensure that supplies of this precious fossil fuel last as long as possible,

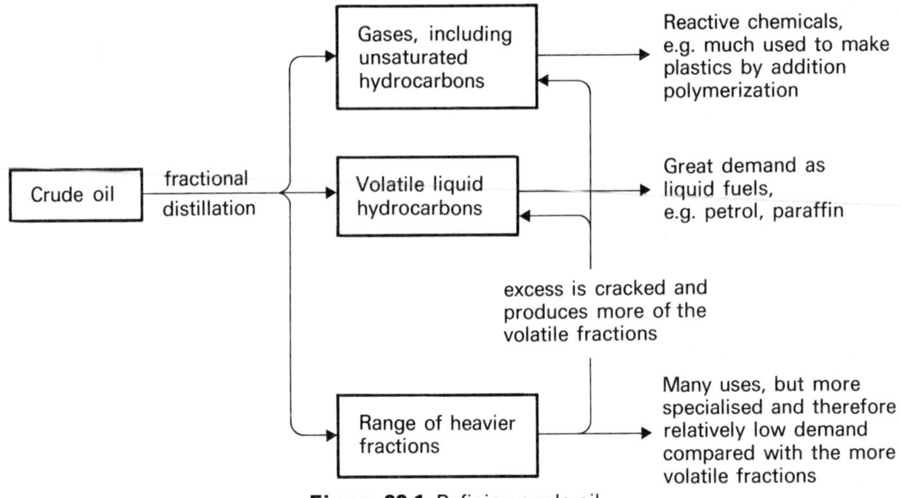

Figure 22.1 Refining crude oil

and that we must develop new (or return to old) sources of liquid fuels and of organic chemicals.

Estimates of fuel reserves should always be treated with caution, as there are so many unknown factors. For example, Table 22.3 shows the known world reserves of oil, and the expected rate of use, as in 1980. It is likely, however, that more oil will be extracted from existing oil fields than is suggested in the table, and the figures do not take into account the oil which can be extracted from tar sands. As the price of oil increases, it will become economic to extract it from such sources. In addition, new oil reserves may be discovered, although their rate of discovery is decreasing.

It is also important to understand how consumption figures are calculated. In Table 22.3, the Middle East supplies may look impressive, but their expected 'life' is calculated on the basis of consumption in the area of origin only, i.e. in the Middle East, where the actual demand for oil is very small. Similarly, in terms of worldwide demand North Sea reserves are very small indeed, and clearly the West must hope for supplies of Middle East oil for some years to come.

In 1980 it was estimated that the total reserves of oil were sufficient to supply worldwide demand only up to the year 2006, and this is the point which must give us serious cause for concern. No doubt this forecast will be modified as consumption patterns change, and new sources are discovered, etc., but no matter how optimistic we are the probability remains that at some fairly early stage in the next century we will have to live without oil.

Table 22.3 World oil consumption and reserves (1980)

	Known reserves (millions of tonnes)	*Consumption in area of origin (millions of tonnes per annum)*	*Years of life if used only in area of origin*
North America	5 500	975	6
Latin America	5 800	202	29
Western Europe	3 300	715	5
Middle East	50 300	83	606
Africa	7 700	60	128
Asia-Pacific	2 700	443	6
Communist countries	12 800	598	21

(Information supplied by Esso Petroleum Company Ltd)

The raw material

The raw material, crude oil, is a complex mixture of hydrocarbons, formed over millions of years by the breakdown, under pressure, of the remains of sea organisms. Our modern society has depended so much on crude oil as a source of fuels and other chemicals, that we are using up in just over a hundred years a resource which has taken millions of years to form, and which has no foreseeable replacement. Until comparatively recently, the attitude of many people was very short-sighted indeed; oil and its products were often not used efficiently, many seemed to imagine that it would last for ever, and a concentrated effort to develop alternative sources of energy was rather late in starting.

Prior to the Second World War, oil refineries were generally built close to the source of crude oil, but since then new refineries have been constructed in regions where the products can be used directly. A modern refinery is a vast, complex site where the various components extracted from crude oil are processed into a whole variety of important chemicals. The crude oil is thus shipped or otherwise transported to the refineries, processed to yield fractions, and then further modified and reacted at the refinery to produce the desired products, which are now more numerous than ever before.

Most students think of petroleum mainly in connection with fuels, and indeed many important fuels *are* produced from crude oil (e.g. petrol, paraffin, fuel oil, and bottled gas). But petroleum is also an important source of many other chemicals. Sulphur and carbon black are two inorganic materials produced from crude oil but, in the main, this mixture of hydrocarbons is the starting point for the production of a large and varied range of important organic chemicals. Some of these, in the past, were produced from other sources. For example, benzene (an important organic chemical with a cyclic structure, of formula C_6H_6) was produced by the destructive distillation of coal (p. 501), and ethyne (acetylene) was made from calcium carbide via coke, but it became more economical (in the short term!) to manufacture them from crude oil. The time is rapidly approaching when we will have either to return to traditional methods of manufacture, or to develop new ones. The chemistry of crude oil, upon which we are still so dependent, is likely to become of purely academic interest in the not too distant future.

The effect of supply and demand on the oil industry

As supplies dwindle, the economics of the petroleum industry will no doubt change, as they have done from time to time. In the very early days of the industry, before the internal combustion engine was invented, 'petrol' was an inconvenient byproduct and it was often burned off. The main fraction required in these early days was paraffin for lamps etc., and so the heavier fractions were also regarded as being relatively unimportant. The invention of the petrol engine then caused a great demand for the petrol fraction and also an increased demand for the lubricants etc. obtained from the heavier fractions. The balance changed yet again when the 'plastics age' started, and the bulk of the heavier fractions were cracked to convert them into more of the smaller, unsaturated molecules which were needed to build up polymer molecules. At the same time, cracking also produced larger quantities of molecules of the type found in the petrol fraction, and this 'extra' petrol was also in demand, being swallowed up by the ever increasing number of cars. Cracking also produced many other simple organic chemicals which had previously been obtained from other sources, and so the petroleum industry replaced traditional sources of organic chemicals. Consequently, a large proportion of the heavier molecules in the less volatile fractions are now broken down into smaller, lighter molecules because of the great demand for organic chemicals with relatively small molecules, especially petrol and the building units for plastics.

Smaller proportions of the heavy fractions are used directly. Gas oil is used to make diesel oil for buses, lorries, and locomotives. Fuel oil is burned to provide the heat energy needed to raise steam in ships, heating plants, and power stations. Lubricating oil is essential wherever moving parts are present, be they very small cog wheels in a watch or the pistons in an internal combustion engine. Bitumen is used on a large scale in road making, and as a water-proofing material for roofing felt and underground pipes. This illustrates how changing technology can alter the demand for a particular fraction.

The future

The oil industry has thus emphasized or modified some of the fractions according to the demand at a particular stage in its history; what is unimportant today may become more important tomorrow. Entirely new priorities will have to be faced in the near future, and crude oil may well be processed in a different way so as to provide only the chemicals which cannot yet be made from other sources.

For example, roughly 55 per cent of the crude oil used in the UK in 1978 was burned to provide heat energy and electricity (Table 22.4). This part of the total oil demand could, in theory, be met by other methods, e.g. an increase in the generation of electricity by nuclear power and by an increase in the use of coal. The other 45 per cent of oil used in the UK was converted into the simpler hydrocarbons which are in turn used as fuels for transport (petrol, diesel oil, aviation spirit, etc.) and in the organic chemicals industry. At the moment, there is no obvious replacement or other source for these simple hydrocarbons. It may well be, therefore, that there will be a move away from the use of oil as a source of heat and electricity. Such a move could make our reserves of oil last for twice as long as previously anticipated. It would also mean that even greater amounts of the heavier fractions would be converted (e.g. by cracking) into simpler hydrocarbons, so as to produce only those materials which cannot yet be obtained from other sources. (This aspect of the 'energy crisis' is considered again in Chapter 26.) Cars use a phenomenal amount of petrol, and this comes either directly from crude oil as one of the fractions or as the result of the modification of other fractions by cracking. If an alternative fuel is devised for transport, this aspect will change again, and even petrol might then be used as a source of chemicals rather than as a fuel.

It is important to make one final point in this context. Natural gas will also be exhausted in the near future, and in any case cannot be considered as an alternative source of organic chemicals because it consists largely of methane. Molecules of methane contain only one carbon atom and do not join together to form larger molecules. Some countries, including Britain, do have greater stocks of the third fossil fuel, coal. It may seem tempting to suggest that when the oil runs out we can solve all of our problems by simply resorting to coal as a source of fuels and organic chemicals, just as we used it before the oil boom. Such a course of action may indeed be necessary, if we have no other alternative, but it can only be a short-term solution; it only buys us more time. Stocks of coal, too, are finite and we could again use up in a short space of time a rich source of energy and chemicals which has taken millions of years to form. The future of civilization as we know it must depend upon new fuels (this is considered further in Chapter 26), new organic processes, more efficient use of our precious resources, and the recycling of materials which are becoming increasingly scarce.

Table 22.4 UK oil consumption, 1978

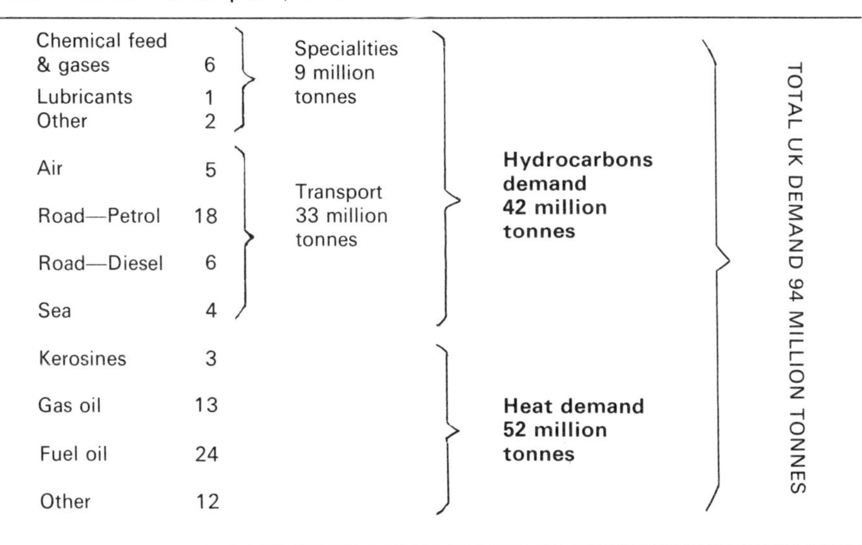

Use	Million tonnes	Group	Demand	Total
Chemical feed & gases	6	Specialities 9 million tonnes	Hydrocarbons demand 42 million tonnes	TOTAL UK DEMAND 94 MILLION TONNES
Lubricants	1			
Other	2			
Air	5	Transport 33 million tonnes		
Road—Petrol	18			
Road—Diesel	6			
Sea	4			
Kerosines	3		Heat demand 52 million tonnes	
Gas oil	13			
Fuel oil	24			
Other	12			

(Information supplied by Esso Petroleum Company Ltd)

Points for discussion

1. How do geologists decide which areas of the Earth's surface may contain oil?
2. What is the only *sure* way of finding out whether oil is present in these areas?
3. What special difficulties are there in drilling for oil at sea?
4. Describe, briefly, three different types of rig used for drilling under the sea.
5. How is crude oil moved from (a) oilwell to port, (b) port to port, and (c) port to refinery?
6. In which countries does the supply of oil exceed the local demand?

Check your understanding

1. The relative amounts of each hydrocarbon component in a crude oil varies depending on the source of the crude oil. The diagram below shows the composition, by mass, of a certain crude oil obtained from West Africa.

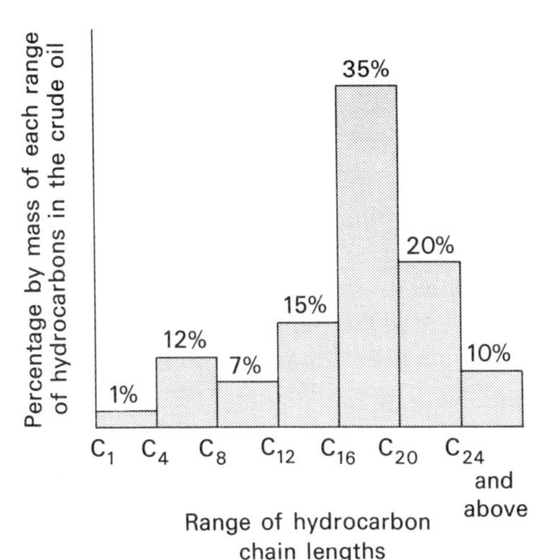

(a) Complete the following table with one suitable example in each case.

Hydrocarbon	*Structure*	*Name*
C_1	H │ H—C—H │ H	
C_2		
C_3		
C_4		

(b) What is the normal physical state of the following hydrocarbon ranges?
(i) C_1 to C_4 (ii) C_4 to C_{12} and (iii) C_{24} and above.

The market demand by mass, for hydrocarbons in the same ranges is given in the diagram on the following page.

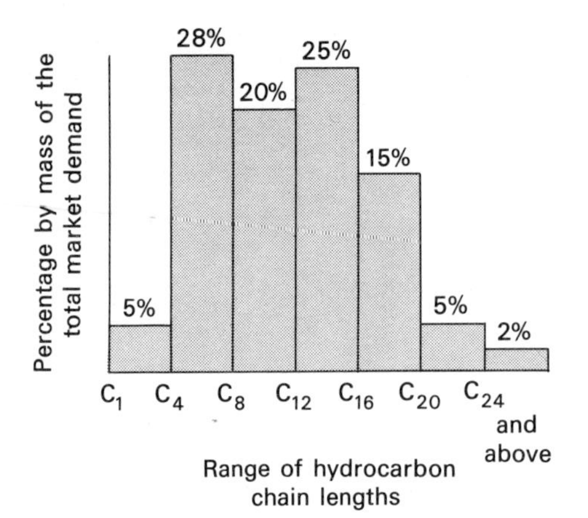

(c) Why is there (i) such a high demand for the C_1 to C_4 range? (ii) such a low demand for the C_{16} and above hydrocarbons?
(d) Comparing the two diagrams, it can be seen that for the crude oil: (i) the demand for the C_1 to C_4 and the C_4 to C_8 ranges is greater than the production; (ii) the total production of hydrocarbons upwards of C_{12} is greater than the market demand. How does the refinery overcome this situation?
(e) Give reasons why the market demand for the C_1 to C_4 range is as high as 5 per cent by mass. (A.E.B. 1977)

22.4 ALCOHOLS, ORGANIC ACIDS, AND ESTERS

Alcohols

Remember that a functional group is an atom or group of atoms present in an organic molecule which tends to dominate the chemical properties of the molecule; the relatively inert carbon and hydrogen skeleton has no major influence on chemical properties.

The alcohols are a series of compounds containing the —OH functional group and having the general formula $C_nH_{2n+1}OH$. The alcohols are named by replacing the –e of the appropriate alkane by the ending –ol, e.g. methanol, CH_3OH (from methane, CH_4), ethanol, C_2H_5OH, and propanol, C_3H_7OH. The only alcohol studied in elementary work is ethanol, a colourless liquid with a characteristic odour, which boils at 78 °C. This liquid is sometimes called ethyl alcohol, or even just 'alcohol', which is particularly unfortunate as it is only one of many alcohols.

The preparation of ethanol by fermentation

The word fermentation comes from the Latin *fevere*, which means to 'boil up'. The process can be summarized as follows.

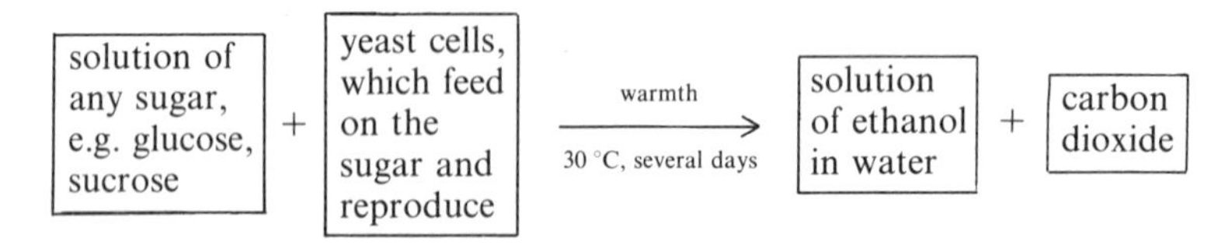

The yeast cells contain enzymes (biological catalysts) which break down the sugars to produce energy needed for the cells to live and reproduce. These enzymes, unlike inorganic catalysts, can be deactivated by over-heating so it is essential to keep the fermenting solution warm but not hot. Ethanol is produced when yeast respires anaerobically, i.e. breaks down sugars to produce energy without using oxygen. This is an alternative version of normal respiration, which is aerobic. Carbon dioxide and energy are formed in both cases, but the sugar is not completely broken down into carbon dioxide in anaerobic respiration.

$$\underset{\text{glucose}}{C_6H_{12}O_6(aq)} \rightarrow \underset{\text{ethanol}}{2C_2H_5OH(aq)} + 2CO_2(g) \text{ (anaerobic)}$$

cf.

$$C_6H_{12}O_6 + 6O_2 \rightarrow 6CO_2 + 6H_2O \text{ (aerobic)}$$

Experiment 22.3
The laboratory preparation of ethanol by fermentation **

Apparatus
250 cm^3 conical flask fitted with bung and delivery tube as in Figure 22.2, two 100 cm^3 beakers, simple distillation apparatus, evaporating basin, Bunsen burner, spatula, splints.
Cane sugar or glucose, yeast, ammonium phosphate, calcium hydroxide solution.

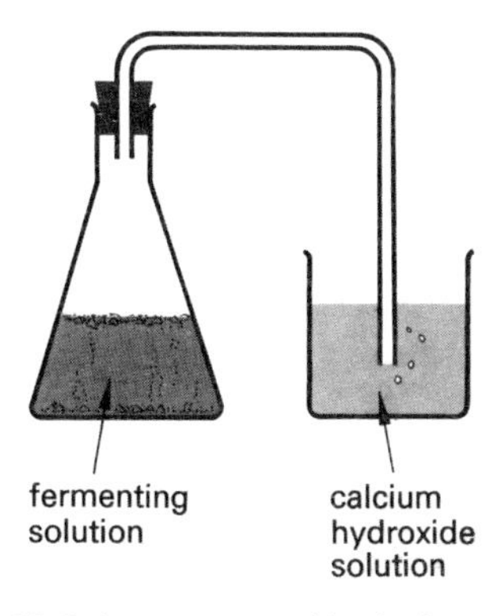

Figure 22.2 Apparatus used in the fermentation of sugar by yeast

Procedure
(a) Dissolve approximately 15 g of cane sugar or glucose in about 50 cm^3 of water in a conical flask. Add to this solution two spatula measures of yeast and a little ammonium phosphate (a yeast nutrient).
(b) Stopper the flask and leave it in a warm room, preferably near a warm radiator.
(c) Observe the flask occasionally and when the solution begins to ferment arrange for the delivery tube to dip into a beaker of calcium hydroxide solution. Record any observations you make. Leave the solution fermenting for several days.
(d) After several days decant (pour off) most of the fermented solution into a round bottom flask and assemble the apparatus for the distillation of this solution.
(e) Distil the solution and record the temperature at which liquid first begins to distil. Collect the first 1 or 2 cm^3 of distillate in an evaporating basin and attempt to ignite this liquid using a lighted splint. Also attempt to ignite a sample of the undistilled liquid from the round bottom flask by transferring a few drops into another evaporating basin and applying a lighted splint.

Results
Describe what you saw during the fermentation, what happened to the calcium hydroxide solution, and what happened when you tried to ignite the two different samples of liquid.

Conclusion
Explain all that happened in the experiment, remembering to include (i) an equation and an explanation about the distillation, (ii) why the distillation was needed, and (iii) why the distillate was richer in ethanol than the mixture being distilled.

Note: The distillate is not *pure* ethanol; it still contains some water. It is not possible to produce pure ethanol from a mixture with water by distillation (even fractional distillation). Final traces of water can be removed by adding a carefully chosen substance which will react with the water but not with the ethanol, and can then be separated (along with the products of its reaction with water) from the ethanol, e.g. by further distillation. Metallic calcium is often used for this purpose.

The manufacture and uses of ethanol
Much of the world's supply of ethanol comes from fermentations of the above type, although the source of sugar or other form of carbohydrate may vary, e.g. potatoes, rice, or sugar cane. The liquid from such distillations is used to produce various types of alcoholic drinks as well as ethanol itself. Ethanol is also manufactured by the hydration of ethene (p. 410) obtained from the cracking of petroleum. Large quantities of ethanol are used as a solvent and in the manufacture of other important organic chemicals.

Alcoholic drinks

Many fermentations are carefully controlled and modified so as to produce solutions which are then used to make alcoholic drinks (Figure 22.3). Starch can be used as a starting material instead of sugar because starch is a polymer of sugar (p. 428). For example, beer is made from the starch in barley, and the resulting alcoholic solution is then boiled with hops to give it a bitter taste. Similarly, a wide variety of wines may be made from substances containing either starches or sugars. In addition to grapes, such substances as elderberries, beetroot, potatoes, rhubarb, etc. can be used. It is estimated that over one million people in Britain make wine in their homes in this way.

Figure 22.3 A malt whisky distillery at Muir-of-Ord, Perthshire

When making wine you must be careful not to over-expose the wine to air in the early stages of fermentation, because oxidizing bacteria, whose spores are always present in air, may oxidize the wine to vinegar, i.e. the ethanol to ethanoic acid. This is how vinegar is made. Fortunately, when the alcoholic content of the wine is greater than about 12 per cent, the bacteria become inactive and so strong wines and all spirits do not turn sour on exposure to air. Beer, which has a lower alcoholic content, would turn sour in air fairly rapidly and so 'stabilizers' are usually added. In contrast, when the alcoholic content of wine reaches

Stabilizers are added to prevent beer going sour

about 17 per cent, the yeast enzymes then cease to function and further increase in the alcoholic content of the solution must be brought about by distillation, or even direct addition of alcohol, as in port.

You have probably noticed that some spirits are said to be 40° proof or 70° proof, etc. This does *not* mean that such solutions contain 40 per cent or 70 per cent alcohol. The term is derived from an old method of determining the alcohol content of a solution, as such solutions were taxed according to the amount of alcohol they contained. The method consisted of pouring the alcoholic liquor over gunpowder and then applying a flame. If the gunpowder was left dry enough to ignite, it was 'proof' that the liquor under test did not contain too much water. The liquor was then said to be *proof*. If the gunpowder was left too damp to ignite, the liquor was 'underproof'. Nowadays the Customs and Excise officer determines the amount of alcohol by the much less exciting method of measuring the density of the solution with a hydrometer.

Some of the chemical properties of ethanol

Combustion Ethanol burns in a plentiful supply of air with a clean, blue flame:

$$C_2H_5OH(l) + 3O_2(g) \rightarrow 2CO_2(g) + 3H_2O(l)$$

Methylated spirit consists largely of ethanol. At the moment, more energy is used in producing methylated spirit (e.g. in distilling the aqueous solution) than can be reobtained by burning it. However, recent developments give hope that an economic method of concentrating a solution produced by fermentation is possible, in which case the use of ethanol as a liquid fuel might increase dramatically.

Reaction with sodium Sodium reacts with ethanol. There is effervescence as hydrogen is given off, and the other product is sodium ethoxide, C_2H_5ONa, which can be obtained as a white solid if the excess ethanol is carefully evaporated off.

$$2C_2H_5OH(l) + 2Na(s) \rightarrow 2C_2H_5ONa \text{ (in ethanolic solution)} + H_2(g)$$

This is one of the few 'safe' ways of disposing of excess sodium, although it still produces a situation where a flammable gas is evolved from a flammable liquid.

Oxidation of ethanol This reaction has been briefly referred to earlier as that used in the 'breathalyser test' (p. 296). Acidified potassium dichromate(VI) is a powerful oxidizing agent and it oxidizes ethanol. The orange solution (or, in the breathalyser test, crystals) changes to a dark green colour as the dichromate(VI) ion is reduced to the green $Cr^{3+}(aq)$ ion, and the ethanol is oxidized to ethanoic (acetic) acid.

$$C_2H_5OH(l) + 2[O] \rightarrow CH_3COOH(aq) + H_2O(l)$$

Reaction with organic acids to make esters When an organic acid is heated with an alcohol in the presence of a little concentrated sulphuric acid as a catalyst, a reaction takes place to form an *ester*. Esters are a series of compounds containing the functional group

$$\begin{array}{c} O \\ \| \\ -C-O-C- \end{array}$$

e.g. $$CH_3COOH(l) + C_2H_5OH(l) \xrightleftharpoons{\text{concentrated } H_2SO_4\text{, heat}} CH_3COOC_2H_5(l) + H_2O(l)$$

$$\text{Organic acid} + \text{alcohol} \rightleftharpoons \text{Ester} + \text{water}$$

Ethyl ethanoate (ethyl acetate) is the organic product, and it can also be shown as

$$CH_3-\overset{\displaystyle O}{\overset{\|}{C}}-O-C_2H_5$$

The simple esters are colourless liquids with strong, pleasant odours. They are immiscible with water, and tend to float on top of the reaction mixtures. They are used in perfumes and flavourings, and as solvents.

In the above reaction, the forward reaction is called *esterification* and the reverse reaction is a hydrolysis. Animal fats and vegetable oils are more complicated esters, formed by the combination of the trihydroxy alcohol glycerol,

$$\begin{array}{l} CH_2OH \\ | \\ CHOH \\ | \\ CH_2OH \end{array}$$

and long chain organic acids such as stearic acid. The hydrolysis of such fats by alkali produces the sodium salts of the acids (i.e. soap, page 182) and free glycerol. This special type of hydrolysis is called *saponification*.

Note: Terylene is a polyester (p. 429).

Experiment 22.4
Some properties of ethanol**

Sodium, and also concentrated sulphuric acid and glacial ethanoic acid, must be used only by the teacher or under direct supervision.

Apparatus

Teat pipettes, watch-glass or evaporating basin, Bunsen burner, splints, rack of test-tubes, sharp knife, filter paper, 50 cm³ beaker, boiling tube.

Supply of ethanol, supply of sodium, fairly concentrated solution of potassium dichromate(VI), concentrated sulphuric acid, glacial ethanoic (acetic) acid.

Procedure

(a) Place a few drops of ethanol on to a watch-glass or evaporating basin and ignite the liquid using a lighted splint. Does ethanol burn with a clean flame? If so, what does this tell you about its structure?

(b) Pour about 2 cm^3 of the ethanol into a test-tube and add one small piece of freshly cut sodium about the size of a rice grain (care). Test any gas evolved with a lighted splint. When all the sodium has 'dissolved', carefully warm the solution, and then place a few drops on a microscope slide using a teat pipette. Allow the ethanol to evaporate. Record all your observations.

(c) Pour about 2 cm^3 of the potassium dichromate(VI) solution into a test-tube and *carefully* add about 1 cm^3 of concentrated sulphuric acid. To this solution add a few drops of ethanol and shake the tube well but carefully. *Gently* warm for a few moments in a small Bunsen flame, and note any changes which occur. What kind of reaction is taking place, e.g. is it an addition, a substitution, a combustion, or a redox reaction? What is happening to the potassium dichromate(VI)?

(d) Place about 1 cm^3 of ethanol and 1 cm^3 of glacial ethanoic acid into a test-tube and carefully add about 0.5 cm^3 of concentrated sulphuric acid. Warm the tube over a small flame for a few minutes, gently shaking the contents during warming. Then pour the contents of the tube into a small beaker half full of cold water. Record your observations and note any detectable odour.

Results and conclusion

Record your results in the usual way. Make sure that you can explain the results, and write equations for the reactions.

Organic acids (carboxylic acids)

These are a series of compounds containing the carboxyl group, —COOH, also shown as

$$-C\begin{matrix} \nearrow O \\ \searrow OH \end{matrix}$$

and having the general formula $C_nH_{2n+1}COOH$. The only one likely to be encountered at elementary level is ethanoic acid, CH_3COOH, commonly called acetic acid. The oxidation of ethanol leads eventually to the formation of this acid, and wines left exposed to the air for some time may have a sour taste due to its presence.

Physical properties of ethanoic acid

The pure acid ('glacial ethanoic acid') is a viscous liquid which freezes at 16 °C and then looks like ice. The acid has a very powerful smell of vinegar. Vinegar is largely dilute ethanoic acid.

Ethanoic acid as a typical acid

The substance is an acid because it can donate a proton to a water molecule (p. 72). It is important to realize that only *one* of the four hydrogen atoms in the molecule is 'acidic', and this is the one in the carboxyl group: CH_3COOⒽ.

$$CH_3COOH(l) + H_2O(l) \rightleftharpoons CH_3COO^-(aq) + H^+(aq).$$

The salts of the acid are called ethanoates (acetates). The acid shows all of the usual properties of an acid, changing blue litmus to red, producing effervescence with a carbonate to give carbon dioxide, and producing effervescence with magnesium ribbon to give hydrogen. It is, however, a weak acid (p. 74).

Check your understanding

1. Write the formula for the seventh member of the series of alcohols.
2. Write a fully balanced equation for the combustion of propanol.
3. Draw the structures of the isomers of butanol.

Points for discussion

1. Why are other substances (e.g. pyridine and a colouring agent) added to ethanol to produce methylated spirit?
2. Why is pure ethanoic acid sometimes called *glacial* ethanoic acid?
3. Ethanoic acid is a monobasic acid, and yet it contains four hydrogen atoms. Explain what this statement means.

22.5 POLYMERS

To an inorganic chemist a 'molecule' of sulphuric acid, H_2SO_4, containing seven atoms is a large molecule. However, in organic chemistry even a molecule of glucose, $C_6H_{12}O_6$, containing twenty-four atoms is considered to be comparatively small, as certain molecules are known which contain tens of thousands of atoms. Such molecules, with a vastly different scale of molecular size, are often *polymers*.

The derivation of the word polymer (from the Greek, *polys*, 'many'; *meros*, 'part') gives a clue as to how such large molecules are formed. They are in fact built up by the linking together of many smaller units, called *monomers*, to form much larger units which may consist of long chains, sheets, or three-dimensional networks. This process is called *polymerization*. Consider a particular monomer M; if two monomers combine together then a dimer M—M (M_2) is formed. If three monomers combine then a trimer M—M—M (M_3) is formed etc. When '*n*' monomers (where *n* usually varies between fifty and 50 000) combine then a polymer (M_n) will be formed.

Some polymers occur naturally and are called *natural polymers* (e.g. starch and cellulose). Others are man-made and are called *synthetic polymers*. Artificial fibres are synthetic polymers which can be made into yarn for clothing. Most synthetic polymers are called plastics, and they have given rise to a large and still expanding industry.

The plastics age

Plastics are used so frequently in everyday life that we tend to take them for granted and to forget that they have only been discovered comparatively recently. When plastics first began to replace traditional materials (e.g. plastic buckets instead of those made of galvanized iron, and plastic guttering instead of the cast iron variety) they were regarded as cheap substitutes and inferior in quality to the metals they replaced. This showed a complete lack of understanding about the advantages and versatility of plastics; cheap they may have been, but inferior they were not. There are no other materials which show so many outstanding practical qualities in the one substance. Plastics technology is now very advanced. Chemists can produce many different plastics with a wide variety of properties. Each particular plastic can also be produced in many different forms, e.g. low density and high density varieties (by altering the packing of the molecules), sheet, film, yarn, etc. Plastics have so revolutionized food packaging, household goods, the clothing industry, and the toy industry (to quote just four examples) that the times in which we live have been described as a plastics age.

Plastics do have their disadvantages, however. We live in a society which seems to generate more and more household refuse, much of which is plastics. Most traditional household refuse will rot down (if it is of vegetable or animal origin) or corrode (if it is metallic), but plastics are not affected by bacteria or corrosion. The disposal of 'waste' plastics can therefore be a problem.

The *source* of the monomers needed to make plastics is also causing concern. At the moment the monomers are produced almost entirely from petroleum, but they are becoming more expensive to manufacture as the crude oil becomes more expensive and difficult to obtain. Alternative sources are possible, e.g. coal, but they are not as convenient as oil and would certainly result in even more expensive products. In addition, coal will also be used up eventually. Each of these factors can be partially solved by recycling as much plastic waste as possible, i.e. by melting it down and remoulding it (p. 515). Unfortunately, some plastics (the thermosetting ones, see page 425) cannot be recycled in this way, and people are still reluctant to save (or separate from other waste material) substances such as paper and plastics. One thing is certain; plastics will continue to increase in cost, and some articles which are *conveniently* (but not essentially) manufactured from plastics will again be made from traditional materials, e.g. plastic carrier bags may be replaced by the traditional paper bags, etc.

More to do

1. Write a list of the advantages of plastics. You should be able to write at least eight important advantages.
2. Find out the difference between moulding and extruding—two methods which are used to shape plastics.
3. Polystyrene cement does not behave like a traditional 'glue'; how does it work?

Two different kinds of polymerization

There are two main types of polymerization.

1. **Addition polymerization** *is the successive linking together of unsaturated monomers. The only product is a single, large molecule. Each polymer normally contains only one kind of monomer molecule.*
2. **Condensation polymerization** *is the linking together of a large number of molecules by a reaction in which two products are formed, one of which is a small molecule (usually water). Such polymers may contain two different kinds of monomer molecule (e.g. in Terylene) or only one kind of monomer molecule (e.g. in starch).*

Two other terms commonly used in describing plastics

A **thermosetting polymer** *softens when first heated but then undergoes a chemical change due to the formation of a network of cross-linkages between polymer chains. This produces a rigid structure which cannot be softened or remoulded by later heating.*

A **thermoplastic (thermosoftening) plastic** *softens on heating and hardens on cooling. This process can be repeated if necessary, and such materials are easily moulded into shape but are not heat resistant.*

Synthetic addition polymers

Many of these have a monomer with the same fundamental structure:

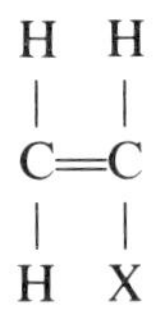

where X varies. For example, if X is a chlorine atom the monomer is chloroethene (vinyl chloride) and if X is a benzene ring (C_6H_5—) the monomer is phenylethene (styrene). See Table 22.5 for examples of addition polymers.

Some uses of these addition polymers

Polythene	Packaging, plastic bags, plastic film
Polyvinylchloride	Insulation for electric cables, raincoats, upholstery, suitcase coverings, gramophone records
Polystyrene	Heat insulator in buildings, packaging, model making. Expanded form made by generating gas in syrup during polymerization
Polymethyl Metracrylate	Used as a glass substitute, e.g. in aircraft windows, reflectors on vehicles, TV guard screens, etc.
Polypropylene	Buckets, washing-up bowls, food containers, and especially useful for moulded boxes with hinged lids—a hinge made of polypropylene can be 'folded' hundreds of times without breaking.

Table 22.5 Some synthetic addition polymers

Name of polymer	*Formula of monomer*	*Name of monomer*	*Reaction conditions*	*Part of polymer chains*
Polyethene (polythene) (high density form)	H H C=C H H	Ethene (gas)	Low temperature and pressure, special catalyst	(H H H H —C—C—C—C— H H H H)
Polyvinylchloride (PVC) or polychloroethene	H H C=C H Cl	Vinyl chloride (chloroethene)	60 °C, high pressure, H_2O_2 catalyst	(H H H H —C—C—C—C— H Cl H Cl)
Polystyrene (polyphenylethene)	H H C=C H C_6H_5	Phenylethene (styrene), syrupy liquid	Catalyst, e.g. dibenzoyl peroxide, heat	(H H H H —C—C——C—C— H C_6H_5 H C_6H_5)
Polymethyl Metracrylate	H CH_3 C=C H $COOCH_3$	'Methyl-methacrylate', syrup	Catalyst, e.g. dibenzoyl peroxide, heat	(H CH_3 H CH_3 —C—C——C—C— H $COOCH_3$ H $COOCH_3$)
Propylene (polypropene)	H H C=C H CH_3	Propene	Similar to polyethene	(H H H H —C—C——C—C— H CH_3 H CH_3)

Experiment 22.5
The laboratory preparation of polystyrene***

Apparatus
100 cm³ round bottom flask fitted with reflux condenser, Bunsen burner, tripod, gauze, oil bath, 100 cm³ and 250 cm³ beakers, measuring cylinder, 0–360 °C thermometer, spatula, filter paper, filter funnel, crucible, pipe clay triangle.
Phenylethene (styrene), methanol, lauroyl peroxide.

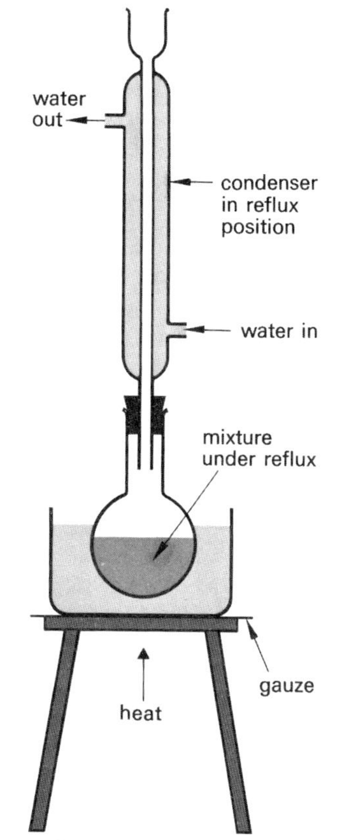

Figure 22.4 Apparatus used for refluxing mixtures

Procedure
(a) Mix together about 10 cm³ of phenylethene and a pinch of lauroyl peroxide in a round bottom flask.
(b) Reflux the mixture over an oil bath (care) for about forty-five minutes, as shown in Figure 22.4. The temperature of reflux should be about 150 °C.
(c) Allow the contents of the flask to cool and pour the mixture into approximately five times its own volume of methanol in a beaker. Note the appearance of the polymer. It can be made to harden to a wax by agitating it with a spatula under the surface of a fresh supply of methanol.
(d) Isolate some of the wax formed in (c) by filtration and dry off the excess methanol using filter paper. Carefully heat a little of this solid wax in a crucible until it becomes pliable and then allow it to cool.

Notes about the experiment
(i) Compare the appearance of the product with that of the monomer. It should be obvious that the former is likely to consist of larger molecules and that at least some polymerization has taken place.
(ii) The lauroyl peroxide is a catalyst for this polymerization.

Results and conclusion
Record your observations in the usual way, and in your conclusion explain the type of polymerization taking place, with a simple 'equation'.

Natural condensation polymers
Proteins from amino acids

Note: Proteins are often called polymers, although strictly speaking they should be called macromolecules and not polymers. Particular amino acids in a protein do not occur at regular intervals, and there may be many different amino acids in one molecule. As a contrast to this, true polymers contain regular repeating units, and usually only one or two different monomers.

Amino acids have the general formula

$$\begin{array}{c} \text{R} \\ | \\ \text{NH}_2\text{—C—COOH,} \\ | \\ \text{H} \end{array}$$

where R is a side chain or hydrogen atom, which varies from one acid to another. NH_2 is the amino functional group. About 20 different amino acids occur in nature. Note that each molecule contains two different functional groups, NH_2 and COOH. (—[A]— and —[B]— are used below to signify the central, variable part of each monomer.)

$$NH_2\text{—[A]—}COOH + NH_2\text{—[B]—}COOH \rightarrow$$
amino acid amino acid

$$NH_2\text{—[A]—}CONH\text{—[B]—}COOH + H_2O$$
dimer

As the product still contains an NH_2 and a COOH group, the process can continue at each end of the molecule, to build up a large molecule called a *protein*. Remember that a protein molecule will contain many different 'monomers', although each of the 'monomers' will be an amino acid, and that each 'monomer' could occur many times in one molecule of the polymer.

Carbohydrates (starch) from sugars

There are many different sugars, but only the polymers of glucose, $C_6H_{12}O_6$, are considered at an elementary level. For convenience, glucose is best considered as a 'block of atoms' (which takes no part in the polymerization) and 'reactive ends' of the molecule, which do.

$$OH\text{—[ ]—}OH + OH\text{—[ ]—}OH \rightarrow OH\text{—[ ]—}O\text{—[ ]—}OH + H_2O$$

two molecules of the monomer (a monosaccharide), glucose — dimer (a disaccharide), maltose

(The 'block' is $C_6H_{10}O_4$.) *Note*: that as the dimer still contains 'reactive ends', the process can continue and a polymer (a polysaccharide) is built up, in this case starch.

Synthetic condensation polymers

Nylon

This is similar to the polymerization of amino acids, but involves two different types of monomer, each of which contains two functional groups of the same type, e.g.

$$COOH\text{—[ ]—}COOH + NH_2\text{—[ ]—}NH_2 \rightarrow$$
$$COOH\text{—[ ]—}CONH\text{—[ ]—}NH_2 + H_2O$$

This process continues, to produce the polymer. The composition of the 'block' varies according to the type of nylon, but typically contains 6 or 10 carbon atoms. It is used mainly as a fibre, e.g. in clothing, ropes, and fishing lines.

Experiment 22.6
Laboratory preparation of nylon**

Apparatus

Two 100 cm³ beakers, 5 cm³ and 50 cm³ measuring cylinders, forceps, glass rod, access to fume cupboard.

Decan-1,10-dioyl chloride (sebacoyl dichloride), 1,6-diaminohexane (hexamethylene diamine), tetrachloromethane, sodium hydroxide solution (concentration 1 mol dm^{-3}). Gloves should be worn during this experiment, and the operations should be conducted in a fume cupboard. Tetrachloromethane is toxic.

Procedure

(a) Dissolve about 1 cm^3 of decan-1,10-dioyl chloride in 10 cm^3 of tetrachloromethane in a 100 cm^3 beaker.
(b) In a separate beaker, make a solution of 1 g of 1,6-diaminohexane in 20 cm^3 of the sodium hydroxide solution.
(c) Carefully pour the solution from (b) on top of the solution made in (a), taking care to avoid mixing of the two solutions.
(d) Using forceps take hold of the thin film of nylon which forms at the interface of the two solutions. Gently pull this film upwards and out of the beaker. Wrap this 'rope' of nylon around a glass rod and continue to remove nylon until the 'rope' breaks (Figure 22.5).

Results and conclusion

Record all of your observations, and in your conclusion explain which type of polymer nylon is, giving simplified formulae to show what happens during the polymerization. The chemical reaction which produces nylon in this experiment takes place at the junction of the two immiscible liquids.

Figure 22.5 Preparation of nylon

Polyethylene terephthalate

This is a polyester.

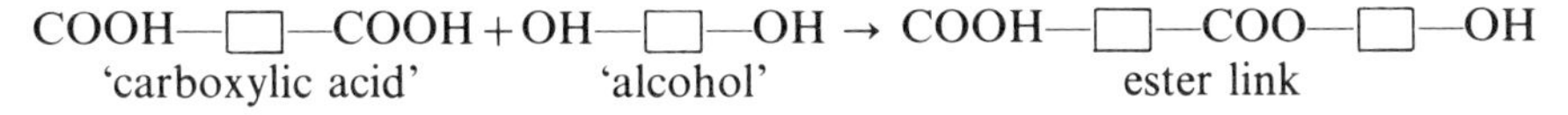

This process continues, to build up the polymer. It is used mainly as a fibre, e.g. in clothing, for boat sails, and in fishing lines. The full formulae of the monomers, and those of some other common condensation polymers, are summarized in Table 22.6.

Check your understanding

1. Plastics molecules can be made by the combination of a number of smaller molecules. These smaller molecules are called
A polymers
B allotropes
C isomers
D dimers
E monomers

2. Plastics are to be found in every home, factory, and type of transport.
(a) Name two common plastics (polymers) which are derived from ethene and give one use for each.
(b) Give one reason why the use of these plastics continues to increase.
(c) Why is the disposal of these plastics a problem for our society?
(d) What natural resource is the basis of these plastic materials?
(e) State one other use of this natural resource.

Table 22.6 Some common condensation polymers

Common name and formula of monomers	*Common name and formula of polymer*	*Some uses of polymer*
(a) HOOC–C_6H_4–COOH + HO—CH_2—CH_2—OH Benzene-1,4-dicarboxylic acid (terephthalic acid) + Ethane-1, 2-diol (ethylene glycol)	$[—OC—C_6H_4—COOCH_2CH_2O—]_n$ Polyethylene terephthalate	Mainly as a fibre (e.g. in clothing, and fishing lines)
(b) C_6H_5OH + $H_2C{=}O$ + C_6H_5OH Hydroxy-benzene (phenol), Methanal (formaldehyde)	$[—CH_2—C_6H_3(OH)—CH_2—C_6H_3(OH)—CH_2—]_n$ Bakelite	To make buttons, knife handles, switches and distributor heads for motor vehicles, cameras, radio and telephone equipment
(c) $O{=}C(NH_2)_2$ + $O{=}CH_2$ + $O{=}CH_2$ Carbamide (urea), Methanal (formaldehyde)	Cross-linked network of —N—CH_2—N—CH_2—N— chains joined through C=O groups, $[\ldots]_n$ Urea–formaldehyde	Domestic kitchenware, e.g. plates, saucers, cups etc., control knobs, bottle caps

3.

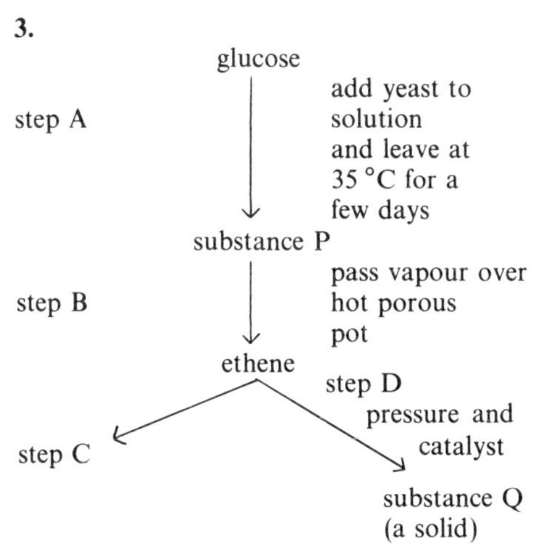

Look at the reaction scheme opposite involving carbon compounds and answer the following questions.
(a) (i) Name the process involved in step A.
(ii) Name the substance P.
(iii) Write an equation for the conversion of glucose to substance P.
(iv) As the reaction in step A is slow give a reason why the reaction is not heated to 100 °C to speed it up.
(b) Name the reagent and conditions necessary for step C.
(c) (i) Name the process involved in step D.
(ii) Name the solid substance Q.
(iii) Write an equation for the conversion of ethene to substance Q.
(iv) Give one use for substance Q.
(J.M.B.)

4. Modern plastics/synthetic fibres tend to be replacing more traditional materials. Complete the following table.

Materials	*TWO advantages of the modern material*	*TWO advantages of the traditional material*
PVC guttering instead of iron guttering		
Polythene bags instead of paper bags		
Synthetic fibres instead of cotton or wool		

(A.E.B. 1978)

5. Ethene is an important material for the petrochemical industry; approximately two million tonnes is used per year.
(a) Explain how ethene is produced industrially.
(b) Complete the following table which is concerned with the products that are manufactured from ethene.
(c) State, with reasons, which one of the polymers in (b) you think is the least expensive to produce.
(d) Explain briefly why 'Polyethylene terephthalate' is a different type of polymer from the others mentioned in (b).
(e) Explain how ethene is converted into ethanol industrially. (A.E.B. 1976)

Product derived from ethene	*One large scale use*	*Reason for this use*
Poly(ethene)		
Poly(vinylchloride)		
Polyethylene terephthalate		
Poly(styrene)		
Ethanol		

Point for discussion

If a piece of bread is retained in the mouth and chewed for some time, it often tastes sweet. Can you explain why?

22.6 A SUMMARY OF CHAPTER 22

The following 'check list' should help you to organize the work for revision.

1. *Definitions*

(a) Organic chemistry (p. 401)
(b) Isomers (p. 403)
(c) Homologous series (p. 405)
(d) Substitution reaction (p. 407)
(e) Addition reaction (p. 409)
(f) Addition polymerization (p. 425)
(g) Condensation polymerization (p. 425)
(h) Thermosetting plastic (p. 425)
(i) Thermoplastic (thermosoftening) plastics (p. 425)

2. *Other points*

Make sure that you understand the following.
(a) The difference between a graphical formula,

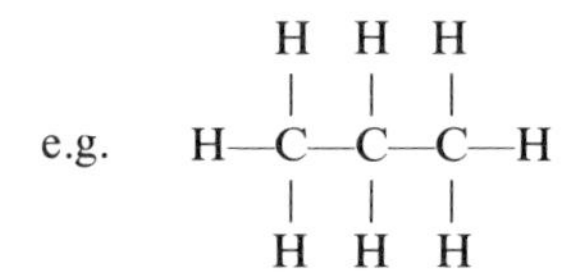

a structural formula (e.g. $CH_3CH_2CH_3$), and a molecular formula (e.g. C_3H_8).
(b) What a functional group is (page 403 and also defined on page 418), e.g. the OH group in an alcohol such as methanol, CH_3OH.
(c) The general features of any homologous series, e.g. (i) members have similar chemical properties, (ii) melting points and boiling points increase 'down' the series, and (iii) members have the same general formulae.
(d) The general formulae for the homologous series you have studied, e.g. the alkanes C_nH_{2n+2}, the alkenes C_nH_{2n}, and the alcohols, $C_nH_{2n+1}OH$.
(e) Some of the principles used in naming *simple* organic compounds.
(f) The differences between saturated and unsaturated compounds. Saturated compounds contain only single covalent C—C bonds and they may take part in substitution reactions (or burn, or be cracked) but they *never* take part in addition reactions. Unsaturated compounds contain double or triple C—C bonds and normally take part in addition reactions.
(g) The physical and chemical properties of methane, as a typical member of the homologous series of alkanes (p. 406).
(h) Carbon *monoxide* is one of the products of combustion when hydrocarbons are burned in a restricted air supply.
(i) The physical and chemical properties of ethene as a typical member of the homologous series of alkenes (p. 409). Compare and contrast these with the reactions of methane, using appropriate scientific words such as substitution, saturated, etc. wherever possible. (See also the table on page 412.) Most of the reactions of methane are substitution reactions, but ethene generally takes part in addition reactions.
(j) Desulphurization of crude oil, and the main fractions produced by the fractional distillation of crude oil (p. 412).
(k) What 'cracking' means, and why most of the heavier fractions are cracked to produce simple alkanes and alkenes.
(l) A general understanding of the problems caused by our dependence on crude oil even though supplies are limited, why refineries are now often situated well away from oil fields, and how refinery operations are affected by economics and especially demand.
(m) How alcoholic drinks are manufactured, with examples.
(n) The simple physical and chemical properties of ethanol (p. 422).
(o) The physical properties of ethanoic acid, and its behaviour as a typical monobasic, weak, acid (p. 423).
(p) The meaning of polymer, dimer, monomer, natural polymer, synthetic polymer, and synthetic fibre.

(q) An appreciation of why the 'plastics revolution' has occurred; the problem of disposal of plastics, the need to recycle the thermosoftening ones, and the problem of continuing to supply suitable monomer molecules from, e.g., dwindling reserves of oil.
(r) How addition polymers are formed; some common examples, their 'structures' and their uses (pp. 425–6).
(s) Examples of some condensation polymers; simple 'equations' to show their formation, and some of their uses (pp. 428–30).

3. Important experiments

The preparation of a solution of ethanol by fermentation, and the concentration of the solution by distillation.

QUESTIONS

Multiple choice questions

1. The empirical formula of ethanoic acid (acetic acid) is
A CH_2O B CH_4O C C_2H_3O
D C_2H_4O E C_2H_6O (C.)

2. Which one of the following is the general formula of an organic acid?
A C_nH_{2n}
B C_nH_{2n+2}
C $C_nH_{2n+1}COOH$
D $C_nH_{2n}O_n$
E $C_nH_{2n+1}OH$ (C.)

3.

```
                          H
                          |
                        H—C—H               H                    H  H  H
  H  H  H  H            H |  H              |                    |  |  |
  |  |  |  |            | |  |            H—C—H  H             H—C—C—C
H—C—C—C=C             H—C—C=C             |      |               |  |  ‖
  |  |     |            |    |            H—C—C=C                H  H  ‖
  H  H     H            H    H              |  |  |                  H—C—H
                                            H  H  H
    (i)                  (ii)                (iii)                  (iv)

                              H  H  H  H
                              |  |  |  |
                            H—C—C=C—C—H
                              |        |
                              H        H
                                 (v)
```

Which of the five structures shown represent identical molecules?
A (i) and (v) only
B (i) and (iii) only
C (ii) and (iii) only
D (i), (iii), and (iv)
E (i), (iii), and (v) (C.)

Structured questions and questions which require longer answers

4. (a) A hydrocarbon X has an empirical formula C_2H_5 and its relative molecular mass (molecular weight) is 58.
(i) Name X and give its molecular formula.
(ii) Write down the structural formulae of two isomers of X.
(iii) How would you expect X to react with chlorine in diffused light?
(iv) Name and give the molecular formulae of 3 other hydrocarbons which are in the same homologous series as X.
(b) Name and give the *full* structural formula of *one organic* compound which is formed by the oxidation of ethanol. (O. & C.)

5. (a) Name *four* fractions obtained by the fractional distillation of petroleum and give *one* major use for each fraction.
(b) At present, most of our ethanol is made from ethene. How is this done on a large scale?
(c) In future, it may be necessary to make

ethene from ethanol. How could this be done industrially?
(d) Ethene and bromine were allowed to react together at a temperature high enough to ensure that all the substances involved in the reaction were gaseous. (i) Write the equation for the reaction. (ii) Give the structural formula for the product. (iii) State what would be seen during the course of the reaction. (iv) If 50 cm^3 of ethene and 50 cm^3 of bromine vapour were used, what volume of gaseous product would be formed?
(J.M.B.)

6. (a) Describe briefly how you would obtain a dilute solution of ethanol in the laboratory starting from glucose ($C_6H_{12}O_6$). Give the equation for the reaction.
(b) A small, almost pure, sample of ethanol can be obtained in the laboratory from a dilute solution of ethanol by fractional distillation. Draw a labelled diagram of the apparatus required if ethanol, boiling point 78 °C, is to be collected.
(c) If ethanol is heated for about thirty minutes with an acidified solution of potassium dichromate(VI) the mixture turns green. From this mixture a pure substance, X, with a melting point of 17 °C can be isolated. Explain (i) what you think is the function of the potassium dichromate(VI) and why the solution turns green; (ii) what would happen to substance X if a bottle containing it were left in an unheated storeroom throughout the year.
(d) X contains 6.67 per cent of hydrogen, 40 per cent carbon, 53.3 per cent oxygen and has a relative molecular mass of 60. Calculate the empirical formula of X and give its molecular formula. (J.M.B.)

7. Explain, with a suitable example in each case, the meaning of each of the following terms: (i) chromatography, (ii) fractional distillation, (iii) an ionic crystal, (iv) a covalent bond, (v) structural isomerism. (J.M.B.)

8. What do you understand by the term *allotrope*? Show how a knowledge of the crystal structures of diamond and graphite can be used to explain the difference in their hardness.

The empirical formula of a gaseous hydrocarbon X is CH_2 and its density is the same as that of nitrogen under the same conditions. When X was passed into bromine water, the latter lost its colour and a sweet-smelling product was obtained. Under suitable conditions X can be converted to a white solid which also has the empirical formula CH_2. Identify X and explain these observations.

State briefly how X may be obtained from ethanol. (A.E.B. 1975)

9. (a) Briefly explain in terms of the electronic theory the bonding between the atoms in a molecule of methane, CH_4.
(b) Methane is said to be a *saturated* hydrocarbon; name one *unsaturated* hydrocarbon. How does the structure of an unsaturated hydrocarbon differ from that of a saturated hydrocarbon? Compare the reactivities of methane and the unsaturated hydrocarbon with bromine, and indicate the structures of the products formed.
(c) State (without describing the apparatus used) (i) how an unsaturated hydrocarbon can be obtained from ethanol in the laboratory, (ii) how an unsaturated hydrocarbon can be converted into a polymer.
(d) Why is it dangerous to use an appliance burning methane (natural gas) in a badly ventilated room? (A.E.B. 1977)

10. (a) Describe, giving the essential reaction conditions, how ethanol may be converted in the laboratory to: (i) ethene, (ii) ethyl ethanoate (ethyl acetate), (iii) sodium ethoxide.
(b) Using appropriate examples, explain briefly how alkanes are used: (i) as industrial fuels, and (ii) as a source for making more useful products by the process of cracking.
(c) (i) Give the structural formulae of ethane and ethene.
(ii) Describe a single chemical test to distinguish between these two gases.
(iii) State and explain *one* important industrial use of ethene. (W.J.E.C.)

11. Explain *four* of the following terms used in organic chemistry giving suitable examples: (a) homologous series, (b) isomerism, (c) unsaturation, (d) polymerization, (e) fermentation.

Give *one* characteristic chemical reaction of (i) ethanol, (ii) ethanoic acid (acetic acid). (W.J.E.C.)

12. A CH_3OH; B CH_3COOH;
C $CH_3CH{=}CH_2$; D CH_3CH_2OH;
E $CH_3CH_2CH_2CH_3$.
(a) For each of the compounds in the list A to E above, give its name, the name and general formula of the homologous series to which it belongs, and state whether the compound is a solid, liquid, or gas at room temperature and pressure.
(b) On complete combustion, 0.100 mole of one of the compounds A to E gave 13.2 g of

carbon dioxide. Identify the compound, explaining your reasoning.
(c) What is meant by an *ester*? Using only compounds chosen from A to E, describe how you would prepare an ester. Write the equation for the preparation. (C.)

13. (a) Which homologous series of organic compounds can be represented by the following general formulae:

Series A $C_nH_{(2n+2)}$;
Series B C_nH_{2n};
Series C $C_nH_{(2n+1)}COOH$;
Series D $C_nH_{(2n+1)}OH$?

(b) Give the name and structural formula of *one* compound in each series.
(c) Describe reactions by which
(i) a *named* compound of series B can be converted to a compound of series A;
(ii) a *named* compound of series D can be converted to a compound of series B.
(d) Name an important natural source of compounds of series A and give *two* industrial uses of such compounds. (C.)

14. Chloroethane can be formed from an alkene by an addition reaction and from an alkane by a substitution reaction. Explain the meaning of the terms alkene, alkane, addition reaction, and substitution reaction and write equations for the two reactions referred to in the first sentence.

Describe (a) *two* other addition reactions of an alkene, and (b) *one* reaction by which chloroethane can be obtained from ethanol. (C.)

15. By drawing their structural formulae show the differences in structure between ethane and ethene. Explain how the bond between a carbon atom and a hydrogen atom in these compounds is formed.

By naming the reagents, stating the conditions, and writing an equation for each reaction, describe how ethene could be converted into (a) ethane, (b) chloroethane, (c) 1,2-dibromoethane, (d) ethanol. (J.M.B.)

23 Gas volume calculations: analysis

23.1 GAS VOLUME CALCULATIONS

What causes the volume of a gas to change?

Many chemical reactions take place between one or more gases, and often the volume of gas changes during such reactions. There are two main reasons why such a change in volume can occur.

Physical changes

Both temperature and pressure changes can cause the volume of a gas to change. For example, in an exothermic reaction or one which is being heated, the temperature of the substances will rise, and the volume of any gas present will increase. We should not measure the volume of gas in such a system. For example, in Experiment 9.1 (p. 134) we only measured the volume of gas after cooling to the original temperature. Gas volumes must always be compared, therefore, under the same conditions of temperature and pressure. Two such sets of conditions are commonly used.

Standard temperature and pressure (s.t.p.) is 0 °C (273 K) and 760 mm mercury (i.e. one atmosphere) pressure.

Room temperature and pressure (r.t.p.) is a much less precise term, being used for calculations where *exact* answers are not important. This set of conditions is assumed to be 25 °C and 760 mm mercury pressure.

Chemical changes

You have met examples where a gas is *formed* during a reaction, and in such cases the volume of gas therefore increases throughout the reaction, e.g.

$$CaCO_3(s) + 2HCl(aq) \rightarrow CaCl_2(aq) + CO_2(g) + H_2O(l)$$
$$Mg(s) + H_2SO_4(aq) \rightarrow MgSO_4(aq) + H_2(g)$$

There are also reactions in which a gas is *used up*, e.g.

$$2Cu(s) + O_2(g) \rightarrow 2CuO(s),$$

or when two gases react to form a gas which takes up a *smaller volume*, e.g.

$$2H_2(g) + O_2(g) \rightarrow 2H_2O(g)$$
$$N_2(g) + 3H_2(g) \rightarrow 2NH_3(g)$$

In these last three cases, the gas volume *decreases* during the reaction. In the last two examples it appears as if we are ending up with less than we started with. In fact the Law of Conservation of Mass still applies, for there are exactly the same number of *atoms* present at the end as were there at the beginning. The 'secret' of these last two volume changes is of course the number of gas *molecules* present, which has decreased and it is this which has brought about the change in volume. This is explained by Avogadro's Law.

Avogadro's Law

In 1811 an Italian, Amadeo Avogadro, proposed the following, which is now accepted as an important law.

Equal volumes of all gases, under the same conditions of temperature and pressure, contain equal numbers of molecules.

This may seem a fairly obvious idea to you, because your chemistry course from its earliest stages has probably emphasized the existence of molecules, and you know that the pressure of a gas is caused by the bombardment of its molecules on the walls of its container. However, at the time of Avogadro the explanation was a completely new idea because, although scientists accepted the idea that atoms existed, they had no idea that atoms of the same element sometimes 'went around' in small groups (e.g. pairs, Cl_2, or fours, P_4) to form *molecules*.

We now have a great deal of evidence in support of Avogadro's Law, e.g. by the answers to the following questions.

Points for discussion

1. The densities of some common gases (measured at 25 °C and 760 mm mercury pressure) are given below. For each gas, calculate the volume (at 25 °C and 760 mm mercury pressure) which would be occupied by a mole of the gas molecules. (Remember that the noble gases have 'molecules' which consist only of single atoms.)

(a) Argon, 1.66 g dm^{-3}
(b) Fluorine, 1.58 g dm^{-3}
(c) Neon, 0.84 g dm^{-3}
(d) Nitrogen, 1.17 g dm^{-3}
(e) Oxygen, 1.33 g dm^{-3}

2. What can you say about the number of molecules present in a mole of each gas?

3. Are the volumes occupied by a mole of each gas at 25 °C and 760 mm pressure about the same?

4. Explain how your answers to questions 1, 2, and 3 help to support Avogadro's Law.

Avogadro's Law and gas pressure

Suppose that we have equal volumes of two gases, A and B, and they are both at the same temperature and pressure. From your work on the kinetic theory you will remember that the pressure in each container is considered to be caused by the 'impact force' of the molecules hitting the walls of the container, and this must be the same in each container.

This 'impact force' is really a combination of two things; (i) how often the molecules hit the walls of the container (which depends upon their speed), and (ii) how hard they hit the walls, which depends upon the speed of the molecules and also upon their mass. If molecules of A have more mass than those of B (i.e. they have a larger relative molecular mass) they may be expected to hit the walls with more 'force' than those of B. On the other hand, for a given temperature, the more massive molecules will move more slowly and therefore the effect of their greater mass is overcome by the fact that they hit the walls less hard and less often. These opposite effects more or less cancel each other out, so that at a given temperature and pressure the 'impact force' is just about the same for a given number of *any* kind of gas molecule.

It must follow, therefore, that if our containers of A and B are of the same volume, temperature, and pressure, the *total* 'impact force' must be the same in each container, and this is because the total number of molecules must be the same in each container (Avogadro's Law).

Gay-Lussac's Law

The ways in which gas volumes change as the result of a chemical reaction were first investigated by a French scientist, Joseph Gay-Lussac. He was trying to find out if air was a mixture or a compound, and collected a great variety of air samples and analysed them. (He even made a perilous journey in a balloon to collect some air at high altitudes.) He analysed his samples by exploding the oxygen present with some added hydrogen in a special piece of apparatus. He found that, in all cases, two volumes of hydrogen combined with one volume of oxygen and this simple but constant ratio led him to experiments with other gases. In 1808 he summarized his findings in a statement which is now known as Gay-Lussac's Law.

When gases combine, they do so in volumes which bear a simple ratio to one another and to the volume of the product(s) if gaseous. All volumes must be compared at the same temperature and pressure.

This statement may not mean much to you when you first read it. In simple terms, it means that if a chemical reaction takes place in which gases are used up or formed, then the volumes of the gases which take part in the reaction will always be in a simple ratio. For example, suppose that there is a reaction in which a gas A is used up and a gas B is formed. 10 cm^3 of gas A could be used up in the reaction, and perhaps 10 cm^3 of gas B would be formed. In this case the ratio of the volume of A used to the volume of B formed is simply 1:1. On the other hand, there are no reactions in which the reacting volumes of gases are in complicated ratios like 10 cm^3 of one gas to 168 cm^3 of another (a ratio of 1:16.8). The ratios between the volumes of reacting gases are usually 1:1, 1:2, 2:3, or 3:1.

Note: (i) These different ratios are not alternatives in any one reaction! The ratio is fixed for each particular reaction involving gases, but it is always a simple ratio.
(ii) The volumes must be compared under equal conditions of temperature and pressure.

These ideas, and the significance of them, will become clearer as you work through the section.

Experiment 23.1
An illustration of Gay-Lussac's Law***

Apparatus
Two glass syringes (100 cm^3), two stands and syringe holders, three-way stopcock, rubber tubing, access to fume cupboard.
Oxygen cylinder, apparatus for producing pure nitrogen monoxide.

Procedure
(a) Working in a fume cupboard, connect the two syringes and the stopcock by rubber tubing as shown in Figure 23.1.
(b) Fill one syringe with nitrogen monoxide and push the gas out again. Do this several times, finally leaving 40 cm^3 of the gas in the syringe. Repeat the procedure using oxygen in the other syringe so that 50 cm^3 of oxygen remain in the syringe after the preliminary 'washing out'.
(c) Turn the stopcock so that the two syringes are connected and push 5 cm^3 of oxygen into the syringe containing the nitrogen monoxide. Close the stopcock,

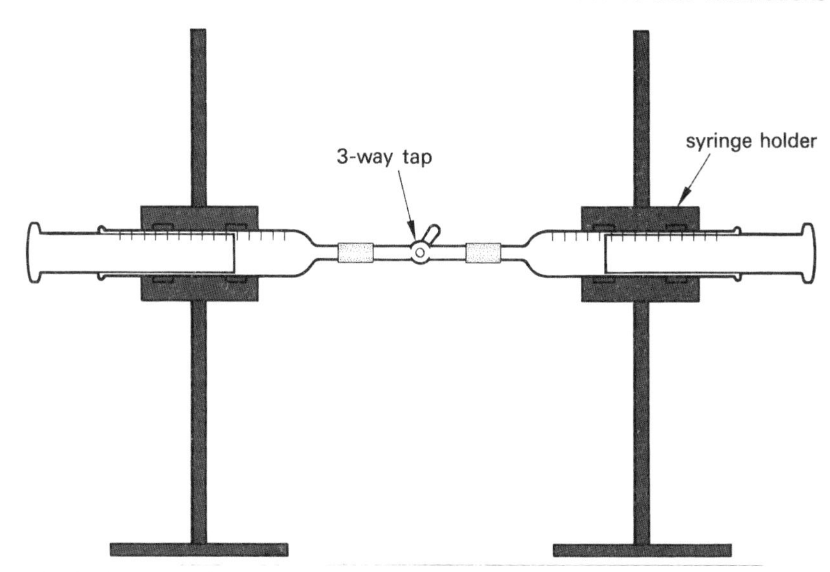

Figure 23.1 Measuring the volume change during the reaction between nitrogen monoxide and oxygen

wait until there is no further change in volume and note the volume of gas in each syringe.
(d) Repeat the process with further 5 cm^3 portions of oxygen until there is no change in the *total* volume of gas in the two syringes.

Results
Record: (i) any colour changes which took place, and name and give the formula of the gas formed during the reaction; (ii) the volume of oxygen used to *complete* the reaction; (iii) the volume of nitrogen monoxide which reacted; (iv) the volume of the gas formed as the product.

Notes about the experiment

(i) In experiments of this type, it is usual to have one of the reacting gases in *excess* so that all of the other gas is used up. In the experiment oxygen was in excess, so that all the nitrogen monoxide was used up but there was some unused oxygen left at the end. In using the volumes to illustrate Gay-Lussac's Law, it is important to use only the *reacting* volumes and/or the volume of the product(s); do not include any *excess*.
(ii) Typical results from some other investigations of the same type are given below. The gases in each reaction were measured under the *same* conditions of temperature and pressure.
(a) 100 cm^3 of hydrogen react with 100 cm^3 of chlorine to produce 200 cm^3 of hydrogen chloride.
(b) 200 cm^3 of hydrogen react with 100 cm^3 of oxygen to produce 200 cm^3 of steam.
(c) 100 cm^3 of carbon monoxide react with 50 cm^3 of oxygen to produce 100 cm^3 of carbon dioxide.
(d) 50 cm^3 of nitrogen react with 150 cm^3 of hydrogen to produce 100 cm^3 of ammonia.

Conclusion
Use your experimental results, and also those given above, to illustrate Gay-Lussac's Law.

Avogadro's Law and volume changes

Avogadro's Law makes it easy to understand why in the reactions

$$2H_2(g) + O_2(g) \rightarrow 2H_2O(g) \quad \text{and}$$
$$N_2(g) + 3H_2(g) \rightarrow 2NH_3(g)$$

there is a decrease in total volume (measured at constant pressure) even though there is no change in the total number of atoms present.

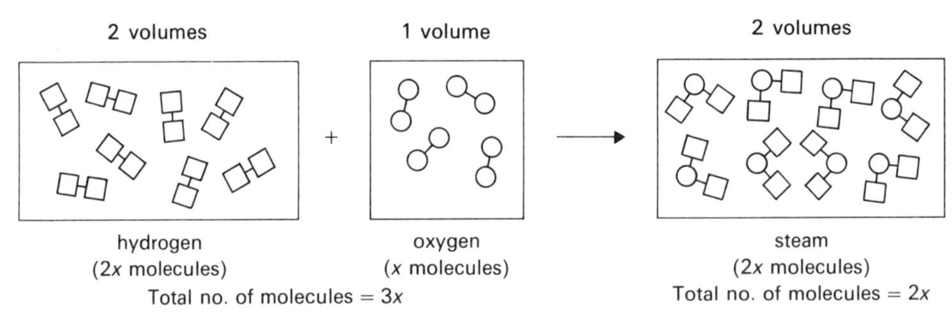

We can see that in the example above the volume of hydrogen is twice the volume of oxygen, because (by Avogadro's Law) at the same temperature and pressure there are twice as many molecules of hydrogen as there are molecules of oxygen. When these molecules combine to form steam, however, the atoms rearrange themselves to form a smaller number of larger molecules and therefore (by Avogadro's Law) the total volume decreases. No atoms have been lost or gained; they have simply rearranged themselves. A similar argument can be applied to other gas reactions.

You may feel that larger molecules must take up more volume than smaller ones (e.g. H_2O molecules take up more volume than a similar number of H_2 molecules), and therefore that the total volume should remain unchanged. However, the *size* of the molecule is practically irrelevant in these arguments. Molecules in a gas are so far apart that their size has almost no effect on the volume of the gas, which is mainly space. The *number* of molecules is far more important than their size.

Calculations involving Gay-Lussac's Law

You could be provided with experimental data and asked to use it to illustrate Gay-Lussac's Law. The essential points to remember in performing these calculations are given below.

1. Use only the volumes of gases which are *used up* or *formed* during the reaction. Some gases may be in excess, and the excess should not enter into the calculation.
2. If steam is one of the products, and the measurements are made at room temperature or below, the steam will condense to a *negligible* volume of liquid water, and this should not enter into the calculations.
3. If more than one gas is formed, it is sometimes necessary to find the volumes of the individual gases by absorbing one of them in a solution (e.g. carbon dioxide can be absorbed in potassium or sodium hydroxide solution). The volumes before and after absorption indicate the volume of the gas which has been absorbed.

The following example should make this clear.

48 cm^3 of methane were mixed with 212 cm^3 (an excess) of oxygen, and the mixture was exploded. The products, after cooling and reverting to room temperature and pressure, occupied 164 cm^3 which became 116 cm^3 after shaking with sodium hydroxide solution. We are asked to show that these results are in accordance with Gay-Lussac's Law.

Methane + oxygen → carbon dioxide + steam (which condenses to a negligible volume of water).
The oxygen is in excess, so all the methane is used up (48 cm^3).
The volume of carbon dioxide formed = (164 − 116) cm^3 = 48 cm^3 (this was absorbed by the sodium hydroxide solution).
The remaining gas was excess oxygen, 116 cm^3.

Volume of oxygen *used* = (212 − 116) cm^3 = 96 cm^3.
∴ 48 cm^3 methane react with 96 cm^3 oxygen to form 48 cm^3 of carbon dioxide.
1 volume methane reacts with 2 volumes of oxygen to form 1 volume of carbon dioxide.
The ratio of reacting volumes is 1:2:1, which is in accordance with Gay-Lussac's Law.

Check your understanding

1. Use a diagram like the one shown on page 440 to explain the change in volume when nitrogen and hydrogen gases combine to form ammonia.
2. 60 cm^3 oxygen were added to 24 cm^3 of carbon monoxide and the mixture was ignited. The product occupied 72 cm^3, of which 48 cm^3 were unused oxygen. Show that this is an illustration of Gay-Lussac's law.
3. 24 cm^3 of methane and 96 cm^3 of oxygen were exploded together. The final volume, measured under the original conditions, was 72 cm^3, neglecting the water formed. 48 cm^3 of this was unused oxygen. Show that this illustrates Gay-Lussac's Law.
4. 60 cm^3 of hydrogen were mixed with 40 cm^3 of chlorine and exposed to bright sunlight. The total volume of gas was unchanged, but when shaken with water the volume was reduced to 20 cm^3 of gas, which was shown to be pure hydrogen. Are these results in accordance with Gay-Lussac's Law?

23.2 MOLAR VOLUMES

Counting molecules by volume

One of the important uses of the mole is that it enables us to 'count' atoms or molecules by weighing, e.g. the relative atomic mass of an element, weighed out in grams, will always contain 6.02×10^{23} atoms of the element. In a similar way, volumes of gases can be used to 'count molecules'. We have seen that equal volumes of gases under the same conditions of temperature and pressure contain an equal number of molecules. If we can find the number of molecules in a certain volume of gas under standard conditions, we can use this to find the number of molecules in any other volume of gas under the same conditions. For example, if there are x molecules of hydrogen in 1000 cm^3 at s.t.p., there will be x molecules of oxygen in 1000 cm^3 of oxygen at s.t.p., $\frac{1}{2}x$ molecules of oxygen in 500 cm^3 of oxygen at s.t.p., etc.

The mole enables us to count atoms by weighing

The conditions chosen as a basis are either at s.t.p. or at 25 °C and 760 mm mercury pressure (which approximates to room temperature and pressure). The data provided on page 437 should have enabled you to show that the volume occupied by a mole of any gas at 25 °C and 760 mm is, to a close approximation, always the same (because it contains the same number of molecules) and that this volume is 24 dm^3. This is known as the *molar volume at 25 °C and 760 mm pressure* (approximately room temperature and pressure). Similarly, the molar volume at s.t.p. can be shown to be 22.4 dm^3. The molar volume in all cases must contain 6.02×10^{23} molecules because it is a mole of gas.

This idea is used in calculations of the type considered under the next few headings.

Calculations involving molar volume

Calculations of this type involve the following steps.

1. The first step *must* be to produce a balanced equation for the reaction.
2. Use the equation and the given masses of reactants to calculate the number of moles of gases being formed.
3. Convert the number of moles of gas into a volume, either at s.t.p. or at 25 °C and 760 mm mercury pressure (approximately room temperature and pressure, r.t.p.) by using the fact that one mole of gas occupies 22.4 dm^3 at s.t.p. or 24 dm^3 at r.t.p. (Alternatively you may be given the mass of a gas which you are asked to convert to a volume. Here you use the fact that the molecular mass of a gas takes up a volume of 22.4 dm^3 at s.t.p. or 24 dm^3 at r.t.p.)

The following example should make this clear.

Calculate the volume of carbon dioxide (measured at s.t.p.) evolved when 10.0 g of potassium hydrogencarbonate are completely decomposed by heating.

Step 1 $2KHCO_3(s) \rightarrow K_2CO_3(s) + CO_2(g) + H_2O(g)$

Step 2 2 moles → 1 mole

200 g → 1 mole

10 g → $\frac{10}{200} \times 1 \text{ mole} = \frac{1}{20} \text{ mole}$

Step 3 $\frac{1}{20}$ mole of carbon dioxide at s.t.p. occupies $\frac{1}{20} \times 22.4\ dm^3 = 1.12\ dm^3$

If the question had asked for the volume at 25 °C and 760 mm mercury pressure (r.t.p.), then step 3 would become:

$\frac{1}{20}$ mole of carbon dioxide at 25 °C and 760 mm mercury pressure occupies $\frac{1}{20} \times 24\ dm^3 = 1.2\ dm^3$.

Check your understanding

Remember that 1 dm^3 = 1 litre = 1000 cm^3

1. What is the volume occupied at s.t.p. by, (a) a mole of hydrogen, (b) 2 moles of chlorine, (c) 0.5 moles of carbon dioxide, (d) 0.1 mole of hydrogen sulphide?

2. At a temperature of 323 K (50 °C) and a pressure of 760 mm, will a mole of oxygen contain (a) more than, (b) less than, (c) the same number as, the Avogadro Number of molecules?

3. What is the volume, measured at s.t.p. of, (a) 16 g of oxygen, (b) 71 g of chlorine, (c) 10 g of hydrogen, (d) 4.4 g of carbon dioxide, (e) 16 g of sulphur dioxide?

4. What is the mass of (a) 22.4 dm^3 of ammonia, (b) 112 cm^3 of carbon monoxide, (c) 5.6 dm^3 of hydrogen sulphide, (d) 4480 dm^3 of nitrogen? (All volumes measured at s.t.p.)

5. What volume of hydrogen (measured at 25 °C and 760 mm mercury pressure) will be produced by the action of excess dilute hydrochloric acid on 6.5 g of zinc?

6. What would be the volume of carbon dioxide produced, at 25 °C and 760 mm mercury pressure, if 2.5 g of pure calcium carbonate reacted completely with dilute nitric acid?

7. 50 cm^3 of a solution containing 3.4 g per litre of hydrogen peroxide was completely decomposed according to the following equation:

$$2H_2O_2(aq) \rightarrow 2H_2O(l) + O_2(g).$$

(One mole of a gas at room temperature and pressure occupies 24 litres. Relative atomic masses: H = 1, O = 16.)

(a) Calculate (i) the mass of 1 mole of hydrogen peroxide, (ii) the volume of oxygen, measured at room temperature and pressure, produced in this experiment.

(b) Hydrogen peroxide is sold by volume strength. Assume that '10 volume hydrogen peroxide' means that, if one litre of hydrogen peroxide were completely decomposed, 10 litres of oxygen, measured at room temperature and pressure, would be evolved. Calculate the volume strength of a solution of hydrogen peroxide containing 3.4 g per litre.

(c) Read the following passage and insert the missing words. Each space can be filled by one word. You may use more than one word if you wish, provided that no errors are introduced thereby.

Some manganese(IV) oxide was placed in a beaker which was then heated to make sure that it was dry. The beaker and contents were allowed to cool and were weighed. 50 cm^3 of the solution containing 3.4 g of hydrogen peroxide per litre were added. The oxygen was evolved faster and at a lower temperature than if the manganese(IV) oxide were absent, and the final volume of gas was found to be ______. When the reaction was complete, the mixture was heated carefully to remove all the ______. The beaker and contents were cooled and weighed. Their mass was found to be ______. Therefore, the manganese(IV) oxide had acted as a ______. (J.M.B.)

8. (a) Find the ratio of the number of atoms in 16 g of sulphur to the number of atoms in 46 g of sodium.
(b) Find the ratio of the number of molecules in 32 g of methane, CH_4, to the number of molecules in 5.6 litres of hydrogen at s.t.p.
(c) One oxide of manganese, Mn, contains 2.4 g of oxygen combined with 0.1 mole of manganese atoms. What is the simplest formula of this oxide? (W.J.E.C.)

Working out equations from volumes

Statements such as those made by Gay-Lussac can be used to work out a chemical equation, if we know the formulae of the molecules concerned.
For example, if 1 volume of nitrogen reacts with 3 volumes of hydrogen to form 2 volumes of ammonia (volumes measured under the same conditions), then x molecules of nitrogen react with $3x$ molecules of hydrogen to form $2x$ molecules of ammonia (where x is the number of molecules contained in one volume, and using Avogadro's Law). Divide throughout by x.
∴ 1 nitrogen molecule reacts with 3 hydrogen molecules to form 2 ammonia molecules,

$$\therefore\ N_2(g)+3H_2(g) \rightarrow 2NH_3(g).$$

These calculations can also be used the other way round, as in the first two below.

Check your understanding

1. A mixture of 500 cm^3 hydrogen and 125 cm^3 chlorine is exploded in bright sunshine. Give the names and volumes of the gases remaining after the reaction, measured at the same temperature and pressure.
2. Give the names and volumes of the remaining gases when a mixture of 50 cm^3 oxygen and 40 cm^3 carbon monoxide is ignited (all volumes measured at room temperature and pressure).
3. State Gay-Lussac's law of combining volumes.
200 cm^3 of a gaseous element X_2 reacted with 650 cm^3 of a gaseous element Y_2 to form 450 cm^3 of a mixture of XY_3 and Y_2. It was later found that 50 cm^3 of excess of Y_2 remained unused. All volumes were measured under the same conditions of temperature and pressure.
(a) What volume of XY_3 was formed in the reaction?
(b) Write a statement to show the relationship between the volumes of X_2 and Y_2 used and the volume of XY_3 formed.
(c) Give a balanced molecular equation for the reaction. (J.M.B.)

Molecular masses from molar volumes

You will remember that it is comparatively easy to calculate an *empirical formula* from reacting masses (p. 308), but in order to find a *molecular formula* we also need to know the relative molecular mass (p. 311). There are two fairly simple ways of finding a molecular mass from a gas volume, so the information needed to calculate a molecular mass is sometimes given in one of these ways.

Using the molar volume directly

If we can weigh a certain volume of gas at either s.t.p. or at 25 °C and 760 mm pressure (r.t.p.), it is a simple matter to determine the relative molecular mass of the gas. For example: 1 dm^3 of chlorine at 25 °C and 760 mm pressure (r.t.p.) has a mass of 2.96 g.

$\therefore$ 24 dm^3 of chlorine at 25 °C and 760 mm pressure has a mass of 24×2.96 g = 71.0 g.

But 24 dm^3 of any gas at 25 °C and 760 mm pressure is its molar volume, i.e. the volume of one mole of the gas.

71.0 g is the mass of one mole of chlorine molecules, and therefore 71.0 is the relative molecular mass of chlorine.

A similar argument can be used for 22.4 dm^3 of a gas at s.t.p. Remember that we do not say that the relative molecular mass of chlorine is 71.0 *grams*; relative molecular (and atomic) masses have no units.

Molecular masses from vapour densities

The vapour density of a gas is the mass of a certain volume of the gas divided by the mass of an equal volume of hydrogen, at the same temperature and pressure.

The vapour density of any gas, G, can therefore be expressed as:

$$\frac{\text{Mass of any volume of G}}{\text{Mass of the same volume of hydrogen}}$$

(volumes measured at the same temperature and pressure).

This can be used to derive a very important relationship. If we consider the special case when the volume is the molar volume (e.g. at s.t.p.), then the mass of the molar volume of G will be its relative molecular mass expressed in grams, and the mass of the molar volume of hydrogen will be 2 g (i.e. the relative molecular mass of hydrogen expressed in grams).

$$\therefore \text{ vapour density of G} = \frac{\text{molar mass of G}}{2}$$

$\therefore$ $2 \times$ vapour density of any gas = molar mass of the gas.

This is a very important relationship because, as it is relatively easy to determine the vapour density of a gas, it provides a simple means of finding its relative molecular mass. You should now know two different ways of finding the relative molecular mass of a gas; the task of comparing the masses of particles too small to be seen is not as difficult as it may have appeared!

The following example shows how a vapour density can be used to calculate a relative molecular mass, and the questions which follow should provide practice in these and other similar calculations.

An evacuated flask has a mass of 80.050 g. When filled with hydrogen its mass is 80.052 g and filled with chlorine at the same temperature and pressure its mass is 80.120 g. Calculate from these results, the relative molecular mass of chlorine.

Mass of hydrogen = (80.052 − 80.050 g) = 0.002 g
Mass of same volume of chlorine = (80.120 − 80.050) g = 0.070 g

$$\text{Vapour density of chlorine} = \frac{0.070}{0.002} = 35$$

Relative molecular mass of chlorine = $35 \times 2 = \underline{70}$

Check your understanding

1. Calculate from the relative molecular mass, the vapour density of each of the following gases: carbon monoxide, hydrogen chloride, nitrogen, sulphur dioxide.
2. What are the relative molecular masses of gases having the following vapour densities: (a) 8.5, (b) 22, (c) 16, (d) 17?
3. A gas cylinder filled with hydrogen holds 10 g of the gas. The same cylinder holds 160 g of gas A or 220 g of gas B under the same temperature and pressure conditions. Calculate the relative molecular masses of gas A and gas B.
4. Determine the molecular formulae of compounds having the following percentage composition by mass:

(a) carbon = 80 per cent, hydrogen = 20 per cent, vapour density = 15.
(b) nitrogen 30.4 per cent, oxygen 69.6 per cent, vapour density = 46.

Note: In all the following cases, the volumes given are measured at the same temperature and pressure.
5. A measured volume of an oxide of nitrogen is decomposed by heated nickel; the volume of the gas is reduced by half and the gas remaining is nitrogen. Deduce the formula for the gas. (Vapour density of gas is 23.)
6. A mixture of 125 cm^3 of hydrogen and 125 cm^3 of chlorine is exposed to bright sunshine in a suitable container. The resulting gas, which is found to be pure hydrogen chloride, has a volume of 250 cm^3. Deduce the formula for hydrogen chloride.
7. When excess charcoal was burnt in a known volume of pure oxygen, and the apparatus cooled, it was found that there was no change in volume but the residual gas was pure carbon dioxide. What is the formula for the gas? (Relative molecular mass = 44.)

23.3 VOLUMETRIC ANALYSIS

Basic calculations involving the concentrations of solutions

The concentrations of solutions are usually measured in mol dm^{-3}, or by the slightly more confusing term 'molarity'.

A solution which contains 1.0 mole of dissolved substance for every cubic decimetre of solution can have its concentration expressed in various ways:
(a) the concentration is 1.0 mol dm^{-3};
(b) the solution is 1.0 molar (usually shortened to 1.0 M);
(c) the solution has a molarity of 1.0.
Similarly a solution of exactly half this concentration could be described as (a) having a concentration of 0.5 mol dm^{-3}, (b) being 0.5 molar (0.5 M), or (c) having a molarity of 0.5.

A molar solution contains one mole of dissolved substance per dm^3 of solution.

In this book, concentrations are normally given as mol dm^{-3}, but the corresponding molarity may be given in brackets. Thus when 58.5 g of sodium chloride (23 + 35.5) is dissolved in water and made up to 1.0 dm^3 of solution, the solution is recorded as having a concentration of 1.0 mol dm^{-3} (1.0 M), with respect to sodium chloride.

An understanding of this principle, and variations on the basic theme, are essential for any understanding of volumetric analysis. Read again the section on 'moles and solutions' starting on page 303, and make sure that you fully understand it.

Check your understanding

1. Calculate the mass of solute dissolved in 1 dm^3 of each of the following solutions.
(a) sulphuric acid of concentration 2.0 mol dm^{-3} (2.0 M).
(b) nitric acid of concentration 0.5 mol dm^{-3} (0.5 M).
(c) potassium hydroxide of concentration 0.1 mol dm^{-3} (0.1 M).

2. Calculate the concentrations of the following solutions, in mol dm^{-3}.
(a) 10.6 g of anhydrous sodium carbonate in 500 cm^3 of solution.
(b) 6.9 g of potassium carbonate in 100 cm^3 of solution.
(c) 9.8 g of sulphuric acid in 250 cm^3 of solution.
(d) 0.7 g of potassium hydroxide in 50 cm^3 of solution.

Preparing a standard solution

A standard solution is one which has been made up to an exact, stated concentration (usually in mol dm^{-3}), using a very pure sample of the chemical and deionized or distilled water.

Suppose that it is required to make up 250 cm^3 of a standard solution containing approximately 0.1 mol dm^{-3} (0.1 M) of anhydrous sodium carbonate.

1. Work out the mass of anhydrous sodium carbonate needed to make up the solution.
The mass of 1 mole of anhydrous sodium carbonate,

$$Na_2CO_3 = (2 \times 23) + 12 + (3 \times 16) = 106 \text{ g}$$

(This would be 286 g if hydrated sodium carbonate, $Na_2CO_3 . 10H_2O$, were used.)
∴ If 106 g of sodium carbonate (anhydrous salt) are dissolved and made up to 1 dm^3 of solution, the solution would have a concentration of 1 mol dm^{-3} (1 M) of sodium carbonate.
∴ If 10.6 g are dissolved and made up to 1 dm^3 of solution, the solution would have a concentration of 0.1 mol dm^{-3} (0.1 M).
∴ 250 cm^3 of a solution having the same concentration would need $\frac{1}{4} \times 10.6$ g $= 2.65$ g.
2. Weigh out accurately, on a watch-glass or in a weighing bottle, a mass of anhydrous sodium carbonate *approximately* equal to the calculated mass.
3. Dissolve *all* of this material, carefully, in deionized water in a beaker. The beaker must be absolutely clean, the volume of water must be *less* than the final volume required, and all traces of solid on the watch-glass (or in the weighing bottle) must be rinsed into the beaker with deionized water from a wash-bottle (Figure 23.2).

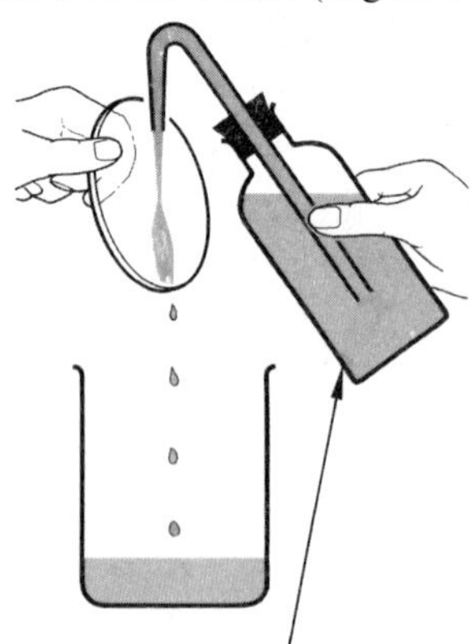

Figure 23.2 Washing the last traces of sodium carbonate from the watch-glass and into the beaker

4. Carefully transfer all of the solution into a 250 cm^3 volumetric flask, using a clean funnel. Carefully rinse out the beaker with deionized water and transfer the washings into the volumetric flask. Make sure that the volume of water used to rinse out the beaker is such that the *total* volume of water used is still less than 250 cm^3.
5. Use a little more deionized water to wash the funnel, again collecting the washings in the volumetric flask.
6. Remove the funnel, and then carefully add deionized water to the flask until the bottom of the meniscus is exactly on the graduation mark. Stopper the flask and mix the solution thoroughly by turning the flask upside down several times with shaking. (Remember to keep your thumb over the stopper.)

7. Calculate the concentration of your final solution. You may have been puzzled when, in step 2, you were asked to weigh out *accurately* an *approximate* mass. This is not a contradiction; it is a perfectly sensible statement. For example, the actual mass may have been 2.92 g instead of the 2.65 g needed to make the solution have a concentration of exactly 0.10 mol dm^{-3}. The actual solution is therefore *more* concentrated than 0.1 mol dm^{-3}. It has a concentration of $\frac{2.92}{2.65} \times 0.1$ mol dm^{-3}, $= 0.110$ mol dm^{-3}. The point of this is that it might have taken some time to weigh out exactly 2.65 g, but by weighing out *accurately* a mass *near to* 2.65 g, we still end up with a standard solution.

Experiment 23.2
To prepare a standard solution of sodium carbonate*

Apparatus
Watch-glass, 250 cm^3 beaker, 250 cm^3 graduated flask, spatula, stirring rod, wash-bottle, filter funnel, access to balance sensitive to 0.001 g.
Pure anhydrous sodium carbonate, de-ionized water.

Procedure
Use the steps outlined earlier to produce 250 cm^3 of a standard solution of sodium carbonate of concentration approximately 0.1 mol dm^{-3}. Keep the solution if you are going to do Experiment 23.4 later.

Make sure that you record *all* of your weighings, and that you clearly show the calculations made in obtaining the standard solution.

Performing a titration

In volumetric analysis, one solution of known concentration is reacted with a second solution of unknown concentration in order to determine the concentration of the latter. A pipette and burette are used, and the solutions are reacted together under controlled conditions in order to find the reacting volumes. An indicator is normally used to decide the 'end point' of the reaction, i.e. the point at which the volumes of the solutions have completely reacted. The whole procedure is called a *titration*. From the volumes of the solutions used in the titration, and the concentration of one solution, the concentration of the other solution can be determined.

Revise the correct use of a pipette and burette, and note the following points which are sometimes forgotten or misunderstood.

1. All readings must be taken at eye level, and with the apparatus vertical.
2. It is not necessary to dry the conical flask between operations.
3. It is always advisable to perform a rough titration first.
4. Shake the flask during addition of the liquid from the burette.
5. Repeat operations until two or more readings agree within 0.1 cm^3.
6. When using a strong acid (sulphuric, hydrochloric, or nitric) with a strong

A rough titration?

alkali (sodium or potassium hydroxides) any indicator is suitable. When titrating a strong acid with sodium or potassium carbonate solution, methyl orange must be used as indicator.

7. This technique can also be used to prepare some soluble salts, as described on page 86.

Calculations using titration results

1. Ensure that you know the concentration of *one* of the two solutions involved. If this is not given directly, it will be necessary to work it out from the mass of solid dissolved in a certain volume of solution.

Suppose that 25.0 cm^3 of a solution of hydrochloric acid of concentration 0.1 mol dm^{-3} (0.1 M) react with 21.5 cm^3 of a solution of sodium hydroxide, and we need to calculate the concentration of the sodium hydroxide solution.

2. Write down the balanced equation for the reaction:

$$HCl(aq) + NaOH(aq) \rightarrow NaCl(aq) + H_2O(l)$$

3. Calculate the number of moles used in the titration for the chemical the concentration of which is known, e.g.
25.0 cm^3 of hydrochloric acid used, of concentration 0.1 mol dm^{-3} (0.1 M).
1000 cm^3 of the hydrochloric acid solution contain 0.1 moles.

$\therefore$ 1 cm^3 of the hydrochloric acid solution contains $\frac{1}{1000} \times 0.1$ moles.

$\therefore$ 25.0 cm^3 of the acid solution contains $25 \times \frac{1}{1000} \times 0.1$ moles.

4. From the equation, calculate how many moles of the *other* substance this has reacted with. In this example, one mole of hydrochloric acid reacts with one mole of sodium hydroxide.

$\therefore$ If $\frac{25 \times 0.1}{1000}$ moles of acid are used, $\frac{25 \times 0.1}{1000}$ moles of alkali will react with it.

5. Calculate how many moles of this second substance would be contained in 1000 cm^3 of its solution; this is the concentration required, e.g.

$\frac{25 \times 0.1}{1000}$ moles of alkali are used. This was in 21.5 cm^3 of solution.

$\therefore$ 1 cm^3 of the alkali solution contains $\frac{25 \times 0.1}{1000 \times 21.5}$ moles.

$\therefore$ 1000 cm^3 of the alkali solution contains

$$\frac{1000 \times 25 \times 0.1}{1000 \times 21.5} \text{ moles} = \frac{25 \times 0.1}{21.5} \text{ moles} = 0.110 \text{ moles.}$$

The alkali solution has a concentration of 0.110 mol dm^{-3} (0.110 M).

Note: It is absolutely vital to base your calculation on a correctly balanced equation. If the same experimental results had been obtained for a reaction between hydrochloric acid and sodium carbonate solution, the final answer (i.e. the concentration of the sodium carbonate solution) would be different, because the equation for this reaction shows that the reagents react in the ratio 1:2 instead of 1:1 as in the first example.

$$2HCl(aq) + Na_2CO_3(aq) \rightarrow 2NaCl(aq) + CO_2(g) + H_2O(l)$$

In step 3, therefore, the number of moles of sodium carbonate reacting would be *half* the number of moles of acid used, i.e.

$\frac{1 \times 25 \times 0.1}{2 \times 1000}$ moles, and this would be used in the final steps rather than $\frac{25 \times 0.1}{1000}$.

Check your understanding

Determine the concentrations (in mol dm^{-3}) of the following solutions:

(a) hydrochloric acid, 25.0 cm^3 of which neutralize 20.0 cm^3 of sodium hydroxide solution of concentration 0.15 mol dm^{-3} (0.15 M);

(b) sulphuric acid, 20.0 cm^3 of which neutralize 30.0 cm^3 of a potassium hydroxide solution of concentration 0.1 mol dm^{-3} (0.1 M);

(c) sodium hydroxide, 10.0 cm^3 of which neutralize 15.0 cm^3 of hydrochloric acid solution of concentration 2.5 mol dm^{-3} (2.5 M);

(d) nitric acid, 10.0 cm^3 of which react with 25.0 cm^3 of a solution of sodium carbonate of concentration 0.2 mol dm^{-3} (0.2 M).

Experiment 23.3
To find the concentration of a solution of hydrochloric acid*

Read again the points on page 447 and on page 448, and then use the apparatus and procedure given for Experiment 6.5 (page 87) to find the volume of sodium hydroxide solution (of a given concentration) which reacts exactly with 25.0 cm^3 (or 10.0 cm^3) of hydrochloric acid of unknown concentration. Use phenolphthalein as indicator, repeat the titrations until two or more readings agree within 0.1 cm^3, and record all of your results in a table similar to the one used in Experiment 6.5. Note that in this case there is no need to repeat the experiment without indicator (as was done in Experiment 6.5), because we are concerned only with calculating the concentration of the acid, not with the need to make a sample of the salt formed by the reaction.

Results and conclusion

Use your results to calculate the concentration of the hydrochloric acid solution, setting out your calculation as shown in the worked example on page 448.

Experiment 23.4
To find the concentration of (a) a sulphuric acid solution and (b) a nitric acid solution*

Apparatus and procedure

(a) Using the same apparatus and procedure as in Experiment 23.3, titrate samples of the sulphuric acid solution with the standard solution of sodium hydroxide, and perform the necessary calculations in order to find the concentration of the sulphuric acid solution.

(b) Titrate the nitric acid of unknown concentration against samples of your standard sodium carbonate solution prepared in Experiment 23.2, so as to determine the concentration of the acid. This time use methyl orange as the indicator.

Check your understanding

1. $Na_2CO_3 + 2HNO_3 \rightarrow 2NaNO_3 + CO_2 + H_2O$

(a) 25.0 cm^3 of a solution of sodium carbonate were found to require 20.0 cm^3 of 1.0 M nitric acid for complete neutralization.

Calculate the concentration of the sodium carbonate solution in grams of anhydrous sodium carbonate (Na_2CO_3) per litre.
(b) The sodium carbonate solution was actually made up by dissolving 49.6 g of a hydrated form of sodium carbonate, $Na_2CO_3.xH_2O$, in water and diluting the solution to 1 litre. Calculate x. (Relative atomic masses, H = 1.0, O = 16, C = 12.0, Na = 23.0.) (J.M.B.)

2. 10.0 cm^3 of a solution of sodium hydroxide required 25.0 cm^3 of a solution of hydrochloric acid of concentration 0.4 mol dm^{-3} for neutralization. Calculate the concentration of the alkali solution in mol dm^{-3}

3. The relative formula mass of lead nitrate is 331. 6.62 g of lead nitrate were dissolved in water and the volume of the solution made up accurately to 500 cm^3. What was the molarity of the solution? (W.J.E.C.)

4. The alkaline effluent from a certain paper-making factory has to be neutralized before it can be released into the normal sewage/drainage system. The alkali in the effluent is sodium hydroxide. Each day a 1000 litre tank of this effluent has to be treated with 120 litres of 2 M hydrochloric acid for neutralization. (a) Give the balanced equation for the reaction between sodium hydroxide and hydrochloric acid. (b) Calculate the mass of sodium hydroxide in the tank in the 1000 litres of effluent solution. (c) If the sodium hydroxide in the effluent were to be re-used it would have to be in a much more concentrated solution. How could this concentrating of the sodium hydroxide be carried out? (d) Suggest why this effluent is neutralized rather than concentrated and re-used. (e) If nitric acid were to be used in the neutralization rather than hydrochloric acid, what effect would this have on the sewage/drainage system? (A.E.B. 1978)

23.4 TESTS FOR COMMON GASES AND IONS

Some of these have been dealt with in other sections of the book, but they are collected and summarized here (Tables 23.1 to 23.4) both for reference and to reinforce your earlier work.

Table 23.1 Tests for some common gases

Gas	*Test*	*Result of test if positive*
Hydrogen	Trap gas in test-tube, apply lighted taper	Squeaky pop
Oxygen	Apply glowing taper to gas sample in test-tube	Taper relights
Carbon dioxide	Pass gas into calcium hydroxide solution (lime water), e.g. by collecting sample in teat pipette and ejecting into small volume of calcium hydroxide solution in test-tube	Calcium hydroxide solution goes 'milky' due to fine precipitate of calcium carbonate
Ammonia	1. Expose gas to damp red litmus paper, or 2. Expose gas to fumes of hydrogen chloride, e.g. from a bottle of concentrated hydrochloric acid	1. Paper goes blue 2. Dense white cloud of ammonium chloride formed
Chlorine	Expose gas to damp blue litmus paper	Paper goes pink and is then bleached
Hydrogen chloride	Expose gas to ammonia fumes, e.g. from bottle of concentrated ammonia solution	Dense white cloud of ammonium chloride
Nitrogen	Apply all other gas tests	If none positive, gas probably nitrogen
Sulphur dioxide	Expose gas to filter paper soaked in acidified potassium dichromate(VI) solution	Paper changes from orange to green
Hydrogen sulphide	Expose gas to filter paper soaked in lead nitrate (or ethanoate) solution	Dark coloured stain of lead sulphide forms on paper

Table 23.2 Tests for some common anions

Anion	*Test*	*Result of test if positive*
Soluble chloride	Acidify solution with dilute nitric acid and then add silver nitrate solution	White precipitate of silver chloride formed which easily dissolves in dilute ammonia solution.
Soluble sulphate	Acidify solution with dilute hydrochloric acid and then add barium chloride solution	White precipitate of barium sulphate formed
Nitrates	Make solution alkaline with dilute sodium hydroxide solution, add Devarda's alloy, and warm. (See extra test to check ammonia gas does not come from ammonium ions, p. 333)	Ammonia gas formed (test in usual way)
Carbonates and hydrogen carbonates	Add a little dilute hydrochloric acid	Effervescence, carbon dioxide gas given off (test in usual way)

Table 23.3 Some flame tests for metal ions

Test	*Result if positive*	
Clean nichrome or platinum wire by repeated dipping in hydrochloric acid and heating in a roaring Bunsen flame. When no colour given to flame by wire, moisten the wire with dilute hydrochloric acid and pick up small sample of compound on it. Hold wire and sample in colourless flame, half way up the flame and to one side of the blue cone	Intense golden yellow:	Na^+
	Apple green:	Ba^{2+}
	Green-blue:	Cu^{2+}
	Brick red:	Ca^{2+}
	Lilac:	K^+
	Scarlet:	Li^+

Table 23.4 Other tests for positive ions in solution

Ion	*Test*	*Result if positive*
Ammonium	Add dilute sodium hydroxide solution, then warm	Ammonia gas produced (test in usual way)
Copper(II)	Add dilute ammonia solution, dropwise, with stirring	Pale blue precipitate of copper(II) hydroxide formed initially, but then dissolves to form deep blue solution
Zinc	Add dilute ammonia solution, dropwise, with stirring to one sample and add dilute sodium hydroxide in a similar way to a second sample	In *both* samples, a white precipitate (zinc hydroxide) forms initially but then dissolves in excess alkali
Iron(II)	Add dilute ammonia or dilute sodium hydroxide solution	Dirty-green precipitate of iron(II) hydroxide
Iron(III)	As with iron(II)	Red-brown precipitate of iron(III) hydroxide
Aluminium and lead(II)	1. Do test as for zinc 2. If necessary (see results column) add dilute hydrochloric acid to a third sample	If white precipitate forms in both cases but only dissolves in excess sodium hydroxide, solution could contain Pb^{2+} or Al^{3+}. Do test (2), no precipitate if Al^{3+}, white precipitate (of lead chloride) if Pb^{2+} present

Note: For an explanation of the reactions involved in the tests for metal ions, see page 329.

Point for discussion

Table 23.5 shows a 'yes/no' procedure by which a solution can be analysed for some common cations. Study this carefully; it will often work only if the solution to be analysed contains only *one* cation.

(a) Explain the chemical reactions involved at each stage of the flow sheet.
(b) Can you produce a different chart of the same kind?
(c) You may be allowed to use Table 23.5, or an *approved* scheme of your own, to analyse a solution which contains only one cation.

Table 23.5 An analysis scheme

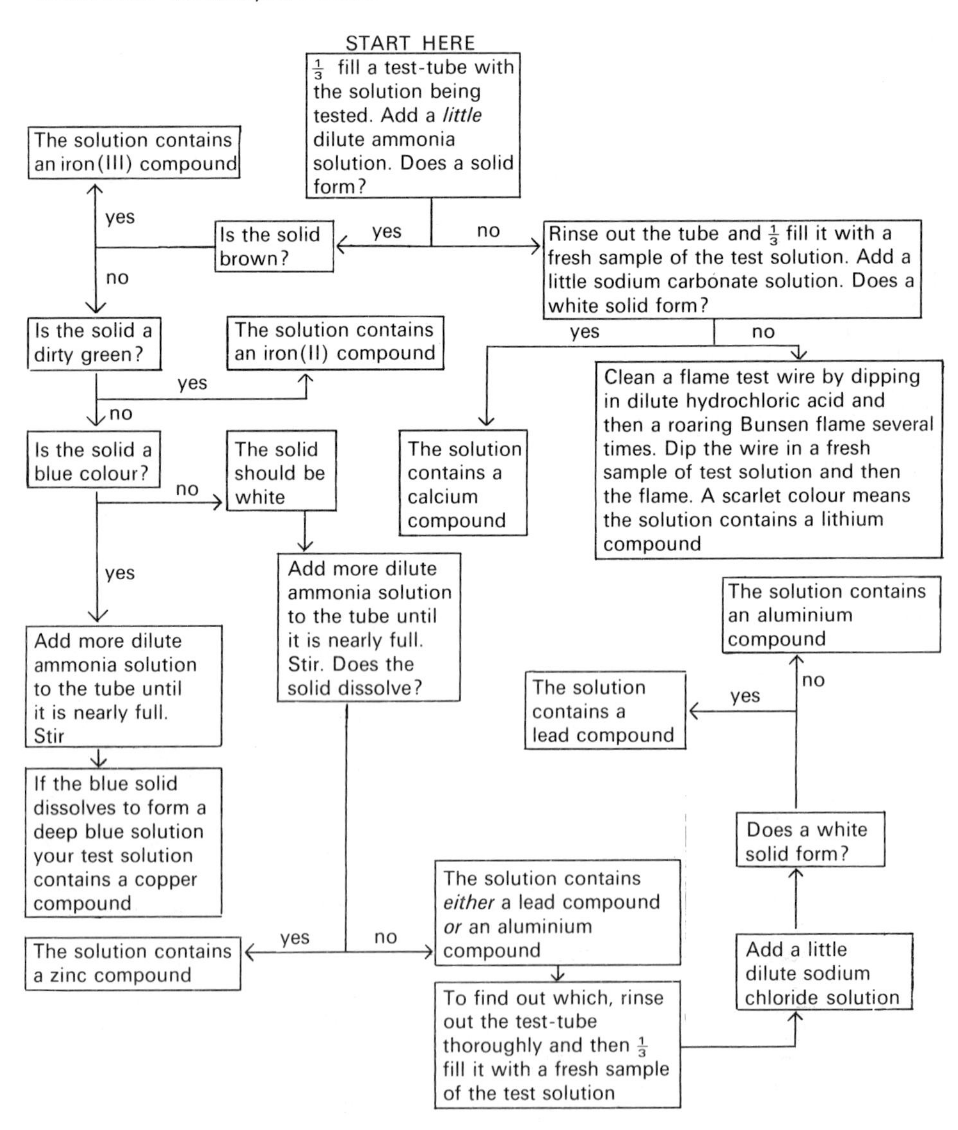

23.5 A SUMMARY OF THE MAIN POINTS IN CHAPTER 23

The following 'check list' should help you to organize the work for revision.

1. *Definitions*

(a) Avogadro's Law (p. 437)
(b) Gay-Lussac's Law (p. 438)
(c) Vapour density (p. 444)
(d) A molar solution (p. 445)
(e) A standard solution (p. 446)

2. *Other points*

(a) Both physical and chemical changes can affect the volume of a gas during a chemical reaction.
(b) Gas volumes must be compared under standard conditions, e.g. at s.t.p. (0 °C and 760 mm) or at 25 °C and 760 mm mercury pressure.
(c) How Avogadro's Law helps to explain changes in gas volume caused by chemical reactions (p. 439).
(d) Calculations involving Gay-Lussac's Law.
(e) The molar volume at s.t.p. is 22.4 dm^3, and at 25 °C and 760 mm pressure is 24 dm^3. The molar volume is the volume of a mole of gas, i.e. of 6.02×10^{23} molecules of gas.
(f) How to calculate the volume of a gas evolved in a reaction (measured under standard conditions) by using an equation and the molar volume (p. 442).
(g) How to determine an equation from reacting volumes (p. 443).
(h) How to calculate a molecular mass from a molar volume.
(This may form part of a calculation to determine the molecular formula of a compound.)
(i) How to calculate a molecular mass from a vapour density.
(j) How to calculate the concentration of a solution in mol dm^{-3}.
(k) How to prepare a standard solution.
(l) How to perform a titration, and the importance of choosing the correct indicator.
(m) Calculations involving titration results.
(n) Tests for common gases and ions.

QUESTIONS

Multiple choice questions

1. In an experiment, 1 cm^3 of a hydrocarbon X required 4 cm^3 of oxygen for complete combustion, to give 3 cm^3 of carbon dioxide, all volumes being measured at s.t.p. Which one of the following formulae represents X?
A CH_4 B C_2H_3 C C_2H_4
D C_3H_4 E C_3H_8 (C.)

2. If a 10 cm^3 pipette were not available, which one of the following procedures would you use to measure a 10 cm^3 portion of a solution for an accurate titration?
A Measure 10 cm^3 from a cylinder
B Measure 10 cm^3 from a burette
C Weigh 10 times the mass of 1 cm^3 of water
D Weigh 10 g of the solution
E Measure 20 cm^3 from a pipette and divide the solution into two equal parts (C.)

3. If, in volumetric analysis between an acid (in the burette) and an alkali, you needed to re-use the same titration flask after the first titration, which one of the following is the best procedure for rinsing the flask?
A rinse with tap water and then with distilled water
B rinse with tap water and then with acid
C rinse with a little of the alkali
D rinse with distilled water and then with a little of the alkali
E rinse with a little of the acid (C.)

4. Which one of the following gives a white precipitate with barium chloride solution, and a brick-red flame test?
A sodium sulphate
B copper(II) sulphate
C calcium sulphate
D ammonium sulphate
E calcium chloride

Structured questions and questions requiring longer answers

5. Aluminium and carbon form a compound, aluminium carbide.
(a) 7.2 g of aluminium carbide contain 5.4 g of aluminium. Find the simplest (or empirical) formula of aluminium carbide.
(b) The relative molecular mass of aluminium carbide is 144. What is its molecular formula?
(c) On reaction with water, aluminium carbide gives aluminium hydroxide and the gaseous hydrocarbon, methane. (i) Write the balanced equation for the reaction. (ii) Find the volume of methane, at room temperature and pressure, produced from 0.2 mol of aluminium carbide, when it is treated with an excess of water. (The molar volume of a gas at room temperature and pressure is 24 litres (dm^3)). (W.J.E.C.)

6. Sodium and oxygen combine under certain conditions to form sodium peroxide (Na_2O_2). Sodium peroxide reacts with water in accordance with the following equation.

$$2Na_2O_2(s) + 2H_2O(l) \rightarrow 4NaOH(aq) + O_2(g)$$

(Molar volume of a gas at room temperature and pressure is 24 dm^3).
(a) What mass of sodium will contain the same number of atoms as 0.8 g of oxygen?
(b) What volume of oxygen would be produced at room temperature and pressure when 3.9 g of sodium peroxide reacts completely with water?
(c) If the sodium hydroxide produced in reaction (b) were made up to 1 dm^3 with distilled water, what would be the concentration of sodium hydroxide in mol dm^{-3} in the solution? (W.J.E.C.)

7. Two experiments were carried out, using in *each* case the same volume of a solution of hydrogen peroxide.
Experiment A: The solution was made alkaline at room temperature.
Experiment B: A small amount of manganese(IV) oxide (manganese dioxide), was added at room temperature.
The volume of oxygen evolved, measured at room temperature and pressure, was plotted against time for each experiment. The graphs obtained were as shown in the diagram opposite:
(a) In which experiment, A or B, did the reaction begin more rapidly? Using the graphs, explain how you arrived at your answer.
(b) State and explain the function of the manganese(IV) oxide in Experiment B.
(c) Both experiments eventually gave the same volume of oxygen. Using the following equation for the decomposition of hydrogen peroxide:

$$2H_2O_2(aq) \rightarrow 2H_2O(l) + O_2(g),$$

calculate the volume of oxygen evolved at room temperature and pressure by the complete decomposition of 1000 cm^3 of hydrogen peroxide solution containing 0.68 g dm^{-3}. (Molar volume of a gas at room temperature and pressure is 24 dm^3.) (W.J.E.C.)

8. (a) Butane, C_4H_{10}, is a *saturated* hydrocarbon. (i) Write down the structural formulae of the two isomers of butane. (ii) Give the equation for the reaction which takes place when butane is burnt in a plentiful supply of air. (iii) Calculate the volume of oxygen which would be used to burn completely 1 litre of butane gas at ordinary laboratory temperature. (Assume that all volumes are measured at the same temperature and pressure.) (iv) Describe in detail why it is dangerous to burn butane in a limited supply of air.
(b) Butene, C_4H_8, is an *unsaturated* hydrocarbon. (i) Describe *one* test which you would use to show that butene is unsaturated. (ii) How could this gas be converted into butane? (iii) Give the equation for this reaction. (iv) Calculate the number of molecules in 0.56 litre of butene gas at s.t.p. (O. & C.)

9. 25 cm^3 of a solution containing 6.0 g of sodium hydroxide per litre, exactly neutralize 30 cm^3 of a solution of nitric acid.
(a) Calculate the molarity of the sodium hydroxide solution.

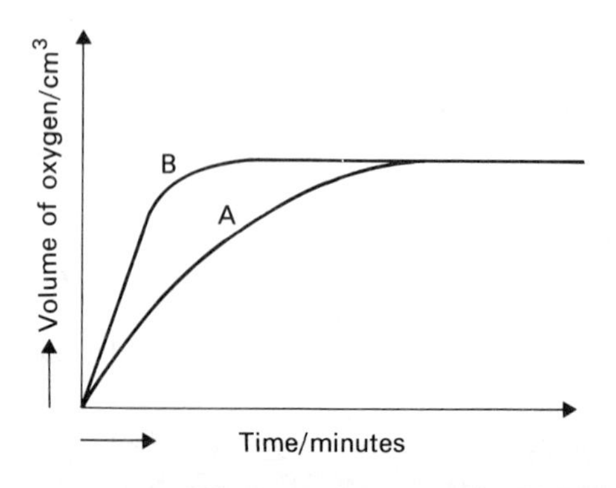

(b) Using the above results, calculate the molarity of the nitric acid and its concentration in grams per litre.
(c) If you were supplied with 200 cm^3 of each of the solutions describe carefully how, by titration, you would arrive at the results given in the first statement.
(d) Having done the titration, describe how you would obtain a pure, dry crystalline sample of sodium nitrate. (J.M.B.)

10. 5 cm^3 of 1.0 M hydrochloric acid were measured into a conical flask. 1 g of powdered zinc (excess) was added to the acid and hydrogen was evolved. The equation for the reaction is $Zn + 2HCl \rightarrow ZnCl_2 + H_2$.
(a) Describe an experiment by which you could study the rate of the reaction.
(b) Sketch a graph indicating the changes in volume of hydrogen observed during the experiment, continuing the graph for some minutes after the reaction has finished. Mark this graph A. Label your axes clearly and explain the reasons for the shape of your graph.
(c) Using the same axes as for part (b), sketch the graph you would expect to obtain if the experiment was repeated using 1 g of granulated zinc. Mark this graph B. Explain the reasons why graphs A and B have different shapes.
(d) Calculate the volume of hydrogen gas which should theoretically be evolved when the reaction is completed at room temperature and pressure.
(e) Copper acts as a catalyst for this reaction. If 0.2 g of copper powder were added to the initial reaction mixture, what effect would this have on the total volume of hydrogen evolved? (J.M.B.)

11. (a) What is meant by (i) an acid, (ii) a molar solution of an acid?
(b) Acetic acid, CH_3COOH, is a monobasic acid. (i) What ions are present in an aqueous solution of this acid? (ii) How many grams of acetic acid are present in 500 cm^3 of a molar solution? (iii) A molar solution of acetic acid conducts electricity but not so easily as a molar solution of hydrochloric acid. What explanation can you offer for this fact?
(c) Calculate the minimum volume of molar hydrochloric acid required to react with 1 gram of calcium carbonate and the volume of gas which would be evolved at standard temperature and pressure. (A.E.B. 1973)

12. (a) Describe a simple experiment which shows that two gases in contact mix thoroughly. What term is used to describe such a process? What evidence concerning the nature of gases is provided by experiments of this kind?
Explain the following as fully as possible. Two unstoppered bottles, one containing concentrated ammonia solution, the other containing concentrated hydrochloric acid, are placed a few centimetres apart in a draught-free enclosure. In a short time dense white smoke is formed near the neck of the hydrochloric acid bottle.
(b) A gas contains by mass 24.0 per cent carbon and 76.0 per cent fluorine (F); its density is 4.167 g per litre at room temperature and pressure. Calculate (i) the simplest formula of the gas; (ii) its relative molecular mass; (iii) the molecular formula. (A.E.B. 1976)

13. When exploded with excess oxygen, ethane (C_2H_6) reacts as in the equation:

$$2C_2H_6(g) + 7O_2(g) \rightarrow 4CO_2(g) + 6H_2O(l)$$

If 20 cm^3 of ethane are exploded with 100 cm^3 of oxygen and the gaseous products are reduced to the original laboratory temperature and pressure, what will be the volume and composition of the resulting mixture of gases? (C.)

14. A mixture of 400 cm^3 of a gas A, and 400 cm^3 of hydrogen reacted completely in the presence of a catalyst to give 400 cm^3 of a new gas, B. When 400 cm^3 of this new gas B, were mixed with 400 cm^3 of hydrogen in the presence of a different catalyst, 400 cm^3 of ethane were produced. (All volumes measured under the same conditions of temperature and pressure.)

A liquid, D, was produced when B and bromine reacted together. Addition of another catalyst to B gave a solid, E, which did not react with bromine and had a molecular mass of about 20 000. Explain these results, identify A, B, D, E and give equations for the reactions.

How might polyvinyl chloride be obtained from A? (J.M.B.)

15. In a reaction between a solution of a metallic hydroxide (formula MOH) and dilute hydrochloric acid, 20.0 cm^3 of the alkali reacted with 25.0 cm^3 of the acid. The

acid concentration was 4.00 g per litre and that of the alkali was 7.67 g per litre. Calculate the formula mass of the alkali and hence the relative atomic mass of the metal.

Describe in detail how you would determine the volume of the acid and alkali which exactly neutralize each other. (J.M.B.)

16. Describe in detail how you would prepare 250 cm^3 of an exactly 0.05 M solution of sodium carbonate and use it to determine the exact concentration of an approximately 0.05 M solution of sulphuric acid. You are provided with the usual apparatus for volumetric analysis. Name the indicator and state the colour change. Explain how you would work out the result.

$$Na_2CO_3(aq) + H_2SO_4(aq) \rightarrow Na_2SO_4(aq) + CO_2(g) + H_2O(l)$$

$$2Na^+ + CO_3^{2-} + 2H^+ + SO_4^{2-} \rightarrow 2Na^+ + SO_4^{2-} + CO_2 + H_2O$$

(C.)

17. When heated in a current of nitrogen, magnesium reacts with it to form magnesium nitride, Mg_3N_2.

Magnesium nitride reacts with water to form magnesium hydroxide and ammonia.

(a) Write the equation for the first reaction and calculate the volume of nitrogen, measured at s.t.p., required to react with 8.0 g of magnesium.

(b) Write the equation for the second reaction and calculate the volume of ammonia produced, measured at s.t.p., if the magnesium nitride formed in (a) reacts completely with water. (Assume that all the ammonia can be driven off and collected.) (J.M.B.)

18. 'Calcite is pure calcium carbonate.' Is this statement borne out by the following experimental results?

A large crystal of calcite weighing 3.753 g was placed in 25.0 cm^3 of a 2 M solution of hydrochloric acid and left there until action ceased. After washing and drying, the crystal was found to weigh 1.253 g.

What volume of carbon dioxide at s.t.p., was formed during the reaction? (C.)

19. Bottles containing three white powders, sodium carbonate, calcium hydroxide and ammonium chloride respectively, have lost their labels. Using only water and dilute hydrochloric acid as additional reagents, what experiments would you perform in order to re-label the bottles correctly? (J.M.B.)

24 Two more important non-metals: sulphur, chlorine and their compounds

24.1 SULPHUR AND ITS COMPOUNDS

Sulphur the element

Extraction

A large proportion of the world supply of sulphur is extracted from deposits of the element in Texas and Louisiana, USA. These deposits are about 30 metres thick, and each consists of a dome-shaped layer of sulphur and limestone. The deposits lie at depths greater than 160 metres, and cannot be mined in the conventional way because they occur under quicksands. Instead, the sulphur is extracted by an ingenious method known as the *Frasch process*. A metal tube about 15 cm in diameter and containing two concentric inner tubes is sunk into the top of the sulphur deposit (Figure 24.1). Water, superheated to 170 °C under pressure, is forced, still under pressure, down the outer tube thus melting the sulphur, which collects in a pool. Hot compressed air is forced down the centre tube, producing a light froth of molten sulphur which rises up the middle tube and is collected in large containers. After the sulphur has solidified, the sides of the containers are dismantled and the blocks of sulphur are broken up. The sulphur obtained by this method is usually over 99.5 per cent pure. Considerable quantities of sulphur are also obtained from crude oil and natural gas. All living organisms contain sulphur, mainly in proteins. Oil, for example, is thought to be produced by the decay of marine life, during which process sulphur is liberated from the proteins. Sulphur must be removed from crude oil before it is processed into fuels etc., because any sulphur or sulphur compounds in fuels will burn to form sulphur dioxide which would rapidly corrode the burners and pollute the atmosphere.

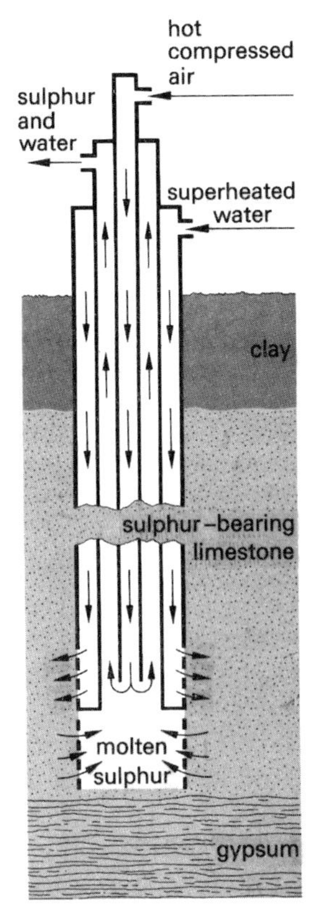

Figure 24.1 Sulphur extraction by the Frasch process

Physical properties of sulphur

Sulphur is a yellowish solid and is used in the laboratory either as roll sulphur (made by casting molten sulphur into 'sticks') or as flowers of sulphur (a yellow powder made by condensing sulphur vapour). It is insoluble in water but soluble in some solvents (e.g. carbon disulphide, Experiment 16.1). It shows the typical physical properties of a non-metal, i.e. it has a low melting point and boiling point and is a poor conductor of heat and electricity. The changes which take place when sulphur is heated have already been discussed, as have the various allotropes of the element (p. 275).

Chemical properties of sulphur

The only reactions which you are likely to encounter are its combustion in air or oxygen (p. 154), and the formation of metal sulphides (e.g. page 28).

Uses of sulphur

Large quantities are used in the manufacture of sulphuric acid (p. 545). Smaller quantities are used to make matches, gunpowder, drugs, vulcanized rubber, and sulphur-based insecticides and fungicides.

Sulphur dioxide, SO_2

Laboratory preparation

The gas can be made by adding a dilute acid to a sulphite, followed by gentle warming, e.g.

$$Na_2SO_3(s)+2HCl(aq) \rightarrow 2NaCl(aq)+H_2O(l)+SO_2(g)$$

Alternatively, the gas can be made by warming copper metal with concentrated sulphuric acid (Figure 24.2). In either case it is collected by downward delivery in a fume cupboard.

$$Cu(s)+2H_2SO_4(aq) \xrightarrow{\text{heat}} CuSO_4(aq)+SO_2(g)+2H_2O(l)$$

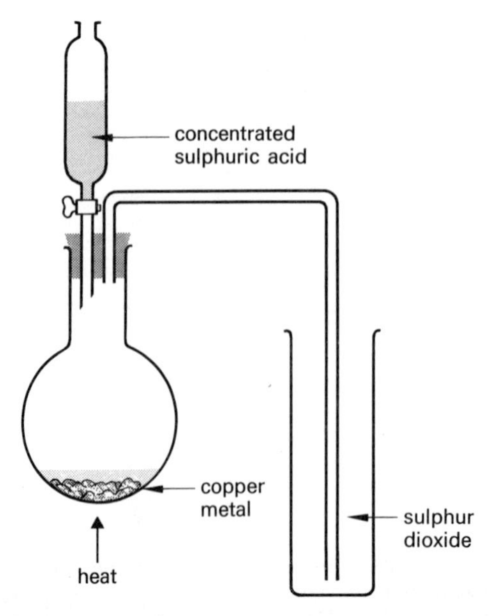

Figure 24.2 Laboratory preparation of sulphur dioxide

Physical properties

Solubility in water	*Colour*	*Odour*	*Density relative to air*	*Toxicity*
Quite soluble. Gives fountain experiment	None	Unpleasant, sharp smell, leaving metallic taste	More dense	Toxic

Note: The gas is easily liquefied under slight pressure.

Chemical properties

As a reducing agent The gas is a good reducing agent if water is present, when it reacts and dissolves to form a solution of sulphurous acid,

$$SO_2(g)+H_2O(l) \rightarrow H_2SO_3(aq)$$

which contains sulphite ions, $SO_3^{2-}(aq)$. The reductions brought about by sulphur dioxide are really reactions of the sulphite ion. A sulphite ion can accept oxygen from the substance being reduced and is itself oxidized to the sulphate ion,

$$SO_3^{2-}(aq)+[O] \rightarrow SO_4^{2-}(aq)$$

Alternatively, you can consider the reaction in terms of electron transfer,

$$SO_3^{2-}(aq)+H_2O(l)-2e^- \rightarrow SO_4^{2-}(aq)+2H^+(aq)$$

The next experiment illustrates the following examples of these reducing properties. Make sure that you can explain what is being oxidized and reduced in each case.

$$H_2O_2(aq)+SO_3^{2-}(aq) \rightarrow SO_4^{2-}(aq)+H_2O(l)$$

(or, if it is easier to remember as two separate half equations,

$$H_2O_2(aq) \rightarrow H_2O(l)+[O], \quad \text{and} \quad SO_3^{2-}(aq)+[O] \rightarrow SO_4^{2-}(aq))$$

$$Br_2(aq)+SO_3^{2-}(aq)+H_2O(l) \rightarrow 2Br^-(aq)+SO_4^{2-}(aq)+2H^+(aq)$$

(in bromine water; brown-yellow) (colourless)

potassium manganate(VII) solution, containing purple MnO_4^- ion $+SO_3^{2-}(aq) \rightarrow$ colourless solution containing Mn^{2+} ion $+SO_4^{2-}(aq)$

potassium dichromate(VI) solution, containing orange $Cr_2O_7^{2-}$ ion $+SO_3^{2-}(aq) \rightarrow$ green solution containing Cr^{3+} ion $+SO_4^{2-}(aq)$

Experiment 24.1

Sulphur dioxide as a reducing agent**

Apparatus

Cylinder of sulphur dioxide or sulphur dioxide generator, two 50 cm³ beakers, teat pipettes, access to fume cupboard.

Dilute hydrochloric acid, barium chloride solution, 'ten volume' hydrogen peroxide solution, small beakers containing solutions of bromine water, acidified potassium manganate(VII) and acidified potassium dichromate(VI).

Procedure

All these experiments should be carried out in a fume cupboard.

(a) Bubble sulphur dioxide into a beaker half full of water in order to form an aqueous 'solution' of the gas.

(b) Add a little of the 'solution of sulphur dioxide' to a few cm^3 of hydrogen peroxide solution. Add a few drops of dilute hydrochloric acid, and then some barium chloride solution.

(c) Add a little of the 'sulphur dioxide solution' to some bromine water.

(d) Repeat (c) using first acidified potassium manganate(VII) and then acidified potassium dichromate(VI) instead of the bromine water.

Results and conclusion

Record all your results, and explain what is happening in each case. Use equations for (b) and (c) only. You should understand the point of adding barium chloride in (b).

As a bleaching agent Some substances are bleached when oxygen is added to them, i.e. when they are oxidized. This is why chlorine and hydrogen peroxide (both good oxidizing agents) are good bleaching agents. It is also possible to decolourize other substances by *removing* oxygen, i.e. by reducing them. Moist sulphur dioxide can bleach some wool and silk dyes, straw, and flowers by reducing them. It is used to bleach wood pulp for making newspapers. Such reactions are not permanent for the substances are reoxidized slowly in the air, which is why old newspapers go yellow.

The next experiment illustrates these bleaching effects.

Experiment 24.2

The bleaching action of sulphur dioxide **

Apparatus

Cylinder of sulphur dioxide or alternative supply, and as required by procedure (a). Coloured articles of wool, silk, and straw. Flowers.

Procedure

(a) Devise an apparatus or system whereby the articles can be left in contact with moist sulphur dioxide for a period of time.

(b) After treatment with sulphur dioxide leave the articles in air and preferably also in sunlight for a few days. Note any changes which take place throughout the experiment.

Results

Record what happened to the coloured articles after they were exposed to moist sulphur dioxide, and also after the treated articles were left in sunlight.

Conclusion

Explain all your observations.

As an acid gas The gas turns damp blue litmus paper red because it is an acidic oxide and forms a solution of 'sulphurous acid' when dissolved in water.

$$SO_2(g) + H_2O(l) \rightarrow H_2SO_3(aq)$$

A solution of sulphurous acid is also formed if a solution of a strong acid such as dilute hydrochloric acid is added to a sulphite (e.g. sodium sulphite, Na_2SO_3). Sulphurous acid is unstable. It readily decomposes on heating, giving off sulphur dioxide, and it is easily oxidized to sulphuric acid.

Uses of sulphur dioxide

Large quantities of sulphur dioxide are manufactured as a first stage in the production of sulphuric acid. It is also used as a preservative for canned fruit and fruit juices; the 'Campden' tablets bought to sterilize containers used for home wine-making liberate the gas when dissolved in water. It is also used to bleach wood pulp before it is made into paper.

Sulphur trioxide (sulphur(VI) oxide), SO_3

Laboratory preparation

This is also an illustration of an important industrial process (p. 545), which is of course done on a larger scale.

Sulphur dioxide and oxygen gases are mixed and passed over a hot catalyst (usually vanadium(V) oxide in the laboratory) as in Figure 24.3. The reactant gases must be dried, and atmospheric moisture must be prevented from reaching the product because sulphur trioxide reacts violently with moisture. If every part of the apparatus is dry, the oxide collects as a white solid, often in the form of silky, needle-like crystals.

$$2SO_2(g) + O_2(g) \rightleftharpoons 2SO_3(g)$$

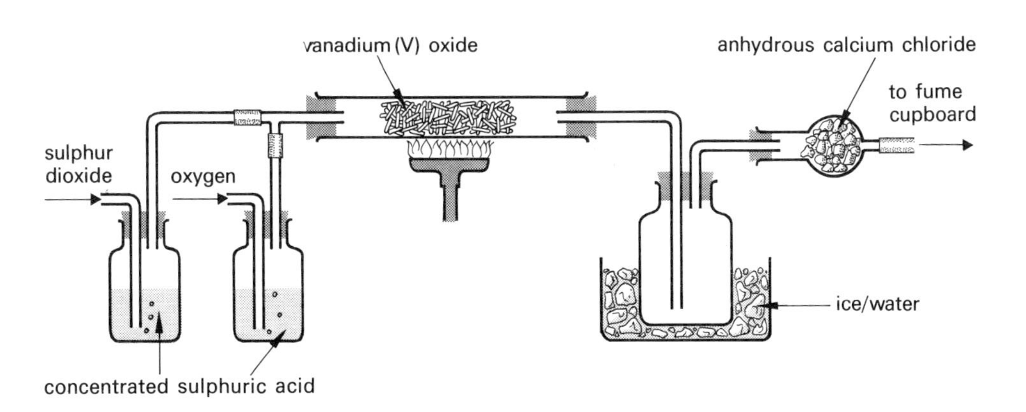

Points for discussion

1. In the preparation, it is usual to pass dry oxygen through the apparatus for a minute or so before starting the reaction. Why is this done?

2. How could you use soap solution and a small paint brush to find out if any of the 'joints' are leaking?

3. What was the purpose of the calcium chloride tube?

Physical properties

The white solid melts at 17 °C and the liquid boils at 44 °C, but because it fumes strongly with moist air it gives the appearance of being a gas at room temperature.

Chemical properties

It is an acidic oxide, reacting violently with water to form sulphuric acid.

$$SO_3(s) + H_2O(l) \rightarrow H_2SO_4(aq)$$

This reaction is so violent that a mist of acid droplets forms rather than a solution, and so when sulphuric acid is prepared or manufactured the sulphur trioxide is dissolved in concentrated sulphuric acid rather than water. Absorption then takes place more efficiently, and a very concentrated acid is produced.

Sulphuric acid, H_2SO_4

Preparation and manufacture

Sulphur trioxide is prepared as just described and dissolved in concentrated sulphuric acid so that the solution becomes even more concentrated. The acid is transported in the concentrated form. The details of the manufacturing process are given on page 545.

The properties of dilute sulphuric acid

This acid is a typical strong acid and its reactions have been described in other chapters.

The properties of concentrated sulphuric acid

Physical properties The pure acid is a colourless, viscous liquid; the concentrated acid contains about 98 per cent of the pure liquid. A great amount of heat is evolved when the concentrated acid comes into contact with water, and when the acid is being diluted it *must* be added to the water. If water is added to the acid, some water is quickly vaporized to steam and 'spits out' with corrosive droplets of the acid.

Chemical properties *Note*: These reactions are not given by the dilute acid (see also Table 5.5, page 75).

1. Its affinity for water makes it ideal for drying gases, but not those which react with it, such as ammonia.
2. The concentrated acid will remove:

(i) chemically combined water such as water of crystallization, e.g. when added to hydrated copper(II) sulphate crystals.

$$\underset{\text{(blue crystals)}}{CuSO_4.5H_2O(s)} \xrightarrow{-5H_2O} \underset{\text{(white powder)}}{CuSO_4(s)} + \text{steam}$$

(ii) hydrogen and oxygen in a 2:1 ratio from some compounds, e.g. when added to crystals of a sugar

$$\underset{\text{(white crystals)}}{C_6H_{12}O_6(s)} \xrightarrow{-6H_2O} \underset{\text{(black solid)}}{6C(s)} + \text{steam}$$

In the above cases the acid is acting as a *dehydrating agent*. Do not confuse this with the ability of concentrated sulphuric acid to act as a drying agent.

3. The concentrated acid is a powerful oxidizing agent, especially when hot, and is itself reduced to sulphur dioxide in the process.

$$H_2SO_4(aq) \rightarrow H_2O(aq) + SO_2(g) + [O]$$

For example, many metals are oxidized by the hot, concentrated acid. Copper is oxidized in the usual laboratory preparation of sulphur dioxide (p. 458), the sulphur dioxide being formed by the reduction of the acid. (See also page 295.)

Experiment 24.3
Some reactions of concentrated sulphuric acid**

Apparatus

Evaporating basins, protective mat, spatula. White sugar, concentrated sulphuric acid, hydrated copper(II) sulphate.

Procedure

(a) Place a few spatula measures of sugar into an evaporating basin and stand the basin on a mat.

(b) *Carefully* pour a small amount of concentrated sulphuric acid on to the sugar and record your observations.
(c) Repeat (a) and (b) but use hydrated copper(II) sulphate instead of the sugar.

Results and conclusion
Use your results to explain what is happening in each case. Give equations.

Uses of sulphuric acid
Sulphuric acid is one of the most important industrial chemicals, and is used in the manufacture of rayon, fertilizers, dyes, plastics, drugs, and explosives. When extra sulphur trioxide is dissolved in it, a very corrosive, fuming liquid called oleum is produced, which is used in the organic chemicals industry.

Check your understanding

1. Dehydration of a substance is best described as
A the removal of moisture
B the addition of water
C the removal of water
D the addition of moisture
E the removal of the elements of water

2. Sulphur
A forms two alkaline oxides
B is spontaneously flammable
C burns with a blue flame
D conducts electricity in the molten state
E is usually stored in the form of sticks in water

3. (a) What do you understand by the term 'allotropy'?
(b) Name two crystalline allotropes of sulphur, (i) and (ii).
(c) Which is the stable allotrope at room temperature?
(d) Describe briefly how you would prepare a sample of the allotrope which is unstable at room temperature.
(e) What would happen to the crystals formed in (d) if they were left at room temperature?
(f) If 3.2 g of allotrope (i) were completely oxidized to sulphur dioxide, what mass of sulphur dioxide would be formed?
(g) If 3.2 g of allotrope (ii) were similarly oxidized, what mass of sulphur dioxide would be formed? (J.M.B.)

4. Write brief notes, and give suitable equations, in support of the statement 'dilute sulphuric acid is a typical, strong acid'. Try to cover the whole range of acidic behaviour in your examples.

5. Describe *briefly* how you would prepare a sample of each of the following, starting from sulphuric acid, and one other *compound* each time. Use a different type of compound for each example.
(a) sodium sulphate crystals,
(b) magnesium sulphate crystals,
(c) barium sulphate,
(d) carbon,
(e) potassium hydrogensulphate solution.

6. Explain why it is important to write '*dilute* sulphuric acid', or '*concentrated* sulphuric acid' instead of just 'sulphuric acid' in describing a property of the acid.

24.2 CHLORINE AND ITS COMPOUNDS

Chlorine the element

Davy first identified chlorine as an element in 1810, and gave it the name of chlorine from the Greek, *chloros*, meaning greenish-yellow.

Laboratory preparation
Chlorine is prepared by the oxidation of concentrated hydrochloric acid:

$$\underset{\text{(from oxidizing agent)}}{[O]} + 2HCl(aq) \rightarrow H_2O(l) + Cl_2(g)$$

The usual oxidizing agent is potassium manganate(VII), and a typical apparatus is shown in Figure 24.4. The potassium manganate(VII) should be covered by a thin layer of water before the acid is added. Another common oxidizing agent used for the purpose is manganese(IV) oxide, which is slightly less convenient as the mixture then needs to be warmed before the reaction will take place. The gas must be prepared in a fume cupboard.

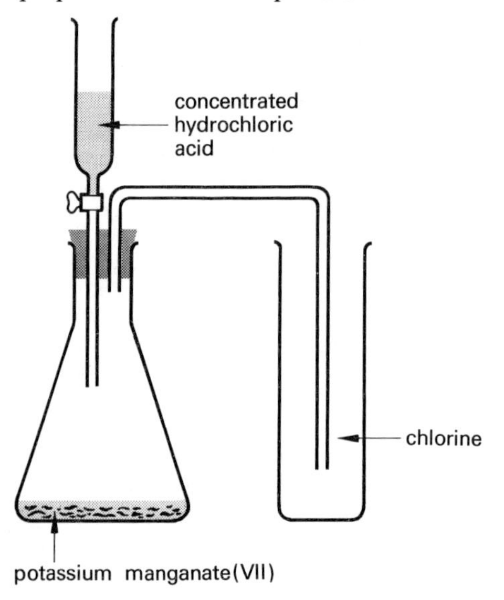

Figure 24.4 Laboratory preparation of chlorine

Point for discussion

How would you modify Figure 24.4 if manganese(IV) oxide is to be used as the oxidizing agent?

Physical properties

Solubility in water	*Colour*	*Odour*	*Density relative to air*	*Toxicity*
Slightly soluble	Green-yellow	Choking, characteristic	More dense	Very toxic

Chemical properties

Direct combination with metals Chlorine is a very reactive element, and readily combines with many other elements, both metals and non-metals. Two common examples are:

1. *sodium*: if a small piece of sodium is ignited on a combustion spoon, then plunged into a gas jar of chlorine, the metal burns with a yellow flame, forming dense, white clouds of solid sodium chloride,

$$2Na(s) + Cl_2(g) \rightarrow 2NaCl(s)$$

2. *magnesium*: if a piece of magnesium ribbon is ignited on the end of a combustion spoon and then placed in a gas jar of chlorine, the metal burns with a white flame, forming a dense white cloud of solid magnesium chloride,

$$Mg(s) + Cl_2(g) \rightarrow MgCl_2(s)$$

3. You may have studied its reaction with iron wool (p. 242), and the alloy 'Dutch metal' reacts very rapidly when placed (unheated) in chlorine. You may see demonstrations of these reactions.

Chlorine as an oxidizing agent Chlorine is a powerful oxidizing agent. Chlorine atoms readily gain electrons (i.e. oxidize another substance by removing electrons from it) to form stable Cl^- ions, as in the halogen displacement reactions (p. 246) and the conversion of iron(II) salts to iron(III) salts (p. 342), all of which are examples of oxidation. Similarly, its affinity for hydrogen means that it removes hydrogen from compounds, and so the reactions with hydrogen, considered below and in Experiment 24.4, are also oxidations.

Point for discussion

A pale green solution X forms a dirty green precipitate when added to sodium hydroxide solution. When chlorine is passed through another sample of X, the solution turns pale yellow-orange. The yellow-orange solution gives a red-brown precipitate when added to sodium hydroxide solution. Explain these facts, and give equations for the three reactions referred to.

The affinity of chlorine for hydrogen The gas has a great affinity for hydrogen, and in addition to reacting explosively with the element in sunlight, it reacts with many compounds containing hydrogen. It is easy to work out what happens in such cases, because the chlorine always removes the hydrogen to form hydrogen chloride, depositing or liberating 'whatever is left' from the compound. Typical experiments include the following:

Hydrocarbons		
a wax taper	*turpentine*	*Hydrogen sulphide*
Taper ignited and placed in gas jar of chlorine. Continues to burn with a smoky flame, depositing large amounts of black carbon and forming steamy fumes of hydrogen chloride	Piece of filter paper soaked in warm turpentine and dropped into gas jar of chlorine. The turpentine bursts into a sheet of red flame, forming large amounts of black carbon and steamy fumes of hydrogen chloride $C_{10}H_{16}(l) + 8Cl_2(g) \rightarrow 10C(s) + 16HCl(g)$	Gas jars of the two gases are placed mouth to mouth. Instant reaction produces steamy fumes of hydrogen chloride and yellow deposits of sulphur form on the sides of the jars $H_2S(g) + Cl_2(g) \rightarrow 2HCl(g) + S(s)$

Experiment 24.4
The affinity of chlorine for hydrogen***

Apparatus
Chlorine generator, combustion spoon, two gas jars and slides, Bunsen burner, filter paper, test-tube and holder, tongs, access to fume cupboard.
Wax taper (or small candle), turpentine.

Procedure
(a) Working in a fume cupboard, collect two gas jars of chlorine.
(b) Light a wax taper, or place a small length of candle on to a combustion spoon and light it. Plunge the taper (or spoon) into a gas jar of chlorine and record your observations.

(c) Pour a little turpentine into a test-tube and warm it gently (care, turpentine is flammable). Fold a piece of filter paper so that it will fit into the test-tube. Hold the paper in the tongs, and dip it into the tube so that it becomes moistened with warm turpentine. Then drop the filter paper into a gas jar of chlorine. Again record your observations.

Results and conclusion
Use your results to explain what happened in each case, giving equations.

Chlorine as a bleaching agent In the presence of moisture, the gas rapidly bleaches coloured flowers, litmus paper, inks, and other materials dyed with vegetable dyes. The chlorine reacts with the moisture to form a mixture of two acids:

$$Cl_2(g) + H_2O(l) \rightarrow HOCl(aq) + HCl(aq)$$

The former, chloric(I) acid (commonly called hypochlorous acid), bleaches substances by oxidizing the colourings in them to a colourless form,

$$HOCl(aq) \rightarrow HCl(aq) + [O]$$

This acidity of chlorine solution, followed by a bleaching action, is used as a test for the gas (p. 450). This reaction with water is a further illustration of the affinity of chlorine for hydrogen.

Points for discussion

1. If a saturated solution of chlorine in water is placed in a burette as shown in Figure 24.5, and left in sunlight for a few days, a gas collects at the top of the burette. What do you think this gas is? (It is not chlorine.) Explain its formation.
2. Compare the ways in which chlorine and sulphur dioxide can act as bleaching agents.
3. Why should substances be washed after they have been bleached by either chlorine or sulphur dioxide?

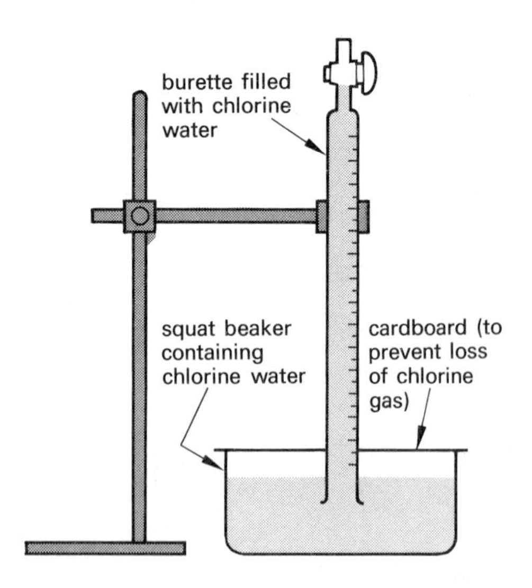

Figure 24.5 Leaving chlorine water in sunlight

The reactions of chlorine with alkalis Chlorine reacts with alkalis in two different ways.

With *cold*, *dilute* alkali,

$$Cl_2(g) + 2NaOH(aq) \rightarrow NaCl(aq) + NaOCl(aq) + H_2O(l)$$

NaOCl, sodium chlorate(I), is commonly called sodium hypochlorite; salts of this kind act like hypochlorous acid, and they are used in preparing bleaches.

With *hot*, *concentrated* alkali,

$$3Cl_2(g) + 6NaOH(aq) \rightarrow 5NaCl(aq) + \underset{\text{sodium chlorate(V)}}{NaClO_3(aq)} + 3H_2O(l)$$

Chlorine as a member of the halogen group The halogen group was considered as part of the study of the Periodic Table (p. 239), when other reactions of chlorine were discussed, e.g. the halogen displacement reactions.

Uses of chlorine

Large quantities are used to introduce chlorine atoms into molecules of organic compounds to make rubbers, plastics (especially PVC), and solvents. It is also used for sterilizing drinking water, water in swimming pools, and sewage, in preparing dry-cleaning fluids and commercial bleaching agents, and in making hydrochloric acid.

A dry cleaning fluid?

Hydrogen chloride, HCl

Preparation

Hydrogen chloride gas is prepared by the action of moderately concentrated sulphuric acid on a solid metal chloride such as sodium chloride, in a fume cupboard.

$$NaCl(s) + H_2SO_4(aq) \rightarrow NaHSO_4(aq) + HCl(g)$$

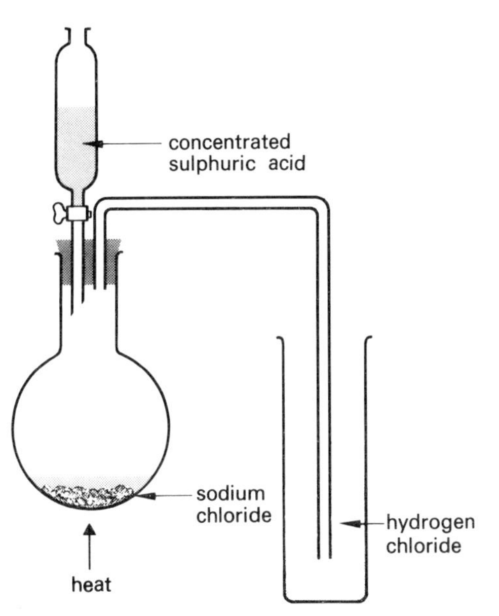

Warming may be needed after the initial vigorous reaction has subsided, and a typical apparatus is shown in Figure 24.6. (Note that this reaction has already been referred to as a possible test for an *insoluble* chloride, page 345.)

Figure 24.6 Laboratory preparation of hydrogen chloride

Physical properties

Solubility in water	*Colour*	*Odour*	*Density relative to air*	*Toxicity*
Extremely soluble—gives fountain experiment	None, but fumes in air	Sharp and characteristic	More dense	Toxic

Chemical properties

These have already been described elsewhere in the book, e.g. its reaction with ammonia (p. 450) is also used as a test.

Note: A solution of hydrogen chloride in a dry, organic solvent (e.g. methylbenzene, as on pp. 71–2) contains hydrogen chloride *molecules* and is not acidic, whereas a solution of hydrogen chloride in water contains $Cl^-(aq)$ and $H^+(aq)$ ions and is acidic.

Hydrochloric acid, HCl(aq)

Preparation

Hydrogen chloride gas is prepared as just described and then passed into water via an inverted funnel, as shown for ammonia in the diagram on page 378.

Properties

Hydrochloric acid is a typical strong acid, and its reactions have been described throughout the book. The concentrated acid behaves as a more reactive form of the dilute acid; there are no additional properties like the oxidizing and dehydrating properties of concentrated sulphuric acid, or the oxidizing properties of concentrated nitric acid (see Table 5.5, p. 75).

Check your understanding

1. A solution of hydrogen chloride in methylbenzene (toluene), an organic solvent, does not conduct electricity and has no apparent reaction with a solid metal oxide or a solid metal carbonate. A solution of hydrogen chloride in water conducts electricity and reacts with both a metal oxide and a metal carbonate. What does this information indicate about the nature of the bonding in hydrogen chloride when it is in solution in (a) methylbenzene (toluene), (b) water?

Write equations, either molecular or ionic, for the reactions of an aqueous solution of hydrogen chloride with (i) magnesium oxide, (ii) zinc carbonate. (J.M.B.)

2. Hydrochloric acid
A is present in the stomach
B can be reduced to chlorine
C is a covalent compound
D bleaches litmus paper
E forms both normal and acid salts

3. For each of the incorrect answers to question 2, explain why it is incorrect.

4. Write brief notes, and give suitable equations, in support of the statement 'dilute hydrochloric acid is a typical strong acid'. Try to cover the whole range of typical acidic behaviour.

5. Name one acidic and one alkaline gas which , in each case, is very soluble in water. Draw a diagram to show how a very soluble gas can be safely dissolved in water contained in a beaker and explain briefly how the apparatus avoids the danger of 'sucking back'. (J.M.B.)

6. Two beakers each contain a dilute acid, but they have lost their labels. One of the beakers contains dilute hydrochloric acid, and the other contains dilute sulphuric acid. What is the simplest way of finding out which acid is in each beaker?

24.3 A SUMMARY OF CHAPTER 24

This 'check list' should help you to organize the work for revision.
(a) The extraction and physical properties of sulphur (pp. 457–8), including the changes which take place when sulphur is heated (p. 276), and the allotropes of sulphur (p. 276).
(b) The combustion of sulphur, and some uses of the element.
(c) The laboratory preparation of sulphur dioxide (by adding a dilute acid to a sulphite, or by warming copper with concentrated sulphuric acid), and the physical properties of the gas (p. 458).

(d) The chemical properties of sulphur dioxide, i.e. examples of its reactions as a reducing agent (p. 459), as a bleaching agent (p. 460), and as an acid (p. 460).
(e) Some uses of sulphur dioxide (p. 460).
(f) The laboratory preparation of sulphur trioxide from sulphur dioxide and oxygen, and its physical properties.
(g) The reaction of sulphur trioxide with water ($SO_3 + H_2O \rightarrow H_2SO_4$), and the need to form a solution of the compound by dissolving it in concentrated sulphuric acid rather than water.
(h) The principle of the preparation of sulphuric acid (industrial details, page 545).
(i) Examples of the behaviour of dilute sulphuric acid as a typical strong acid (acid + base, acid + carbonate, acid + fairly reactive metals), and in the preparation of salts.
(j) The physical properties of concentrated sulphuric acid, and its reactions as a dehydrating agent and as an oxidizing agent (see also page 295). (Dehydrating is not the same as drying. Dehydrating is the removing of chemically combined water.)
(k) Some uses of sulphuric acid (p. 463).
(l) The laboratory preparation of chlorine (by oxidizing concentrated hydrochloric acid), and its physical properties.
(m) The reactions of chlorine with metals, as an oxidizing agent by the removal of electrons, as an oxidizing agent by the removal of hydrogen, as a bleaching agent, with alkalis, and in displacement reactions with other members of the halogen group (pages 240–1, and also page 246).
(n) Some uses of chlorine.
(o) The laboratory preparation of hydrogen chloride from sodium chloride and concentrated sulphuric acid, and the physical properties of the gas.
(p) Examples of the behaviour of dilute hydrochloric acid as a typical strong acid (acid + base, acid + carbonate, acid + fairly reactive metals).
(q) Concentrated hydrochloric acid has no *special* properties other than those of the dilute acid (unlike concentrated sulphuric and nitric acids).
(r) It is particularly important to use state symbols when referring to hydrogen chloride or hydrochloric acid. HCl(g) is a covalently bonded, molecular, gas but HCl(aq) is an ionic solution. However, a solution of hydrogen chloride in a dry organic solvent still contains *molecules* of hydrogen chloride, and therefore does not act as an acid.

QUESTIONS

1. (a) Draw a labelled diagram of the apparatus used in the laboratory to convert sulphur dioxide to sulphur(VI) oxide, SO_3, and to collect the product. Write an equation for the reaction.
(b) Describe what you would observe if sulphur(VI) oxide were cautiously treated with water and thus explain why this method of formation of sulphuric acid from sulphur(VI) oxide is unsuitable for a large scale process. What is the industrial alternative?
(c) Pure sulphuric acid is a liquid which boils at 330 °C at standard atmospheric pressure. What type of bonding is present in the pure liquid? How do you therefore justify the word 'acid' in its name?
(d) Name the gas given off when concentrated sulphuric acid is added to sodium chloride at room temperature and name the product formed when this gas is passed into water. Write an equation for one of these reactions. (J.M.B.)

2. Describe briefly how you would separate a pure sample of the first named substance from the impurity in each of the following mixtures.
(a) Magnesium sulphate from a mixture with glass.
(b) A solution of the orange dye which, together with a blue dye, forms black ink.
(c) Copper filings from a mixture with magnesium filings.
(d) Oxygen from a mixture with chlorine.
(e) Ammonium chloride from a mixture with sodium chloride. (J.M.B.)

3. Give the reactions by which the following gases may be obtained in the laboratory: (a) hydrogen chloride from sodium chloride; (b) ammonia from ammonium sulphate. (*Note*: Diagrams and details of collection are *not* required.)
What precaution would you take when passing these gases into water?
An acid is a proton donor: a base is a proton acceptor.

Illustrate this statement by reference to the changes that take place when hydrogen chloride and ammonia are separately dissolved in water. Explain why hydrochloric acid is said to be a *strong* acid whereas ammonia solution is considered to be a *weak* base.

Describe and explain what you would expect to see when ammonium chloride is gently heated in the apparatus below.

(A.E.B. 1975)

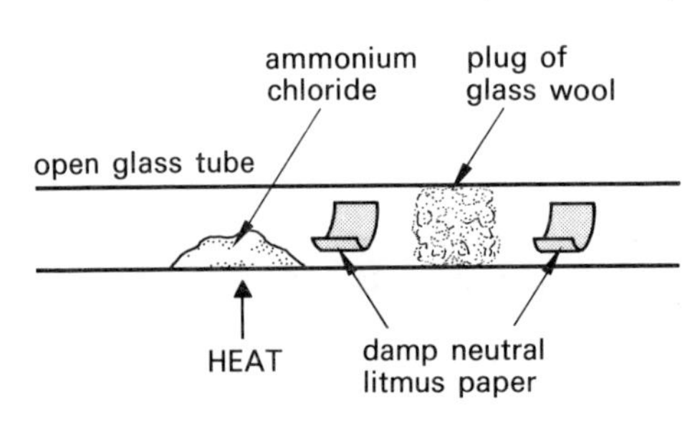

4. (a) Chlorine and bromine are members of the same chemical family of elements. The atomic number of chlorine is 17.

(i) What name is given to this family of elements? Name one other member and state its physical appearance at room temperature.

(ii) Give electronic diagrams of a chlorine atom and of a hydrogen chloride molecule.

(iii) Why do the elements in this chemical family show similarities in their chemical reactions?

(iv) State the conditions under which each of the elements chlorine and bromine reacts with hydrogen. Indicate how these are related to the relative reactivities of the two elements.

(b) Hydrogen chloride dissolves readily in water. The change is exothermic and the solution produced is strongly acidic.

(i) What precautions would you take when dissolving hydrogen chloride in water?

(ii) What is meant by the term exothermic?

(iii) Explain why the solution is strongly acidic. (A.E.B. 1976)

5. (a) Chlorine may be prepared by oxidizing hydrochloric acid.

(i) Name *one* suitable oxidizing agent.

(ii) Write the equation for the reaction and say why you regard this to be an oxidation reaction.

(iii) State how you would react the oxidizing agent with hydrochloric acid so as to produce a reasonable quantity of chlorine for use in the laboratory (a diagram is *not* required).

(iv) Why is it desirable to wash the chlorine with a little water before collecting it? Why would it be incorrect to use sodium hydroxide solution to wash the gas?

(v) State how you would collect the chlorine. Why is the method of collection suitable?

(vi) Give *one* chemical test for chlorine.

(b) Describe *one* experiment you have seen in which chlorine reacts with a named metallic element.

(c) Warming a gaseous oxide of chlorine, Cl_xO_y, causes it to decompose: $2Cl_xO_y \rightarrow xCl_2 + yO_2$. In an experiment 20 cm^3 of the oxide gave 20 cm^3 of chlorine and 10 cm^3 of oxygen (all gas volumes being measured at the same temperature and pressure). Calculate the formula of this oxide showing your reasoning clearly. (A.E.B. 1977)

6. (a) (i) Sketch and label an apparatus by which, starting from sodium chloride, hydrogen chloride may be generated and safely dissolved in water.

(ii) Write the equation for the reaction by which the gas is obtained.

(iii) State clearly the changes that take place when hydrogen chloride molecules dissolve in water.

(b) State why a solution of hydrogen chloride in water acts as a *strong* acid whilst a solution containing ethanoic (acetic) acid, CH_3COOH, is a *weak* acid.

(c) What do you understand by a *molar solution* of an acid? Calculate the volume of molar hydrochloric acid required to neutralize 0.1 mol of sodium carbonate by the reaction:

$$Na_2CO_3 + 2HCl \rightarrow 2NaCl + CO_2 + H_2O$$

(d) Give, with reasons, *one* chemical test that would enable you to distinguish between dilute hydrochloric acid and dilute sulphuric acid. (A.E.B. 1977)

7. Describe and explain laboratory experiments to illustrate the following.

(a) The preparation of a reasonably pure sample of copper(II) sulphate from metallic copper or some suitable compound of the metal.

(b) Chlorine is a more powerful oxidizing agent than iodine.

(c) Concentrated sulphuric acid can act as:

(i) a dehydrating agent,

(ii) an oxidizing agent.

(d) Sulphur dioxide can act as a reducing agent. (W.J.E.C.)

8. (a) Starting from the *element* sulphur, give an account of one method for the manufacture of sulphuric acid.
(b) How does the atmosphere become polluted by sulphur compounds? What undesirable effects does this pollution have? How can it be kept to a minimum?
(c) Describe the reaction of concentrated sulphuric acid with:
(i) sucrose, $C_{12}H_{22}O_{11}$,
(ii) potassium nitrate. (O. & C.)

9. (a) Describe how chlorine can be obtained from sodium chloride:
(i) using electrolysis (the sign and nature of the electrodes must be stated),
(ii) without using electrolysis (the names of the other chemicals required and the equations for the reactions must be given).
(b) Describe the reaction of chlorine with:
(i) a *named* hydrocarbon,
(ii) aqueous potassium iodide.
Using (ii) as an example, explain, in terms of electron transfer, why chlorine is classified as an oxidizing agent. (O. & C.)

10. (a) Neither water nor dry sulphur trioxide crystals alone turn dry blue litmus paper red. However, the litmus paper does turn red when it is added to a solution of sulphur trioxide crystals in water. Explain this observation.
Write an ionic equation to show what happens when sulphur trioxide dissolves in water.
(b) Describe what is seen when blue copper(II) sulphate crystals are added to
(i) dilute sulphuric acid,
(ii) concentrated sulphuric acid.
(J.M.B.)

25 Energy in chemistry

25.1 SOME BASIC IDEAS

Introduction

Energy is easy to recognize but difficult to define. The nearest we can get to a simple definition is to say that energy is something that can do work, and that all the many forms of energy are interconvertible.

Heat and electricity are familiar forms of energy. Other examples are chemical, kinetic, and potential energy. Each of these types of energy can be changed into any of the other forms. Thus the energy stored in coal (chemical energy) can be used when the coal burns (heat energy) to change water into steam and so drive the turbines in a dynamo (kinetic energy) to produce electricity (electrical energy). This in turn can be used to move a cable car up a mountain (kinetic energy) until it reaches the summit where it has energy due to its position (potential energy).

From earliest times man has used the stored chemical energy in fuels such as wood and peat to provide warmth, but until comparatively recently muscular effort was the main source of energy for any work that had to be done. (The energy used by all the slaves during the twenty years required to build an Egyptian pyramid is approximately equal to the energy needed for two minutes of a space rocket launch.) The Industrial Revolution was due to the discovery that the chemical energy stored in coal could be transformed to kinetic energy in machines. Today we use the energy stored in coal and oil (either directly or via electricity) to do all the heavy work in factories, to transport us in trains, ships, and aeroplanes and to provide heat and light for our work and leisure.

Much of the energy from fossil fuels is converted into electrical energy which is vitally important for the modern way of life. Increasing use of electrical appliances, the needs of the developing countries, and the population explosion are producing an accelerating demand for electrical power. This, added to the fact that fossil fuels are also the main sources of many widely used organic chemicals, has resulted in a tremendous increase in the rate at which coal, natural gas, and oil are being used up. Supplies of fossil fuels once thought to be limitless are in danger of running out in the near future. A great deal of research is being done to find alternative sources of energy. Nuclear power stations use the energy released by nuclear fission (p. 109) to produce electricity, but other ways of using the tremendous stores of energy locked up in atomic nuclei (e.g. fusion power, page 503) are still in their infancy. Other sources now

A new source of energy?

being used to supply energy include fuel cells, the sun, the wind, and the waves. Although most of our present sources of energy, from the food we eat to fossil fuels such as crude oil, are derived from the sun's energy by photosynthesis, the *direct* use of sunlight as a source of energy is still insignificant. The problem of maintaining a supply of energy is immediate and pressing; some new sources of energy, or more efficient methods of using known sources, will be needed in the near future. This is discussed in more detail in the next chapter, but for the moment you need to have some understanding of how we measure energy changes, and how chemicals can give up some of the energy they contain.

Units

Most chemical reactions produce a heat change, and such energy changes are normally measured in joules. As quite often a *considerable* amount of energy is given out (or taken in) during a chemical reaction, it is usually more convenient to use the kilojoule (kJ) as the unit of energy. 1000 J = 1 kJ.

It is also important to state what *quantity* of chemicals react to produce (or absorb) a certain amount of energy. Chemists usually calculate heat changes per mole of chemical, because then we are always making fair comparisons between a fixed number of particles of different substances. A typical statement of an energy change might be: 29.3 kJ mol^{-1}, which is the same as 29 300 J mol^{-1}.

Note: (i) In some cases, we note the energy change for the quantities of chemicals contained in a balanced *equation* for the reaction, and not per mole of a particular substance.

(ii) The energy contained in foods is sometimes still measured in units called calories or kilocalories. To convert calories to joules, multiply by 4.18. Industrial units of energy are considered on page 492.

Where does energy come from?

Experiment 25.1
Some simple energy changes**

Apparatus

Two 100 cm^3 beakers, stirring rod, −10 to 110 °C thermometer, spatula, small measuring cylinder.

Concentrated sulphuric acid, ammonium nitrate, or ammonium chloride.

Procedure

(a) Half fill one of the beakers with water, and note the temperature of the water.

(b) Carefully pour 15 cm^3 of concentrated sulphuric acid into the measuring cylinder. Wipe the thermometer so that it is absolutely dry, and then note the temperature of the acid.

(c) Carefully and slowly pour the acid into the water, stirring constantly with the thermometer, and note the temperature of the solution formed.

(d) Rinse the thermometer with water and wipe it dry. Half fill the other beaker with water, and again note its temperature.

(e) Add spatula measures of ammonium nitrate (or ammonium chloride) to the water, and stir to dissolve the solid. Note the temperature of the solution after about 10 spatula measures have been dissolved.

Results

Record all of your observations.

Notes about the experiment

The purpose of the experiment is not to find out *why* energy changes occur, nor to measure them accurately, but rather to demonstrate that:

(i) even simple processes such as these can cause considerable energy changes; and

(ii) energy can be *released* (in which case the chemicals and their container increase in temperature) or *absorbed* (in which case the temperature of the chemicals and their container falls). (This last point may puzzle you. When ammonium nitrate (or chloride) dissolves in water, energy is needed for it to do so; energy is taken from the water, and the temperature of the water therefore goes down.)

Points for discussion

1. Why was it important, in the experiment, to add the concentrated acid to the water instead of adding the water to the acid?

2. Can you remember any other chemical reactions (other than combustions) which release or absorb energy?

3. Some people are surprised by the statement that, in order for ammonium nitrate to dissolve in water, it takes energy from the water. How can water which is cold contain energy?

How do chemicals 'contain' energy?

It seems strange that fairly cold water and fairly cold concentrated sulphuric acid can release so much heat energy when they are mixed together. It shows that even 'cold' substances contain energy, and that some of this energy can ,be released in certain circumstances. There are several ways in which chemicals 'contain energy'.

Kinetic energy The atoms within a molecule vibrate and rotate, and the molecules as a whole vibrate, rotate, and in some cases move from place to place, i.e. they have kinetic energy. When a chemical reaction takes place, the kinetic energies of the reacting molecules will always be different from those of the product molecules. If the product molecules have less kinetic energy than the molecules of the reactants, energy will be given out during the reaction. The opposite can also happen.

Let us consider what happens when steam condenses to liquid water. The molecules of steam have much more kinetic energy than the molecules of liquid water. When we 'slow down' the steam molecules, i.e. when we change the steam to water, the 'extra' kinetic energy which the molecules had as steam is released as heat energy. This energy is called the latent heat of condensation. (Latent = 'hidden', because at the point where steam changes to water a thermometer would show no change in temperature.)

It is important to remember that terms such as 'cold' and 'hot' are only *relative*. Ice may seem cold to us, but it would be very 'hot' when compared with the temperature of liquid air (about $-195\,°C$). Indeed, a container of liquid air would be brought rapidly to the boil if it was placed on a block of ice; the ice would be a source of heat energy to the liquid air. At all temperatures above $-273\,°C$ molecules have kinetic energy. Water at room temperature (say $20\,°C$) is thus about 293 degrees above absolute zero, and so although it may appear fairly cold to us, in actual fact it contains a great deal of kinetic energy. Some of this kinetic energy was used up in dissolving the ammonium salt in Experiment 25.1, which is why the temperature of the solution went down.

The energy locked in chemical bonds Substances also contain energy in the bonds which hold the particles together. Each molecule of water, for example, has two covalent bonds between oxygen and hydrogen atoms,

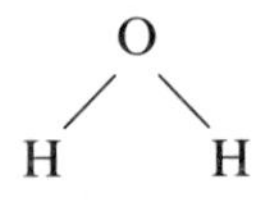

A large amount of energy is needed to break open a water molecule into separate particles of hydrogen and oxygen. It must follow, therefore, that the reverse is true;

when the two O—H bonds are *formed* in a molecule of water, a considerable amount of energy is released.

There are many kinds of bonding between atoms, molecules, and ions. Some bonds are stronger than others. The strong bonds include covalent bonds, ionic bonds, and metallic bonds (see page 285). When any such strong bonds are made, a considerable amount of energy is released. Conversely, a considerable amount of energy is needed to break such bonds. There is also a range of weaker bonds, such as the intermolecular forces between water molecules (p. 281), sulphur molecules, and iodine molecules (p. 275). The making and breaking of these weaker bonds has the same general effect as with stronger ones, but the energy changes are smaller. For example, when substances dissolve in water, bond-making and bond-breaking processes take place, but these involve the weaker *intermolecular* forces between water molecules and not the strong O—H covalent bonds within water molecules (p. 283). Even 'cold' water, therefore, also contains another source of energy, i.e. the ability to react with something else and form stronger bonds (intermolecular or intramolecular) than are present already. Energy is released in the process.

Other sources of energy within atoms and molecules Some of the other sources of energy within atoms and molecules are more complicated. One source of particular interest is the great amount of energy locked up in holding the protons together (i.e. in opposing their self-repulsive forces) in the nuclei of atoms. Such enormous energy factors are not affected by ordinary chemical reactions, because only the 'outer parts' of an atom (i.e. the electrons) change in the making and breaking of bonds. They are, however, used in nuclear power stations where the colossal amounts of energy from 'nuclear reactions' are produced under controlled conditions.

What determines whether a reaction gives out energy or takes in energy?

As there are so many forms of energy within a substance, and some of these forms are not altered during a reaction (e.g. the nuclear energy), we never measure the *total* amount of energy a substance contains. We are only concerned with measuring a *change* in energy when substances react to become something else, i.e. we measure the overall release or gain in energy which takes place in a reaction.

The important energy changes which take place during a chemical reaction are caused by kinetic energy and the energy of chemical bonding, mentioned in the previous section, and the more important of these is the energy change due to the making and breaking of bonds. When chemicals react, some bonds are broken *and* some new ones are made. Whether energy is given out or taken in (considering the whole reaction) depends largely on the number and 'strength' of the various bonds which are being broken and made. *All* chemical reactions involve both bond-making and bond-breaking (energy releasing and energy absorbing) processes. If, overall, it requires less energy to break existing bonds (in the reactants) than is released by forming new ones (in the products), the measured energy change will be a release of energy from the reacting substances.

The more energy something contains, the more unstable it is when compared to something which contains less energy. Most of the reactions you study are the type which give out more energy than they take in, and so the products contain less energy than the reactants, i.e. they are more stable.

$$\underbrace{\underset{\text{break}}{A{-}A} + \underset{\text{break}}{B{-}B}}_{\text{use energy (1)}} \rightarrow \underbrace{\underset{\text{make}}{A{-}B} + \underset{\text{make}}{A{-}B}}_{\text{release energy (2)}}$$

If the energy (1) is less than the energy (2), the total reaction will release energy. However considerations such as these are not the only ones which determine whether a reaction actually takes place, or how quickly it reacts, as you will learn if you study chemistry to a more advanced level.

These ideas are summarized in Figure 25.1, which also shows that the energy change due to the making and breaking of bonds also causes an energy change (usually less important) because of the differences in kinetic energy between the reactant molecules and the product molecules (as discussed earlier). It is not important that you should learn by heart the ideas under discussion; they are used only to give you a simple understanding of where energy comes from (or goes to) during a chemical reaction.

Note that the bonds within hydrogen molecules (H—H) and oxygen molecules (O═O) have to be *broken* during the course of the reaction illustrated in Figure 25.1, because molecules of steam

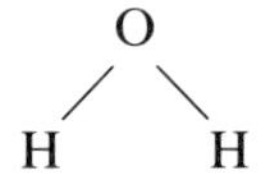

contain neither H—H bonds nor O═O bonds. However, new bonds are also *made* during the reaction. The 'separated' hydrogen and oxygen atoms join to form two new bonds inside each molecule of steam. The energy needed to break 2 moles of H—H bonds and 1 mole of O═O bonds is less than the energy given out when H—O bonds are made in forming 2 moles of steam. There is also a further release of energy when steam changes to liquid water because more bonds are then made, this time intermolecular bonds (bonds between one molecule and the next). This particular reaction thus takes place with an overall release of energy, even though some of the steps absorb energy.

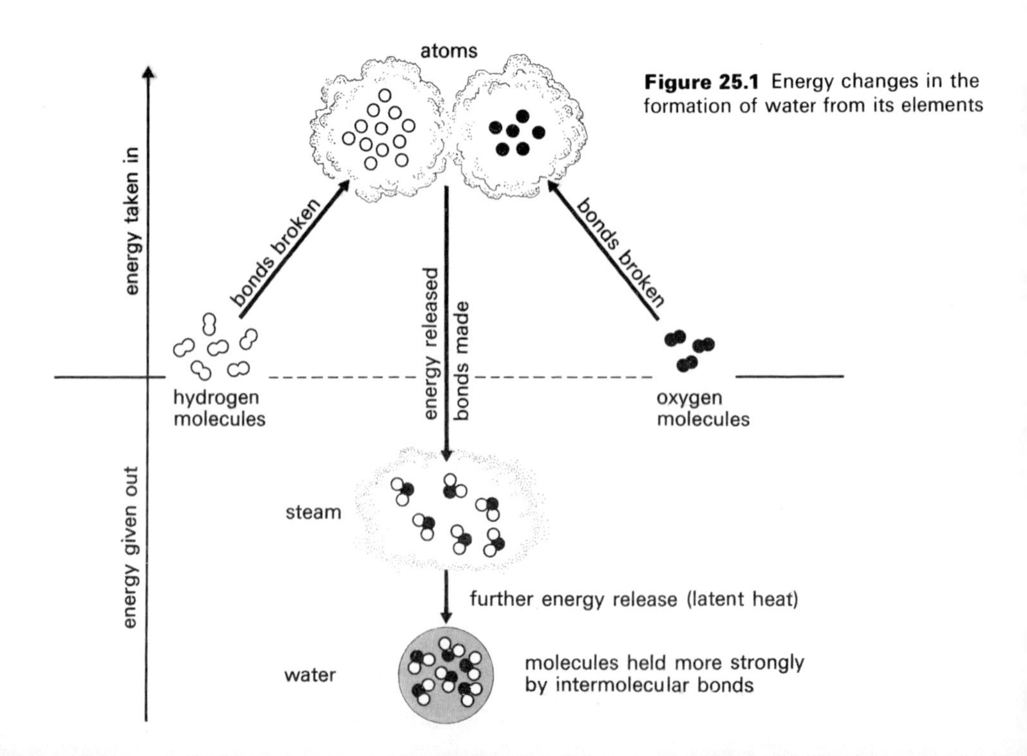

Figure 25.1 Energy changes in the formation of water from its elements

Some important points about energy changes

The following terms and ideas are essential for any understanding of energy changes in chemistry.

1. When chemicals react and give out energy (i.e. the container and its contents become hotter because of the reaction), the reaction is said to be *exothermic*.
2. When energy is absorbed during a reaction (i.e. the container and its contents fall in temperature because of the reaction), the reaction is said to be *endothermic*.
3. When chemical bonds are broken, energy is required to do this and this stage in a reaction is endothermic.
4. When chemical bonds are made, energy is released and such a stage in a reaction is exothermic.
5. A measured energy change is the *difference* between the total energy content of the starting materials (reactants) and that of the final materials (products), and is given the symbol ΔH.

The energy or heat content of the reacting substances (sometimes called the enthalpy) is given the symbol H. As explained earlier, we never measure the actual heat content of a system, only the *change* in H. This change is given the symbol ΔH, because Δ is the symbol normally used when a change in some property or quantity is being measured. (Δ is the Greek capital 'delta', and so ΔH is spoken as 'delta H'.) Thus $\Delta H =$ energy of products − energy of reactants. If the reactants contain more energy than the products, heat is given out during the reaction (i.e. it is exothermic) and ΔH will have a negative value. In an endothermic reaction, the energy of the products will be greater than the energy of the reactants and ΔH will have a positive value. These ideas are summarized in Figure 25.2. The overall energy change in a familiar exothermic reaction (that between magnesium and oxygen to form magnesium oxide) is summarized in Figure 25.3. Remember that

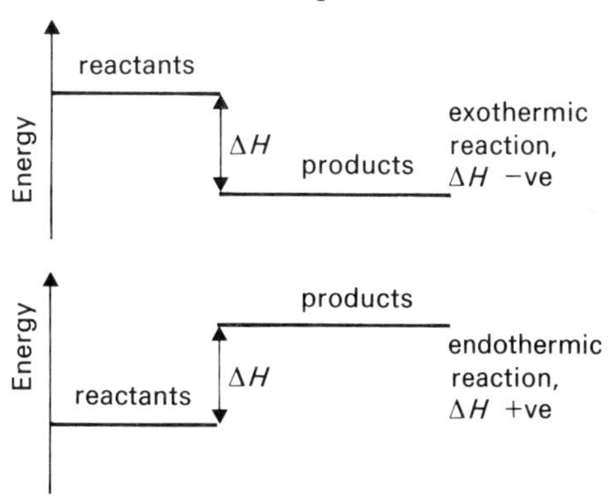

Figure 25.2 The ΔH convention

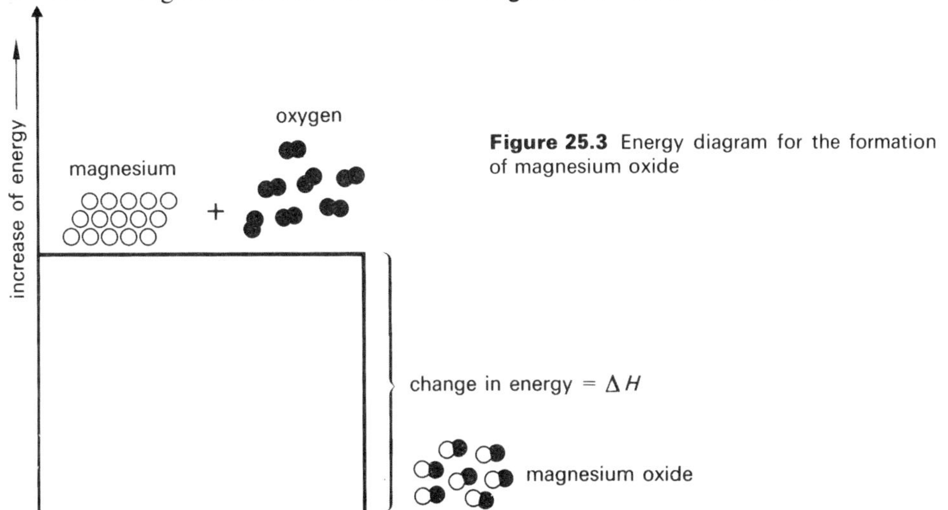

Figure 25.3 Energy diagram for the formation of magnesium oxide

when magnesium burns in oxygen, a considerable amount of energy is given out as heat and light. The reactants, magnesium and oxygen, thus clearly contain more energy than the product, magnesium oxide, and this 'extra' energy is given out during the reaction. As ΔH = energy of products − energy of reactants, its value in exothermic reactions (such as the combustion of magnesium) must be negative.

Remember: *endothermic reactions, ΔH positive,*
exothermic reactions, ΔH negative.

6. Most chemical reactions involve both bond-making (exothermic) and bond-breaking (endothermic) processes, and the overall ΔH for the reaction depends upon the difference in magnitude of each of the steps.

Common types of energy change

Latent heat of vaporization

When water is boiled, its temperature does not change even though it is still being heated. This is because the molecules are being 'pulled apart from each other' in the liquid state and changed into free molecules in the gas state. This involves breaking intermolecular bonds (bonds between molecules), and the process is thus endothermic, i.e. it requires heat energy. The energy needed to change liquid to vapour at the same temperature is called the *latent heat of vaporization*, and is often expressed in kJ per mole, i.e. kJ mol^{-1}. If the latent heats of vaporization of different liquids are compared (in kJ mol^{-1}), the figures are proportional to the strength of the intermolecular forces present. For example, the molar heat of vaporization of water is 41.2 kJ mol^{-1}, and that for tetrachloromethane, CCl_4, is 30.5 kJ mol^{-1}. This means that more energy is needed to separate 6.02×10^{23} molecules of water at its boiling point than is needed to separate 6.02×10^{23} molecules of tetrachloromethane at its boiling point. Water has a high latent heat of vaporization because it has relatively strong intermolecular forces (see page 281), which require a lot of energy to break them. Similar considerations apply to the *latent heat of fusion*, when a liquid solidifies or a solid melts.

Heat of reaction

The 'heat of reaction' is the heat change associated with a particular chemical reaction, the reaction normally being stated in the form of an equation. Always look at the units of ΔH carefully. The heat change is normally stated as that for the molar quantities shown in the equation, e.g.

$$N_2(g) + 3H_2(g) \rightarrow 2NH_3(g); \qquad \Delta H = -92.4 \text{ kJ}.$$

This means that when one mole of nitrogen reacts with three moles of hydrogen to make two moles of ammonia, 92.4 kJ of heat energy are given out.

Heat of precipitation

This is the heat change which occurs when one mole of a substance is precipitated from solution, e.g.

$$AgNO_3(aq) + NaCl(aq) \rightarrow AgCl(s) + NaNO_3(aq); \qquad \Delta H = -50.16 \text{ kJ mol}^{-1}$$

50.16 kJ is the heat of precipitation (of one mole) of silver chloride.

Such reactions are exothermic because strong bonds are being made (an ionic lattice) and all bond-breaking processes have occurred before reaction, when the reagents were dissolved in water. This is an example of where an *ionic equation* shows the only *chemical* reaction actually taking place,

$$Ag^+(aq)+Cl^-(aq) \rightarrow AgCl(s)$$

Heat of combustion

The 'heat of combustion' is the heat change when one mole of a substance (in its normal state, i.e. in the form in which it is normally found at room temperature and pressure) is completely burned in oxygen. This information is of particular importance because we rely upon the combustion of fossil fuels for most of our energy needs.

Heat of solution

The heat of solution is the energy change which occurs when a solid dissolves in water, but such energy changes are not measured precisely at an elementary level because the heat change depends upon not only the amount of solid being dissolved, but also on the volume of solvent it is dissolved in. It is important, however, to understand the processes which occur when a solid dissolves, and how these can give rise to endothermic and exothermic heats of solution. This is considered in the next section.

Heat of neutralization

This is the heat change which occurs when a mole of water is made by the neutralization of either hydroxide ions or oxide ions (from a base) or hydrogen ions from an acid, e.g.

$$H^+(aq)+OH^-(aq) \rightarrow H_2O(l)$$

or

$$2H^+(aq)+O^{2-}(aq) \rightarrow H_2O(l)$$

Check your understanding

1. What is meant by the terms exothermic and endothermic reactions? Illustrate your answer by means of energy diagrams.

2. Name six forms of energy. For each of the forms you give, mention one example.

3. For the reaction

$H_2(g)+Cl_2(g) \rightarrow 2HCl(g)$, $\Delta H = -182$ kJ

This means

A the heat absorbed when 1 mole of hydrogen chloride is formed from its elements is 91 kJ

B the heat evolved when 1 mole of hydrogen chloride is formed from its elements is 91 kJ

C the heat of reaction of hydrogen and chlorine is 91 kJ

D the heat energy of hydrogen chloride is greater than for the heat energies of one molecule of hydrogen and one molecule of chlorine

4. Draw energy level diagrams for the following reactions.

(a) $P_4(s)+5O_2(g) \rightarrow P_4O_{10}(s)$ $\Delta H = -3005$ kJ

(b) $B_2H_6(g)+3O_2(g) \rightarrow B_2O_3(s)+3H_2O(g)$ $\Delta H = -2040$ kJ

(c) $N_2(g)+2O_2(g) \rightarrow 2NO_2(g)$ $\Delta H = +66$ kJ

5. State which of the following reactions are exothermic and which are endothermic.

(a) $H_2(g)+I_2(g) \rightarrow 2HI(g)$ $\Delta H = +50.1$ kJ

(b) $CO(g)+H_2(g) \rightarrow H_2O(g)+C(s)$ $\Delta H = -129$ kJ

(c) $Ag(s)+\frac{1}{2}Cl_2(g)+aq \rightarrow AgCl(s)$ $\Delta H = -61$ kJ

(d) $N_2(g) \rightarrow 2N(g)$ $\Delta H = +470$ kJ

If these reactions were carried out in heat insulated containers would the temperature in each container, i.e. of the contents, be higher or lower after the reaction was complete?

6. Explain the following. The heat of neutralization per mole of water formed is the same for many neutralizations, but the heat of neutralization per mole of hydrochloric acid is only half of the value per mole of sulphuric acid.

25.2 A MORE DETAILED LOOK AT SOME ENERGY CHANGES

Some general experimental points

1. When a temperature change takes place during a chemical reaction, both the chemicals *and* the container are affected. In calculating the energy given out (or absorbed) during the reaction, we must therefore take into account the energy given out (or absorbed) by both the reacting substances and the container.
2. The heat energy given out (or absorbed) by a substance when it changes temperature is given by:

Heat energy lost or absorbed = mass × specific heat capacity × temperature change (often abbreviated $mc\theta$)

The mass is measured in grams, the specific heat capacity in J g^{-1} $°C^{-1}$, and temperature in °C.
3. The specific heat capacity of dilute solutions in water (e.g. a dilute solution of copper(II) sulphate) is usually assumed to be the same as that of pure water, i.e. 4.18 J g^{-1} $°C^{-1}$. This introduces a slight error into the calculations, but the error is insignificant when compared to those arising from other practical procedures, which are summarized under the next heading.
4. If a polythene bottle or waxed-paper cup is used as a container, it acts as an effective insulator (if closed) against heat losses to the surroundings. In addition, containers like these have a negligible heat capacity. This means that we can ignore any heat energy which such a container will lose or gain during a reaction. In such cases we can restrict the calculations to the heat energy absorbed or lost by the solution only, which greatly simplifies the calculations.
5. The mass of 1 cm^3 of water is 1.0 g, and in elementary calculations it is assumed that the mass of 1 cm^3 of solution is also 1.0 g. This again introduces a slight but insignificant error into the calculations but makes the experimental work easier.

Some other errors in simple practical work on energy changes

1. As all energy changes obey the Law of Conservation of Energy, then theoretically the energy content of the system at the start of the reaction should equal the energy content of the system at the end, but in practice there will always be some energy lost to (or gained from) the surroundings. This factor has been ignored in the simple calculations considered in this chapter, because heat transfers to or from the surroundings have been kept to a minimum. This has been achieved by using containers which are themselves good insulators or which are lagged by insulating material such as cotton wool, and by choosing reactions in which the temperature of the system does not rise far above room temperature. Reactions which produce temperatures well above room temperature, for example, will obviously result in high heat losses to the surroundings.
2. Thermometers are often read inaccurately, especially if graduated in intervals of 1 °C only.
3. When solutions are mixed, each solution dilutes the other. In some cases this dilution effect causes a temperature change. This effect has been ignored in the practical work and calculations in this chapter.

Check your understanding

(The specific heat capacity of water is 4.18 J $g^{-1}\,°C^{-1}$.)

1. How many joules of heat are required to heat
(a) 100 cm^3 of water from 10 °C to 20 °C,
(b) 500 cm^3 of water from 20 °C to 50 °C,
(c) 1.0 dm^3 of water from 20 °C to 100 °C?

2. How many joules of heat are lost when
(a) 100 cm^3 of water cool from 50 °C to 20 °C,
(b) 500 cm^3 of water cool from 70 °C to 30 °C?

3. How many joules of heat energy are required to heat 100 cm^3 of water in a glass beaker weighing 50 g from 20 °C to 35 °C? The specific heat capacity of the beaker is 0.84 J $g^{-1}\,°C^{-1}$. (Assume that there are no heat losses to the atmosphere.)

A sample calculation to determine a heat of precipitation

Experiment 25.2 provides a way of finding the heat of precipitation of silver chloride. The following data are taken from such an experiment, and are used to show how the calculation is performed.

Suppose that 25.0 cm^3 of silver nitrate solution (concentration 0.5 mol dm^{-3}) are added to 25.0 cm^3 of sodium chloride solution (concentration 0.5 mol dm^{-3}) in a polythene bottle, and that there is a temperature rise of 3.0 °C. (The polythene bottle can be considered to have a negligible heat capacity.)

25 cm^3 + 25 cm^3 = 50 cm^3 solution.

Heat energy gained by solution in rising through 3 °C $= mc\theta = 50 \times 4.18 \times 3$ J (assuming that the heat capacity of the solution is equal to that of water, and that 1 cm^3 of solution has a mass of 1.0 g).

Heat energy gained by container is negligible as polythene has a very low specific heat capacity.

∴ Total heat energy gained by system $= 50 \times 4.18 \times 3$ J (ignoring heat losses to the surroundings).

The solution contains $\frac{25}{1000} \times 0.5$ mol of Ag^+ (or Cl^-), and therefore $\frac{25 \times 0.5}{1000}$ mol of silver chloride are precipitated = 0.0125 mol

∴ precipitation of 1 mole of silver chloride would produce

$$\frac{1}{0.0125} \times 50 \times 4.18 \times 3\ \text{J} = 50\,160\ \text{J} \quad \text{or} \quad 50.16\ \text{kJ}$$

The heat of precipitation of silver chloride = -50.16 kJ mol^{-1}.

Note: If the container was made of a material which has a significant heat capacity (e.g. glass), then the heat absorbed by the container must be included in the calculation.

Experiment 25.2
To determine the heat of precipitation of silver chloride***

Apparatus

Small polythene bottle of about 60 cm^3 capacity fitted with a bung carrying a 0–50 °C thermometer, 25 cm^3 measuring cylinder. Stoppered bottles of solutions of silver nitrate, ammonium chloride, potassium chloride, and sodium chloride, each of concentration 0.5 mol dm^{-3}. (Allow these solutions to stand overnight in the laboratory to attain room temperature.)

Procedure

(a) Make out a table for your results as shown at the top of p. 482.

(b) Measure out 25 cm^3 of silver nitrate solution and pour it into the polythene bottle. Insert the bung fitted with the thermometer and note the temperature of the solution. It may be necessary to invert the bottle in order to cover the bulb of the thermometer with liquid. Handle the bottle

	With KCl	*With NH_4Cl*	*With NaCl*
Total volume of solution (cm^3)	50	50	50
Final temperature (°C)			
Initial temperature (°C)			
Temperature difference			

so that the minimum quantity of heat is transferred from the hand to the liquid.
(c) Measure out 25.0 cm^3 of ammonium chloride solution and add it to the silver nitrate solution in the polythene bottle.

Quickly replace the bung in the bottle and shake *gently* to allow mixing of the solutions. Note the maximum steady temperature.
(d) Repeat procedures (b) and (c) using (i) potassium chloride and (ii) sodium chloride in place of the ammonium chloride.

Results
Record the temperatures in the table of results.

Conclusion
Using the worked example as a guide, use your results to calculate a value for the heat of precipitation of silver chloride. You should be able to do this for each set of results.

Points for discussion

1. The value for the heat of precipitation of silver chloride, as calculated from the experimental results, should be approximately the same whether you used sodium chloride, ammonium chloride, or potassium chloride. Explain why this is so.
2. Explain why ΔH (for the precipitation of silver chloride) is given a *negative* value.

Heats of combustion

Heats of combustion are particularly important, because they represent one way of comparing the heat energy produced when different fuels are burned. A more detailed discussion of this is given in Chapter 26. It is important to realize that not all the heat energy liberated by a combustion process can be made into *useful* energy, e.g. only about 40 per cent of the heat energy liberated by the combustion of a primary fuel in a power station is converted into useful electricity.

You may do or see an experiment in which you find the heat of combustion of an alcohol such as ethanol, using the apparatus shown in Figure 25.4. The burner containing the alcohol is weighed before and after burning so that the mass of alcohol which has been burned can be calculated. The rise in temperature of the apparatus enables us to calculate the heat output from the burning alcohol. In order to perform the calculation it is essential to know the specific heat capacities of the water, copper, and glass which form the apparatus, all of which absorb heat energy from the burning alcohol. This is normally simplified by calibrating the apparatus so that the heat capacity of the *apparatus as a whole* can be determined, and this single heat capacity is used in the calculation. It then remains to calculate the heat output when a mole of the alcohol is burned, and this is the heat of combustion of the alcohol.

Another much simpler (and less accurate) technique is to use the apparatus shown in Figure 25.5. The following results were obtained by using such an apparatus, and the calculation which follows shows how the results can be used to find the heat of combustion of the alcohol.

Mass of bottle + ethanol (initially) = 24.630 g
Mass of bottle + ethanol (finally) = 24.400 g
Mass of ethanol burnt = 0.230 g
Final temperature of water = 33.0 °C
Initial temperature of water = 18.0 °C

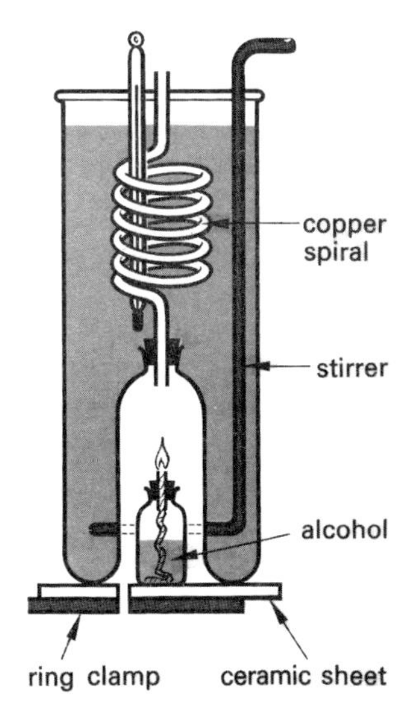

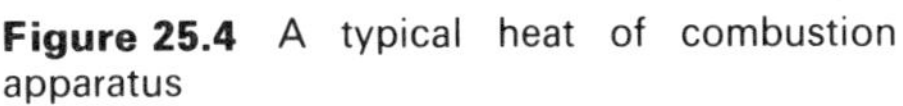

Figure 25.4 A typical heat of combustion apparatus

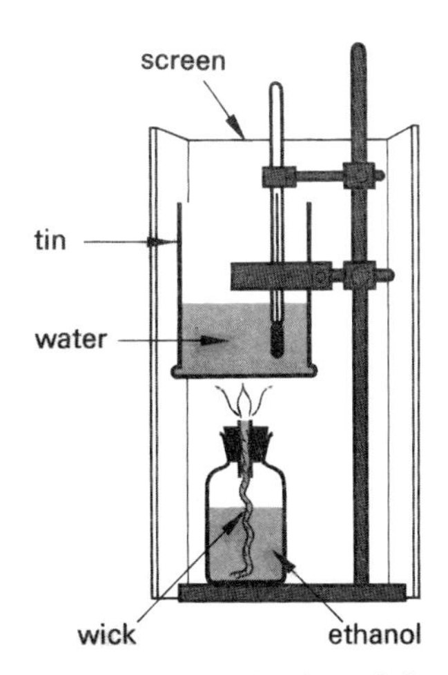

Figure 25.5 The determination of the heat of combustion of ethanol

Rise of temperature = 15.0 °C
Mass of water = 100 g
Heat evolved = Mass of water × specific heat capacity of water × rise of temperature
= 100 × 4.18 × 15 joules
= 6.27 kJ

Thus when 0.23 g of ethanol is completely burnt the heat evolved is 6.27 kJ. Therefore if 46 g ethanol (one mole) is completely burnt the heat evolved is

$$\frac{6.27 \times 46}{0.23} = 1254 \text{ kJ mol}^{-1}$$

The heat of combustion of ethanol is

$$\Delta H = -1254 \text{ kJ mol}^{-1}$$

Points for discussion

1. What is the main error involved in using the apparatus shown in Figure 25.5?
2. Why was the heat capacity of the tin can ignored in the calculation?
3. The heat of combustion is the heat change when one mole of a substance is completely burned in oxygen. Why is it not necessary to burn one mole of substance in order to find this value?
4. It was suggested earlier that the heat capacity of the complete apparatus as used in Figure 25.4 can be determined by first 'calibrating the apparatus'. Can you suggest how this might be done?

Check your understanding

1. When 4.0 g of sulphur is burned in a special calorimeter, the heat evolved raises the temperature of 500 g of water by 17 °C. Calculate the heat of combustion per mole of sulphur atoms.
2. The combination of sulphur dioxide and oxygen to produce sulphur trioxide is exothermic.

$$2SO_2(g) + O_2(g) \rightarrow 2SO_3(g), \ \Delta H = -187 \text{ kJ}$$

If 0.2 mole of sulphur trioxide is formed, what would be the heat energy evolved?
3. The formula for methanol is CH_3OH. It was found in an experiment to determine the latent heat of vaporization that to turn 8.0 g of methanol at its boiling point into vapour required 187 kJ of heat energy. What is the molar heat of vaporization of methanol?

Heats of neutralization

At an elementary level, it is usual to consider only the neutralization reactions between strong acids and strong alkalis in making these calculations. As strong acids and alkalis are completely dissociated in water (i.e. their molecules split up completely into hydrated ions when added to water), all of the *bond-breaking* processes have occurred before the chemicals react. The only change which then occurs during the neutralization is the *bond-making* process by which $H^+(aq)$ ions from the acid react with $OH^-(aq)$ ions from the alkali to form water molecules. The other ions present are merely spectator ions.

Note: In calculations be wary of reactions involving sulphuric acid, e.g.

$$H_2SO_4(aq) + 2NaOH(aq) \rightarrow Na_2SO_4(aq) + 2H_2O(l)$$

In this reaction, *two* moles of $OH^-(aq)$ ions are neutralized by *each* mole of sulphuric acid. Make sure that ΔH of neutralization is expressed *per mole* of $OH^-(aq)$ (or $H^+(aq)$ or H_2O). The heat change for the quantities shown in the equation above is -115.0 kJ, but the heat of neutralization per mole of $OH^-(aq)$ or $H^+(aq)$ or H_2O is -57.5 kJ.

A typical calculation on heat of neutralization can be summarized as follows.

Suppose that 50 cm^3 of hydrochloric acid has been neutralized by 50 cm^3 of sodium hydroxide solution, both having a concentration of 2.0 mol dm^{-3}. The temperature *change* in the reaction is θ °C, and the container has negligible heat capacity and is well insulated.

Total volume of solution = 100 cm^3.

Total heat change (ignoring heat capacity of container and heat losses) = $mc\theta$ (where m is the mass, c the specific heat capacity and θ the temperature change, of the reactants) $= 100 \times 4.18 \times \theta$ joules $= 4180\theta$ joules.

50 cm^3 of hydrochloric acid of concentration 2.0 mol dm^{-3} contain $\frac{2 \times 50}{1000}$ mol $= 0.1$ mol.

Each mole of hydrochloric acid provides 1 mole of $H^+(aq)$ ions for neutralization.

$\therefore$ neutralization of 0.1 mol of $H^+(aq)$ liberate 4180θ joules.

$\therefore$ neutralization of 1.0 mol of $H^+(aq)$ liberate $\frac{4180\theta}{0.1}$ joules $= 41800\theta$ joules.

Heat of neutralization of hydrochloric acid, ΔH, $= -41800\theta$ J mol^{-1}.

Experiment 25.3
To determine some heats of neutralization*

Apparatus

Wax cup 'calorimeter' in suitably sized beaker, cotton wool, thermometer 0–100 °C, copper wire stirrer, 50 cm^3 measuring cylinder.

Solutions of sodium hydroxide, potassium hydroxide, hydrochloric acid, and sulphuric acid, each of concentration 2.0 mol dm^{-3}.

Procedure

(a) Set out a table for your results as shown at the top of the next page.

(b) Pad the outside of the cup with cotton wool, and place it in the beaker.

(c) Measure out 50 cm^3 of sodium hydroxide solution into the calorimeter. Place the thermometer in the calorimeter and note the temperature.

(d) Measure 50 cm^3 of hydrochloric acid into the measuring cylinder. (If the cylinder is used to measure out the sodium hydroxide it should be washed out with water and then rinsed wtih hydrochloric acid before measuring the 50 cm^3 of acid.)

	Hydrochloric acid and sodium hydroxide	*Hydrochloric acid and potassium hydroxide*	*Sulphuric acid and sodium hydroxide*	*Sulphuric acid and potassium hydroxide*
Total volume of solution (cm^3)				
Final temperature (°C)				
Initial temperature (°C)				
Temperature change				

(e) Pour the acid into the calorimeter, stir and note the highest steady temperature.
(f) Repeat the experiment using other combinations of acid and alkali as shown in the table headings in (a), but using 25 cm^3 of sulphuric acid instead of 50 cm^3.

Results
Complete the table of results and use your results to calculate the heat of neutralization in each experiment, in kJ per mole of water formed.

Conclusion
Make your own conclusion, based on the following points.
1. Your answers for each of the experiments should be roughly the same. Can you explain why the heat of neutralization in these experiments should be the same, no matter which combination of acid and base is used?
2. The accepted value for the heat of neutralization of any strong acid by any strong base is -57.5 kJ mol^{-1}. How close were your answers? Can you explain any discrepancies?
3. Why was it important to be careful when calculating the results from experiments involving sulphuric acid?

Point for discussion
Can you suggest why, if you spill a little concentrated sulphuric acid on your hand, you should immediately dilute it with a large quantity of running water rather than neutralize it with sodium hydroxide solution?

Heats of neutralization involving weak acids and bases
When a *weak* acid or base is involved in a neutralization, the heat of neutralization is not necessarily -57.5 kJ mol^{-1} of water formed. This may surprise you, because we are still measuring the heat change when a mole of water is formed by the reaction between $H^+(aq)$ ions and $OH^-(aq)$ ions, just as we are in the calculations with strong acids and bases. However, there are two additional factors which contribute to the overall heat change when a weak acid or alkali is used.

1. A weak acid or alkali is *not* fully dissociated in solution; some of its molecules remain as dissolved but undissociated molecules. In order for a mole of $H^+(aq)$ or $OH^-(aq)$ ions to be made available for neutralization, the undissociated molecules must 'break up' to release more ions, i.e. further bond breaking must take place, and this requires energy. This does not occur when solutions of strong acids or alkalis are used, because they are already fully dissociated.
2. The ions released by the bond-breaking process in 1 then react with water molecules to become hydrated, i.e. some bonds are then *made*, and this step is exothermic. Again, this step does not affect the calculations involving only strong acids and bases because it has already occurred when the solutions are provided.

Thus the energy change involving, for example, the neutralization of the weak acid ethanoic acid by sodium hydroxide solution, is not just the energy change associated with the neutralization reaction

$$H^+(aq) + OH^-(aq) \rightarrow H_2O(l)$$

This basic reaction is further complicated by additional bond-breaking and bond-making processes. Bonds in the undissociated molecules of ethanoic acid have to be broken, and then the ions so formed become hydrated. It happens in this particular example that the energy changes involved in these two additional processes more or less cancel each other out. The heat of neutralization is -55.2 kJ mol^{-1}, which is very close to the value for strong acids and bases. This is summarized in Figure 25.6.

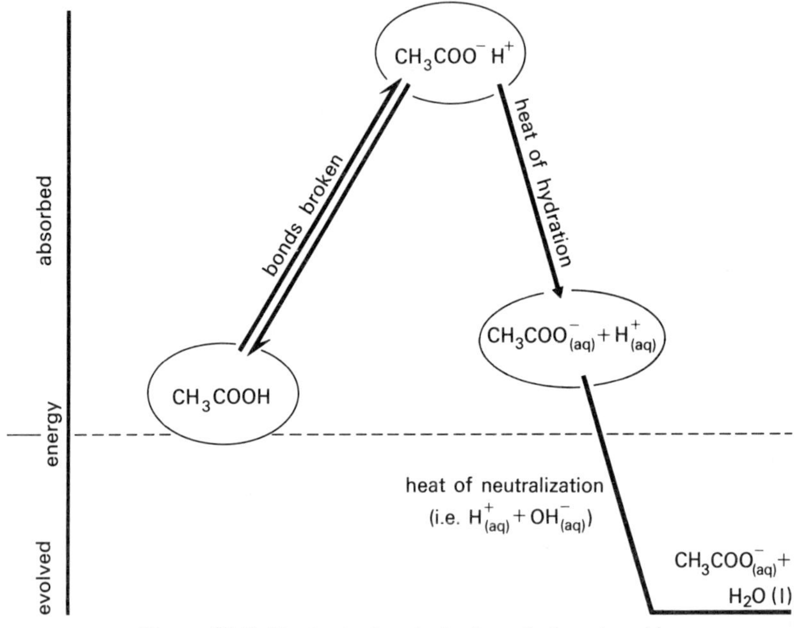

Figure 25.6 The heat of neutralization of ethanoic acid

Heats of solution

As explained earlier, it is not usual to express heats of solution in a quantitative way at an elementary level. It is important, however, to *understand* the processes which occur when a solid dissolves in water, and the *type* of heat change which each stage produces. The following experiment provides information with which to discuss these points.

Experiment 25.4
The energy changes which occur when a solid dissolves in water*

Apparatus
Small polythene bottle of about 60 cm^3 capacity fitted with a rubber bung carrying a thermometer (0–50 °C), spatula.
Anhydrous and hydrated copper(II) sulphate, ammonium nitrate or ammonium chloride, deionized water.

Procedure
(a) Make out a table for your results as shown at the top of the next page.

	Anhydrous copper(II) sulphate	Hydrated copper(II) sulphate	Ammonium nitrate (or chloride)
Final temperature (°C)			
Initial temperature (°C)			
Temperature change (rise or fall)			
Type of overall energy change (endothermic or exothermic)			

(b) Half fill the polythene bottle with deionized water.
(c) Replace the bung and record the temperature of the water.
(d) Remove the bung, quickly add about four spatula measures of anhydrous copper(II) sulphate. Replace the bung and after gently shaking the bottle, record the temperature.
(e) Repeat the procedure with (i) hydrated copper(II) sulphate and (ii) the ammonium salt as the solute.

Results
Complete the table of results.

Notes about the experiment
When a solid dissolves in water, two opposing energy changes always take place.
(i) The particles in the solid separate from each other; this separation involves the breaking of bonds and so is endothermic.
(ii) The free ions or molecules formed in (i) then *react* with water molecules to become hydrated. This is a bond-making process, which is therefore exothermic.
Note: Most ions are hydrated in solution, and this is what the state symbol (aq) means. The $Cu^{2+}(aq)$ ion is really $[Cu(H_2O)_6]^{2+}$, because each copper(II) ion usually bonds with six water molecules when it becomes hydrated.

Some substances will dissolve in water endothermically, because the energy change in (i) is greater than that in (ii). On the other hand, some solids dissolve exothermically in water because the energy change in step (ii) is greater than that in step (i). Make sure that you understand these points.

Conclusion
Use your results and the information above to explain what happened in the experiment. Note in particular that you have to explain why two different forms of copper(II) sulphate behave differently (in terms of heat changes) when they dissolve in water. (Do not make the mistake of thinking that the ions in the crystals are completely hydrated. In the crystal lattice the ratio of $Cu^{2+}:H_2O$ is 1:4, whereas in free hydrated ions of copper the ratio is 1:6. It is not possible to get any more water into the crystals without them dissolving, but the ions are further hydrated when the solid dissolves.)

Points for discussion
1. Why do some solids dissolve in water with almost no temperature change?
2. What can you conclude about the processes which take place when concentrated sulphuric acid dissolves in water (Experiment 25.1, page 473)?

25.3 A SUMMARY OF CHAPTER 25

1. Some of the important points
The following 'check list' should help you to organize the work for revision.
(a) All forms of energy are interchangeable. Many chemical reactions produce (or absorb) heat energy, but they may also produce electrical energy (p. 261) or light energy.
(b) Heat changes are measured in units of kJ mol^{-1} or kJ for the molar quantities given in an equation.
(c) When heat energy is given out in a chemical reaction, the substances and their container rise in temperature and the reaction is said to be exothermic.

(d) When heat energy is absorbed in a chemical reaction, the substances and their container fall in temperature and the reaction is said to be endothermic.
(e) Even 'cold' substances contain energy, in several ways, e.g. in the bonds which hold particles together and in the kinetic energy of the particles.
(f) Energy is required to break bonds; bond breaking is endothermic.
(g) Energy is released when bonds are made; bond making is exothermic.
(h) We only measure energy *changes* during a reaction, not the actual energy content of a particular chemical.
(i) Energy changes are given the symbol ΔH. ΔH = energy of products − energy of reactants; the sign is negative if the overall reaction is exothermic, and positive if the overall reaction is endothermic.
(j) You should be familiar with the names given to different kinds of energy changes, as on page 478.
(k) You should understand the assumptions which are made in calculations on energy changes (p. 480), and also the other sources of error in the practical work.
(l) How to calculate a heat of precipitation, a heat of combustion, and a heat of neutralization from experimental data. Note that heat of neutralization is measured per mole of $H^+(aq)$ or $OH^-(aq)$ used, or per mole of H_2O molecules formed, and this is particularly important when sulphuric acid is involved in a calculation.
(m) You should understand that the heat of neutralization for any reaction between a strong base and a strong acid is always the same.
(n) You should understand the types of reaction (both bond-making and bond-breaking) which occur when a solid dissolves in water, and how these affect the overall heat change which takes place.

2. *Important experiments*

(a) An experiment to determine a heat of precipitation, e.g. Experiment 25.2.
(b) An experiment to show the energy changes which take place when a solid dissolves in water, e.g. Experiment 25.4.

QUESTIONS

1. Explain, with examples, the meaning of (a) kinetic energy, (b) potential energy, (c) chemical energy.

2. Explain clearly why ΔH is given a *negative* sign for an exothermic reaction.

3. Why is more energy evolved when hydrogen and oxygen combine to form a mole of liquid water, than when they form a mole of steam?

4. The equation for the reaction in which magnesium burns to form magnesium oxide suggests that the reaction takes place in one step. In fact, several steps are involved. Explain these statements.

5. Substance A (which consists of A_2 molecules) reacts with a substance B (which consists of B_2 molecules) to form molecules of the substance AB. A—A bonds are about as strong as B—B bonds, but A—B bonds are stronger than either. For the reaction

$$A_2(g) + B_2(g) \rightarrow 2AB(g)$$

(a) state whether the reaction will be exothermic or endothermic, and explain your reasoning.
(b) draw an energy level diagram for the reaction.

6. How would you expect the heat of solution of anhydrous sodium carbonate, Na_2CO_3, to compare with that of hydrated sodium carbonate, $Na_2CO_3.10H_2O$?

7. Why is the heat change due to the following reaction not called a standard heat of combustion?

$$2C(s) + O_2(g) \rightarrow 2CO(g)$$

8. Define the term heat of neutralization. When a dilute solution of a strong acid is neutralized by a solution of a strong base, the heat of neutralization is found to be nearly the same in all cases. How can you explain this? How can you explain the cases where there is a difference?

9. (a) How does (i) the motion of the molecules, and (ii) the spacing between the molecules differ in the three states of matter, viz. solid, liquid, and vapour?

(b) How and why does atmospheric pressure affect the boiling point of a liquid?
(c) (i) An ionic solid $(X^+Y^-)_n$ may have its crystal lattice broken down by two different physical processes. Give the names of these two processes and describe, for each of them, how the necessary energy to break down the crystal lattice is obtained.
(ii) For each of the processes named in (c) (i), state with a reason, whether you would expect that process always to be endothermic. (J.M.B.)

10. Propane, C_3H_8, is a gas at room temperature. It burns in excess air or oxygen forming carbon dioxide and water vapour and releasing 2200 kJ mol^{-1} of propane.
(a) Write the equation for the reaction and calculate the energy released when 1 gram of propane burns in this way.
(b) Sketch an apparatus by which you could burn propane (obtained from a cylinder) and collect the water formed. State *two* tests (one of them chemical) by which water may be identified.
(c) What do you understand by the term *diffusion* of gases? Briefly describe one experiment that illustrates gaseous diffusion.
(d) Under the same experimental conditions, why do propane and carbon dioxide diffuse at the same rate? (A.E.B. 1977)

26 Fuels

26.1 SOME BASIC IDEAS ABOUT FUELS

What are fuels?

A fuel is a substance used to provide energy. In the previous chapters, we have stressed the increasing demand for energy, in the home, in industry, and for transport. Whether the increasing demand is really *necessary* is another matter, as is discussed in section 26.2.

Most fuels are burned, e.g. coal, natural gas, and paraffin. The energy produced in the combustion process is used either directly as heat, converted into more convenient forms of energy such as electricity, or used to drive machinery, e.g. in the internal combustion engine by 'exploding' petrol and air. The carbohydrates and fats that we eat as part of our diet are fuels because they provide us with energy. Most fuels contain compounds of carbon and hydrogen, and when they burn carbon dioxide and steam are usually produced, as well as heat energy. The production of this heat energy by the making and breaking of chemical bonds has been explained on page 475.

The chemical energy within these fuels has normally been obtained from the energy of the sun (e.g. coal and oil have been produced indirectly by photosynthesis) and we are re-using this energy when we burn the fuels. Some of the relationships between fuels are summarized in Figure 26.1.

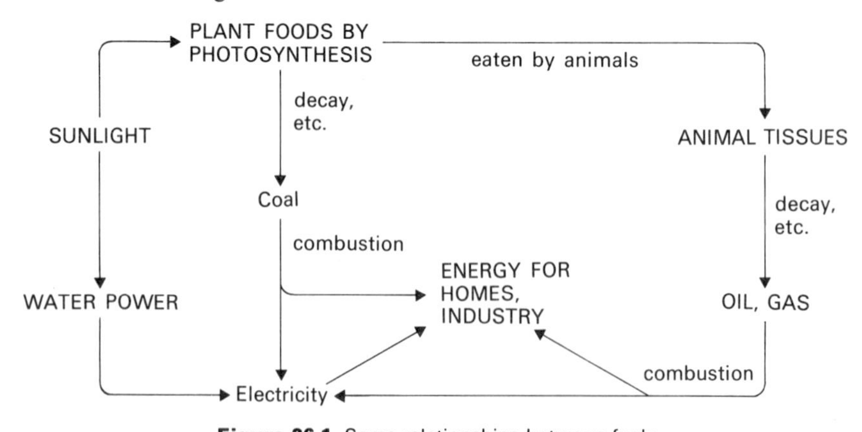

Figure 26.1 Some relationships between fuels

Choosing a fuel

Any chemical reaction which gives out energy is a *potential* fuel system. There are thousands of exothermic reactions which could in theory be used in this way. For example, many substances can burn in air and when they do so they give out energy in the form of heat and light, i.e. the heat of combustion has a negative value (p. 477).

However, very few substances are actually used as fuels. There is more to choosing a fuel than simply deciding whether it is capable of burning in air.

Some considerations in choosing a fuel

These are some of the questions we should ask about a substance when considering it for use as a fuel.

1. Does the substance occur naturally or is it made artificially? (If the latter, is it expensive to make, and do we use up energy and important materials in making it?)
2. If it occurs naturally, does it occur in large deposits (which are easy to develop) or is it scattered in small, well-separated deposits (which may be more difficult to exploit)?
3. If it occurs naturally, is there a plentiful supply?
4. Can it be transported and stored easily?
5. Does it contain impurities which, if burned, might produce toxic chemicals and pollute the atmosphere? If the answer is yes, would it be costly to purify the fuel before it is burned? Alternatively, can the impurity (once removed) be sold as a useful byproduct, in order to reduce overall costs?
6. Does the fuel produce *solid* waste when it burns? If so, could this 'waste' be used or would it be difficult to dispose of?
7. Does the potential fuel have a high heat output (i.e. a high 'calorific value')? In other words, does a given mass (or volume) of it produce as much energy as (or more energy than) an equal mass or volume of other fuels when it is completely burned? (This point is discussed in more detail on page 492.)
8. Is the fuel itself toxic or difficult to handle? (For example, the fuel used in nuclear power stations is toxic and radioactive. In spite of this 'handicap', nuclear power is used because it is considered to have other factors in its favour.)
9. If the fuel occurs naturally, does it occur in areas which are easy to develop, e.g. near to transport services? (Deposits of a fuel discovered in the Antarctic would be expensive to exploit.)
10. Is the fuel easy to ignite, or does it need to be heated before it is ignited?
11. How widely can the fuel be used? Can it be used only in special circumstances (e.g. a rocket fuel) or are its applications many and varied?

Additional points about choosing a fuel

If a source of energy is to be used on a domestic scale, it should be possible to transport it to every house which needs it (point 4 above). Compare, for example, the convenience of gas, electricity, coal, and oil. Supplies of the first two may be connected directly to a house and are instantly available at the turn of a switch or tap. The other two have to be ordered, delivered, and then stored. However, inconveniences of this type may be outweighed by other advantages.

Similarly, it is more convenient to be able to feed electricity and gas round the country by a system of pipes or cables, whereas other sources of energy are dependent upon rail or road transport, which adds to their cost. The network of pipes and cables is 'permanent'. Although additions or modifications may be required to meet increases in demand, the cost of this is small compared to the cost of new vehicles, depots etc. needed to deal with an increased demand for coal.

Point 5 above is clearly very important. All of the fossil fuels contain sulphur formed from the decay of living cells. This must be removed from the fuels if possible, so as to reduce the quantities of sulphur dioxide liberated on combustion. The desulphurization of oil has been explained on page 144, and that of coal is referred to on page 502.

Some quite spectacular illustrations of point 6 (p. 491) can be made in the laboratory. For example, if a *small* quantity of ammonium dichromate(VI), $(NH_4)_2Cr_2O_7$, is heated gently in a crucible in a fume cupboard, a reaction takes place in which a considerable amount of energy is given out. However, the reaction also produces a green solid which takes up more volume than the original chemical. Both the solid product (chromium(III) oxide, Cr_2O_3) and the reactant are toxic, and the reaction takes place very rapidly; it cannot be controlled. Ammonium dichromate(VI) is therefore not suitable for use as a fuel.

Point 7 (p. 491) can be confusing. The heat energy produced by the burning of a certain quantity of fuel is still sometimes described as its calorific value, even though chemists no longer use the calorie as a unit of energy. We measure heat changes in kJ per mole of fuel, because we are interested in comparing an equal number of particles of fuel each time. Such units are not convenient in industry, however, because fuels have to be transported and the cost of transport depends upon how dense they are. It would not be appropriate in industry to compare the heat given out by a mole of natural gas (which would take up a volume of approximately 24 000 cm^3 at room temperature and pressure) with that given out by a mole of carbon (in coal) which would take up a tiny volume. For this reason the 'calorific' values of solid fuels are usually compared per unit mass, e.g. per tonne, and those of gases by unit volume, e.g. per m^3. Note also that the units quoted by industry involve much greater quantities than simple moles, because fuels are being handled on a much larger scale than in the laboratory. Another reason for using these units is that industrial fuels are not usually pure substances. For example, coal is not pure carbon and so we cannot refer to a mole of ordinary coal.

Although industrial units of energy are not the same as the ones we use in the laboratory, we should be familiar with some of them because they are used in daily life. When we buy a gas fire or greenhouse heater its heat output is normally rated in B.T.U. per hour. A British Thermal Unit (B.T.U.) is the quantity of heat required to raise the temperature of one pound of water by 1 °F, and one B.T.U. is approximately equal to 1.05 kJ. Similarly, gas bills are calculated according to the price of a Therm (1 Therm = 100 000 B.T.U.).

As an illustration of point 11 (p. 491), technological and safety factors restrict the use of nuclear fuels to large scale applications (power stations) except in limited cases like nuclear submarines, satellites, and similar very specialized items. Similarly, fuel oil is burned in power stations but is quite unsuitable as a domestic fuel because it is difficult to ignite. Many fuels are converted into electricity, which is almost universal in its applications, but it is at the moment impractical for large scale or long distance road transport because of difficulties in providing suitable storage (batteries).

Most of these arguments really add up to one single factor, i.e. cost. It is this which has the biggest impact because it not only influences our fuel bills but has an effect on the price of almost anything we buy. We can expect fuel costs to continue to rise, and there will also be fluctuations in the relative cost of individual fuels according to supply and demand. When a house is built, very careful consideration has to be given to the kind of heating system which is to be used; a method which is economical at the present time may prove to be very expensive in ten years' time. Similarly, it is becoming increasingly important that houses should be well insulated to reduce heating costs and to conserve energy.

Points for discussion

1. Name three chemicals you have seen burn in the laboratory, and two substances (not common fuels) which you have seen burn in daily life. Compare and contrast their suitability as fuels.

2. Imagine asking the 11 questions listed earlier (p. 491), about the suitability of *oil* as a supply of fuel, first in 1900 and then today. What changes have taken place since 1900 which have resulted in different answers being given to some of the questions today?

3. Which of the questions referred to earlier might not have been asked in 1900?

4. Can you think of a common solid fuel which is comparatively difficult to ignite (see question 10, p. 491)?

5. Can you think of an *advantage* of extracting a newly discovered reserve of fuel in the Antarctic?

6. List some fuels which only have highly specialized uses and name these uses.

7. In what units is the consumption of electricity measured? Is there such a quantity as a 'mole of electricity'?

8. Up to 1960, most homes were heated by solid fuel, even when they had central heating. Since then there has been a gradual move away from solid fuel to oil and gas.

(a) Suggest reasons for the change.

(b) Do you think that the trend will continue? Give reasons for your answer.

Check your understanding

1. Both gaseous methane (CH_4) and gaseous propane (C_3H_8) can be used as fuels.

(a) Write equations for the complete combustion of (i) methane and (ii) propane in oxygen.

(b) Indicate whether these reactions would be

A both endothermic

B both exothermic

C (i) endothermic and (ii) exothermic

D (i) exothermic and (ii) endothermic

(c) What volume of oxygen will just react completely with 10 litres of propane? (Assume that all volumes are measured at the same temperature and pressure.) (J.M.B.)

2. The following equations represent the combustion of hydrocarbon gases. The enthalpy changes (ΔH values) are those for the equations as written.

$$CH_4 + 2O_2 \rightarrow CO_2 + 2H_2O \quad \Delta H = -880 \text{ kJ}$$

$$C_3H_8 + 5O_2 \rightarrow 3CO_2 + 4H_2O \quad \Delta H = -2200 \text{ kJ}$$

$$2C_4H_{10} + 13O_2 \rightarrow 8CO_2 + 10H_2O \quad \Delta H = -5750 \text{ kJ}$$

(Relative atomic masses: A_r (H) = 1, A_r (C) = 12. One mole of gas at room temperature and pressure occupies 24 dm^3.)

(a) Calculate the amount of heat liberated by the complete combustion of 1 g of methane.

(b) Calculate the heat liberated by the complete combustion of 1 dm^3 of methane, measured at room temperature and pressure.

(c) Explain briefly why these gases are often described as 'clean' fuels.

(d) Large quantities of carbon dioxide enter the atmosphere from the burning of carbon fuels. Explain briefly why the proportion by volume of carbon dioxide in the air is generally less than 0.05 per cent.

(e) Give one reason why ethene, C_2H_4, would be unsuitable for use as a fuel.

26.2 PROBLEMS OF THE PAST, PRESENT, AND FUTURE

In this chapter we will discuss some of the ways in which energy is currently provided, some of the steps which are being taken to preserve our stocks of fossil fuels, and some sources of energy which are being developed for the future. Before the more scientific aspects are considered, we must appreciate that wasteful use of energy cannot be allowed to continue and that personal attitudes are extremely important.

The increasing demand for energy

Until the Industrial Revolution, man relied upon the horse and sailing ship as the main modes of transport. The materials he worked with and used in everyday life were made by hand or by simple machines. Nowadays, almost everything we in the developed countries use requires some form of external energy, and rapid personalized transport is considered to be essential.

Just as important is the fact that almost everything used in the developed countries also requires energy to *make* it. Think of the uses of steel, for example, and remember that to make the iron in a blast furnace and then to convert it into steel machined to a particular specification, a large amount of energy is required. It has been calculated that on average each person in the UK uses every year an amount of energy equivalent to that contained in 6 tonnes of coal, and this does not include the energy needed to make the material things we buy.

It is difficult to appreciate the great changes which have taken place in a comparatively short time in the history of mankind; we have moved from dependence on the horse to being able to land on the moon. In doing so our appetite for energy has increased enormously. Some people have come to rely on energy so much that they are prepared to pay a high price for it, and steep rises in fuel prices have had little effect on the often wasteful and prolific way in which energy is used in some parts of the world. We cannot and should not continue to use energy for non-essential purposes.

It is often assumed that the consumption of energy will continue to increase. This need not be the case, but if it is to be avoided there must be a change in attitude towards the use of energy. As we have stressed in earlier chapters, currently usable supplies of energy have their limits, and we must consider the problem as being very serious. A better attitude towards energy conservation, particularly in the developed countries, would help to preserve our stocks of fossil fuels and give us more time for the development of alternatives.

Personal attitudes

In the past, fuels have not been used as efficiently as they might have been. People living 100 years ago did not appreciate how the demand for energy would increase, nor did they imagine that the apparently enormous stocks of coal and oil could be used up in a comparatively short time. It did not seem important that steam engines were not particularly efficient in their use of energy (early ones were only about 1 per cent efficient!), nor that most of the heat produced by the burning of coal in a domestic open fire was wasted up the chimney.

Most people now appreciate that every time we burn some fossil fuel we are using up in seconds a substance which has taken millions of years to form, and that such precious materials must be used in a responsible way.

Figure 26.2 shows how energy was used in the UK in 1978. The details may change slightly over the years, but the important factor illustrated by the data is that 26 per cent of the energy was used for domestic purposes. In other words, an important proportion of the total energy demand is that used by individuals in their homes and daily life. Our own contributions to the problem may *seem* insignificant, but the efforts of many individuals can make a massive contribution to energy saving. It has been estimated that we still waste so much energy in Britain that it would be possible to cut

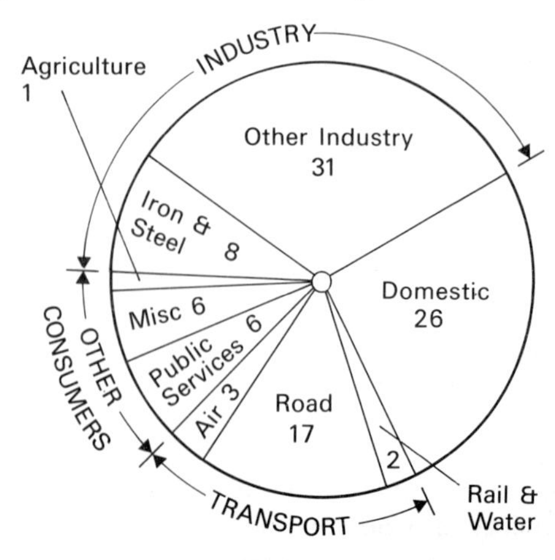

Figure 26.2 How energy was used in the UK in 1978 (*Information supplied by the Department of Energy*)

our overall consumption of energy by one-third without suffering any noticeable loss in living standards. This could be achieved by insisting on much higher standards for the insulation of houses, on more efficient use of energy in the home and industry, and making more use of 'waste' heat (e.g. warm water from power stations). Money spent on improving such factors is soon repaid by the saving in fuel bills. Those of us who live in the UK must be particularly wary about complacency, because for a short period of time we will be self-sufficient in energy, i.e. we will produce more (from North Sea oil, natural gas etc.) than we need. This luxurious state of affairs is misleading: before long, unless our attitudes change, the demand for energy in the UK will almost certainly exceed that which we can then provide ourselves, and it will become increasingly difficult to buy imported sources of energy.

Points for discussion

1. Make a list of ways in which you could save energy in your own home, in daily life, and in school.
2. Write an essay about a 'normal day' in the UK without coal, oil, or gas.
3. How can the fuel consumption of a car be improved by better driving habits?

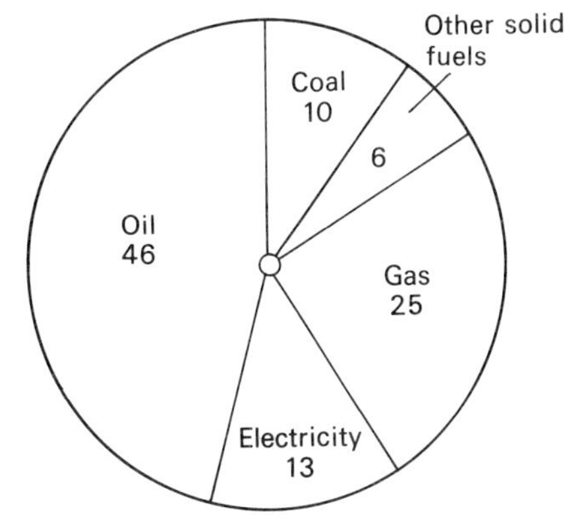

Figure 26.3 The consumption of individual fuels in the UK (1978) (*Information supplied by the Department of Energy*)

The future

Some basic facts about energy

Known reserves of oil and natural gas will not last for long into the next century, and yet in 1980 as much as 71 per cent of the energy consumed in Britain was provided by oil and gas (see Figure 26.3), and in addition some electricity was also produced from these fuels. At the present rate of use, Britain's coal reserves are expected to last until the year 2300 if used in Britain alone, and the total known reserves of coal will supply world-wide demand until the year 2500. However, if coal has to be used as an alternative to the rapidly diminishing supplies of oil and gas, as would seem to be inevitable, then coal stocks will be used up by a much earlier date. In addition, it is important to realize that although coal can be used to generate electricity and to provide heat, it is not a convenient source of other forms of energy needed for transport, and its use creates more pollution problems than are encountered with oil and gas. Other sources of energy are clearly needed.

Energy losses in generating electricity

Electricity is a classic and wonderfully versatile form of energy. It can produce heat, light, and power; it is clean, instant, convenient, produces no waste, is easily transported, and automatically available to almost every home. It allows appliances to be portable because of the use of sockets or batteries. Most of the sources of energy likely to be developed in the future are in fact alternative methods of generating electricity, and so it is important to understand a problem involved in the conversion of energy into electricity.

At the moment, most electricity is generated by using the energy of combustion of a fossil fuel to heat water and create steam, which in turn drives a turbine to make electricity. An important proportion of the heat energy obtained by burning a fuel is lost in its conversion into electricity and in supplying electrical energy to our homes and industry. One tonne of coal, 0.6 tonnes of oil, and 24 500 ft^3 of natural gas provide roughly the same amount of heat energy when burned at a power station, but only 40 per cent of that energy 'reappears' in the electricity they produce. As alternative ways of generating electricity are developed, therefore, it is important to ensure that the conversion of energy into electricity is efficient. Alternative methods will reduce the need to use our precious fossil fuels in this way, particularly when they can be used more efficiently for other purposes, such as producing the chemical raw materials for many industries.

The need to develop an alternative fuel for transport

There is another problem in connection with the use of energy in the future. We can regard electricity as the most likely source of heat in the future, and also as the source of many other forms of energy. Unfortunately, for all its versatility electricity is not at the present time suitable as a transport fuel except for short journeys (e.g. by battery driven cars and milk floats) and for rail networks. This may change in the future as better batteries or fuel cells (p. 506) are developed, but in the short term at least we must develop other fuels for transport to replace oil and petrol.

To illustrate the problem, in 1979 more than 35 per cent of the crude oil processed in the UK was used for transport. Even if all the diesel locomotives on the rail network were changed to electric locomotives that 35 per cent would only become 34 per cent! This is used entirely by cars, lorries, ships, and aircraft, and electricity cannot replace oil in these cases at present.

Much research at the moment is based on the development of hydrogen as a fuel for transport. One way of producing hydrogen is by electrolysis of 'water', but this is expensive because we are using energy (and an expensive form of energy) in order to produce the hydrogen. If electricity can be generated more efficiently, there is hope that its use to make hydrogen for transport would then be an economical proposition. It could also be piped to towns using the existing gas network. (Coal gas is about 50 per cent hydrogen, page 501). Fuel cells could then generate electricity and at the same time produce pure water (p. 506), which could be very useful in some parts of the world.

Research is also being conducted in the use of ethanol as a fuel for transport. An ethanol and water mixture can be produced fairly cheaply by fermentation, but until recently distilling the mixture so as to concentrate the ethanol for use as a fuel cost more in energy than was gained from the product. There is now hope that ethanol can be concentrated by a more efficient method.

Check your understanding

The American re-usable Orbiter Shuttle Spacecraft is powered for its final stage by three high pressure liquid oxygen–liquid hydrogen engines. The hydrogen and oxygen are burned together and this reaction gives the final thrust to boost the Orbiter to 18 000 m.p.h. and place it into orbit.

(a) Give the balanced equation for the burning together of hydrogen and oxygen.

(b) Calculate the mass of oxygen needed, and the mass of combustion products formed, when 20 tonnes of liquid hydrogen is combusted in an Orbiter Shuttle.

(c) Give two problems that would be encountered using liquid oxygen and liquid hydrogen in this situation.

(d) Why does this engine cause little atmospheric pollution? (A.E.B. 1979)

Using coal stocks in an attempt to preserve oil and natural gas

As already explained, we are still highly dependent on oil and gas and we are making a comparatively small demand on coal, of which we have greater stocks. The large proportion of oil used directly to provide heat energy could, in theory, be replaced by an increased use of coal and by the development of nuclear power and other sources of electricity. This would enable us to preserve our supplies of oil and to use them more efficiently. Also, if coal is processed as on page 500, some of the byproducts will help to meet the demand for transport fuels and organic chemicals.

Such a policy is fine in theory but not so easy in practice. It is almost impossible to extract coal at a rate sufficient for it to replace oil as a source of bulk heat. An increase in the rate of extraction of coal causes environmental problems, e.g. in developing new reserves in populated areas, and in dealing with the vast amounts of smoke involved in its processing and with the large volumes of sulphur dioxide formed by its combustion. (Coal contains, on average, more sulphur than either oil or natural gas.) A change over to a coal-based economy will also cause transport problems; it is far more difficult to transport a solid than it is to transport a liquid or a gas.

Points for discussion

1. Can you explain why it is 'almost impossible to extract coal at the same rate as oil is currently used to provide bulk heat'?

2. What advantages are there in processing coal in sites developed underground?

3. The transportation of large amounts of chemicals and raw materials is commonplace. However, safe handling and choice of method of transport are essential.

Complete the following table.

Chemical to be transported	*Method of transport needed*	*Two problems associated with transporting the chemical by this means*
Crude oil from Middle East to Europe		
A continuous supply of ethene gas from refinery to chemical plant 100 miles distant		
Liquid oxygen from a central depot to small users in a 50 mile radius		
100 tonne batches daily, of liquid chlorine to a plant 30 miles distant		

(A.E.B. 1977)

Discovering new stocks of oil and gas

The rate of discovery of new oil and gas fields has slowed down in recent years, but large sums of money are still being spent on attempts to discover new supplies, and there may well be further discoveries. As new reserves become more and more difficult to find, there is a distinct possibility that more energy could be spent on trying to make the discoveries than can be recovered from them!

26.3 FOSSIL FUELS

Fossil fuels are those which have been formed by the decay of once-living materials over a period of millions of years. They include crude oil, natural gas, and coal.

Oil

The origin of crude oil, and the petrochemicals industry, have been explained in Chapter 22, and throughout Chapters 22 and 26 we have stressed our dependence on oil and the need to provide alternatives.

Gas

Crude oil and natural gas are frequently found together, and this is because they are very similar in origin. Gas and oil are formed from the remains of marine creatures (e.g. plankton, Figure 26.4) which lived millions of years ago. These remains have been covered with layers of sediment which have been converted to rock. The remains went through a series of complicated changes produced by high temperatures, high pressures, and bacterial action. Natural gas consists mainly of one hydrocarbon, methane, whereas crude oil contains hundreds of hydrocarbons, from simple ones like ethane and propane (dissolved in the liquid) to those with much larger molecules. We can readily understand that methane is formed from the decay of animal material when we consider that a cow produces on average 500 000 cm^3 of methane per day and that suitable treatment of sewage produces large volumes of the gas.

The word gas is used to describe several different kinds of fuel, and so we should always make it clear as to which particular 'gas fuel' we are referring to. For example, in America 'gas' is used as an abbreviated form of 'gasoline', which in Britain is called petrol. In Britain gas might mean natural gas (North Sea gas), coal gas, or the bottled gas supplied in cylinders. Bottled gas can contain any of several hydrocarbons, e.g. propane, butane, or a mixture of the two, which are liquefied under pressure and stored in pressurized containers. These hydrocarbons are produced by the refining

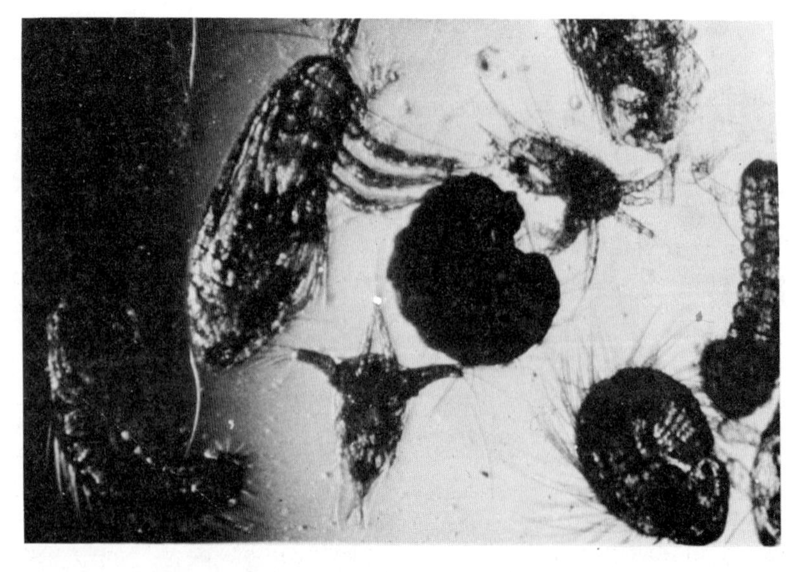

Figure 26.4 Plankton from the North Sea (*Shell*)

of crude oil. Although more expensive, volume for volume, than piped gas, bottled gas is easily portable and is used for heating and lighting in camping, caravans, portable blow lamps, etc.

'But you said fill it with gas'

Nowadays, the gas supplied to homes in Britain is natural gas, whereas until the early 1970s it was coal gas. Coal gas was the first gas used on a large scale as a fuel, and it is produced from coal (p. 501). Natural gas has several advantages over coal gas, e.g. it is not poisonous (coal gas is) and it has a higher 'calorific' value. In addition, natural gas does not have to be manufactured and it requires very little processing; it is only treated to remove sulphur compounds (their concentration is usually very low) and a special substance is added to give it a smell so that leaks are more easily detected. On the other hand, it cost a great deal of money to convert gas appliances so that they were suitable to burn natural gas instead of coal gas.

At the moment piped gas is regarded as a *premium* fuel, i.e. an important and clean fuel which is very convenient to use for domestic purposes. When supplies of natural gas are used up there will still be a demand for some form of piped gas because it is such a convenient fuel. This could be hydrogen (p. 496), or we may again have to produce piped gas from coal. However, to avoid spending enormous sums of money to convert gas appliances back to burning coal gas, much research is taking place into making *synthetic natural gas* (SNG) from coal. This has a composition similar to natural gas, and can be burned in the same appliances as natural gas. At the moment, Britain leads the field in the development of SNG.

Unfortunately, all of these forms of gas are (or are produced from) fossil fuels, and they are likely to become more and more expensive as supplies of the resources are reduced. The convenience of maintaining continuing supplies of piped gas will have to be paid for by the consumer.

Coal

Coal has been known to mankind for thousands of years. It is found in layers or *seams*, formed from plant material which died millions of years ago. Analysis shows that coal is a kind of 'modified wood', and fragments of plants are frequently found embedded in it. Each layer of decaying vegetable material (perhaps dead trees, leaves, and branches in swampy areas) became covered with a layer of sand and mud as the land sank slightly over a long period of time. In turn, this was covered by another layer of vegetable material from later growth. As the process continued, the layers of sand and mud hardened to form rock. The deeper layers of decaying vegetation gradually changed to coal as they went through a series of changes caused by enormous pressures from the weight of rock above, bacterial action, and a high temperature.

As coal is the end product of gradually changing vegetable matter, we would expect several stages on the way. Peat is the first stage. This can usually be found near the surface of the earth, and so has not gone through the temperature and pressure changes necessary for its conversion into coal. The next stage produces a crumbly, soft kind of coal called *lignite*. The most common kind of coal is *bituminous coal*, which is harder than lignite but not as hard as the *anthracite* produced by the final stage.

The history of coal mining in Britain is closely connected with very significant changes. Coal played a vital part in making Britain a world power during the Industrial

Revolution. It provided energy to drive locomotives and machines which were to revolutionize our whole way of life. It provided one of the raw materials for the conversion of iron ore into iron and steel, which in turn gave rise to tremendous developments in engineering. It provided coal gas (by destructive distillation) and electricity (from coal-fired power stations)—which revolutionized life at home and in industry. You have probably learned of the great changes in the working conditions in the mines and how these reflected the changing opinions of the times. You may know something of the dangers of mining; the risks of falling rock, and of poisonous or explosive gases. You might like to find out more about conditions in a modern coal mine. Your library may contain books which will provide very interesting information about coal and coal mining (perhaps with particular reference to the area in which you live), and your school may have copies of some of the excellent booklets produced by the National Coal Board.

More to do

Try to find answers to the following questions.
1. What is an open cast mine?
2. How do engineers know where to sink a pit?
3. Who invented the miner's safety lamp? How does it work?
4. Why were canaries taken down into pits?
5. What is 'fire damp'? What is 'after damp' and why is it so dangerous?
6. What are the following men noted for: Newcomen, Murdoch, and Abraham Darby?

Experiment 26.1
'Improving' coal*

Apparatus
As in Figure 26.5. Rack of test-tubes, clamp stand, Bunsen burner, test-tube holder.
Supply of powdered coal, dilute sodium hydroxide solution, red litmus paper.

Procedure
(a) Half fill the smaller hard glass tube with powdered coal, and assemble the apparatus as in Figure 26.5.
(b) Heat the coal, gently at first, and apply a flame to the end of tube A at frequent intervals.
(c) Heat the coal strongly until no further reaction takes place. Remove the bung from the large test-tube to prevent 'sucking back'. Allow the apparatus to cool.
(d) Cautiously smell the liquid in the larger tube.
(e) Transfer a little of the aqueous solution from the larger test-tube to a clean test-tube. Add a few cm³ of dilute sodium hydroxide solution and gently heat the mixture. Remove the tube from the flame and test any evolved gas with damp red litmus paper.
(f) Remove the solid from the cool smaller tube and examine it carefully.

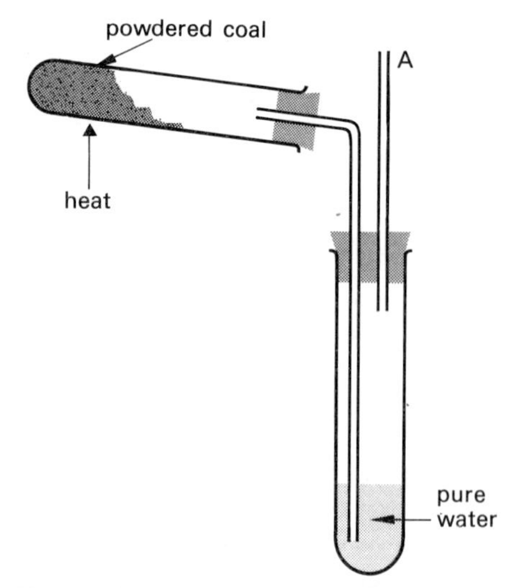

Figure 26.5 The destructive distillation of coal

Results
Record all your observations.

Conclusion
Note that the coal was not *burned* during the experiment. It was broken down by heat in the presence of a little air, and the

products were then separated from each other by distillation. This process is called the *destructive distillation* of coal. There are high temperature and low temperature forms of the process, as is explained later.

Make your own conclusion, based upon the information above, your observations, and answers to the following questions.

1. What happened when a flame was applied to tube A? What do you think had been formed so that this could happen?
2. What substances were condensed or dissolved in the water?
3. How did the solid residue compare with the original powdered coal? Can it still burn and so be used as a fuel? If so, does it have any advantages or disadvantages as a fuel when compared with coal?
4. List at least four substances produced by the destructive distillation of coal.

The advantages of the destructive distillation of coal

Coal contains a wide variety of chemicals, most of which are not used efficiently when raw coal is used directly as a fuel. At the moment, the main reason for carrying out the destructive distillation of coal is to provide coke and other solid smokeless fuels. In addition however, another clean fuel is produced, coal gas, together with important chemicals. The tar, oil, and ammoniacal products are potentially extremely valuable, and can be processed into a wide variety of chemicals which can be used to produce such substances as fertilizers, nylon, other plastics, paints, explosives, disinfectants, cleaning agents, printing inks, perfumes, pesticides, and detergents. At the moment it is uneconomic to process some of these other byproducts, as their derivatives can be obtained more cheaply from other sources, but this is likely to change. It is obvious that this type of process is a very efficient way of using the wealth to be found in 'black diamonds'.

Coal gas

Coal gas is made by heating coal without burning it, as in Experiment 26.1. Its composition varies, because the operating conditions used to make it also vary, and different kinds of coal are used. For example, coke is normally made by the destructive distillation of coal at temperatures around 1000 °C and for a period varying from 12 hours to 30 hours. The actual operating conditions chosen depend upon the type of coke (or other smokeless fuel) which is required; the coke used in blast furnaces is very different from that used as a domestic smokeless fuel. The coal gas obtained during the process also varies, therefore, according to the operating conditions.

A modern coke oven may take a charge of over 20 tonnes of coal, and typical yields per tonne of coal processed at a high temperature (1350 °C) are:

Coke	0.7 tonnes
Tar	35.0 dm^3
Benzole	12 dm^3
Ammonium sulphate	12 kg
Coal gas	500 m^3

(Benzole is a mixture of many, fairly volatile hydrocarbons from which many other substances can be made, including petrol.)

A typical composition of a coal gas sample is:

Hydrogen	50 per cent
Methane	30 per cent
Ethene	4 per cent
Carbon monoxide	8 per cent

Smaller quantities of carbon dioxide, nitrogen, and oxygen.

The disadvantages of coal gas are that carbon monoxide (which it always contains) is poisonous, and the explosive nature of certain gas–air mixtures makes leaks or fractures in pipes potentially dangerous. (This latter fact is also true of natural gas.) It also requires heat energy to produce it, although the value of other byproducts also obtained by the process is some compensation for this.

Smokeless fuels

The discharge of smoke from industry and the home resulted in a very serious form of air pollution, as explained on page 145. The Clean Air Act was introduced in Britain in 1956, and since then many regions have become smokeless zones. It is possible to burn solid fuels in such areas without breaking the act because bituminous coal can be processed to produce solid smokeless fuels. These tend to be more expensive than raw coal because they have to be processed.

Coke is normally produced by a high temperature destructive distillation, as explained earlier, but other solid smokeless fuels are produced by a low temperature process at about 500 °C. In all of these cases the smoke and other volatile compounds are driven off from the coal. The solid products of such reactions vary according to the actual process but they are usually given trade names such as Homefire, Roomheat, Coalite, Rexco, Sunbrite, and so on. The gas and liquid products are similar to those from the high temperature process but vary in detailed composition and relative proportions.

How coal is used today

There was a time when large quantities of coal were treated to produce coal gas, were used to raise steam in locomotives etc., and were burnt as solid fuel in the home. Coal is no longer treated with the main intention of providing coal gas, because natural gas at the moment is readily available. It is still destructively distilled, however, in order to produce coke (for industry and for use as a domestic smokeless fuel) and to provide other solid smokeless fuels. The gas produced during these processes has been relegated in importance to being a useful side product, and is often used on the site.

By far the greatest proportion of the coal produced in Britain today is burned in power stations to generate electricity. There is little doubt that coal will play an increasingly important part in our lives in the future, and the economics of the industry are likely to show many changes. We have already mentioned that coal can be processed to produce Synthetic Natural Gas. The direct combustion of coal makes it an important fuel, and as its processing can also produce a wide variety of chemicals and smokeless fuels we can appreciate that the coal industry has a secure future—at least until the coal seams are exhausted!

Points for discussion

1. In a coal fired power station 1000 tonnes of coal is burned in a day. The average sulphur content of the coal is 1.6 per cent by mass.
(a) Calculate the mass of sulphur dioxide produced by the power station in one day, assuming that all the sulphur is oxidized according to the equation:

$$S + O_2 \rightarrow SO_2.$$

(b) What will happen to this sulphur dioxide in the atmosphere?
(c) Name three other substances released into the atmosphere from the combustion of coal.
(d) The technology exists for the removal of sulphur dioxide from the flue gases of power stations; explain why the sulphur dioxide is not removed. (A.E.B. 1977)

2. It is not always appreciated that a gas leak can be very dangerous, whether it is of natural gas or coal gas. The combustion of a methane/air mixture can give rise to a flame temperature of over 1700 °C, and this can lead to a pressure of between 6 and 7 atmospheres in an enclosed space such as a house. Explosions caused by gas leaks can be very serious.

List the various things you should do (and not do) if you suspect that there is a leak of gas in your home.

26.4 A BRIEF CONSIDERATION OF OTHER SOURCES OF ENERGY

Fusion

Nuclear fusion is a process in which light atomic nuclei join (fuse) together to form heavier nuclei, and give out energy in the process. The sun's energy is derived from nuclear fusion, and if we can reproduce this process in a controlled way on Earth we will have a source of energy which is almost limitless. Note that the term fusion must not be confused with atomic *fission*, which is the source of energy in a nuclear power station (p. 107). In fission, the nuclei of 'heavy' elements split up to form lighter nuclei and release energy in the process. Both processes are sources of enormous amounts of energy, but the energy available from a single fusion far exceeds that from a single fission.

Fusion:

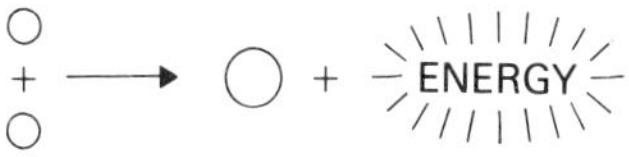

○ + ○ ⟶ ○ + ENERGY

Fission:

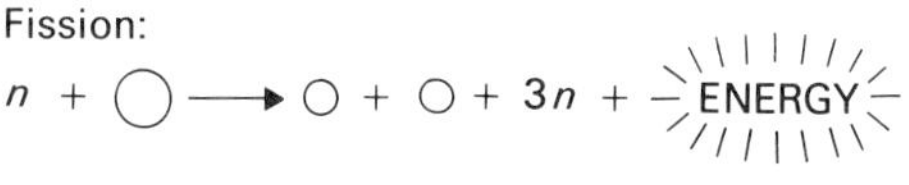

n + ○ ⟶ ○ + ○ + 3n + ENERGY

The fusion fuel would be two different isotopes of hydrogen, deuterium (hydrogen atoms containing one neutron, page 159), and tritium (hydrogen atoms containing two neutrons). In the fusion process one atom of each kind fuse together to make an atom of helium. Enough deuterium exists in the oceans of the earth to provide fusion fuel for 100 million million years. Tritium is derived from lithium, of which there is also a plentiful supply. In theory, therefore, fusion would provide almost limitless energy; one tonne of fusion fuel would produce the same energy as 5 million tonnes of coal. The waste products of a fusion process, unlike those of a fission process, would be non-toxic and comparatively easy to dispose of.

Unfortunately, there are enormous technological problems to be overcome before fusion can provide usable power, as distinct from that generated in an H-bomb. For example, the fuel would have to be kept in a reactor for just one second or so at a temperature hotter than that at the centre of the sun, i.e. over a hundred million °C! This temperature can be achieved in an H-bomb by triggering the reaction with an 'ordinary' atomic bomb, and attempts are being made to use laser beams to make atoms fuse together. In addition there is the problem of using, in a safe way, the colossal amounts of energy that such a process would produce. The technological problems are so great that few people can see fusion making any significant contribution to the world's energy supplies until at least the early part of the next century, if ever it can be controlled at all.

Points for discussion

1. Draw diagrams to represent structures of atoms of (a) deuterium and (b) tritium.
2. If an atom of deuterium combines with an atom of tritium to make an atom of 'ordinary' helium, what else would be liberated apart from energy?
3. What is heavy water? What mass would a mole of heavy water have?

Nuclear power from atomic fission

A brief description of fission and nuclear power is given on page 109. Britain has had nuclear power stations since 1955, and in 1980 approximately 14 per cent of our electricity was generated by nuclear power. However, nuclear power stations also have their problems. We are still trying to solve technological difficulties caused by the extreme conditions under which they operate, and the radioactive waste which they produce is potentially a very serious health risk. The waste from the early reactors includes a considerable amount of uranium which could be used to provide more energy, but this further conversion cannot be achieved in a conventional nuclear power station (which is estimated to make use of only about one-hundredth of the potential energy in uranium). This is particularly important, because, as with fossil fuels, there is only a limited supply of uranium in the world. The so-called 'fast' reactors (p. 110) use uranium sixty times more efficiently than earlier nuclear reactors, but they present even greater technological problems.

Points for discussion

1. Suppose that an atom of $^{235}_{92}U$ absorbs a neutron and then splits into an atom of $^{212}_{82}Pb$, two more neutrons, and one other atom. What would be the mass number and atomic number of the other atom? To which element would this other atom belong? Would it be the most abundant nuclide of the element?
2. Some of the advantages of nuclear power have been discussed earlier in the book, and some further points in favour of an expansion in nuclear power are listed below. Nevertheless, some people are concerned about the possible dangers of such an expansion. Read the following points carefully. Can you suggest counter-arguments? Can you think of any other arguments against further development of nuclear power? (It would be as well to revise the work on pages 107–15, and in particular to make sure you understand the potential of nuclear power, before you answer the questions.)
(a) Nuclear power uses fuel which is not required for any other purpose (cf. oil).
(b) All the 'highly active' radioactive waste produced so far by Britain's nuclear reactors would, in total, take up less volume than two average-sized semi-detached houses.
(c) No smoke or soot is produced by a nuclear power station.
(d) The public is exposed to background radiation all the time, and the extra radiation from nuclear power is less than 0.2 per cent of the natural background level of radiation.
(e) It is much cheaper to generate electricity by nuclear power than by burning oil, coal, or gas.
(f) An advanced gas-cooled reactor can make as much electricity from 1 kg of enriched uranium fuel as could be produced from about 60 tonnes of coal.
(g) Development of fast reactors would enable current stocks of uranium in Britain to provide as much energy as our entire coal reserves.
(h) The statistical risk of a member of the public dying as the result of an accident at a nuclear power station is less than that of being struck by lightning.
(i) There has been more research into possible risks from using nuclear power than for any other fuel.
(j) Used nuclear fuel is transported in containers which are designed to withstand extreme accidents. Over a twenty year period these containers have travelled over 3 million miles without causing harm to any member of the public.

(The information above is modified from data supplied by British Nuclear Fuels Ltd.)

Geothermal energy

The Earth's core consists of molten material at a very high temperature, and the temperature gradually decreases from the core to the crust. Even in Britain it is possible to extract heat energy from under the Earth's crust, although we do not have any natural volcanic activity. There are hot water springs at Bath, for example, and in several areas it is possible to extract water at a temperature of about 70 °C from depths of about 1750 metres. Natural hot water is used in several parts of the world to supplement domestic heating systems.

As technology improves, it may become possible to extract heat energy from underground rocks rather than underground water. This is an attractive proposition, for it has been estimated that there is enough heat energy in the top 10 km of the Earth's crust in Britain alone to be equivalent to the energy obtainable from 50 000 million million tonnes of coal. Unfortunately, the energy is 'spread thinly' and will be difficult to extract, but even a tiny fraction of that which is theoretically available would prove to be of great value, and research is being conducted into this possibility.

Solar energy

Man frequently makes indirect use of the fusion process occurring in the sun, e.g. by trapping some of the energy of sunlight in greenhouses. Similarly, plants are extremely efficient at utilizing sunlight as a source of energy to build up foods such as sugars. The total annual input of solar energy on the earth is more than 5000 times greater than the annual consumption of energy. Solar energy is thus an attractive source of energy; it is clean, pollution-free, and does not consume any of the Earth's natural resources.

Photoelectric cells can convert solar energy into electricity (e.g. in some types of photographic light meter), but at the moment they are very expensive and can convert only a small proportion of the available energy into electricity. Solar panels (e.g. on the roof of a house) use solar energy as a supplementary source of heat for domestic hot water. Research is being conducted into establishing suitable orbiting stations in space which would be capable of trapping some of the sun's energy.

Unfortunately, the utilization of solar energy also has its problems. The intensity of sunlight at ground level is irregular and is particularly weak in winter, the time when we make our greatest demand on electricity, and of course there is no sunlight at night. Nevertheless, solar energy remains an attractive supplementary source of energy, particularly for domestic heating of various kinds.

Points for discussion

1. Figure 26.6 is a simple diagram of a solar still (i.e. a distillation system that depends on heat from the sun's rays). Carefully study the diagram.

(a) Explain how the still works.

(b) To what natural cycle is this distillation system similar?

(c) What are the differences between the processes in this still and those in the natural cycle?

(d) Suggest *three* improvements in design, or in materials used in the construction of this still.

(e) If water contaminated with methanol was put into the still instead of sea water, then pure water could not be obtained from the still. Why is this? (A.E.B. 1977)

2. What are the advantages of siting solar cells in space rather than on Earth? What are the disadvantages of doing so?

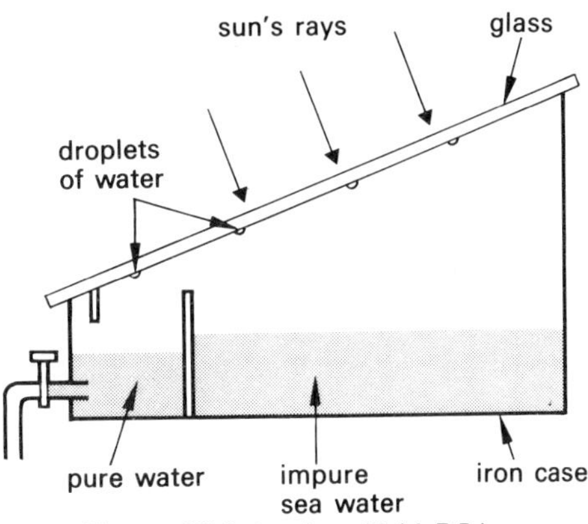

Figure 26.6 A solar still (*A.E.B.*)

Small scale sources of electricity. Fuel cells

Apart from the methods discussed in this chapter, there are other ways of 'generating' electricity, e.g. by voltaic cells (like the common 'dry cell' battery and other systems described on page 261) and by fuel cells. Voltaic cells and fuel cells have certain features in common. In both types of cell chemical energy is changed *directly* into electrical energy, and this is therefore a more efficient conversion than that which takes place in a conventional power station. They both need an electrolyte to complete the flow of electrons between the electrodes. They have useful applications for small scale power generation, but unfortunately they cannot be considered as sources of power on a large scale. The voltaic cell, in particular, has other disadvantages too.

In a voltaic cell, the production of electricity stops as soon as the electrolyte and/or one of the electrodes is used up, and this limits its use. A fuel cell overcomes this disadvantage by feeding gases or liquids into the cell so that it continues to work as long as the 'fuels' are added to it, and the materials of the cell are unchanged. Unfortunately, this results in a bulky and non-portable cell, unlike the conventional battery.

Fuel cells

The first attempt to make a fuel cell was by William Grove who, in 1840, constructed a cell which he called a gaseous voltaic battery. It consisted of two well-separated platinum electrodes in dilute sulphuric acid as an electrolyte. Electricity did not flow at first because both electrodes were made of the same material, but when hydrogen gas was bubbled over one electrode and oxygen gas over the other, an electric current was generated. This current ceased as soon as the gas flow was cut off. Unlike a voltaic cell, this cell did not depend upon the materials from which it was constructed but upon the materials supplied to it. The electrodes were not eaten away or chemically involved, and the cell provided the possibility of a continuous output of electrical energy.

The simple fuel cell produced by Grove works as follows. Platinum is a catalyst and will adsorb gases on to its surface, where they are able to react more easily.

At the anode: $2H_2(g) \rightarrow 4H^+(aq) + 4e^-$

At the cathode: $O_2(g) + 2H_2O(l) + 4e^- \rightarrow 4OH^-(aq)$

Electrical energy is therefore produced, the only two substances consumed are hydrogen and oxygen (supplied externally), and water is the only chemical product, formed by the combination of $H^+(aq)$ and $OH^-(aq)$ ions.

Hydrox fuel cells

In the simple fuel cell considered above, hydrogen is the fuel and it is oxidized by oxygen. Such cells are called hydrox cells, and a typical arrangement is shown in Figure 26.7. Note that hydrogen can be *burned* in oxygen to produce heat energy and that this can then be used to produce electricity, but in the case of the fuel cell the electricity is obtained directly, and this saves the wasteful loss of energy during heat transfer.

Disadvantages of the early hydrox cells

Early cells of the Grove type suffered from various disadvantages, such as the small surface area of the platinum electrodes used and the fact that the catalyst was 'poisoned' by impurities in the gases entering the cell. Also, as water is produced by the operation of the cell, prolonged use would result in a diluted electrolyte. It has also been found that a porous electrode enables gas, liquid, and solid to come into contact more easily, and so electrodes have been modified to improve contact between the reactants.

Research into fuel cells was intensified by the development of the USA space programme, and the modern type of hydrox cell is now very efficient. Many other fuels have been developed for these cells, e.g. hydrazine (N_2H_4), methanol (CH_3OH), and various hydrocarbons. The cheapest oxidant is still oxygen. Further developments will depend upon the purposes for which fuel cells are required and their cost relative to other sources of energy. Cylinders of hydrogen, for example, are relatively expensive (see page 540), and the use of hydrazine is restricted because of its expense and toxicity. Higher temperature cells (operating at say 500 °C) are still experimental and use molten electrolytes; this gives some indication of the potential of fuel cells.

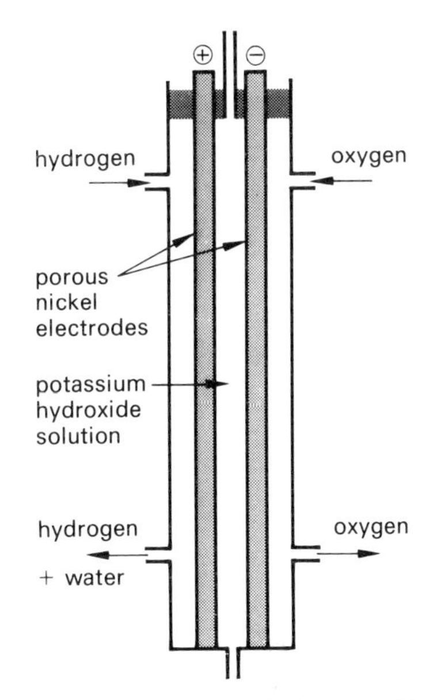

Figure 26.7 A simple hydrox cell

Wave power

In theory this is one of the more promising untapped sources. Wave power (not to be confused with tidal power) would provide a *constant* source of energy, and it is estimated that wave power from a 1000 km stretch of Atlantic coastline in the north-west of Britain could provide half of our present demand for electricity. Once again, however, there are formidable technological problems to be overcome before this method can be used on a large scale.

Tidal power

Britain is fortunate to have a long coastline, and in theory both the incoming and outgoing tidal flows can generate electricity. If a suitable barrage is built across an estuary it could have built-in water-powered turbines which could generate electricity. As the tide flowed in, the turbines would be turned one way, and their movement would be reversed when the tide went out; electricity would be generated by both processes. Ideally, such a barrage would be built across an estuary in which there is a considerable tidal range, and the Severn Estuary is one of the most promising sites in Britain for this purpose. The Rance tidal barrage has been working in France for some time.

Tidal power also has its problems as a potential source of energy. The cost of building the barrage and of maintaining it would be very high. Also, in the simplest type of scheme, the generation of electricity would not be constant. It would rise and fall according to the tidal flow, and peak output would vary from day to day. Most important of all, a single scheme of this kind could provide only a small proportion of our electricity demand.

Wind power

Man has long used the wind as a source of energy in windmills and sailing ships, but this traditional source of power has taken on a new importance because of the energy crisis. Wind power is attractive, because it is more readily available in winter when we need more electricity, and it would not use up any of our natural resources, but it has many disadvantages. For example, it is not a constant source; there are many calm days, even in Britain.

In Britain, the coastal areas and hill tops in the west and north-west are the most suitable areas for wind machines. To give some idea of the potential of the method, it has been estimated that 10 000 wind machines with rotors about 45 m in diameter could provide the same energy as eight million tonnes of coal per year. Unfortunately, even this apparently vast amount of energy is no more than 2.5 per cent of our present needs.

26.5 A SUMMARY OF THE MAIN POINTS IN CHAPTER 26

The following 'check list' should help you to organize the work for revision.
(a) Factors to be considered in deciding whether a substance is suitable as a fuel (pp. 491–2).
(b) A general understanding of the various factors which have led to the energy crisis, and why the problem might intensify.
(c) An appreciation of the fact that the generation of electricity in a conventional power station converts only about one-third of the energy available from the primary fuel into electricity.
(d) An understanding of the need to conserve our stocks of fossil fuels and to develop alternative ways of generating electricity, and of developing an alternative fuel for transport.
(e) How natural gas is formed, its advantages, its composition, its limited life, and the fact that a Synthetic Natural Gas can be made from coal.
(f) How coal is formed, and the main kinds of coal (lignite, bituminous, anthracite).
(g) The destructive distillation of coal, in the laboratory and on a large scale, and the advantages of processing coal in this way.
(h) Coal gas, smokeless fuels, and the other products from the destructive distillation of coal.
(i) A general understanding of other sources of energy, and their advantages and disadvantages.

Important experiment

The destructive distillation of coal, e.g. Experiment 26.1.

QUESTIONS

Multiple choice questions

The following information concerns questions 1–3. The letters A–E represent gases which will burn upon ignition in air. The equations for their combustion are given together with the enthalpies (heats) of reaction as represented by the equation as written. Answer the questions by choosing the correct gas.

A hydrogen sulphide
$2H_2S + 3O_2 \rightarrow 2H_2O + 2SO_2$
$\Delta H = -1064$ kJ

B methane
$CH_4 + 2O_2 \rightarrow CO_2 + 2H_2O$
$\Delta H = -810$ kJ

C carbon monoxide
$2CO + O_2 \rightarrow 2CO_2$
$\Delta H = -566$ kJ

D hydrogen
$2H_2 + O_2 \rightarrow 2H_2O$
$\Delta H = -572$ kJ

E ethyne
$2C_2H_2 + 5O_2 \rightarrow 4CO_2 + 2H_2O$
$\Delta H = -2600$ kJ

1. Which gas will release the greatest quantity of heat per mole if it could be completely burned in air?

2. Which gas, if used as a fuel, would cause the greatest pollution hazard?

3. Which gas would create the greatest difficulty in supplying the volume of oxygen necessary for complete combustion?

Structured questions and questions requiring longer answers

4. (a) What do you understand by the term 'fossil fuel'?
(b) Name three fossil fuels in common use at the present time.
(c) Give the names of the two elements predominating in each of these fuels which are largely responsible for the combustible nature of the fuel.
(d) Give two reasons, excluding pollution effects, why these resources should be used sparingly as fuels.

5. Name two examples of (a) a gaseous fuel (b) a liquid fuel and (c) a solid fuel. Compare the advantages and disadvantages of each type.

6. (a) How do the processes of respiration and the burning of fuel (i) resemble one another, (ii) differ from one another?
(b) When we burn coal, we are making use of stored energy that came originally from the sun. Explain this statement. (C.)

7. Articles on fuels appear quite frequently in newspapers bearing such headlines as:
'Nuclear power—the price to be paid is "super long-lived" radioactive waste'
'Energy crisis—hydrogen could be wonder fuel of the future'
'Waste plastics and paper to heat towns'
Critically discuss two of these headlines, giving the relevant chemical facts and explaining the problems involved and the solutions possible. (A.E.B. 1976)

8. Crude oil and natural gas are often found together. Natural gas consists mainly of methane, with smaller quantities of ethane and other hydrocarbons. Sulphur compounds are removed and a substance with a characteristic smell is added before the gas is supplied to the consumer. Natural gas requires more oxygen to burn it than does town gas (mainly a mixture of hydrogen and methane): hence any appliance designed for town gas has to be converted in order to burn natural gas.

When crude oil is fractionally distilled, one of the first fractions is petrol and one of the last fractions is heavy gas oil.
(a) Why are crude oil and natural gas often found together?
(b) Why are sulphur compounds removed before the gas is supplied to the consumer?
(c) Why is a substance added to give natural gas a characteristic smell?
(d) What do you understand by the word hydrocarbon?
(e) Write separate equations for the combustion (i) of methane, (ii) of hydrogen, and hence show why more oxygen is required to burn natural gas than an equal volume of town gas.
(f) How can it be shown that petrol is a mixture and not a single compound?
(g) How do petrol and heavy gas oil differ in (i) colour, (ii) viscosity, (iii) their behaviour on burning? (C.)

27 The chemical industry

We usually regard the chemical industry as that which extracts and processes chemical raw materials. Raw materials may include unprocessed substances obtained directly from the environment, such as sea water, air, nitrates, oil, metal ores, limestone, sulphur, and coal. Raw materials can also include chemicals obtained by processing naturally occurring substances, which are then used as the basis of other manufacturing processes. Thus sulphuric acid, chlorine, and many organic chemicals are important raw materials for other industries although the materials do not occur naturally. The products of these chemical industries are thus great in number and very wide in their range of applications. They include fertilizers, soaps and detergents, metals and alloys, drugs, solvents, dyes, explosives, adhesives, plastics, insecticides, and paints. To some people the chemical industry brings to mind smelly factories pouring out pollutants and making substances with strange names and complicated formulae which have little to do with daily life; factories where the health of the workers is harmed by the conditions in which they work, and where valuable materials are constantly removed from the environment and replaced by heaps of solid waste. These images of industry are not objective ones.

As we have stressed in section 9.4 there are many areas of the world where natural activities pour more pollutants into the air than industry does. In the past, the largest source of air pollution in Britain was the burning of raw coal in domestic fires. Standards of pollution control in industry are now much better than they were, and are constantly being improved as are safety standards. Many more people are killed and injured on the roads and in aircraft accidents than are killed as the result of industrial activity, but the latter often receive far more bad publicity.

Our present standard of living owes much to the chemical industry. The chemicals produced and processed by the industry *do* affect all of us in our daily lives, no matter how unlikely this may seem. We now need to strike the correct balance between factors such as conservation, recycling, maintaining our standard of living, improving our environment, replacing some of our resources and not wasting others, and developing new sources of energy. In the last chapter we gave a brief outline of some of the possible alternative sources of energy, and of the serious technological problems involved with their development. This factor alone illustrates that the industrial world presents a challenge and a stimulus which no doubt some of you will one day enjoy.

27.1 THE EARTH'S RESOURCES: RECYCLING

The Earth's resources

When we talk about the Earth's resources we mean those naturally occurring substances which can be processed to provide all the materials for modern living. The main resources include the air (from which we abstract oxygen and nitrogen), fossil fuels (from which we obtain organic chemicals as well as energy), the seas (which as yet are relatively unexploited), and minerals, which may exist as pure elements, pure compounds, or mixtures. Some of these individual resources are considered in more detail later in the chapter, and the fossil fuels (Chapters 22 and 26) and air (Chapter 9) have already been discussed.

Minerals

Some common minerals?

Table 27.1 The percentage abundance of some of the elements in the Earth's crust (by mass)

Metals		*Non-metals*	
		Oxygen	46.5%
		Silicon	27.6%
Aluminium	8.1%		
Iron	5.1%		
Calcium	3.6%		
Sodium	2.8%		
Potassium	2.6%		
Magnesium	2.1%		
Titanium	0.6%		
		Hydrogen	0.1%
		Phosphorus	0.1%
		Carbon	0.01%
Manganese	0.01%		
		Sulphur	0.005%

Table 27.1 shows the percentage abundance of some of the more common elements in the Earth's crust. Some of these elements occur *native*, i.e. as the uncombined element. Elements found native include the non-metal sulphur, and the unreactive metals gold and silver. Some elements occur native in sea water, but as yet it is not economical to extract them. The German chemist Fritz Haber once attempted to obtain gold from sea water on a large scale. There are millions of pounds worth of gold in each km^3 of sea water, but it would cost more than the gold is worth to extract it.

Most metals and non-metals occur as minerals, i.e. in combined form as compounds with other elements. Some common minerals are listed in Table 27.2, and others are found in sea water. Some of the dissolved compounds in sea water have been extracted commercially to produce elements such as bromine, magnesium, and iodine.

Most minerals that we use are solids, and they are rarely found in the pure state. They are normally mixed with rock and other unwanted material, and the whole mixture is known as an ore. The stages which take place in extracting an element from an ore can be summarized as follows.

1. Obtain the ore (e.g. usually mine it).
2. Purify the ore to obtain the pure mineral (i.e. remove unwanted rock, etc).
3. Chemical reaction to 'break open' the mineral and obtain the pure element.

Step 2 is often difficult because some ores contain only a very small proportion of mineral. Step 3 is often expensive, particularly for the more reactive elements which require large quantities of energy (e.g. electricity for electrolysis) to separate them from their compounds. Aluminium, for example, forms about one-twelfth by mass of the Earth's crust, but many of its 'ores' are not economical sources of the metal; clay contains about 25 per cent by mass of aluminium, but step 3 would be very difficult if clay were to be used as a starting material for the manufacture of aluminium.

The kind of chemistry involved in step 3 is discussed in more detail later in this chapter (e.g. in the extractions of sodium, aluminium, and iron), and in general terms on page 523.

Table 27.2 Some important minerals

Mineral	*Element extracted from mineral*	*Name and formula of important compound in mineral*
Bauxite	Aluminium	Hydrated aluminium oxide, $Al_2O_3.H_2O$ or $Al_2O_3.3H_2O$
Copper pyrites	Copper	Copper(II)iron(II) sulphide, $CuFeS_2$
Fluorspar	Fluorine	Calcium fluoride, CaF_2
Haematite	Iron	Iron(III) oxide, Fe_2O_3
Galena	Lead	Lead(II) sulphide, PbS
Carnallite	Magnesium and potassium	Hydrated magnesium potassium chloride, $KCl.MgCl_2.6H_2O$
Cinnabar	Mercury	Mercury(II) sulphide, HgS
Pentlandite	Nickel	A sulphide of nickel, copper, and iron
Rock salt	Sodium and chlorine	Sodium chloride, NaCl
Argentite	Silver	Silver sulphide, Ag_2S
Cassiterite	Tin	Tin(IV) oxide, SnO_2
Rutile	Titanium	Titanium(IV) oxide, TiO_2
Zinc blende	Zinc	Zinc sulphide, ZnS
Calamine	Zinc	Zinc carbonate, $ZnCO_3$
Uraninite	Uranium	Triuranium octoxide, U_3O_8

Resources are not limitless

It is important to realize that our resources cannot last for ever, and that we cannot afford to continue with the inefficient ways of using them which have been common in the past. This problem has already been emphasized with respect to fossil fuels, and this example is particularly important because fossil fuels cannot be recycled; once they have been used to provide energy, they have gone for ever. Resources of oil were at one time exploited in a very convenient way, i.e. by drilling on land. Such is the demand for oil that we have been forced to seek it in places where it is becoming increasingly difficult to extract, e.g. beneath the gale-struck waters of the North Sea. In future years we will probably be forced to extract and process materials which are at present too inconvenient or too uneconomic to consider.

The problems of dwindling resources are already emerging in the case of metals. It became accepted that we should build cars and other machinery which were only used for a few years and were then regarded as useless scrap which was allowed to corrode away. We now realize that we cannot afford to treat our metals in this way. Research is taking place to discover new ways in which we can recycle metals and other materials in limited supply, i.e. ways in which they can be 'used again' when their use as a particular product comes to an end. Unfortunately, one of the main difficulties to be faced is our attitude towards material things; we have become accustomed to replacing goods long before their useful lives are over because of fashion changes or because more 'up to date' versions are available.

The one resource which seems to remain unchanged in spite of all our demands on it is the oxygen in the air. Oxygen is constantly being released by photosynthesis and this roughly balances the amount which is used up. Some resources can be replaced by efficient planning. Timber, for example, is used to make paper and for constructional purposes, and this can be replaced by growing new trees. This is a long-term target, however, for trees grow very slowly and we are still felling them at a rate much faster than that at which new ones are being planted.

Synthetic materials have helped to solve the problem of excessive demand for particular natural substances. Natural rubber alone, for example, could not supply the present demand for rubber, but there are many forms of synthetic rubber which help to solve the problem. Again, the demand for textiles is so great that natural materials such as cotton and wool could almost certainly not meet the demand. Synthetic fibres such as nylon and terylene can be used in place of natural materials. Unfortunately, although synthetic materials are in some ways better than the natural ones they have replaced, they also consume resources in their manufacture (e.g. synthetic fibres are made from products of the oil industry). The importance of recycling as a method of conserving our resources becomes more and more apparent.

Point for discussion

1. How was native gold extracted in the days of the 'Wild West'? How did the process work?

2. Twenty-five elements occur native. How many of these can you list?

Recycling

Recycling is essentially re-using, and it includes the reclamation of waste material and its reprocessing. Recycling is not a new idea. The 'rag and bone man' used to be a familiar figure, there have been scrap-metal merchants for many years, and waste vegetable and animal materials have long been composted by gardeners and farmers to return important chemicals to the soil. Perfect recycling occurs in nature, e.g. in the carbon, nitrogen, and water cycles (pages 373, 383, and 163). In natural cycles like these the material passes endlessly through the cycle. We have 'interfered' with the nitrogen cycle (p. 384) but fortunately there is a large reserve of nitrogen in the air which can be fixed and added to the cycle as fertilizers. Unfortunately, there are not large reserves of copper, tin, etc., and therefore it is important that we recycle as much of these materials as possible, so as to preserve our resources for future generations.

Some are always lost from the cycle

Industrial cycles can never return *all* the waste material for reprocessing; some is always lost from the cycle so there is still some overall loss of material. However, this is an improvement on the 'open' system in which raw materials are continuously consumed and 'lost', i.e.

raw material → extraction → processing → used and waste material.

Recycling is an attempt to save some of the waste by returning it to the processing stage, i.e.

raw material → extraction → processing → used and waste material
recycling (used and waste material → recycling → processing)

Note that material which has been recycled does not need to go through the extraction stage. The extraction stage often consumes large amounts of energy, e.g.

electricity, and so recycling also saves on energy in addition to saving raw materials. It is less expensive to recycle steel from tin cans than to produce it from imported iron ore. Similarly, the production of copper from the ore may take up to ten times the quantity of electricity needed to produce the same quantity of the metal from scrap. Aluminium requires 15 000 kW hours of electricity per tonne from bauxite (the ore) but only 3000 kW hours per tonne of scrap.

In the future, the recycling of both industrial and domestic waste will become increasingly important as the technology for separating and treating such waste improves, and the processes become more economic (i.e. the price of recycled material compares favourably with that for obtaining the same substance from raw materials). In order to achieve this aim, positive pressure, co-operation from industry (particularly the manufacturing industries), and a better attitude from the individual are all needed. For example, products could be designed and manufactured so as to be more easily recycled. Consider a simple metal container, which might typically be made from a mixture of products such as a body of tinned steel with a soldered seam (solder contains lead and tin) and aluminium ends. Such a container is difficult to recycle because the different components have to be separated. If the same container were to be made entirely of aluminium, it would be easy to recycle it.

Point for discussion

Even if it is more expensive to recycle a product than to obtain it from raw materials, there is still a case for recycling. Do you agree with this statement? Give reasons for your answer.

Reprocessing waste materials

The main classes which are treated in this way are metals, paper, glass, plastics, oils, and rubber. We will consider the reprocessing of some metals and plastics in this section.

Metals Most reclaimed metals are from manufacturer's scrap. This is uncontaminated material, i.e. it is not mixed with any other substances, and is fed back into the smelter. Used scrap, such as that from old cables, motor vehicles, alloy wastes, and metal scrap from domestic waste, usually contains several metals, and so the reprocessing is more complicated. Great advances have been made in recycling the metals in scrap cars, and modern methods can now separate individual metals from a mixture of metals at a reasonably economical rate.

'Ferrous' metals Large amounts of 'ferrous metals' (i.e. those containing iron or steel) are now recycled, and a typical 'charge' in a steel furnace may include up to 70 per cent of scrap. One tonne of scrap steel can produce the same amount of new steel as would be obtained by using 1.5 tonnes of iron ore, 1.0 tonne of coke, and 0.5 tonne of limestone, all of which are valuable resources. In addition, there is an energy saving. As scrap steel is likely to contain a wide variety of elements (in addition to carbon and iron) you may imagine that it would be extremely difficult to produce a new steel of an exact specification for a particular purpose. This is not as difficult as it may appear, because the molten mixture can be rapidly analysed by instruments which record the exact percentages of all the elements present. It is then comparatively simple to add further amounts of iron, carbon, or other elements so as to bring the new steel up to the required specification.

Tin cans are potentially an important source of scrap ferrous metal and tin. It is now possible to extract tin cans from domestic rubbish very efficiently, but the recycling process is not easy because tin cans contain several materials. In addition to steel they may contain aluminium, they may have joints soldered with a tin–lead alloy, and the inside of the can may be lacquered.

Aluminium Large amounts of aluminium are recycled. Many 'aluminium' items are in fact not made of the pure element, but are aluminium alloys. When scrap aluminium is melted it therefore contains other elements as well as aluminium, and it is not easy to extract the pure metal from the mixture. As explained on page 514, this is not really a problem for the molten mixture can be analysed quickly and precisely and appropriate amounts of other materials can be added to bring the mixture to the required specification for a particular alloy. The production of aluminium ingots is now based almost entirely on recycled scrap aluminium. The electrolytic production of aluminium from bauxite (p. 534) is important where supplies of the *pure* metal are required, e.g. in adding to molten scrap in alloy production, for replacing inevitable 'losses' from the cycle, and to expand the aluminium industry.

Aluminium cans (which are coming increasingly into use), saucepans, and milk bottle tops all form potential sources of aluminium scrap. Unfortunately, 50 000 cans are needed to produce one tonne of scrap metal and at least 20 tonnes must be processed at a time to make the undertaking worthwhile.

Copper Many copper alloys are recycled. Scrap dealers sort out the copper-based scrap into alloy types such as gun metal and the various brasses, and these are delivered to the smelter. Here, as with aluminium reprocessing, appropriate amounts of scrap alloy, virgin metal, and other elements are added to bring the mixture to a particular specification, and it is then cast into ingots which are called 'certified ingots'. If a fairly cheap alloy is being produced, e.g. a yellow brass which does not have a precise specification, the molten mixture may contain 60 per cent of scrap.

Lead The lead industry has for many years recycled large amounts of lead scrap from roofing material, plumbing, and especially car batteries. Recycled lead now contributes a high proportion of total lead 'production'.

Summary The recycling of metals is becoming more efficient all the time. Most foundries now collect even the grinding dust and slag which they produce. These can be reprocessed by vacuum melting so that all but 5 per cent of the metal content can be extracted. To put this in perspective, capital expenditure on machinery capable of extracting foundry dust will pay for itself (in the value of extracted metal) in less than a year!

Plastics A large range of packaging materials is made of plastics, as are many common articles which have a fairly short life, such as toys. Much of this waste material is thrown away with household refuse, and plastics waste now forms about 6 per cent of the world's rubbish. The price of plastics is tied to the price of the crude oil from which they are obtained. As crude oil becomes more difficult to obtain and its price rises, so the recycling of plastics becomes more and more worthwhile.

Only thermosoftening plastics can be completely recycled, because they can be softened and remoulded by heating (p. 425). Unfortunately, plastic is not one substance; there are over 30 different types of thermosoftening plastics in common use, each with its own particular properties and advantages. Moreover, a 'plastic' article will contain several other materials in addition to the basic plastic. To convert plastic waste back into high quality plastics, each with a particular specification, it is essential that each type of plastic is separated from the waste and then processed independently. This involves hand sorting or the use of expensive machinery. One solution to this problem is to use indirect cycling, i.e. to convert mixed plastics into materials which do not need a high degree of purity or uniformity, e.g. shoe soles, bicycle saddles, toys, building boards, pipes, and plant pots.

Even if we cannot recycle some plastics (e.g. the thermosetting varieties), it is important that we should find a use for them rather than abandon them to spoil the environment, for waste plastics of a traditional kind will not rot or corrode; they simply cause litter. Some plastics are now being made *biodegradable* so that they can be acted upon by bacteria after use, and they will eventually rot away. These new plastics are not suitable for all purposes, however, and the problem of disposal of the thermosetting varieties will still continue. All plastics have high calorific values and so in theory waste plastic could be burned to provide energy. In a typical process which makes use of this, waste plastic is mixed with paper and put through a mill. The resulting pellets, known as RDF (refuse-derived fuel) are used as a substitute for coal, but the combustion products may be toxic.

Points for discussion on the recycling of other substances

Try to 'guess' the correct answers to questions 1 to 4, which concern other materials which are recycled. You may get a few surprises!

1. On average, how many newspapers are printed and then unsold each day in Britain? (a) 100 000 (b) 200 000 (c) 1 million

2. A tonne of waste paper can save the lives of a number of mature trees. How many? (a) 5 (b) 10 (c) 17

3. What proportion of the glass produced in the UK is used by the container industry, mostly in the form of *non-returnable* bottles? (a) 10 per cent (b) 40 per cent (c) 70 per cent

4. What proportion of the rubber used in Britain is for tyre manufacture? (a) 10 per cent (b) 30 per cent (c) 50 per cent

5. Make your own comments about the issues raised by the previous questions.

6. Try to find out more information about the recycling of paper and glass.

A summary of the main advantages of recycling waste

1. A saving of resources: trees, metals, oil, energy.
2. Improvement of the environment: no unsightly tips, less use of valuable land.
3. Saving money: smaller bills for taxpayer.

Point for discussion

After copper has been extracted from its ore it is cast into thick plates of impure copper. These plates are set up in electrolysis cells, as shown in Figure 27.1, to refine the metal electrolytically. In a typical electrolysis plant, there are some 1500 tanks and some 100 000 plates undergoing refining. The electrolysis proceeds for approximately 14 days during which time the pure cathode will increase in mass from 5 kg to over 120 kg.

(a) Give an ionic equation for the reaction at (i) the cathode, (ii) the anode.

(b) What happens to the impurities in the copper?

(c) Why is the electrolysis carried out over a long period of time?

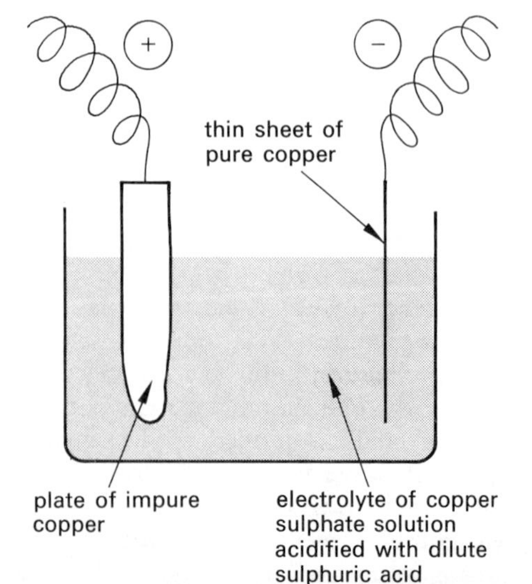

Figure 27.1 Purification of copper (*A.E.B.*)

(d) State *two* properties of copper and for *each* property give an industrial application.
(e) World reserves of copper are being used up rapidly. Comment on the feasibility of (i) recycling copper, (ii) the use of alternative materials for copper. (A.E.B. 1979)

27.2 THE DEVELOPMENT OF A FACTORY OR INDUSTRY

The various stages involved

A typical sequence of events which takes place in the development of an industry is summarized in Table 27.3. Remember that a 'raw material' is something used to produce a final product; it may be a natural resource, or it could be a product from another industry.

The first stage shown in Table 27.3 can be an expensive and frustrating one. Millions of pounds can be spent on trying to find a new oil field, for example, but there is no guarantee that the search will be successful. Industries which depend on raw materials other than natural resources do not have this particular problem. Their difficulties arise from their dependence upon other industries. Any problems in these industries will be beyond the control of the dependent firm, but will nevertheless affect them.

Table 27.3 A summary of the main steps involved in the development of a factory or industry

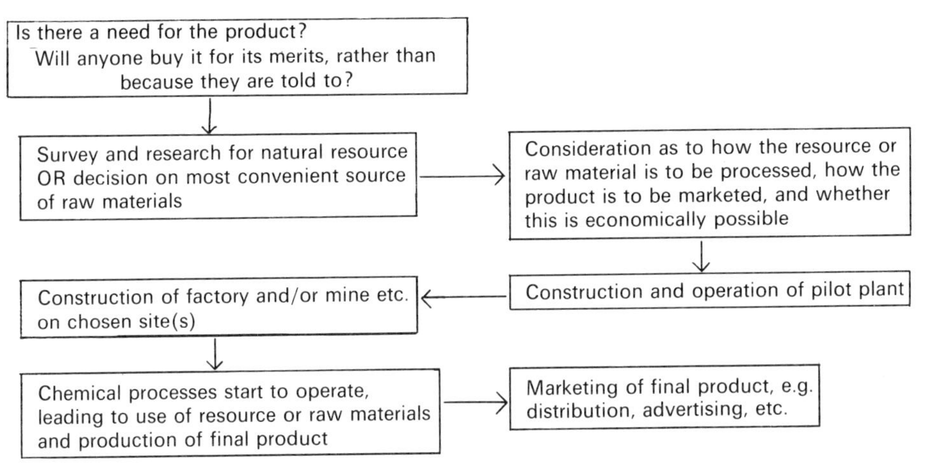

Note: *Complete* developments of this kind, 'from scratch', are rare. Usually, a firm is exploiting an established source of supply, or a known market, or developing an existing site.

Choosing a site for a factory

You have now learned how it is possible to influence the rates of chemical reactions and you have learned important considerations about the use of natural resources. You will need to consider many more factors, however, before you can suggest a suitable site for a factory and an appropriate set of working conditions for the process. We can only deal with some general aspects here, and you must not imagine that you will be competent to choose a suitable site and process when you have read this unit! Remember in all these considerations that the basic aim is to produce the desired substances as cheaply as possible.

Sources of materials

A factory uses raw materials, and it is desirable that these should be easily obtainable. If raw materials are obtained directly from the environment it is much more economical to have a fairly substantial supply in one place than to have smaller deposits in several places. Similarly, a substantial deposit of raw material may be of little use if it is in a very inaccessible part of the world or if it can only be extracted with great difficulty. (See the problem of extracting gold from sea water, mentioned on page 511.) Nevertheless, we are now having to consider both *areas* (e.g. the North Sea) and *materials* (e.g. low grade ores) which were previously uneconomic propositions, because the more accessible or cheaply obtainable raw materials are rapidly being used up.

The method by which a raw material is to be obtained or extracted may also influence a choice of site. For instance, it is more economical to extract a substance by surface mining than it is to extract it from below ground.

A typical example of the way in which an industrial complex can grow up around an important source of easily obtained raw materials is the alkali industry centred upon Cheshire and Merseyside, where deposits of common salt, coal, and limestone are readily available, together with the import and export facilities of a large port, and where there is a ready market for bleaches and dyes in the Lancashire cotton industry. The salt is electrolysed, either in solution to produce chlorine, hydrogen, and sodium hydroxide, or when fused to obtain sodium and chlorine. A whole complex of industries, such as the manufacture of organic chemicals, plastics, dyes, solvents, and detergents, has grown up around these primary and secondary raw materials.

Fuels

Fuel supplies are an important consideration, and very expensive. The fuel to be used should be readily available. So much electricity is used at some factories, for example, that generating stations are often included on the site. Occasionally, the site may be chosen to make use of natural sources of energy, e.g. in the past it may have been important to draw power from a water wheel situated in a suitable flow of water. The production of one tonne of aluminium by electrolysis requires approximately 15 000 kilowatt hours of electricity. This is why there is a large scale production of aluminium at Kinlochleven in Scotland, where there is hydroelectric power available.

Transport

Closely connected with all these factors is the problem of the transport of the raw materials and the finished products. Rail, road, sea, and air links may be important, depending on the materials involved, and pipelines are used for transporting liquids and powders over land and under water. You should be able to suggest advantages and disadvantages of these various forms of transport (see also the *Point for discussion* on page 497).

The actual form in which materials are moved is also important. For example, it is more economical to transport sulphuric acid as the concentrated acid than as the dilute acid, for in the latter case we would be paying for the transport of something which was mainly water. Similarly, oxygen is normally transported as a liquid, which is far more concentrated (about 1700 times) than the gas form at normal pressure.

Other factors

Some developments may be planned for the district which could have important consequences in the future. The cost of the land may be prohibitive. It may be possible to obtain a Government grant if the site is in a development area. Some regions have a tradition for a particular type of work, and local expertise can be very useful. Large volumes of water are often needed for cooling, and washing, and a site in a hard water area could be undesirable. There is also the problem of the disposal of waste; it may be necessary to choose a site near to a large river for the removal of suitably treated waste. Industrial development could cause traffic congestion problems or have an adverse effect on a tourist industry. The preservation of National Parks may have to take precedence over short-term and non-essential economic considerations. Even the direction of the prevailing wind may be important, particularly if gases are emitted by the factory and it is situated near a populated area.

Points for discussion

1. Why do you think that a considerable quantity of aluminium is produced in Norway, even though the country has no natural source of aluminium ore?

2. Why do you think that sodium carbonate is normally transported as 'soda ash' (the anhydrous form) rather than the more commonly used crystals?

Technological problems

The large scale

It is not easy for young chemists to appreciate the differences between a school experiment and the same process on an industrial scale, yet the essential chemistry of the reaction is probably the same. If you compare the apparatus you used for the distillation of crude oil and that for the preparation of sulphur trioxide with photographs of similar processes on an industrial scale (e.g. Figures 4.6 and 27.21), you will begin to realize some of the technological problems involved.

Try to imagine the complications in scaling up even a simple operation such as filtration. Large quantities of suspension may have to be filtered per hour on an industrial scale, and the process needs to be continuous. If the filtration has to be stopped to clean the filter (comparable to replacing the filter paper in a funnel) the process becomes inefficient. An example of filtration on a large scale is shown in Figure 27.2.

Figure 27.2 A filter used in a phosphoric acid plant (*Albright and Wilson*)

Making the best use of available energy

Energy requirements are often very great. In a school laboratory you probably do not worry about making efficient use of a Bunsen burner. For one thing you are not paying for the gas. Imagine a simple distillation on a large scale. The heat losses from the surfaces of the vessels could be very high indeed, and this would represent a most uneconomical use of the apparatus. A very important problem in industry is how to make maximum use of a given amount of heat energy.

Suppose some reactants have to be heated. Waste gases from the reaction may be hot, and so some arrangement is made whereby these gases pass on their heat to incoming reactants before they are allowed to escape. There are many variations on this theme, but every effort is made to utilize any 'waste' heat, or the heat evolved from exothermic reactions, in spite of the technological problems this may create. (See page 548, and Figure 27.3.)

Figure 27.3 Heat exchangers at Fawley refinery (*An Esso photograph*). Extra heat exchangers at Fawley refinery have produced an energy saving of 10% on the unit

Construction materials

At school level there are rarely any serious problems concerning the materials we use for a particular apparatus. Glass tubing and vessels are readily obtainable in a variety of shapes and sizes, and these easily withstand the temperatures used in school laboratories. It is clearly impossible to use glass so extensively in industry where, apart from the obvious problems, very high temperatures and pressures may be used. Again, if a piece of tubing breaks in a laboratory experiment it is an easy matter to replace it, but industry simply cannot afford to have problems of this kind. If a plant has to close down to make any repairs or replacements to pipelines or vessels, the loss of revenue for every minute of non-operation could be enormous, and there is also the additional problem that dangerous chemicals may escape through the fracture. Remember, too, that there may be hundreds of miles of pipeline even in a small chemical plant.

Metals are often used to overcome problems of this kind but they create their own difficulties. Metals are easily corroded, particularly by acids and alkalis, the very substances they are often expected to transport. Chemists are always searching for new materials and ways to improve upon older ones in an effort to overcome such problems. There are now metals available which can withstand corrosion, high temperatures, and high pressures, but they are very expensive, e.g. stainless steel. It is often cheaper in the long term, however, to utilize such components in spite of the high initial outlay. Plastic tubing is being used more widely because plastics are so versatile that they can be produced to have the particular properties required, and of course the search continues for new alloys which can easily be worked, retain a high degree of strength, resist corrosion, and are cheap to produce.

Catalysts

Catalysts are widely used in industry and these again produce problems which are rarely encountered in school. The catalytic chambers must be designed so that a large surface area of catalyst is exposed to the reactants. Catalysts are very easily poisoned by impurities and if this happens the process ceases to be efficient. This means that the purity of the reactants may have to be carefully controlled, and this in turn may necessitate banks of purifiers and complex analysis schemes. For example, the iron catalyst used in the synthesis of ammonia is easily poisoned by sulphur, arsenic, water, carbon monoxide, and carbon dioxide; as the nitrogen and hydrogen used as reactants may contain these substances initially, they must be carefully purified. This factor is rarely encountered at school level, partly because of the small quantities used and also because the reactants are generally obtained in a high degree of purity from reagent bottles, instead of being the comparatively impure raw materials used in industry. Indeed, the reagent bottle is often the end product of an industrial process which has had to battle with very impure substances.

Waste disposal

The disposal of the waste materials produced on an industrial scale is another problem. It is a simple matter to pour a few cubic centimetres of effluent down a school sink or into a 'residue' bottle, and rare indeed are the school experiments which produce large quantities of smoke and dirt. There has been growing concern about the waste products of industrial processes which have been discharged into the air, rivers, and seas, and which pollute our environment. Industry is faced with the task of making effluent harmless by removing dust and toxic substances from waste gases and liquids. As always such processes are very expensive and pose considerable technological problems.

Safety

Although simple safety precautions must always be observed for any practical process, be it in school or in industry, the dangers are potentially much greater on the industrial scale, where large amounts of materials are used and often at higher temperatures and pressures. This imposes further restrictions on the design of the plant.

Summary

We have tried to confine these factors to a few generalizations. There are many more aspects we could consider, such as the complicated machinery needed to perform a multitude of operations, the systems used to carry materials around the plant from one stage to another, the packing of products, and the maintenance of the whole plant. No doubt you can think of others.

Economics

General

The problem of economics is connected with all the ideas we have considered so far, and indeed it is almost impossible to separate economic considerations from the other factors. We have already mentioned fuel, heat conservation, transport, and choice of materials.

It is important to realize that a chemical factory can rarely be considered in isolation. The raw materials it consumes are often the products of other chemical factories and the materials it produces are quite likely to be the feedstock of other industries. The whole picture is thus very complicated. A strike in any one factory, or an increase in the cost of just one raw material, may have repercussions throughout the whole of the chemical industry because so many of the processes are interdependent.

Use of byproducts

Another important factor is that whenever possible any chemicals produced as byproducts should be utilized in some way so that they are not uneconomic waste. If the byproducts can be easily converted into useful materials they are often used on the site, so that a chemical factory which produces one substance only is a rarity. Alternatively, the byproducts may be sold to other companies for use in their manufacturing processes.

Examples of the economic use of byproducts have already been given, and a particularly good example is the production of sodium carbonate (p. 542). See also how calcium sulphate was used to make sulphuric acid, page 548.

In the past many byproducts were regarded as waste, but we are gradually realizing that almost any byproduct has some use, and that recycling even part of it may be very important.

Supply and demand

As the technological age changes so does the demand for particular substances. There are often many ways of producing a specific chemical, particularly if it is organic, and one method may well be the most important one for several years because the byproducts are also utilized. But the demand for a certain substance, or the price and availability of the raw materials, may suddenly change and so the basic chemical process has to be altered. Industry is geared all the time to this problem of supply and demand. For example, the most important chemicals from crude oil are at the moment obtained from the petrol fraction and from the alkane and alkene gases. Because of this the petrochemical industry is orientated towards the demand by cracking the bulk of the heavier fractions into simpler molecules. This has not always been the case, and there was a time when the petrol fraction was almost regarded as a waste product. In the years to come, this situation is likely to change again, as we discussed on page 416.

Points for discussion

1. The first commercial process for making sodium metal was the Castner process, in which sodium hydroxide was melted and electrolysed. The reactions taking place were:

At the anode:

$$4OH^- - 4e^- \rightarrow 2H_2O + O_2$$

At the cathode:

$$4Na^+ + 4e^- \rightarrow 4Na$$

This has been replaced by the Downs process (p. 533) in which molten sodium chloride is used as the electrolyte, and the reactions taking place are:

At the anode:

$$2Cl^- - 2e^- \rightarrow Cl_2$$

At the cathode:

$$2Na^+ + 2e^- \rightarrow Na$$

Suggest reasons why the Downs Cell replaced the Castner process. (The Downs process has several advantages over the Castner process, but it also has one disadvantage. This disadvantage is partly overcome by modifying the electrolyte. Think carefully before you answer the question; you may find a book of data useful.)

2. The following article was written in 1970. It is written in fairly technical language, and involves chemicals which are not familiar to you. Read the article carefully and then answer the following two questions, using your own words.
(a) Why have there been several different ways of manufacturing phenol since the First World War?
(b) What changes have taken place since the article was written which may result in yet another different process for the manufacture of phenol?

A fairly common chemical which has had a varied career in terms of supply and demand is phenol, which was originally obtained from coal tar. In World War I there was a great demand for phenol because it can be converted into picric acid, an explosive. This sudden increase in demand could not be met by the supply from coal tar, and so a new process was developed involving the sulphonation of benzene. When the war was over the demand for phenol decreased just as dramatically as it had earlier increased, but before industry could adjust to the new situation large stocks of unused phenol built up, and the market price of the chemical dropped sharply.

Unexpectedly, a new situation arose. Phenol-formaldehyde plastics were developed and the demand for phenol increased again. The surplus war stocks were soon used up but, as the price of the surplus raw material was so low, the first phenol-formaldehyde plastics were unrealistically cheap. The demand for the plastics would probably have decreased if the price had suddenly increased. To meet this new demand for phenol, industry could not revert to the previous methods of manufacture, as the cost of the product would have been very much higher than that of the stockpiled supplies, and thus the price of the plastic would have risen alarmingly. To overcome this difficulty yet another scheme was put into operation to produce phenol more cheaply than ever before, and so maintain the comparatively low prices for the plastics.

Following the changing pattern of supply and demand other changes have taken place. The latest method of manufacture, the Cumene process, is well established because the chief raw material, propene, is now readily obtained from the petroleum fractions which the petrochemical industry itself is geared to produce, and one of the byproducts, propanone (acetone), is of great commercial value for other manufacturing processes.

The chemistry of the process

There may be several different ways of manufacturing a certain chemical, and the actual choice is governed by the balance between supply and demand (which may itself be continually changing).

Once a decision has been made with regard to the site and the basic process, we then have to work out how we can obtain the highest yield of the product by controlling the conditions of the reaction by the methods we discussed earlier (Chapter 20). The essential idea is to obtain a high yield in the shortest possible time but only within the limits of reasonable economy. For example, a higher yield obtained by an increase in pressure may not be economical because of increased maintenance or technological problems involved in using the high pressure, in which case suitable compromise conditions are chosen.

It is a useful exercise for you to devise a simple costing operation at school. The class can study several ways of making a sample of the same chemical, perform the experimental work, calculate the yield, and then account for the price of raw materials, energy costs, labour costs, value of the product, life of the apparatus used, time factors, and so on.

Points for discussion

1. The scheme outlined on p. 254 could represent some of the events which must take place in the exploitation of any natural resource. Study the diagram carefully and then answer the questions.

Stage I		Stage II		Stage III
Search and survey for the resource	→	Decisions to exploit the resource	→	Building of plant and/or mine
				↓
		Stage V		Stage IV
		Commercial use of resource	←	Purification of resource

(a) Which stage could cost large sums of money and yet yield no commercial return?
(b) Research shows that a particular ore is contaminated with large amounts of a useless clay-like material and with sulphur. State one way in which Stage IV could have an adverse effect upon the environment.
(c) If the resource is a metal oxide, what chemical process must it undergo in order that the metal can be separated?
(d) If this metal occupied a position above zinc in the activity series, how would you attempt to achieve the process mentioned in (c)?
(e) State two factors which would influence the decision made as to where the chemical plant necessary for such a process (as in c) might be built.
(f) In most extraction processes, the chief running costs arise from the cost of the raw material and one other factor. Name this second factor.
(g) Give two reasons why the recycling of materials is likely to increase in the future.

2. Read the section on 'Sources of materials' (p. 518) and then answer the following questions.
(a) Name three unprocessed substances which the chemical industry may obtain from the environment.
(b) Which compound mentioned in the passage has given the chemical industry centred on Cheshire and Merseyside its name?
(c) Which substance mentioned in the passage would be used as a bleach in the Lancashire cotton industry?
(d) Name two secondary raw materials.
(e) Give three reasons, each of a different nature, why the alkali industry has developed in the Cheshire/Merseyside area.

3. Read the following paragraph concerning the extraction of zinc on a commercial scale and then answer the questions below.

The concentrated ore, zinc blend (ZnS), is roasted in air. The sulphur dioxide evolved is removed and used in the manufacture of sulphuric acid. The zinc oxide is then mixed with coke in a furnace and heated to 1400 °C. The zinc vapour evolved is condensed and the liquid run into moulds. The carbon monoxide produced is used to heat the furnace.

(a) Give the *chemical* name for the ore of zinc.
(b) Balance the following equation:

$$ZnS(s)+O_2(g) \rightarrow ZnO(s)+SO_2(g)$$

(c) Why is the sulphur dioxide evolved in this reaction not allowed to escape into the air?
(d) State *one* function of the coke used in the process.
(e) Write the equation for the chemical change which the zinc oxide undergoes in the furnace.
(f) Give *two* major uses of zinc.

4. Baren Island is a fictitious island close to the South Polar ice cap. It is known for its penguin, seal, and whale populations and its steaming 'hot spring' area. The island is mainly hills of approximately 1500 metre which are blanketed in snow and ice. It is not inhabited, the only relic of civilization being the ghost town of St. Wellington once used by whalers and polar explorers.

Extensive geological surveys have shown that raw materials are present in large quantities (see map, Figure 27.4). Coal is three to six metres below the surface and in thick

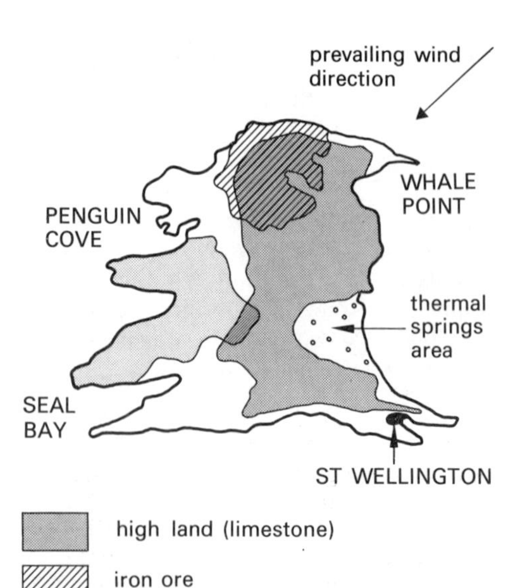

Figure 27.4 Developing Baren Island (*A.E.B.*)

Figure 27.5 Developing Treepool (*A.E.B.*)

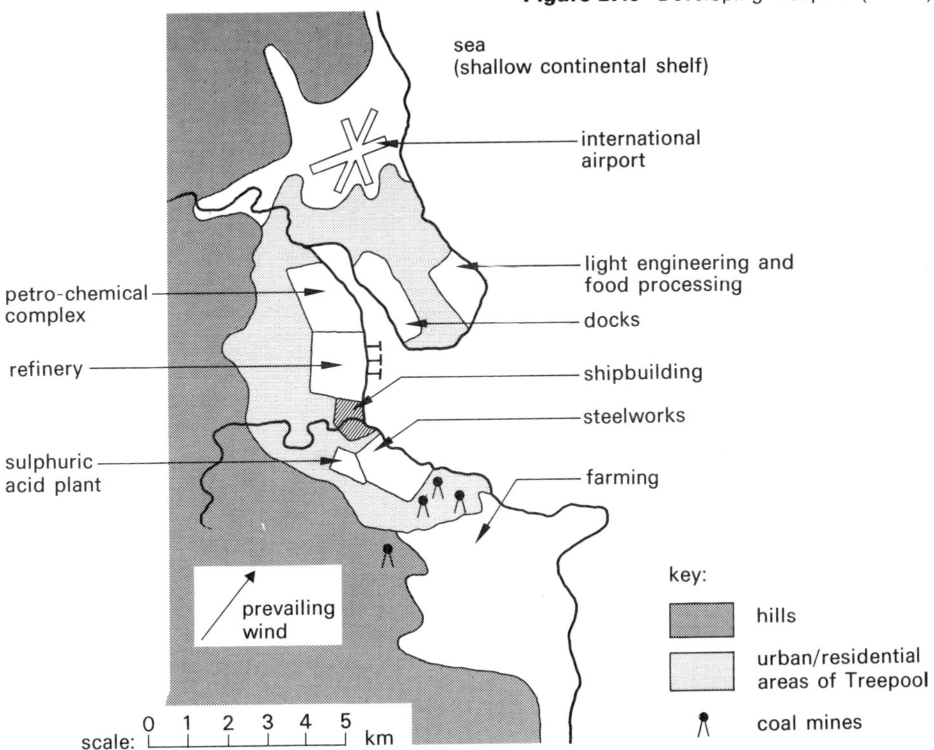

seams; iron ore is high grade and whole mountains appear to be solid ore; the surface rock is limestone.

It has been decided to develop these materials and it will only be economic if all the metal extraction processes are carried out on the island.

Discuss the problems which might arise, and suggest possible solutions for the following:
(a) Developing the ghost town of St. Wellington to house a work force of 1500 people in terms of (i) buildings, (ii) water supply and sewage, (iii) heating and lighting, (iv) food, (v) entertainments and social life.
(b) Developing the metal extraction complex with reference to (i) mining the raw materials, (ii) best site, (iii) disturbance of wild life and pollution to the environment.
(c) Developing transport systems for raw materials and for the refined metal.

(A.E.B. 1977)

5. The fictitious industrial area of Treepool is shown on the map (Figure 27.5). The development of Treepool over the last 120 years from a small fishing port to its present state has been largely historical and without any overall planning.

(a) Study the map of Treepool carefully and then answer the following questions. (*Note*: In the actual examination question, an outline map was provided on which to indicate some of your answers. You may be asked to prepare answers for a class discussion, or to draw your own outline map before answering the question.)

(i) What pollution problems might you expect to find in present-day Treepool?

(ii) What health problems might you expect to find in Treepool?

(b) Assuming that you could have had knowledge, both of the industries that were to be developed in the Treepool area and of their environmental problems, indicate clearly on the outline map where you would have sited each of the following:

(i) the chemical plant and refinery,
(ii) the steel works,
(iii) the sulphuric acid plant,
(iv) residential areas,
(v) the docks and shipbuilding,
(vi) farming areas,
(vii) the airport,
(viii) communications (roads, railways and airport).

(c) For each of the items (i) to (viii) in (b), explain briefly why you have sited them as you have on your replanned map.
(d) State with reasons *one* other industry which you could foresee developing in the area in the future.
(e) State, with reasons, how pollution could be controlled in the new Treepool which you have planned. (A.E.B. 1976)

6. Complete the following table which concerns common symbols (pictograms) used to give information about hazardous chemicals.

Symbol (pictogram)	Meaning	One example of a chemical of this type
	Poisonous (toxic)	
		Hexane

Give *two* advantages of the use of this type of symbol (pictogram) on chemicals.
(A.E.B. 1979)

27.3 SOME EXAMPLES OF INDUSTRIAL PROCESSES

The extraction of iron and its conversion into steel

In 1740 Britain produced less than 20 000 tonnes of iron per year, though we produced more metals than any other country. It is interesting to note that Britain now produces more than three times this amount of iron in one day.

Steel has been made on a large scale since the middle of the nineteenth century. The key step was Bessemer's invention of a process for converting pig iron into steel in 1856. Now there are many specialized steels which are made for particular applications. Steel is the most commonly used structural metal, and nearly all the iron extracted is converted into some kind of steel alloy. There are springy steels, hard steels, stainless steels, and many others, and almost everything we use in daily life has depended on steel at some stage of its manufacture.

The production of iron

The fundamental chemical reaction in the production of the metal is the *reduction* of iron(III) oxide (the essential constituent of iron ore) to the metal itself. In addition, other reactions remove some of the impurities from the ore and provide the heat energy needed for the process.

The overall reduction process can be represented in simple form as

$$Fe_2O_3 - 3[O] \rightarrow 2Fe$$

In theory various reducing agents can be used, e.g. carbon, carbon monoxide (CO), and hydrogen. The process is carried out in a large tower called a blast furnace (because of the 'blast' of hot air forced into the base of the tower) which is illustrated in Figure 27.6 and in diagrammatic form in Figure 27.7.

The traditional blast furnace uses coke as an important part of the *charge* (i.e. the solid materials which are added to the furnace). The coke has two main functions; it burns to provide heat energy, and in doing so it forms carbon monoxide, which is the main reducing agent.

Figure 27.6 The blast furnaces and stoves (*British Steel Corporation*)

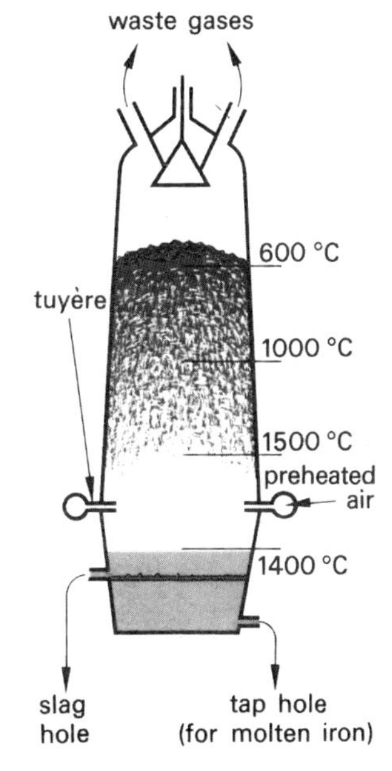

Figure 27.7 The blast furnace

The main reactions which occur inside a traditional blast furnace are summarized below. Pollution and energy aspects are summarized along with those of the steel making process on page 530.

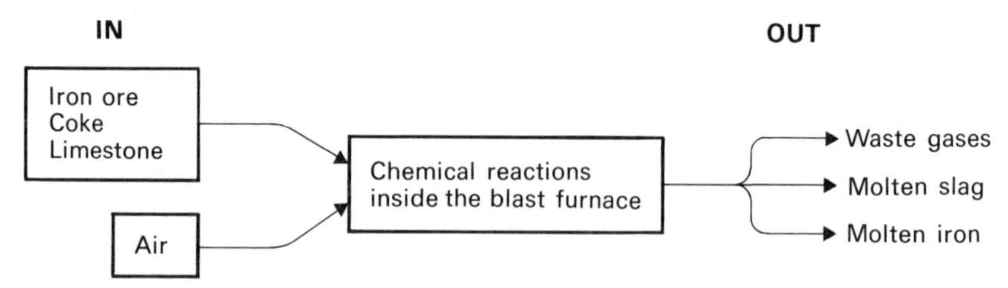

Raw materials Iron ore (impure haematite, iron(III) oxide, Fe_2O_3), coke, and limestone are added as the charge at the top of the furnace, and oxygen is used up from air 'blasted' in at the bottom of the furnace through the tuyères (French—nozzle for air blast).

Chemical reactions producing iron The coke burns in the preheated air:

$$C(s)+O_2(g) \rightarrow CO_2(g) \quad \Delta H = -352 \text{ kJ mol}^{-1}$$

The carbon dioxide reacts with more hot coke to form carbon monoxide:

$$CO_2(g)+C(s) \rightarrow 2CO(g) \quad \Delta H = +171 \text{ kJ mol}^{-1}$$

The carbon monoxide then reduces the iron(III) oxide in stages to iron in the upper part of the furnace at a temperature of about 700 °C. The *overall* equation for the reduction process is:

$$Fe_2O_3(s)+3CO(g) \rightarrow 2Fe(l)+3CO_2(g)$$

This reaction is also exothermic.

Removal of waste products Waste gases (mainly nitrogen and oxides of carbon) escape at the top. The limestone decomposes to form calcium oxide,

$$CaCO_3(s) \rightarrow CaO(s)+CO_2(g)$$

which reacts with the silicon impurities present in the ore to produce slag (molten calcium silicate) which floats on top of the molten iron.

$$CaO(s)+SiO_2(s) \rightarrow CaSiO_3(l)$$

The slag is 'tapped off' from time to time.

Removal of the iron The dense molten iron forms a layer at the bottom of the furnace and is tapped from the furnace at the appropriate time (Figure 27.8). The product (cast iron or pig iron) still contains a significant proportion of carbon and is extremely brittle. Most of the iron produced is immediately converted into steel.

Modifications to the traditional blast furnace

There have been many technological modifications to the traditional blast furnace. In some parts of the world a fuel which is readily available (e.g. natural gas or lignite, page 499) is used to replace some or all of the coke. The essential reduction process in the furnace remains the same, but the proportions of the individual reducing agents (carbon, carbon monoxide, and hydrogen) will vary according to the particular method being used.

Figure 27.8 Tapping the blast furnace (*British Steel Corporation*)

Approximately 40 per cent of the cost of the iron produced by a coke-fuelled blast furnace is due to the cost of the coke itself. In Britain the fuel used in the blast furnaces is still based on coke (and therefore the main reducing agent is carbon monoxide), but additional materials are now injected into the tuyères along with the preheated air. These additional materials may be natural gas, fuel oil, and poor quality coals and cokes. They act as both a fuel and a source of carbon and carbon monoxide.

The conversion of iron into steel

Principle This involves *removing* impurities such as sulphur, silicon, phosphorus, and excess carbon, and may involve the *addition* of controlled quantities of other elements such as manganese, cobalt, and tungsten. These additions give the different varieties of steel their special properties. Note that the excess carbon in pig iron is mainly responsible for its brittle nature, but a small controlled proportion of carbon is left in the steel because it is an essential constituent of the alloy.

The essential chemical reaction in the blast furnace is the *reduction* of the iron ore, but in steel making the impurities are *oxidized* and then removed from the mixture. Excess carbon is oxidized to gaseous carbon monoxide which escapes, and similarly sulphur is removed as oxides of sulphur. The other oxidized impurities combine with calcium oxide (which is added in calculated quantities to the molten mixture) to form a slag, e.g. phosphorus as phosphates and silicon as silicates in the slag.

Although the basic chemistry of the steel-making process remains unchanged, the exact way in which it is carried out changes rapidly as technology improves. The Bessemer process, which used air to oxidize the impurities, is now only a historical curiosity and the Open Hearth Process is rapidly disappearing from use. Apart from the Electric Arc Process (p. 530), all of the modern methods can be described as Basic Oxygen processes, in which oxygen (rather than air) is used directly to oxidize the impurities.

The Basic Oxygen Process There are several variations of the method, the latest versions being called the LD process (named after the steel-producing towns of Linz and Donawitz) and the OBM (Oxy-Bottom Maxhutte) process. What matters at this stage, however, is the chemistry of the process (which has just been summarized) and not the precise technological details, which are likely to continue to change. Figure 27.9 should thus be regarded as a *typical* convertor using the Basic Oxygen Process.

An LD furnace (Figure 27.9) can convert a charge of 350 tonnes into steel in forty minutes, and more recently introduced processes can convert a smaller charge (say 200 tonnes) in only twenty-five minutes. The Basic Oxygen Process is always used on the same site as the iron-making process, because it needs a molten charge and it is more economical to use molten pig iron straight from a blast furnace than to waste energy in remelting solidified iron. The furnace is tilted to receive the charge, which might be 70 per cent molten iron and 30 per cent scrap steel. A water-cooled steel lance blows oxygen into the melt at about 5–15 atmospheres pressure. As explained earlier, the oxygen 'burns out' some of the impurities as volatile oxides and oxidizes other compounds so that they form a slag with the calcium oxide, which is added as required.

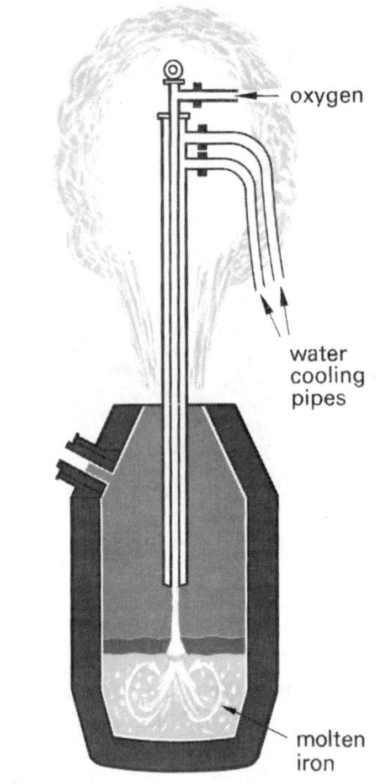

Figure 27.9 The LD process (*British Iron and Steel Federation*)

The Electric Arc Furnace Recycling is likely to become increasingly important, and the Electric Arc Furnace is particularly useful for scrap as its charge consists of *cold* metal. An Electric Arc Furnace can thus be situated away from a blast furnace complex, and in this sense is more versatile. It is also useful for making special steels. On the other hand it is very expensive on fuel (see page 531) and is likely to remain a relatively small contributor to steel making.

A modern Electric Arc Furnace can produce 150 tonnes of steel in four hours. The furnace (Figure 27.10) has a circular bath in which the charge is placed. To charge the furnace, the carbon electrodes are withdrawn, the movable roof is swung back, and the charge is lowered onto the circular bath from an overhead crane. After the electrodes and roof have been replaced in position, a powerful electric current is passed through the charge. An arc is struck between the charge and the electrodes, and a great deal of heat is generated which melts the charge. Samples of the molten steel are tested at frequent intervals, and appropriate amounts of calcium oxide, fluorspar, and iron(III) oxide are added. These combine with most of the impurities in the metal to form a molten slag. Other elements are added if necessary to bring the steel to the required specification. The slag forms a layer above the molten steel, from where it is raked off from time to time.

Environmental factors and energy considerations

The iron and steel industry has had to battle with potentially serious pollution problems, because the processes are 'dirty' ones. The blast furnace uses vast quantities

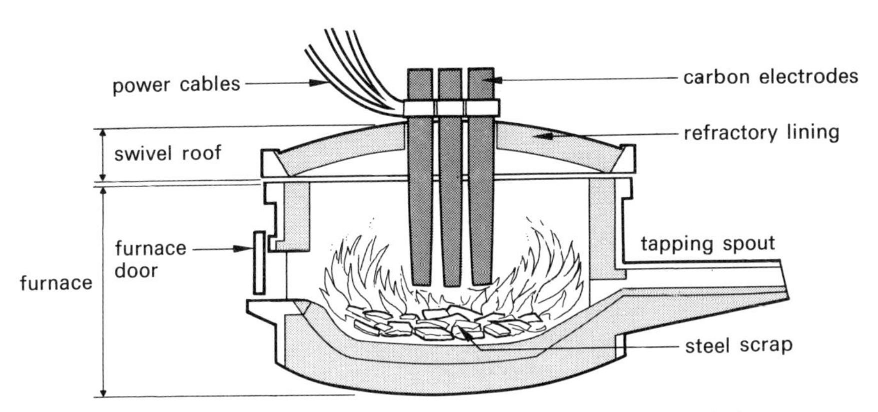

Figure 27.10 The Electric Arc Furnace (*British Steel Corporation*)

of solids, and dust is a problem. When coal is treated to produce coke for the process, many other chemicals are formed (p. 501) which must be carefully removed and used. The reactions inside the furnaces produce vast quantities of smoke and gases, some of which are toxic (e.g. sulphur dioxide), and similar problems occur during the conversion to steel. There is also the problem of disposing of large quantities of solid slag.

Modern iron and steel processes are much more efficient than they used to be. The dust and smoke are almost completely removed, and the waste gases are thoroughly 'scrubbed' to clean them before they are allowed to enter the atmosphere. Similarly, slag is no longer indiscriminately dumped; it is used in the construction of roads, for filling in quarries which are then 'planted out', and so on.

The use of a double bell charging system (Figure 27.11) prevents escape of waste gases and heat energy as the blast furnace is charged. In part (a) of the figure both bell hoppers are in the raised (closed) position. When the furnace is charged, the charge is placed on top of the (still closed) small hopper. In (b) the small hopper is lowered so that the charge falls on to the large hopper. In (c) the small hopper is closed and the large hopper is lowered to allow the charge to fall into the furnace. At all stages during this sequence of operations at least one of the hoppers is closed, thus ensuring that waste gases cannot enter the atmosphere. (The shape of the bell also helps in a more even distribution of the charge in the furnace.)

Energy demands are minimized in the blast furnace because most of the reactions which occur in the furnace are overall exothermic and are self-sustaining once started. For this reason, a blast furnace normally operates *continuously* for up to two years or so, at which stage it has to be temporarily closed down so that the refractory (heat-resistant) lining can be replaced. The heat energy of the hot waste gases is used to preheat the oxygen in the blast, so that every possible use is made of the available energy.

In the conversion into steel, the Electric Arc Furnace consumes large amounts of electrical energy and the carbon electrodes are rapidly burned away. Each of these factors results in high running costs, and the Basic Oxygen Process is likely to remain the most important steel-manufacturing process.

Check your understanding

1. A large blast furnace can use up to 4000 tonnes of air per day.

(a) State which gas from the air is used in the furnace and give the equation for the reaction involving this gas.

(b) Give the equation for the reduction of iron ore (Fe_2O_3) to iron in the furnace.

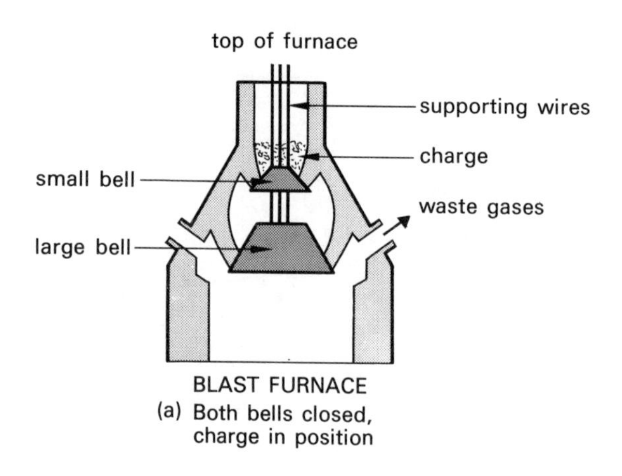

(a) Both bells closed, charge in position

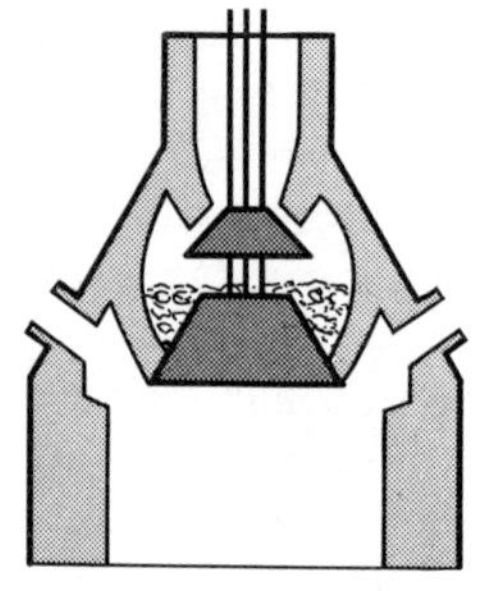

(b) Small bell open, charge falls on to second bell

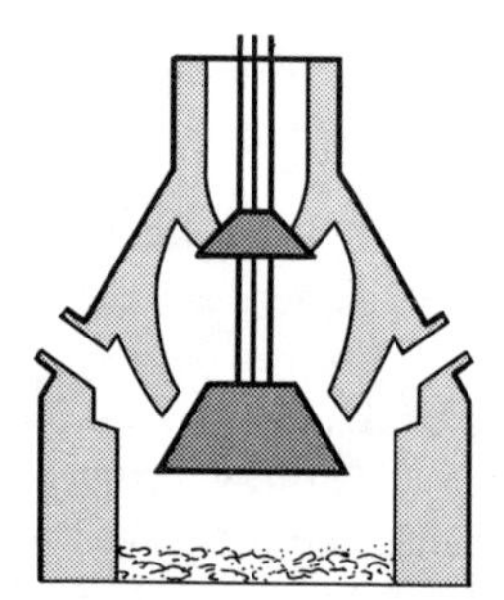

(c) Small bell closed, large bell open. Charge falls into furnace

Figure 27.11 The double bell system

(c) Why is limestone added to the charge for the blast furnace?
(d) The iron produced by a blast furnace is brittle and is known as pig iron. Name the main impurity which causes pig iron to be brittle.
(e) Describe briefly how pig iron is converted into steel.
(f) Name two metals which are added to steel to give it greater strength and hardness.
(g) What atmospheric pollution problems are associated with the conversion of iron into steel? (A.E.B. 1978)

2. The corrosion of iron is estimated to cost several hundred million pounds per year in this country.
(a) Explain why this corrosion, or rusting, costs so much.
(b) Give two factors which help rusting to occur.
(c) Give two different methods of preventing rusting.
(d) With very large ships several kilograms of rust can form on a ship in a few days. Calculate how many kg of oxygen and how many kg of iron would be used up if 8 kg of rust accumulated on a ship in one day. Assume the formula of rust is Fe_2O_3. (A.E.B. 1976)

Points for discussion

1. Can you suggest a property which silicon oxide, phosphorus oxides, and sulphur oxides have in common, and which results in them reacting with calcium oxide to form a slag during the steel-making process?

2. Can you suggest why the main steel-making processes are referred to as *Basic* Oxygen Processes?

The extraction of sodium and aluminium

Metals high in the activity series are so reactive that they are always found combined in nature as ionic compounds. Such is the readiness of these metals to lose electrons that it is very difficult to change their ions back into the metal atoms by making them gain electrons, and their compounds are difficult to decompose. It is comparatively easy to reduce iron(III) oxide to the metal by using carbon monoxide in a blast furnace, but this method cannot be used for compounds of sodium and aluminium which have to be decomposed by electrolysis in order to produce the metals.

The extraction of sodium in the Downs Cell

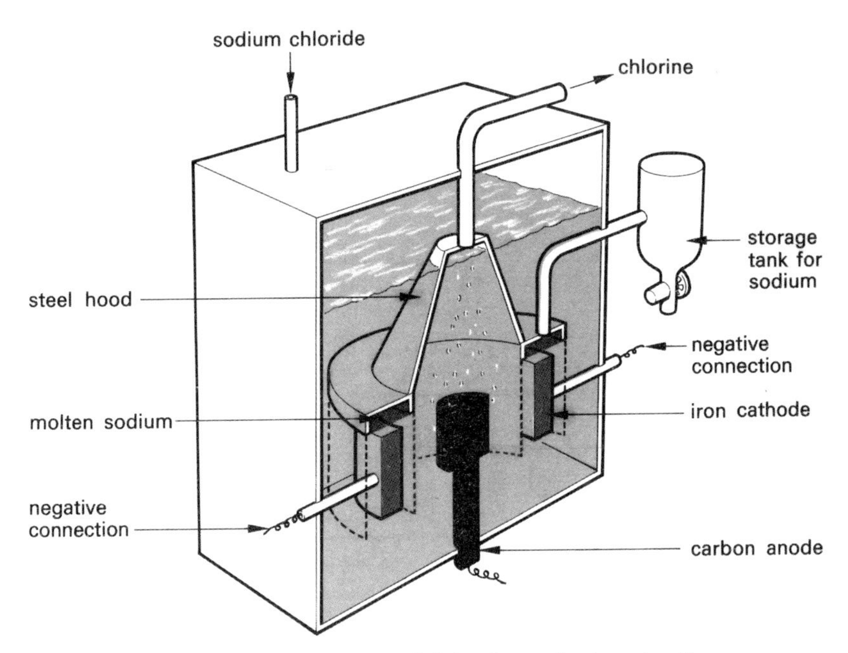

Figure 27.12 The Downs Cell for the production of sodium

A typical Downs cell is illustrated in Figure 27.12. A molten ionic compound of sodium has to be used as the electrolyte because if an aqueous solution was used, hydrogen would be discharged at the cathode in preference to sodium. Molten sodium chloride is the important part of the electrolyte, and sodium ions are discharged at the cathode to form sodium metal. Chlorine is formed at the anode, and is a valuable byproduct.

At the cathode: $2Na^+ + 2e^- \rightarrow 2Na$
At the anode: $2Cl^- - 2e^- \rightarrow Cl_2$

The molten sodium is collected in an inverted circular trough from where it runs into an outer storage tank. The steel hood prevents the chlorine coming into contact with the sodium (the two would react together) and also prevents the sodium from reacting with air.

Environmental factors and energy considerations The melting point of sodium chloride is high (801 °C), and this prevented its use in earlier cells. The operating temperature is reduced to 600 °C by adding up to 60 per cent of calcium chloride to the electrolyte. The calcium chloride acts as an 'impurity' in the sodium chloride and therefore lowers its melting point (p. 54). This allows a considerable saving on the energy needed to melt the electroyte. Note that the fused electrolyte contains both sodium ions and calcium ions as cations, but the product at the cathode is mainly sodium. The small proportion of calcium which is present in the molten alloy crystallizes out on cooling, leaving fairly pure molten sodium.

Both products (sodium and chlorine) are of commercial importance, which helps to make the process economical, as does the comparatively cheap and apparently plentiful supply of the electrolyte, sodium chloride, which is readily available in Britain. There is no waste material.

In spite of the factors quoted, the process still has high running costs because of its large demand on electrical energy, some of which helps to keep the electrolyte molten and the rest of which is used for the actual electrolysis.

The extraction of aluminium

Aluminium is a very important metal; between 1969 and 1980 the demand for aluminium doubled, and world consumption in 1980 ran at 15 million tonnes.

Sources of raw materials and initial purification Bauxite, the main ore of aluminium, is the impure hydrated oxide, $Al_2O_3 . 3H_2O$. Supplies of bauxite are obtained mainly from Australia (Queensland) and Jamaica. The ore has to be purified before it is suitable for use in electrolysis, because it contains up to 50 per cent by mass of impurities, mainly iron(III) oxide and silicon compounds. The purification process is normally carried out near to the bauxite mines, and the purified oxide is then shipped to areas where aluminium is extracted from it by electrolysis.

The purification of the bauxite can be summarized as follows. The ore is ground up and treated with hot sodium hydroxide solution under pressure. This dissolves the (amphoteric) aluminium oxide but not the impurities, which are filtered off as a 'red mud'. Pure aluminium hydroxide is crystallized from the solution and then heated to form the pure oxide.

$$2Al(OH)_3(s) \rightarrow Al_2O_3(s) + 3H_2O(g)$$

The electrolytic process Reducing agents such as carbon cannot reduce the oxide to the metal, and the electrolysis of an aqueous solution of an aluminium compound would result in the discharge of hydrogen at the cathode in preference to aluminium. The oxide itself has a very high melting point (2050 °C) and the molten compound is a very poor conductor. For technical reasons other compounds of aluminium are not suitable as electrolytes. The commercial extraction of aluminium only became feasible, therefore, when Hall (in the USA) and Héroult (in France) independently discovered in 1886 that the mineral cryolite, when molten, can act as a 'solvent' for aluminium oxide at a temperature of 900 °C, and that the solution is a good conductor and yields aluminium metal at the cathode on electrolysis. (Cryolite has the formula Na_3AlF_6.) The electrolyte usually contains a 10 per cent solution of the oxide in cryolite.

The important details of the electrolysis A typical cell used for the production of aluminium is illustrated in Figure 27.13, and the process is often called the Hall–Héroult process. The cell is a rectangular steel tank lined with heat-resistant bricks, on the inside of which is a carbon lining which initially acts as the cathode. The anodes are carbon blocks made of very pure graphite.

The reactions taking place in the cell are complicated (more than 20 different ions are known to exist in the solution!), and the ionic equations listed below are only a *summary* of the reactions which actually take place.

At the cathode: $2Al^{3+} + 6e^- \rightarrow 2Al$
At the anode: $3O^{2-} - 6e^- \rightarrow 1\frac{1}{2}O_2$

The aluminium is molten at the operating temperature of the cell, and is drawn off from time to time. The molten metal becomes the actual cathode surface during the electrolysis.

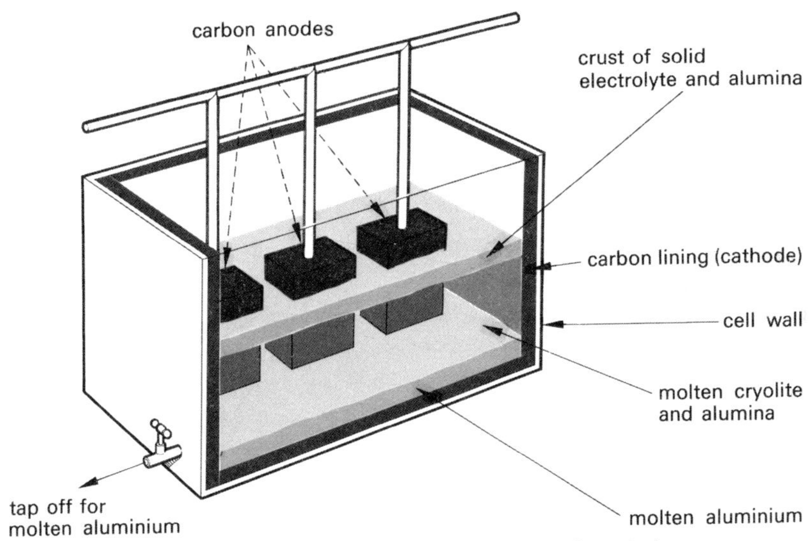

Figure 27.13 The extraction of aluminium by electrolysis

Environmental factors and energy considerations During the purification of the bauxite, very large quantities of 'red mud' are produced. This consists mainly of iron(III) oxide and silicon compounds. It is not economic to extract iron from the mud, so it is dumped. This occurs in the regions of the bauxite mines and not where the extraction plants are situated, so causing environmental problems in the bauxite-producing countries rather than the UK. It compares with the production of slag in the blast furnace, in the sense that a solid material is produced as a byproduct of the main process. Note that 4 tonnes of ore and 0.15 tonnes of sodium hydroxide produce 2 tonnes of the pure oxide, which eventually yields 1 tonne of the metal. It is estimated that known reserves of bauxite should last until the year 2010.

In theory the cryolite used to dissolve the purified oxide is not decomposed during the electrolysis, but in practice some cryolite is 'lost' and it has to be replaced. Natural supplies of cryolite are declining because of this, but fortunately the material can be made artificially and most of the cryolite now used in the process is synthesized.

Much of the oxygen produced at the anodes oxidizes the carbon to carbon dioxide, and so the oxygen produced is not collected as a byproduct. The anodes are fairly rapidly 'worn away' by this reaction, and need to be replaced regularly. (The production of 1 tonne of metal results in the loss of 0.6 tonnes of graphite from the anodes.) Replacement of the anodes is expensive, and since the cell produces only one product of commercial importance there is no sale of byproducts to help offset the running costs.

The main cost of the process is the very large amount of electricity it uses. The electrolysis process is demanding in electricity because each Al^{3+} ion which is discharged requires three electrons (cf. one electron per Na^+ ion in the Downs cell), i.e. 3 faradays per mole of aluminium ions, but not all of the electricity consumed is used in the actual electrolysis; 75 per cent of the electrical energy is used in keeping the electrolyte molten.

Points for discussion

1. The energy required for the production of aluminium is 14.9 kWh kg^{-1}, which corresponds to 53.6 MJ kg^{-1}. As explained on page 495, the conversion of energy from a primary fuel into electricity results in a loss of about two-thirds of the energy which is theoretically available. In terms of a *primary* fuel, therefore, the production of aluminium requires even more energy, i.e. 215 MJ kg^{-1}. Compare this with the primary fuel demands of 30 MJ kg^{-1} for pig iron and 85 MJ kg^{-1} for copper.
2. Aluminium is expensive to produce, and it is likely to become even more expensive as the price of energy soars and the bauxite ores become more difficult to obtain. In spite of this, the use of aluminium doubled between 1969 and 1980. Why is the metal so important and in such great demand?
3. Why is bauxite purified and then exported, rather than exported as bauxite ore?
4. What other substances could be used as a source of the metal, apart from bauxite and recycled scrap?
5. Why are aluminium extraction plants often situated in areas with a low density of population?
6. Modern man's needs include transport, electricity, and aluminium. In obtaining these he also causes pollution to the atmosphere. Complete the following table.

Activity of man	*Main gaseous pollutant produced (name and formula)*	*Effect of this gas on living material*
Transportation using the internal combustion engine		
Generation of electricity in coal fired power stations		
Aluminium extraction		

(A.E.B. 1979)

The production of chlorine and sodium hydroxide by the electrolysis of sodium chloride solution (brine)

The electrolysis of sodium chloride solution yields three products, sodium hydroxide, chlorine, and hydrogen. (Compare this with the Downs Cell, page 533, which uses *molten* sodium chloride.) Until the 1950s the electrolysis of brine was carried out mainly to produce sodium hydroxide, which is the most important alkali. The chlorine and hydrogen also produced were 'inconvenient' byproducts because they could not be used commercially at the same rate as they were produced. Figure 27.14 shows how the demand for chlorine has grown since 1950, largely because of the production of PVC. The electrolysis of brine has therefore increased in importance, and all three products now find ready markets. The process is now operated on a very large scale, and in a typical year 30 million tonnes of each product may be manufactured.

The principle of the electrolysis

The ions present in sodium chloride solution are $Na^+(aq)$, $Cl^-(aq)$, $H^+(aq)$, and $OH^-(aq)$. On electrolysis, hydrogen is normally discharged at the cathode (this is done indirectly in the mercury cell) and chlorine at the anode, and the solution thus becomes richer in sodium hydroxide.

discharged at cathode ← $H^+(aq)$ $Na^+(aq)$ $Cl^-(aq)$ → discharged at anode $OH^-(aq)$

Chlorine reacts with sodium hydroxide and with hydrogen, and so the cell has to be designed to make sure that the products are kept separate from one another. Two types of cell are in current use.

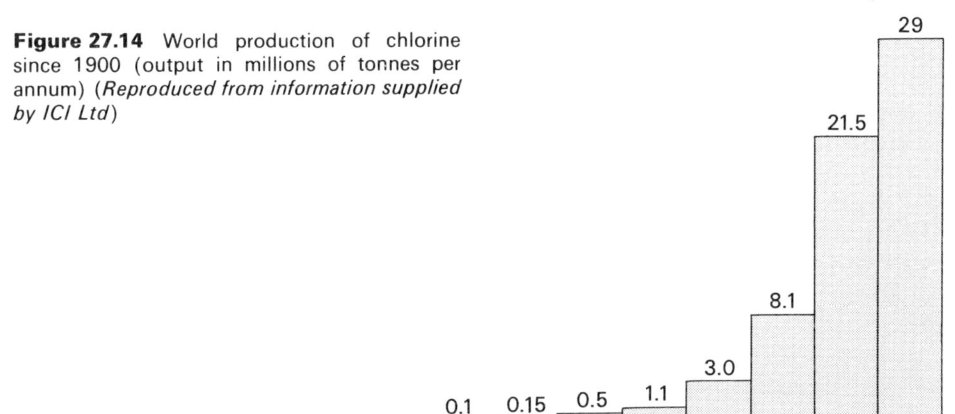

Figure 27.14 World production of chlorine since 1900 (output in millions of tonnes per annum) (*Reproduced from information supplied by ICI Ltd*)

The Diaphragm Cell

The essential chemistry which takes place in the Diaphragm Cell is shown in Figure 27.15. The titanium anodes and steel gauze cathodes are separated from each other by a porous asbestos diaphragm. Saturated brine (containing about 25 per cent by mass of sodium chloride) is added to the anode compartment, and the level of solution is kept higher than that in the cathode compartment so that liquid slowly seeps through the diaphragm from anode compartment to cathode compartment. However, the liquid passing through the diaphragm from the anode is less concentrated in chloride ions than the initial solution, because while it is in the anode compartment chloride ions are continuously discharged at the anode to produce chlorine gas, which is piped off.

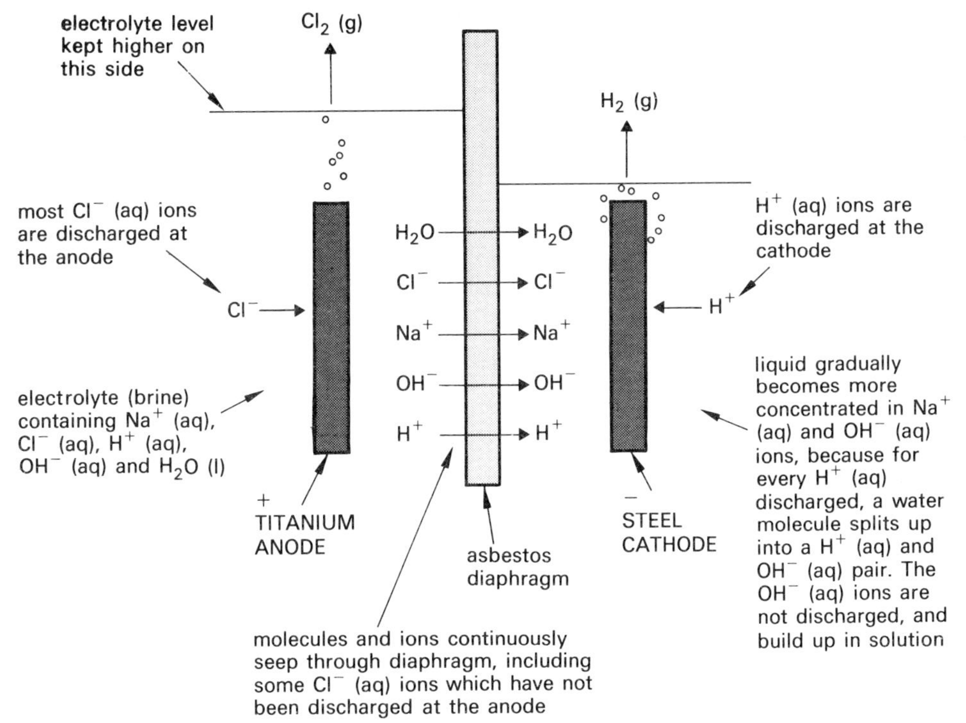

Figure 27.15 Simple illustration of reactions in the Diaphragm Cell (*from information supplied by ICI Ltd*)

At the anode: $2Cl^-(aq) - 2e^- \rightarrow Cl_2(g)$

The liquid 'arriving' in the cathode compartment has thus already 'lost' chloride ions (about 50 per cent of those present initially), and it also loses hydrogen ions because these are discharged at the cathode to form hydrogen gas, which is piped off.

At the cathode: $2H^+(aq) + 2e^- \rightarrow H_2(g)$

These reactions are summarized in Figure 27.15.

The solution in the cathode compartments is drained off as necessary. It contains about 10 per cent by mass of sodium hydroxide and 15 per cent by mass of sodium chloride. It is heated until the volume has been reduced by about 80 per cent, and the less soluble sodium chloride crystallizes out of the concentrated solution, from which it can be separated. The liquid so produced contains about 50 per cent by mass of sodium hydroxide and only a trace (about 1 per cent) of sodium chloride. This step is a fractional crystallization (see page 173).

An actual cell contains several anodes and cathodes (Figure 27.16) and up to 60 cells may be joined in series. The outward appearance and actual design details of the cells change from time to time as improvements are made, but the essential chemistry of the process remains unchanged. It is thus more important to understand Figure 27.15 than Figure 27.16.

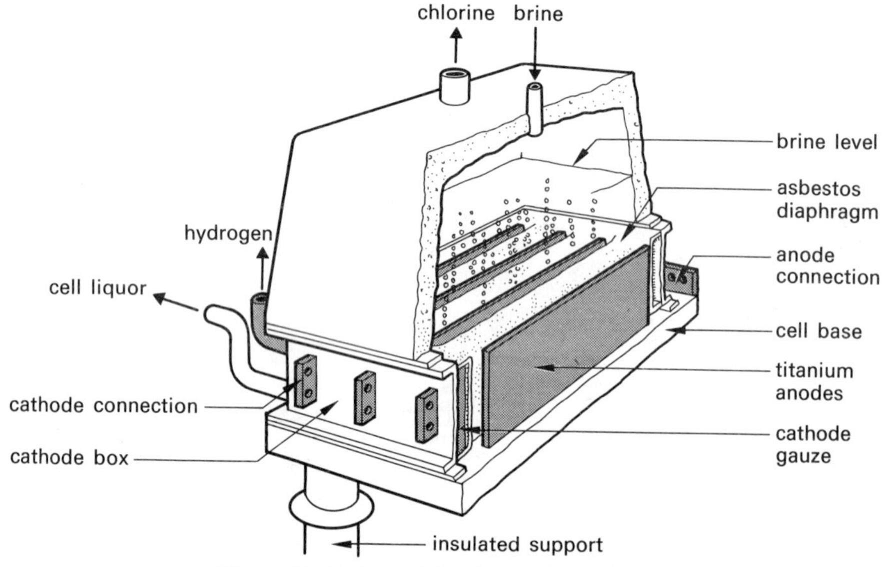

Figure 27.16 Typical Diaphragm Cell (*ICI Ltd*)

The Mercury Cell

This usually consists of two troughs of opposing slopes (about 1 in 80) which are illustrated in Figure 27.17. The electrolyte is sodium chloride solution and the anodes are made of titanium, as in the Diaphragm Cell, but the cathode is a film of mercury which flows slowly across the floor of the cell (in both compartments).

The anode reaction is exactly the same as that in the Diaphragm Cell, and the discharged chlorine gas is piped off.

At the anode: $2Cl^-(aq) - 2e^- \rightarrow Cl_2(g)$

The cathode reaction is entirely different. When the cathode is mercury, sodium ions are discharged in preference to hydrogen ions. The discharged sodium metal does not react with the water in the solution because it immediately forms a liquid alloy (amalgam) with the mercury, and it is carried with the mercury into the second compartment.

At the cathode: $2Na^+(aq) + 2e^- \rightarrow 2Na(s)$

$Na(s) + Hg(l) \rightarrow Na/Hg(l)$

The sodium amalgam is decomposed in the second compartment (which is often called the 'decomposer') when it makes contact with the graphite blocks. The essential reaction which takes place can be summarized as:

$$2Na/Hg(l) + 2H_2O(l) \rightarrow 2NaOH(aq) + H_2(g) + 2Hg(l)$$

The mercury is recycled to the first compartment, the hydrogen is piped off. The liquid removed from the decomposer contains 50 per cent of sodium hydroxide by mass, i.e. it is equivalent to that produced by the Diaphragm Cell, but it is of higher purity.

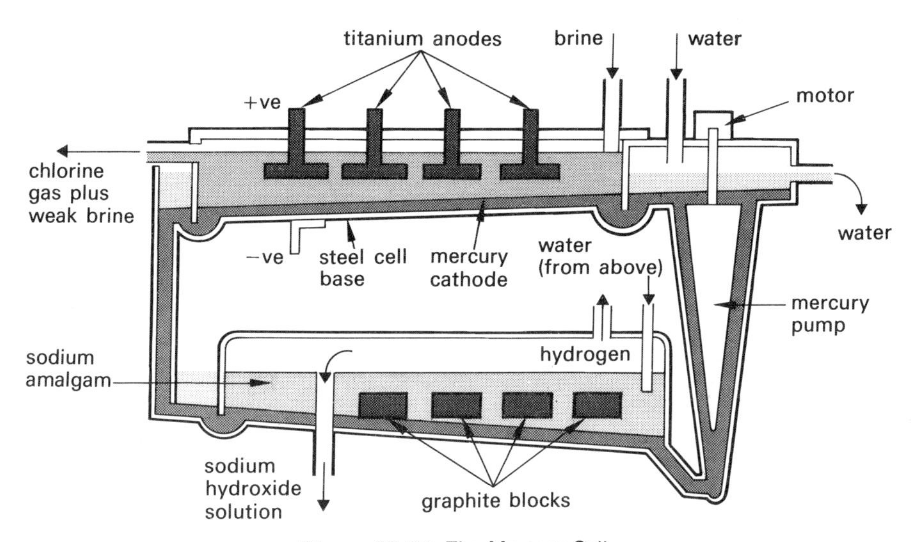

Figure 27.17 The Mercury Cell

Environmental factors and energy considerations

The starting material, sodium chloride, is readily available and plentiful in Britain, and all three products are of commercial importance. Even more important, demand for them is roughly in proportion to the quantities produced by the process, and so there is no excess of any of them. These factors all contribute to the economy of the process. (Remember that the Downs Cell also produces chlorine, but it is operated on a much smaller scale than the electrolysis of sodium chloride solution. The Downs process is operated mainly to produce sodium, and its contribution to the global supply of chlorine is not very important.)

There is no solid waste produced by either Diaphragm or Mercury Cells, and no pollutant gases should enter the atmosphere. The choice between the two types of cell thus depends upon the energy they each consume, the capital costs required to build them, the purity of the sodium hydroxide they produce, and the cost of combating the potential hazard of mercury pollution in the Mercury Cell. These factors are compared in Table 27.4.

Energy demands again represent a high proportion of the cost of the process. The power consumed by a typical plant in order to electrolyse sodium chloride solution is equivalent to the output of a large modern power station or the requirements of a large town.

Table 27.4 A comparison between the Diaphragm and Mercury Cells

	Diaphragm cell	*Mercury cell*
Construction	Simple and inexpensive	Expensive (especially cost of mercury)
Energy demands	Run at lower voltage (3.8 V) than Mercury cells, but saving in electricity offset by heat energy needed to evaporate solution and crystallize out sodium chloride	Run at higher voltage (4.5 V) but no other energy demands
Purity of sodium hydroxide	Reasonably pure for most purposes, but does contain some sodium chloride	Very high purity
Other factors	None	Mercury is toxic and must be removed from effluent

(From data supplied by ICI Ltd)

Points for discussion

1. The electrical energy consumed by the Mercury Cell in order to produce a given quantity of sodium hydroxide is almost exactly the same as the combined electrical and heat energy needed to produce the same quantity of product by the Diaphragm Cell. In terms of *primary* fuel demand, however, the Diaphragm Cell is cheaper to operate. Can you explain this?

2. Two of the products of the process are transported in 'liquid' form. Chlorine gas is easy to liquefy and is normally transported as liquid chlorine in tankers. Sodium hydroxide is normally transported as the 50 per cent aqueous solution produced by the process, again in tankers. However, it is not practical to liquefy hydrogen and this is normally transported under pressure in steel cylinders.

(a) A typical lorry and trailer carrying cylinders of hydrogen may weigh 32 000 kg, of which 100 kg is actual gas. Such a lorry is said to have a *payload* of 0.3 per cent. A liquid chlorine tanker normally carries 16 000 kg of liquid chlorine and has a typical total mass of 23 000 kg. What is the payload (as a per cent of total mass) of the typical chlorine tanker?

(b) The electrolysis of brine produces equal volumes of chlorine and hydrogen, but the consumer has to pay a much higher price for hydrogen than for chlorine. Why is this so?

The fractional distillation of liquid air

Commercially, oxygen is extremely important. Very large quantities are needed in order to change pig iron into steel (p. 529). On such a large scale, the only economic source of oxygen is the air. The air is a mixture, and thus it can be separated into its various components by one or more of the techniques discussed in Chapter 4. It is essential to obtain oxygen with a high degree of purity, and so the air is liquefied and then fractionally distilled. Each component 'boils off' at its own boiling point (e.g. liquid nitrogen at −195 °C and liquid oxygen at −183 °C) and is collected separately.

The steps involved in the process (*see Figure 27.18*)

1. The first step is to remove both carbon dioxide and water vapour from the air, because at the low temperatures used in the process both of these would solidify and block the pipes.

2. The remaining gases (nitrogen, oxygen, and noble gases) are compressed to about 200 atmospheres pressure. This produces heat (in the same way that the pumping up of a car tyre or bicycle tyre also causes heat) and so the compressed gases must be immediately cooled again. This is achieved by allowing them to pass through a central

pipe surrounded by another pipe which contains the cold gases produced by a later stage in the process (Figure 27.18).
3. The cooled, compressed gases are then allowed to expand by emerging from a fine hole into a chamber at a lower pressure (Figure 27.18). The molecules are close together in the compressed gas, and when they expand into a 'low pressure chamber' they move further apart. To do this they need to overcome the weak intermolecular forces between them. As there is no 'outside' source of heat, the energy needed to overcome the intermolecular forces must come from the energy of the molecules themselves. This causes a reduction in temperature. A similar situation occurs when air under pressure escapes from a bicycle tyre or car tyre valve; it feels cold.

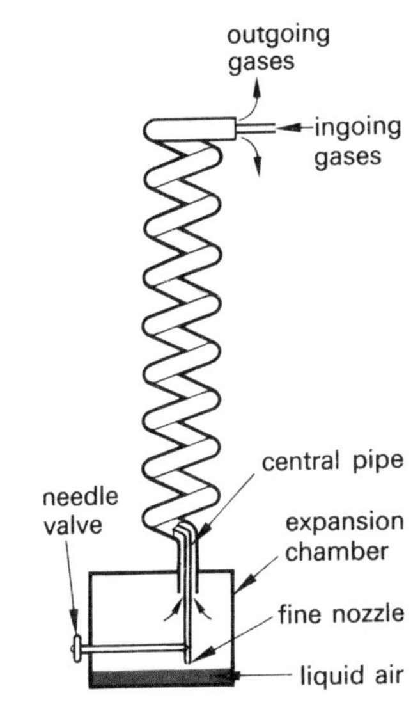

Figure 27.18 The liquefaction of air

4. The cold gases are taken back to the compressor, cooling incoming gases on the way, and are then compressed again to 200 atmospheres and repeatedly taken through the same cycle. The gases become colder after each cycle, and eventually they liquefy.
5. The liquid air is then allowed to evaporate in a fractionating column. The gas which boils off first is mainly nitrogen (b.p. −195 °C) and this is compressed into cylinders. The liquid left behind consists finally of liquid oxygen, which is transported as such or is allowed to evaporate and then compressed into cylinders. The individual noble gases can also be obtained in a high degree of purity by carefully controlling the evaporation of the liquid air, but for many purposes they are allowed to remain in the 'liquid oxygen' for they do not interfere with its normal applications.

The main steps in the process are summarized in Figure 27.19.

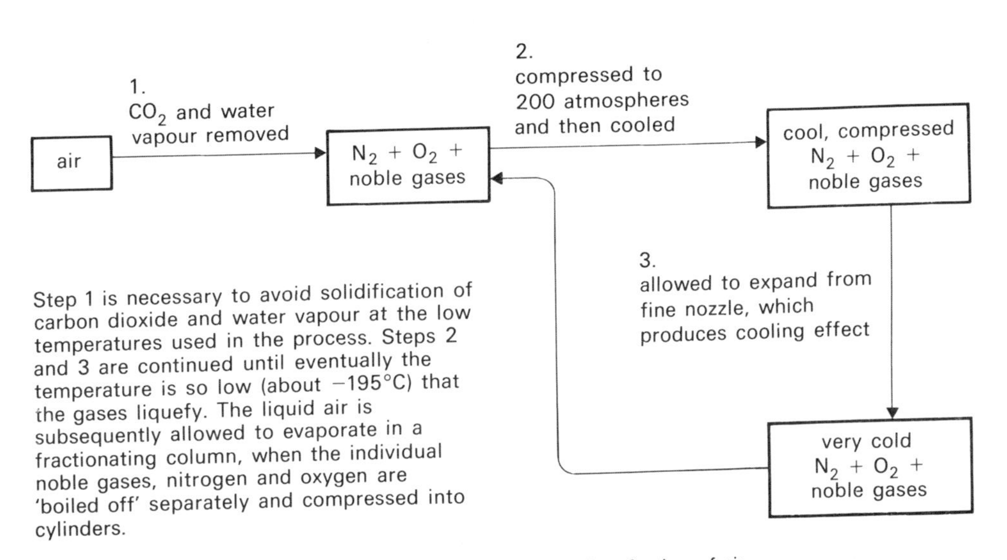

Figure 27.19 A summary of the liquefaction of air

Environmental factors and energy considerations

This is a very clean process; there is no pollution, no waste, and the oxygen removed from the atmosphere is soon replaced through photosynthesis.

The raw material is obviously cheap(!) and readily available, and the process is conducted mainly to produce oxygen. The nitrogen and noble gases do have special uses, but are not required on the same scale as oxygen.

Energy demands are fairly low. The gases are continuously compressed (electrical energy needed) but the main cooling effect is self-contained, and no external energy is needed for the fractional distillation.

If the gases are transported in cylinders, their cost is quite high (cf. hydrogen, page 540) but oxygen is usually liquefied and transported in tankers, and many steel-making sites have a direct supply of liquid oxygen by pipeline.

Point for discussion

A domestic refrigerator uses the same principle as that used to liquefy air. How does it work?

The manufacture of sodium carbonate by the Ammonia–Soda (Solvay) Process

This is another very important industrial process, which is utilized on a large scale throughout the world. Much of the sodium carbonate produced is used in the manufacture of glass, but its other applications are many and varied. They include steel processing, enamelling, soap additives, the heavy chemicals industry, textiles, dyes, food and drink, and the processing of oils, fats, waxes, and sugars. In addition, some sodium hydrogencarbonate is produced in the manufacturing sequence, and this also has important uses, e.g. in the food trade. Such is the success of the process, that sodium hydrogencarbonate is one of the purest industrial chemicals, containing less than 30 g of impurities (mainly sodium chloride) per tonne.

The chief raw materials for the process are sodium chloride and calcium carbonate, and the *theoretical* overall equation can be considered as

$$2NaCl + CaCO_3 \rightarrow CaCl_2 + Na_2CO_3$$

However, as calcium carbonate does not react directly with sodium chloride, the operating process has to involve several steps.

The operating process

Figure 27.20(a) is a flow diagram of the process. The limestone is heated in a kiln where it dissociates into calcium oxide and carbon dioxide:

$$CaCO_3(s) \rightarrow CaO(s) + CO_2(g) \qquad (1)$$

The carbon dioxide is passed into the Solvay tower (Figure 27.21 (b)). Ammonia is used to saturate the incoming sodium chloride solution, and the resulting 'ammoniacal brine' must be cooled because of the considerable amount of heat liberated during this solution. This 'ammoniacal brine' next passes into the Solvay tower up which the carbon dioxide is blown. The reactions occurring may be summarized as:

$$\underbrace{Na^+(aq) + Cl^-(aq)}_{\text{Brine}} + CO_2(g) + NH_3(g) + H_2O(l) \rightarrow$$

$$NaHCO_3(s) + NH_4^+(aq) + Cl^-(aq) \qquad (2)$$

The precipitated sodium hydrogencarbonate is filtered off on a rotating filter, washed

Figure 27.20 Manufacture of sodium carbonate

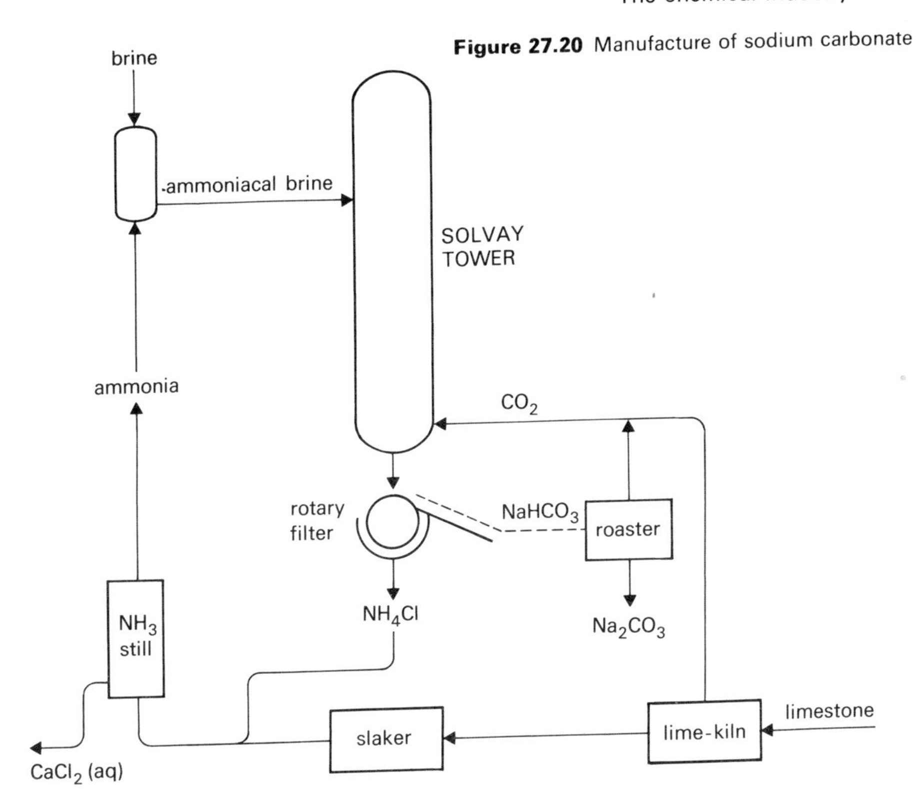

(a) A flow diagram of the Ammonia–Soda or Solvay Process

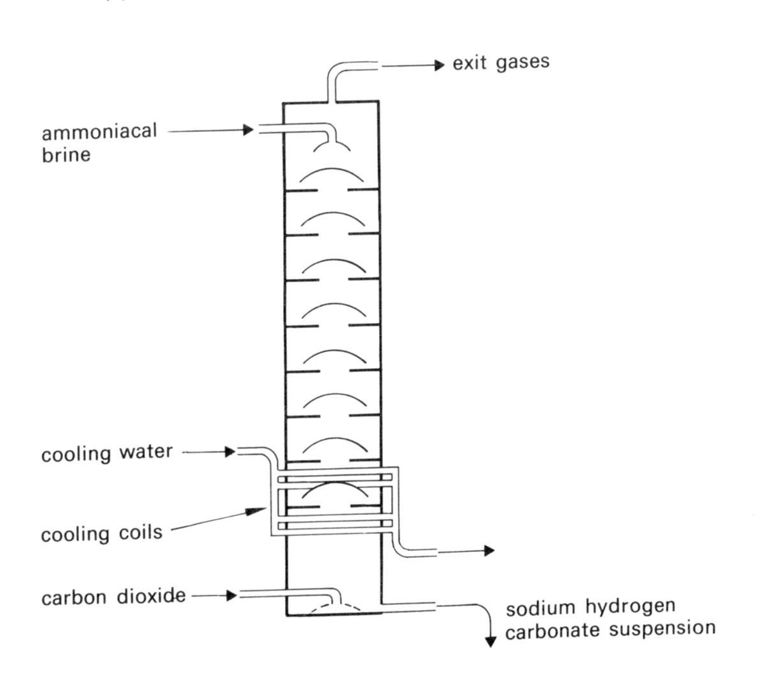

(b) Cross sectional diagram of Solvay tower

to remove any ammonium chloride, and then heated in a furnace to give sodium carbonate and carbon dioxide:

$$2NaHCO_3(s) \rightarrow Na_2CO_3(s) + CO_2(g) + H_2O(g) \quad (3)$$

The latter gas is passed back into the Solvay tower, and this is only one of the ways in which the gases used in the process are regenerated.

The calcium oxide from (1) is 'slaked' by addition of water:

$$CaO(s) + H_2O(l) \rightarrow Ca(OH)_2(s)$$

The suspension of calcium hydroxide is heated with the filtrate from the Solvay tower (equation (2)), and ammonia is produced:

$$2NH_4Cl(aq) + Ca(OH)_2(s) \rightarrow 2NH_3(g) + CaCl_2(aq) + 2H_2O(l)$$

If you look carefully at the flow diagram you will see that theoretically there is only one 'waste' product and that is calcium chloride from the ammonia still. The carbon dioxide and the ammonia are recycled into the process, and unavoidable small losses (such as ammonia gas) are made good. This is an excellent illustration of the economical use of byproducts.

Environmental factors and energy considerations

The starting materials (sodium chloride and calcium carbonate) are readily available and relatively cheap. The sodium chloride is normally obtained by pumping water into salt deposits and then pumping out saturated brine. For this reason factories are often situated near to salt deposits, especially where limestone quarries are also in the vicinity. The multi-step process also involves coke (to fuel the lime-kilns) and very large supplies of water for cooling etc., and these factors will also be important in considering a suitable site.

The process results in some solid waste (mainly calcium chloride), dust from quarrying operations, some pollutant gases, and large quantities of liquid effluent. The calcium chloride is not dumped; it is used, e.g. to hasten the curing of concrete, and so it is neither a waste product in the true sense, nor a product of commercial importance.

The main pollutant gas is sulphur dioxide, produced by burning the coke in the lime-kilns, but the gases from the kilns are thoroughly scrubbed before being returned to the atmosphere.

The liquid effluent is mainly water which has been used for cooling, and as this contains no pollutant material it can be discharged directly into a river. The temperature of the river water will be increased slightly, which may be important (see page 203). Some liquid effluent is also produced directly by the chemical reactions in the plant; it contains dissolved sodium and calcium chlorides, and insoluble impurities derived from the limestone. This chemical effluent cannot be discharged directly into a river for it would cause serious silting problems. The majority of the dissolved and suspended solids are returned below ground by pouring the treated effluent back down the bore holes used originally to extract brine. The remainder of the chemical effluent is made harmless and cooled before being discharged into a river.

The main energy demand is for the fuel used in the lime-kilns, because the decomposition of limestone is a highly endothermic process which is carried out at a temperature in excess of 1100 °C. Ironically, many of the subsequent chemical reactions are exothermic and yet the heat energy so produced is 'inconvenient', i.e. it cannot be used to help the efficiency of the process. The process is thus demanding in energy, but in terms of the use and recycling of materials it is extremely efficient.

Points for discussion

1. It is interesting to note that brine contains impurities such as calcium salts which would interfere with the Ammonia–Soda process. In order to remove these, some of the final product, sodium carbonate, is added to the brine to precipitate out calcium ions as calcium carbonate.

$$Ca^{2+}(aq) + CO_3^{2-}(aq) \rightarrow CaCO_3(s)$$

This is another example of making a factory pay for itself!

2. Before permission is granted for a factory to discharge waste into a river, the inspectors have to be satisfied that the effluent does not have a high *biological oxygen demand* (BOD). (a) What does this mean, and why is it important? (See page 197 if in doubt.) (b) There are minute traces of a certain inorganic ion in the effluent from the Ammonia–Soda process which does increase the BOD of the effluent. The ion is present in very low concentrations and is not a significant pollutant. Which ion do you think this is?

The manufacture of sulphuric acid by the Contact Process

J. von Liebig once said that it was possible to judge the commercial prosperity of a country by the amount of sulphuric acid it produces, and this remains a good indicator of industrial activity even today. Sulphuric acid is one of the most important industrial chemicals, with world production exceeding 110 million tonnes per year and with applications which involve almost all manufacturing processes (see Tables 27.5 and 27.6).

The history of the commercial production of sulphuric acid shows interesting variations. The basic variables have been the source of sulphur dioxide and the actual process by which the sulphur dioxide is converted into sulphur trioxide, which is then used to make the acid.

Most of the world's supply of sulphuric acid is now produced by the Contact process, a small version of which is used in Figure 24.3 (p. 461). See also the summary 'flow chart' on page 554.

The source of sulphur dioxide

The most important source of sulphur dioxide is sulphur, which is obtained from natural deposits (see page 457) or from the desulphurization of certain crude oils and natural gas (e.g. the 'sour' natural gas found in parts of France and Canada). The sulphur is then burned to produce sulphur dioxide:

$$S(l) + O_2(g) \rightarrow SO_2(g) \quad \Delta H = -297 \text{ kJ mol}^{-1} \qquad (1)$$

Some sulphur dioxide is also produced when sulphide minerals are roasted as part of a process to extract certain metals (e.g. zinc), but in Britain this contribution is very small. As North Sea oil and natural gas are also low in sulphur, most of our supply of sulphur dioxide comes from imported sulphur.

The conversion of sulphur dioxide into sulphur trioxide

Typical converters used for this stage of the process are illustrated in Figure 27.21. The chemical reaction taking place in the converters is:

$$2SO_2(g) + O_2(g) \rightleftharpoons 2SO_3(g) \quad \Delta H = -98 \text{ kJ mol}^{-1} \qquad (2)$$

The gases must be pure and dry. The reaction is reversible, producing an equilibrium mixture of sulphur dioxide, sulphur trioxide, and oxygen. The following conditions are chosen to ensure a good and economic yield of sulphur trioxide.

Catalyst A catalyst containing vanadium(V) oxide is used. This is not easily 'poisoned' by impurities in the sulphur dioxide, and so can remain in use for a reasonable length of time (e.g. five or six years).

Table 27.5 World production of sulphuric acid, 1976 (113 million tonnes in total)

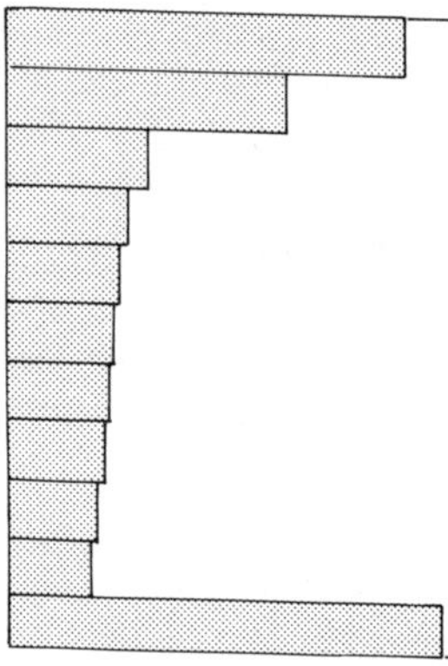

USA	25.6%
USSR	17.7%
Japan	5.4%
West Germany	4.1%
France	3.5%
Poland	3.2%
United Kingdom	2.9%
Canada	2.8%
Spain	2.5%
Italy	2.4%
Rest of the world	29.9%

(Information from *Sulphur*—journal published by the British Sulphur Corporation, London)

Table 27.6 Uses of sulphuric acid in the UK, 1976. (Total 3.37 million tonnes)

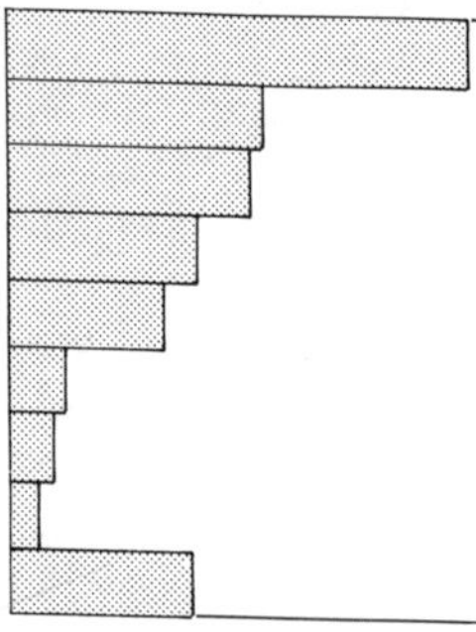

Fertilizers/Agricultural	30.4%
Chemicals	16.4%
Paints/Pigments	15.7%
Detergents/Soaps	11.7%
Natural/Man-made fibres	9.4%
Metallurgy	2.3%
Dyestuffs/Intermediates	2.0%
Oil/Petrol	1.0%
Miscellaneous	11.1%

(Information reproduced from figures published by the National Sulphuric Acid Association, London)

Temperature of about 500 °C Application of the principle of Le Chatelier (p. 360) shows that low temperatures produce the best yields of sulphur trioxide, but under such conditions the rate of reaction is too slow and so an operating temperature of about 500 °C is used. This is a compromise temperature, and the process of arriving at such a compromise in order to obtain the maximum yield of a product in a given time at an economical rate is called *optimization*; it is frequently met with in industry, and the temperature chosen for a particular Contact process is the *optimum* temperature for that plant.

Atmospheric pressure Application of the principle of Le Chatelier shows that high pressures increase the yield of sulphur trioxide. However, a good yield is obtained even at atmospheric pressure and so the use of high pressures (and the consequent increase in energy demand and also in the cost of plant built to contain the pressure) is not justified. (In practice, a pressure slightly above atmospheric pressure is used to push gases through the plant.)

Use of excess oxygen and the removal of sulphur trioxide Excess oxygen is used (from the air) and sulphur trioxide is continuously removed from the equilibrium mixture, both of which help to improve the efficiency of the process, by driving the reversible reaction to the right.

Figure 27.21 Convertors, sulphuric acid Contact plant (*Albright and Wilson*)

The conversion of sulphur trioxide into sulphuric acid

As explained on page 461, sulphur trioxide reacts violently with water, forming a stable mist of sulphuric acid droplets which is difficult to absorb. (The problem occurs when the gaseous sulphur trioxide meets water vapour above the surface of the water, and before it makes contact with liquid water.) The gas is therefore passed into 98 per cent sulphuric acid (from which virtually no water vapour escapes), and the reaction with the small amount of water present in the acid takes place without the formation of a mist. Some acid is continuously removed as product, and water is added to the remainder of the acid to keep its concentration at around 98 per cent. The net result is a steady production of sulphuric acid (98 per cent) from sulphur trioxide and water:

$$SO_3(g) + H_2O(l) \rightarrow H_2SO_4(l) \quad \Delta H = -130 \text{ kJ mol}^{-1} \qquad (3)$$

The absorption vessels have to be cooled because the absorption is exothermic. See page 554 for a summary of the overall process.

Note: Some industrial processes require the use of *oleum*. This is concentrated sulphuric acid which contains sulphur trioxide.

Environmental factors and energy considerations

The conversion of sulphur dioxide into sulphur trioxide, and the absorption of the latter, are now so efficient that the waste gases entering the atmosphere contain only between 0.03 and 0.05 per cent of the sulphur dioxide formed in the first stage. All new sulphur-burning sulphuric acid plants in the UK have to work to 99.5 per cent efficiency, and it is unlikely that this can be improved. This is a remarkable achievement when we remember that the second stage in the process is reversible, is not carried out under the theoretically ideal conditions, and produces a mixture of sulphur dioxide and sulphur trioxide. The amount of sulphur dioxide released into the atmosphere from a modern sulphuric acid plant is negligible when compared with that from a power station, but nevertheless it still represents a significant amount of pollutant gas per annum.

The Contact process shows a remarkable balance sheet in terms of energy. If you look back at the three reactions involved in the process (labelled (1), (2), and (3)) you will see that they are all exothermic. The first two are particularly important, because the extra heat energy given out by these reactions is produced at an already high temperature (above the boiling point of water) and so can be used to produce steam, which in turn can generate electricity and provide heating. Unfortunately, reaction (3) takes place at low temperatures (below the boiling point of water) and the heat energy given out cannot be used to raise steam; it has to be regarded as 'waste' heat, for if the temperature in the absorbers were to rise it would interfere with the absorption process. The other two reactions, however, produce enough energy to operate and heat the whole plant. They also produce *extra* energy for use in other factories or by other processes, i.e. energy can be 'sold'. This makes the production of sulphuric acid very attractive in terms of energy, and the price of the acid is very low indeed when compared with most other industrial chemicals.

In some sites which operate a variety of processes, a sulphuric acid plant is deliberately incorporated because the sale of the product (sulphuric acid) pays for the importation of sulphur, and the energy produced is not only sufficient to operate the whole process, it is also enough to 'pay' for maintenance and labour, and to 'power' other processes on the same site.

Points for discussion

1. The Contact process provides a useful example of a chemical reaction which you have studied in the laboratory and which is also conducted on a large scale. Read again the information on 'problems of the large scale' etc. (pages 518 to 521), and then make out a table comparing and contrasting the industrial and laboratory processes.

2. The only sulphur-containing mineral found in quantity in Britain is calcium sulphate, which occurs as gypsum and anhydrite (p. 186). Until recently, it was economical to use this mineral as a source of sulphur dioxide because the high energy demand for mining the rock and converting it into sulphur dioxide was offset by the fact that two commercially important materials could be manufactured by the one basic process, i.e. sulphuric acid and cement. As the price of energy has increased rapidly, and because the demand for cement is variable, both products are now made by separate, more economical processes. This is another example of the way in which changing circumstances can result in a different manufacturing process (see page 522).

Check your understanding

Certain deposits of natural gas extracted in France are found to contain 17 per cent by mass of a contaminant gas called hydrogen sulphide (H_2S). This contaminant gas is removed by mixing sulphur dioxide with the natural gas and passing the mixture over a heated catalyst. The natural gas is not affected but the contaminant gas reacts according to the equation

$$2H_2S + SO_2 \xrightarrow[\text{heat}]{\text{catalyst}} 2H_2O + 3S$$

(a) In 100 tonnes of this contaminated natural gas what mass of hydrogen sulphide is present?

(b) Calculate the mass of sulphur dioxide required to react with this amount of hydrogen sulphide.

(c) What total mass of sulphur will be formed?

(d) How much of this total mass of sulphur formed will have come from the contaminant gas, hydrogen sulphide?

(e) Indicate the stages, including equations, by which sulphur is converted into sulphuric acid.

(f) Explain how sulphur can be used to vulcanize rubber. (A.E.B. 1979)

The synthesis of ammonia by the Haber Process

In section 21.3 (p. 383) we stressed the problem of feeding the increasing world population, the need to fix atmospheric nitrogen, and the importance of fertilizers.

Nearly all of the nitrogenous fertilizers are made from ammonia, and indeed over 80 per cent of the ammonia produced is converted into fertilizers. Much of the remaining 20 per cent is converted into nitric acid (see page 552) which in turn leads to other fertilizers, dyestuffs, synthetic fibres, and explosives. There are eight ammonia plants in Britain (Figure 27.22), collectively producing 7000 tonnes of liquid ammonia each day.

Virtually all of the world's supply of ammonia is made by the Haber process. The process is named after the German chemist L. F. Haber, who developed a catalyst which enabled nitrogen and hydrogen to be reacted directly together to give economic yields of ammonia. The process was developed rapidly between 1911 and 1920, and since then the synthesis reaction has remained almost unchanged, but the source of hydrogen has varied from time to time and from place to place, according to economic conditions. The production of hydrogen is the most expensive stage in the process, and so it is understandable that this step fluctuates according to other factors. (This corresponds to the way in which sources of sulphur dioxide have varied in the Contact process.) See the summary 'flow chart' of the process on page 554.

Figure 27.22 A 1200 tonne per day ammonia plant at Billingham (*ICI*)

The source of hydrogen

At the moment, the ammonia plants operating in Britain obtain their hydrogen from natural gas. A series of steps is involved, but the details need not concern us here.

In the early days of the process, the hydrogen was produced by the electrolysis of 'water'. This requires a large consumption of expensive electricity, and is no longer economic. It has an important advantage, however, in that the hydrogen so produced is very pure. This is an important consideration because catalysts are easily poisoned by impurities. Catalysts are normally used in finely divided form (e.g. as pellets) so that they have a large surface area upon which the gases can meet and react. (A selection of industrial catalysts is shown in Figure 27.23.) This unfortunately also means that they are easily poisoned by impurities in the gases. As sources of natural gas decline, and if relatively cheap sources of electricity become available, this method of producing hydrogen could again become important (cf. the production of hydrogen as a fuel for transport, page 496).

The electrolytic method was replaced by a series of reactions involving coke and steam. Where natural gas and oil are not readily available, hydrogen is still produced by this method, and we may well revert to it in Britain when supplies of natural gas decline.

Figure 27.23 A selection of industrial catalysts (*ICI*)

The synthesis step

The reaction taking place is:

$$N_2(g) + 3H_2(g) \rightleftharpoons 2NH_3(g) \quad \Delta H = -46.2 \text{ kJ mol}^{-1}$$

A mixture of one volume of nitrogen (from the air) and three volumes of hydrogen is used. The reaction is reversible and so an equilibrium mixture containing nitrogen, hydrogen, and ammonia is produced. The ammonia is continuously removed from the mixture (as liquid ammonia), which drives the reaction to the right (principle of Le Chatelier, page 360). The mixture leaving the catalyst beds contains about 15 per cent ammonia, and the unreacted gases are recycled. The following operating conditions are found to be the most economic.

Pressure of 200 atmospheres The principle of Le Chatelier shows that high pressures produce good yields of ammonia. This is indeed the case as is shown by the following data (by courtesy of ICI). The percentages are the proportions of ammonia at equilibrium.

Temperature (°C)	Per cent ammonia at equilibrium at pressure (kg cm^{-2}): 155	259	362
350	46.21	57.46	65.16
450	22.32	32.87	39.31
550	9.91	15.58	20.76

Some plants have operated at pressures much higher than 200 atmospheres (to improve the yield of ammonia), but working conditions of this kind are expensive to operate and maintain.

Temperature of 380 °C–450 °C Application of the principle of Le Chatelier shows that low temperatures will improve the yield of ammonia (see data above), but unfortunately low temperatures would result in the reaction rate being too slow to be economical. An optimum temperature is around 400 °C.

Iron catalyst The catalyst increases the rate of the reaction but does not affect the yield of ammonia. It is easily poisoned by impurities (especially carbon monoxide and sulphur) and breaks up if the temperature is allowed to exceed 660 °C. See page 554 for a summary of the overall process.

Environmental factors and energy considerations

The Haber process is fairly 'clean', and operates with a high degree of efficiency. Solid waste is confined to spent catalysts. Large quantities of water are used for cooling, and so there is much liquid effluent but this is mainly 'warm water'. The process is efficient in its use of energy because a considerable amount of heat energy is liberated by the exothermic reactions in the production of hydrogen from natural gas. This energy is used to raise high pressure steam which in turn drives the turbines needed to compress the gas to 200 atmospheres pressure. The synthesis reaction is also exothermic, and so is self-sustaining; in fact the reacting gases have to be cooled to keep them at the required temperature. The production of hydrogen is expensive, however, for it is obtained from natural gas which is at the moment regarded as a premium fuel. Nevertheless, we can argue that it is more sensible to use natural gas as a source of materials rather than merely burn it.

Points for discussion

1. Fuels which are to be burned are often desulphurized so that sulphur dioxide is not formed on combustion. The methane used in the Haber process is not 'burned', so why is it necessary to desulphurize it?
2. Four of the eight ammonia plants in Britain are situated at Billingham on the River Tees. Remembering that they were built many years ago, suggest reasons for the choice of site.
3. Circumstances have changed since the construction of the plants, but the Billingham site has proved to be a very convenient one in spite of these changes. Can you explain this statement?
4. Liquid effluent discharged from a Haber plant must be monitored in case it has a high BOD (p. 197). What could cause it to have a high BOD, and why would it do so?
5. It is correct to describe the formation of ammonia in the Haber process as a *synthesis*, but this term should not be applied to the production of sulphuric acid in a Contact process. What exactly does the term synthesis mean, when used to describe a chemical reaction?
6. It is important to cool the reacting gases in the final stage of the Haber process, because otherwise the exothermic reaction would produce a high temperature. Suggest two reasons why a high temperature would be undesirable.

The industrial manufacture of nitric acid

As explained earlier, some of the ammonia produced in the Haber process is immediately converted into nitric acid (usually on the same site), and nitric acid is an important industrial chemical with many uses. The raw materials required are ammonia, air, and water, and the steps in the process are summarized below. See also the summary 'flow chart' on page 554.

1. Ammonia gas is mixed with about eight times its volume of air and preheated to about 300 °C by passage through a heat interchanger (Figure 27.24).

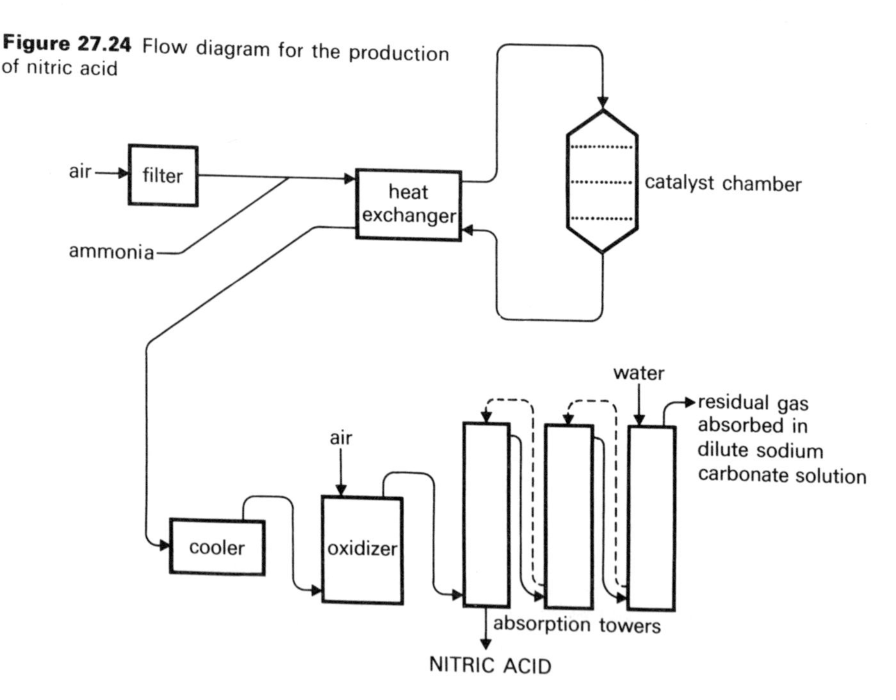

Figure 27.24 Flow diagram for the production of nitric acid

2. The mixture of gases then passes through a catalyst chamber (Figure 27.25) containing several layers of a platinum/rhodium gauze (90 per cent platinum) which acts as the catalyst. The reaction is started by an electrically heated wire just above the catalyst, and under these conditions the following reaction takes place:

$$4NH_3(g) + 5O_2(g) \rightarrow 4NO(g) + 6H_2O(g) \quad \Delta H = -1087 \text{ kJ}$$

The reaction is very exothermic, and once started the heater is turned off and the temperature of the reacting mixture is controlled at about 900 °C.

3. The gases leaving the catalyst chamber include excess oxygen and also nitrogen from the air. They pass through a heat interchanger where they give up some of their heat energy in warming up the incoming air/ammonia mixture. Further cooling brings the temperature of the gases down to a level at which the nitrogen monoxide can react with the excess oxygen to form nitrogen dioxide. (This reaction does not take place at high temperatures.)

$$2NO(g) + O_2(g) \rightarrow 2NO_2(g)$$

4. The mixture of gases finally passes into a series of absorption towers where it meets counter-currents of nitric acid of gradually increasing dilution until only water flows

through the last tower (Figure 27.24). The nitrogen dioxide, an acidic oxide, reacts with the water in the towers to form a mixture of nitric and nitrous acids

$$2NO_2(g)+H_2O(l) \rightarrow HNO_2(aq)+HNO_3(aq)$$

The nitrous acid is oxidized to nitric acid by the excess oxygen present,

$$HNO_2(aq)+[O] \rightarrow HNO_3(aq)$$

Any unreacted oxides of nitrogen are extracted and returned to the first absorption tower, and the concentration of the acid obtained by the process is between 55 and 65 per cent, i.e. ordinary concentrated nitric acid. See page 554 for a summary of the overall process.

Point for discussion

Make sure that you understand the process used for the manufacture of nitric acid, study Figure 27.24 carefully, and then make comments on the environmental and energy aspects of the process in the usual way.

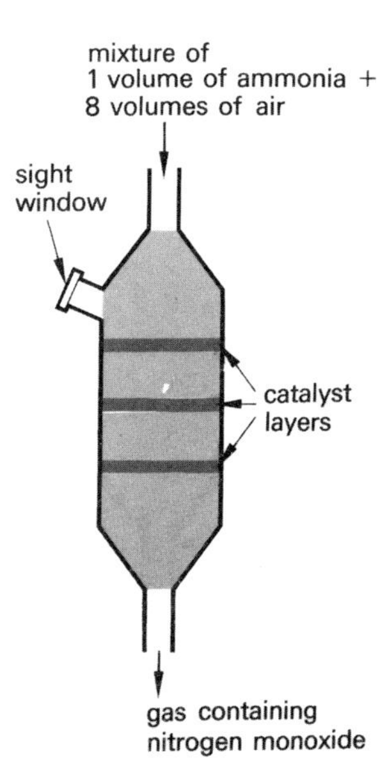

Figure 27.25 The catalyst chamber for the production of nitric acid

27.4 A SUMMARY OF THE MAIN POINTS IN CHAPTER 27

The following 'check list' should help you to organize the work for revision.

(a) The meaning of the terms raw materials, resource, native element, mineral, and ore.

(b) Minerals usually need to undergo a chemical reaction in order to obtain an element from them, and this is one of the most important stages in the extraction process because it usually demands energy.

(c) An appreciation of the fact that our resources cannot last for ever, that some are already in short supply, and that our attitude towards the problem should be more positive.

(d) The need to recycle materials with a limited source. A *general* appreciation of the kind of problems encountered with recycling, and of the advantages to be gained by the process, e.g. a saving of energy, resources, and valuable land.

(e) A *general* understanding of the factors which are considered when choosing a site for an industry, so that you can make sensible suggestions about choosing a site (when given appropriate data) or justify the fact that a particular industry has been developed in a certain region.

Note: For the industrial processes, there is no need to learn all of the details or facts given in the book, some of which are only used to provide background information and to illustrate principles frequently used in industry. Your teacher will give you guidance as to what you need to know.

(f) The raw materials used, and the chemical changes which take place, when pig iron is made in a blast furnace (p. 527).

(g) The reactions which take place when molten iron is converted into steel, using the Basic Oxygen Process and/or the Electric Arc method (pp. 529–32).
(h) A general understanding of the environmental and energy factors involved in making iron and steel (p. 530).
(i) The manufacture of sodium by the Downs Cell, and the environmental and energy factors associated with the process (p. 533).
(j) The manufacture of aluminium, and the environmental and energy factors associated with the process (p. 534).
(k) The production of chlorine, sodium hydroxide, and hydrogen by the electrolysis of brine using the Diaphragm and/or Mercury Cells (p. 536). The environmental and energy factors associated with the process, and a comparison between the two cells (p. 539).
(l) The fractional distillation of liquid air (p. 540).
(m) The manufacture of sodium carbonate by the Ammonia–Soda process, the way in which materials are recycled in the process, and the environmental and energy considerations (pp. 542–5).
(n) The manufacture of sulphuric acid by the Contact process, and the environmental and energy considerations (pp. 545–8).

$$\text{source of } SO_2 + O_2 \text{ (air)} \xrightarrow[\text{500 °C, atmospheric pressure}]{\text{vanadium(V) oxide}} SO_3 \xrightarrow[\text{water in concentrated sulphuric acid}]{} H_2SO_4$$

(o) The synthesis of ammonia by the Haber process, and the environmental and energy factors involved (pp. 548–51).

$$\text{source of } H_2 + N_2\text{(air)} \xrightarrow[\text{200 atmospheres, 450 °C}]{\text{iron catalyst}} NH_3 \longrightarrow \text{liquefied and removed}$$

(p) The industrial manufacture of nitric acid (p. 552).

$$NH_3 + O_2 \xrightarrow[\text{900 °C}]{\text{platinum catalyst}} NO \xrightarrow[\text{oxidize}]{\text{cool,}} NO_2 \xrightarrow[\text{with water}]{\text{react}} HNO_3 + HNO_2$$

$$HNO_2 \xrightarrow{\text{oxidize}} HNO_3$$

QUESTIONS

1. (a) Historically man was able to extract metals from their ores in the following order.
Copper—impure in the form of bronze—Bronze Age
Iron—Iron Age
Aluminium } only in the past 150 years
Sodium }
Give chemical explanations for this chronological order of extraction.
(b) For either iron or aluminium
(i) describe the extraction of the metal, giving chemical and ionic equations wherever possible;
(ii) discuss the importance of the metal to modern man. (A.E.B. 1979)

2. To generate wealth and maintain good living standards man has to exploit and use the Earth's resources. In so doing many problems may arise. For two of the following situations
(a) building and commissioning a nuclear power station,
(b) opening new limestone quarries and production plant close to a National Park,
(c) building and operating a sulphuric acid plant on a river estuary close to a densely populated area,

discuss the problems that may arise, with special reference to the following main points.

Location and facilities of the site
Nature of the raw materials and transport
Economics—demand for product etc.
Social effects—work force required
Pollution effects—or other possible dangers

Wherever possible relate your answer to chemical facts, reasons and solutions. (A.E.B. 1979)

3. (a) Crude iron is produced in a blast furnace from a mixture of iron ore, coke, and limestone.

(i) Give the common name and formula of one iron ore used.

(ii) What happens chemically to the coke?

(iii) Give the equation for one reaction by which iron ore is converted to iron.

(iv) Why is limestone used? What happens to it?

(v) What are the essential differences between the composition of crude iron and that of steel?

(b) State, without further description, one reaction in each case by which you could:

(i) obtain a solution containing iron(II) ions (Fe^{2+}) from metallic iron;

(ii) convert this solution to one containing iron(III) ions (Fe^{3+}).

Briefly describe a chemical test to confirm that the change in (ii) has taken place.

(A.E.B. 1976)

4. (a) Iron is smelted by heating an iron ore with coke and limestone at a high temperature in a blast furnace. Hot air is blown in at the base of the furnace.

(i) Give the common name of one iron ore used.

(ii) What happens to the coke?

(iii) How is the high temperature of the furnace maintained?

(iv) How is the iron ore converted to iron?

(v) Why is limestone added, and what happens to it?

(vi) State how the air is heated before being blown into the furnace.

(vii) Name two impurities usually present in pig iron.

(viii) Why is this method used for producing pig iron but not for producing aluminium?

(b) (i) Under identical conditions, why does iron corrode much faster than aluminium?

(ii) Why does zinc plating protect iron from corrosion even when the zinc layer is broken?

(A.E.B. 1977)

5. (a) Discuss the social and environmental effects which have followed the successful extraction of iron and aluminium from their ores.

(b) What is meant by 'recycling of metals', and why is recycling of such importance?

(A.E.B. 1976)

6. (a) Discuss the social, environmental, and economic problems associated with the large scale generation of electricity from either fossil fuels or nuclear fuels.

(b) The extraction and refining of certain metals is dependent on electrolysis.

(i) For one such metal describe carefully either its electrolytic extraction or its electrolytic refining.

(ii) Discuss the uses of the metal you have selected in (b) (i). (A.E.B. 1978)

7. (a) Describe the essential features of an industrial process by which chlorine and sodium hydroxide are manufactured electrolytically. Name the electrolyte, give the polarity of the electrodes, state the materials of which they are made, and write equations for the discharge of the ions at the electrodes. Your account should state clearly how and where each product is obtained and should indicate how these two products are kept separate.

(b) Name the products and write an equation for a reaction between chlorine and sodium hydroxide solution.

(c) State two large scale uses of chlorine.

(J.M.B.)

8. Describe briefly how each one of the following substances is manufactured from the given starting material.

(a) Nitric acid from ammonia.

(b) Sodium from sodium chloride.

(c) Sulphuric acid from sulphur dioxide.

(d) Ethanol from ethene. (J.M.B.)

9. (a) Draw a labelled diagram to show how gas jars of dry ammonia can be prepared and collected. Write an equation for the reaction.

(b) Describe a method of showing that ammonia is very soluble in water.

(c) Under what conditions does ammonia react with copper(II) oxide? State what would be seen during the reaction and either write an equation for the reaction or name the products.

(d) State briefly how crystals of ammonium sulphate can be made in the laboratory starting from ammonia solution. (J.M.B.)

10. (a) Draw a labelled diagram and write an equation to show how sulphur(VI) oxide can be prepared in the laboratory from sulphur dioxide.

(b) (i) In industry, sulphuric acid is made by passing sulphur(VI) oxide into 98 per cent sulphuric acid and reacting the product with water. Give one reason why sulphuric acid is not manufactured by passing sulphur(VI) oxide directly into water. (ii) Give two large scale industrial uses of sulphuric acid.

(c) Concentrated sulphuric acid can act as an oxidizing agent and as a dehydrating agent. Give one example of each and state the products of each reaction.
(d) In 1843, Justus von Liebig wrote: 'The commercial prosperity of a country may be judged from the amount of sulphuric acid it consumes.' Explain very briefly what he meant and state, with a reason, whether you think that this statement still applies today.
(J.M.B.)

11. Describe briefly the manufacture of nitric acid from ammonia, giving the essential chemistry but no technical details. Give two uses of nitric acid and one physical property. Describe two experiments that illustrate the oxidizing power of concentrated nitric acid. Include in your accounts evidence that oxidation has occurred. (C.)

12. Name two important ores of iron. Outline the extraction of iron by the blast furnace process, explaining the essential chemistry. (No diagram is required, nor are technical details.) What differences are there in composition between cast iron and steel? Describe, giving any necessary conditions, the action of (a) hydrochloric acid, (b) nitric acid (one reaction only), and (c) steam on iron. (C.)

13. Answer the following questions about the manufacture of iron and steel (no diagrams are required).
(a) Give the name and formula of one mineral from which iron is extracted.
(b) Explain how carbon monoxide is formed in the blast furnace.
(c) Write the equation for one reaction by which metallic iron is formed in the furnace.
(d) Explain clearly why limestone (calcium carbonate) is used in the blast furnace and suggest what you think would happen if the limestone were not present.
(e) Name three impurities likely to be present in the 'pig iron' formed in the blast furnace. Give one effect of these impurities on the physical properties of the iron.
(f) Explain how these impurities are removed during the conversion of pig iron into steel. (C.)

14. Titanium(Ti) is the seventh most abundant metal in the Earth's crust. One of the most important ores of titanium is rutile(TiO_2).
Titanium is manufactured from rutile in two stages as follows:
(i) Chlorine is passed over a heated mixture of rutile and carbon giving titanium(IV) chloride ($TiCl_4$) and carbon monoxide. The titanium(IV) chloride (melting point −23 °C; boiling point 136 °C) is condensed and purified by fractional distillation.
(ii) Titanium(IV) chloride is reduced to titanium by heating it with a reactive metal such as magnesium in an atmosphere of argon. The solid titanium separates from the molten magnesium chloride (melting point 714 °C) which is tapped off.

Titanium has a high mechanical strength and a low density and is very resistant to corrosion.

(a) Explain what is meant by (i) an ore, (ii) condensed, (iii) fractional distillation.
(b) In each case, suggest an element which could have been used in the manufacture of titanium (i) in place of magnesium, (ii) in place of argon.
(c) In each case, construct the equation for (i) the preparation of titanium(IV) chloride, (ii) the reduction of the chloride to the metal.
(d) Justify the use of the term reduction in (c) (ii).
(e) In each case, give the name and formula of any compound which has the same type of bonding as (i) titanium(IV) chloride, (ii) magnesium chloride.
(f) Suggest a commercial use for titanium metal. (C.)

15. (a) Give an account of the chemical principles and the main reactions involved in the blast furnace, which uses the raw materials iron(III) oxide, coke, and limestone. (No diagram of industrial plant is required.)
(b) Explain, in terms of electron transfer, the conversion of an Fe(III) compound into an Fe(II) compound. Use one particular reaction in which this conversion takes place to explain why this is a redox reaction.
(c) Describe how you would distinguish between solutions containing Fe^{2+} and Fe^{3+} ions. Give one chemical test in each case.
(O. & C.)

16. (a) Give an account of the manufacture of aluminium, starting from the purified oxide.
(b) (i) Aqueous ammonia is added in excess to aqueous aluminium sulphate. Describe what is seen and give the ionic equation for the reaction.
(ii) What difference would be observed if aqueous sodium hydroxide were used instead of aqueous ammonia?
(c) Explain what you understand by anodized aluminium.

(d) State one other important use of aluminium. Give the properties of the metal on which this use depends. (O. & C.)

17. (a) Describe the industrial preparation of sodium hydroxide by the electrolysis of brine (sodium chloride solution). Include in your answer an explanation of the reactions occurring at the electrodes.
(b) How and under what conditions does sodium hydroxide react with the following? (i) carbon dioxide, (ii) ammonium sulphate, (iii) ethanoic acid (acetic acid), (iv) zinc sulphate. (W.J.E.C.)

18. The atmosphere is a relatively thin layer of gases approximately 40 km thick. Man puts many unwanted materials into the atmosphere and also uses it as a source of chemicals. Discuss critically (a) the various causes of pollution (gases, particles, radiation) in the atmosphere, (b) the importance of the atmosphere in providing raw materials for various industries. (A.E.B. 1977)

19. Describe the manufacturing process by which chlorine is made in the Diaphragm Cell. Your answer should include details of the following: (i) the nature of the electrolyte, the electrodes and the essential cell construction, (ii) the reactions occurring at the electrodes. State three factors which will influence the siting of the factory. Name two other direct products of the process, stating one use of each.

Index